Macrolide Antibiotics
Chemistry, Biology, and Practice

Macrolide Antibiotics

Chemistry, Biology, and Practice
Second Edition

Edited by
SATOSHI ŌMURA
*Kitasato Institute for Life Sciences
Kitasato University
and The Kitasato Institute
Tokyo, Japan*

ACADEMIC PRESS

An imprint of Elsevier Science

Amsterdam   Boston   London   New York   Oxford   Paris   San Diego
San Francisco   Singapore   Sydney   Tokyo

This book is printed on acid-free paper. ∞

Copyright 2002, Elsevier Science (USA).

All rights reserved.
No part of this publication may be reproduced or transmitted in any form or by any means, electronic or mechanical, including photocopy, recording, or any information storage and retrieval system, without permission in writing from the publisher.

Requests for permission to make copies of any part of the work should be mailed to: Permissions Department, Harcourt, Inc., 6277 Sea Harbor Drive, Orlando, Florida 32887-6777.

Academic Press
An imprint of Elsevier Science
525 B Street, Suite 1900, San Diego, California 92101-4495, USA
http://www.academicpress.com

Academic Press
Harcourt Place, 32 Jamestown Road, London NW1 7BY, UK
http://www.academicpress.com

Library of Congress Catalog Card Number: 2001097961

International Standard Book Number: 0-12-526451-8

Printed in the United States of America

02 03 04 05 06 07 MB 9 8 7 6 5 4 3 2 1

Contents

Contributors ix
Preface xi

1. Discovery of New Macrolides
   KAZURO SHIOMI AND SATOSHI ŌMURA

   I. Introduction 2
   II. Macrolides from Actinomycetes 8
   III. Macrolides from Bacteria Including Myxobacteria 22
   IV. Macrolides from Fungi 27
   V. Macrolides from Plants and Lichens 31
   VI. Macrolides from Insects 37
   VII. Other Macrolides 40
   VIII. Concluding Remarks 45
         References 46

2. Discovery of New Macrolides from Marine Organisms
   MASAMI ISHIBASHI

   I. Introduction 57
   II. Macrocyclic Lactones of Marine Organism Origin 58
   III. Concluding Remarks 89
         References 89

3. Chemical Modification of Macrolides
   TOSHIAKI SUNAZUKA, SADAFUMI OMURA, SHIGEO IWASAKI,
   AND SATOSHI ŌMURA

   I. Introduction 99
   II. Fourteen-Membered Macrolides 100
   III. Sixteen-Membered Macrolide Antibiotics and the Avermectin Family 145
   IV. Concluding Remarks 164
         References 165

4. Total Synthesis of Macrolides
   TADASHI NAKATA

   I. Introduction 181
   II. Synthetic Strategy for Macrolide Synthesis 182
   III. Total Synthesis of Selected Macrolides..................................... 210
   IV. Concluding Remarks 271
        References 275

5. Biosynthesis, Regulation, and Genetics of Macrolide Production
   HARUO IKEDA AND SATOSHI ŌMURA

   I. Introduction 286
   II. Reaction Mechanism of Polyketide Biosynthesis 287
   III. Polyketide Synthase.. 289
   IV. Genes Encoding Modular Polyketide Synthase......................... 295
   V. Sugar Biosynthesis 314
   VI. Genetic Manipulation of PKS Genes... 319
        References 320

6. Pharmacokinetics and Metabolism of Macrolides
   YOSHIRO KOHNO

   I. Introduction 327
   II. Pharmacokinetics and Metabolism.. 328
   III. Drug Interaction... 350
   IV. Concluding Remarks 354
        References 354

7. Antimicrobial Macrolides in Clinical Practice
   SALVADOR ALVAREZ-ELCORO AND JOSEPH D. C. YAO

   I. Introduction 363
   II. Fourteen- and Fifteen-Membered Macrolides............................. 364
   III. Sixteen-Membered Macrolides.. 380
   IV. Concluding Remarks 382
        References 382

8. Ivermectin in Clinical Practice
   OSAMU ZAHA, TETSUO HIRATA, FUKUNORI KINJO, AND ATSUSHI SAITO

   I. Introduction 403
   II. Novel Activity of Ivermectin in Clinical Practice 405
   III. Concluding Remarks 414
        References 416

Contents

9. Tacrolimus and Other Immunosuppressive Macrolides in Clinical Practice
   TADAHIRO AMAYA, JUN HIROI, AND IRA D. LAWRENCE

   I. Introduction 421
   II. Tacrolimus, a Brief Developmental History 424
   III. Novel Activity of Tacrolimus and Other Immunosuppressive Macrolides in Clinical Practice 425
   IV. Concluding Remarks 442
   References 444

10. Mode of Action and Resistance Mechanisms of Antimicrobial Macrolides
    YOSHINORI NAKAJIMA

    I. Introduction 453
    II. Mode of Action of Macrolide Antibiotics 454
    III. Mechanisms of Resistance to Antimicrobial Macrolides 472
    IV. Important Developments in Macrolide Antibiotics 485
    V. Concluding Remarks 486
    VI. Addendum 487
    References 488

11. Mode of Action of Macrolides with Motilin Agonistic Activity—Motilides
    NOBUHIRO INATOMI, FUMIHIKO SATO, ZEN ITOH, AND SATOSHI ŌMURA

    I. Introduction 501
    II. Mode of Action of Motilin 504
    III. Invention of Motilides 507
    IV. Biological Activity of Motilides 510
    V. Clinical Trials of Motilides 526
    VI. Concluding Remarks 527
    References 528

12. Novel Activity of Erythromycin and Its Derivatives
    SHOJI KUDOH, ARATA AZUMA, JYUN TAMAOKI, HAJIME TAKIZAWA, KOH NAKATA, AND HAJIME GOTO

    I. Erythromycin Treatment in Diffuse Panbronchiolitis................ 534
    II. Inhibition of Chloride Channel 541
    III. Effects of Macrolides on Cytokine/Chemokine Expression 546
    IV. Modulation of Bacterial Function 553
    V. New Challenge for Novel Action 557
    References 564

13. Mode of Action of Avermectin
SATOSHI ŌMURA

   I. Introduction 571
  II. Target of Avermectin Action 571
 III. Cloning and Structure of Avermectin Binding Protein 572
 IV. Concluding Remarks 575
      References 575

14. Mode of Action of FK506 and Rapamycin
NOBUHIRO TAKAHASHI

   I. Introduction 577
  II. Initial Cellular Target for FK506 and Rapamycin; Peptidyl
      Prolyl *cis-trans* Isomerases (Rotamases, Immunophilins) 586
 III. Target of FK506–FKBP12 Complex: Calcineurin 599
 IV. Target of Rapamycin–FKBP12 Complex: mTOR/FRAP/RFAT 604
  V. Intervention of Intracellular Signaling Pathways by FK506
      and Rapamycin 607
      References 611

Index 623

Contributors

Numbers in parentheses indicate the pages on which the authors' contributions begin.

SALVADOR ALVAREZ-ELCORO (363), Division of Infectious Diseases, Mayo Clinic, Jacksonville, Florida 32224

TADAHIRO AMAYA (421), Development Division, Fujisawa Pharmaceutical Company, Osaka, Japan

ARATA AZUMA (533), Fourth Department of Internal Medicine, Nippon Medical School, Tokyo, Japan

HAJIMI GOTO (533), School of Medicine, Kyorin University, Tokyo, Japan

TETSUO HIRATA (403), First Department of Internal Medicine, Faculty of Medicine, University of the Ryukyus, Okinawa, Japan

JUN HIROI (421), Medical Science Research, Fujisawa Pharmaceutical Company, Osaka, Japan

HARUO IKEDA (285), The Kitasato Institute for Life Sciences, Kitasato University, Tokyo, Japan

NOBUHIRO INATOMI (501), Discovery Research Laboratory III, Takeda Chemical Industries, Ltd., Osaka, Japan

MASAMI ISHIBASHI (57), Graduate School of Pharmaceutical Sciences, Chiba University, Chiba, Japan

ZEN ITOH (501), 5-10, Chiyodamachi 1-chome, Maebashi, Gunma, Japan

SHIGEO IWASAKI (99), The Kitasato Institute, Tokyo, Japan

FUKUNORI KINJO (403), First Department of Internal Medicine, Faculty of Medicine, University of the Ryukyus, Okinawa, Japan

YOSHIRO KOHNO (327), Research Center, Taisho Pharmaceutical Company, Ltd., Saitama, Japan

SHOJI KUDOH (533), Fourth Department of Internal Medicine, Nippon Medical School, Tokyo, Japan

IRA D. LAWRENCE (421), Research and Development, Fujisawa Healthcare, Inc., Chicago, Illinois 60015-2548

YOSHINORI NAKAJIMA (453), Division of Microbiology, Hokkaido College of Pharmacy, Hokkaido, Japan

KOH NAKATA (533), Research Institute, International Medical Center of Japan, Tokyo, Japan

TADASHI NAKATA (181), RIKEN, Synthetic Organic Chemistry Laboratory, Saitama, Japan

SADAFUMI OMURA (99), The Kitasato Institute, Tokyo, Japan

SATOSHI ŌMURA (1, 99, 285, 501, 571), Kitasato Institute for Life Sciences, Kitasato University, and The Kitasato Institute, Tokyo, Japan

ATSUSHI SAITO (403), First Department of Internal Medicine, Faculty of Medicine, University of the Ryukyus, Okinawa, Japan

FUMIHIKO SATO (501), Pharmacology Laboratory II, Takeda Chemical Industries, Ltd., Osaka, Japan

KAZURO SHIOMI (1), Kitasato Institute for Life Sciences, Kitasato University, and The Kitasato Institute, Tokyo, Japan

TOSHIAKI SUNAZUKA (99), School of Pharmaceutical Sciences, Kitasato University, Tokyo, Japan

NOBUHIRO TAKAHASHI (577), Tokyo University of Agriculture and Technology, Tokyo, Japan

HAJIME TAKIZAWA (533), Graduate School of Medicine, University of Tokyo, Tokyo, Japan

JYUN TAMAOKI (533), Tokyo Women's Medical University, Tokyo, Japan

JOSEPH D. C. YAO (363), Division of Infectious Diseases, Mayo Clinic, Jacksonville, Florida

OSAMU ZAHA (403), First Department of Internal Medicine, Faculty of Medicine, University of the Ryukyus, Okinawa, Japan

Preface

Since the first edition of this book was issued in 1984, a number of new macrolides with a variety of characteristics involving not only the structure but also the biological activity have been discovered. Many of these compounds are the results of research achievements on marine natural products that have seen remarkable development during these 18 years. And macrolides having peculiar structures have been chosen as targets for organic syntheses and have contributed greatly to the development of this field.

The analyses of biosynthetic genes on naturally occurring organic compounds is a research field that has recently achieved conspicuous progress. The reader should note in particular that this research has been carried out primarily on the "polyketide," which is the backbone structure of macrolide antibiotics.

Since the first edition was published, semisynthetic macrolide antibiotics such as clarithromycin and roxithromycin and immunosuppressants such as FK-506 (tacrolimus) and rapamycin have been introduced in clinical practice(s) and have shown positive results.

Erythromycin, a representative macrolide, and its derivatives have been found to show excellent anti-diffuse panbronchiolitis activity, and they now receive clinical attention beyond their original use as antibiotics. Moreover, the use of ivermectin, which was initially used as an antinematode agent for animals in 1981, procedures for exterminating onchocerciasis has been undertaken on a large scale. Consequently, in 1999 34 million people in an endemic area centered in Africa were saved from this disease with only a single annual administration. Thus, application of macrolides for clinical use has been a surprising development.

The action mechanism of macrolides with particular activities has been actively investigated. More than ever, macrolides have become interesting compounds in both basic and applied sciences.

In this edition, readers will learn of the development of macrolide research during the past 18 years and its perspective at hand. Each chapter is written by a specialist in that field to explore our fascination with macrolides to its full extent. I am grateful to the author of each chapter.

Finally, I offer hearty thanks to Dr. Shigeo Isawaki and Dr. Noelle Gracy for their valiant efforts and hope that this book will come to the attention of many people and prove useful in their own research fields.

Satoshi Ōmura, Ph.D.
Kitasato University and
The Kitasato Institute
Tokyo, Japan

1. Discovery of New Macrolides 5

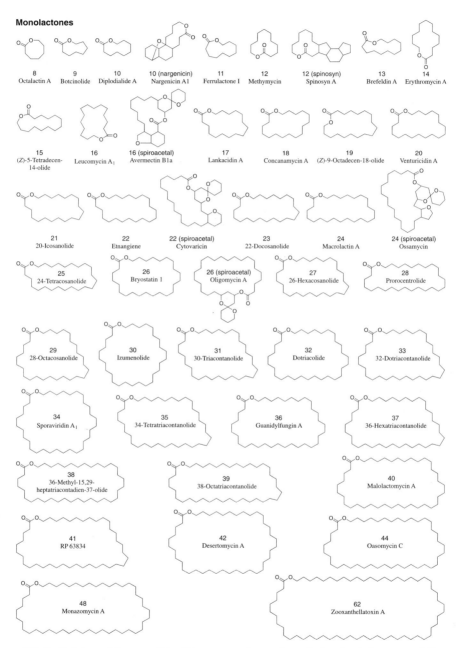

Fig. 1. Skeletons of the macrolides. The compound name for each ring size is the representative compound for each class. These compounds are described in the text.

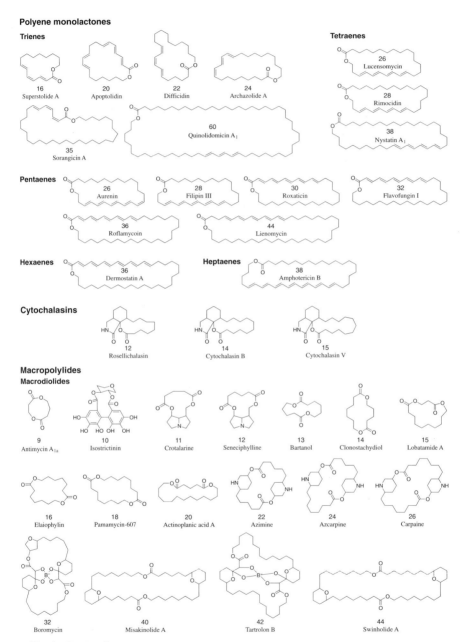

Fig. 1. *Continued.*

1. Discovery of New Macrolides

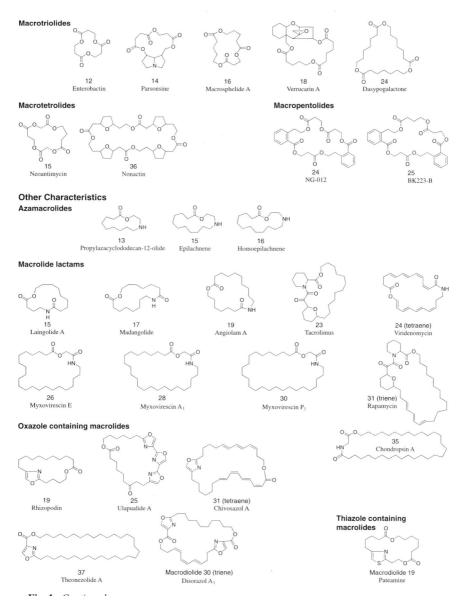

Fig. 1. *Continued.*

TABLE II
Macrolides Classified by Their Producing Organisms

Producing organism	Reported number[a]
Actinomycetes	826
Myxobacteria	125
Other bacteria	37
Fungi	218
Lichens	3
Algae	69
Plants	683
Invertebrates	190
Insects	51
Vertebrates	27
Total[a]	2212

[a]Sixteen macrolides are produced by two or three kinds of organisms and, hence, are counted in duplicate or triplicate.

II. Macrolides from Actinomycetes

Actinomycetes are the largest source of natural macrolides. They produce more than three hundred 16-membered macrolides (including spiroacetals) and nearly one hundred 14-membered macrolides (Table III). They also produce various sizes of macrolides up to a 60-membered ring, including polyene macrolides, macrodiolides, macrotetrolides, and immunosuppressive macrolide lactams. Compounds already described in the first edition of this book [1] are mentioned here briefly.

A. Medium-Ring Macrolides (Eight- to Ten-Membered Rings)

Two 8-membered ring macrolides, octalactins A (Fig. 2, **1**) and B, are known to be produced by marine *Streptomyces* (see Chapter 2, Section II.A.1) [7]. Only one 9-membered ring macrolide, juglorubin (**2**), has been reported to be produced by actinomycetes [8].

All 10-membered ring macrolides have skeletons similar to nargenicin A_1 (**3**) [9]. The skeleton of luminamicin (= coloradocin, **4**) has an additional two fused rings, and the position of the epoxide bridge on decalin differs from that of **3** [10, 11]. Compound **4** showed antibacterial activity against anaerobic bacteria, especially against *Clostridium* sp.

TABLE III
Macrolides Produced by Actinomycetes

Ring size	Structural characteristic	Number of reported compounds	Example[a]
Monolactone			
8		2	Octalactin A (**1**)
9		1	Juglorubin (**2**)
10	Nargenicin group	10	Nargenicin A_1 (**3**)
12		9	Methymycin (**5**)
12	Spinosyn group	23	Spinosyn A (**9**)
14		87	Erythromycin A (**11**)
15		1	SCH 23831 (**23**)
16		211	Leucomycin A_1 (**17**)
16	Spiroacetal	119	Avermectin B1a (**18**)
17		5	Lankacidin A (**25**)
18		30	Concanamycin A (**27**)
20		8	Venturicidin A (**29**)
22	Spiroacetal	7	Cytovaricin (**31**)
24		2	Maduralide (**30**)
24	Spiroacetal	11	Ossamycin (**32**)
26	Spiroacetal	22	Oligomycin A (**33**)
30		2	Izumenolide (**35**)
32		14	Dotriacolide (**36**)
34		18	Sporaviridin A_1 (**41**)
36		36	Guanidylfungin A (**38**)
40		6	Malolactomycin A (**39**)
41		1	RP 63834 (**40**)
42		10	Desertomycin A (**42**)
44		2	Oasomycin C (**44**)
48		2	Monazomycin A (**43**)
20	Triene	1	Apoptolidin (**45**)
22	Triene	1	Labilomycin (**46**)
60	Triene	3	Quinolidomicin A_1 (**47**)
26	Tetraene	9	Lucensomycin (**48**)
28	Tetraene	2	Rimocidin (**49**)
38	Tetraene	4	Nystatin A_1 (**50**)
26	Pentaene	1	Aurenin (**51**)
28	Pentaene	16	Filipin III (**52**)
30	Pentaene	3	Roxaticin (**53**)
32	Pentaene	5	Flavofungin I (**54**)
36	Pentaene	3	Roflamycoin (**55**)
44	Pentaene	2	Lienomycin (**56**)
36	Hexaene	2	Dermostatin A (**57**)
38	Heptaene	24	Amphotericin B (**58**)

(*continued*)

TABLE III (continued)

Ring size	Structural characteristic	Number of reported compounds	Example[a]
Macropolylide			
9	Macrodiolide	33	Antimycin A_{1a} (**59**)
16	Macrodiolide	9	Elaiophylin (**60**)
18	Macrodiolide	18	Pamamycin-607 (**61**)
20	Macrodiolide	1	Actinoplanic acid A (**62**)
32	Macrodiolide	9	Boromycin (**63**)
15	Macrotetrolide	1	Neoantimycin (**64**)
36	Macrotetrolide	9	Nonactin (**65**)
Macrolide lactam			
23		18	Tacrolimus (**67**)
24		1	Viridenomycin (**66**)
31	Triene	12	Rapamycin (**68**)

[a] Structures of these compounds are depicted in Figs. 2–13.

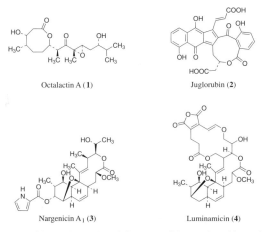

Fig. 2. Structures of 8- to 10-membered ring macrolides produced by actinomycetes.

B. Twelve-Membered Ring Macrolides

Methymycin (Fig. 3, **5**) [12] represents glycosidic 12-membered ring macrolides. The skeletons of PA-46101 A (**6**) and B are similar to those of the chlorothricin group (see Section II.C), but the former have 12-membered rings [13]. A few nonglycosidic compounds have been also reported. Lactimidomycin (**7**) and FD-895 (**8**) have glutarimide and epoxyolefin side chains, respectively, and showed cytotoxic activity [14, 15]. One large group of macrolides has a long side chain at the carbon that bears a ring lactone oxygen. Such compounds have various

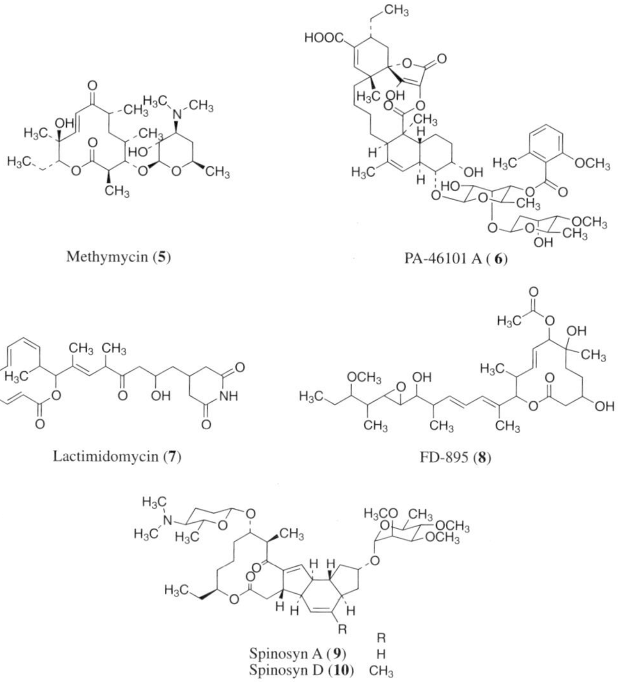

Fig. 3. Structures of 12-membered ring macrolides produced by actinomycetes.

biological activity and are antifungal or cytotoxic. Compounds **7** and **8** are considered to belong to this group. Some macrolides in this group have a functional group at the end of the side chain; for example, amines for desertomycins and monazomycins, guanidines for guanidylfungins and malolactomycins, and sugars for bafilomycins and concanamycins.

Recently, spinosyns A–Y have been isolated from a new species *Saccharopolyspora spinosa* by mosquito larvicide assay [16]. Spinosyns possess the core structure of a tricyclic *as*-indacene fused to a 12-membered ring lactone. In addition, an aminosugar (forosamine) and a neutral sugar (2,3,4-tri-*O*-methylrhamnose) are glycosidically linked to the tetracyclic framework. Spinosad (Tracer), a naturally occurring mixture of spinosyns A (**9**) and D (**10**), is used commercially as an insecticide [17]. Spinosad has been found to provide effective control of pests in the insect orders Lepidoptera, Diptera, and Thysanoptera, and some species of Coleoptera and Orthoptera. It demonstrates both rapid contact and ingestion activity in insects. Spinosad is relatively low in toxicity to mammals and birds and is only slightly to moderately toxic to aquatic organisms. The mode of action of spinosad is characterized by excitation of the insect nervous system, but no cross resistance to spinosad has been demonstrated by known insecticides.

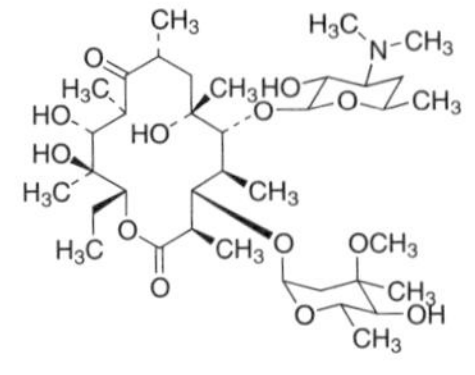

Erythromycin A (**11**)

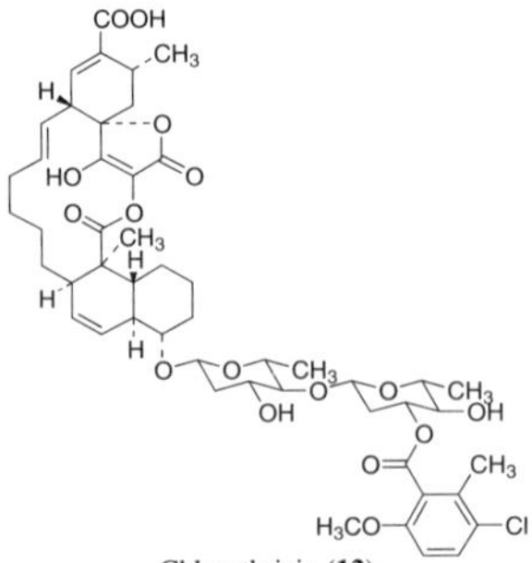

Chlorothricin (**12**)

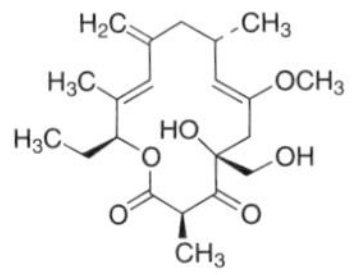

Rustmicin (**13**)

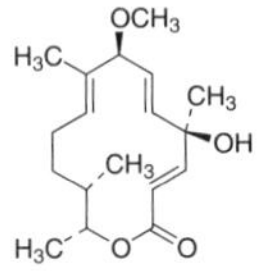

Albocycline (**14**)

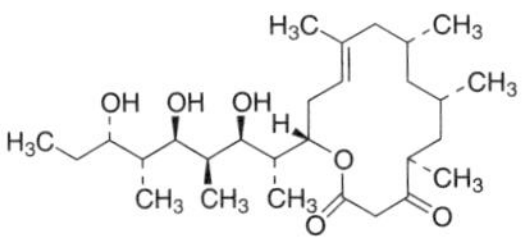

Sekothrixide (**15**)

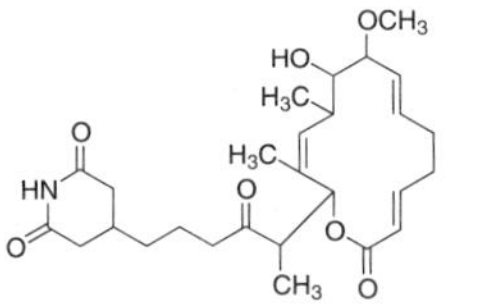

Migrastatin (**16**)

Fig. 4. Structures of 14-membered ring macrolides produced by actinomycetes.

C. Fourteen-Membered Ring Macrolides

Erythromycin A (Fig. 4, **11**) [18] and some other glycosidic 14-membered ring macrolides are widely used as antibacterial antibiotics. They also have various different activities (see Chapters 11 and 12). Chlorothricin (**12**) [19] and its analogues have fused pentacyclic skeletons. As for nonglycosidic compounds, rustmicin (= galbonolide A, **13**) group compounds showed antifungal activity [20]. Compound **13** inhibited inositol phosphoceramide synthase, resulting in the accumulation of ceramide and the loss of all of the complex sphingolipids [21]. Albocycline (**14**) is another nonglycosidic group compound [22]. Sekothrixide (**15**) and migrastatin (**16**) belong to the group having long side chains [23, 24]. The latter compound (**16**) inhibited spontaneous migration of human cancer cells.

D. Sixteen-Membered Ring Macrolides

Leucomycins (leucomycin A_1, Fig. 5, **17**) [25] and related glycosidic 16-membered ring macrolides are used clinically. Spiroacetal-fused 16-membered ring macrolides, such as avermectins (avermectin B1a, **18**) [26] and milbemycins [27],

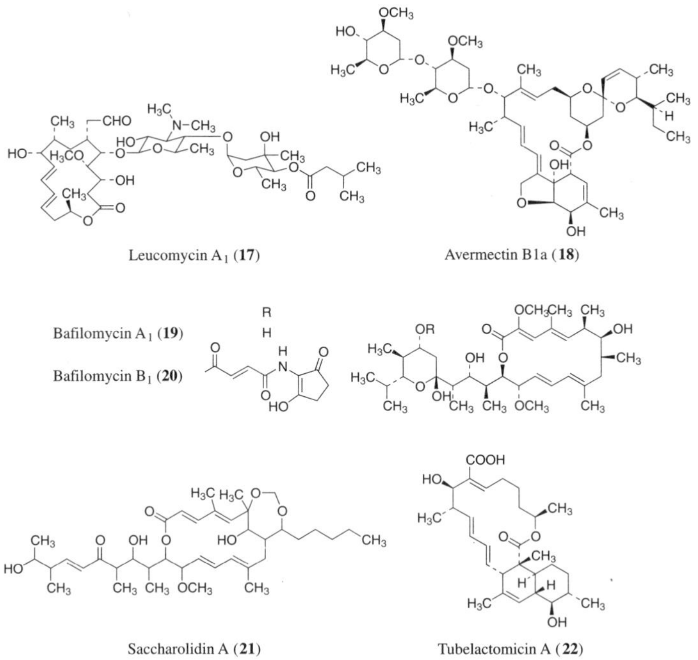

Fig. 5. Structures of 16-membered ring macrolides produced by actinomycetes.

are widely used as anthelmintics and insecticides (see Chapters 8 and 13). Bafilomycins (bafilomycin A_1, **19**) belong to the group having long side chains, and they are used as vacuolar-type H^+-ATPase inhibitors [28, 29]. Bafilomycin B_1 (**20**) was originally isolated as an antimicrobial and nematocidal antibiotic named setamycin [30]. An antiparasitic macrolide, saccharolidin A (**21**), also has a long side chain, but its skeleton is fused by 1,3-dioxepane [31]. Tubelactomicin A (**22**) showed strong activity against acid-fast bacteria. It is a nonglycosidic macrolide with fused decalin, and its skeleton is similar to that of **12** [32].

E. Fifteen-, Seventeen-, and Eighteen-Membered Ring Macrolides

Only one 15-membered ring macrolide, SCH 23831 (Fig. 6, **23**), has been reported to be a product of actinomycetes [33]. The structure is closely related to rosamicin (**24**), a 16-membered ring macrolide [34], but the skeleton of **23** formed a cyclopentanopyridine ring and thus it became a 15-membered ring. All 17-membered ring macrolides are similar to lankacidin A (= bundlin B, **25**) [35]. Glycosidic 18-membered ring macrolides are lipiarmycin A_3 (= clostomicin B_1, **26**)

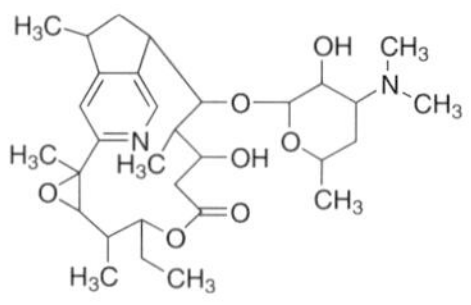

SCH 23831 (**23**)

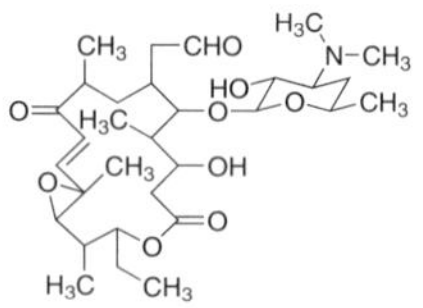

Rosamicin (**24**)

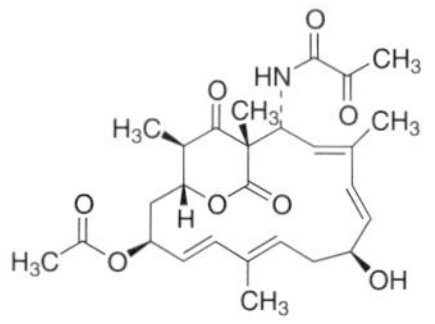

Lankacidin A (**25**)

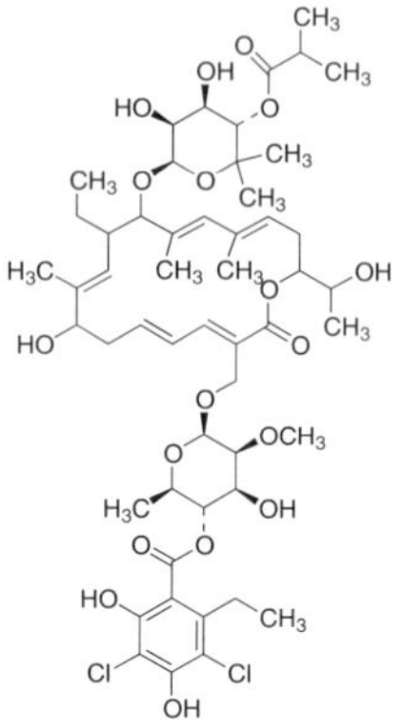

Lipiarmycin A_3 (**26**)

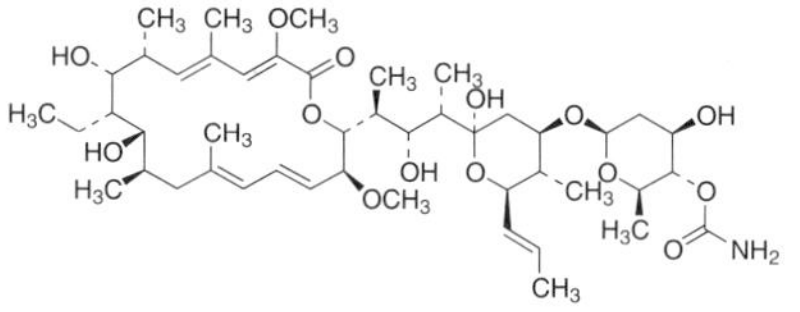

Concanamycin A (**27**)

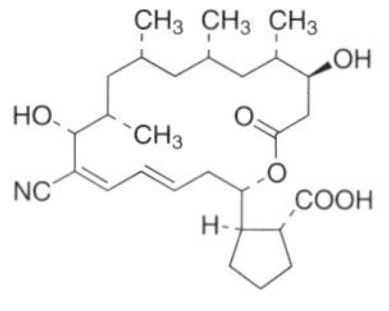

Borrelidin (**28**)

Fig. 6. Structures of 15- to 18-membered ring macrolides produced by actinomycetes.

and related antibiotics [36]. Long side chains of concanamycins (concanamycin A, **27**) resemble those of bafilomycins (**19** and **20**), and concanamycins are also used as vacuolar-type H^+-ATPase inhibitors [37, 38]. Bindseil and Zeeck proposed the term *plecomacrolides* for macrolides bearing such side chains [39]. Elaiophylin-group macrodiolides are also included in plecomacrolides (see Section II.H). Borrelidin (**28**) is a nonglycosidic 18-membered ring macrolide with a side chain of cyclopentanecarboxylic acid [40].

F. Twenty- to Forty-Eight-Membered Ring Macrolides

Macrolides possessing more than 20-membered rings usually have long side chains or spiroacetals at the adjacent position of ring lactone oxygen. An antifungal macrolide venturicidin A (Fig. 7, **29**) has a 20-membered ring lactone [41],

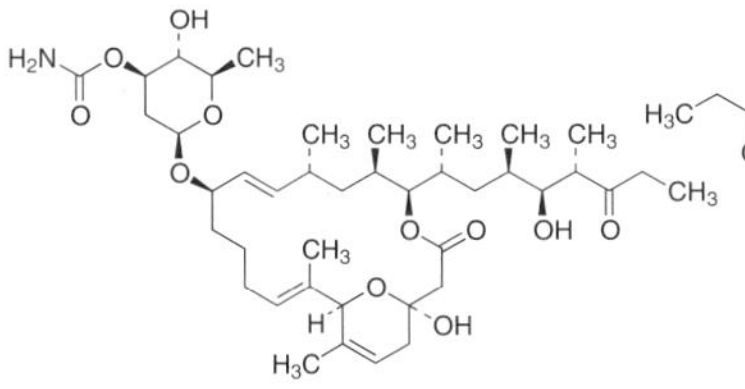

Venturicidin A (**29**)

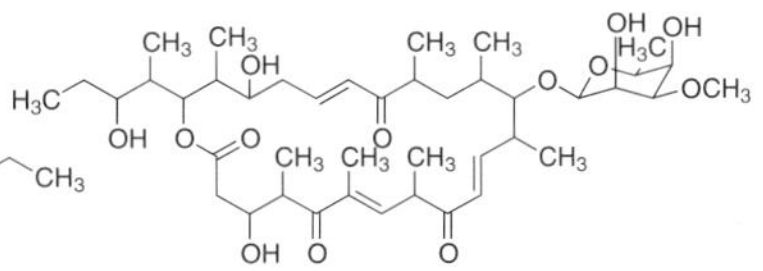

Maduralide (**30**)

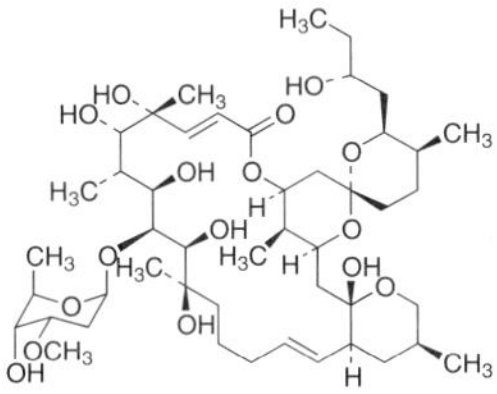

Cytovaricin (**31**)

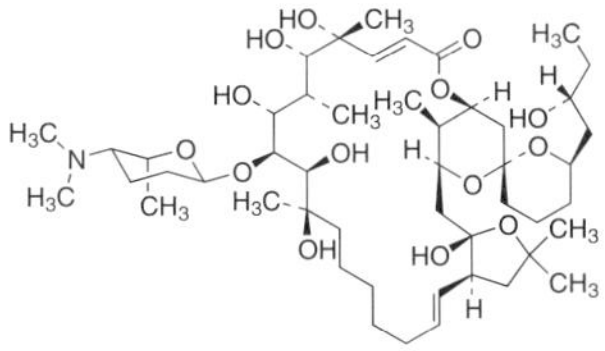

Ossamycin (**32**)

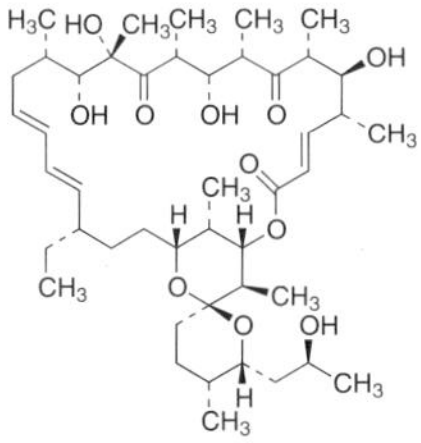

Oligomycin A (**33**)

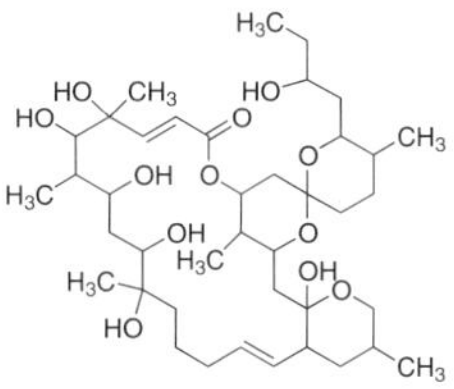

Phthoramycin (**34**)

Fig. 7. Structures of 20- to 26-membered ring macrolides produced by actinomycetes.

and maduralide (**30**) is a 24-membered ring macrolide produced by marine actinomycetes (see Chapter 2, Section II.A.1) [42]. Spiroacetal macrolides of 22-, 24-, and 26-membered rings such as cytovaricin (**31**), ossamycin (**32**), and oligomycins, respectively, have been isolated [43–45]. Oligomycin A (**33**) is widely used as a H^+-transporting ATP synthase (F_OF_1) inhibitor [46]. H^+-transporting ATP synthase is also inhibited by **29**, **31**, and **32** [46–48]. Phthoramycin (**34**), another 22-membered ring spiroacetal macrolide, inhibited cellulose biosynthesis [49].

Macrolides of 30-membered ring izumenolide (Fig. 8, **35**) and 32-membered ring dotriacolide (**36**) have a few sulfonates, and they inhibited β-lactamase [50, 51]. Macrolides of 32-membered ring copiamycin (= niphithricin A, **37**), 36-membered ring guanidylfungins (guanidylfungin A, **38**), 40-membered ring malolactomycins (malolactomycin A, **39**), and 41-membered ring RP 63834 (**40**) have a guanidino residue at the end of each side chain, and they are antifungal antibiotics [52–55]. Sporaviridins (sporaviridin A_1, Fig. 9, **41**) are 34-membered

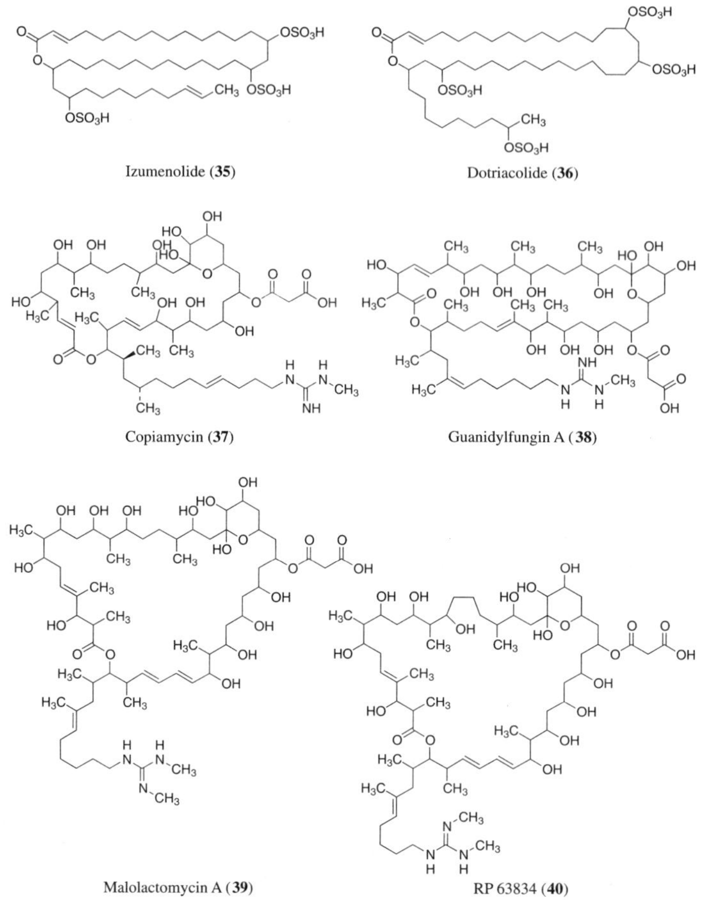

Fig. 8. Structures of 30-, 32-, 36-, 40-, and 41-membered ring macrolides produced by actinomycetes.

ring glycoside antibiotics, and they have seven sugars including amino sugars [56]. Macrolide antibiotics of 42-membered ring desertomycin A (**42**) and 48-membered monazomycin A (**43**) have an amine residue at the end of each side chain [57–59]. The structure of **42** was elucidated by one of the earliest NMR studies of ^{1}H-detected heteronuclear multiple-bond ^{1}H-^{13}C correlation (HMBC) and homonuclear Hartmann–Hahn (HOHAHA) experiments, which are generally used for the structure elucidation of natural products [58]. Oasomycins (oasomycin C, **44**) have 42- or 44-membered ring lactones, and the end of each side chain is a carboxylic acid or a γ-lactone [60]. Zerlin and Thiericke proposed the term *marginolactone* for a macrolide of ring size larger than a 32-membered

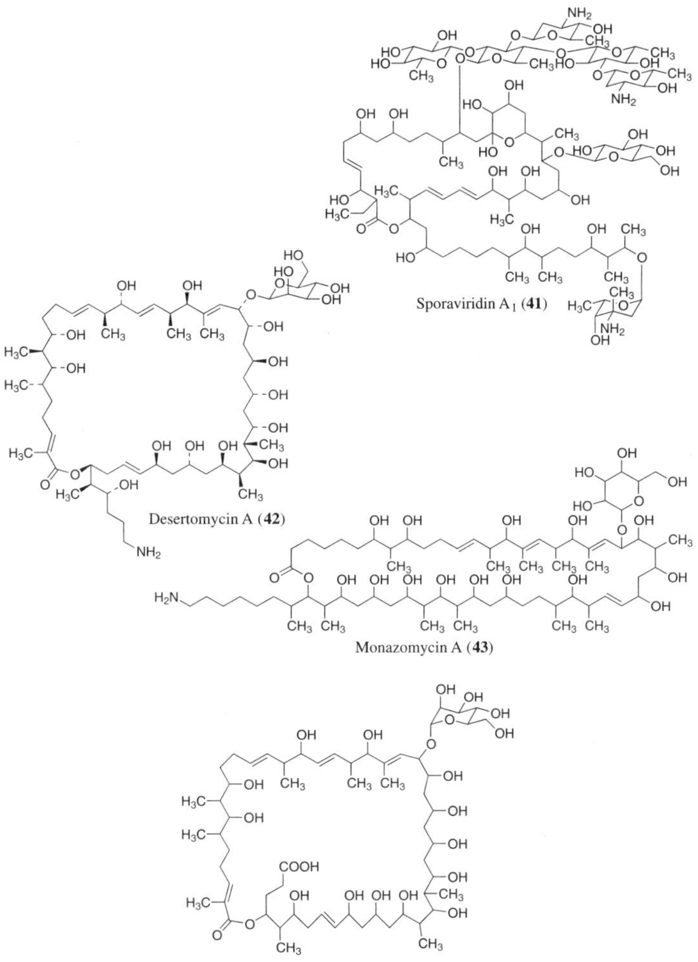

Sporaviridin A_1 (**41**)

Desertomycin A (**42**)

Monazomycin A (**43**)

Oasomycin C (**44**)

Fig. 9. Structures of 34-, 42-, 44-, and 48-membered ring macrolides produced by actinomycetes.

ring bearing a side chain with a terminal amino or guanidino (or carboxylic acid) functionality [61].

G. Polyene Macrolides

Most polyene macrolides show antifungal activity, and they are produced by actinomycetes. Polyene macrolides are precisely described in Chapters 9 and 12 of the first edition of this book [1]. Though a few octaene macrolides have been reported, their structures have not been elucidated.

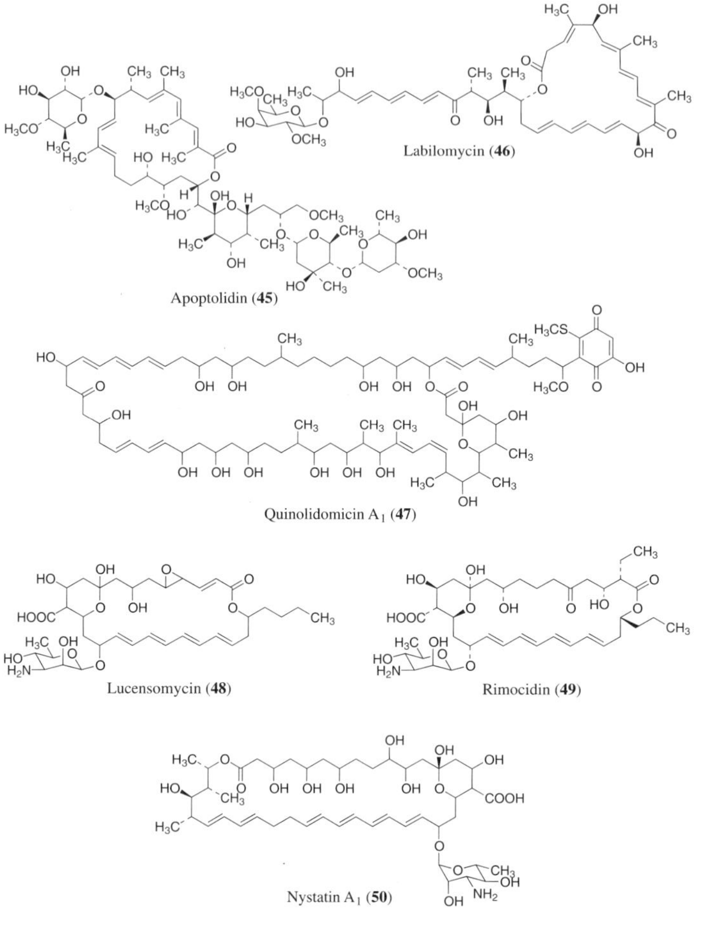

Fig. 10. Structures of triene and tetraene macrolides produced by actinomycetes.

Apoptolidin (Fig. 10, **45**) is a 20-membered ring triene macrolide, and it induced apoptosis in transformed cells [62]. It is an F_0F_1 inhibitor like **33**. Labilomycin (= pulvomycin, **46**) is a 22-membered ring triene macrolide [63]. Other triene macrolides, quinolidomicins (quinolidomicin A_1, **47**), have the largest ring sizes (a 60-membered ring) among macrolides produced by actinomycetes [64]. Compound **47** inhibited the growth of various tumor cells including multidrug-resistant cells. Tetraene macrolides with three different ring sizes have been reported as 26-membered ring lucensomycin (**48**), 28-membered ring rimocidin (**49**), and 38-membered ring nystatin A_1 (**50**) [65–67]. Pentaene macrolides have

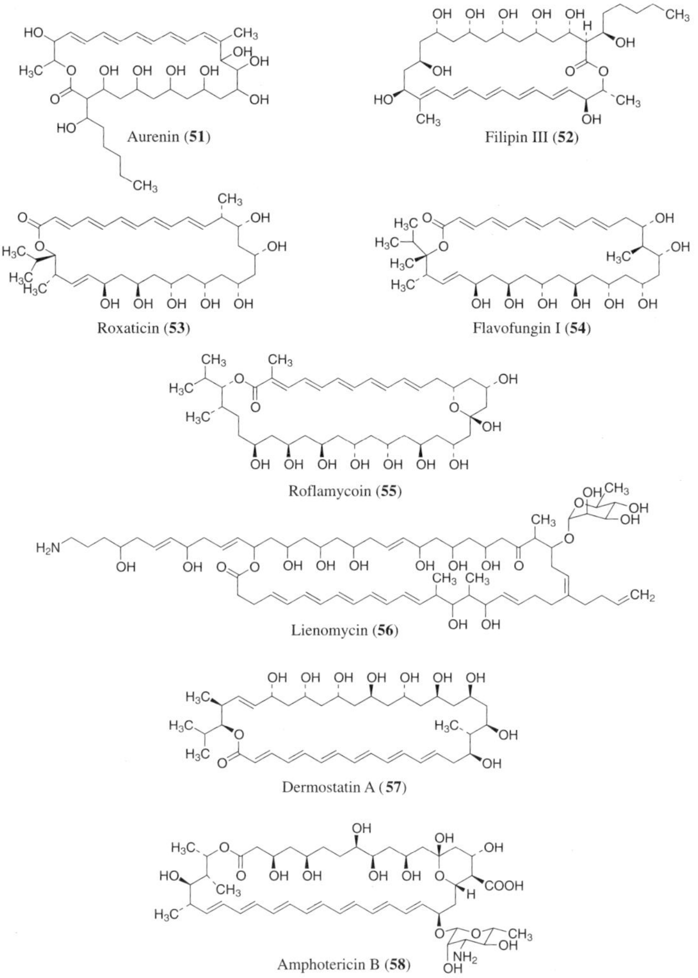

1. Discovery of New Macrolides

Fig. 11. Structures of pentaene, hexaene, and heptaene macrolides produced by actinomycetes.

a variety of ring sizes: 26-membered ring aurenin (Fig. 11, **51**), 28-membered ring filipin III (**52**), 30-membered ring roxaticin (**53**), 32-membered ring flavofungin I (**54**), 36-membered ring roflamycoin (= flavomycoin, **55**), and 44-membered ring lienomycin (**56**) [68–73]. Nothing but dermostatins A (**57**) and B were 36-membered ring hexaene macrolides whose structures have been already elucidated [74]. Amphotericin B (**58**) is a widely used antifungal heptaene macrolide having a 38-membered ring [75].

H. Macrodiolides and Macrotetrolides

Some actinomycetes produce macrolides with two and four ester linkages, which are called macrodiolides (dilactone macrolides) and macrotetrolides (tetralactone macrolides), respectively. A 9-membered ring macrodiolide antimycin A_{1a} (Fig. 12, **59**) is well known as an inhibitor of ubiquinol-cytochrome c reductase (complex III) [76]. Elaiophylin (**60**) is a 16-membered ring macrodiolide having antifungal and cytotoxic activity [77]. Although **60** is included in plecomacrolides (see Section II.E), it does not inhibit vacuolar-type H^+-ATPase, and it is a plasma-proton ATPase inhibitor [38, 39]. An 18-membered ring macrodiolide pamamycin-607 (**61**) is an aerial mycelium inducer [78]. A 20-membered ring

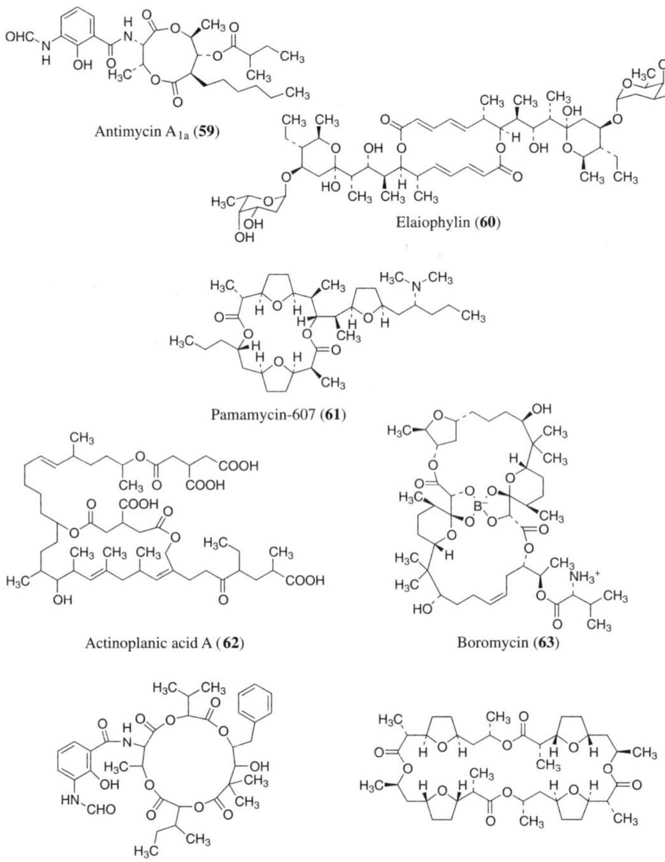

Fig. 12. Structures of macropolylides produced by actinomycetes.

macrodiolide actinoplanic acid A (**62**) was isolated as an inhibitor of Ras farnesyl-protein transferase [79]. Boromycin (**63**) is a 32-membered ring macrodiolide antibiotic, and it is the first natural product in which boron has been found [80]. Neoantimycin (**64**) belongs to the group of antimycin, but it is a 15-membered ring macrotetrolide [81]. Nonactin (**65**) is typical of 36-membered ring macrotetrolides [82].

I. Macrolide Lactams

Some macrocyclic lactones each contain one amide linkage in their skeletons, and they are called macrolide lactams (or macrolactam lactones). Twenty-four-membered ring viridenomycin (Fig. 13, **66**) is a tetraene macrolide lactam and showed antibacterial, antifungal, and antitumor activities [83, 84]. Tacrolimus (= FK-506, **67**) and rapamycin (**68**) are a 24-membered ring macrolide lactam and a 31-membered ring triene macrolide lactam, respectively [85, 86]. They are used as immunosuppressants (see Chapters 9 and 14). Recently, sanglifehrins (sanglifehrin A, **69**) were isolated by screening of cyclophilin-binding assay [87]. They showed immunosuppressive activity in a mixed-lymphocyte reaction. Although their structures are somewhat similar to **67** and **68**, sanglifehrins are 24-membered ring macrocyclic lactones with three amide linkages.

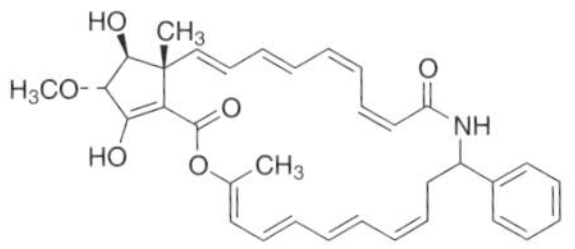

Viridenomycin (**66**)

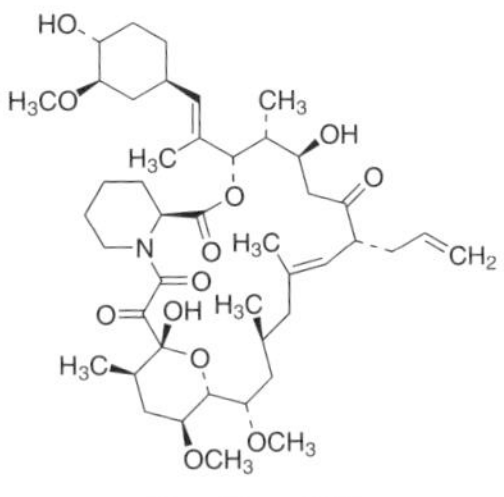

Tacrolimus (**67**)

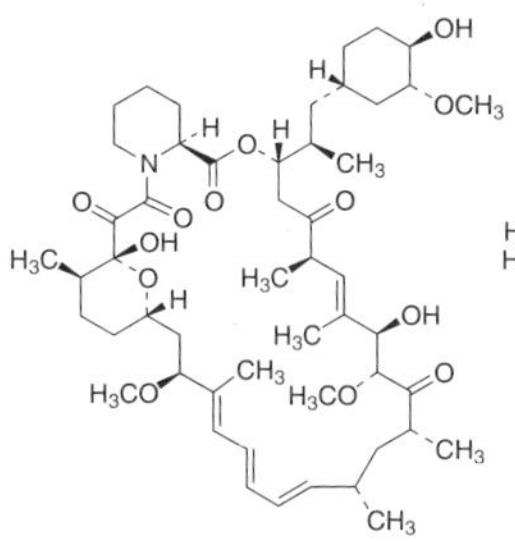

Rapamycin (**68**)

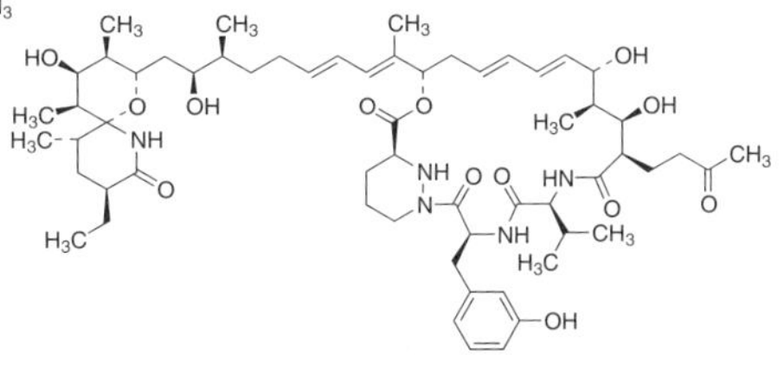

Sanglifehrin A (**69**)

Fig. 13. Structures of macrolide lactams and a related compound produced by actinomycetes.

TABLE IV
Macrolides Produced by Myxobacteria

Ring size	Structural characteristic	Number of reported compounds	Example[a]
Monolactone			
12		4	Mycolactone A (**71**)
14		2	Ripostatin A (**73**)
16		6	Epothilone A (**75**)
18		27	Soraphen $A_{1\alpha}$ (**76**)
22		1	Etnangiene (**77**)
24	Triene	1	Archazolide A (**83**)
35	Triene	14	Sorangicin A (**84**)
Macrodiolide			
42		4	Tartrolon B (**86**)
Macrolide lactam			
19		3	Angiolam A (**88**)
26		3	Myxovirescin E (**90**)
28		29	Myxovirescin A_1 (**89**)
30		2	Myxovirescin P_1 (**91**)
Oxazole containing			
19		1	Rhizopodin (**92**)
31	Tetraene	7	Chivosazol A (**93**)
30	Triene macrodiolide	21	Disorazol A_1 (**94**)

[a]Structures of these compounds are depicted in Figs. 14–17.

III. Macrolides from Bacteria Including Myxobacteria

Bacteria other than actinomycetes also produce macrolides. About 100 macrolides have been produced by myxobacteria, and about 40 by other bacteria (Tables IV and V). Myxobacteria produce macrolides with up to 42-membered rings including polyene macrolides, macrodiolides, macrolide lactams, and oxazole-containing macrolides. The other bacteria produce macrolides with up to 26-membered rings including polyene macrolides, macrodiolides, and macrotriolides.

A. Eight- to Thirty-Five-Membered Ring Macrolides

Aurantinins A (Fig. 14, **70**) and B have been isolated from *Bacillus aurantinus*. They are 8-membered ring macrolides and are active against gram-positive bacteria [88].

Mycolactones A (**71**) and B are 12-membered ring macrolides produced by a myxobacterium, *Mycobacterium ulcerans* [89, 90]. *M. ulcerans* causes Buruli ulcer, a severe human skin disease that occurs primarily in Africa and Australia. Mycolactones are suggested to play a major role in the pathology of the disease

1. Discovery of New Macrolides   23

TABLE V
Macrolides Produced by Bacteria Other Than Actinomycetes and Myxobacteria

Ring size	Structural characteristic	Number of reported compounds	Example[a]
Monolactone			
8		2	Aurantinin A (**70**)
12		2	Oximidine I (**72**)
14		3	Aerocavin (**74**)
20		1	Macrolactin L (**80**)
22		3	Macrolactin H (**79**)
24		13	Macrolactin A (**78**)
26	Spiroacetal	2	Oligomycin SC-2 (**81**)
22	Triene	6	Difficidin (**82**)
Macropolylide			
15	Macrodiolide	4	Lobatamide A (**85**)
12	Macrotriolide	1	Enterobactin (**87**)

[a]Structures of these compounds are depicted in Figs. 14–16.

by causing necrosis and immunosuppression. The isolation of mycolactones is the first report of macrolides deriving from the genus *Mycobacterium* as well as the first identification of a macrolide from any bacterial pathogen.

Oximidines I (**72**) and II are produced by *Pseudomonas* sp., and they are 12-membered macrolides containing *O*-methyloxime moieties [91]. They inhibited the growth of oncogene-transformed cell lines and showed the inhibition against mammalian vacuolar-type H^+-ATPase with unprecedented selectivity [92].

Ripostatins A (**73**) and B are 14-membered ring macrolides produced by myxobacterium *Sorangium cellulosum* [93]. They acted especially on *Staphylococcus aureus* and inhibited eubacterial RNA polymerase. Compound **73** exists as an equilibrium mixture of ketone form and hemiacetal form in methanol. An unpigmented strain of *Chromobacterium violaceum* produces another 14-membered ring macrolide, aerocavin (**74**), which is active against gram-positive and gram-negative bacteria [94].

Epothilones (epothilone A, **75**) are 16-membered ring macrolides produced by *Sorangium cellulosum* [95]. Epothilones represent a novel structural class of compounds, the first to be described since the original discovery of taxol, which not only mimic the biological effects of taxol but also appear to bind to the same microtubule-binding site as taxol [96]. In contrast to taxol, epothilones retained a much greater toxicity against P-glycoprotein-expressing multidrug-resistant cells.

Soraphen $A_{1\alpha}$ (= soraphen A, **76**) and etnangiene (**77**) are 18- and 22-membered ring macrolides, respectively, also produced by *S. cellulosum* [97–99]. Compound **76** is a potent inhibitor of fungal growth, and its primary target was suggested to be acetyl-CoA carboxylase [98]. Compound **77** is a nucleic acid polymerase inhibitor.

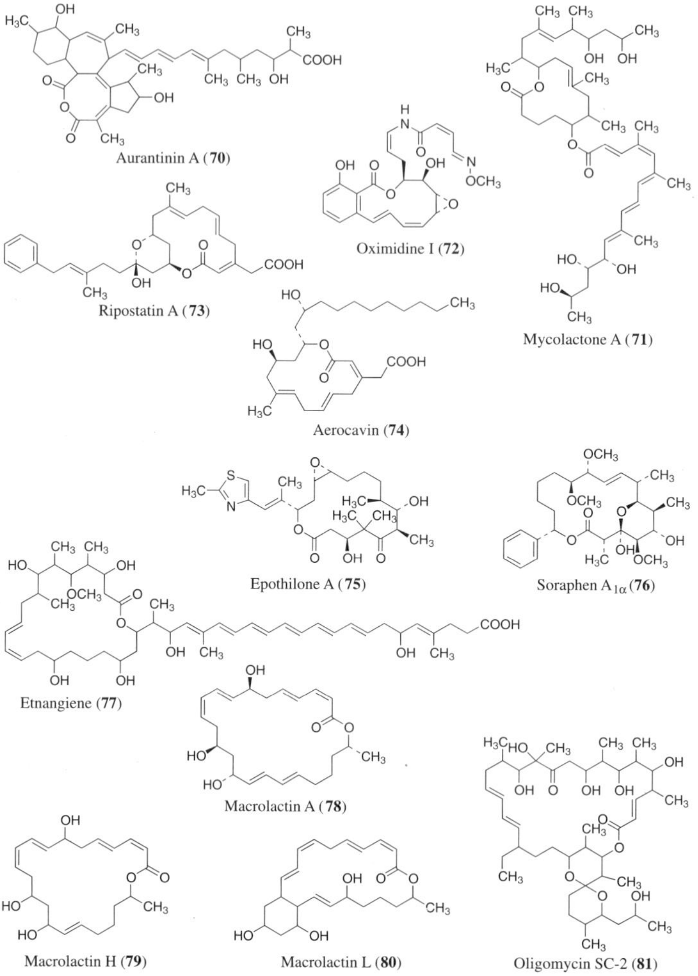

Fig. 14. Structures of 8- to 26-membered ring macrolides produced by bacteria.

Macrolactins A (**78**)–F are 24-membered ring macrolides isolated from a taxonomically undefinable deep-sea bacterium [100] (see Chapter 2, Section II.A.1). Macrolactins G, H (**79**), and I–L (**80**) are isolated from *Bacillus* sp. [101]. Compounds **79** and **80** are 22- and 20-membered ring macrolides, respectively, while the others are 24-membered ring macrolides.

Spiroacetal-fused 26-membered ring macrolides, oligomycins, are generally produced by actinomycetes (see Section II.F), but oligomycins SC-1 and 2 (**81**) are produced by a bacterium *Pantoea agglomerans* [102]. BE-56384, isolated from *Streptomyces* sp., has a structure identical to that of **81** [103].

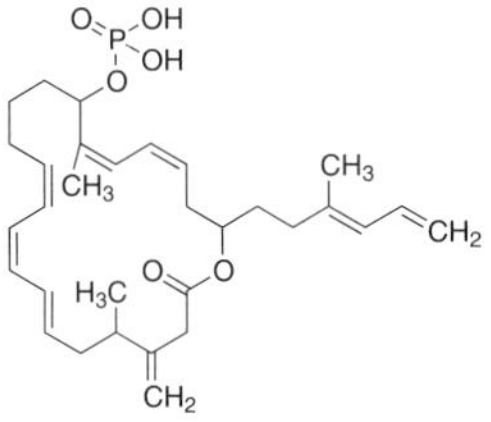

Difficidin (**82**)

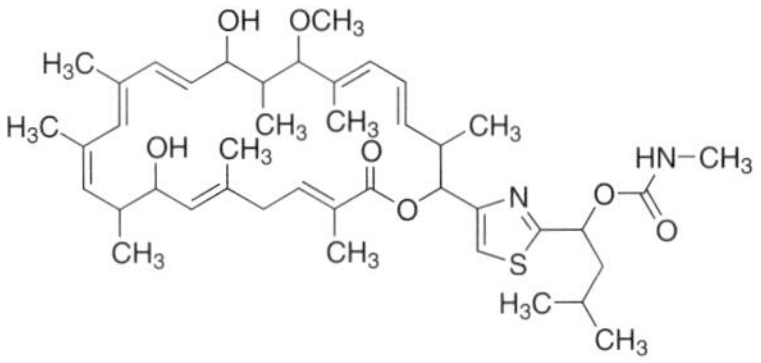

Archazolide A (**83**)

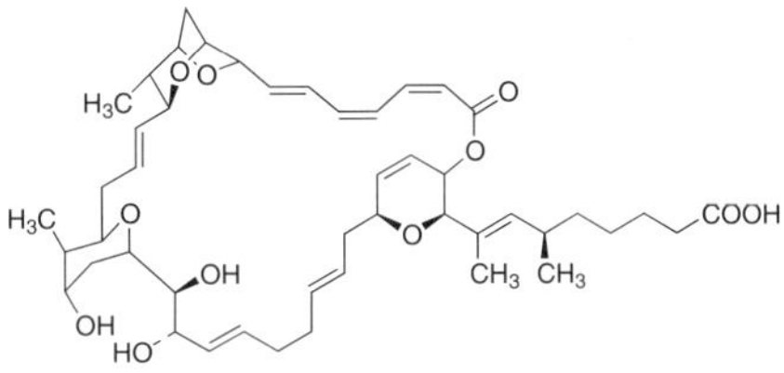

Sorangicin A (**84**)

Fig. 15. Structures of triene macrolides produced by bacteria.

B. Triene Macrolides

Difficidin (Fig. 15, **82**) is a 22-membered ring triene macrolide with a phosphate ester produced by *Bacillus subtilis* [104]. It showed broad activity against aerobic and anaerobic bacteria. Archazolide A (**83**) is a 24-membered ring triene macrolide isolated from a myxobacterium, *Archangium gephyra* [105].

Sorangicins (sorangicin A, **84**) are 35-membered ring triene macrolides produced by *S. cellulosum* [106]. They acted mainly against gram-positive bacteria, including mycobacteria, and inhibited eubacterial RNA polymerase [107].

C. Macrodiolides and Macrotriolides

A 15-membered ring macrodiolide, lobatamide A (Fig. 16, **85**), was isolated from tunicates (see Chapter 2, Section II.D.4). The structurally identical compound YM-75518A has been reported to be produced by *Pseudomonas* sp. [108]. It showed the inhibition against mammalian vacuolar-type H^+-ATPase in a manner similar to that of oximidines (oximidine I, **72**) [82].

Tartrolons (tartrolon B, **86**) are 42-membered ring macrodiolides produced by *S. cellulosum* [109]. Compound **86** contains a boron atom, and the substructure and configuration of the boron-binding site are identical to those of boromycin (**63**).

Enterobactin (**87**) is a 12-membered ring macrotriolide initially isolated from *Salmonella typhimurium* [110]. It is involved in microbial transport and metabolism of iron. It is an iron-sequestering agent of the phenolate group and is overproduced by *Escherichia coli* and related enteric bacteria under low-iron stress.

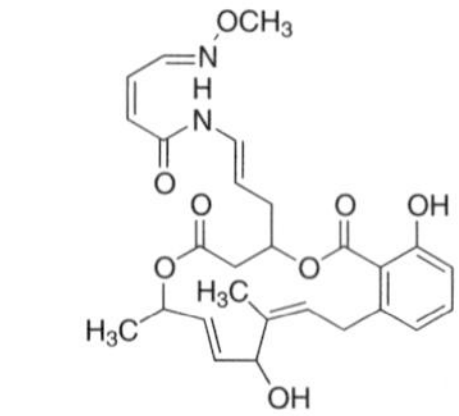

Lobatamide A (**85**)

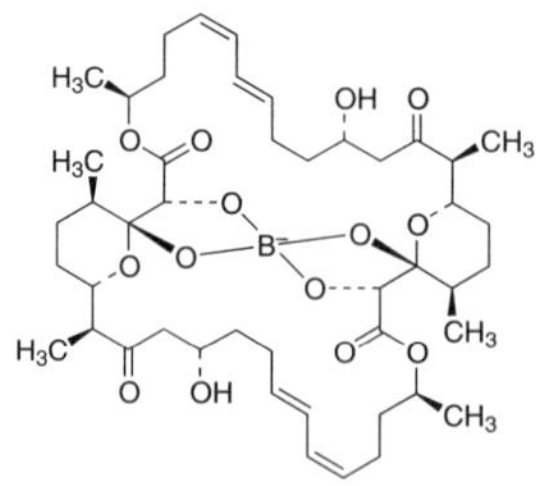

Tartrolon B (**86**)

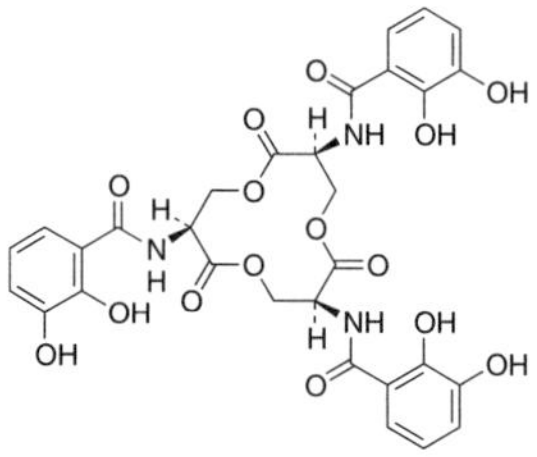

Enterobactin (**87**)

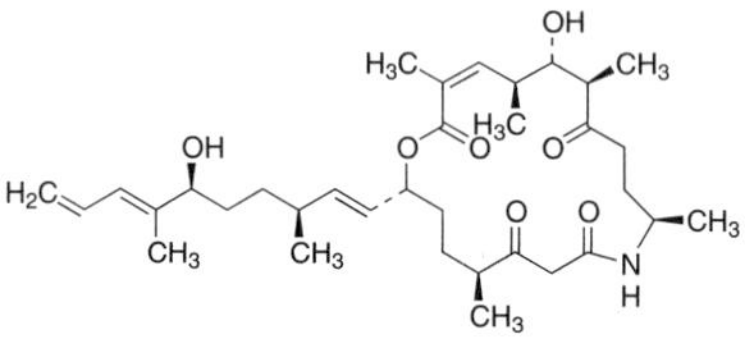

Angiolam A (**88**)

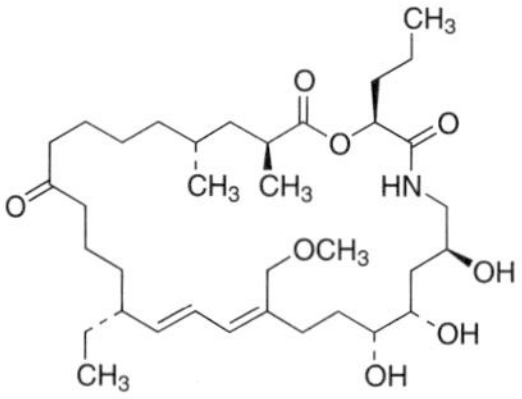

Myxovirescin A_1 (**89**)

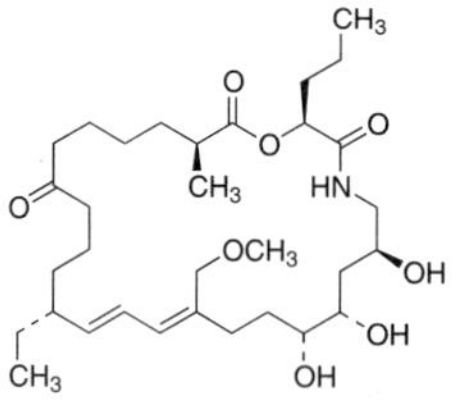

Myxovirescin E (**90**)

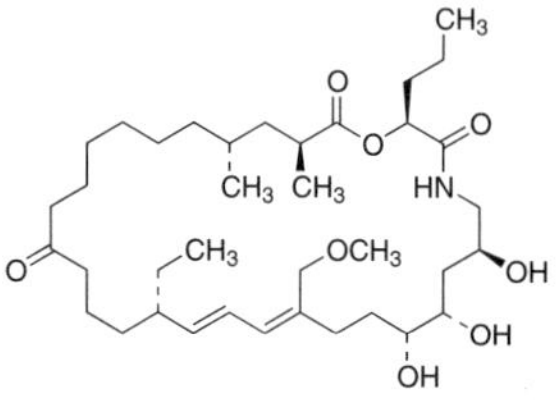

Myxovirescin P_1 (**91**)

Fig. 16. Structures of macropolylides and macrolide lactams produced by bacteria.

D. Macrolide Lactams and Oxazole-Containing Macrolides

Angiolam A (**88**) is a 19-membered ring macrolide lactam antibiotic produced by a myxobacterium, *Angiococcus disciformis* [111]. It inhibited protein synthesis.

Myxovirescins (myxovirescin A_1 = megovalicin C, **89**) are macrolide lactam antibiotics isolated from myxobacteria of *Myxococcus* sp. [112, 113]. Myxovirescin E (**90**) has a 26-membered ring, and myxovirescin P_1 (**91**) and P_2 have

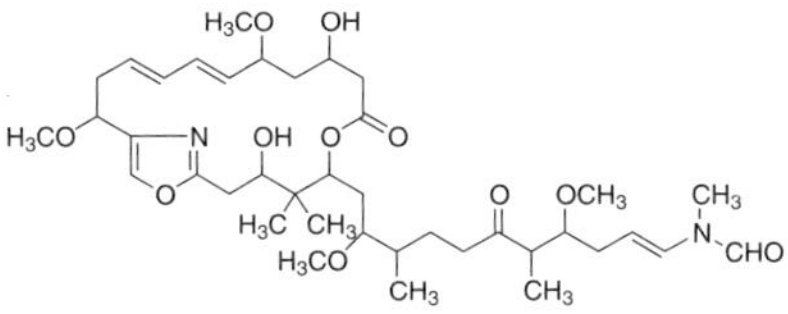

1. Discovery of New Macrolides 27

Rhizopodin (**92**)

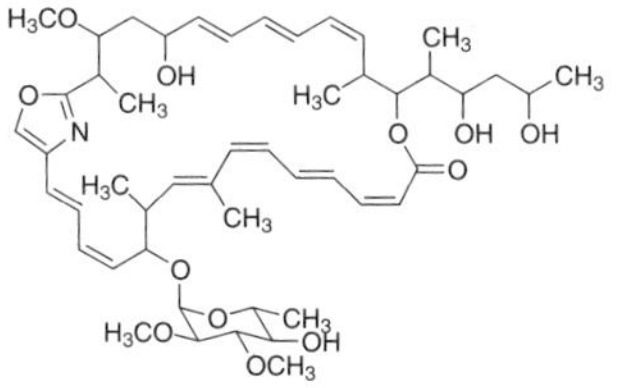

Chivosazol A (**93**)

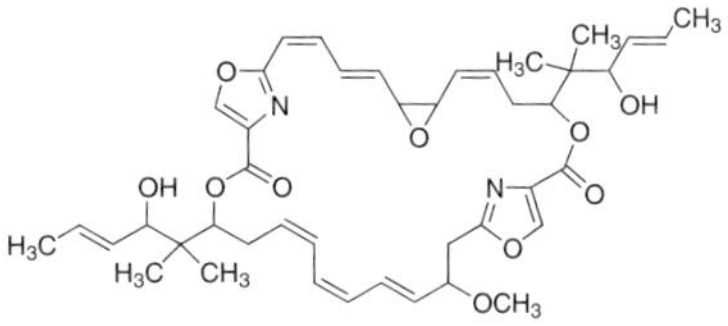

Disorazol A₁ (**94**)

Fig. 17. Structures of oxazole-containing macrolides produced by bacteria.

30-membered rings [114], while the other myxovirescins have 28-membered rings.

A cytotoxic compound, rhizopodin (Fig. 17, **92**), is a 19-membered ring macrolide whose skeleton contains an oxazole [115]. It is produced by *Myxococcus stipitatus*. It caused formation of rhizopodia-like structures in animal cells and caused reorganization of the actin cytoskeleton [116]. Chivosazols (chivosazol A, **93**) are oxazole-containing macrolides having 31-membered ring tetraene [117]. They are produced by *S. cellulosum*.

Disorazols (disorazol A₁, **94**) are oxazole-containing macrodiolides having a 30-membered ring [118]. They are produced by *S. cellulosum* and are highly cytotoxic and active against fungi.

IV. Macrolides from Fungi

About 200 macrolides have been isolated from fungi (Table VI). The ring sizes of macrolides produced by fungi are rather small. They produce monolactone macrolides and macrodiolides with up to 16-membered rings, macrotriolides with up to 18-membered rings, and macrotetrolides with up to 25-membered rings. Mycotoxins are produced by fungi, and a part of them such as curvularin and zearalenone belong to macrolides. Some cytochalasins and trichothecenes also have macrocyclic lactones. Some fungal macrolides have been already described in the Chapter 13 of the first edition of this book [1].

A. Eight- to Sixteen-Membered Ring Macrolides

Botcinolide (Fig. 18, **95**) is a 9-membered ring phytotoxic macrolide produced by *Botrytis cinerea* [119]. Dermocanarins (dermocanarin 1, **96**) are

TABLE VI
Macrolides Produced by Fungi

Ring size	Structural characteristic	Number of reported compounds	Example[a]
Monolactone			
8		1	Decarestrictine M (**98**)
9		16	Botcinolide (**95**)
10		48	Decarestrictine D (**97**)
12		37	Patulolide A (**100**)
13		7	Brefeldin A (**102**)
14		32	Zearalenone (**103**)
16		9	Rhizoxin (**107**)
12	Cytochalasin	4	Rosellichalasin (**108**)
14	Cytochalasin	7	Cytochalasin B (**109**)
15	Cytochalasin	1	Cytochalasin V (**110**)
Macropolylide			
13	Macrodiolide	2	Bartanol (**111**)
14	Macrodiolide	8	Clonostachydiol (**112**)
16	Macrodiolide	22	Roritoxin A (**113**)
16	Macrotriolide	10	Macrosphelide A (**115**)
18	Macrotriolide	8	Verrucarin A (**114**)
24	Macropentolide	6	NG-012 (**116**)
25	Macropentolide	1	BK223-B (**117**)

[a]Structures of these compounds are depicted in Figs. 18–20.

naphthylanthraquinones containing a 9-membered ring lactone bridge [120]. They are the principal pigments of *Dermocybe canaria*.

Decarestrictine D (**97**) is a 10-membered ring macrolide isolated from *Penicillium* spp. [121]. Compound **97** showed an inhibitory effect on cholesterol biosynthesis *in vitro* and *in vivo*. The other decarestrictines also have 10-membered ring lactones, but only decarestrictine M (**98**) has an 8-membered ring lactone [122]. Pinolidoxin (**99**) is a main phytotoxin isolated from a phytopathogenic fungi, *Ascochyta pinodes* [123]. It is also a 10-membered ring macrolide. Patulolides (patulolide A, **100**) are 12-membered ring antifungal macrolides isolated from *Penicillium urticae* [124]. CJ-12,950 (**101**) and its geometrical isomer CJ-13,357 are 12-membered ring macrolides produced by *Mortierella verticillata* [125]. They are potent inducers of the low-density lipoprotein (LDL) receptor gene and enhance LDL receptor expression in human hepatocyte cells.

Brefeldin A (**102**) is a 13-membered ring macrolide initially isolated from *Penicillium decumbens* [126]. It caused rapid and extensive disruption of Golgi morphology. The molecular target of **102** has not been defined, but its effect has been suggested to be due to the activation of ADP-ribosylation [127].

Zearalenone (**103**) is produced by *Fusarium* spp. having a benzene ring to which is fused a 14-membered ring macrolide [128]. It exhibits estrogen-like

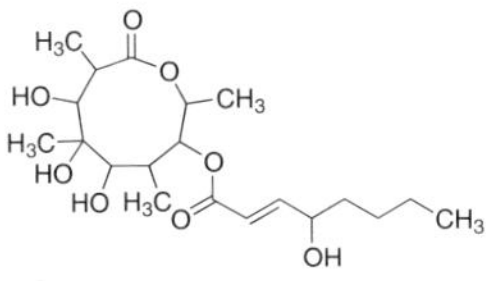

Botcinolide (**95**)

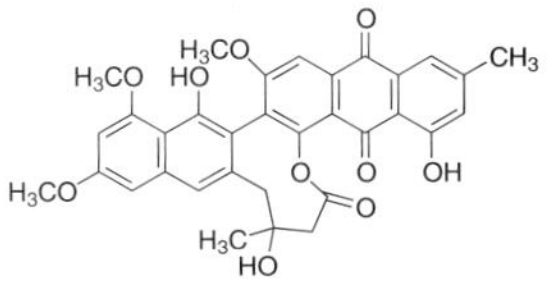

Dermocanarin 1 (**96**)

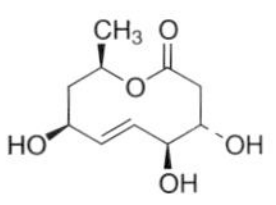

Decarestrictine D (**97**)

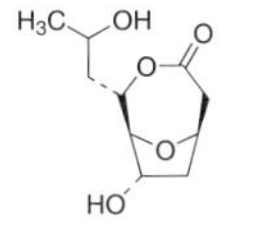

Decarestrictine M (**98**)

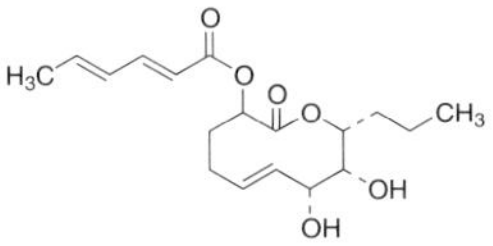

Pinolidoxin (**99**)

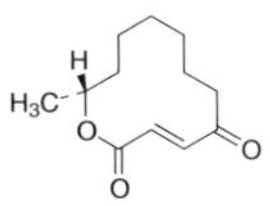

Patulolide A (**100**)

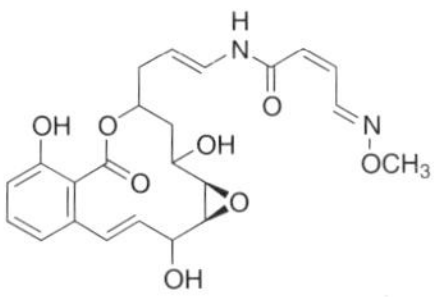

CJ-12,950 (**101**)

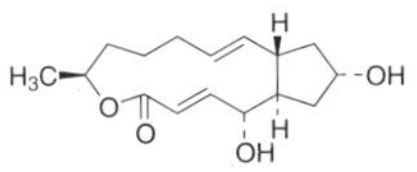

Brefeldin A (**102**)

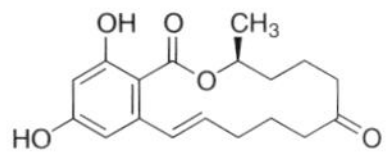

Zearalenone (**103**)

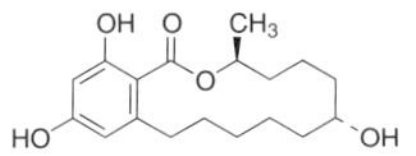

Zeranol (**104**)

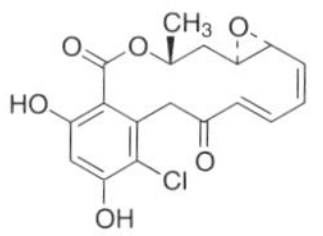

Radicicol (**105**)

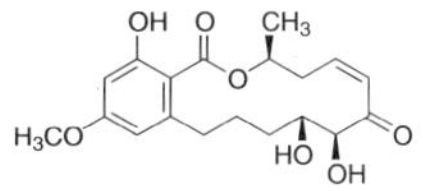

L-783,277 (**106**)

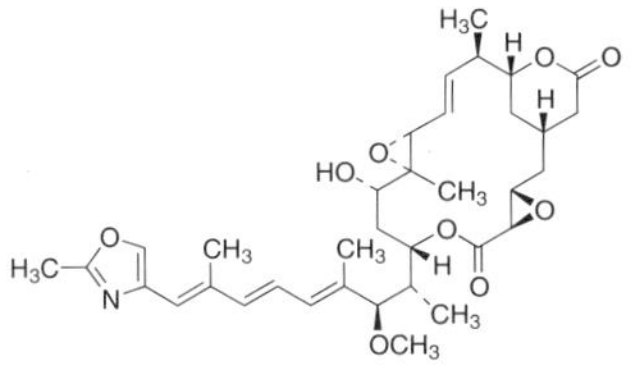

Rhizoxin (**107**)

Fig. 18. Structures of 8- to 16-membered ring macrolides produced by fungi.

effects, and its synthetic analogue, zeranol (**104**), is used as a growth promoter of livestock [129]. Radicicol (= monorden, **105**) was initially isolated from *Monosporium bonorden*, and the skeleton of the structure is the same as that of **103** [130]. It induced the reversal of transformed phenotypes of Rous sarcoma virus-transformed fibroblasts, and **105** is a specific protein-tyrosine kinase inhibitor [131]. L-783,277 (**106**) is another zearalenone-type macrolide produced by *Phoma* sp. and a specific inhibitor of MAP kinase kinase [132].

Rhizoxin (**107**) is a 16-membered ring phytotoxin macrolide produced by *Rhizopus chinensis* [133]. It showed potent antifungal and antitumor activities by inhibiting tubulin polymerization [134].

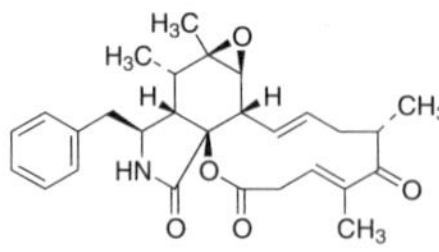

Rosellichalasin (**108**)

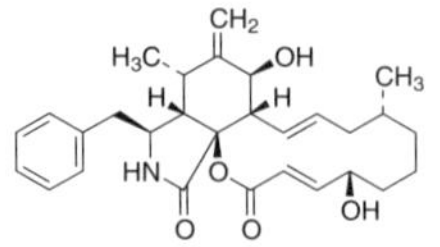

Cytochalasin B (**109**)

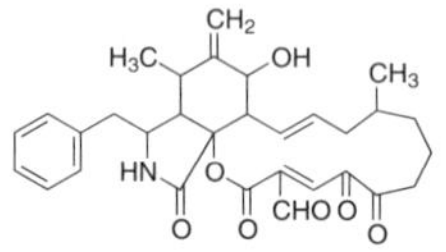

Cytochalasin V (**110**)

Fig. 19. Structures of cytochalasins produced by fungi.

B. Cytochalasins

Cytochalasins are produced by fungi and show various effects on eukaryotic cells. They bound to actin and induced actin depolymerization [135]. Their structural skeletons are each composed of a macrocyclic ring with a fused hydrogenated isoindole. Several compounds having 12- or 14- or 15-membered ring macrocyclic lactones are numbered among the cytochalasins.

Rosellichalasin (Fig. 19, **108**) is a 12-membered ring macrolide isolated from *Rosellinia necatrix* [136]. Cytochalasin B (= phomin, **109**) was the first isolated cytochalasin. It is produced by *Phoma* sp. and has a 14-membered ring macrolide [137]. Cytochalasin V (**110**) is a 15-membered ring macrolide bearing a formyl group on the macrocyclic ring produced by *Phoma exigua* [138].

C. Macrodiolides, Macrotriolides, and Macropentolides

Bartanol (Fig. 20, **111**) is a 13-membered ring macrodiolide isolated from *Cytospora* sp. [139]. Clonostachydiol (**112**) is a 14-membered ring macrodiolide produced by *Clonostachys cylindrospora* and exhibited anthelmintic activity [140].

Some trichothecenes, a group of mycotoxins, have macrodiolide or macrotriolide skeletons. Trichothecenes inhibited protein synthesis by binding to the ribosomal peptidyltransferase site [141]. Roritoxins (roritoxin A, **113**) are 16-membered ring macrodiolides isolated from *Myrothecium roridum* [142]. Verrucarin A (**114**) is an 18-membered ring macrotriolide produced by *Myrothecium* spp. [143].

Macrosphelides (macrosphelide A, **115**) are 16-membered ring macrotriolides isolated by *Microsphaeropsis* sp. [144]. Compound **115** showed potent inhibitory activity against cell adhesion of human leukemia cells (HL-60) to LPS-activated

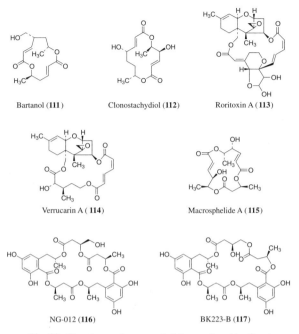

Fig. 20. Structures of macropolylides produced by fungi.

human umbilical vein endothelial cells (HUVEC) and showed an antimetastatic effect *in vivo*.

NG-011 and NG-012 (= BK223-A, **116**) are 24-membered ring macropentolides produced by *Penicillium verruculosum* [145, 146]. They potentiated the neurite outgrowth induced by nerve growth factor in rat PC12 cells. BK223-B (**117**), a 25-membered ring macropentolide, is also produced by *P. verruculosum* and inhibited the growth of plant pathogenic fungi [146].

V. Macrolides from Plants and Lichens

Only three macrolides have been isolated from lichens (Table VII). On the other hand, plants produce about 700 macrolides (Table VIII), but most of them are macrodiolides or macrotriolides such as tannins and alkaloids (e.g., pyrrolizidines). Their biosynthetic pathways are quite different from polyketide macrolides. A small group of monolactone macrolides has up to 19-membered ring lactones. Some of them have a musky odor.

TABLE VII
Macrolides Produced by Lichens

Ring size	Structural characteristic	Number of reported compounds	Example[a]
Monolactone			
18		1	Aspicilin (**118**)
Macropolylide			
16	Macrodiolide	1	Lepranthin (**119**)
24	Macrotriolide	1	Dasypogalactone (**120**)

[a]Structures of these compounds are depicted in Fig. 21.

TABLE VIII
Macrolides Produced by Plants

Ring size	Structural characteristic	Number of reported compounds	Example[a]
Monolactone			
8		10	Almuheptolide A (**121**)
9		1	Salvifoliolide (**122**)
10		1	Jasmine ketolactone (**124**)
12		1	Lasiodiplodin (**123**)
13		2	Yuzu lactone (**125**)
14		5	13-Tridecanolide (**126**)
15		2	(Z)-5-Tetradecen-14-olide (**127**)
16		2	Exaltolide (**128**)
17		1	Ambrettolide (**130**)
18		1	17-Heptadecanolide (**131**)
19		1	(Z)-9-Octadecen-18-olide (**132**)
Macropolylide			
9	Macrodiolide	1	1,3-Di-O-[3,4-bis-(3,4-dihydroxyphenyl)-cyclobutane-1,2-dicarbonyl]-4,5-di-O-caffeoylquinic acid (**133**)
10	Macrodiolide	38	Isostrictinin (**138**)
11	Macrodiolide	290	Crotalarine (**141**)
12	Macrodiolide	180	Seneciphylline (**142**)
13	Macrodiolide	19	Doronenine (**143**)
15	Macrodiolide	67	Evonine (**146**)
16	Macrodiolide	41	Chaksine (**148**)
22	Macrodiolide	1	Azimine (**149**)
24	Macrodiolide	1	Azcarpine (**150**)
26	Macrodiolide	1	Carpaine (**151**)
14	Macrotriolide	12	Parsonsine (**152**)
18	Macrotriolide	4	Shizukaol B (**153**)

[a]Structures of these compounds are depicted in Figs. 22–26.

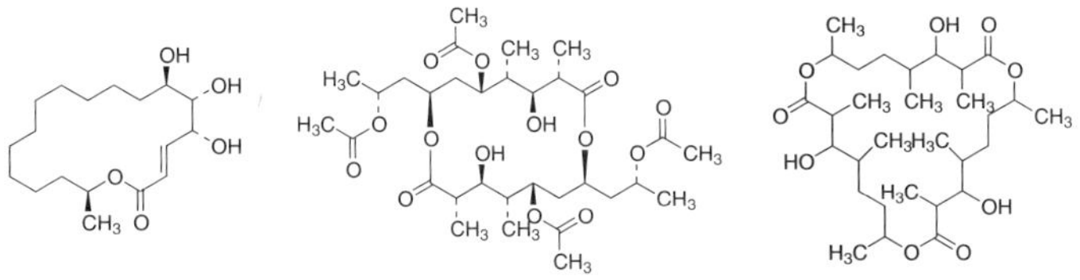

Fig. 21. Structures of macrolides produced by lichens.

A. Macrolides from Lichens

Aspicilin (Fig. 21, **118**) was the first macrolide isolated from lichens. It is an 18-membered ring macrolide and is produced by *Aspicilia gibbosa* [147]. Lepranthin (**119**) is a 16-membered ring macrodiolide produced by *Arthonia impolita* [148]. Dasypogalactone (**120**) is a 24-membered ring macrotriolide isolated from *Usnea dasypoga* [149].

B. Eight- to Nineteen-Membered Ring Macrolides from Plants

Almuheptolide A (Fig. 22, **121**) is an 8-membered ring macrolide isolated from *Goniothalamus arvensis* [150]. It showed potent and selective inhibitory activity toward mammalian mitochondrial respiratory chain complex I. Salvifoliolide (**122**) is a 9-membered ring macrolide produced by *Trichogonia* sp. [151]. Its skeleton may be derived from sesquiterpene. Lasiodiplodin (**123**) is a 12-membered ring macrolide isolated from *Euphorbia splendens* [152]. It showed a potent antileukemic activity.

Some simple monolactones produced by plants have good flavor such as musky odor, and some are contained in essential oils and used as perfume [153]. Jasmine ketolactone (**124**) is a 10-membered ring macrolide isolated from jasmine oil (*Jasminum grandiflorum*) [154]. Yuzu lactone (**125**) is a 13-membered ring macrolide and considered to be responsible for the characteristic flavor of Yuzu fruit (*Citrus junos*) [155]. A 14-membered ring macrolide, 13-tridecanolide (**126**), was obtained as a volatile constituent of angelica (*Angelica archangelica*) root oil [156]. (*Z*)-5-Tetradecen-14-olide (**127**) is a 15-membered ring macrolide isolated from the seed oil of ambrette (*Abelmoschus moschatus*) [157]. Exaltolide (**128**) is a 16-membered ring macrolide isolated from angelica root oil, and it is well known to have a musklike odor [158]. (*R*)-15-Hexadecanolide (**129**) is also a 16-membered ring macrolide isolated from galbanum (*Ferula galbaniflua*) resin [159]. A 17-membered ring macrolide, ambrettolide (**130**), was isolated from the seed oil of ambrette [158], and an 18-membered ring macrolide, 17-heptadecanolide (**131**), was isolated from angelica root oil [156]. (*Z*)-9-Octadecen-18-olide (**132**) is a 19-membered ring macrolide, and it is a constituent of oriental tobacco (*Nicotiana tabacum*) flavor [160]. It is interesting that **129**, **130**, and **132** have been also isolated from insects (see Section VI.A).

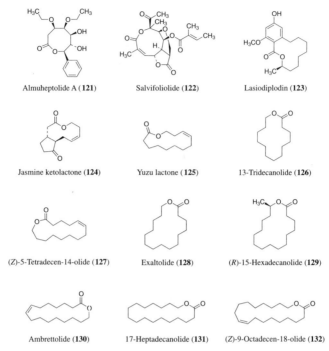

Fig. 22. Structures of 8- to 19-membered ring macrolides produced by plants.

C. Macrodiolides and Macrotriolides from Plants

1,3-Di-*O*-[3,4-bis-(3,4-dihydroxyphenyl)-cyclobutane-1,2-dicarbonyl]-4,5-di-*O*-caffeoylquinic acid (Fig. 23, **133**) is a 9-membered ring macrodiolide isolated from the aerial parts of *Pluchea symphytifolia* [161]. Scapaundulins (scapaundulin A, **134**) are dimeric labdane diterpenoids and have 10-membered ring macrodiolides [162]. They were isolated from a Japanese liverwort *Scapania undulata*. Jasminin (**135**) is an 11-membered ring macrodiolide isolated from the leaves of *Jasminum primulinum* as a bitter component [163]. Pycnocomolide (**136**) is a 12-membered ring macrodiolide isolated from the dried leaves of *Pycnocoma cornuta* [164]. It is a phorbol ester derivative and agonizes protein kinase C. A diterpene, corymbi-7,13*E*-dienolide (**137**), was isolated from the roots of *Corymbium villosum*, and it has a 15-membered ring macrodiolide [165].

Some ellagitannins have macrocyclic dilactones with ring sizes of 10- to 12-membered rings. Isostrictinin (Fig. 24, **138**) was isolated from *Casuarina stricta* and has a 10-membered ring macrodiolide [166]. Pedunculagin (**139**), isolated from *Quercus* spp., has one 11-membered and one 10-membered ring macrodiolides [167]. One 12-membered and one 11-membered ring macrodiolides are contained in the structure of geraniin (**140**), which was isolated from *Geranium thunbergii* [168].

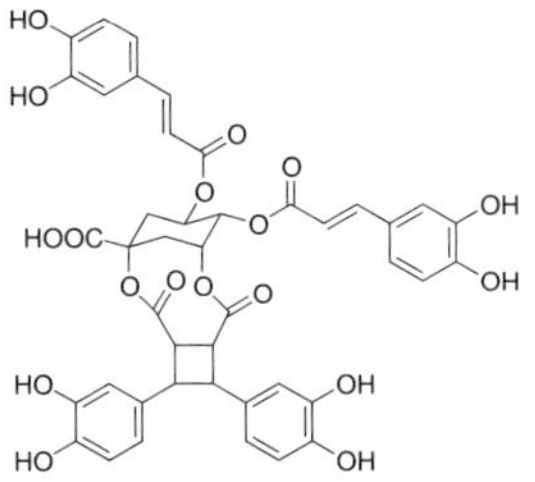

Fig. 23. Structures of 9- to 15-membered ring macrodiolides produced by plants.

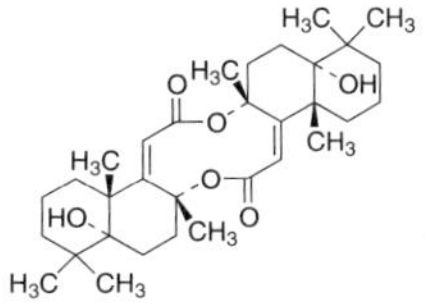

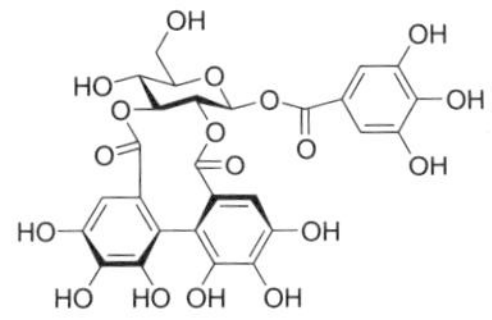

Fig. 24. Structures of macrodiolide tannins produced by plants.

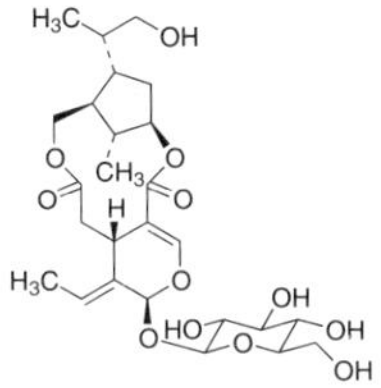

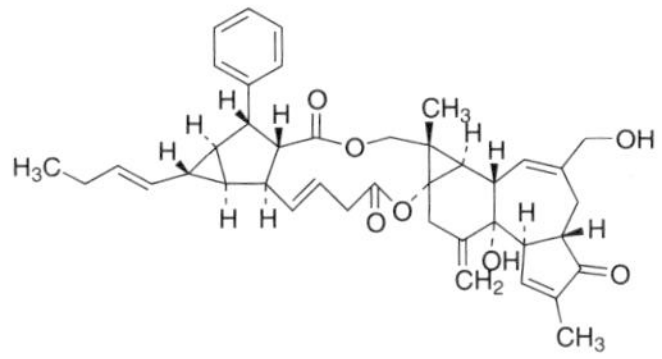

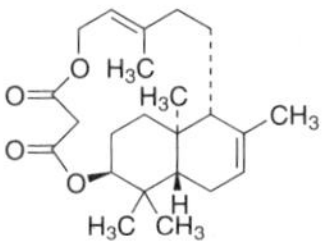

Some alkaloids have macropolylide skeletons in their structures; in particular, many pyrrolizidine alkaloids have structures with each pyrrolizidine skeleton fused to a macrocyclic dilactone or trilactone [169]. Crotalarine (Fig. 25, **141**) is a 10-membered ring macrodiolide isolated from *Crotalaria burhia* [170]. Seneciphylline (**142**) is a 12-membered ring macrodiolide isolated from *Senecio* spp.

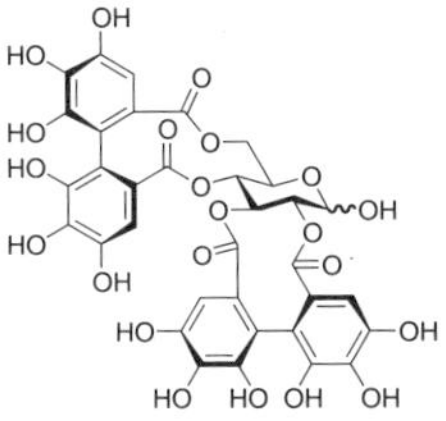

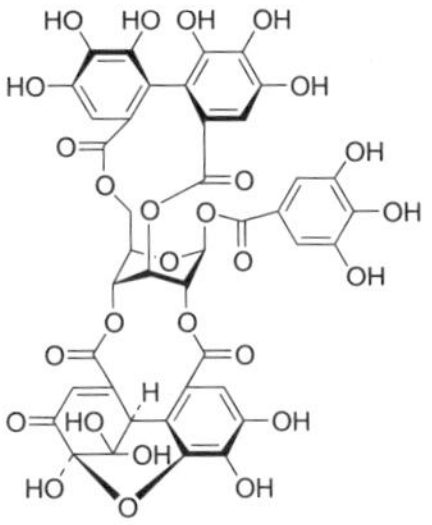

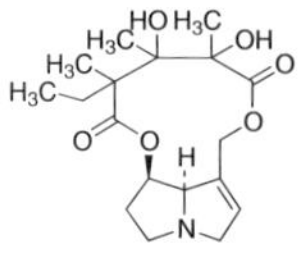

Crotalarine (**141**)

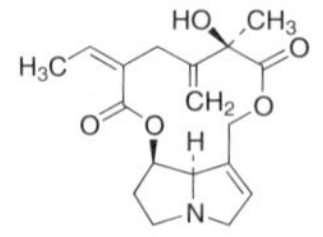

Seneciphylline (**142**)

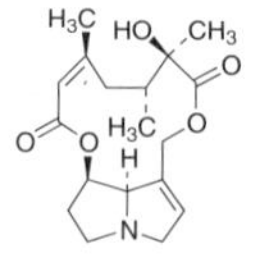

Doronenine (**143**)

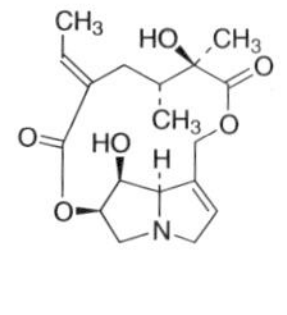

Madurensine (**144**)

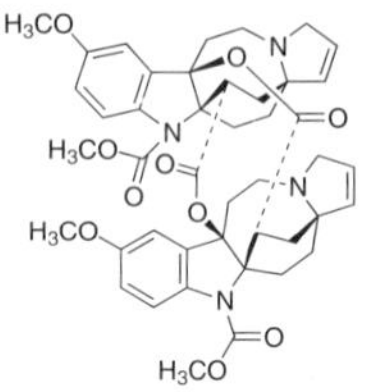

Tenuisine A (**145**)

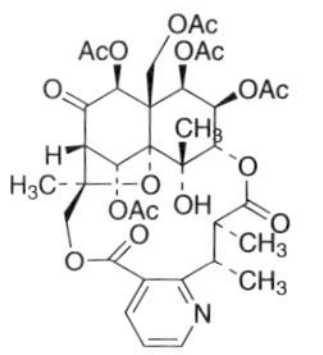

Evonine (**146**)

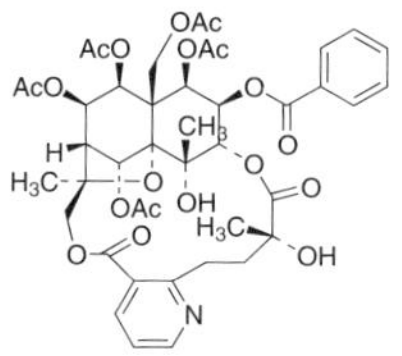

Wilfordine (**147**)

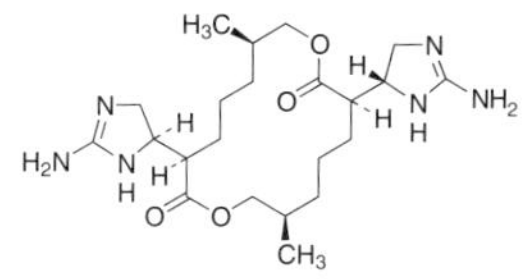

Chaksine (**148**)

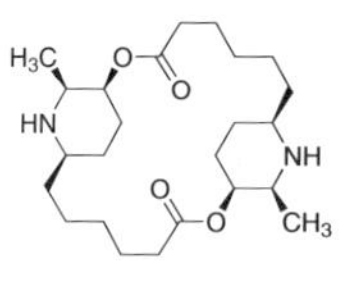

Azimine (**149**)

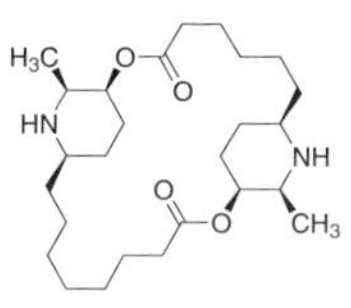

Azcarpine (**150**)

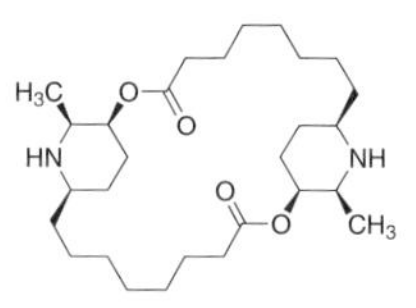

Carpaine (**151**)

Fig. 25. Structures of macrodiolide alkaloids produced by plants.

[171]. Doronenine (**143**), produced by *Senecio doronicum*, and madurensine (**144**), produced by *Crotalaria* spp., are both 13-membered ring macrodiolides, but the ester oxygen of **144** bonds to C-6, which is unusual [172, 173].

Tenuisine A (**145**) is a bisindole 10-membered ring macrodiolide isolated from *Kopsia tenuis* [174]. Evonine (**146**) is a 15-membered ring macrodiolide containing a pyridine and a decalin [175]. It was isolated from the seeds of the shrub *Euonymus europaeus*. Wilfordine (**147**) was isolated from the whole roots of *Tripterygium wilfordii* [176]. Its structure is similar to that of **146**, but it is a 16-membered ring macrodiolide. It showed insecticidal activity. Chaksine (**148**) is an alkaloid of the seeds of *Cassia absus* and has a 16-membered ring macrodiolide with two 2-amino-2-imidazoline side chains [177]. Three piperidine alkaloids, azimine (**149**), azcarpine (**150**), and carpaine (**151**), were isolated from the leaves of *Azima tetracantha* [178]. Compounds **149**, **150**, and **151** are macrodiolides with ring sizes of 22, 24, and 26, respectively.

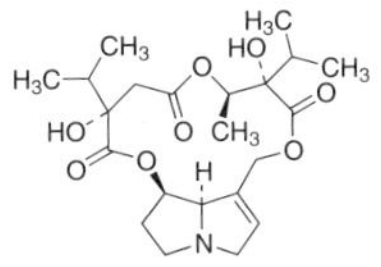

Parsonsine (152) 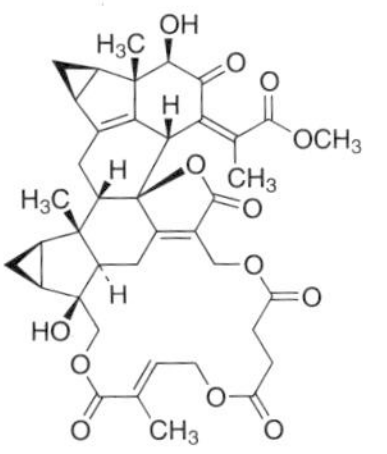   Shizukaol B (153)

Fig. 26. Structures of macrotriolides produced by plants.

Two groups of macrotriolides have been reported. One is 14-membered ring parsonsine-type pyrrolizidine alkaloids, and the other is 18-membered ring dimeric sesquiterpenes, shizukaols. Parsonsine (Fig. 26, **152**) was isolated from *Parsonia heterophylla* [179]. Shizukaols (shizukaol B, **153**) were isolated from the roots of *Chloranthus serratus* [180].

VI. Macrolides from Insects

Insects produce monolactone macrolides having from 10- to 39-membered rings and azamacrolides having 13-, 15-, and 16-membered rings (Table IX). They are used as sex pheromones, aggregation pheromones, and defensive compounds. All of them have simple cyclic structures with or without alkyl side chains. It is interesting that all large macrolides having more than a 16-membered ring size produced by insects have ring sizes of odd numbers except one example (a 22-membered ring).

A. Ten- to Thirty-Nine-Membered Ring Macrolides

Phoracantholide I (Fig. 27, **154**) is a defense secretion of the eucarypt longicorn, *Phoracantha synonyma*, and it is a 10-membered ring macrolide [181].

Some macrolides are used as aggregation pheromones. Ferrulactones I (= cucujolide I, **155**) and II (**156**) are 11- and 12-membered ring macrolides, respectively, produced by the rusty grain beetle *Cryptolestes ferrugineus* [182]. (Z)-3-Dodecen-12-olide (**157**) and cucujolide III (**158**) are 13- and 14-membered ring macrolides, respectively, produced by *C. pusillus* and *C. turcicus*, respectively [183, 184].

(R)-15-Hexadecanolide (Fig. 22, **129**), produced by stinkbug *Piezodorous hybneri*, ambrettolide (**130**), secreted from Dufour's gland of solitary bees, and (Z)-9-octadecen-18-olide (**132**), produced by social wasp *Polybia sericea*, are 16-, 17- and 19-membered ring macrolides, respectively, and they are also produced by plants (see Section V.B) [185–187]. Compound **130** has a strong musklike odor.

TABLE IX
Macrolides Produced by Insects

Ring size	Number of reported compounds	Example[a]
Monolactone		
10	2	Phoracantholide I (**154**)
11	2	Ferrulactone I (**155**)
12	4	Ferrulactone II (**156**)
13	2	(Z)-3-Dodecen-12-olide (**157**)
14	5	Cucujolide III (**158**)
16	1	(R)-15-Hexadecanolide (**129**)
17	2	Ambrettolide (**130**)
19	4	(Z)-9-Octadecen-18-olide (**132**)
21	4	20-Icosanolide (**159**)
22	1	21-Henicosanolide (**162**)
23	3	22-Docosanolide (**160**)
25	5	24-Tetracosanolide (**161**)
27	4	26-Hexacosanolide (**163**)
29	2	28-Octacosanolide (**164**)
31	1	30-Triacontanolide (**165**)
33	1	32-Dotriacontanolide (**166**)
35	1	34-Tetratriacontanolide (**167**)
37	1	36-Hexatriacontanolide (**168**)
39	1	38-Octatriacontanolide (**169**)
Azamacrolide		
13	1	9-Propyl-10-azacyclododecan-12-olide (**170**)
15	4	Epilachnene (**171**)
16	1	Homoepilachnene (**172**)

[a]Structures of these compounds are depicted in Figs. 27 and 28.

Compound **129** is a sex pheromone and **132** is a trail pheromone. A 21-membered ring macrolide, 20-icosanolide (Fig. 27, **159**), a 23-membered ring macrolide, 22-docosanolide (**160**), and a 25-membered ring macrolide, 24-tetracosanolide (**161**), are produced by various bee species and are female sex pheromones [188]. Compounds **160** and **161** are also contained in defense secretions of termites, *Armitermes* spp. [189].

21-Henicosanolide (**162**) is a 22-membered ring macrolide, and it is secreted from Dufour's gland of bee, *Colletes* sp. [190].

Large ring size macrolides are produced by termites as defense secretions [189]. 26-Hexacosanolide (**163**), 28-octacosanolide (**164**), 30-triacontanolide (**165**), 32-dotriacontanolide (**166**), 34-tetratriacontanolide (**167**), 36-hexatriacontanolide (**168**), and 38-octatriacontanolide (**169**) are 27-, 29-, 31-, 33-, 35-, 37-, and 39-membered ring macrolides, respectively, produced by termite *A. teevani*. Interestingly, **164**, **165**, and **166** are also produced by horses (see Section VII.B).

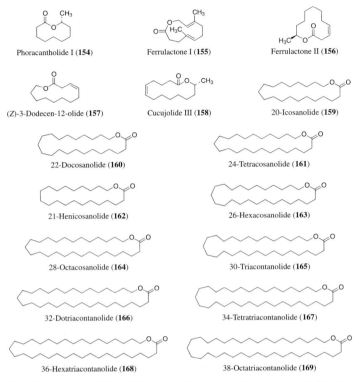

Fig. 27. Structures of 10- to 39-membered ring macrolides produced by insects.

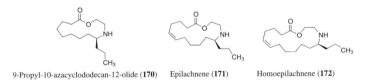

Fig. 28. Structures of azamacrolides produced by insects.

B. Azamacrolides

A ladybird beetle, *Epilachna varivestis*, produces azamacrolides as pupal defensive secretion [191]. 9-Propyl-10-azacyclododecan-12-olide (Fig. 28, **170**) is a 13-membered ring azamacrolide, epilachnene (**171**) is a 15-membered ring azamacrolide, and homoepilachnene (**172**) is a 16-membered ring azamacrolide.

VII. Other Macrolides

A. Macrolides from Algae and Invertebrates

Macrolides from algae and invertebrates, except for insects, are precisely described in Chapter 2. Therefore, compounds having unique skeletons produced only by algae or invertebrates are merely mentioned here. The lists of macrolides produced by algae and invertebrates are shown in Tables X and XI, respectively.

Prorocentrolide (Fig. 29, **173**) is a 28-membered ring macrolide produced by dinoflagellate *Prorocentrum lima* [192]. Zooxanthellatoxins A (**174**) and B are 62-membered ring macrolides isolated from dinoflagellate *Symbiodinium* sp.,

TABLE X
Macrolides Produced by Algae

Ring size	Structural characteristic	Number of reported compounds	Example[a]
Monolactone			
9		1	Neohalicholactone
12		1	Amphidinolide Q
13		1	Hybridalactone
14		5	Oscillariolide
15		3	Amphidinolide J
16		7	Lyngbyaloside
18		2	Acutiphycin
19		4	Hoffmanniolide
20		2	Amphidinolide A
22		9	Scytophycin B
25		2	Amphidinolide C
26		7	Amphidinolide B_1
27		2	Amphidinolide G
28		3	Prorocentrolide (**173**)
29		1	Amphidinolide M
32		1	Goniodomin A
34		2	Pectenotoxin 2
62		2	Zooxanthellatoxin A (**174**)
Macropolylide			
14	Macrodiolide	10	Aplysiatoxin
12	Macrotriolide	1	Pinnatifolide
Macrolide lactam			
15		2	Laingolide (**175**)
17		1	Madangolide (**176**)

[a]Structures of numbered compounds are depicted in Fig. 29. Structures of unnumbered compounds are depicted in Chapter 2.

TABLE XI
Macrolides Produced by Invetebrates

Ring size	Structural characteristic	Number of reported compounds	Example[a]
Monolactone			
8		9	Solandelactone A
9		2	Halicholactone
10		4	Didemnilactone A
12		3	Salicylihalamide A
13		10	Prostaglandin A_2 1,15-lactone
14		14	Callipeltoside A
16		8	Latrunculin A
17		2	Aplyolide B
18		7	Tedanolide
19		1	Halichlorine
20		4	Laulimalide
21		1	Neolaulimalide
22		5	Lasonolide A
24		10	Iejimalide A
25		3	Lituarine A
26		23	Bryostatin 1 (**177**)
27		1	Neristatin 1
31		7	Halichondrin B
34		6	Pectenotoxin 1
42		10	Altohyrtin A
16	Triene	2	Superstolide A (**178**)
Macrodiolide			
10		1	Singardin
14		2	Aplysiatoxin
15		6	Lobatamide A
40		3	Misakinolide A (**179**)
42		1	Isobistheonellide A
44		9	Swinholide A (**180**)
Macrolide lactam			
35		2	Chondropsin A (**181**)
Oxazole containing			
25		30	Ulapualide A (**182**)
37		3	Theonezolide A (**183**)
Thiazole containing			
19	Macrodiolide	1	Pateamine (**184**)

[a]Structures of numbered compounds are depicted in Figs. 30 and 31. Structures of unnumbered compounds are depicted in Chapter 2.

and their 62-membered rings are the largest ones in natural macrolides [193]. Laingolide (**175**) and madangolide (**176**) are macrolide lactams produced by blue-green alga *Lyngbya bouillonii*, and they have 15- and 17-membered rings, respectively [194, 195].

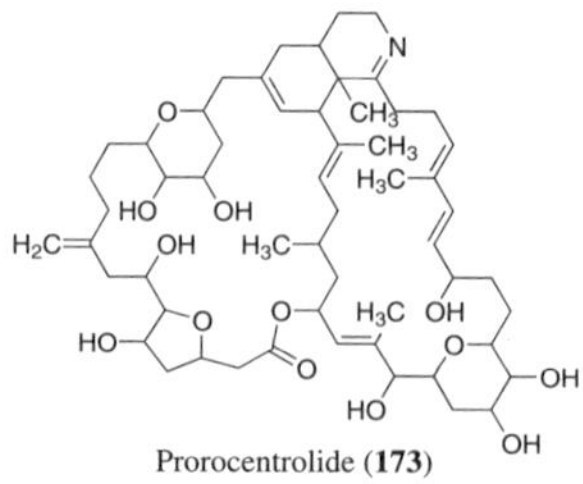

Prorocentrolide (**173**)

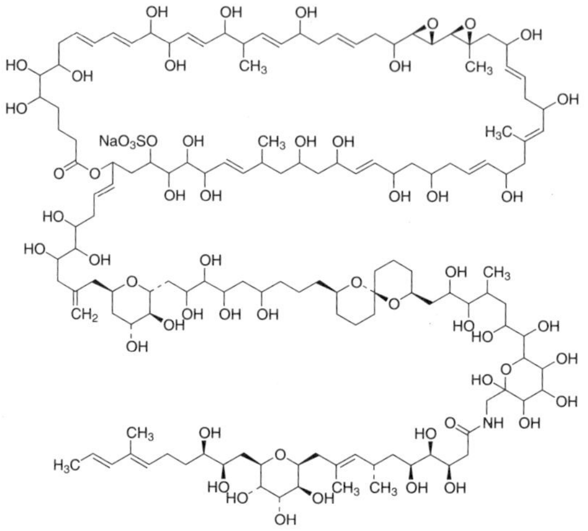

Zooxanthellatoxin A (**174**)

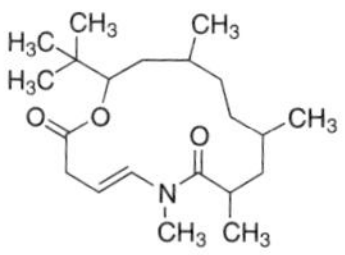

Laingolide (**175**) 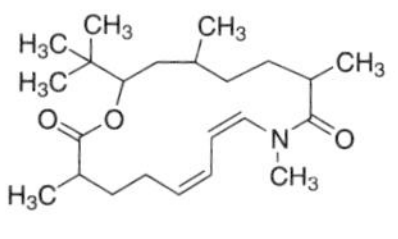 Madangolide (**176**)

Fig. 29. Structures of macrolides produced by algae.

Bryostatin 1 (Fig. 30, **177**) is a 26-membered ring macrolide produced by the sea-mat *Bugula neritina* [196]. Superstolide A (**178**) is a 16-membered ring triene macrolide isolated from the sponge *Neosiphonia superstes* [197]. Misakinolide A (**179**) and swinholide A (**180**) are 40- and 44-membered ring macrodiolides, respectively, produced by sponge *Theonella* spp. [198, 199]. A 35-membered ring macrolide lactam, chondropsin A (Fig. 31, **181**), was isolated from a sponge *Chondropsis* sp. [200]. Ulapualide A (**182**) and theonezolide A (**183**) are 25- and 37-membered ring macrodiolides, respectively, each containing

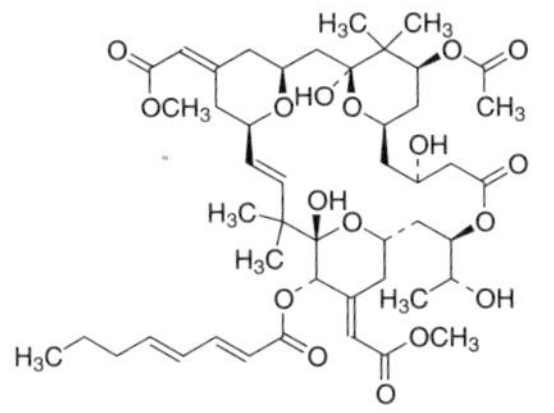

Bryostatin 1 (**177**)

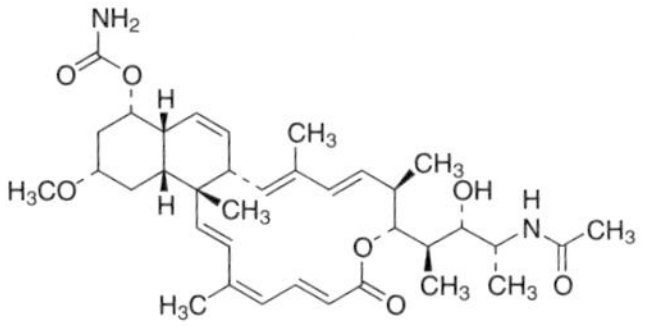

Superstolide A (**178**)

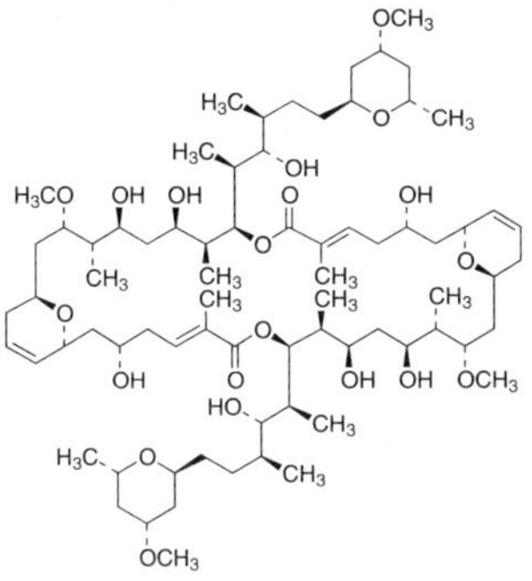

Misakinolide A (**179**)

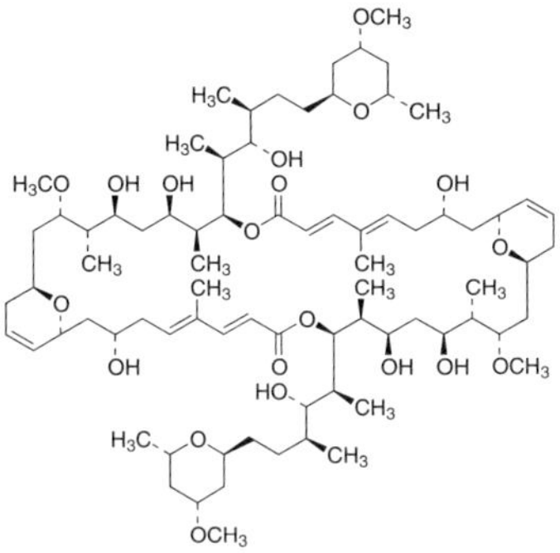

Swinholide A (**180**)

Fig. 30. Structures of monolactone macrolides and macrodiolides produced by invertebrates.

an oxazole ring [201, 202]. Compound **182** was isolated from a nudibranch *Hexabranchus sanguineus*, and **183** was isolated from a sponge *Theonella* sp. Pateamine (**184**) is only one macrocyclic lactone in natural products that contains a thiazole ring in its skeleton [203]. It is a 19-membered ring macrodiolide with a thiazole isolated from a sponge *Mycale* sp.

B. Macrolides from Vertebrates

It is well known that macrocyclic ketones like muscone are produced by some mammals such as musk deer, civet cat, and Louisiana muskrat [153]. Some mammals also produce simple monolactones as shown in Table XII. They are rather large in size—up to 38-membered rings.

A 15-membered ring macrolide, 4-tetradecen-14-olide (Fig. 32, **185**), has been reported as an anal sac secretion of striped hyena *Hyaena hyaena*, and it may be used to mark territorial boundaries [204]. Ambrettolide (**130**) is a 17-membered ring macrolide produced by plants and insects (see Sections V.B and VI.A). It is also detected in the lipid from musk of muskrat *Ondatra zibethica* [205].

All other mammal macrolides are isolated or detected from the skin surface lipid of horse and other *Equus* spp. (donkey, zebra, ass, and mule) and named

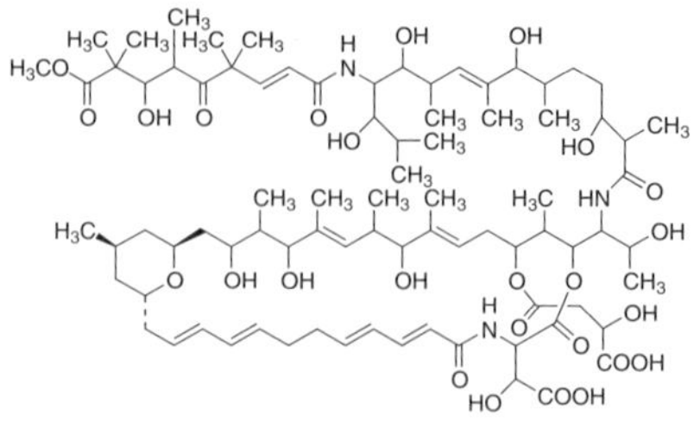

Chondropsin A (**181**)

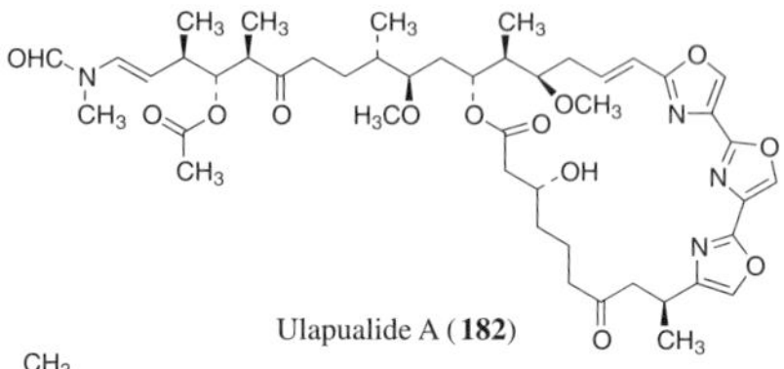

Ulapualide A (**182**)

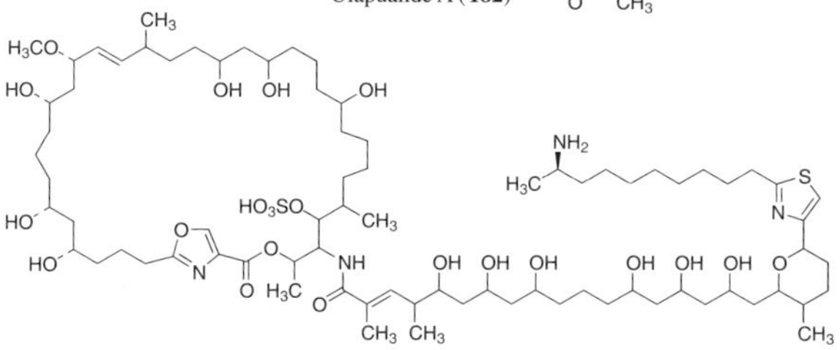

Theonezolide A (**183**)

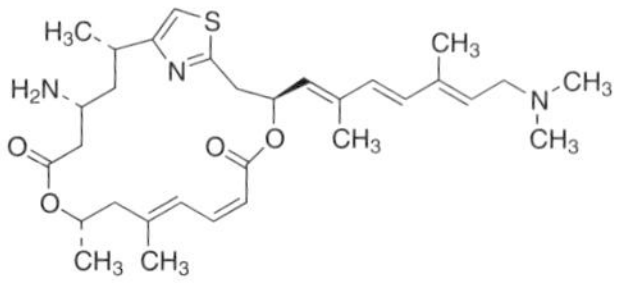

Pateamine (**184**)

Fig. 31. Structures of nitrogen-containing macrolides produced by invertebrates.

equolides [206, 207]. The representative macrolides of each ring size are as follows: 29-membered ring 28-octacosanolide (**164**), 30-membered ring 28-methyl-29-nonacosanolide (**186**), 31-membered ring 30-triacontanolide (**165**), 32-membered ring 30-methyl-31-hentriacontanolide (**187**), 33-membered ring 32-dotriacontanolide (**166**), 34-membered ring (Z)-32-methyl-25-tritriaconten-33-olide (**188**), 36-membered ring (Z,Z)-34-methyl-13,27-pentatriacontadien-35-olide (**189**), and 38-membered ring (Z,Z)-36-methyl-15,29-heptatriacontadien-37-olide (**190**). Compounds **164**, **165**, and **166** are also produced by insects (see Section VI.A).

TABLE XII
Macrolides Produced by Vertebrates

Ring size	Number of reported compounds	Example[a]
Monolactone		
15	1	4-Tetradecen-14-olide (**185**)
17	1	Ambrettolide (**130**)
29	1	28-Octacosanolide (**164**)
30	1	28-Methyl-29-nonacosanolide (**186**)
31	2	30-Triacontanolide (**165**)
32	1	30-Methyl-31-hentriacontanolide (**187**)
33	2	32-Dotriacontanolide (**166**)
34	7	(Z)-32-Methyl-25-tritriaconten-33-olide (**188**)
36	5	(Z,Z)-34-Methyl-13,27-pentatriacontadien-35-olide (**189**)
38	5	(Z,Z)-36-Methyl-15,29-heptatriacontadien-37-olide (**190**)

[a] Structures of these compounds are depicted in Fig. 32.

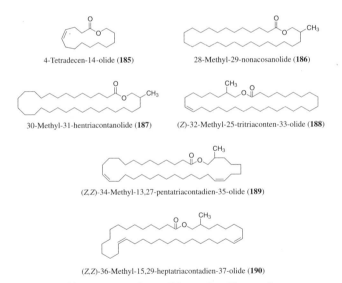

Fig. 32. Structures of macrolides produced by vertebrates.

VIII. Concluding Remarks

A large variety of macrolides have been isolated from natural sources. Macrolides have various ring sizes up to 62 and exhibit various characters: spiroacetal-fused macrolides, polyene macrolides, macropolylides, and macrolides containing amino nitrogen, amide nitrogen, oxazole rings, or thiazole rings in their skeletons.

Many macrolides have interesting biological activities such as antiviral, antibacterial, antifungal, anthelmintic, phytotoxic, insecticidal, antitumor, and immunosuppressive activities. Those activities are due to their various biochemical activities and thus some macrolides are important tools as biochemical reagents. Some macrolides have a musklike odor, and they are used as perfume ingredients. Some pheromones produced by insects are also macrolides. It is interesting that mammals also produce macrolides.

We hope different and interesting macrolides will continue to be isolated from natural sources in the near future.

References

1. Ōmura, S. (1984). "Macrolide Antibiotics. Chemistry, Biology, and Practice." Academic Press, London.
2. Woodward, R. B. (1957). Struktur und Biogenese der Makrolide. Eine neue Klasse von Naturstoffen. *Angew. Chem.* **69**, 50–58.
3. Masamune, S., Bates, G. S., and Corcoran, J. W. (1977). Macrolides. Recent progress in chemistry and biochemistry. *Angew. Chem. Int. Edit. Engl.* **16**, 585–607.
4. Laatsch, H. (1998). "AntiBase 3.0, A Database for Rapid Structure Identification of Microbial Metabolites." Chemical Concepts GmbH, Weinheim, Germany.
5. Buckingham, J. (2000). "Dictionary of Natural Products on CD-ROM. Version 9:1." Chapman & Hall/CRC Press, Boca Raton, FL.
6. Chemical Abstracts Service (2001). "CAS Database." American Chemical Society, Washington, DC.
7. Tapiolas, D. M., Roman, M., Fenical, W., Stout, T. J., and Clardy, J. (1991). Octalactins A and B: Cytotoxic eight-membered-ring lactones from a marine bacterium, *Streptomyces* sp. *J. Am. Chem. Soc.* **113**, 4682–4683.
8. Lessmann, H., Krupa, J., and Lackner, H. (1993). Juglorubin. *Z. Naturforsch. B: Chem. Sci.* **48**, 672–682.
9. Celmer, W. D., Chmurny, G. N., Moppett, C. E., Ware, R. S., Watts, P. C., and Whipple, E. B. (1980). Structure of natural antibiotic CP-47,444. *J. Am. Chem. Soc.* **102**, 4203–4209.
10. Ōmura, S., Iwata, R., Iwai, Y., Taga, S., Tanaka, Y., and Tomoda, H. (1985). Luminamicin, a new antibiotic. Production, isolation and physicochemical and biological properties. *J. Antibiot.* **38**, 1322–1326.
11. Rausmussen, R. R., Scherr, M. H., Whittern, D. N., Buko, A. M., and McAlpine, J. B. (1987). Coloradocin, an antibiotic from a new *Actinoplanes*. II. Identity with luminamicin and elucidation of structure. *J. Antibiot.* **40**, 1383–1393.
12. Donin, M. N., Pagano, J., Dutcher, J. D., and McKee, C. M. (1953–1954). Methymycin, a new crystalline antibiotic. *Antibiot. Ann.* 179–185.
13. Matsumoto, M., Kawamura, Y., Yoshimura, Y., Terui, Y., Nakai, H., Yoshida, T., and Shoji, J. (1990). Isolation, characterization and structures of PA-46101 A and B. *J. Antibiot.* **43**, 739–747.
14. Sugawara, K., Nishiyama, Y., Toda, S., Komiyama, N., Hatori, M., Moriyama, T., Sawada, Y., Kamei, H., Konishi, M., and Oki, T. (1992). Lactimidomycin, a new glutarimide group antibiotic. Production, isolation, structure and biological activity. *J. Antibiot.* **45**, 1433–1441.
15. Seki-Asano, M., Okazaki, T., Yamagishi, M., Sakai, N., Takayama, Y., Hanada, K., Morimoto, S., Takatsuki, A., and Mizoue, K. (1994). Isolation and characterization of a new 12-membered macrolide FD-895. *J. Antibiot.* **47**, 1395–1401.

16. Kirst, H. A., Michel, K. H., Mynderase, J. S., Chio, E. H., Yao, R. C., Nakatsukasa, W. M., Boeck, L. D., Occlowitz, J. L., Paschal, J. W., Deeter, J. B., and Thompson, G. D. (1992). Discovery, isolation, and structure elucidation of a family of structurally unique, fermentation-derived tetracyclic macrolides. In "Synthesis and Chemistry of Agrochemicals III" (D. R. Baker, J. G. Fenyes, and J. J. Steffens, Eds.), pp. 214–225. American Chemical Society, Washington, DC.
17. Thompson, G., and Scott, H. (1999). Spinosad. *Pestic. Outlook*, **10**, 78–81.
18. McGuire, J. M., Bunch, R. L., Anderson, R. C., Boaz, H. E., Flynn, E. H., Powell, H. M., and Smith, J. W. (1952). Ilotycin, a new antibiotic. *Antibiot. Chemother.* **2**, 281–283.
19. Keller-Schierlein, W., Muntwyler, R., Pache, W., and Zaehner, H. (1969). Metabolic products of microorganisms. LXXIII. Chlorothricin and dechlorothricin. *Helv. Chim. Acta* **52**, 127–142.
20. Takatsu, T., Nakayama, H., Shimazu, A., Furihata, K., Ikeda, K., Furihata, K., Seto, H., and Otake, N. (1985). Rustmicin, a new macrolide antibiotic active against wheat stem rust fungus. *J. Antibiot.* **38**, 1806–1809.
21. Mandala, S. M., Thornton, R. A., Milligan, J., Rosenbach, M., Garcia-Calvo, M., Bull, H. G., Harris, G., Abruzzo, G. K., Flattery, A. M., Gill, C. J., Bartizal, K., Dreikorn, S., and Kurtz, M. B. (1998). Rustmicin, a potent antifungal agent, inhibits sphingolipid synthesis at inositol phosphoceramide synthase. *J. Biol. Chem.* **273**, 14942–14949.
22. Nagahama, N., Suzuki, M., Awataguchi, S., and Okuda, T. (1967). New antibiotic, albocycline. I. Isolation, purification, and properties. *J. Antibiot. Ser. A* **20**, 261–266.
23. Kim, Y. J., Furihata, K., Shimazu, A., Furihata, K., and Seto, H. (1991). Isolation and structural elucidation of sekothrixide, a new macrolide effective to overcome drug-resistance of cancer cell. *J. Antibiot.* **44**, 1280–1282.
24. Takemoto, Y., Nakae, K., Kawatani, M., Takahashi, Y., Naganawa, H., and Imoto, M. (2001). Migrastatin, a novel 14-membered ring macrolide, inhibits anchorage-independent growth of human small cell lung carcinoma Ms-1 cells. *J. Antibiot.* **54**, 1104–1107.
25. Hata, T., Sano, Y., Ohki, N., Yokoyama, Y., Matsumae, A., and Ito, S. (1953). Leucomycin, a new antibiotic. *J. Antibiot. Ser. A* **6**, 87–89.
26. Burg, R. W., Miller, B. M., Baker, E. E., Birnbaum, J., Currie, S. A., Hartman, R., Kong, Y.-L., Monaghan, R. L., Olson, G., Putter, I., Tunac, J. B., Wallick, H., Stapley, E. O., Ōiwa, R., and Ōmura, S. (1979). Avermectins, new family of potent anthelmintic agents: Producing organism and fermentation. *Antimicrob. Agents Chemother.* **15**, 361–367.
27. Mishima, H., Kurabayashi, M., Tamura, C., Sato, S., Kuwano, H., Saito, A., and Aoki, A. (1975). Structures of milbemycin β1, β2, and β3. *Tetrahedron Lett.* 711–714.
28. Werner, G., Hagenmaier, H., Drautz, H., Baumgartner, A., and Zaehner, H. (1984). Metabolic products of microorganisms. 224. Bafilomycins, a new group of macrolide antibiotics. Production, isolation, chemical structure and biological activity. *J. Antibiot.* **37**, 110–117.
29. Bowman, E. J., Siebers, A., and Altendorf, K. (1988). Bafilomycins: A class of inhibitors of membrane ATPases from microorganisms, animal cells, and plant cells. *Proc. Natl. Acad. Sci. USA* **85**, 7972–7976.
30. Ōmura, S., Otoguro, K., Nishikiori, T., Ōiwa, R., and Iwai, Y. (1981). Setamycin, a new antibiotic. *J. Antibiot.* **34**, 1253–1256.
31. Swanson, A. G., Monday, R. A., and Perry, D. A. (1992). Structure elucidation by NMR spectroscopy of saccharolidin A, a novel antiparasitic macrolide from *Saccharothrix aerocolonigenes*. *Magn. Reson. Chem.* **30**, S87–S95.
32. Igarashi, M., Hayashi, C., Homma, Y., Hattori, S., Kinoshita, N., Hamada, M., and Takeuchi, T. (2000). Tubelactomicin A, a novel 16-membered lactone antibiotic, from *Nocardia* sp. I. Taxonomy, production, isolation and biological properties. *J. Antibiot.* **53**, 1096–1101.
33. Puar, M. S., Brambilla, R., Bartner, P., Schumacher, D., and Jaret, R. S. (1979). SCH 23831, a novel macrolide from *Micromonospora rosaria*. *Tetrahedron Lett.* 2767–2770.

34. Wagman, G. H., Waitz, J. A., Marquez, J., Murawski, A., Oden, E. M., Testa, R. T., and Weinstein, M. J. (1972). New *Micromonospora*-produced macrolide antibiotic, rosamicin. *J. Antibiot.* **25**, 641–646.
35. Sakamoto, J. M. J., Kondo, S., Yumoto, H., and Arishima, M. (1962). Bundlins A and B, two antibiotics produced by *Streptomyces griseofuscus* nov. sp. *J. Antibiot. Ser. A* **15**, 98–102.
36. Coronelli, C., White, R. J., Lancini, G. C., and Parenti, F. (1975). Lipiarmycin, a new antibiotic from *Actinoplanes*. II. Isolation, chemical, biological, and biochemical characterization. *J. Antibiot.* **28**, 253–259.
37. Kinashi, H., Someno, K., and Sakaguchi, K. (1984). Isolation and characterization of concanamycins A, B and C. *J. Antibiot.* **37**, 1333–1343.
38. Dröse, S., Bindseil, K. U., Bowman, E. J., Siebers, A., Zeeck, A., and Altendorf, K. (1993). Inhibitory effect of modified bafilomycins and concanamycins on P- and V-type adenosinetriphosphatases. *Biochemistry* **32**, 3902–3906.
39. Bindseil, K. U., and Zeeck, A. (1994). The chemistry of unusual macrolides, 2. Spectroscopic and biosynthetic investigations of the V-type ATPase inhibitor concanamycin A. *Liebigs Ann. Chem.* 305–312.
40. Berger, J., Jampolsky, L. M., and Goldberg, M. W. (1949). Borrelidin, a new antibiotic with antiborrelia activity and penicillin-enhancement properties. *Arch. Biochem.* **22**, 476–478.
41. Rhodes, A., Fantes, K. H., Boothroyd, B., McGonagle, M. P., and Crosse, R. (1961). Venturicidin: A new antifungal antibiotic of potential use in agriculture. *Nature* **192**, 952–954.
42. Pathirana, C., Tapiolas, D., Jensen, P. R., Dwight, R., and Fenical, W. (1991). Structure determination of maduralide: A new 24-membered ring macrolide glycoside produced by a marine bacterium (Actinomycetales). *Tetrahedron Lett.* **32**, 2323–2326.
43. Kihara, T., Kusakabe, H., Nakamura, G., Sakurai, T., and Isono, K. (1981). Cytovaricin, a novel antibiotic. *J. Antibiot.* **34**, 1073–1074.
44. Schmitz, H., Jubinski, S. D., Hooper, I. R., Crook Jr. K. E., Price, K. E., and Lein, J. (1965). Ossamycin, a new cytotoxic agent. *J. Antibiot. Ser. A* **18**, 82–88.
45. Smith, R. M., Peterson, W. H., and McCoy, E. (1954). Oligomycin, a new antifungal antibiotic. *Antibiot. Chemother.* **4**, 962–970.
46. Matsuno-Yagi, A., and Hatefi, Y. (1993). Studies on the mechanism of oxidative phosphorylation. ATP synthesis by submitochondrial particles inhibited at F_O by venturicidin and organotin compounds. *J. Biol. Chem.* **268**, 6168–6173.
47. Salomon, A. R., Zhang, Y., Seto, H., and Khosla, C. (2001). Structure–activity relationships within a family of selectively cytotoxic macrolide natural products. *Org. Lett.* **3**, 57–59.
48. Walter, P., Lardy, H. A., and Johnson, D. (1967). Antibiotics as tools for metabolic studies. X. Inhibition of phosphoryl transfer reactions in mitochondria by peliomycin, ossamycin, and venturicidin. *J. Biol. Chem.* **242**, 5014–5018.
49. Tanaka, Y., Sugoh, M., Ji, W., Iwabuchi, J., Yoshida, H., and Ōmura, S. (1995). Screening method for cellulose biosynthesis inhibitors with herbicidal activity. *J. Antibiot.* **48**, 720–724.
50. Bush, K., Bonner, D. P., and Sykes, R. B. (1980). Izumenolide—a novel β-lactamase inhibitor produced by *Micromonospora*. II. Biological properties. *J. Antibiot.* **33**, 1262–1269.
51. Ikeda, Y., Kondo, S., Sawa, T., Tsuchiya, M., Ikeda, D., Hamada, M., Takeuchi, T., and Umezawa, H. (1981). Dotriacolide, a new β-lactamase inhibitor. *J. Antibiot.* **34**, 1628–1630.
52. Arai, T., Kuroda, S., Ohara, H., Katoh, Y., and Kaji, H. (1965). Copiamycin, a new antifungal antibiotic derived from *S. hygroscopicus* var. *chrystallogenes*. *J. Antibiot. Ser. A* **18**, 63–67.
53. Takesako, K., and Beppu, T. (1984). Studies on new antifungal antibiotics. Guanidylfungins A and B. I. Taxonomy, fermentation, isolation and characterization. *J. Antibiot.* **37**, 1161–1169.
54. Kobinata, K., Koshino, H., Kusakabe, H., Kobayashi, Y., Yamaguchi, I., Isono, K., and Osada, H. (1993). Isolation and characterization of a new antibiotic, malolactomycin A. *J. Antibiot.* **46**, 1912–1915.

55. Frechet, D., Danzer, M., Debu, F., Monegier du Sorbier, B., Reisdorf, D., Snozzi, C., and Vuilhorgne, M. (1991). Structure elucidation of RP 63834, a new macrocyclic lactone antibiotic. *Tetrahedron* **47**, 61–70.
56. Harada, K., Kimura, I., Sakazaki, T., Murata, H., and Suzuki, M. (1989). Structural investigation of the antibiotic sporaviridin. XIV. Isolation of components of intact sporaviridin. *J. Antibiot.* **42**, 1056–1062.
57. Uri, J., Bognar, R., Bekesi, I., and Varga, B. (1958). Desertomycin, a new crystalline antibiotic with antibacterial and cytostatic action. *Nature* **182**, 401.
58. Bax, A., Aszalos, A., Dinya, Z., and Sudo, K. (1986). Structure elucidation of the antibiotic desertomycin through the use of new two-dimensional NMR techniques. *J. Am. Chem. Soc.* **108**, 8056–8063.
59. Akasaki, K., Karasawa, K., Watanabe, M., Yonehara, H., and Umezawa, H. (1963). Monazomycin, a new antibiotic produced by a *Streptomyces*. *J. Antibiot. Ser. A* **16**, 127–131.
60. Grabley, S., Kretzschmar, G., Mayer, M., Philipps, S., Thiericke, R., Wink, J., and Zeeck, A. (1993). Oasomycins, new macrolactones of the desertomycin family. *Liebigs Ann. Chem.* 573–579.
61. Zerlin, M., and Thiericke, R. (1994). Common principles in macrolactone (marginolactone) biosynthesis. Studies on the desertomycin family. *J. Org. Chem.* **59**, 6986–6993.
62. Kim, J. W., Adachi, H., Shin-ya, K., Hayakawa, Y., and Seto, H. (1997). Apoptolidin, a new apoptosis inducer in transformed cells from *Nocardiopsis* sp. *J. Antibiot.* **50**, 628–630.
63. Akita, E., Maeda, K., and Umezawa, H. (1963). Isolation and characterization of labilomycin, a new antibiotic. *J. Antibiot. Ser. A* **16**, 147–151.
64. Hayakawa, Y., Matsuoka, M., Shin-ya, K., and Seto, H. (1993). Quinolidomycins A_1, A_2 and B_1, novel 60-membered macrolide antibiotics. I. Taxonomy, fermentation, isolation, physico-chemical properties and biological activity. *J. Antibiot.* **46**, 1557–1562.
65. Arcamone, F., Bertazzoli, C., Canevazzi, G., DiMarco, A., Ghione, M., and Grein, A. (1957). La estruscomycina, nuovo antibiotico prodotto dallo *Streptomyces lucensis* n. sp. *Giorn. Microbiol.* **4**, 119–128.
66. Davisson, J. W., Tanner, Jr., F. W., Finlay, A. C., and Solomons, I. A. (1951). Rimocidin, a new antibiotic. *Antibiot. Chemother.* **1**, 289–290.
67. Hazen, E. L., and Brown, R. (1951). Fungicidin, an antibiotic produced by a soil actinomycete. *Proc. Soc. Exp. Biol. Med.* **76**, 93–97.
68. Frolova, V. I., Kuzovkov, A. D., Ushakova, T. A., and Elizarova, R. N. (1967). Isolation of the new pentaenic antibiotic aurenin. *Khim. Farm. Zh.* **1**, 36–39.
69. Whitfield, G. B., Brock, T. D., Ammann, A., Gottlieb, D., and Carter, H. E. (1955). Filipin, an antifungal antibiotic: Isolation and properties. *J. Am. Chem. Soc.* **77**, 4799–4801.
70. Maehr, H., Yang, R., Hong, L. N., Liu, C. M., Hatada, M. H., and Todaro, L. J. (1989). Microbial products. 9. Roxaticin, a new oxo pentaene antibiotic. *J. Org. Chem.* **54**, 3816–3819.
71. Uri, J., and Békési, I. (1958). Flavofungin, a new crystalline antifungal antibiotic: Origin and biological properties. *Nature* **181**, 908.
72. Schlegel, R., and Thrum, H. (1968). Flavomycoine, a new antifungal polyene antibiotic. *Experientia* **24**, 11–12.
73. Brazhnikova, M. G., Kudinova, M. K., Lavrova, M. F., Borisova, V. N., Kruglyak, E. B., Kovsharova, I. N., and Proshlyakova, V. V. (1971). Isolation, purification, and physicochemical properties of lienomycin. *Antibiotiki* **16**, 483–487.
74. Narasimhachari, N., Ramachandran, S., and Swami, M. B. (1966). Purification of dermostatin, an antifungal antibiotic from *Streptomyces*. *Hind. Antibiot. Bull.* **8**, 111–112.
75. Gold, W., Stout, H. A., Pagano, J. F., and Donovick, R. (1955–1956). Amphotericins A and B, antifungal antibiotics produced by a streptomycete. I. *In vitro* studies. *Antibiot. Ann.* 579–586.
76. Leben, C., and Keitt, G. W. (1948). An antibiotic substance active against certain phytopathogens. *Phytopathology* **38**, 899–906.

77. Arcamone, F., Bertazzoli, C., Ghione, M., and Scotti, T. (1959). Melanosporin and elaiophylin, new antibiotics from *Streptomyces melanosporus*. *Giorn. Microbiol.* **7**, 207–216.
78. Kondo, S., Yasu, K., Natsume, M., Katayama, M., and Marumo, S. (1988). Isolation, physicochemical properties and biological activity of pamamycin-607, an aerial mycelium-inducing substance from *Streptomyces alboniger*. *J. Antibiot.* **41**, 1196–1204.
79. Singh, S. B., Liesch, J. M., Lingham, R, B., Goetz, M, A., and Gibbs, J. B. (1994). Actinoplanic acid A: A macrocyclic polycarboxylic acid which is a potent inhibitor of Ras farnesyl-protein transferase. *J. Am. Chem. Soc.* **116**, 11606–11607.
80. Huetter, R., Keller-Schierlein, W., Knuesel, F., Prelog, V., Rodgers, G. C., Jr., Suter, P., Vogel, G., Voser, W., and Zaehner, H. (1967). Metabolic products of microorganisms. LVII. Boromycin. *Helv. Chim. Acta* **50**, 1533–1539.
81. Cassinelli, G., Grein, A., Orezzi, P., Pennella, P., and Sanfilippo, A. (1966). New antibiotics produced by *Streptoverticillium orinoci*. *Arch. Mikrobiol.* **55**, 358–368.
82. Corbaz, R., Ettlinger, L., Gäumann, E., Keller-Schierlein, W., Kradolfer, F., Neipp, L., Prelog, V., and Zähner, H. (1955). Stoffwechselprodukte von Actinomyceten. 3. Mitteilung. Nonactin. *Helv. Chim. Acta* **38**, 1445–1448.
83. Hasegawa, T., Kamiya, T., Henmi, T., Iwasaki, H., and Yamatodani, S. (1975). Viridenomycin, a new antibiotic. *J. Antibiot.* **28**, 167–175.
84. Nakagawa, M., Toda, Y., Furihata, K., Hayakawa, Y., and Seto, H. (1992). Studies on viridenomycin, a novel 24-membered macrocyclic polyene lactam antibiotic. *J. Antibiot.* **45**, 1133–1138.
85. Kino, T., Hatanaka, H., Hashimoto, M., Nishiyama, M., Goto, T., Okuhara, M., Kosaka, M., Aoki, H., and Imanaka, H. (1987). FK-506, a novel immunosuppressant isolated from a *Streptomyces*. I. Fermentation, isolation, and physico-chemical and biological characteristics. *J. Antibiot.* **40**, 1249–1255.
86. Vezina, C., Kudelski, A., and Sehgal, S. N. (1975). Rapamycin (AY-22,989), a new antifungal antibiotic. I. Taxonomy of the producing streptomycete and isolation of the active principle. *J. Antibiot.* **28**, 721–726.
87. Sanglier, J. J., Quesniaux, V., Fehr, T., Hofmann, H., Mahnke, M., Memmert, K., Schuler, W., Zenke, G., Gschwind, L., Maurer, C., and Schilling, W. (1999). Sanglifehrins A, B, C and D, novel cyclophilin-binding compounds isolated from *Streptomyces* sp. A92-308110. I. Taxonomy, fermentation, isolation and biological activity. *J. Antibiot.* **52**, 466–473.
88. Nishikiori, T., Masuma, R., Ōiwa, R., Katagiri, M., Awaya, J., Iwai, Y., and Ōmura, S. (1978). Aurantinin, a new antibiotic of bacterial origin. *J. Antibiot.* **31**, 525–532.
89. George, K. M., Chatterjee, D., Gunawardana, G., Welty, D., Hayman, J., Lee, R., and Small, P. L. C. (1999). Mycolactone: A polyketide toxin from *Mycobacterium ulcerans* required for virulence. *Science* **283**, 854–857.
90. Gunawardana, G., Chatterjee, D., George, K. M., Brennan, P., Whittern, D., and Small, P. L. C. (1999). Characterization of novel macrolide toxins, mycolactones A and B, from a human pathogen *Mycobacterium ulcerans*. *J. Am. Chem. Soc.* **121**, 6092–6093.
91. Kim, J. W., Shin-ya, K., Furihata, K., Hayakawa, Y., and Seto, H. (1999). Oximidines I and II: Novel antitumor macrolides from *Pseudomonas* sp. *J. Org. Chem.* **64**, 153–155.
92. Boyd, M. R., Farina, C., Belfiore, P., Gagliardi, S., Kim, J. W., Hayakawa, Y., Beutler, J. A., McKee, T. C., Bowman, B. J., and Bowman, E. J. (2001). Discovery of a novel antitumor benzolactone enamide class that selectively inhibits mammalian vacuolar-type (H^+)-ATPases. *J. Pharmacol. Exp. Ther.* **297**, 114–120.
93. Irschik, H., Augustiniak, H., Gerth, K., Höfle, G., and Reichenbach, H. (1995). The ripostatins, novel inhibitors of eubacterial RNA polymerase isolated from myxobacteria. *J. Antibiot.* **48**, 787–792
94. Singh, P. D., Liu, W. C., Gougoutas, J. Z., Malley, M. F., Porubcan, M. A., Trejo, W. H., Wells, J. S., and Sykes, R. B. (1988). Aerocavin, a new antibiotic produced by *Chromobacterium violaceum*. *J. Antibiot.* **41**, 446–453.

1. Discovery of New Macrolides 51

95. Gerth, K., Bedorf, N., Höfle, G., Irschik, H., and Reichenbach, H. (1996). Epothilons A and B: Antifungal and cytotoxic compounds from *Sorangium cellulosum* (Myxobacteria). Production, physico-chemical and biological properties. *J. Antibiot.* **49**, 560–563.
96. Bollag, D. M., McQueney, P. A., Zhu, J., Hensens, O., Koupal, L., Liesch, J., Goetz, M., Lazarides, E., and Woods, C. M. (1995). Epothilones, a new class of microtubule-stabilizing agents with a taxol-like mechanism of action. *Cancer Res.* **55**, 2325–2333.
97. Bedorf, N., Schomburg, D., Gerth, K., Reichenbach, H., and Hoefle, G. (1993). Antibiotics from gliding bacteria. LIV. Isolation and structure elucidation of soraphen $A_{1\alpha}$, a novel antifungal macrolide from *Sorangium cellulosum*. *Liebigs Ann. Chem.* 1017–1021.
98. Vahlensieck, H. F., Pridzun, L., Reichenbach, H., and Hinnen, A. (1994). Identification of the yeast ACC1 gene product (acetyl-CoA carboxylase) as the target of the polyketide fungicide soraphen A. *Curr. Genet.* **25**, 95–100.
99. Hoefle, G., Reichenbach, H., Irschik, H., and Schummer, D. (1998). Etnangiene manufacture with *Sorangium*. German Offen. 19630980.
100. Gustafson, K., Roman, M., and Fenical, W. (1989). The macrolactins, a novel class of antiviral and cytotoxic macrolides from a deep-sea marine bacterium. *J. Am. Chem. Soc.* **111**, 7519–7524.
101. Nagao, T., Adachi, K., Araki, M., Nishijima, M., Chin, Y., and Sano, M. (1997). New macrolactins manufacturing with *Bacillus* sp. *Jpn. Kokai Tokkyo Koho* 09-255677.
102. Kamimura, D., Kuramoto, M., Yamada, K., Yazawa, K., and Kano, M. (1997). Oligomycin SC compounds of *Pantoea agglomerans* as anticancer agents. *Jpn. Kokai Tokkyo Koho* 09-208587.
103. Yamauchi, T., Nakajima, S., Hirayama, M., Ojiri, K., and Suda, H. (1998). Antitumor substance BE-56384 and its production. *Jpn. Kokai Tokkyo Koho* 10-101676.
104. Zimmerman, S. B., Schwartz, C. D., Monaghan, R. L., Pelak, B. A., Weissberger, B., Gilfillan, E. C., Mochales, S., Hernandez, S., Currie, S. A., Tejera, E., and Stapley, E. O. (1987). Difficidin and oxydifficidin: Novel broad spectrum antibacterial antibiotics produced by *Bacillus subtilis*. I. Production, taxonomy and antibacterial activity. *J. Antibiot.* **40**, 1677–1681.
105. Hoefle, G., Reichenbach, H., Sasse, F., and Steinmetz, H. (1993). Archazolide A, its manufacture with *Archangium gephyra*, and its use as neoplasm inhibitor and fungicide. German Patent. 4142951.
106. Jansen, R., Wray, V., Irschik, H., Reichenbach, H., and Hoefle, G. (1985). Antibiotics from gliding bacteria. XXX. Isolation and spectroscopic structure elucidation of sorangicin A, a new type of macrolide-polyether antibiotic from gliding bacteria. *Tetrahedron Lett.* **26**, 6031–6034.
107. Irschik, H., Jansen, R., Gerth, K., Hoefle, G., and Reichenbach, H. (1987). Antibiotics from gliding bacteria. 32. The sorangicins, novel and powerful inhibitors of eubacterial RNA polymerase isolated from myxobacteria. *J. Antibiot.* **40**, 7–13.
108. Suzumura, K., Takahashi, I., Matsumoto, H., Nagai, K., Setiawan, B., Rantiatmodjo, R. M., Suzuki, K., and Nagano, N. (1997). Structural elucidation of YM-75518, a novel antifungal antibiotic isolated from *Pseudomonas* sp. Q38009. *Tetrahedron Lett.* **38**, 7573–7576.
109. Schummer, D., Irschik, H., Reichenbach, H., and Hoefle, G. (1994). Tartrolons: New boron-containing macrodiolides from *Sorangium cellulosum*. *Liebigs Ann. Chem.* 283–289.
110. Pollack, J. R., and Neilands, J. B. (1970). Enterobactin, an iron transport compound from *Salmonella typhimurium*. *Biochem. Biophys. Res. Commun.* **38**, 989–992.
111. Kohl, W., Witte, B., Kunze, B., Wray, V., Schomburg, D., Reichenbach, H., and Hoefle, G. (1985). Antibiotics from gliding bacteria. XXVII. Angiolam A. A novel antibiotic from *Angiococcus disciformis* (Myxobacterales). *Liebigs Ann. Chem.* 2088–2097.
112. Gerth, K., Irschik, H., Reichenbach, H., and Trowitzsch, W. (1982). The myxovirescins, a family of antibiotics from *Myxococcus virescens* (Myxobacterales). *J. Antibiot.* **35**, 1454–1459.
113. Miyashiro, S., Yamanaka, S., Takayama, S., and Shibai, H. (1988). Novel macrocyclic antibiotics: Megovalicins A, B, C, D, G and H. I. Screening of antibiotic-producing myxobacteria and production of megovalicins. *J. Antibiot.* **41**, 433–438.

114. Trowitzsch-Kienast, W., Schober, K., Wray, V., Gerth, K., Reichenbach, H., and Höfle, G. (1989). Antibiotika aus Gleitenden Bakterien, XLI. Zur Konstitution der Myxovirescine B-T und Biogenese des Myxovirescins A. *Liebigs Ann. Chem.* 345–355.
115. Sasse, F., Steinmetz, H., Hoefle, G., and Reichenbach, H. (1993). Antibiotics from gliding bacteria. 49. Rhizopodin, a new compound from *Myxococcus stipitatus* (myxobacteria) causes formation of rhizopodia-like structures in animal cell cultures: Production, isolation, physico-chemical and biological properties. *J. Antibiot.* **46**, 741–748.
116. Gronewold, T. M. A., Sasse, F., Liinsdorf, H., and Reichenbach, H. (1999). Effects of rhizopodin and latrunculin B on the morphology and on the actin cytoskeleton of mammalian cells. *Cell Tissue Res.* **295**, 121–129.
117. Irschik, H., Jansen, R., Gerth, K., Hoefle, G., and Reichenbach, H. (1995). Antibiotics from gliding bacteria. 66. Chivosazol A, a new inhibitor of eukaryotic organisms isolated from myxobacteria. *J. Antibiot.* **48**, 962–966.
118. Jansen, R., Irschik, H., Reichenbach, H., Wray, V., and Hoefle, G. (1994). Antibiotics from gliding bacteria. LIX. Disorazoles, highly cytotoxic metabolites from the sorangicin-producing bacterium *Sorangium cellulosum*, strain So ce12. *Liebigs Ann. Chem.* 759–773.
119. Cutler, H. G., Jacyno, J. M., Harwood, J. S., Dulik, D., Goodrich, P. D., and Roberts, R. G. (1993). Botcinolide: A biologically active natural product from *Botrytis cinerea*. *Biosci. Biotechnol. Biochem.* **57**, 1980–1982.
120. Gill, M., and Gimenez, A. (1990). Pigments of fungi. 18. The dermocanarins, unique macrocyclic lactones from the fungus *Dermocybe canaria*. *Tetrahedron Lett.* **31**, 3505–3508.
121. Grabley, S., Granzer, E., Hütter, K., Ludwig, D., Mayer, M., Thiericke, R., Till, G., Wink, J., Philipps, S., and Zeeck, A. (1992). Secondary metabolites by chemical screening. 8. Decarestrictines, a new family of inhibitors of cholesterol biosynthesis from *Penicillium*. I. Strain description, fermentation, isolation and properties. *J. Antibiot.* **45**, 56–65.
122. Grabley, S., Hammann, P., Hütter, K., Kirsch, R., Kluge, H., Thiericke, R., Mayer, M., and Zeeck, A. (1992). Secondary metabolites by chemical screening. 20. Decarestrictines, a new family of inhibitors of cholesterol biosynthesis from *Penicillium*. III. Decarestrictines E to M. *J. Antibiot.* **45**, 1176–1181.
123. Evidente, A., Lanzetta, R., Capasso, R., Vurro, M., and Bottalico, A. (1993). Pinolidoxin, a phytotoxic nonenolide from *Ascochyta pinodes*. *Phytochemistry* **34**, 999–1003.
124. Sekiguchi, J., Kuroda, H., Yamada, Y., and Okada, H. (1985). Structure of patulolide A, a new macrolide from *Penicillium urticae* mutants. *Tetrahedron Lett.* **26**, 2341–2342.
125. Dekker, K. A., Aiello, R. J., Hirai, H., Inagaki, T., Sakakibara, T., Suzuki, Y., Thompson, J. F., Yamauchi, Y., and Kojima, N. (1998). Novel lactone compounds from *Mortierella verticillata* that induce the human low density lipoprotein receptor gene: Fermentation, isolation, structural elucidation and biological activities. *J. Antibiot.* **51**, 14–20.
126. Singleton, V. I., Bohonos, N., and Ullstrup, A. J. (1958). Decumbin, a new compound from a species of *Penicillium*. *Nature* **181**, 1072–1073.
127. Weigert, R., Colanzi, A., Limina, C., Cericola, C., Di Tullio, G., Mironov, A., Santini, G., Sciulli, G., Corda, D., De Matteis, M. A., and Luini, A. (1997). Characterization of the endogenous mono-ADP-ribosylation stimulated by brefeldin A. *Adv. Exp. Med. Biol.* **419** (ADP-Ribosylation in Animal Tissues), 337–342.
128. Stob, M., Baldwin, R. S., Tuite, J., Andrews, F. N., and Gillette, K. G. (1962). Isolation of an anabolic, uterotrophic compound from corn infected with *Gibberella zeae*. *Nature* **196**, 1318.
129. Hodge, E. B., Hidy, P. H., and Wehrmeister, H. J. (1966). Estrogenic compounds and animal growth promoters. U.S. Patent 3239345.
130. Delmotte, P., and Delmotte-Plaquee, J. (1953). A new antifungal substance of fungal origin. *Nature* **171**, 344.

131. Kwon, H. J., Yoshida, M., Fukui, Y., Horinouchi, S., and Beppu, T. (1992). Potent and specific inhibition of p60v-*src* protein kinase both *in vivo* and *in vitro* by radicicol. *Cancer Res.* **52**, 6926–6930.
132. Zhao, A., Lee, S. H., Mojena, M., Jenkins, R. G., Patrick, D. R., Huber, H. E., Goetz, M. A., Hensens, O. D., Zink, D. L., Vilella, D., Dombrowski, A. W., Lingham, R. B., and Huang, L. (1999). Resorcylic acid lactones: Naturally occurring potent and selective inhibitors of MEK. *J. Antibiot.* **52**, 1086–1094.
133. Iwasaki, S., Kobayashi, H., Furukawa, J., Namikoshi, M., Okuda, S., Sato, Z., Matsuda, I., and Noda, T. (1984). Studies on macrocyclic lactone antibiotics. VII. Structure of a phytotoxin "rhizoxin" produced by *Rhizopus chinensis*. *J. Antibiot.* **37**, 354–362.
134. Takahashi, M., Iwasaki, S., Kobayashi, H., Okuda, S., Murai, T., Sato, Y., Haraguchi-Hiraoka, T., and Nagano, H. (1987). Studies on macrocyclic lactone antibiotics. XI. Antimitotic and antitubulin activity of new antitumor antibiotics, rhizoxin and its homologs. *J. Antibiot.* **40**, 66–72.
135. Brown, S. S., and Spudich, J. A. (1979). Cytochalasin inhibits the rate of elongation of actin filament fragments. *J. Cell Biol.* **83**, 657–662.
136. Kimura, Y., Nakajima, H., and Hamasaki, T. (1989). Structure of rosellichalasin, a new metabolite produced by *Rosellinia necatrix*. *Agric. Biol. Chem.* **53**, 1699–1701.
137. Rothweiler, W., and Tamm, C. (1966). Isolation and structure of phomin. *Experientia* **22**, 750–752.
138. Evidente, A., Lanzetta, R., Capasso, R., Vurro, M., and Bottalico, A. (1992). Cytochalasins U and V, two new cytochalasans, from *Phoma exigua* var. *heteromorpha*. *Tetrahedron* **48**, 6317–6324.
139. Hanson, K., O'Neill, J. A., Simpson, T. J., and Willis, C. L. (1994). Bartanol and bartallol, novel macrodiolides from *Cytospora* sp. ATCC 20502. *J. Chem. Soc. Perkin Trans.* **1**, 2493–2497.
140. Grabley, S., Hammann, P., Thiericke, R., Wink, J., Philipps, S., and Zeeck, A. (1993). Secondary metabolites by chemical screening. 21. Clonostachydiol, a novel anthelmintic macrodiolide from the fungus *Clonostachys cylindrospora* (strain FH-A 6607). *J. Antibiot.* **46**, 343–345.
141. Shifrin, V. I., and Anderson, P. (1999). Trichothecene mycotoxins trigger a ribotoxic stress response that activates c-Jun *N*-terminal kinase and p38 mitogen-activated protein kinase and induces apoptosis. *J. Biol. Chem.* **274**, 13985–13992.
142. Jarvis, B. B., and Yatawara, C. S. (1986). Roritoxins, new macrocyclic trichothecenes from *Myrothecium roridum*. *J. Org. Chem.* **51**, 2906–2910.
143. Haerri, E., Loeffler, W., Sigg, H. P., Staehelin, H., Stoll, C., Tamm, C., and Wiesinger, D. (1962). On the verrucarins and roridins, a group of highly cytostatically active antibiotics from *Myrothecium* species. *Helv. Chim. Acta* **45**, 839–853.
144. Hayashi, M., Kim, Y.-P., Hiraoka, H., Natori, M., Takamatsu, S., Kawakubo, T., Masuma, R., Komiyama, K., and Ōmura, S. (1995). Macrosphelide, a novel inhibitor of cell-cell adhesion molecule. I. Taxonomy, fermentation, isolation and biological activities. *J. Antibiot.* **48**, 1435–1439.
145. Ito, M., Maruhashi, M., Sakai, N., Mizoue, K., and Hanada, K. (1992). NG-011 and NG-012, novel potentiators of nerve growth factor. I. Taxonomy, isolation, and physico-chemical and biological properties. *J. Antibiot.* **45**, 1559–1565.
146. Breinholt, J., Jensen, G. W., Nielsen, R. I., Olsen, C. E., and Frisvad, J. C. (1993). Antifungal macrocyclic polylactones from *Penicillium verruculosum*. *J. Antibiot.* **46**, 1101–1108.
147. Hesse, O. (1900). Beitrag zur Kenntniss der Flechten und ihrer charakteristischen Bestandtheile. *J. Prakt. Chem.* **62**, 430–480.
148. Polborn, K., Steglich, W., Connolly, J. D., and Huneck, S. (1995). Structure of the macrocyclic bis-lactone lepranthin from the lichen *Arthonia impolita;* an X-ray analysis. *Z. Naturforsch. B: Chem. Sci.* **50**, 1111–1114.
149. Suwarso, W., Priyono, Gani, R. L., Krohn, K., and John, M. (1999). Dasypogalactone, a new C3-symmetric macrolactone from the Indonesian lichen *Usnea dasypoga*. *Eur. J. Org. Chem.* 1719–1721.

150. Bermejo, A., Tormo, J. R., Cabedo, N., Estornell, E., Figadere, B., and Cortes, D. (1998). Enantiospecific semisynthesis of (+)-almuheptolide A, a novel natural heptolide inhibitor of the mammalian mitochondrial respiratory chain. *J. Med. Chem.* **41**, 5158–5166.
151. Bohlmann, F., Zdero, C., Jakupovic, J., Gerke, T., Wallmeyer, M., King, R. M., and Robinson, H. (1984). New sesquiterpene lactones and rosane derivatives from *Trichogonia* species. *Liebigs Ann. Chem.* 162–185.
152. Lee, K. H., Hayashi, N., Okano, M., Hall, I. H., Wu, R. Y., and McPhail, A. T. (1982). Antitumor agents. Part 51. Lasiodiplodin, a potent antileukemic macrolide from *Euphorbia splendens*. *Phytochemistry* **21**, 1119–1121.
153. Mookherjee, B. D., and Wilson, R. A. (1982). The chemistry and fragrance of natural musk compounds. *In* "Fragrance Chemistry. The Science of the Sense of Smell" (E. T. Theimer, Ed.), pp. 433–494. Academic Press, London.
154. Naves, Y. R., and Grampoloff, A. V. (1942). Etudes sur les matieres vegetales volatiles XX. Sur la composition de l'extrit etheropetrolique (essence concrete) de la fleur de jasmin. *Helv. Chim. Acta* **25**, 1500–1514.
155. Matsuura, Y., Hata, G., Abe, S., Sakai, I., and Abe, A. (1980). Tech. Data of the 8th International Congress of Essential Oils, Cannes, France, pp. 497–499.
156. Taskinen, J., and Nykanen, L. (1975). Chemical composition of angelica root oil. *Acta Chem. Scand. Ser. B* **29**, 757–764.
157. Maurer, B., and Grieder, A. (1977). (Z)-5-Tetradecen-14-olide, a new macrocyclic lactone, and two unsaturated straight chain acetates from ambrette seed absolute. *Helv. Chim. Acta* **60**, 1155–1160.
158. Kerschbaum, M. (1927). Über Lactone mit großen Ringen—die Träger des vegetabilischen Moschus-Duftes. *Ber. Dtsch. Chem. Ges.* **60**, 902–909.
159. Kaiser, R., and Lamparsky, D. (1978). New macrolides and some sesquiterpenoid derivatives from Galbanum resin. *Helv. Chim. Acta* **61**, 2671–80.
160. Demole, E., and Enggist, P. (1978). Identification of twenty-one novel constituents of oriental tobacco flavor (*Nicotiana tabacum* L.) including (E)-3-methyl-non-2-en-4-one, pentadecan-15-olide, 8α,13:9α,13-diepoxy-15,16-dinorlabdane, (Z)-octadec-9-en-18-olide, and (E)-2-ethylidene-6,10,14-trimethylpentadecanal. *Helv. Chim. Acta* **61**, 2318–2327.
161. Scholz, E., Heinrich, M., and Hunkler, D. (1994). Caffeoylquinic acids and some biological activities of *Pluchea symphytifolia*. *Planta Med.* **60**, 360–364.
162. Yoshida, T., Toyota, M., and Asakawa, Y. (1997). Scapaundulins A and B, two novel dimeric labdane diterpenoids, and related compounds from the Japanese liverwort *Scapania undulata* (L.) Dum. *Tetrahedron Lett.* **38**, 1975–1978.
163. Kubota, T., Ichikawa, N., and Kamikawa, T. (1968). Isolation and characteristics of jasminin, a bitter component from the leaves of *Jasminum primulinum*. *Nippon Kagaku Zasshi* **89**, 62–65.
164. Bergquist, K. E., Obianwu, H., and Wickberg, B. (1989). Isolation and structure determination of a novel phorbol derivative in an intramolecular diester macrolide. *J. Chem. Soc. Chem. Commun.* 183–184.
165. Zdero, C., and Bohlmann, F. (1988). Macrolide diterpenes and other *ent*-labdanes from *Corymbium villosum*. *Phytochemistry* **27**, 227–231.
166. Okuda, T., Yoshida, T., Hatano, T., Yazaki, K., and Ashida, M. (1982). Ellagitannins of the *Casuarinaceae, Stachyuraceae*, and *Myrtaceae*. *Phytochemistry* **21**, 2871–2874.
167. Schmidt, O. T., Würtele, L., and Harréus, A. (1965). Über natürliche Gerbstoffe, XXXIII. Pedunculagin, eine 2.3;4.6-Di-[(−)-hexahydroxy-diphenoyl]-glucose aus Knoppern. *Ann. Chem.* **690**, 150–162.
168. Okuda, T., Yoshida, T., and Nayeshiro, H. (1976). Geraniin, a new ellagitannin from *Geranium thunbergii*. *Tetrahedron Lett.* 3721–3722.
169. Rizk, A.-F. M. (Ed.) (1990). "Naturally Occurring Pyrrolizidine Alkaloids," CRC Press, Boca Raton, FL.

1. Discovery of New Macrolides   55

170. Ali, M. A., and Adil, G. A. (1973). Isolation and structure of crotalarine, a new alkaloid from *Crotalaria burhia*. *Pak. J. Sci. Ind. Res.* **16**, 227–229.
171. Orekhov, A. (1935). Über *Senecio* Alkaloide. I. Mitteil: Die Alkaloide von *Senicio platyphyllus*. D.C. *Chem. Ber.* **68**, 650.
172. Roeder, E., Wiedenfeld, H., and Frisse, M. (1980). Pyrrolizidinalkaloid aus *Senecio doronicum*. *Phytochemistry* **19**, 1275–1277.
173. Atal, C. K., Kapur, K. K., Culvenor, C. C. J., and Smith, L. W. (1966). A new pyrrolizidine amino alcohol in alkaloids from *Crotalaria* species. *Tetrahedron Lett.* 537–544.
174. Kam, T.-S., Yoganathan, K., and Li, H.-Y. (1996). Tenuisines A, B and C, novel bisindoles with C2 symmetry from *Kopsia tenuis*. *Tetrahedron Lett.* **37**, 8811–8814.
175. Pailer, M., and Libiseller, R. (1962). *Euonymus* alkaloids. I. Purification and constitution of evonine from *E. europaeus*. *Monatsh. Chem.* **93**, 403–16.
176. Acree, F. Jr., and Haller, H. L. (1950). Wilfordine, an insecticidal alkaloid from *Tripterygium wilfordie* Hook. *J. Am. Chem. Soc.* **72**, 1608–1611.
177. Aggarwal, M. L., Ray, J. N., and Sen, D. C. (1946). Chaksine. *Sci. Culture* **12**, 201.
178. Rall, G. J. H., Smalberger, T. M., De Waal, H. L., and Arndt, R. R. (1967). Dimeric piperdine alkaloids from *Azima tetracantha* azimine, azcarpine, and carpaine. *Tetrahedron Lett.* 3465–3469.
179. Eggers, N. J., and Gainsford, G. J. (1979). Parsonsine ($C_{22}H_{33}NO_8$): A pyrrolizidine alkaloid from *Parsonsia heterophylla* A. Cunn. *Cryst. Struct. Commun.* **8**, 597–603.
180. Kawabata, J., and Mizutani, J. (1992). Studies on the chemical constituents of Chloranthaceae plants. Part 8. Dimeric sesquiterpenoid esters from *Chloranthus serratus*. *Phytochemistry* **31**, 1293–1296.
181. Moore, B. P., and Brown, W. V. (1976). The chemistry of the metasternal gland secretion of the eucalypt longicorn *Phoracantha synonyma* (Coleoptera: Cerambycidae). *Aust. J. Chem.* **29**, 1365–1374.
182. Wong, J. W., Verigin, V., Oehlschlager, A. C., Borden, J. H., Pierce, H. D., Jr., Pierce, A. M., and Chong, L. (1983). *Cryptolestes ferrugineus* (Coleoptera: Cucujidae). *J. Chem. Ecol.* **9**, 451–474.
183. Millar, J. G., Pierce, H. D., Jr., Pierce, A. M., Oehlschlager, A. C., Borden, J. H., and Barak, A. V. (1985). Aggregation pheromones of the flat grain beetle, *Cryptolestes pusillus* (Coleoptera: Cucujidae). *J. Chem. Ecol.* **11**, 1053–1070.
184. Millar, J. G., Pierce, H. D., Jr., Pierce, A. M., Oehlschlager, A. C., and Borden, J. H. (1985). Aggregation pheromones of the grain beetle, *Cryptolestes turcicus* (Coleoptera: Cucujidae). *J. Chem. Ecol.* **11**, 1071–1081.
185. Leal, W. S., Soares, Kuwahara, S., Shi, X., Higuchi, H., Marino, C. E. B., Ono, M., and Meinwald, J. (1998). Male-released sex pheromone of the stink bug *Piezodorus hybneri*. *J. Chem. Ecol.* **24**, 1817–1829.
186. Bergstrom, G. (1974). Natural odoriferous compounds. X. Macrocyclic lactones in the Dufour gland secretion of the solitary bees *Colletes cunicularius* and *Halictus calceatus* (Hymenoptera, Apidae). *Chem. Scr.* **5**, 39–46.
187. Clarke, S. R., Dani, F. R., Jones, G. R., Morgan, E. D., and Turillazzi, S. (1999). Chemical analysis of the swarming trail pheromone of the social wasp *Polybia sericea* (hymenoptera: vespidae). *J. Insect Physiol.* **45**, 877–883.
188. Smith, B. H., Carlson, R. G., and Frazier, J. (1985). Identification and bioassay of macrocyclic lactone sex pheromone of the halictine bee *Lasioglossum zephyrum*. *J. Chem. Ecol.* **11**, 1447–1456.
189. Prestwich, G. D. (1982). From tetracycles to macrocycles. Chemical diversity in the defense secretions of nasute termites. *Tetrahedron* **38**, 1911–1919.
190. Albans, K. R., Aplin, R. T., Brehcist, J., Moore, J. F., and O'Toole, C. (1980). Dufour's gland and its role in secretion of nest cell lining in bees of the genus *Colletes* (Hymenoptera: Colletidae). *J. Chem. Ecol.* **6**, 549–564.

191. Attygalle, A. B., McCormick, K. D., Blankespoor, C. L., Eisner, T., and Meinwald, J. (1993). Defense mechanisms of arthropods. 116. Azamacrolides: A family of alkaloids from the pupal defensive secretion of a ladybird beetle (*Epilachna varivestis*). *Proc. Natl. Acad. Sci. USA* **90**, 5204–5208.
192. Torigoe, K., Murata, M., Yasumoto, T., and Iwashita, T. (1988). Prorocentrolide, a toxic nitrogenous macrocycle from a marine dinoflagellate, *Prorocentrum lima*. *J. Am. Chem. Soc.* **110**, 7876–7877.
193. Nakamura, H., Asari, T., Ohizumi, Y., Kobayashi, J., Yamasu, T., and Murai, A. (1993). Isolation of zooxanthellatoxins, novel vasoconstrictive substances from the zooxanthella *Symbiodinium* sp. *Toxicon* **31**, 371–376.
194. Klein, D., Braekman, J.-C., Daloze, D., Hoffmann, L., and Demoulin, V. (1996). Laingolide, a novel 15-membered macrolide from *Lyngbya bouillonii* (Cyanophyceae). *Tetrahedron Lett.* **37**, 7519–7520.
195. Klein, D., Braekman, J. C., Daloze, D., Hoffmann, L., Castillo, G., and Demoulin, V. (1999). Madangolide and laingolide A, two novel macrolides from *Lyngbya bouillonii* (Cyanobacteria). *J. Nat. Prod.* **62**, 934–936.
196. Pettit, G. R., Herald, C. L., Doubek, D. L., Herald, D. L., Arnold, E., and Clardy, J. (1982). Isolation and structure of bryostatin 1. *J. Am. Chem. Soc.* **104**, 6846–6848.
197. D'Auria, M. V., Debitus, C., Paloma, L. G., Minale, L., and Zampella, A. (1994). Superstolide A: A potent cytotoxic macrolide of a new type from the New Caledonian deep water marine sponge *Neosiphonia superstes*. *J. Am. Chem. Soc.* **116**, 6658–6663.
198. Sakai, R., Higa, T., and Kashman, Y. (1986). Misakinolide A, an antitumor macrolide from the marine sponge *Theonella* sp. *Chem. Lett.* 1499–1502.
199. Carmely, S., and Kashman, Y. (1985). Structure of swinholide-A, a new macrolide from the marine sponge *Theonella swinhoei*. *Tetrahedron Lett.* **26**, 511–514.
200. Cantrell, C. L., Gustafson, K. R., Cecere, M. R., Pannell, L. K., and Boyd, M. R. (2000). Chondropsins A and B: Novel tumor cell growth-inhibitory macrolide lactams from the marine sponge *Chondropsis* sp. *J. Am. Chem. Soc.* **122**, 8825–8829.
201. Roesener, J. A., and Scheuer, P. J. (1986). Ulapualide A and B, extraordinary antitumor macrolides from nudibranch eggmasses. *J. Am. Chem. Soc.* **108**, 846–847.
202. Kobayashi, J., Kondo, K., Ishibashi, M., Walchli, M. R., and Nakamura, T. (1993). Theonezolide A: A novel polyketide natural product from the Okinawan marine sponge *Theonella* sp. *J. Am. Chem. Soc.* **115**, 6661–6665.
203. Northcote, P. T., Blunt, J. W., and Munro, M. H. G. (1991). Pateamine: A potent cytotoxin from the New Zealand marine sponge, *Mycale* sp. *Tetrahedron Lett.* **32**, 6411–6414.
204. Buglass, A. J., Darling, F. M. C., and Waterhouse, J. S. (1990). Analysis of the anal sac secretion of the Hyaenidae. *Chem. Signals Vertebr.* **5**, 65–69.
205. Chen, Y., Liu, Z., Song, F., and Liu, S. (1998). Analysis of chemical constituents in lipid from musk-rat musk by gas chromatography/mass spectrometry. *Fenxi Huaxue* **26**, 1142–1145.
206. Colton VI, S. W., and Downing, D. T. (1983). Variation in skin surface lipid composition among the Equidae. *Comp. Biochem. Physiol.* **75B**, 429–433.
207. Frost, M. L., Colton VI, S. W., Wertz, P. W., and Downing, D. T. (1984). Structures of the dienoic lactones of horse sebum. *Comp. Biochem. Physiol.* **78B**, 549–552.

Chapter 2

Discovery of New Macrolides from Marine Organisms

MASAMI ISHIBASHI

Graduate School of Pharmaceutical Sciences
Chiba University
Chiba, Japan

I.	Introduction	57
II.	Macrocyclic Lactones of Marine Organism Origin	58
	A. Marine Microorganisms	58
	B. Algae	61
	C. Sponges	67
	D. Other Invertebrates	80
III.	Concluding Remarks	89
	References	89

I. Introduction

Since 1984, when the precursor to this book, *Macrolide Antibiotics—Chemistry, Biology, and Practice*, edited by S. Ōmura, was published, a great number of new macrolides from marine organisms have been discovered. Most of these marine macrolides exhibit interesting biological activity and some of them play important roles as potential molecules for drug development or as tools for basic biological sciences. This chapter deals with recent studies on the isolation and structure determination of macrolides from marine organisms together with their unique biological activities. In this chapter, natural products possessing a macrocyclic lactone moiety are considered to be "macrolides," and most of them are likely to belong to polyketides from a biogenetic viewpoint.

Because so many marine macrolides have been reported in recent years, the selection of compounds here may seem arbitrary and depends on the interest of the reviewer to some extent. The compounds chosen in this chapter are classified on the basis of organisms from which the compounds were isolated, and this chapter contains sections on marine microorganisms, algae, sponges, and other invertebrates. Microalgae were assigned to the algae section in this chapter, though they may sometimes be assigned to microorganisms. For very excellent reviews on marine natural products, a continuous series by D. J. Faulkner has appeared in *Natural Products Reports* [1], covering all aspects of the literature on marine natural products. Many other reviews and books dealing with general or specialized topics in marine natural products research have been described recently [for example, 2–5].

II. Macrocyclic Lactones of Marine Organism Origin

A. Marine Microorganisms

1. Bacteria

Although defining "marine" microorganisms may not be easy because many isolates tolerate a wide range of salinity, usually those bacteria and fungi that are isolated from seawater media, marine sediments, or marine macroorganisms such as the surfaces of algae are called "marine" microorganisms. Therefore metabolites from marine microorganisms are sometimes found to have structures very similar to those obtained from terrestrial bacteria or fungi [6].

Macrolactins, represented by macrolactin A (**1**), were isolated from a taxonomically undefinable gram-positive marine bacterium, which was isolated from sediment at a depth of −980 m in the North Pacific, and the cultures were grown in a seawater-containing liquid culture. This macrolide (**1**) showed selective antibacterial activity against *Bacillus subtilis* and *Staphylococcus aureus* and inhibited B16-F10 murine melanoma cancer cells in *in vitro* assays (IC_{50} = 3.5 μg/ml). Compound **1** also showed significant inhibition of mammalian herpes simplex viruses (types I and II) (IC_{50} = 5.0 μg/ml) and protected T-lymphoblast cells against human HIV viral replication [7]. The stereochemistry of macrolactins was determined by a combination of oxidative degradation, chemical correlation, and ^{13}C-acetonide analysis of 1,3-diol moieties using isotopically enriched acetone [8]. Unfortunately, fermentation of this deep-sea bacterium was unreliable and macrolactins were no longer available in significant yields. Recently, two Asian groups reported isolation of macrolactins from cultured broths of marine bacteria of *Actinomadura* sp. and *Bacillus* sp. [9, 10].

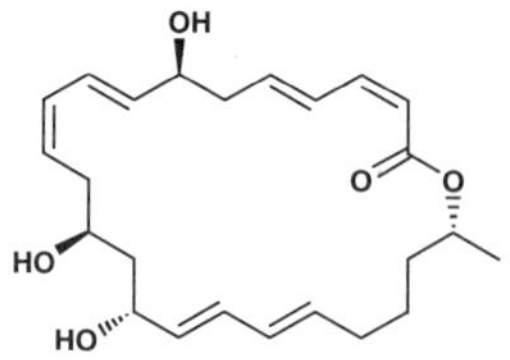

1: macrolactin A

Two 8-membered-ring lactones, octalactins A (**2**) and B (**3**), were isolated from a marine-derived actinomycete of the genus *Streptomyces*, collected from the surface of a gorgonian octocoral *Pacifigorgia* sp. [11], and the structure of **2** was firmly established by X-ray crystallographic analysis. The lactone ring of **2** adopted a boat-chair conformation with a *cis* lactone in the solid state, which might be stabilized by strong intermolecular hydrogen bonding between the lactone carbonyl and two hydroxyl groups. Octalactin A (**2**) showed strong cytotoxic activity toward B16-F10 murine melanoma and HCT-116 human colon tumor cell lines with IC_{50} values of 0.0072 and 0.5 µg/ml, respectively, whereas, interestingly, octalactin B (**3**) was completely inactive in the cytotoxicity assays.

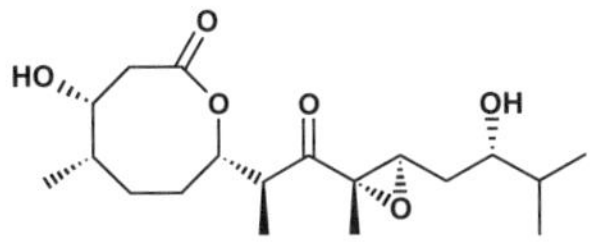

2: octalactin A

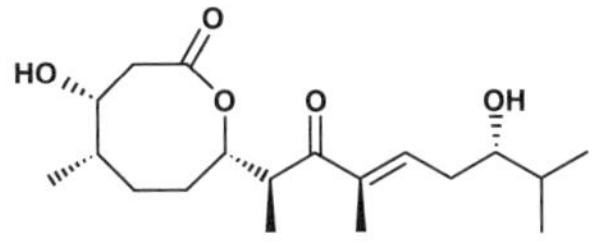

3: octalactin B

Although access to this natural product (**2**) is severely limited, total synthesis of **2** was achieved [for example, 12, 13], and several of its structural and stereochemical analogues were also prepared to test their ability to prevent murine L1210 leukemic cells from synthesizing macromolecules and growing *in vitro* [14].

From a liquid culture of a sediment-derived actinomycete found in the shallow waters of Bodega Bay, California, a unique 24-membered-ring macrolide glycoside, maduralide (**4**), was isolated. This producing organism was suggested to be a member of the suprageneric group Maduramycetes from its cell wall constituents and whole-cell sugar composition [15]. Maduralide (**4**) possesses 6-deoxy-3-*O*-methyl talose pyranoside, which was unprecedented among macrolide glycosides. This macrolide (**4**) showed weak antibiotic activity against *B. subtilis*, but further biological properties of **4** are not yet known.

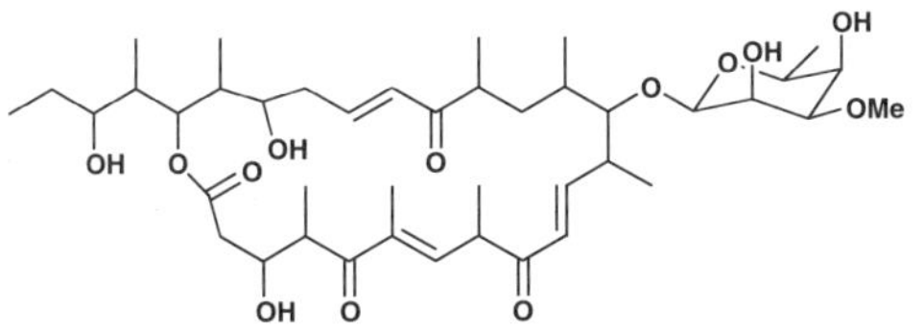

4: maduralide

Lobophorins A and B (**5** and **6**) were isolated from fermentation broths of a marine actinomycete isolated from the surface of the Caribbean brown alga *Lobophora variegata* [16]. Although lobophorins A and B (**5** and **6**) can be considered macrolides because they are related to the kijanimicin class of microbial antibiotics [17], they do not have significant antibiotic properties. However, they show potent anti-inflammatory activities in the PMA (phorbol-myristate-acetate)-induced mouse ear edema model. Related tetrocarcin-class antibiotics [18], arisostatins A and B (**7** and **8**), were isolated from the cultured broth of an actinomycete strain identified as *Micromonospora* sp., which was obtained from a seawater sample collected in Toyama Bay, Japan. These compounds (**7** and **8**) showed antibiotic activity against gram-positive bacteria and cytotoxicity against the human myeloid leukemia U937 cell line (IC_{50} for **7**, 0.4 μg/ml) as well as other breast, brain, colon, and lung cancer cell lines (IC_{50} for **7**, 0.059–0.26 μM). Arisostatin A (**7**) also inhibited the neuritogenesis of NGF-stimulated PC12 cells at less than 1 μM by inhibiting tubulin polymerization [19, 20].

5: lobophorin A, R = NH$_2$
6: lobophorin B, R = NO$_2$

7: arisostatin A, R = NO$_2$
8: arisostatin B, R = NH$_2$

From the mycelia extracts of a marine actinomycete of the genus *Micromonospora*, a macrolide, IB-96212 (**9**), was isolated and its structure was revealed to be a spiroketal-containing 26-membered macrocyclic lactone related to oligomycins [21]. IB-96212 (**9**) exhibited antimicrobial activity only against *Micrococcus luteus* (MIC, 0.4 μg/ml) out of six strains tested for that activity. The compound (**9**) showed a strong cytotoxic activity against the P-388 cell line (IC_{50}, 0.0001 μg/ml), and the activity on P-388 was about four orders of magnitude higher than that of the other three cell lines tested (A-549, HT-29, and MEL-28) [22, 23].

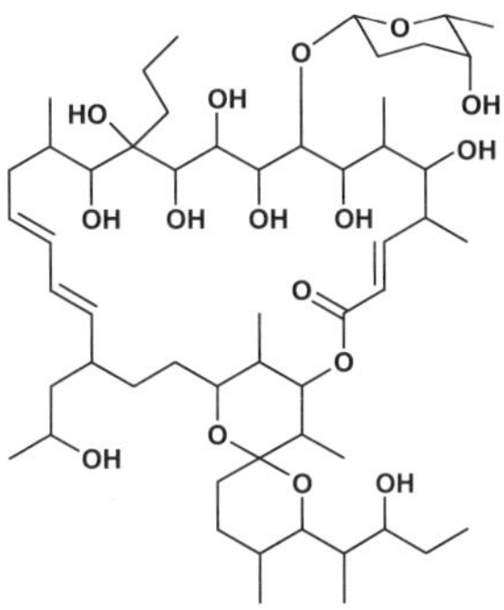

9: IB-96212

2. Fungi

A strain of the fungus *Periconia byssoides*, which was originally isolated from the gastrointestinal tract of the sea hare *Aplysia kurodai*, produced macrosphelide E (**10**) and its congeners [24]. Macrosphelides were first discovered in the fermentation broth of the terrestrial fungus *Macrosphaeropsis* sp. as a novel cell adhesion inhibitor [25].

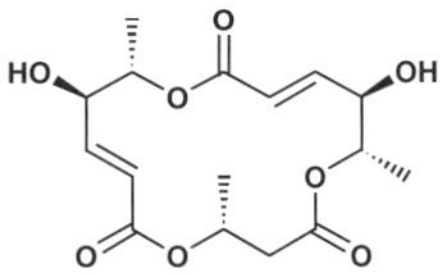

10: macrosphelide E

B. Algae

1. Red Algae

Although seafood poisoning due to macroalgae is rare, sudden and fatal intoxication due to ingestion of the red alga *Polycavernosa tsudai* (formerly *Gracilaria edulis*; Japanese name, ogonori) occurred in Guam in 1991. Thirteen people became ill, and three of them died. The alga had been widely eaten, but no potential risk had been previously recorded. Identification of the causative toxin was therefore imperative, and a novel glycosidic macrolide, polycavernoside A (**11**, 400 μg), was obtained from the alga (2.6 kg) and deemed the toxin responsible for the poisoning [26]. Its LD_{99} value in mice was 200–400 μg/kg, and the toxin (**11**) caused diarrhea, hypersalivation, lacrymation, muscle spasms, and cynosis.

The carbon skeleton of **11** was an unprecedented molecular entity, and its stereochemistry, including absolute configurations, was established by total synthesis [27]. Analogues of polycavernoside A (**11**) possessing a side chain at C-15 different from that of the natural toxin have been synthesized to evaluate their *in vivo* toxicities [28]. Previous outbreaks of fatal poisoning caused by other *Gracilaria* algae such as *G. chorda* and *G. verrucosa* remained unexplained as to the nature of the toxin.

11: polycavernoside A

Poisoning from another *Gracilaria* red alga, *Gracilaria coronopifolia*, in Hawaii occurred in succession in September 1994, and the causative agents of these poisoning incidents were studied to identify two major toxins, aplysiatoxin (**12**) and debromoaplysiatoxin (**13**) [29], which were previously obtained from the sea hare [30] and also from blue-green algae [31]. The existence of these toxins in the residue of algae washed in saline was confirmed by HPLC analysis. Furthermore, blue-green algal parasitism was observed on the surface of the toxic *G. coronopifolia*, suggesting that epiphytic organisms such as blue-green algae might be the true origin of the toxins in *G. coronopifolia*.

12: aplysiatoxin, R = Br
13: debromoaplysiatoxin, R = H

2. Blue-Green Algae

Blue-green algae (cyanophytes) are known to produce some potent toxins as well as valuable bioactive compounds [32], among which are some unique macrolides as discussed here. From a terrestrial scytonematacean alga, namely, *Tolypothrix conglutinata* var. *colorata*, collected at Fanning Island, a toxic 22-membered macrolide, tolytoxin (**14**), was isolated [33]. Tolytoxin (**14**) was also isolated from three species of *Scytonema*, namely, *S. mirabile*, *S. buranicum*, and *S. ocellatum*, and its structure was determined to be 6-hydroxy-7-*O*-methylscytophycin B [34]. Scytophycin B (**15**) was a major cytotoxin obtained from cultured *S. pseudohofmanni*, and its structural assignment including absolute stereochemistry was based on X-ray analysis and CD studies [35]. Tolytoxin (**14**) was highly toxic to mice, exhibiting an LD_{50} (ip) of 1.5 mg/kg, and it also showed potent antifungal activity with MIC values in the range of 0.25 to 8 n*M*. Tolytoxin (**14**) inhibited the growth of a variety of mammalian cells at similar doses, without specific inhibition of macromolecular synthesis [36].

14: tolytoxin, R_1 = OH, R_2 = Me
15: scytophycin B, R_1 = R_2 = H

A 14-membered macrocyclic lactone with a side chain containing unique poly-1,3-diols and bromoolefin units, named oscillariolide (**16**), was isolated from a marine reddish blue-green alga *Oscillatoria* sp., which was collected from Gokashowan Bay, Mie Prefecture, Japan, and was cultured in an inorganic medium. Oscillariolide (**16**) showed inhibition activity of the cell division of fertilized starfish eggs at a concentration of 0.5 µg/ml [37].

16: oscillariolide

A 16-membered macrolide glycoside, lyngbyaloside (**17**), was isolated from the blue-green alga *Lyngbya bouillonii*, collected at a depth of 1–10 m on the coral reef of Laing Island, Papua New Guinea. The relative configurations of 8 out of the 13 stereogenic carbons were proposed from NOE analysis [38]. Lyngbyaloside (**17**), possessing a 2,3,4-tri-*O*-methyl-6-deoxy-α-mannopyranoside, had a kind of structural similarity to the sponge-derived 14-membered macrolide glycoside, callipeltoside A (see Section II.C.7). From the same organism, laingolide (**18**), possessing a 15-membered lactone-lactam structure, was obtained and its planar structure was reported [39]. No bioactivity was described for compounds **17** and **18**.

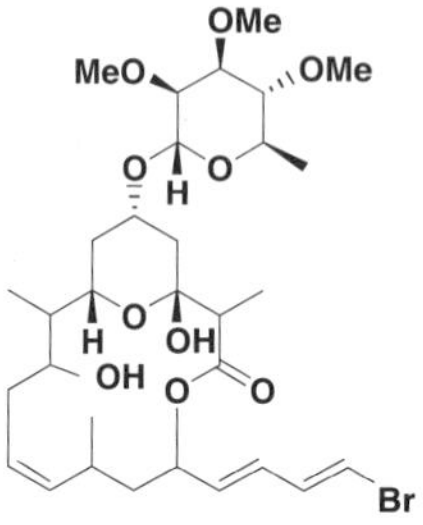

17: lyngbyaloside

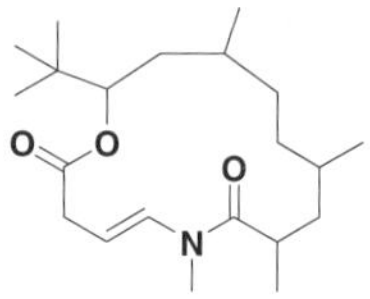

18: laingolide

3. Dinoflagellates

Extensive studies have been conducted on the bioactive metabolites of marine dinoflagellates during the past 10–20 years, revealing many interesting compounds having macrocyclic-lactone structures.

In 1968, a new antibiotic, goniodomin A (**19**), possessing strong antifungal activity, was isolated from a marine dinoflagellate *Goniodoma* sp. [40]. Its structure remained unknown for a long time, and in 1988 it was finally elucidated by extensive spectral studies using a sample reisolated from *G. pseudogoniaulax*, collected in a rock pool at Jogashima, Kanagawa Prefecture, Japan [41]. Pharmacological studies on goniodomin A (**19**) revealed that it induced potent stimulation of the actomyosin ATPase activities, and this activation may be sensitive to the regulatory protein system, the troponin–tropomyosin complex [42].

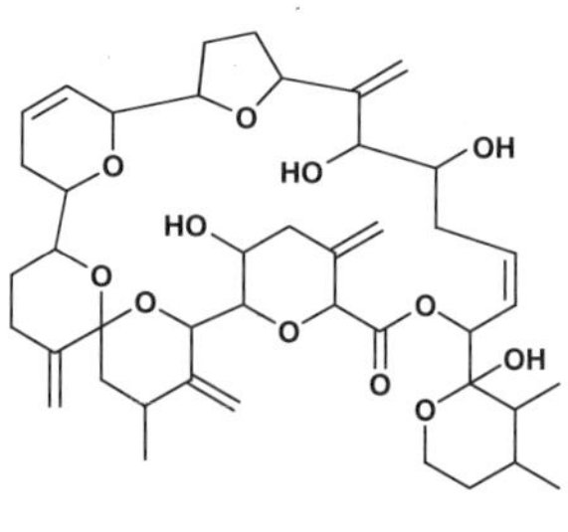

19: goniodomin A

From a benthic dinoflagellate, *Prorocentrum lima*, a toxic macrocycle named prorocentrolide (**20**) was isolated [43]. This dinoflagellate was known to sometimes contain diarrhetic shelfish poisoning (DSP) toxins such as okadaic acid and its esters. This macrocyclic compound (**20**) possessed a unique structure constructed from a C_{49} fatty acid incorporating a C_{27} macrolide and a hexahydroisoquinoline moiety. Prorocentrolide (**20**) had a mouse lethality of 0.4 mg/kg (ip), cytotoxicity against L1210 cells (IC_{50}, 20 µg/ml), and negative antimicrobial activity.

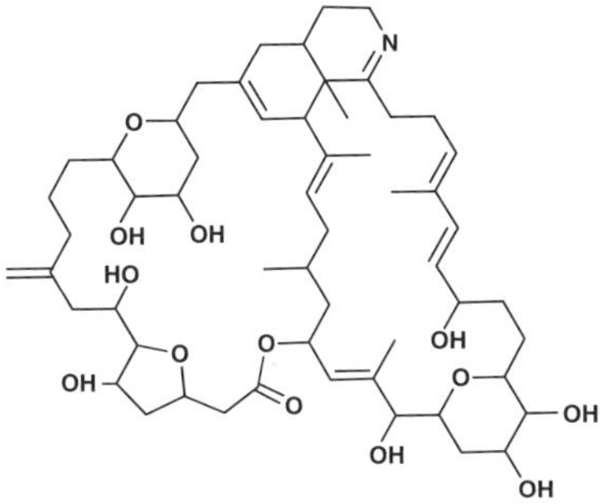

20: prorocentrolide

From another DSP toxin-producing marine dinoflagellate, *Prorocentrum hoffmannianum*, a 19-membered macrolide, hoffmanniolide (**21**), was isolated from a nontoxic fraction. This macrolide (**21**) showed no evidence of cytotoxicity up to 100 µg/ml [44].

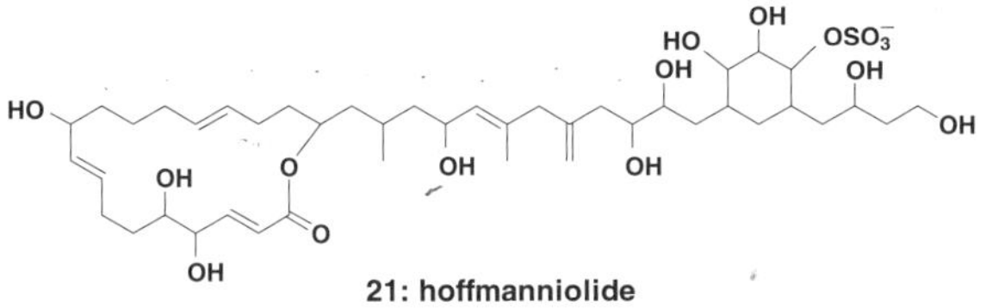

21: hoffmanniolide

An unprecedented class of polyhydroxypolyene macrolide, named zooxanthellatoxin A (**22**), was isolated from a symbiotic marine dinoflagellate *Symbiodinium* sp. belonging to zooxanthellae, which are well-known symbionts distributed in a wide range of marine invertebrates [45]. This toxin (**22**) exhibited potent vasoconstrictive activity, causing sustained contractions of isolated rabbit aorta at concentrations above 7×10^{-7} M. Further pharmacological study revealed that this toxin (**22**) caused thromboxane A_2-dependent and genistein-sensitive aggregation in rabbit platelets [46]. This macrolide (**22**) contains the largest 62-membered ring in natural products, and its relative and absolute stereochemistries are extensively investigated on the basis of modern NMR spectroscopic methods [47] and synthesis of degradation products [48].

22: zooxanthellatoxin A

The amphidinolides are a family of biologically active macrolides isolated from marine dinoflagellates of the genus *Amphidinium*, which are symbionts of Okinawan marine acoel flatworms *Amphiscolops* spp. [49]. This family of macrolides has remarkable structural diversity with 21 reported examples of amphidinolides A through V that are illustrative of macrocyclic formation ranging from 12-membered to 27-membered systems [50]. These macrolides exhibited potent cytotoxic activities against murine leukemia L1210 cells *in vitro* (IC$_{50}$, 0.05 ng/ml for the most potent one), and amphidinolide B (**23**) was shown to initiate rabbit skeletal actomyosin ATPase activity [51]. Among the 21 reported examples of amphidinolides, all of which contained many chiral centers in the molecule, absolute stereochemistries of amphidinolides B (**23**), G (**24**), and H (**25**) have been determined thus far on the basis of X-ray analyses in combination with syntheses of degradation products [49, 52], and achievement of total syntheses of amphidinolides J (**26**), K (**27**), and P (**28**) also established their absolute stereochemistries [53–55].

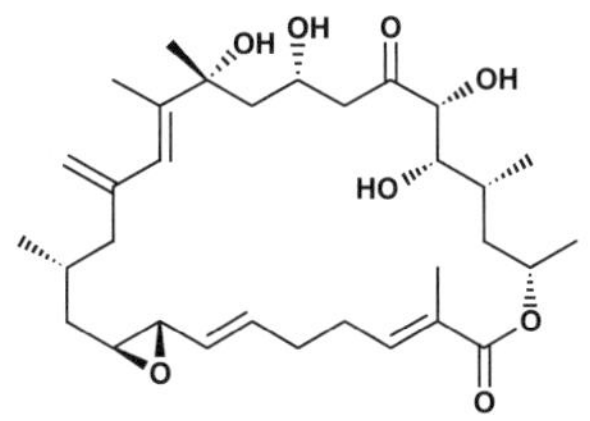

23: amphidinolide B
24: amphidinolide G 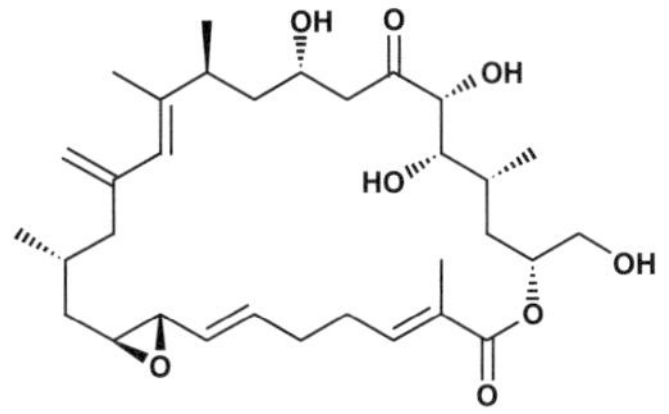
25: amphidinolide H 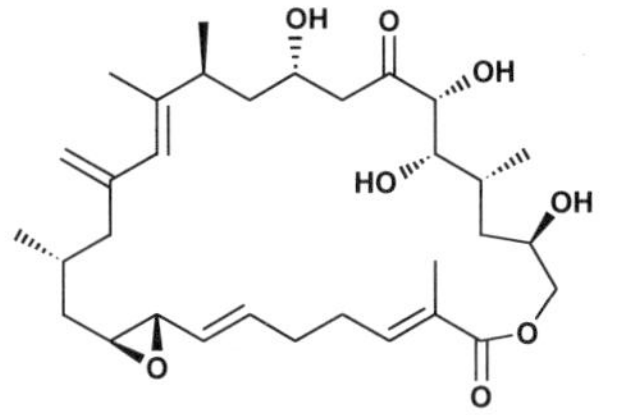
26: amphidinolide J 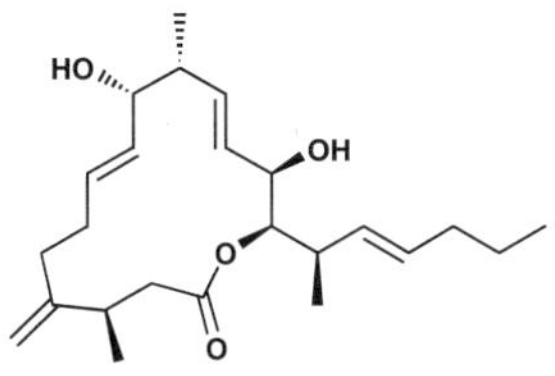
27: amphidinolide K 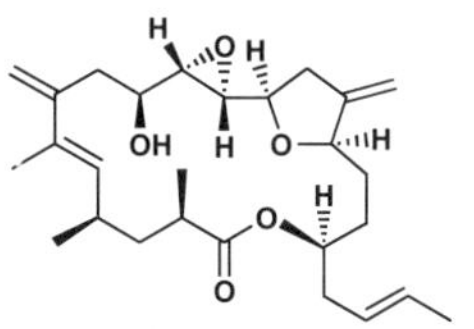
28: amphidinolide P 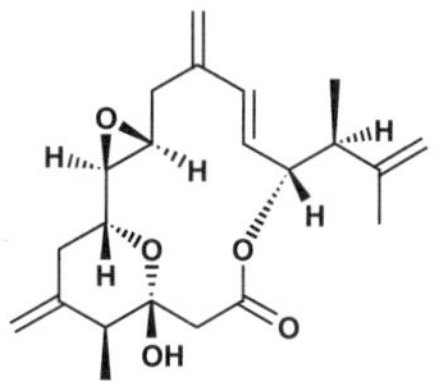

C. Sponges

1. Halichondrins

Halichondrins are a family of potently cytotoxic polyether macrolides, first isolated from *Halichondria okadai*, a common widely distributed sponge on the Pacific coast of Japan. They contain okadaic acid as a cytotoxic constituent as well [56]. The most potent cytotoxic member of this series was halichondrin B (**29**), which exhibited an IC_{50} value of 0.3 nM against L1210 murine leukemia cells *in vitro*, and also displayed potent *in vivo* activity against a number of chemoresistant human solid tumor xenografts [57]. Halichondrin B (**29**) was shown to bind to the vinca domain of tubulin, thereby inhibiting tubulin polymerization and tubulin-dependent GTP hydrolysis [58]. Although the potential for halichondrin B (**29**) to become a clinically important anticancer drug may be high, its yields from natural sources are quite low (1.8×10^{-8} to 4.0×10^{-5} %) and large quantities are required to support drug development research. Production of halichondrin B (**29**) by aquaculture of sponges was investigated, but the generation of 5 kg/year for clinical use would require the aquacultural production of sponge in estimated quantities of more than 5000 ton/year [59]. Synthetic chemists have extensively studied total synthesis of natural products as well as preparation of structurally simplified analogues of halichondrin B (**29**) [for example, 60].

29: halichondrin B

2. Tedanolides

Tedanolide (**30**), a structurally complex 18-membered macrolide, was isolated in 1984 from *Tedania ignis*, which was a widely distributed sponge in the Caribbean commonly referred to as "fire sponge" because contact with the skin induced a localized burning sensation and varying degrees of dermatitis for some individuals [61]. More recently, in 1991, a deoxy congener named 13-deoxytedanolide (**31**) was isolated from the Japanese sponge *Mycale adhaerens* [62]. Tedanolide (**30**) was revealed to be highly cytotoxic with ED_{50} values of 0.25 ng/ml and 16 pg/ml against KB and PS cell lines, respectively, and cell-flow cytofluorometry showed that tedanolide (**30**) caused accumulation of cells in the S phase at concentrations as low as 0.01 µg/ml. Tedanolide (**30**) also exhibited *in vivo* antitumor activity, increasing the life span of mice implanted with lymphocytic leukemia cells (23% at 1.56 µg/ml). 13-Deoxytedanolide (**31**) showed significant cytotoxicity against P388 murine leukemia cells with IC_{50} of 94 pg/ml, and also displays good antitumor activity against P388: *T/C* = 189% at a dose of 0.125 mg/kg.

30: tedanolide, R = OH
31: 13-deoxytedanolide, R = H

3. Spongistatins and Related Macrolides

In 1993 three independent research groups reported a new class of remarkably cytotoxic marine macrolides from marine sponges: spongistatins from *Spongia* sp. of the Eastern Indian Ocean [63], altohyrtins from *Hyrtios altum* of Aragusuku-jima, Okinawa [64], and cinachyrolide A from *Cinachyra* sp. of Hachijo-jima Island, Tokyo [65]. Although these macrolides were available from nature in only trace amounts (0.5–13.8 mg), these extremely potent cytotoxic compounds (GI_{50} 10^{-10} to 10^{-12} mol/liter) were identified as "probably the best to date in the National Cancer Institute's (NCI) evaluation programs." Spongistatin 1 and altohyrtin A had identical planar structures, and the planar structure of cinachyrolide A corresponded to the 15-deacetyl derivative of spongistatin 1 (= altohyrtin A). The relative stereochemistry of the six oxane rings in cinachyrolide A was proposed on the basis of NOESY data [65], while the absolute configurations of altohyrtin A (**32**) and its congeners were elucidated on the basis of detailed NMR analysis using their α-methoxy-α-(trifluoromethyl)phenylacetic acid (MTPA) esters and application of CD exciton chirality methods [66]. In 1998, total syntheses of altohyrtins C (**33**) [67] and A (**32**) [68] were accomplished by two independent groups, thereby establishing that the absolute stereochemistries assigned to altohyrtins were correct, and that altohyrtins A (**32**) and C (**33**) were identical to spongistatins 1 and 2, respectively. These structurally related macrolides exhibited outstandingly potent cytotoxicity (Table I), and *in vitro* studies revealed that spongistatins efficiently inhibit mitosis by blocking microtubule assembly through binding to tubulin in the vinca-alkaloid binding site [69]. To obtain more detailed information, especially in view of the selectivity for tumor cells and other cells, *in vivo* tests and possible medical applications, larger quantities of these compounds are required [70].

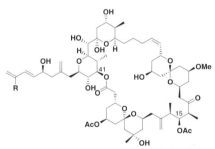

32: altohyrtin A (=spongistatin 1), R = Cl
33: altohyrtin C (=spongistatin 2), R = H

4. Swinholides, Mycalolides, and Related Macrolides

Swinholide A (**34**), a potent cytotoxic macrolide, was first isolated from the Red Sea sponge *Theonella swinhoei* in 1985 and its structure was first proposed as a monomeric 22-membered macrocyclic lactone [71]. In 1990, swinholide A

TABLE I
Cytotoxicity Data and Isolation Yields of Spongistatin, Altohyrtin, and Cinachyrolide

Compound	Yield	Cytotoxicity cell line		
Spongistatin 1	3.5×10^{-6} % [13.8 mg from 400 kg (wet weight) of sponge]	P388	GI_{50}	$2.5–3.5 \times 10^{-11}$ M
Altohyrtin A	6.9×10^{-6} % [7.6 mg from 112 kg of sponge]	L1210 KB	IC_{50} IC_{50}	3×10^{-11} g/ml 1×10^{-11} g/ml
Cinachyrolide A	1.7×10^{-5} % [1.1 mg from 6.6 kg of sponge]	L1210	IC_{50}	$<6 \times 10^{-10}$ g/ml

Source: Data from Pietruszka [70].

(**34**) was reisolated from the Okinawan sponge *T. swinhoei* and its structure, including absolute stereochemistry, was revealed to be 44-membered bislactone on the basis of X-ray analysis [72]. The configurations of each asymmetric carbon in swinholide A (**34**) were consistent with those of scytophycin B (**15**) isolated from blue-green alga *Scytonema pseudohofmanni* (see Section II.B.2). A structurally related macrolide, misakinolide A (**35**), was obtained from the sponge *Theonella* sp., collected at Maeda-misaki, Okinawa, in 1986, and its structure was first proposed as a monomeric 20-membered lactone [73]. In 1987 an identical compound, named bistheonellide A, was isolated from the sponge *Theonella* from Hachijo-jima, and its planar structure was revised to the dimeric macrodiolide structure [74]. In 1990 the absolute stereostructure of **35** was established on the basis of chemical correlation with swinholide A (**34**) [75]. Misakinolide A (= bistheonellide A) (**35**) possesses a 40-membered dilactone structure, and its absolute stereochemistry was the same as that of swinholide A (**34**) except that **35** lacks one conjugated ethylenic moiety in each monomeric unit of **34**. Swinholide A (**34**) had potent cytotoxicity against L1210 and KB cell lines with IC_{50} values of 0.03 and 0.04 μg/ml, respectively, and misakinolide A (**35**) also exhibited potent cytotoxicity with IC_{50} values of 0.035 μg/ml against L1210, 0.01 μg/ml against P388, and 0.0005–0.005 μg/ml against human tumor cells (HCT-8, A-549, and MDA-MB-231). With regard to the *in vivo* antitumor test, misakinolide A (**35**) was positive [T/C = 140% at the dosage of 0.31 mg/kg (mice) against P388 leukemia], while swinholide A (**34**) was toxic and did not show promising antitumor activity [76]. Investigations on the biochemical properties of these macrolides revealed that swinholide A (**34**) disrupts the actin cytoskeleton of cells grown in culture, sequesters actin dimers *in vitro* with a binding stoichiometry of one swinholide A molecule per actin dimer, and rapidly severs F-actin *in vitro* [77].

Misakinolide A (**35**) also binds simultaneously to two actin subunits with virtually the same affinity as swinholide A (**34**), but it does not sever actin filaments like swinholide A (**34**); rather it caps the barbed end of F-actin [78]. The presence of microbial symbionts in *T. swinhoei* was reported and the production of these natural products such as swinholide A (**34**) was believed to be responsible for symbiotic cyanobacteria. Studies by transmission electron microscopy, however, suggested that production of these macrolides was attributable to heterotrophic unicellular bacteria rather than cyanobacteria [79].

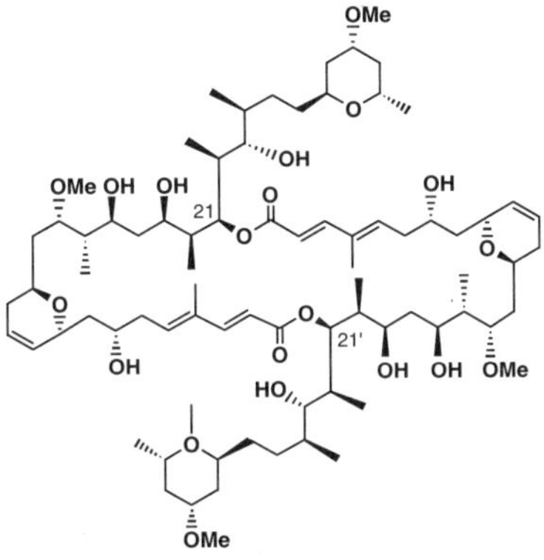

34: swinholide A

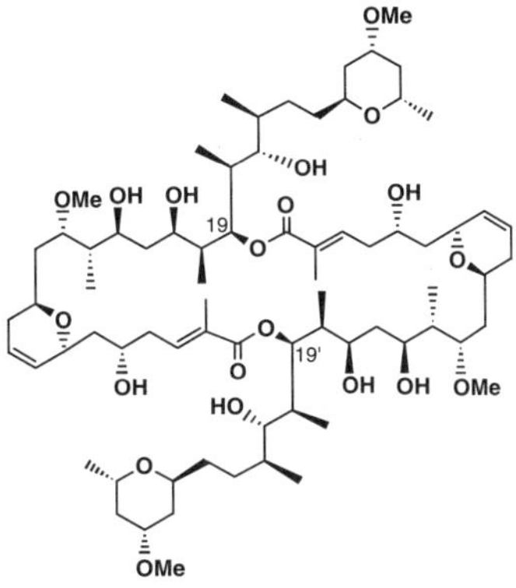

35: misakinolide A

Among bioactive metabolites isolated from sponges and other marine organisms, unusual trisoxazole-containing macrolides such as ulapualide A (**36**) [80], kabiramide C (**37**) [81], halichondramide (**38**) [82], and mycalolide A (**39**) [83] are known, and those macrolides exhibit structural similarities with swinholide A (**34**) and scytophycin B (**15**) such as substitution patterns, particularly in side-chain moieties, and terminal functionality (enamine *N*-methyl formamide group), which were also found in aplyronine A (see Section II.D.2). Ulapualides and kabiramides were contained in the egg masses of nudibranch such as *Hexabranchus sanguineus*, while halichondramides and mycalolides were isolated from marine sponges of the genus of *Halichondria* and *Mycale*, respectively. Interestingly, kabiramide C (**37**) was also isolated from a sponge *Halichondria* sp. [82]. These trisoxazole-containing macrolides exhibited potent cytotoxicity (for example, **36**: IC_{50}, 0.01–0.03 μg/ml against L1210) or antifungal activity (for example, **37**: active against *Candida albicans, Aspergillus niger, Penicillium citrium,* and *Trichophyton interdigitae*), and they were found to be potent actin-depolymerizing agents [84]. The stereochemistry of ulapualide A (**36**) was assigned on the basis of molecular mechanics studies and total synthesis [85], whereas the relative and absolute stereochemistry of mycalolide A (**39**) and its congeners was established through the combined use of chemical synthesis, degradation, and careful analysis of 1D and 2D ^{1}H NMR data [86]; further proof was provided by total synthesis of mycalolide A (**39**) [87].

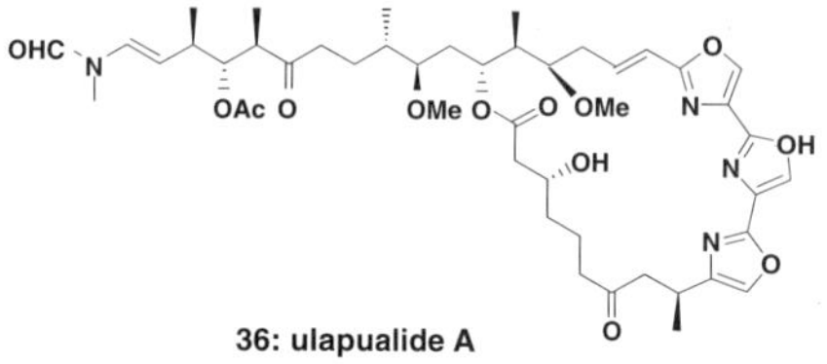

36: ulapualide A

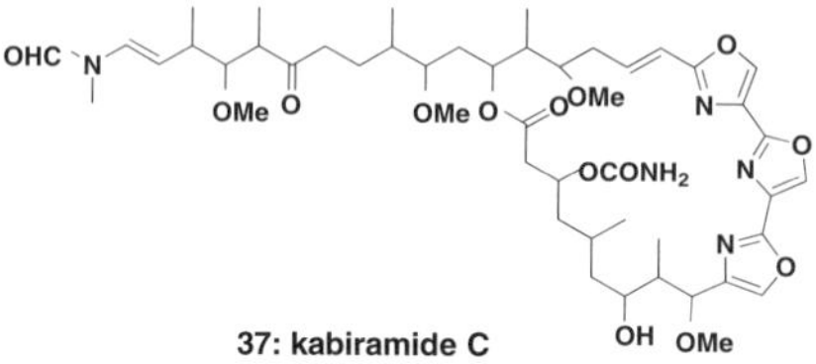

37: kabiramide C

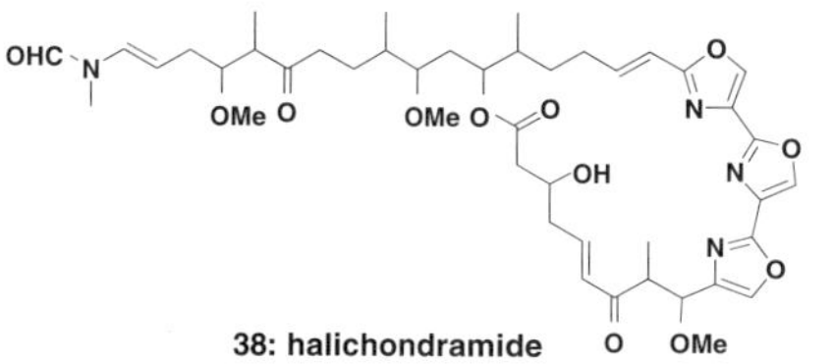

38: halichondramide

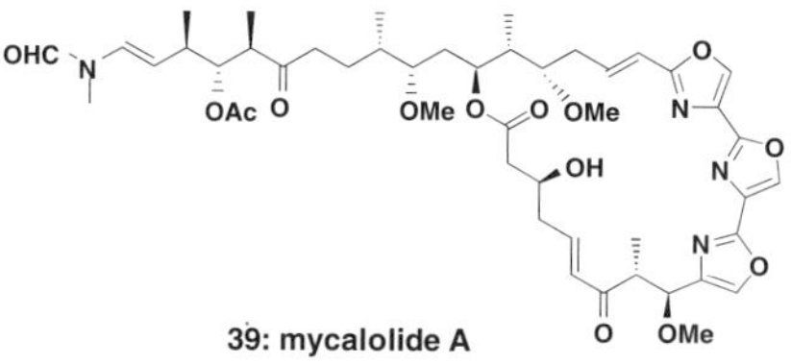

39: mycalolide A

Sphinxolide (**40**) and its congeners were isolated from deep-sea sponges, *Neosiphonia superstes* [88] and *Reidispongia coerulea* [89], that were collected in the south of New Caledonia at depths of 500–515 m. Sphinxolide (**40**) has also been isolated from an unidentified nudibranch of Hawaiian waters, and proved active against the KB cell line (IC$_{50}$ 35 pg/ml) [90]. Sphinxolide (**40**) is a 26-membered macrolide with a side chain bearing an unsaturated amide moiety, which had two slowly interconverting geometrical forms. This phenomenon was already encountered with tolytoxin (**14**) and related macrolides (see Section II.B.2). Sphinxolide (**40**) was revealed, like scytophycins, to cause rapid loss of microfilaments in cultured cells and potently inhibit actin polymerization *in vitro*. Sphinxolide (**40**) was also shown to circumvent multidrug resistance mediated by overexpression of either P-glycoprotein or MRP [91].

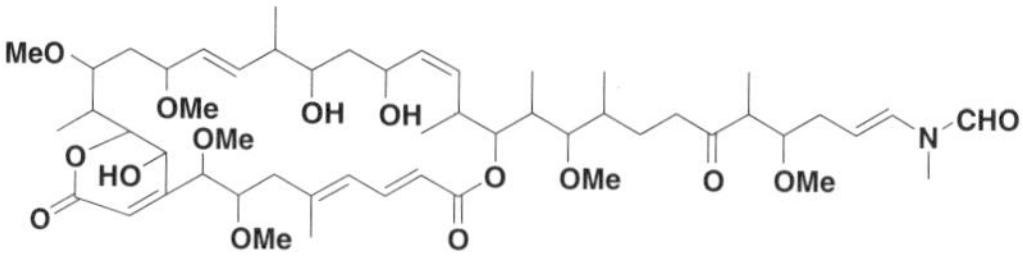

40: sphinxolide

5. Discodermolide

From the sponge *Discodermia dissoluta* collected at Lucay, Grand Bahama Island, at a depth of 33 m, a polyhyrdoxylated lactone, named discodermolide (**41**), was isolated as an immunosuppressive and cytotoxic agent in 1990, and its structure was determined by X-ray crystallographic analysis [92]. Although this polypropionate compound embracing a 6-membered lactone may not be classified as a macrolide, recent studies have revealed discodermolide (**41**) to be a useful tool for investigating microtubule binding and stabilization [93], because discodermolide (**41**) was shown to be a potent antimitotic agent, possessing a mode of action similar to that of a clinical anticancer agent paclitaxel (Taxol). Discodermolide (**41**) and paclitaxel were both found to arrest the cell cycle at the M phase, promote microtubule formation, and have similar inhibitory effects against breast cancer carcinoma [IC$_{50}$, 2.4 nM (**41**) and 2.1 nM (paclitaxel)] [94]. Interestingly, paclitaxel and discodermolide (**41**) represented a synergistic drug combination in human carcinoma cell lines [95].

41: discodermolide

6. Latrunculins and Related Macrolides

Two 2-thiazolidione-bearing marine macrolides, latrunculins A (**42**) and B (**43**), were isolated as the first major ichthyotoxins from the Red Sea sponge *Latrunculia magnifica*. This sponge had never been observed to sustain damage from fish, and on manually squeezing this sponge, it exuded a reddish fluid accompanied by a strong odor and caused fish poisoning and death within minutes. Purification of the toxins of the sponge, monitored by the toxicity to fish, resulted in isolation of two toxins (**42** and **43**) possessing 16- and 14-membered lactone rings, respectively [96]. The relative stereochemistry of latrunculin A (**42**) was solved by direct methods of X-ray crystallographic analysis [96]. The effects of latrunculins A (**42**) and B (**43**) on cultured mouse neuroblastoma and fibroblast cells were examined to reveal that these toxins rapidly induced striking changes in cell morphology that were reversible on removal of the toxin. The toxins also caused major alterations in the organization of microfilaments without obvious effects on the organization of the microtubular system [97]. Interestingly, latrunculin A (**42**) was also isolated from a Pacific nudibranch, *Chromodoris elisabethina*, for which **42** may serve as a defense allomone [98].

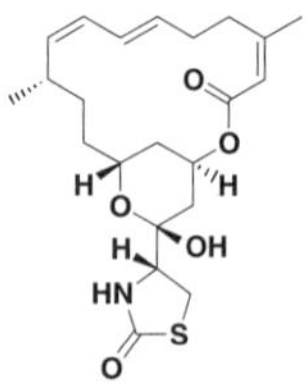

42: latrunculin A

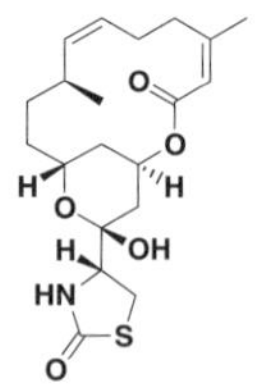

43: latrunculin B

Isolation of latrunculin A (**42**) was also reported from the Fijian sponge *Spongia mycofijiensis* and an associated nudibranch *Chromodoris lochi* [99] as well as from an Indonesian sponge *Hyattella* sp. and a nudibranch predator *Chromodoris lochi* that was found grazing on the sponge [100]. Furthermore, from this sponge–nudibranch association, a 20-membered macrolide, laulimalide (fijianolide B) (**44**), having strong cytotoxicity (IC$_{50}$, 15 ng/ml against KB cell line) was isolated [100, 101]. Latrunculin A (**42**) and laulimalide (**44**) were later isolated again from the Okinawan sponge *Fasciospongia rimosa*, and their absolute configurations were confirmed by X-ray analyses [102].

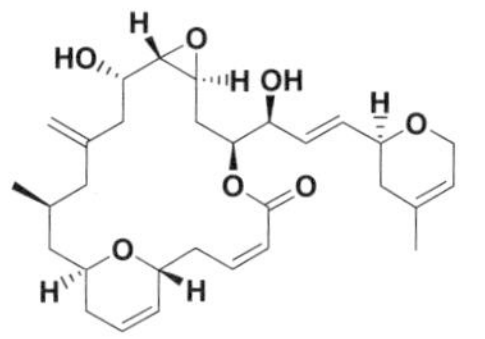

44: laulimalide

From the same Okinawan sponge (*F. rimosa*) that contained latrunculin A (**42**) and laulimalide (**44**), another 20-membered macrolide of a different class, named zampanolide (**45**), was isolated, and its gross structure was determined by 2D NMR and partial stereochemistry by NOE analysis. Zampanolide (**45**), containing high unsaturation and an uncommon carbinol amine functionality, exhibited potent cytotoxicity (IC$_{50}$ 1–5 ng/ml) against the P388, A549, HT29, and MEL 28 cell lines [103].

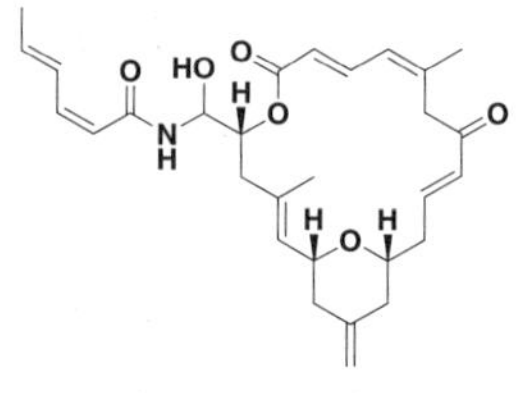

45: zampanolide

7. Other Sponge Metabolites

Theonezolide A (**46**) was isolated in 1993 from the Okinawan marine sponge *Theonella* sp. collected off Ie Island, Okinawa, and the planar structure was elucidated on the basis of extensive spectroscopic analyses [104]. This macrolide (**46**) was the first member of a new class of polyketide natural products consisting of two principal fatty-acid chains with various functionalities such as a sulfate ester, an oxazole, and a thiazole group, constituting a 37-membered macrocyclic lactone ring bearing a long side chain attached through an amide linkage. The absolute configurations of the terminal chiral center at C-75 as well as the two 1,3-diol-type moieties at C-8, C-10, C-14, and C-16 were determined by syntheses of its ozonolysis products [105]. Theonezolide A (**46**) exhibited cytotoxicity against murine lymphoma L1210 and human epidermoid carcinoma KB cells *in vitro* (IC_{50} values of 0.75 and 0.75 µg/ml, respectively). Additionally, theonezolide A (**46**) caused a marked platelet shape change at low concentration (0.2–0.6 µ*M*), and increasing the concentration of **46** to 6 µ*M* or more caused a shape change followed by a small but sustained aggregation [106].

46: theonezolide A

From the Okinawan sponge *Polyfibrospongia* sp. collected off the island of Miyako, a unique macrolide, miyakolide (**47**), was isolated and its structure was determined by X-ray analysis. Although miyakolide (**47**) was structurally similar to the bryostatins, an important class of anticancer compounds (see Section II.D), compound **47** only displayed weak *in vitro* (IC_{50} 17.5 µg/ml) and *in vivo* antitumor activity (*T/C* = 127% at 800 µg/kg) against P388 mouse leukemia [107]. Incidentally, the same Okinawan sponge *Polyfibrospongia* sp. contained a noticeable natural product of a different class from "macrolides," named hennoxazole A (**48**), embracing bis-oxazoles and C-glycoside linkage and displaying potency against herpes simplex virus type 1 and peripheral analgesic activity [108].

Pateamine (**49**), a thiazole-containing macrolide with a 19-membered dilactone functionality, was isolated from the New Zealand marine sponge *Mycale* sp., collected on the southwest coast of the South Island of New Zealand. Pateamine (**49**) had potent *in vitro* antifungal (MIC values: *Candida albicans*, 1 µg/ml; *Trichophyton mentagrophytes*, 20 ng/ml; *Cladosporium resinae*, 0.4 µg/ml) and cytotoxic

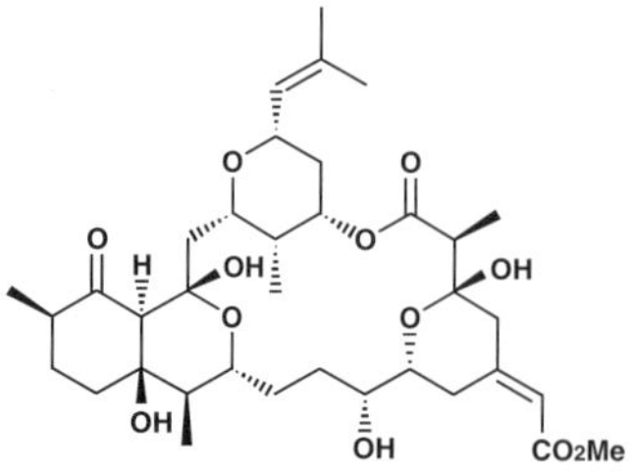

47: miyakolide

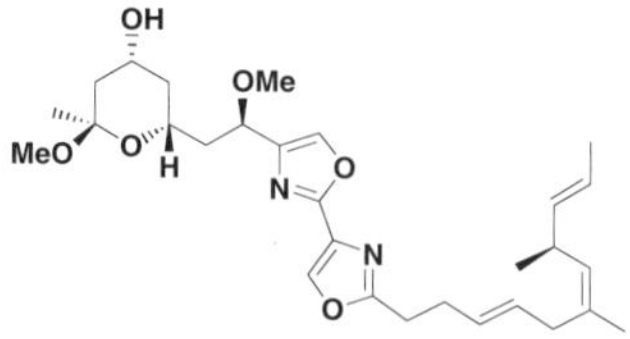

48: hennoxazole A

activity (IC$_{50}$ 0.15 ng/ml against P388) [109]. The relative and absolute stereochemistry of **49** was established by total synthesis [110]. Further studies revealed that pateamine (**49**) displayed immunosuppressive properties [mixed lymphocyte reaction (MLR), IC$_{50}$ = 2.6 nM] and specifically inhibited an intracellular step of the T-cell-receptor-mediated IL-2 transcription factor, indicating that its mode of action was similar to that of FK506 and cyclosporin A but distinct from that of rapamycin [110]. Note that the unrelated, but also strongly bioactive (antiviral and antitumor) nonmacrolide metabolite, mycalamide A (**50**), had been isolated from a sponge of the same species, collected at a different location, the southeast coast of the South Island of New Zealand [111].

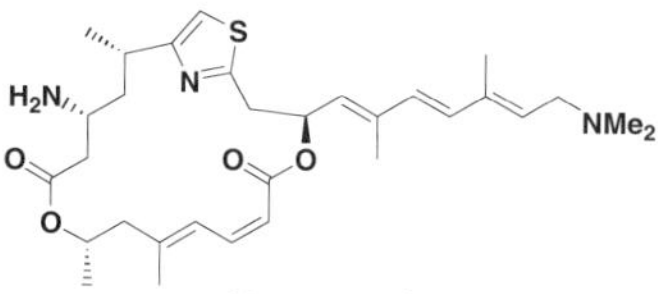

49: pateamine

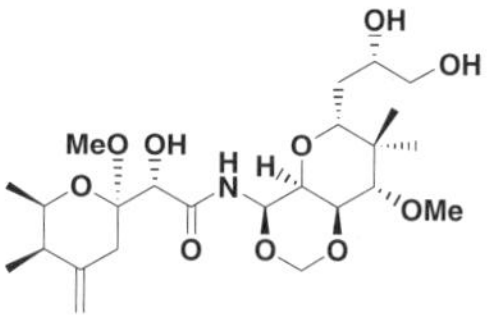

50: mycalamide A

From another collection of the New Zealand sponge *Mycale* sp. from the north coast of the South Island, a polyoxygenated, pyranose-ring containing 16-membered macrolide peloruside A (**51**) was isolated along with pateamine (**49**) and mycalamide A (**50**). Peloruoside A (**51**) was found to be cytotoxic to P388 murine leukemia cells at approximately 10 ng/ml (18 nM). The relative stereochemistry of the 10 stereogenic centers contained in **51** was elucidated using a variety of spectroscopic methods [112].

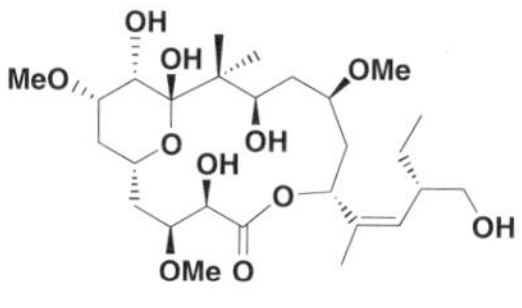

51: peloruside A

From the Indian Ocean marine sponge *Phorbas* sp., potently cytotoxic and antifungal macrolides, phorboxazoles A (**52**) and B (**53**), were isolated. These bis-oxazole-containing macrolides exhibited *in vitro* antifungal activity against *Candida albicans* at 0.1 µg/disk and extraordinary cytostatic activity (mean $GI_{50} < 7.9 \times 10^{-10}\,M$ in the NCI 60 tumor cell line panel) [113]. These compounds (**52** and **53**) thus showed activity on the same order of magnitude reported for spongistatin 1 (**32**) (see Section II.C.3), and were "among the most potent cytostatic agents yet discovered." The configurations of 15 stereocenters in phorboxazoles A (**52**) and B (**53**) were defined by a combination of extensive NMR studies, derivatizations, degradations, and synthesis of model compounds [114] and then corroborated by total synthesis [115].

52: phorboxazole A, R_1 = OH, R_2 = H
53: phorboxazole B, R_1 = H, R_2 = OH

From the shallow-water Caribbean marine sponge *Forcepia* sp., collected in the British Virgin Islands, a bis-tetrahydropyran-containing macrolide, lasonolide A (**54**), was isolated as a potent cytotoxin with IC_{50} values against the A-549 human lung carcinoma and P388 murine leukemia cell lines of 40 and 2 ng/ml, respectively. Compound **54** also showed inhibition activity of cell adhesion in the EL-4.IL-2 cell line with an IC_{50} of 19 ng/ml; however, toxicity against this cell line was weak (>25 µg/ml) [116]. Regio- and stereochemistries for hydroxyl- and ether-bearing stereocenters in the macrolide-ring portion of **54** were assigned on the basis of ROESY and deuterium exchange ^{13}C experiments [116].

54: lasonolide A

A specimen of marine sponge of *Dysidea* sp. collected from Palau contained an unstable 14-membered macrolide, arenolide (**55**), which showed modest cytotoxicity

[IC$_{50}$, 21 mM (HCT116) and 9.8 mM (A2780)]. The relative stereochemistry about the 14-membered ring was determined by interpretation of the NOESY data. Isolation of macrolides such as arenolide (**55**) had never been accomplished with *Dysidea* species; therefore, it was proposed that the sponge may have absorbed the compounds released by organisms in the vicinity [117].

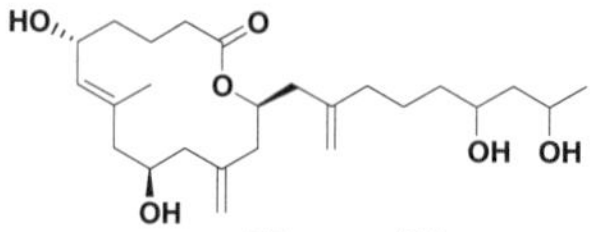

55: arenolide

The extracts of the deep-water New Caledonian sponge *Neosiphonia superstes*, which contained sphinxolide (**40**) (see Section II.C.4), were subjected to continuing bioassay-guided fractionation to result in the isolation of a different structural class of macrolide, superstolide A (**56**), made up of a decalin system fused with a 16-membered highly unsaturated lactone ring. This compound (**56**) was highly cytotoxic against NSCLC-N6-L16 (human bronchopulmonary non-small-cell lung carcinoma) cells with IC$_{50}$ of 0.04 µg/ml and murine leukemia cells expressing resistance toward doxorubicine P388 Dox with an IC$_{50}$ of 0.02 µg/ml. The relative and absolute configurations of stereocenters in **56** were elucidated by a combination of NMR data, acetonide analysis, GLC-modified Horeau's methodology, modified Mosher's method, and molecular dynamics and mechanics calculations [118].

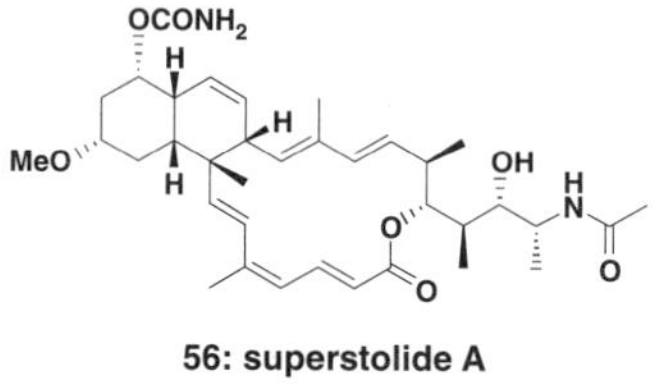

56: superstolide A

Another New Caledonian sponge, *Callipelta* sp., an aminodeoxy sugar-containing macrolide, callipeltoside A (**57**), was isolated. This material was a lithistid sponge collected from the shallow waters off the east coast of New Caledonia. Its extracts showed antifungal and antiviral activity. The relative stereochemistry of **57** was proposed from the analysis of ROESY and NOE difference experiments. Callipeltoside A (**57**) was revealed to be a cytotoxin, even if it had a moderate activity with IC$_{50}$ values against the NSCLC-N6 and P388 cells of 11.26 and 15.26 µg/ml, respectively. Cell cycle analysis by flow cytometry assays of the NSCLC-N6 cell line treated with **57** revealed a cell cycle-dependent effect, involving a dependent G$_1$ blockage [119].

57: callipeltoside A

Salicylihalamide A (**58**), a highly cytotoxic macrolide incorporating salicyclic acid, a 12-membered lactone ring, and an enamide side chain, was isolated from the marine sponge *Halichlona* sp., collected in southwestern Australia. In the NCI 60-cell line human tumor screen, the mean activity GI_{50} concentration level of salicylihalamide A (**58**) was approximately 15 n*M*, and COMPARE pattern recognition analyses of the mean graph profiles of **58** did not reveal any significant correlations with the profiles of known antitumor compounds in the NCI's database [120]. A chemically closely related compound, apicularen A (**59**), was obtained from culture extracts of myxobacterium *Chondromyces robustus* as a highly cytotoxic agent against cultivated human and animal cells (IC_{50}, 0.1–3 ng/ml) with no antimicrobial activity [121].

58: salicylihalamide A   **59: apicularen A**

A calcareous sponge of a new genus, *Leucascandra caveolata*, collected along the east coasts of New Caledonia, contained a new type of macrolide, leucascandrolide A (**60**), which consisted of a doubly *O*-bridged 18-membered macrolide, showing little C_1-branching versus extensive 1,3-dioxygenation and a peculiar side chain. This macrolide (**60**) displayed strong *in vitro* cytotoxicity against KB and P388 cells with IC_{50} values of 0.05 and 0.25 μg/ml, respectively, and was also antifungal, inhibiting growth of *Candida albicans* [122].

From the sponge *Chondropsis* sp., collected at Bass Island, Australia, two structurally novel macrolides, chondropsins A (**61**) and B (**62**), were isolated by antiproliferative bioassay-guided fractionation. These compounds (**61** and **62**) were polyunsaturated, polyhydroxylated, 35-membered macrocycles incorporating both

60: leucascandrolide A

lactone and lactam functionalities with a complex, amide-linked, polyketide side chain. Testing of chondropsin A (**61**) in the NCI 60-cell antitumor screen revealed that this compound showed potent, differential growth inhibition of the 60 cell lines, yielding a mean panel GI_{50} value of 2.4×10^{-8} M. COMPARE algorithm analyses of the mean graph profile of **61** revealed no significant correlation to any mean graph profiles contained in the NCI's standard agents database, implying that the mechanism of tumor growth inhibition by chondropsin A (**61**) is different from that of conventional antitumor agents [123].

61: chondropsin A, R =
62: chondropsin B, R = H

D. Other Invertebrates

1. Bryozoan

The bryostatins are a family of potent antitumor agents, among which bryostatin 1 (**63**) was first isolated in 1968 from the bryozoan (sea-mat) *Bugula neritina*, an innocuous filter feeder that habitually attaches to ship hulls [124, 125]. The structure of bryostatin 1 (**63**) was determined through X-ray crystallographic analysis and its absolute stereochemistry was defined by heavy-atom dipersion analysis in the X-ray structure of a *p*-bromobenzoate derivative [124]. Bryostatin 1 (**63**) bore a distinctive structure constructed from a 20-membered lactone, three pyran rings, two exocyclic unsaturated esters, and an octadienyl ester side chain. Bryostatin 1 (**63**) had significant *in vivo* antitumor activity against murine leukemia, B-cell lymphoma, reticulum cell sarcoma, ovarian carcinoma, and melanoma [125]. In addition, bryostatin 1 (**63**) showed immunopotentiating, hematopotentiating, and protein kinase C (PKC) activating effects. Bryostatin 1 (**63**) showed promising Phase I clinical results, and was moved into Phase II trials against non-Hodgkin's lymphoma, melanoma, and renal cancer by the NCI [126, 127]. Bryostatin 1 (**63**) was therefore demonstrated to be clinically important, but the extractive isolation yield was very low (1.3×10^{-6}%; 18 g of bryostatin from 14 tons of animal) and the current market price for bryostatin 1 (**63**) is expensive ($1156 U.S./500 μg). Attempts to solve the supply problem are currently under way by means of aquaculture or chemical synthesis [for example, 128, 129].

63: bryostatin 1

2. Sea Hare

A potent antitumor compound, aplyronine A (**64**), was first isolated from the sea hare *Aplysia kurodai* off the Pacific coast of Mie Prefecture, Japan. Aplyronine A (**64**) exhibited a high degree of antitumor activity *in vivo* against P388 murine leukemia (*T/C* = 545%, 0.08 mg/kg), Lewis lung carcinoma (*T/C* = 556%, 0.04 mg/kg), Ehrlich carcinoma (*T/C* = 398%, 0.04 mg/kg), colon 26 carcinoma (*T/C* = 255%, 0.08 mg/kg), and B16 melanoma (*T/C* = 201%, 0.04 mg/kg) [130]. The mode of action was shown to be related to the inhibition of actin polymerization. Aplyronine A (**64**) also quickly depolymerized F-actin. The kinetics of depolymerization suggested that aplyronine A (**64**) severs F-actin. The relationship between the concentration of total actin and F-actin at different concentrations of aplyronine A (**64**) suggested that aplyronine A (**64**) forms a 1:1 complex with G-actin. Comparison of the chemical structure of aplyronine A (**64**) and another actin-depolymerizing macrolide, mycalolide (**39**), suggested that the side chain but not the macrolide ring of aplyronine A (**64**) may account for its actin binding and severing activity [131]. The absolute stereochemistry of aplyronine A (**64**) was fully established by chemical degradations and synthetic methods [132], and total synthesis of aplyronine A (**64**) was also achieved [133].

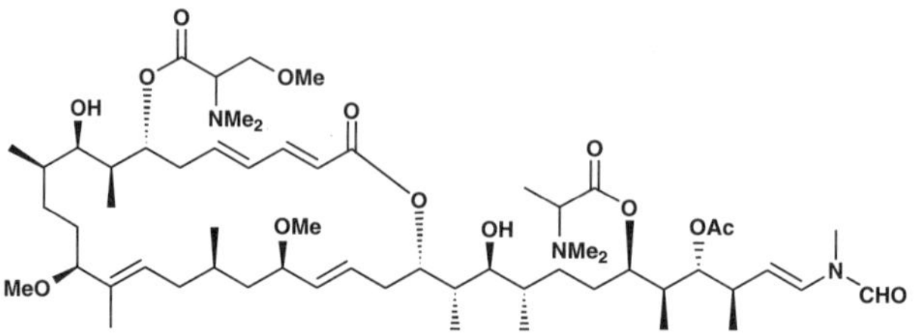

64: aplyronine A

The sea hare *Dolabella auricularia* (Aplysiidae) has been extensively investigated as a rich source of bioactive natural products, and a number of bioactive peptides and depsipeptides such as dolastatin 10 were isolated from both Western Indian Ocean specimens and Japanese specimens of this animal [for example, 134, 135]. This mollusk was also found to contain macrolide-type natural products. From *D. auricularia*, collected by hand at a depth of 0–1 m off the coast of the Shima Peninsula, Mie Prefecture, Japan, a 22-membered macrolide, dolabelide A (**65**), was isolated as a cytotoxin with an IC_{50} value of 6.3 µg/ml against HeLa-S_3 cells. The gross structure of **65** was determined by spectroscopic analysis, and its absolute stereochemistry was elucidated by a combination of chemical means and the NMR spectroscopic method [136].

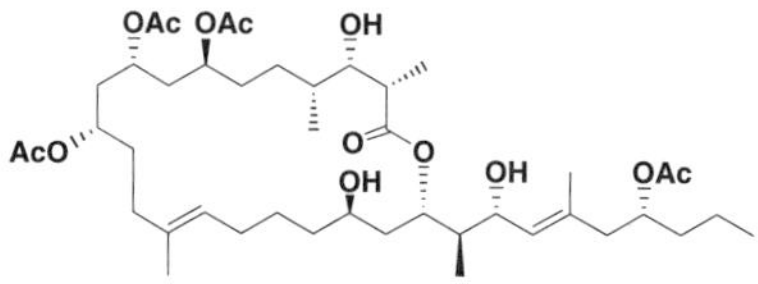

65: dolabelide

Further investigations on the bioassay-guided fractionation of the extracts of the Japanese specimens of *D. auricularia* resulted in isolation of a different class of macrolide, auriside A (**66**), possessing cytotoxicity against HeLa-S_3 cells with an IC_{50} value of 0.17 µg/ml. On the basis of NOESY spectral analysis and degradation experiments, its total structure with absolute configurations was determined to be 14-membered macrolide glycoside containing a bromine-substituted conjugated diene structure, a cyclic hemiacetal moiety, and a 2,4-di-*O*-methyl-L-rhamnopyranoside residue [137]. The structure of auriside A (**66**) appears to be similar to that of the cyanobacterial 16-membered bromo-diene-containing macrolide glycoside, lyngbyaloside (**17**) (see Section II.B.2), and the sponge-derived 14-membered macrolide glycoside, callipeltoside A (**57**) (see Section II.C.7).

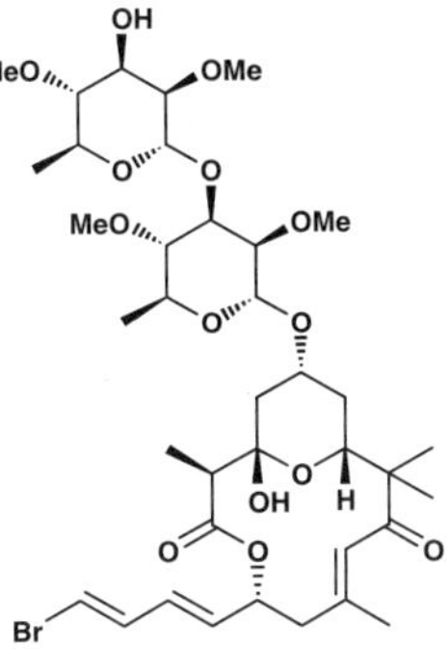

66: auriside A

3. Shellfish

The pectenotoxins [for example, pectenotoxins 1 (**67**), 2 (**68**), 3 (**69**), and 6 (**70**)], a family of cyclic polyether macrolide toxins, were first found in the digestive glands of toxic scallops, *Patinopecten yessoensis*, which were involved in an episode of DSP [138]. The principal toxins of DSP were nonmacrolide polyethers such as okadaic acid (**71**) and dinophysistoxin 1 (**72**), which were also isolated from *P. yessoensis*. Pectenotoxin 2 (**68**) was toxic [minimum lethal dose to mouse, 260 µg/ml (ip)] and induced diarrhea and liver damages, but the mode of action remains unknown, whereas okadaic acid (**71**) and its analogues are known as powerful phosphatase inhibitors. The structure of pectenotoxin 1 (**67**) was determined by X-ray crystallography [138], and the absolute configuration of pectenotoxin 6 (**70**) was elucidated by NMR methods using a chiral anisotropic reagent, phenylglycine methyl ester (PGME) [139]. As is often the case with families of shellfish toxins, the toxic compounds are not produced by the shellfish themselves, but rather are products of toxigenic microalgae or phytoplankton ingested by the shellfish. Strains of the dinoflagellate *Dinophysis fortii*, which produce okadaic acid (**71**) and its analogues, were also found to produce pectenotoxin 2 (**68**). It was notable that pectenotoxin 2 (**68**), bearing no oxygen on the C-47 position, was the only pectenotoxin congener found in the dinoflagellate strains, thus suggesting that pectenotoxin 2 (**68**) is the parent toxin that is metabolized in scallops to yield the oxygenated congeners (**67**, **69**, and **70**) [140]. In addition, pectenotoxin 2 (**68**) was also isolated from a two-marine-sponge association of *Poecillastra* sp. and *Jaspis* sp. collected off Cheju and Komun Islands in the South Sea of Korea as a causative compound for the brine shrimp lethality. Compound **68** displayed potent cytotoxicities in the NCI 60-cell line panel antitumor assay [141]. However, significant *in vitro* antitumor activity of **68** could not be affirmed in the *in vivo* antitumor assay at the NCI.

67: pectenotoxin 1, R = CH$_2$OH
68: pectenotoxin 2, R = CH$_3$
69: pectenotoxin 3, R = CHO
70: pectenotoxin 6, R = CO$_2$H

71: okadaic acid, R = H
72: dinophysistoxin 1, R = Me

Shellfish of the genus *Pinna*, living mainly in shallow areas of the temperate and tropical zones of the Indian and Pacific Oceans, are commonly eaten in Japan and China, and food poisoning resulting from its ingestion occurs frequently. From the viscera of the bivalve *Pinna muricata*, collected in Okinawa, a nonmacrolide toxin, pinnatoxin A (**73**), was isolated (LD$_{99}$ 180 μg/ml), and its structure was elucidated to have an amphoteric macrocycle structure [142]. From this bivalve, pinnaic acid (**74**) and tauropinnaic acid (**75**) were also isolated and they were found to inhibit cytozolic phospholipase A$_2$ (cPLA$_2$) with IC$_{50}$ values of 0.2 and 0.09 mM, respectively [143]. Although **74** and **75** did not have a macrolide structure either, a closely related compound, halichlorine (**76**), which contains a macrocyclic-lactone structure, was isolated from the marine sponge *Halichondoria okadai* collected at a mediolittoral zone in Kanagawa Prefecture, Japan [144]. Halichlorine (**76**) was revealed to significantly inhibit the induction of vascular cell adhesion molecule 1 (VCAM-1) at an IC$_{50}$ of 7 μg/ml. Its absolute stereochemistry was established by synthesis of a degradation product [145].

4. Tunicates

Tunicates (ascidians) are an excellent source of novel biologically active compounds of varied biosynthetic origin, particularly nitrogen-containing amino-acid-derived metabolites [146]. The most important compounds discovered to date from tunicates may be the didemnins, potent cytotoxic and antiviral cyclic

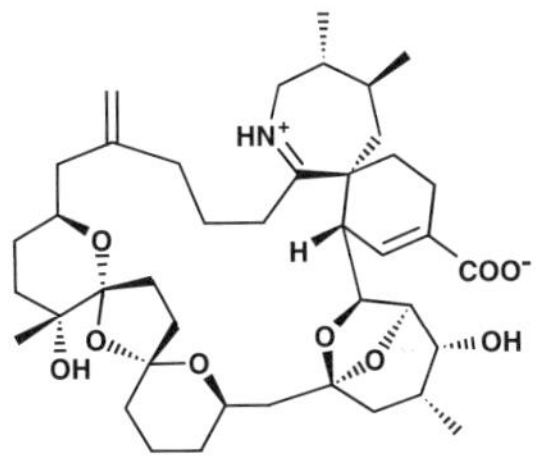

73: pinnatoxin A

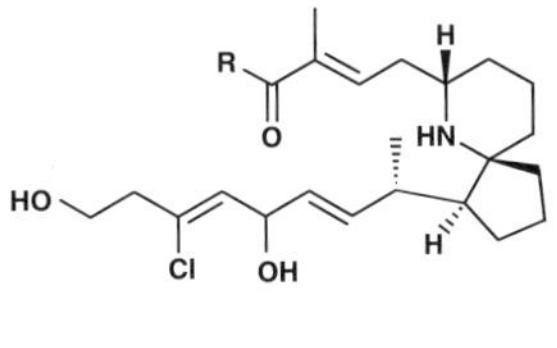

74: pinnaic acid, R = OH
75: tauropinnaic acid, R = NHCH$_2$CH$_2$SO$_3$H

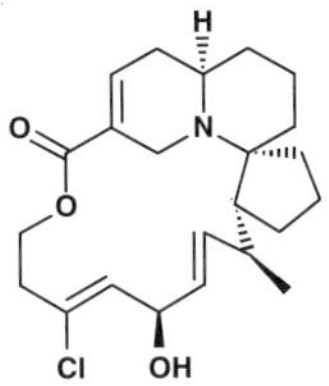

76: halichlorine

depsipeptides from Caribbean *Trididemnum solidum*. The Phase II studies of didemnin B indicated complete or partial remission with non-Hodgkin's lymphoma, but cardiotoxicity caused didemnin B to be dropped from further studies [147]. Not as many, but unique, bioactive macrolides have been isolated from tunicates as described below.

Patellazoles A–C (**77–79**), 24-membered thiazole-containing macrolides, were isolated from two specimens of didemnid tunicate, *Lissoclinum patella*, which were collected from the Fiji Islands [148] and Piti Bomb Holdes, Guam [149]. The patellazoles were potent cytotoxins in the NCI human cell line protocol with mean IC$_{50}$ values of 10^{-3} to 10^{-6} μg/ml and antifungal activity against *Candida albicans*.

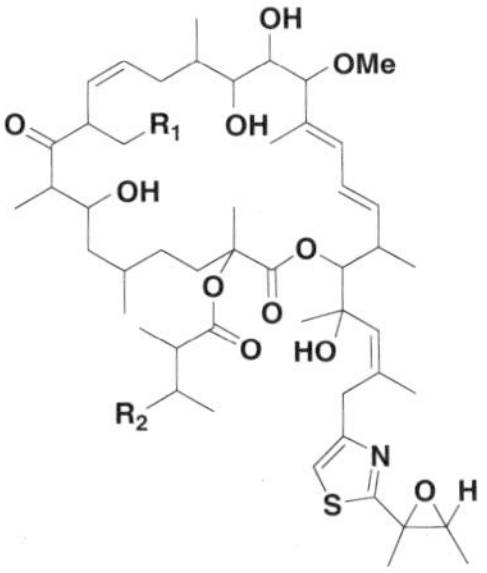

77: patellazole A, R^1 = H, R^2 = H
78: patellazole B, R^1 = H, R^2 = OH
79: patellazole C, R^1 = OH, R^2 = OH

Iejimalides A (**80**) and B (**81**), and their sulfates, iejimalides C (**82**) and D (**83**), respectively, were isolated from the compound tunicate *Eudistioma* cf. *rigida*, collected at Ie Island, Okinawa. These macrolides had a 24-membered macrocyclic lactone with high unsaturation and an *N*-formyl-L-serine residue at the terminal. Iejimalide A (**80**) showed significant cytotoxicity against L1210 (IC_{50}, 62 ng/ml) and L5178Y (IC_{50}, 22 ng/ml), whereas the cytotoxicity of the sulfates was relatively weak (for **82**: IC_{50}, 10 µg/ml against L1210) [150, 151].

80: iejimalide A, R^1 = H, R^2 = H
81: iejimalide B, R^1 = CH_3, R^2 = H
82: iejimalide C, R^1 = H, R^2 = SO_3Na
83: iejimalide D, R^1 = CH_3, R^2 = SO_3Na

From the colonial marine ascidian *Didemnum candidum*, collected near Big Pine Key, Florida, lactonized eicosanoids, named ascidiatrienolide A and its congeners, were isolated. Ascidiatrienolide A was not active in inducing or inhibiting inflammation in several standard mouse ear assays [152]. Although the structure of ascidiatrienolide A was first proposed as **84**, a C_{20} fatty acid having a 9-membered ring lactone, it was later revised to the structure of **85**, having a 10-membered ring lactone, on the basis of synthesis of enantiomers of **84** and **85** [153].

84: ascidiatrienolide A
(initially proposed structure)

85: ascidiatrienolide A
(revised structure)

Three 10-membered lactones with a conjugated triene and tetraene system, named didemnilactones A (**86**) and B (**87**) and neodidemnilactone (**88**), were isolated from the colonial tunicate *Didemnum noseleyi*, collected in Ago Bay, Mie Prefecture, Japan. Their structures, including absolute stereochemistry, were established on the basis of spectral studies and chemical synthesis. Didemnilactones exhibited moderate inhibitory activities against 5-lipoxygenase and 15-lipoxygenase of human polymorphonuclear leukocytes (IC_{50} of **86**, 9.4 and 41 µM, respectively), and didemnilactone A (**86**) showed weak binding activity to leukotriene B_4 receptors of human polymorphonuclear leukocyte membrane fractions (IC_{50}, 38 µM) [154, 155].

Ascidiatrienolide A (**85**) and neodidemnilactone (**88**) proved to be geometrical isomers of the triene moiety with the same absolute configurations of the stereocenters on the lactone ring part.

86: didemnilactone A

87: didemnilactone B

88: neodideminilactone

Lobatamide A (**89**) and its congeners, salycilate-encompassing macrolides, were isolated from Southwestern Pacific tunicates of different collections: (i) from shallow-water (<6 m) Australian collections of the tunicate *Aplidium lobatum* [156, 157] and (ii) from a shallow-water collection of an unidentified Philippine ascidian [157]. Lobatamide A (**89**) may belong to a structurally unique class of macrolides composed of a 15-membered bislactone ring containing a salicylate moiety and an enamide-bearing side chain terminated by an oxime methyl ether structure. Lobatamide A (**89**) and three other congeners were described to be identical with those (named aplidite A and its congeners) originally isolated from a tunicate *Aplidium* sp. of a deep-water collection during a trawling expedition at the Great Australian Bight, though the structures of aplidites were initially proposed differently as those incorporating an orthonitrite functionality [158]. Lobatamide A (**89**) was tested in the NCI's 60-cell line human tumor screen and showed a characteristic pattern of differential cytotoxicity with mean panel GI_{50} values of ~1.6 nM. COMPARE pattern recognition analyses of the mean graph profiles of **89** did not reveal any significant correlations with the profiles of reference antitumor compounds of the NCI's database, suggesting that the lobatamides might act by an unknown mechanism of action. However, the mean graph profiles of **89** showed high (≥0.7) COMPARE correlations with salicylihalamide A (**58**) obtained from a sponge *Halichlona* sp. (see Section II.C.7). This result may be ascribable to the fact that lobatamide A (**89**) and salicylihalamide A (**58**) share common core structural components such as a salicylate moiety with an enamide side chain. In addition, lobatamide A (**89**) proved to be identical to the structure of YM-75518 isolated from a terrestrial *Pseudomonas* sp. cultured from an Indonesian soil sample, and compound **89** was shown to have weak antifungal activity against *Rhodotorula acuta* alone among the tested microorganisms [159]. The occurrence of this identical compound in a terrestrial pseudomonad may suggest the possibility of the production of lobatamides in tunicates by associated microorganisms or by accumulation through dietary intake [157].

89: lobatamide A

The phenomenon described above, that is, that related compounds were obtained from marine tunicates, sponges, and terrestrial bacteria, was also observed in the following compounds of the haterumalide series. From an Okinawan ascidian *Lissoclinum* sp., a cytotoxic chlorinated 14-membered macrolide, haterumalide B (**90**), was isolated, and this macrolide (**90**) completely inhibited the first cleavage of fertilized sea urchin eggs at a concentration of 0.01 µg/ml [160]. A closely related macrolide, haterumalide NA (**91**), and its congeners were isolated from an Okinawan sponge *Ircinia* sp., and the absolute stereochemistry of **91** was detemined by spectroscopic analysis and modified Mosher's method [161]. Haterumalide NA (**91**) exhibited cytotoxicity against P388 cells with an IC$_{50}$ value of 0.32 µg/ml and moderate acute toxicity against mice with an LD$_{99}$ of 0.24 g/kg. A chlorinated macrolide, named oocydin A, was isolated from a red-pigmented antimycotic-producing bacterial strain, *Serratia marcescens*, growing as a epiphyte on *Rhyncholacis pedicillata*, an aquatic plant native to the Carrao river of the Venezuelan–Guyanan region of South America [162]. Oocydin A had an identical planar structure with haterumalide NA (**91**), but the stereochemistry of the C-15 hydroxyl group of oocydin A was assigned differently from that of **91**. Oocydin A demonstrated selective toxicity (antimycotic) toward such phytopathogenic oomycetes as *Pythium ultimum, Phytophthora parasitica, Phytophthora cinnamomi*, and *Phytophthora citrophora* with extremely low MIC values of approximately 0.03 µg/ml), but showed little or no activity against certain fungi imperfecti including several important fungal pathogens of humans such as *Candida albicans, Cryptococcus neoformans*, and *Aspergillus fumigatus*. This macrolide, therefore, may have potential as an antimycotic in agricultural applications for crop protection.

90: haterumalide B   **91: haterumalide NA**

III. Concluding Remarks

The systematic investigation of marine environments as sources of new biologically active substances began only in the mid-1970s, and since then more than 2500 new natural products have been isolated from various marine organisms [163]. These studies have clearly revealed that the marine organisms are a rich source of bioactive compounds having totally novel classes of chemical structures, which were not found in terrestrial resources. Although no compound isolated from a marine source has advanced to commercial use as a chemotherapeutic agent, several such as bryostatin 1 (**63**) are in some phase of clinical development as anticancer drugs. The sea covers 70% of the earth's surface and there could be more than 10 million species of marine macrofauna [164]. Each milliliter of seawater contains up to a million free-living cells of microorganisms such as bacteria, fungi, or microalgae. Very few species of the world's biodiversity in marine environments have been used for biological activity tests for drug discovery purposes or as materials of natural products research. Many more useful natural lead compounds are awaiting discovery, and advances in separation and analytical methods suggest that active substances can be discovered and identified rapidly from a rather small quantity of materials by microscale experiments.

References

1. Faulkner, D. J. (2000). Marine natural products. *Nat. Prod. Rep.* **17**, 7–55, and preceding papers of this series.
2. Faulkner, D. J. (2000). Highlights of marine natural products chemistry (1972–1999). *Nat. Prod. Rep.* **17**, 1–6.
3. Kobayashi, J., and Ishibashi, M. (1999). Marine natural products and marine chemical ecology. *In* "Comprehensive Natural Products Chemistry" (K. Mori, Ed.), Vol. 8, pp. 415–649. Elsevier Science, Oxford.
4. Watanabe, Y., and Fusetani, N. (Eds.) (1997). "Sponge Sciences: Multidisciplinary Perspectives." Springer-Verlag, Tokyo.
5. Higa, T., and Tanaka, J. (1997). Bioactive marine macrolides. *In* "Studies in Natural Products Chemistry" (A. Rahman, Ed.), Vol. 19, pp. 549–626. Elsevier Science, Amsterdam.
6. Fenical, W. (1993). Chemical studies of marine bacteria: Developing a new resource. *Chem. Rev.* **93**, 1673–1683.
7. Gustafson, K., Roman, M., and Fenical, W. (1989). The macrolactins, a novel class of antiviral and cytotoxic macrolides from a deep-sea marine bacterium. *J. Am. Chem. Soc.* **111**, 7519–7524.
8. Rychnovsky, S. D., Skalitzky, D. J., Pathirana, C., Jensen, P. R., and Fenical, W. (1992). Stereochemistry of the macrolactins. *J. Am. Chem. Soc.* **114**, 671–677.
9. Kim, H.-H., Kim, W.-G., Ryoo, I.-J., Kim, C.-J., Suk, J.-E., Han, K.-H., Hwang, S.-Y., and Yoo, I.-D. (1997). Neuronal cell protection activity of macrolactin A produced by *Actinomadura* sp. *J. Microbiol. Biotechnol.* **7**, 429–434.
10. Jaruchoktaweechai, C., Suwanborirux, K., Tanasupawatt, S., Kittakoop, P., and Menasveta, P. (2000). New macrolactins from a marine *Bacillus* sp. Sc026. *J. Nat. Prod.* **63**, 984–986.

11. Tapiolas, D. M., Roman, M., and Fenical, W. (1991). Octalactins A and B: Cytotoxic eight-membered-ring lactones from a marine bacterium, *Streptomyces* sp. *J. Am. Chem. Soc.* **113**, 4682–4683.
12. Buszek, K. R., Sato, N., and Jeong, Y. (1994). Synthesis of octalactin A and B. *J. Am. Chem. Soc.* **116**, 5511–5512.
13. McWilliams, J. C., and Clardy, J. (1994). Total synthesis of (+)-octalactins A and B: Unusual metabolites from a marine microbe. *J. Am. Chem. Soc.* **116**, 8378–8379.
14. Perchellet, J.-P., Perchellet, E. M., Newell, S. W., Freeman, J. A., Ladesich, J. B., Jeong, Y., Sato, N., and Buszek, K. (1998). Antitumor activity of novel octalactin A analogs in murine leukemia cells *in vitro*. *Anticancer Res.* **18**, 97–106.
15. Pathirana, C., Tapiolas, D., Jensen, P. R., Dwight, R., and Fenical, W. (1991). Structure determination of maduralide: A new 24-membered macrolide glycoside produced by a marine bacterium (Actinomycetales). *Tetrahedron Lett.* **32**, 2323–2326.
16. Jiang, Z.-D., Jensen, P. R., and Fenical, W. (1999). Lobophorins A and B, new anti-inflammatory macrolides produced by a tropical marine bacterium. *Bioorg. Med. Chem. Lett.* **9**, 2003–2006.
17. Waitz, J. A., Horan, A. C., Kalyanpur, M., Lee, B. K., Loebenberg, D., Marquez, J. A., Miller, G., and Patel, M. G. (1981). Kijanimicin (Sch 25663), a novel antibiotic produced by *Actinomadura kijaniata* SCC 1256. Fermentation, isolation, characterization and biological properties. *J. Antibiot.* **34**, 1101–1106.
18. Tomita, F., Tamaoki, T., Shirahata, K., Morimoto, M., Ohkubo, S., Minemura, K., and Ishii, S. (1980). Novel antitumor antibiotics, tetrocarcins. *J. Antibiot.* **33**, 668–670.
19. Furumai, T., Takagi, K., Igarashi, Y., Saito, N., and Oki, T. (2000). Arisostatins A and B, new members of tetrocarcin class of antibiotics from *Micromonospora* sp. TP-A0316 I. Taxonomy, fermentation, isolation and biological properties. *J. Antibiot.* **53**, 227–232.
20. Igarashi, Y., Takagi, K., Kan, Y., Fujii, K., Harada, K.-I., Furumai, T., and Oki, T. (2000). Arisostatins A and B, new members of tetrocarcin class of antibiotics from *Micromonospora* sp. TP-A0316 II. Structure determination. *J. Antibiot.* **53**, 233–240.
21. Carter, G. T. (1986). Structure determination of oligomycins A and C. *J. Org. Chem.* **51**, 4264–4271.
22. Fernández-Chimeno, I. R., Cañedo, L., Espliego, F., Grávalos, D., De La Calle, F., Fernández-Puentes, J. L., and Romeo, F. (2000). IB-96212, a novel cytotoxic macrolide produced by a marine *Micromonospora*. I. Taxonomy, fermentation, isolation and biological activities. *J. Antibiot.* **53**, 474–478.
23. Cañedo, L. M., Fernández-Puentes, J. L., Baz, J. P., Huang, X.-H., and Rinehart, K. L. (2000). IB-96212, a novel cytotoxic macrolide produced by a marine *Micromonospora*. II. Physicochemical properties and structure determination. *J. Antibiot.* **53**, 479–483.
24. Numata, A., Iritani, M., Yamada, T., Minoura, K., Matsumura, E., Yamori, T., and Tsuruo, T. (1997). Novel antitumor metabolites produced by a fungal strain from a sea hare. *Tetrahedron Lett.* **38**, 8215–8218.
25. Hayashi, M., Kim, Y. P., Hiraoka, H., Natori, M., Takamatsu, S., Kawakubo, T., Masuma, R., Komiyama, K., and Ōmura, S. (1995). Macrosphelide, a novel inhibitor of cell–cell adhesion molecule. I. Taxonomy, fermentation, isolation and biological activities. *J. Antibiot.* **48**, 1435–1439.
26. Yotsu-Yamashita, M., Haddock, R. L., and Yasumoto, T. (1993). Polycavernoside A: A novel glycosidic macrolide from the red alga *Polycavernosa tsudai* (*Gracilaria edulis*). *J. Am. Chem. Soc.* **115**, 1147–1148.
27. Fujiwara, K., Murai, A., Yotsu-Yamashita, M., and Yasumoto, T. (1998). Total synthesis and absolute configuration of polycavernoside A. *J. Am. Chem. Soc.* **120**, 10770–10771.
28. Barriault, L., Boulet, S. L., Fujiwara, K., Murai, A., Paquette, L. A., and Yotsu-Yamashita, M. (1999). Synthesis and biological evaluation of analogs of the marine toxin polycavernoside A. *Bioorg. Med. Chem. Lett.* **9**, 2069–2072.

29. Nagai, H., Yasumoto, T., and Hokama, Y. (1996). Aplysiatoxin and debromoaplysiatoxin as the causative agents of a red alga *Gracilaria coronopifolia* poisoning in Hawaii. *Toxicon* **34**, 753–761.
30. Kato, Y., and Scheuer, P. J. (1974). Aplysiatoxin and debromoaplysiatoxin, constituents of the marine mollusk *Stylocheilus longicauda* (Quoy and Gaimard, 1824). *J. Am. Chem. Soc.* **96**, 2245–2246.
31. Mynderse, J. S., Moore, R. E., Kashiwagi, M., and Norton, T. R. (1977). Antileukemia activity in the Oscillatoriaceae: Isolation of debromoaplysiatoxin from *Lyngbya*. *Science* **196**, 538–540.
32. Patterson, G. M. L., Larsen, L. K., and Moore, R. E. (1994). Bioactive natural products from blue-green algae. *J. Appl. Phycol.* **6**, 151–157.
33. Moore, R. E., Patterson, G. M. L., Mynderse, J. S., Barch, Jr., J., Norton, T. R., Furusawa, E., and Furusawa, S. (1986). Toxins from cyanophytes belonging to the scytonemataceae. *Pure Appl. Chem.* **58**, 263–271.
34. Carmeli, S., Moore, R. E., and Patterson, G. M. L. (1990). Tolytoxin and new scytophycins from three species of *Scytonema*. *J. Nat. Prod.* **53**, 1533–1542.
35. Ishibashi, M., Moore, R. E., Patterson, G. M. L., Xu, C., and Clardy, J. (1986). Scytophycins: Cytotoxic and antimycotic agents from the cyanophyte *Scytonema pseudohofmanni*. *J. Org. Chem.* **51**, 5300–5306.
36. Patterson, G. M., and Carmeli, S. (1992). Biological effects of tolytoxin (6-hydroxy-7-O-methyl-scytophycin B), a potent bioactive metabolite from cyanobacteria. *Arch. Microbiol.* **157**, 406–410.
37. Murakami, M., Matsuda, H., Makabe, K., and Yamaguchi, K. (1991). Oscillariolide, a novel macrolide from a blue-green alga *Oscillatoria* sp. *Tetrahedron Lett.* **32**, 2391–2394.
38. Klein, D., Braekman, J. C., Daloze, D., Hoffmann, L., and Demoulin, V. (1997). Lyngbyaloside, a novel 2,3,4-tri-O-methyl-6-deoxy-α-mannopyranoside macrolide from *Lyngbya bouillonii* (Cyanobacteria). *J. Nat. Prod.* **60**, 1057–1059.
39. Klein, D., Braekman, J. C., Daloze, D., Hoffmann, L., and Demoulin, V. (1996). Laingolide, a novel 15-membered macrolide from *Lyngbya bouillonii* (cyanobacteria). *Tetrahedron Lett.* **37**, 7519–7520.
40. Sharma, G. M., Michaels, L., and Burkholder, P. R. (1968). Goniodomin, a new antibiotic from a dinoflagellate. *J. Antibiot.* **21**, 659–664.
41. Murakami, M., Makabe, K., Yamaguchi, K., Konosu, S., and Walchli, M. R. (1988). Goniodomin A, a novel polyether macrolide from the dinoflagellate *Goniodoma pseudogoniaulax*. *Tetrahedron Lett.* **29**, 1149–1152.
42. Matsunaga, K., Nakatani, K., Murakami, M., Yamaguchi, K., and Ohizumi, Y. (1999). Powerful activation of skeletal muscle actomyosin ATPase by goniodomin A is highly sensitive to troponin/tropomyosin complex. *J. Pharmacol. Exp. Ther.* **291**, 1121–1126.
43. Torigoe, K., Murata, M., Yasumoto, T., and Iwashita, T. (1988). Prorocentrolide, a toxic nitrogenous macrocycle from a marine dinoflagellate, *Prorocentrum lima*. *J. Am. Chem. Soc.* **110**, 7876–7877.
44. Hu, T., Curtis, J. M., Walter, J. A., and Wright, J. L. C. (1999). Hoffmanniolide: A novel macrolide from *Prorocentrum hoffmannianum*. *Tetrahedron Lett.* **40**, 3977–3980.
45. Nakamura, H., Asari, T., Murai, A., Kan, Y., Kondo, T., Yoshida, K., and Ohizumi, Y. (1995). Zooxanthellatoxin-A, a potent vasoconstrictive 62-membered lactone from a symbiotic dinoflagellate. *J. Am. Chem. Soc.* **117**, 550–551.
46. Rho, M. C., Nakahata, N., Nakamura, H., Murai, A., and Ohizumi, Y. (1997). Involvement of phospholipase C-gamma2 in activation of mitogen-activated protein kinase and phospholipase A2 by zooxanthellatoxin-A in rabbit platelets. *J. Pharmacol. Exp. Ther.* **282**, 496–504.
47. Matsumori, N., Kanono, D., Murata, M., Nakamura, H., and Tachibana, K. (1999). Stereochemical determination of acyclic structures based on carbon-proton spin-coupling constants. A method of configuration analysis for natural products. *J. Org. Chem.* **64**, 866–876.

48. Nakamura, H., Maruyama, K., Fujimaki, K., and Murai, A. (2000). Absolute configuration of the common terminal acid portion of zooxanthellatoxins from a symbiotic dinoflagellate *Symbiodinium* sp. established by the synthesis of its ozonolysis product. *Tetrahedron Lett.* **41**, 1927–1930.
49. Ishibashi, M., and Kobayashi, J. (1997). Amphidinolides: Unique macrolides from marine dinoflagellates. *Heterocycles* **44**, 543–572.
50. Kubota, T., Tsuda, M., and Kobayashi, J. (2000). Amphidinolide V, novel 14-membered macrolide from marine dinoflagellate *Amphidinium* sp. *Tetrahedron Lett.* **41**, 713–716.
51. Matsunaga, K., Nakatani, K., Ishibashi, M., Kobayashi, J., and Ohizumi, Y. (1999). Amphidinolide B, a powerful activator of actomyosin ATPase enhances skeletal muscle contraction. *Biochim. Biophys. Acta* **1427**, 24–32.
52. Kobayashi, J., Shimbo, K., Sato, M., Shiro, M., and Tsuda, M. (2000). Absolute stereochemistry of amphidinolides G and H. *Org. Lett.* **2**, 2805–2807.
53. Williams, D. R., and Kissel, W. S. (1998). Total synthesis of (+)-amphidinolide J. *J. Am. Chem. Soc.* **120**, 11,198–11,199.
54. Williams, D. R., and Meyer, K. G. (1999). Total synthesis of (+)-amphidinolide K. In "Abstracts of Papers, 218th National Meeting of the American Chemical Society, 1999," p. 578-ORGN. American Chemical Society, Washington, DC.
55. Williams, D. R., Myers, B. J., and Mi, L. (2000). Total synthesis of (−)-amphidinolide P. *Org. Lett.* **2**, 945–948.
56. Uemura, D., Takahashi, K., Yamamoto, T., Katayama, C., Tanaka, J., Okumura, Y., and Hirata, Y. (1985). Norhalichondrin A: An antitumor polyether macrolide from a marine sponge. *J. Am. Chem. Soc.* **107**, 4796–4798.
57. Fodstad, O., Breistol, K., Pettit, G. R., Shoemaker, R. H., and Boyd, M. R. (1996). Comparative antitumor activities of halichondrins and vinblastine against human tumor xenografts. *J. Exp. Ther. Oncol.* **1**, 119–125.
58. Bai, R. L., Paull, K. D., Herald, C. L., Malspeis, L., Pettit, G. R., and Hamel, E. (1991). Halichondrin B and homohalichondrin B, marine natural products binding in the vinca domain of tubulin. Discovery of tubulin-based mechanism of action by analysis of differential cytotoxicity data. *J. Biol. Chem.* **266**, 15,882–15,889.
59. Munro, M. H., Blunt, J. W., Dumdei, E. J., Hickford, S. J., Lill, R. E., Li, S., Battershill, C. N., and Duckworth, A. R. (1999). The discovery and development of marine compounds with pharmaceutical potential. *J. Biotechnol.* **70**, 15–25.
60. Wang, Y., Habgood, G. J., Christ, W. J., Kishi, Y., Littlefield, B. A., and Yu, M. J. (2000). Structure–activity relationships of halichondrin B analogues: Modifications at C.30–C.38. *Bioorg. Med. Chem. Lett.* **10**, 1029–1032.
61. Schmitz, F. J., Gunasekera, S. P., Yalamanchili, G., Hossain, M. B., and van der Helm, D. (1984). Tedanolide: A potent cytotoxic macrolide from the Caribbean sponge *Tedania ignis*. *J. Am. Chem. Soc.* **106**, 7251–7252.
62. Fusetani, N., Sugawara, T., and Matsunaga, S. (1991). Cytotoxic metabolites of the marine sponge *Mycale adhaerens* Lambe. *J. Org. Chem.* **56**, 4971–4974.
63. Pettit, G. R., Cichacz, Z. A., Gao, F., Herald, C. L., Boyd, M. R., Schmidt, J. M., and Hooper, J. N. A. (1993). Isolation and structure of spongistatin 1. *J. Org. Chem.* **58**, 1302–1304.
64. Kobayashi, M., Aoki, S., Sakai, H., Kawazoe, K., Kihara, N., Sasaki, T., and Kitagawa, I. (1993). Altohyrtin A, a potent anti-tumor macrolide from the Okinawan marine sponge *Hyrtios altum*. *Tetrahedron Lett.* **34**, 2795–2798.
65. Fusetani, N., Shinoda, K., and Matsunaga, S. (1993). Cinachyrolide A: A potent cytotoxic macrolide possessing two spiro ketals from marine sponge *Cinachyra* sp. *J. Am. Chem. Soc.* **115**, 3977–3981.

66. Kobayashi, M., Aoki, S., Gato, K., and Kitagawa, I. (1996). Absolute stereostructures of altohyrtins A, B, and C and 5-desacetylaltohyrtin A, potent cytotoxic macrolides, from the Okinawan marine sponge *Hyrtios altum*. *Chem. Pharm. Bull.* **44**, 2142–2149.
67. Evans, D. A., Trotter, B. W., Cote, B., Coleman, P. J., Dias, L. C., and Tyler, A. N. (1998). Enantioselective synthesis of altohyrtin C (spongistatin 2): Fragment assembly and revision of the spongistatin 2 stereochemical assignment. *Angew. Chem. Int. Edit. Engl.* **36**, 2744–2747.
68. Hayward, M. M., Roth, R. M., Duffy, K. J., Dalko, P. I., Stevens, K. L., Guo, J., and Kishi, Y. (1998). Total synthesis of altohyrtin A (spongistatin 1): Part 2. *Angew. Chem. Int. Edit. Engl.* **37**, 192–196.
69. Bai, R., Taylor, G. F., Cichacz, Z. A., Herald, C. L., Kepler, J. A., Pettit, G. R., and Hamel, E. (1995). The spongistatins, potently cytotoxic inhibitors of tubulin polymerization, bind in a distinct region of the vinca domain. *Biochemistry* **34**, 9714–9721.
70. Pietruszka, J. (1998). Spongistatins, cynachyrolides, or altohyrtins? Marine macrolides in cancer therapy. *Angew. Chem. Int. Edit. Engl.* **37**, 2629–2635.
71. Carmely, S., and Kashman, Y. (1985). Structure of swinholide-A, a new macrolide from the marine sponge *Theonella swinhoei*. *Tetrahedron Lett.* **26**, 511–514.
72. Kitagawa, I., Kobayashi, M., Katori, T., Yamashita, M., Tanaka, J., Doi, M., and Ishida, T. (1990). Absolute stereostructure of swinholide A, a potent cytotoxic macrolide from the Okinawan marine sponge *Theonella swinhoei*. *J. Am. Chem. Soc.* **112**, 3710–3712.
73. Sakai, R., Higa, T., and Kashman, Y. (1986). Misakinolide-A, an antitumor macrolide from the marine sponge *Theonella* sp. *Chem. Lett.* 1499–1502.
74. Kato, Y., Fusetani, N., Matsunaga, S., Hashimoto, K., Sakai, R., Higa, T., and Kashman, Y. (1987). Antitumor macrodiolides isolated from a marine sponge *Theonella* sp.: Structure revision of misakinolide A. *Tetrahedron Lett.* **28**, 6225–6228.
75. Tanaka, J., Higa, T., Kobayashi, M., and Kitagawa, I. (1990). Marine natural products. XXIV. The absolute stereostructure of misakinolide A, a potent cytotoxic dimeric macrolide from an Okinawan marine sponge *Theonella* sp. *Chem. Pharm. Bull.* **38**, 2967–2970.
76. Kobayashi, M., Kawazoe, K., Okamoto, T., Sasaki, T., and Kitagawa, I. (1994). Marine natural products. XXXI. Structure–activity correlation of a potent cytotoxic dimeric macrolide swinholide A, from the Okinawan marine sponge *Theonella swinhoei*, and its isomers. *Chem. Pharm. Bull.* **42**, 19–26.
77. Bubb, M. R., Spector, I., Bershadsky, A. D., and Korn, E. D. (1995). Swinholide A is a microfilament disrupting marine toxin that stabilizes actin dimers and severs actin filaments. *J. Biol. Chem.* **270**, 3463–3466.
78. Terry, D. R., Spector, I., Higa, T., and Bubb, M. R. (1997). Misakinolide A is a marine macrolide that caps but does not sever filamentous actin. *J. Biol. Chem.* **272**, 7841–7845.
79. Bewley, C. A., Holland, N. D., and Faulkner, D. J. (1996). Two classes of metabolites from *Theonella swinhoei* are localized in distinct populations of bacterial symbionts. *Experientia* **52**, 716–722.
80. Roesener, J. A., and Scheuer, P. J. (1986). Ulapualide A and B, extraordinary antitumor macrolides from nudibranch eggmasses. *J. Am. Chem. Soc.* **108**, 846–847.
81. Matsunaga, S., Fusetani, N., Hashimoto, K., Koseki, K., and Noma, M. (1986). Kabiramide C, a novel antifungal macrolide from nudibranch eggmasses. *J. Am. Chem. Soc.* **108**, 847–849.
82. Kernan, M. R., Molinski, T. F., and Faulkner, D. J. (1988). Macrocyclic antifungal metabolites from the Spanish dancer nudibranch *Hexabranchus sanguineus* and sponges of the genus *Halichondria*. *J. Org. Chem.* **53**, 5014–5020.
83. Fusetani, N., Yasumuro, K., Matsunaga, S., and Hashimoto, K. (1989). Mycalolides A–C, hybrid macrolides of ulapualides and halichondramide, from a sponge of the genus *Mycale*. *Tetrahedron Lett.* **30**, 2809–2812.

84. Saito, S., Watabe, S., Ozaki, H., Fusetani, N., and Karaki, H. (1994). Mycalolide B, a novel actin depolymerizing agent. *J. Biol. Chem.* **269**, 29,710–29,714.
85. Chattopadhyay, S. K., and Pattenden, G. (1998). Total synthesis of ulapualide A, a novel trisoxazole containing macrolide from the marine nudibranch *Hexabranchus sanguineus*. *Tetrahedron Lett.* **39**, 6095–6098.
86. Matsunaga, S., Liu, P., Celatka, C. A., Panek, J. S., and Fusetani, N. (1999). Relative and absolute stereochemistry of mycalolides, bioactive macrolides from the marine sponge *Mycale magellanica*. *J. Am. Chem. Soc.* **121**, 5605–5606.
87. Liu, P., and Panek, J. S. (2000). Total synthesis of (−)-mycalolide A. *J. Am. Chem. Soc.* **122**, 1235–1236.
88. D'Auria, M. V., Paloma, L. G., Minale, L., Zampella, A., Verbist, J.-F., Roussakis, C., and Debitus, C. (1993). Three new potent cytotoxic macrolides closely related to sphinxolide from the New Caledonian sponge *Neosiphonia superstes*. *Tetrahedron* **49**, 8657–8664.
89. D'Auria, M. V., Paloma, L. G., Minale, L., Zampella, A., Verbist, J.-F., Roussakis, C., Debitus, C., and Patissou, J. (1994). Reidispongiolide A and B, two new potent cytotoxic macrolides from the New Caledonian sponge *Reidispongia coerulea*. *Tetrahedron* **50**, 4829–4834.
90. Guella, G., Mancini, I., Chiasera, G., and Pietra, F. (1989). Sphinxolide, a 26-membered antitumoral macrolide isolated from an unidentified Pacific nudibranch. *Helv. Chim. Acta* **72**, 237–246.
91. Zhang, X., Minale, L., Zampella, A., and Smith, C. D. (1997). Microfilament depletion and circumvention of multidrug resistance by sphinxolides. *Cancer Res.* **57**, 3751–3758.
92. Gunasekera, S. P., Gunasekera, M., Longley, R. E., and Schulte, G. K. (1990). Discodermolide: A new bioactive polyhydroxylated lactone from the marine sponge *Discodermia dissoluta*. *J. Org. Chem.* **55**, 4912–4915.
93. Hung, D. T., Nerenberg, J. B., and Schreiber, S. L. (1996). Syntheses of discodermolides useful for investigating microtubule binding and stabilization. *J. Am. Chem. Soc.* **118**, 11054–11080.
94. ter Haar, E., Kowalski, R. J., Hamel, E., Lin, C. M., Longley, R. E., Gunasekera, S. P., Rosenkranz, H. S., and Day, B. W. (1996). Discodermolide, a cytotoxic marine agent that stabilizes microtubules more potently than taxol. *Biochemistry* **35**, 243–250.
95. Martello, L. A., McDaid, M. H., Regl, D. L, Yang, C. P., Meng, D., Pettus, T. R., Kaufman, M. D., Arimoto, H., Danishefsky, S. J., Smith III, A. B., and Horwitz, S. B. (2000). Taxol and discodermolide represent a synergistic drug combination in human carcinoma cell lines. *Clin. Cancer Res.* **6**, 1978–1987.
96. Groweiss, A., Shmueli, U., and Kashman, Y. (1983). Marine toxins of *Latrunculia magnifica*. *J. Org. Chem.* **48**, 3512–3516.
97. Spector, I., Shochet, N. R., Kashman, Y., and Groweiss, A. (1983). Latrunculins: Novel marine toxins that disrupt microfilament organization in cultured cells. *Science* **219**, 493–495.
98. Okuda, R. K., and Scheuer, P. J. (1985). Latrunculin A, ichthyotoxic constituent of the nudibranch *Chromodoris elisabethina*. *Experientia* **41**, 1355–1356.
99. Kakou, Y., Crews, P., and Bakus, G. J. (1987). Dendrolasin and latrunculin A from the Fijian sponge *Spongia mycofijiensis* and an associated nudibranch *Chromodoris lochi*. *J. Nat. Prod.* **50**, 482–484.
100. Corley, D. G., Herb, R., Moore, R. E., Scheuer, P. J., and Paul, V. J. (1988). Laulimalides: New potent cytotoxic macrolides from a marine sponge and a nudibranch predator. *J. Org. Chem.* **53**, 3644–3646.
101. Quiñoà, E., Kakou, Y., and Crews, P. (1988). Fijianolides, polyketide heterocycles from a marine sponge. *J. Org. Chem.* **53**, 3642–3644.
102. Jefford, C. W., Bernardinelli, G., Tanaka, J., and Higa, T. (1996). Structures and absolute configurations of the marine toxins latrunculin A and laulimalide. *Tetrahedron Lett.* **37**, 159–162.
103. Tanaka, J., and Higa, T. (1996). Zampanolide, a new potent cytotoxic macrolide from a marine sponge. *Tetrahedron Lett.* **37**, 5535–5538.

104. Kobayashi, J., Kondo, K., Ishibashi, M., Wälchli, M. R., and Nakamura, T. (1993). Theonezolide A, a novel polyketide natural product from the Okinawan marine sponge *Theonella* sp. *J. Am. Chem. Soc.* **115**, 6661–6665.
105. Sato, M., Takeuchi, S., Ishibashi, M., and Kobayashi, J. (1998). Studies on the stereochemistry of theonezolides A–C: Determination of the absolute stereochemistry of the C-4–C-17 fragment. *Tetrahedron* **54**, 4819–4826.
106. Rho, M.-C., Park, Y.-H., Sasaki, S., Ishibashi, M., Kondo, K., Kobayashi, J., and Ohizumi, Y. (1996). The mode of rabbit platelet shape change and aggregation induced by theonezolide-A, a novel polyketide macrolide, isolated from the Okinawan marine sponge *Theonella* sp. *Can. J. Physiol. Pharmacol.* **74**, 193–199.
107. Higa, T., Tanaka, J., Komesu, M., Garcia Gravalos, D., Puentes, J. L. F., Bernardinelli, G., and Jefford, C. W. (1992). Miyakolide: A bryostatin-like macrolide from a sponge, *Polyfibrospongia* sp. *J. Am. Chem. Soc.* **114**, 7587–7588.
108. Ichiba, T., Yoshida, W. Y., Scheuer, P. J., Higa, T., and Garcia Gravalos, D. (1991). Hennoxazoles, bioactive bisoxazoles from a marine sponge. *J. Am. Chem. Soc.* **113**, 3173–3174.
109. Northcote, P. T., Blunt, J. W., and Munro, M. H. G. (1991). Pateamine: A potent cytotoxin from the New Zealand marine sponge *Mycale* sp. *Tetrahedron Lett.* **32**, 6411–6414.
110. Romo, D., Rzasa, R. M., Shea, H. A., Park, K., Langenhan, J. M., Sun, L., Akhiezer, A., and Liu, J. O. (1998). Total synthesis and immunosuppressive activity of (−)-pateamine A and related compounds: Implementation of a β-lactam-based macrocyclization. *J. Am. Chem. Soc.* **120**, 12,237–12,254.
111. Perry, N. B., Blunt, J. W., Munro, M. H. G., and Pannell, L. K. (1988). Mycalamide A, an antiviral compound from a New Zealand sponge of the genus *Mycale*. *J. Am. Chem. Soc.* **110**, 4850–4851.
112. West, L. M., Northcote, P. T., and Battershill, C. N. (2000). Peloruside A: A potent cytotoxic macrolide isolated from the New Zealand marine sponge *Mycale* sp. *J. Org. Chem.* **65**, 445–449.
113. Searle, P. A., and Molinski, T. F. (1995). Phorboxazoles A and B: Potent cytostatic macrolides from marine sponge *Phorbas* sp. *J. Am. Chem. Soc.* **117**, 8126–8131.
114. Molinski, T. F. (1996). Absolute configuration of phorboxazoles A and B from the marine sponge *Phorbas* sp. 2. C43 and complete stereochemistry. *Tetrahedron Lett.* **37**, 7879–7880.
115. Forsyth, C. J., Ahmed, F., Cink, R. D., and Lee, C. S. (1998). Total synthesis of phorboxazole A. *J. Am. Chem. Soc.* **120**, 5597–5598.
116. Horton, P. A., Koehn, F. E., Longley, R. E., and McConnell, O. J. (1994). Lasonolide A, a new cytotoxic macrolide from the marine sponge *Forcepia* sp. *J. Am. Chem. Soc.* **116**, 6015–6016.
117. Lu, Q., and Faulkner, D. J. (1998). Three dolabellanes and a macrolide from the sponge *Dysidea* sp. from Palau. *J. Nat. Prod.* **61**, 1096–1100.
118. D'Auria, M. V., Debitus, C., Paloma, L. G., Minale, L., and Zampella, A. (1994). Superstolide A: A potent cytotoxic macrolide of a new type from the New Caledonian deep water sponge *Neosiphonia superstes*. *J. Am. Chem. Soc.* **116**, 6658–6663.
119. Zampella, A., D'Auria, M. V., Minale, L., Debitus, C., and Roussakis, C. (1996). Callipeltoside A: A cytotoxic aminodeoxy sugar-containing macrolide of a new type from the marine lithistida sponge *Callipelta* sp. *J. Am. Chem. Soc.* **118**, 11,085–11,088.
120. Erickson, K. L., Beutler, J. A., Cardellina II, J. H., and Boyd, M. R. (1997). Salicylihalamides A and B, novel cytotoxic macrolides from the marine sponge *Haliclona* sp. *J. Org. Chem.* **62**, 8188–8192.
121. Kunze, B., Jansen, R., Sasse, F., Höfle, G., and Reichenbach, H. (1998). Apicularens A and B, new cytostatic macrolides from *Chondromyces* species (Myxobacteria): Production, physicochemical and biological properties. *J. Antibiot.* **51**, 1075–1080.
122. D'Ambrosio, M., Guerriero, A., Debitus, C., and Pietra, F. (1996). Leucascandrolide A, a new type of macrolide: The first powerfully bioactive metabolite of calcareous sponges (*Leucascandra caveolata*, a new genus from the Coral Sea). *Helv. Chim. Acta* **79**, 51–60.

123. Cantrell, C. L., Gustafson, K. R., Cecere, M. R., Pannell, L. K., and Boyd, M. R. (2000). Chondropsins A and B: Novel tumor cell growth inhibitory macrolide lactams from the marine sponge *Chondropsis* sp. *J. Am. Chem. Soc.* **122**, 8825–8829.
124. Pettit, G. R. (1991). The bryostatins. *In* "Progress in the Chemistry of Organic Natural Products" (W. Herz, Ed.), Vol. 57, pp. 153–195. Springer-Verlag, New York.
125. Pettit, G. R. (1996). Progress in the discovery of biosynthetic anticancer drugs. *J. Nat. Prod.* **59**, 812–821.
126. Pagliaro, L., Daliani, D., Amato, R., Tu, S. M., Jones, D., Smith, T., Logothetis, C., and Millikan, R. (2000). A phase II trial of bryostatin-1 for patients with metastatic renal cell carcinoma. *Cancer* **89**, 615–618.
127. Varterasian, M. L., Mohammad, R. M., Shurafa, M. S., Hulburd, K., Pemberton, P. A., Rodriguez, D. H., Spadoni, V., Eilender, D. S., Murgo, A., Wall, N., Dan, M., and Al-Katib, A. M. (2000). Phase II trial of bryostatin 1 in patients with relapsed low-grade non-Hodgkin's lymphoma and chronic lymphocytic leukemia. *Clin. Cancer Res.* **6**, 825–828.
128. Rouhi, A. M. (1995). Supply issues complicate trek of chemicals from sea to market. *Chem. Eng. News,* Nov. 20, 42–44.
129. Evans, D. A., Carter, P. H., Carreira, E. M., Charette, A. B., Prunet, J. A., and Lautens, M. (1999). Total synthesis of bryostatin 2. *J. Am. Chem. Soc.* **121**, 7540–7552.
130. Yamada, K., Ojika, M., Ishigaki, T., Yoshida, Y., Ekimoto, H., and Arakawa, M. (1993). Aplyronine A, a potent antitumor substance, and the congeners aplyronines B and C isolated from the sea hare *Aplysia kurodai. J. Am. Chem. Soc.* **115**, 11020–11021.
131. Saito, S., Watabe, S., Ozaki, H., Kigoshi, H., Yamada, K., Fusetani, N., and Karaki, H. (1996). Novel actin depolymerizing macrolide aplyronine A. *J. Biochem.* **120**, 552–555.
132. Ojika, M., Kigoshi, H., Ishigaki, T., Tsukada, I., Tsuboi, T., Ogawa, T., and Yamada, K. (1994). Absolute stereochemistry of aplyronine A, a potent antitumor substance of marine origin. *J. Am. Chem. Soc.* **116**, 7441–7442.
133. Kigoshi, H., Ojika, M., Ishigaki, T., Suenaga, K., Mutou, T., Sakakura, A., Ogawa, T., and Yamada, K. (1994). Total synthesis of aplyronine A, a potent antitumor substance of marine origin. *J. Am. Chem. Soc.* **116**, 7443–7444.
134. Pettit, G. R., Kamano, Y., Herald, C. L., Fujii, Y., Kizu, H., Boyd, M. R., Boettner, F. E., Doubek, D. L., Schmidt, J. M., Chapuis, J.-C., and Michel, C. (1993). Antineoplastic agents. Part 247. The dolastatins. Part 18. Isolation of dolastatins 10–15 from the marine mollusk *Dolabella auricularia. Tetrahedron* **49**, 9151–9170.
135. Sone, H., Kondo, T., Kiryu, M., Ishiwata, H., Ojika, M., and Yamada, K. (1995). Dolabellin, a cytotoxic bisthiazole metabolite from the sea hare *Dolabella auricularia:* Structural determination and synthesis. *J. Org. Chem.* **60**, 4774–4781.
136. Ojika, M., Nagoya, T., and Yamada, K. (1995). Dolabelides A and B, cytotoxic 22-membered macrolides isolated from the sea hare *Dolabella auricularia. Tetrahedron Lett.* **36**, 7491–7494.
137. Sone, H., Kigoshi, H., and Yamada, K. (1996). Aurisides A and B, cytotoxic macrolide glycosides from the Japanese sea hare *Dolabella auricularia. J. Org. Chem.* **61**, 8956–8960.
138. Yasumoto, T., Murata, M., Oshima, Y., Sano, M., Matsumoto, G. K., and Clardy, J. (1985). Diarrhetic shellfish toxins. *Tetrahedron* **41**, 1019–1025.
139. Sasaki, K., Satake, M., and Yasumoto, T. (1997). Identification of the absolute configuration of pectenotoxin-6, a polyether macrolide compound, by NMR spectroscopic method using a chiral anisotropic reagent, phenylglycine methyl ester. *Biosci. Biotechnol. Biochem.* **61**, 1783–1785.
140. Sasaki, K., Wright, J. L.C., and Yasumoto, T. (1998). Identification and characterization of pectenotoxin (PTX) 4 and PTX7 as spiroketal stereoisomers of two previously reported pectenotoxins. *J. Org. Chem.* **63**, 2475–2480.

141. Jung, J. H., Sim, C. J., and Lee, C.-O. (1995). Cytotoxic compounds from a two-sponge association. *J. Nat. Prod.* **58**, 1722–1726.
142. Uemura, D., Chou, T., Haino, T., Nagatsu, A., Fukuzawa, S., Zheng, S.-Z., and Chen, H.-S. (1995). Pinnatoxin A: A toxic amphoteric macrocycle from the Okinawan bivalve *Pinna muricata. J. Am. Chem. Soc.* **117**, 1155–1156.
143. Chou, T., Kuramoto, M., Otani, Y., Shikano, M., Yazawa, K., and Uemura, D. (1996). Pinnaic acid and tauropinnaic acid: Two novel fatty acids composing a 6-azaspiro[4.5]decane unit from the Okinawan bivalve *Pinna muricata. Tetrahedron Lett.* **37**, 3871–3874.
144. Kuramoto, M., Tong, C., Yamada, K., Chiba, T., Hayashi, Y., and Uemura, D. (1996). Halichlorine, an inhibitor of VCAM-1 induction from the marine sponge *Halichondria okadai* Kadota. *Tetrahedron Lett.* **37**, 3867–3870.
145. Arimoto, H., Hayakawa, I., Kuramoto, M., and Uemura, D. (1998). Absolute stereochemistry of halichlorine: A potent inhibitor of VCAM-1 induction. *Tetrahedron Lett.* **39**, 861–862.
146. Davidson, B. S. (1993). Ascidians: Producers of amino acid derived metabolites. *Chem. Rev.* **93**, 1771–1792.
147. Rinehart, K. L. (2000). Antitumor compounds from tunicates. *Med. Res. Rev.* **20**, 1–27.
148. Zabriskie, T. M., Mayne, C. L., and Ireland, C. M. (1988). Patellazole C: A novel cytotoxic macrolide from *Lissoclinum patella. J. Am. Chem. Soc.* **110**, 7919–7920.
149. Corley, D. G., Moore, R. E., and Paul, V. J. (1988). Patellazole B: A novel cytotoxic thiazole-containing macrolide from the marine tunicate *Lissoclinum patella. J. Am. Chem. Soc.* **110**, 7920–7922.
150. Kobayashi, J., Cheng, J.-F., Ohta, T., Nakamura, H., Nozoe, S., Hirata, Y., Ohizumi, Y., and Sasaki, T. (1988). Iejimalides A and B, novel 24-membered macrolides with potent antileukemic activity from the Okinawan tunicate *Eudistoma* cf. *rigida. J. Org. Chem.* **53**, 6147–6150.
151. Kikuchi, Y., Ishibashi, M., Sasaki, T., and Kobayashi, J. (1991). Iejimalides C and D, new antineoplastic 24-membered macrolide sulfates from the Okinawan marine tunicate *Eudistoma* cf. *rigida. Tetrahedron Lett.* **32**, 797–798.
152. Lindquist, N., and Fenical, W. (1989). Ascidiatrienolides A–C, novel lactonized eicosanoids from the colonial marine ascidian *Didemnum candidum. Tetrahedron Lett.* **30**, 2735–2738.
153. Congreve, M. S., Holmes, A. B., Hughes, A. B., and Looney, M. G. (1993). Ascidiatrienolide A is a 10-membered lactone. *J. Am. Chem. Soc.* **115**, 5815–5816.
154. Niwa, H., Inagaki, H., and Yamada, K. (1991). Didemnilactone and neodidemnilactone, two new fatty acid metabolites possessing a 10-membered lactone from the tunicate *Didemnum moseleyi* (Herdman). *Tetrahedron Lett.* **32**, 5127–5128.
155. Niwa, H., Watanabe, M., Inagaki, H., and Yamada, K. (1994). Didemnilactones A and B and neodidemnilactone, three new fatty acid metabolites isolated from the tunicate *Didemnum moseleyi* (Herdman). *Tetrahedron* **50**, 7835–7400.
156. Galinis, D. L., Mckee, T. C., Pannell, L. K., Cardellina II, J. H., and Boyd, M. R. (1997). Lobatamides A and B, novel cytotoxic macrolides from the tunicate *Aplidium lobatum. J. Org. Chem.* **62**, 8968–8969.
157. Mckee, T. C., Galinis, D. L., Pannell, J. K., Cardellina II, J. H., Laakso, J., Ireland, C. M., Murray, L., Capon, R. J., and Boyd, M. R. (1998). The lobatamides, novel cytotoxic macrolides from southwestern Pacific tunicates. *J. Org. Chem.* **63**, 7805–7810.
158. Murray, L., Lim, T. K., Currie, G., and Capon, R. J. (1995). Aplidites (A–G): Macrocyclic orthonitrites from an Australian tunicate, *Aplidium* sp. *Aust. J. Chem.* **48**, 1253–1266.
159. Suzumura, K., Takahashi, I., Matsumoto, H., Nagai, K., Setiawan, B., Rantiatmodjo, R. M., Suzuki, K., and Nagao, N. (1997). Structural elucidation of YM-75518, a novel antifungal antibiotic isolated from *Pseudomonas* sp. Q38009. *Tetrahedron Lett.* **38**, 7573–7576.
160. Ueda, K., and Hu, Y. (1999). Hateramalide B: A new cytotoxic macrolide from an Okinawan ascidian *Lissoclinum* sp. *Tetrahedron Lett.* **40**, 6305–6308.

161. Takada, N., Sato, H., Suenaga, K., Arimoto, H., Yamada, K., Ueda, K., and Uemura, D. (1999). Isolation and structures of haterumalides NA, NB, NC, ND, and NE, novel macrolides from an Okinawan sponge *Ircinia* sp. *Tetrahedron Lett.* **40**, 6309–6312.
162. Strobel, G., Li, J.-Y., Sugawara, F., Koshino, H., Harper, J., and Hess, W. M. (1999). Oocydin A, a chlorinated macrocyclic lactone with potent anti-oomycete activity from *Serratia marcescens*. *Microbiology* **145**, 3557–3564.
163. Newman, D. J., Cragg, G. M., and Snader, K. M. (2000). The influence of natural products upon drug discovery. *Nat. Prod. Rep.* **17**, 215–234.
164. Harvey, A. (2000). Strategies for discovering drugs from previously unexplored natural products. *Drug Discovery Today* **5**, 294–300.

Chapter 3

Chemical Modification of Macrolides

TOSHIAKI SUNAZUKA
School of Pharmaceutical Sciences
Kitasato University
Tokyo, Japan

SADAFUMI OMURA
SHIGEO IWASAKI
SATOSHI ŌMURA
The Kitasato Institute
Kitasato University
Tokyo, Japan

I. Introduction 99
II. Fourteen-Membered Macrolides 100
   A. Chemical Modification of Fourteen-Membered
      Macrolides before 1984 101
   B. Newer Macrolides: The New Generations 105
   C. Recent Modifications of Erythromycins 117
   D. Ketolides 127
   E. Discussion 142
III. Sixteen-Membered Macrolide Antibiotics
    and the Avermectin Family 145
   A. Chemical Modification of Sixteen-Membered
      Macrolides before 1984 146
   B. Newer Macrolides 147
   C. The Avermectin Family 153
IV. Concluding Remarks 164
    References 165

I. Introduction

Since the discovery of pikromycin in 1950, a number of macrolides have been found from various microorganisms. They include useful antibiotics such as 14-, 15-, and 16-membered macrolides, polyene macrolides, avermectin, FK-506, and so on. Macrolide antibiotics have chemical characteristics different from other antibiotics, such as having macrocyclic lactone, peculiar molecular size and conformation, and high lipophilicity. In addition, some of their glycosidic linkage often gives the local polarity in a molecule, which relates closely with the affinity for a target such as a ribosome or other binding site. On the other hand, the unique structures and physicochemical properties of macrolides show diversity in biological activity, such as antibacterial, antifungal, antiprotozoal, and immunomodulating activities.

Naturally occurring macrolides, however, do not always have the most desirable chemical characteristics and biological properties. Therefore, improvements to their chemical and biological properties by chemical modification and microbial transformation of the lactone ring or sugar moiety have been investigated. In particular, many efforts have been made to resolve such problems as potency, spectrum of activity, drug resistance, pharmacokinetics, and toxicity. Coupled with these efforts were rapid advances in technology, such as nuclear magnetic resonance (NMR) analysis, mass spectroscopy, X-ray analysis, computational analysis, chemical synthesis, and biotechnology, such as molecular biology and gene engineering. Consequently, much progress in the research and development of macrolides has been made in recent decades.

This chapter deals with recent progress in the chemical modification and structure–activity relationships of 14- and 15-membered macrolides (mainly erythromycin derivatives), 16-membered macrolides (mainly the leucomycin and tylosin families), and the avermectin family of macrolides, showing nematocidal, insecticidal, and arachnidicidal activities. Previous reviews of these macrolides were given by Sakakibara and Ōmura in the first edition of this book in 1984 [1].

II. Fourteen-Membered Macrolides

14-Membered macrolides have always been a target of research for the chemical modification and structure–activity relationships. In particular, erythromycin (EM) A (**1**) and oleandomycin have been extensively studied to improve their antimicrobial activity and bioavailability. However, no significant improvement by chemical modification was attained in this regard before 1980 because of their structural complexity, such as the large size of the lactone ring bearing hydroxy, the keto groups, and the amino and/or neutral sugars, accompanied by the structural instability in acidic media, particularly of erythromycins [2]. In practice, acid salts and esters of **1** and oleandomycin (Fig. 1) were mainly synthesized as derivatives to improve their acid stability, intestinal absorption, and serum concentration in order to use them for the clinical treatment of respiratory tract infections [1].

Compound **1**, a representative 14-membered macrolide, originally exhibited unique and superior biological properties *in vitro* and *in vivo*, such as its activity, safety, and mode of action in inhibiting selective bacterial protein synthesis (its favorable antimicrobial spectrum being active also against *Mycoplasma* and *Chlamydia*), as well as good tissue penetration compared with other antibiotics. However, **1** showed several disadvantages, such as weak activity against Gram-negative bacteria, induction of macrolide resistance, a bitter taste, and a tendency to produce gastrointestinal disorder, in addition to an instability in gastric juices

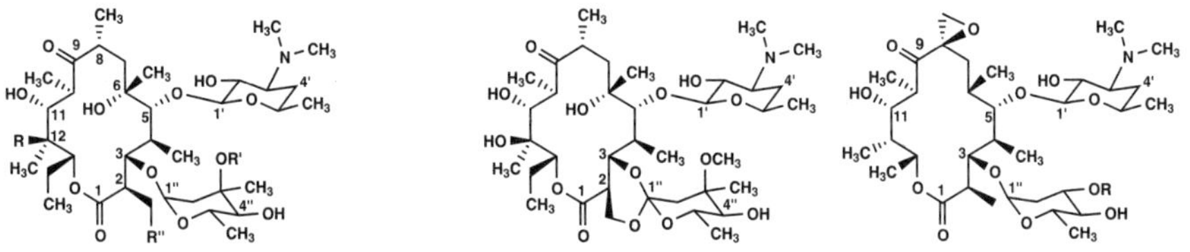

Erythromycin A: R = OH, R' = CH₃, R'' = H
Erythromycin B: R = R'' = H, R' = CH₃
Erythromycin C: R = OH, R' = R'' = H
Erythromycin D: R = R' = R'' = H
Erythromycin F: R = R'' = OH, R' = CH₃

Erythromycin E

Oleandomycin: R = CH₃
Oleandomycin Y: R = H

Fig. 1. Naturally occurring erythromycins and oleandomycins.

and a low serum concentration. These have been the major problems to be conquered by chemical modification.

Since Kurath *et al.* [2] reported the acid degradation pathway of **1** (Scheme 1) in 1971, the roles of the functional groups of its lactone ring and its sugar moieties have gradually been clarified. During these three decades, study of the chemical modifications of **1** has accelerated and has been directed toward the rational approach, resulting in a renaissance of semisynthetic macrolides such as the so-called newer macrolides and ketolides. The progress of these chemical modifications and the structure–activity relationships are descibed here.

A. Chemical Modification of Fourteen-Membered Macrolides before 1984

Since the previous review by Sakakibara and Ōmura [1], progress has been made in the study of the modification and the structure–activity relationships of **1**, and over the past 15 years various derivatives have been reported, as described in Sections II.B through II.D. First, however, we summarize the chemical modification of 14-membered macrolides (mainly **1**) (Figs. 2 through 5) described in the first edition. The modification of other 14-membered macrolides, such as oleandomycins and megalomicins, have been omitted because less progress has been made on them since 1985.

1. Modification of the Sugar Moieties

Concerning the sugar moieties of **1**, the presence of desosamine and cladinose is essential for retaining the original activity (later revised). The presence of the 3′-dimethylamino and 2′-hydroxyl groups on desosamine are essential and their orientations are important for retention of the antibacterial activity (except neutral macrolides). Any changes and modifications of these groups resulted in decreased activity. However, various 2′-esters such as propionate and ethylsuccinate,

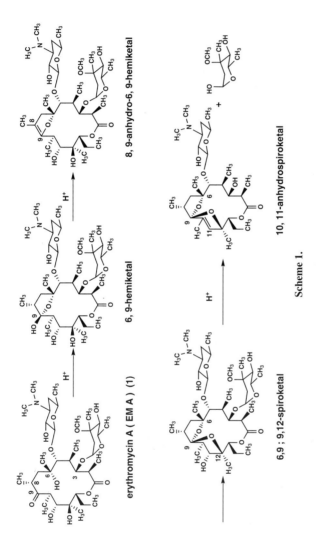

Scheme 1.

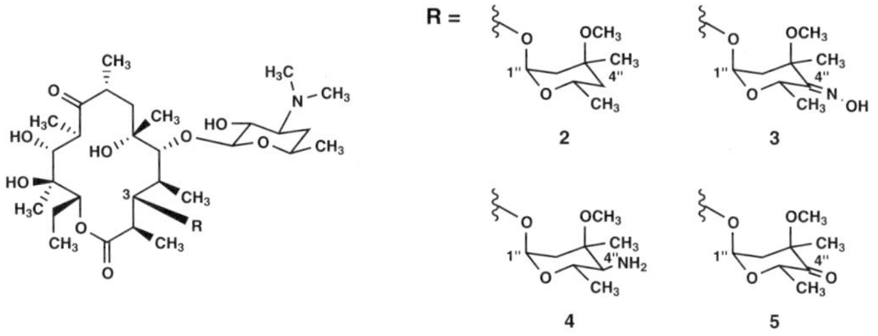

3. Chemical Modification of Macrolides

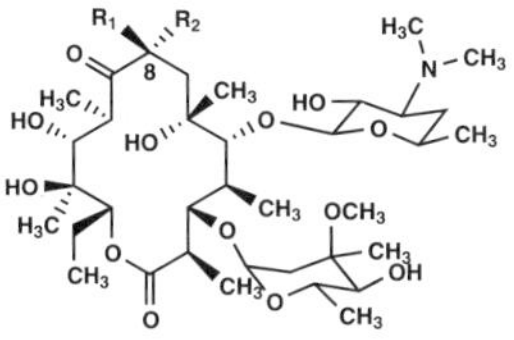

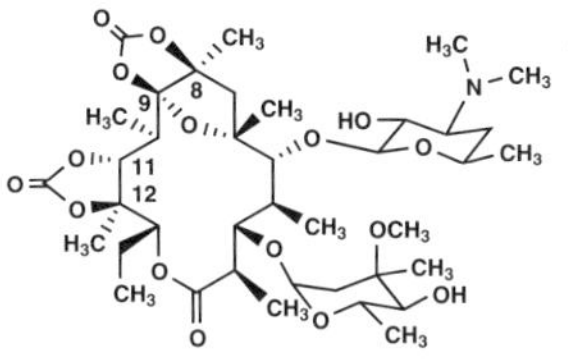

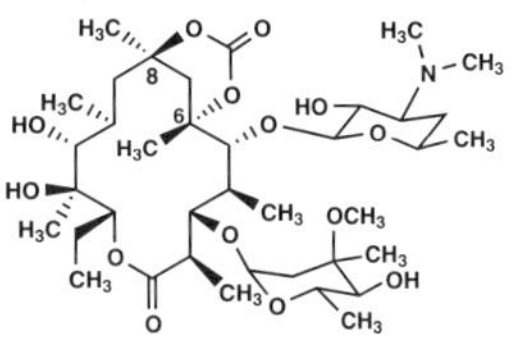

Fig. 2.

Fig. 3.

	R₁	R₂
6a:	OH	CH₃
6b:	CH₃	OH

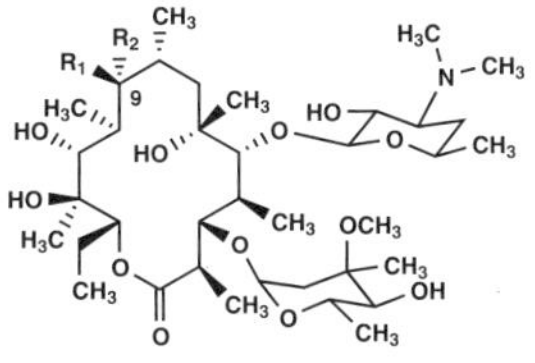

	R₁	R₂
9a :	NH₂	H
9b :	H	NH₂
10a :	OH	H
10b :	H	OH

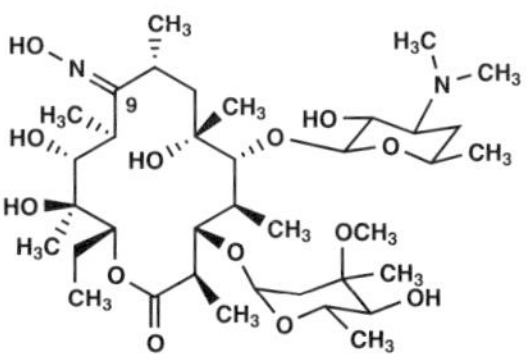

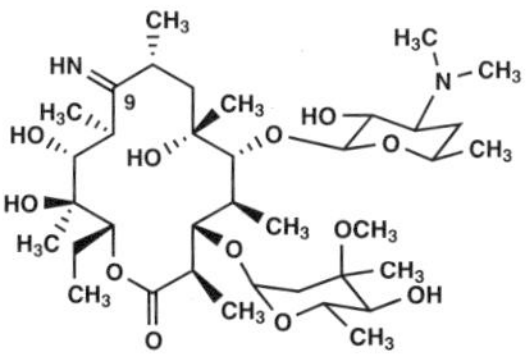

Fig. 4.

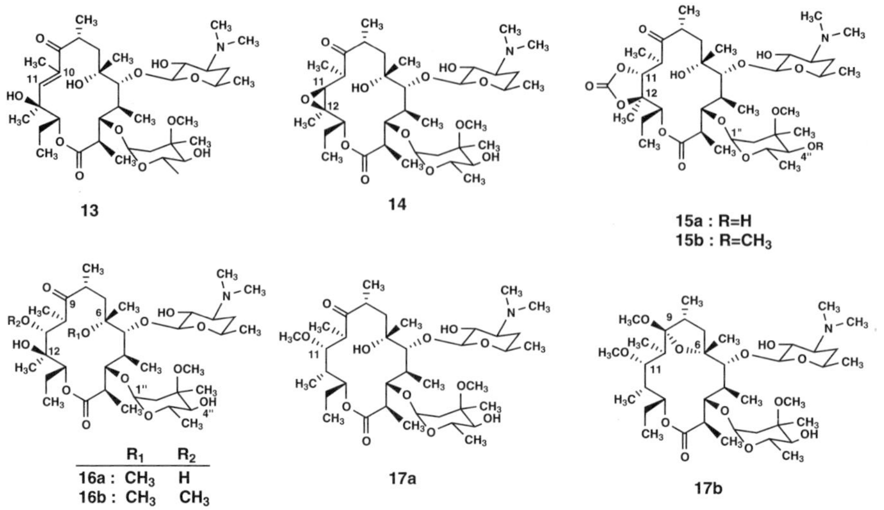

Fig. 5.

which are inactive *in vitro*, are activated by hydrolysis of the esters after oral administration, and their *in vivo* activities were higher than those of **1** due to the increased blood level of **1**.

Regarding the cladinose of **1**, 3″-demethylation of the 3″-methoxyl group diminishes the activity by the effect of the polar 3″-hydroxyl group on the 3′-dimethylamino or 2′-hydroxyl group. The 4″-hydroxy epimer of **1** and the 4″-deoxy derivative (**2**) are comparable to **1** in their antibacterial activity and resistance inducibility to Gram-positive bacteria. EM derivatives having 4″-oxime (**3**) and 4″-amino (**4**) or their 9-oximes produce the same or less activity and diminish resistance inducibility compared with those of **1**. 4″-Didehydro EM (**5**) also reduces the activity. Thus, modification of the 4″-hydroxyl group changes the activity and the macrolide resistance inducibility.

2. Modification of the Lactone Ring

By acylation of the C-11, C-2′, or C-4″ hydroxyl group, the activities of all monoesters decrease and are on the order of C-11 > C-4″ > C-2′. The decrease in activity caused by acylation is in general on the order of the number of esterified hydroxy groups as well as of the acyl chain lengths; the more, the weaker, and the longer, the weaker. Antibacterial activities of (8S)-8-hydroxy EM (**6a**) and (8R)-8-hydroxy EM (**6b**) were 1/8 and 1/30 of **1**, respectively. However, (8S)-8-hydroxy EM 6,9-hemiacetal 8,9-:11,12-dicyclic dicarbonate (**7**) was more active than **6a**.

The 9-keto group of **1** is not essential for its antibacterial activity. It can be replaced by carbonyl groups of 8,9-cyclic carbonate in **7** and of 6,8-cyclic carbonate in 9-deoxo EM 6,8-cyclic carbonate (**8**) without loss of parent activity. The capability of 9-keto replacement was also confirmed by the fact that (9*S*)-9-amino (**9a**) and its *N*-substituted derivatives not only retained their *in vitro* activity but showed enhanced *in vivo* activity compared with **1**. (9*S*)-9-Dihydro EM (**10a**) was less active than **1** *in vitro* but reasonably inhibited cell-free protein synthesis and the binding of **1** to ribosome. In addition, 9-oxime (**11**) and 9-imino (**12**) of **1** were obtained to prepare (9*S*)-9-amino (**9a**) and (9*R*)-9-amino (**9b**) EMs. Compound **9b** and (9*R*)-9-dihydro EM (**10b**) derivatives were less active than their corresponding epimers (**9a** and **10a**).

10,11-Anhydro EM (**13**) and 11,12-epoxy EM (**14**) showed a significant reduction of activity. However, EM 11,12-cyclic carbonate (**15a**) has been considered to stabilize the molecule and enhance the activity. 4″-*O*-Methyl EM 11,12-cyclic carbonate (**15b**), indeed, exhibited much higher activity than **1** against inducibly EM-resistant strains (MLS-i) of staphylococci. 6-*O*-Methyl EM (**16a**) and 6,11-*O*-dimethyl EM (**16b**) also showed higher *in vitro* and *in vivo* activity, caused by their acid stability and good bioavailability. 11-*O*-Methyl erythromycin B (**17a**) and 9,11-di-*O*-methyl erythromycin B 6,9-hemiketal (**17b**) exhibited an MIC value of 0.78 µg/ml against *Staphylococcus aureus*, which was a lower value than that of erythromycin B.

As the previous authors concluded, EM derivatives having (9*S*)-amino (**9**), 6-*O*-methyl (**16a**), 11,12-cyclic carbonate (**15a**), and 4″ hydroxyl analogues are promising substituents for further study.

B. Newer Macrolides: The New Generations

In studies on the effect of functional groups in EM (**1**) derivatives, compounds having 9-oxime (**11**), (9*S*)-9-amino (**9a**), 6-*O*-methyl (**16a**), (8*S*)-8-hydroxy (**6a**), and 11,12-cyclic carbonate (**15a**) substituents were found to stabilize the lactone ring from decomposition in acidic media [2]. Therefore, further improvement of activity and bioavailability in oral administration has been attempted.

Roxithromycin (RXM) (**18**), clarithromycin (CAM) (**16a**), azithromycin (AZM) (**19**), dirithromycin (DRM) (**20**), and flurithromycin (FLM) (**21**), so-called newer macrolides or new generation macrolides, were synthesized by the procedures illustrated in Schemes 2 through 6, respectively. These products, showing broader antibacterial spectra and/or good bioavailability through acid stability, were developed and clinically evaluated in the late 1980s.

1. Roxithromycin

EM 9-oxime (**11**), a key compound of roxithromycin synthesis, is known to increase acid stability by reducing intermolecular ketalization but also diminishes the antimicrobial activity compared with EM (**1**) [3–5]. Compound **11** was obtained by the reaction of EM with hydroxylamine hydrochloride in a methanol solution in the presence of triethylamine (Scheme 2) [6]. The isolated EM oxime

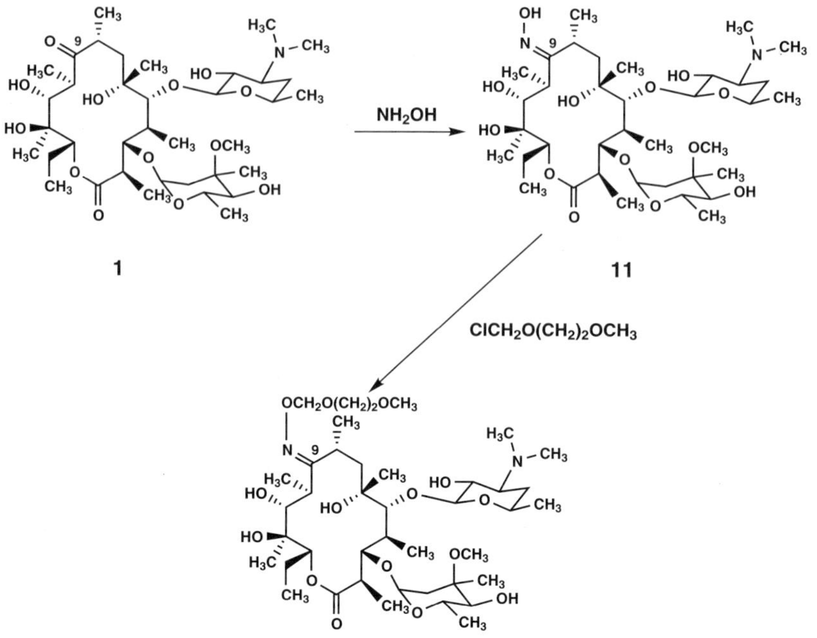

Roxithromycin (18)

Scheme 2.

has the *E* configuration but a small amount of the *Z* isomer, which was less active than the *E* isomer *in vitro* and *in vivo*, was present in the crude product. Further, EM 9-oxime ether derivatives were synthesized and found to show almost the same level of activity as **1** [6]. The reactivity of the hydroxyl function of the oxime group enables selective alkylation by means of alkyl halide in the presence of a base in a solvent such as acetone, THF, or ethyl ether without the protection of other hydroxyl groups. In some cases, *O*-alkyl oxime derivatives were prepared as main products by the reaction of **1** with a variety of *O*-alkyl hydroxylamine. In some preparations, a diether derivative was obtained as a minor product, substituted on the hydroxyl group of the oxime and on the 4″-hydroxyl group [7].

Among a number of oxime ether derivatives of **1** with aromatic and aliphatic substitutions, the compounds bearing heteroatoms in substituents, such as nitrogen, oxygen, and sulfur, exhibited a significant increase in activity *in vivo*. Based on their *in vitro* and *in vivo* activities against staphylococci and streptococci as well as their pharmacokinetics, five compounds **18** and **23** through **26** have been selected (Table I) [6].

TABLE I
Biological Feature of the Selected Derivatives

No.	9-Oxime ether derivative R	In vitro activity[a] MIC (μg/ml)	In vivo activity[b] (mouse) Staphylococcal infection	In vivo activity[b] (mouse) Streptococcal infection	Pharmacokinetic data (rat) T_{max} (hours)	Pharmacokinetic data (rat) C_{max} (hours)	Pharmacokinetic data (rat) C (7 hr) (μg/ml)
	EM (1)	0.15	1	1	N.D.	0.25	N.D.
23	—CH$_2$—CH$_2$—N(CH$_3$)$_2$	0.13	2.36	20	2	1.9	1
24	—CH$_2$—CH$_2$—N(C$_2$H$_5$)$_2$	0.12	1.55	12	1	3.8	0.3
25	—CH$_2$—O—CH$_3$	0.12	2.78	5	0.5	1.6	0.15
26	—CH$_2$—S(→O)—C$_6$H$_5$	0.33	2.5	1.98	1	4.2	0.47
18	—CH$_2$—O—CH$_2$—CH$_2$—O—CH$_3$ (RXM)	0.30	3.84	2.3	0.25	3.8	0.44

Source: Data from Gasc *et al.* [6].
[a]Mean MICs of seven selected sensitive bacteria (3 staphylococci, 4 streptococci).
[b]*In vivo* activity shown as the ratio between ED$_{50}$ of EM/ED$_{50}$ of test compound.

Roxithromycin (**18**), a methoxyethoxymethyl ether oxime derivative of **11**, is more stable than **1** in acidic media. The *E* orientation of the oxime and the chain length that bears oxygen atoms, such as methoxyethoxymethyl, are important for increased *in vitro* and *in vivo* activity. The structural characteristics of **18** obtained by X-ray and NMR analyses have been compared with those of **1** [8, 9]. The modification of **1** at the 9-oxime decreased its affinity for *P*-450, which reflects reduced interaction with hepatic mono-oxygenase *in vitro* and *in vivo* [10]. The change in physicochemical properties such as the octanol–water partition coefficient of **18** ($P = 408$, log $P = 2.61$, 37°C, pH 7.40) indicated a more lipophilic character than that of **1** ($P = 50$, log $P = 1.70$), which can be linked with its good tissue penetration [11].

Although **18** was slightly less active than **1** in antibacterial activity *in vitro*, it exhibited higher tissue distribution and a longer half-life [12]. From the good balance of *in vitro* and *in vivo* activity, pharmacokinetics, and toxicity, RXM (**18**) has been developed and marketed for clinical use [13, 14]. The synthesis procedure for **18** is shown in Scheme 2 [6].

2. Clarithromycin

The alkylation of the C-6 tertiary hydroxyl group of **1** was considered to be difficult due to its expected low reactivity. However, 6-*O*-methylation of **1** could be accomplished initially by the reaction of 2'-*O*, 3'-*N*-bis(bezyloxycarbonyl)-*N*-demethyl EM (**22**) (Fig. 6) with NaH (or KOH) in the presence of CH$_3$I in DMSO-dimethoxyethane. The 2'-*O*, 3'-*N*-bis-(bezyloxycarbonyl) groups were

TABLE II
In Vitro and In Vivo Antibacterial Activities of O-Alkylerythromycins

Compound	MIC (µg/ml) S. aureus Smith	ED_{50} (mg/mouse)[a]
Erythromycin (EM) (**1**)	0.20	0.977[b]
6-O-Methyl EM (**16a**)	0.10	0.273[b]
6-O-Ethyl EM (**27**)	0.3	
6,11-Di-O-methyl EM (**16b**)	0.39	0.446[b]
6,12-Di-O-methyl EM (**28**)	0.20	0.216[c]
6,4″-Di-O-methyl EM (**29**)	3.13	
6,11,12,4″-Tetra-O-methyl EM (**30**)	6.25	
11-O-Methyl EM (**31**)	0.39	0.738[b]
6-O-Methyl EM-B (**32**)	0.20	0.358[c]

Source: Data from Omura et al. [16].
[a]Compounds were administered orally to ICR mouse (4 weeks, male) 1 hr after infection.
[b]Number of test animals is 20.
[c]Number of test animals is 15.

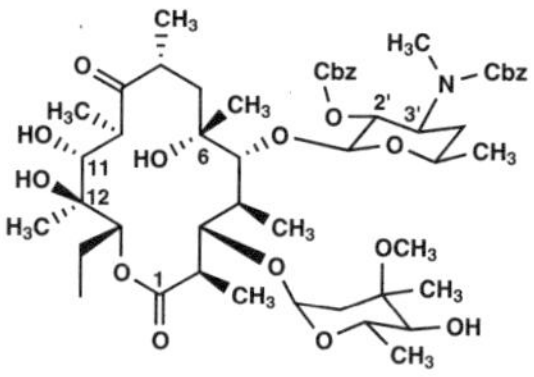

Fig. 6.

(Cbz : -$CO_2CH_2C_6H_5$)

then deprotected from the 6-O-methylated product by hydrogenation with H_2 over Pd/C in ethanol–acetic acid, affording 6-O-methyl-N-demethyl EM. Finally, the 3′-monomethylamino group was methylated with formaldehyde under reductive conditions (H_2-Pd/C). Selective O-methylation of each hydroxyl group in **1** could be performed by a combination of different solvents and reagents. The presence of CAM (**16a**) and 11-O-methylated CAM (**16b**) as minor components was detected and their acid stability confirmed [7, 15, 16]. A number of O-alkyl derivatives of erythromycins A and B (**27** through **32**) were then prepared (Table II). From their chemical and biological characteristics, **16a** was selected as the best compound to be developed [15–17].

To improve the selectivity and yield of **16a**, various synthetic methods were proposed, and as an improved method, a procedure using compound **11** as a key intermediate was performed (Scheme 3). The 2′- and 4″-hydroxy groups and the 9-oxime group in **11** were protected as trimethylsilyl and acetal groups, respectively,

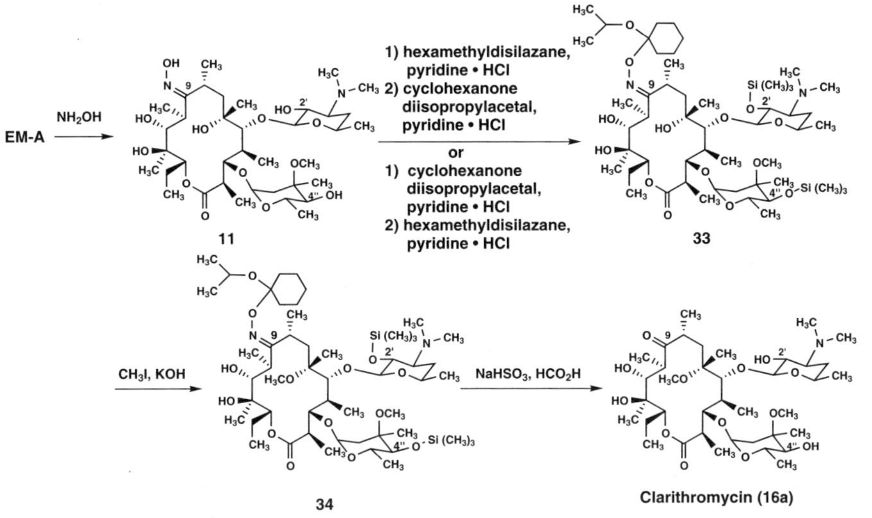

Scheme 3.

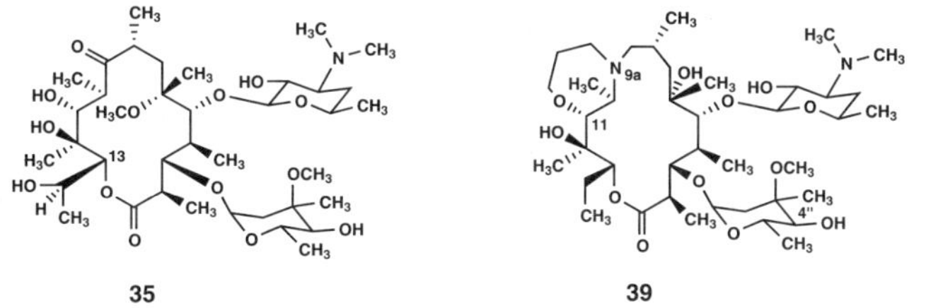

Fig. 7.

to give 2′-4″-O-bis(trimethylsilyl) EM 9-oxime acetal (**33**). An industrial-scale method for the production of **16a** has been established using methylation of **33** to obtain the 6-O-methyl derivative (**34**) followed by deprotection for a more than 45% overall yield [16, 18–20].

A pharmacokinetic study of **16a** revealed that one of the main metabolites in human urine is (14R)-14-hydroxyl CAM (**35**) [21] (Fig. 7). Compound **35** showed almost the same level of *in vitro* activity as **1** and greatly increased *in vivo* activity [16]. Because the excretion amount of **35** in urine was more than 35%, its pharmacokinetic behavior and *in vivo* activity needed to be investigated. Compound **35** was finally obtained by bioconversion of **16a** using a fungus, *Mucor circinelloides* [22].

Compound **16a** is chemically more stable than **1** in gastric juices, as expected, and is well absorbed after oral administration. The serum concentration, tissue distribution, and urinary excretion in animals are also significantly higher than those of **1**. In particular, the concentration of **16a** in lungs was remarkably higher than that in the liver [23]. It seems to be of advantage for the treatment of respiratory tract infection with less liver toxicity. Thereby, **16a** extended the range of clinical applications against various pathogens including *Haemophilus influenzae, Legionella, Chlamydia* [a sexually transmitted disease (STD)] [16], *Helicobacter pylori* [24], and *Mycobacterium avium* [25].

3. Azithromycin

Another direction for the improvement of EM was extension of its antimicrobial spectrum to gram-negative bacteria. 9-Deoxo-9a-aza-9a-methyl-9a-homo EM A (azithromycin) (**19**) was synthesized for this purpose as well as for improvement of acid stability and bioavailability of the drug. Chemical synthesis of **19** was achieved by Beckmann rearrangement of **11**, through imino ether, followed by reduction (Scheme 4). The ring expansion of EM due to the introduction of amine into the lactone ring resulted in a change in the physicochemical and biological properties. Derivatives of this group are named azalides for their structural characteristics of 15-membered aza-macrolides derived from **1** [26].

The reaction of **11** with tosyl chlorides gave imino ether (**36**). The reduction of **36** yielded 9-deoxo-9a-aza-9a-homo-EM A (**37**). The synthesis of its *N*-, *O*-acyl, and *N,O*-diacyl azalide derivatives, their 11, 12-cyclic carbonates, and their *in vitro* antimicrobial activities were studied, but 9a-*N* monoacyl (acetyl, propionyl) compounds exhibited *in vitro* activity 10 to 50 times lower than that of **37** [26].

Subsequently, 9a-*N*-methyl and 9a-*N*-ethyl derivatives were synthesized by a reductive alkylation of **37** with formaldehyde and acetoaldehyde, respectively (route **37** to **19**). Other 9a-*N*-alkyl analogues were also prepared by reaction of alkyl halide with 3'-*N*-oxide (**38**), followed by catalytic hydrogenolysis and tributylphosphine-induced deoxygenation (route **38** to **19**). Compound **19** and related 9a-*N*-alkyl analogues were obtained by the latter route. Compound **19** was further investigated for the role of functionality at C-4″ such as epihydroxy, keto, oxime, or amino groups [27]. Then, a number of derivatives having 9a-*N* substituents or 4″ hydroxyl substituents and their combined substituents of **37** were prepared. The 9a-*N*,11-*O*-cyclic compound (**39**) (Fig. 7) has a unique structure, having an additional ring, and its activity was comparable to that of **19** [27].

These azalides generally exhibited reduced activity *in vitro* than that of **1** against Gram-positive bacteria but showed better activity against Gram-negative bacteria such as *Escherichia coli* and *H. influenzae*. Among them, compound **19** and its analogues of 4″-epimer (**40**) and 4″-amino (**41**) clearly exhibited increased activity against *H. influenzae* compared with EM (Table III) [27, 28].

Compound **19** showed excellent *in vivo* activity (Table IV) and good tissue distribution compared with those of **1** [29]. Additionally, the pharmacokinetic profile of **19** indicated excellent features in several animals, as shown in Table V [27].

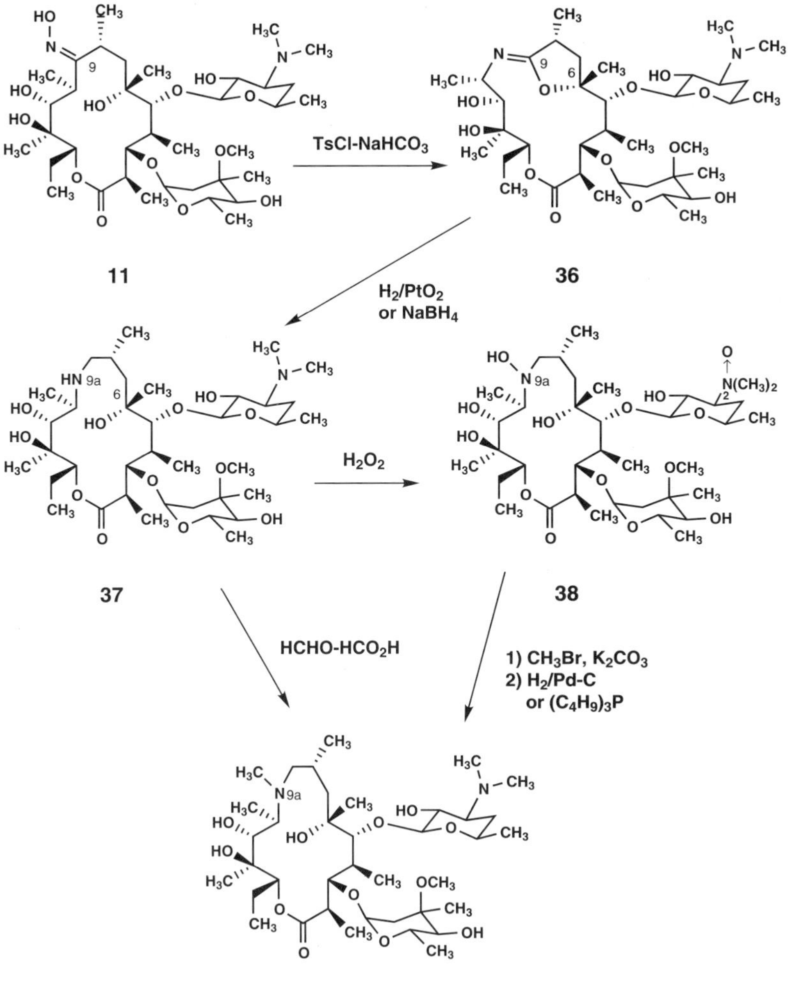

3. Chemical Modification of Macrolides 111

Azithromycin (19)

Scheme 4.

Thus, azithromycin advanced to clinical study and was developed as a newer macrolide [30].

The conformation of lactone ring of **19** was proved to be very similar to that of **1** by NMR and X-ray crystallographic analyses [31, 32]. This fact implies that the ring expansion and the introduction of tertiary amine do not extremely alter the structure–activity relationships between **19** and **1**, but the alteration of hydrophilicity and pKs by two ionizable nitrogen groups of **19** seem to have an effect on the antibacterial spectrum and pharmacokinetics [33].

TABLE III
In vitro Activity of Selected Azalides

Compound	MIC (µg/ml)						
	S. pyogenes	S. aureus			E. coli		H. influenzae[a]
		MLS-s	MLS-i	MLS-c	EM-s	EM-r	
EM (**1**)	≦0.025	0.10	6.25	>50	1.56	100	3.12
AZM (**19**)	≦0.025	0.39	25	>50	0.78	6.25	0.78
4″-Epimer AZM (**40**)	≦0.025	0.39	3.12	>50	0.78	3.12	0.39
4″-Amino AZM (**41**)	≦0.025	0.78	12.5	>50	0.05	1.56	0.20

Source: Data from Bright et al. [27].

MIC_{90} against *Neisseria gonorrhoeae* is 0.12 µg/ml. MLS-s, EM susceptible; MLS-i, inducibly EM resistant; Ery-c, constitutively EM resistant; EM-s, EM sensitive; EM-r, EM resistant.

[a]Average MIC of eight strains.

TABLE IV
In Vivo Activity of Azalides against EM
Susceptible S. aureus in Mice Infection

Azalide	PD_{50} (mg/kg dose)	
	po	sc
EM (**1**)	>200	9.0
AZM (**19**)	71	8.5
4″-Epimer AZM (**40**)	69	22.4
4″-Amino AZM (**41**)	86	10.3

Source: Data from Bright et al. [27].

TABLE V
Oral Pharmacokinetics of AZM in Animals

Compound	Animal	Oral dose (mg/kg)	C_{max}[a] (µg/ml)	AUC (µg · hr/ml)	Half-life (hr)
EM (**1**)	Rat	50	0.6	1.54	2.1
	Mouse	50	0.9	1.5	1.2
	Monkey	10	0.2	0.2	0.6
AZM (**19**)	Rat	50	1.3	9.5	7.7
	Mouse	50	1.6	10.0	6.4
	Beagle	10	1.4	23.4	21.0
	Monkey	10	0.43	2.9	8.7

Source: Data from Girard et al. [29].

[a]C_{max}, peak concentration of the antibiotic in serum/plasma.

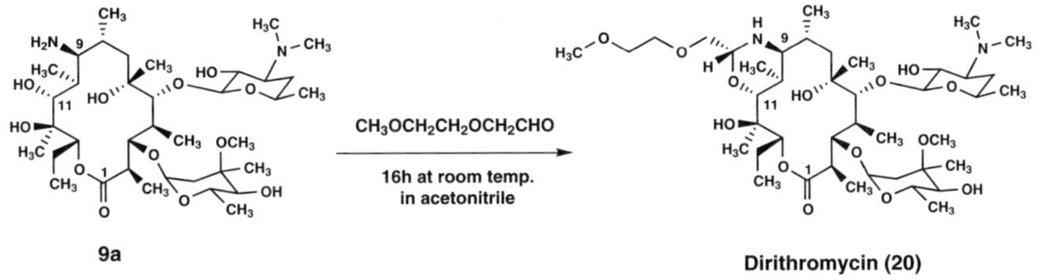

Scheme 5.

4. Dirithromycin

(9S)-9-Erythromycylamine (**9a**), the key intermediate for dirithromycin (**20**) synthesis, could be prepared from any of the EM 9-oxime (**11**), EM 9-hydrazone (**42**), or EM 9-imino (**12**) derivatives [34–36]. However, compound **9a** was not efficiently absorbed by oral administration. Therefore, a number of derivatives of **9a** were synthesized to improve the activity and oral absorbability. They include various Schiff bases of 9-amino obtained by condensation with ketones or aromatic aldehydes [37–39]. EM 9-*N*-11-*O*-oxazines (including **20**, Scheme 5) are also products condensed with aliphatic aldehydes [40, 41]. 9-*N*-Alkylated **9a** was prepared by the ring opening reductive alkylation of oxazine derivatives or by direct reductive alkylation of **9a** with aliphatic aldehydes. A number of 9-*N*-alkyl, 9-*N*-11-*O*-oxazine, 9-*N*-amide, and 9-*N*-sulfonamide derivatives of **9a** have been synthesized and their *in vitro* and *in vivo* activities evaluated, together with bioavailability. Some derivatives showed excellent *in vitro* activity but low blood levels after oral administration [41, 42].

Among these, 1-propyl derivative of **9a** (LY281389) (**43**) as well as three other analogues, having 2-methoxyethyl (**44**), 3-methoxy-1-propyl (**45**), and 2-2-methoxyethoxymethyl (**20**) substituents, were selected and their ED_{50} values after oral and subcutaneous administration in mice were compared (Table VI). Although **45** was consistently more effective than the others after parenteral and oral administration, it showed an unexpected adverse effect [43].

On the other hand, compound **20** was superior in its tissue concentration and serum half-life in addition to its antibacterial activity in comparison with the other candidates [44]. Compound **20** was thus selected as the next candidate for development, and it could be easily synthesized by condensing **9a** with 2-methoxyethoxyacetaldehyde in acetonitrile [43–45] (Scheme 5). Compound **20** has an *R* configuration at the asymmetric carbon atom of 9-*N*-11-*O*-oxazine bridge [46–48]. Compound **20** is readily hydrolyzed to **9a** both *in vitro* and *in vivo* [44]. The *in vitro* antibacterial spectrum and potency of **20** were similar to those of **1** and **9a** [42], but it showed greater *in vivo* activity because of its increased tissue concentration and prolonged *in vivo* half-life, which allow a reduction in dosage and frequency [44, 59].

TABLE VI
Comparative *In Vivo* Evaluation of 9-*N*-Substituted
Erythromycylamine Derivatives

Compound (R)	ED_{50} values in mice (mg/kg × 2)		
	S. aureus[a]	*S. pneumoniae*[a]	*S. pyogenes*[a]
(**1**) EM	15	14	18
(**43**)	19	14	9
(**44**)	44	25	16
(**20**) (DRM)	78	43	37
(**45**)[b]	36	17	19

Source: Data from Kirst *et al.* [41].
9-*N*-Alkyl (R) erythromycylamine; **43**, R = 1-Propyl-;
44, R = 2-Methoxyethyl-; **20**, R = 2-(2-Methoxyethoxy) ethyl; **45**, R = 3-Methoxy-1-propyl-.
[a]MLS-susceptible strain.
[b]Other experiment. Route: po.

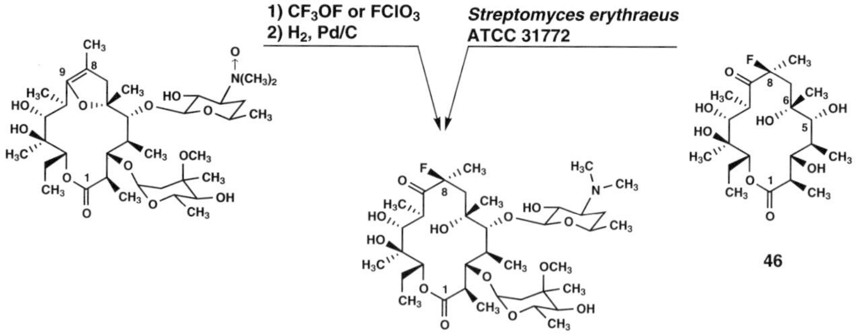

Scheme 6.

5. Flurithromycin

8-Fluorination of enol ether group in 8,9-anhydro erythronolide A 6,9-hemiketal (Scheme 6) was achieved by mild reaction with trifluoromethylhypofluorite (CF_3OF) to give flurithromycin (**21**). The (8*S*)-8-fluorinated products, including **21**, were found to be considerably stable in an acidic solution, although both erythronolide A and **1** were unstable and proceeded to degrade along similar pathways [49].

The microbial glycosidation of (8*S*)-8-fluoro erythronolides were investigated by the same research group. Fluorinated erythronolides A (**46**) and B (**47**) were added into the fermentation broth of *Streptomyces erythraeus* ATCC 31772, a blocked mutant in EM biosynthesis (Scheme 6). In this procedure, (8*S*)-8-fluoro

EM A (**21**) was successfully obtained, together with (8S)-8-fluoro EM C, by the addition of (8S)-8-fluoro erythronolide A. Similarly, (8S)-8-fluoro EM B and D were obtained by the addition of (8S)-8-fluoro erythronolide B [50].

Among these fluorinated derivatives, **21** showed the most promising features of antimicrobial activity *in vivo*, pharmacokinetics, and clinical efficacy [51–54]. The *in vitro* activity of **21** against pathogens of respiratory tract infections was compared with that of **1**. The activity of **21** and **1** were tested against *S. aureus* (16 strains), *Streptococcus pneumoniae* (24), *Streptococcus pyogenes* (24), *Moraxella catarrhalis* (22), and *H. influenzae* (22), and the observed MIC_{50} values (μg/ml) were 0.5 and 0.25, 0.03 and 0.03, 0.125 and 0.03, 0.125 and 0.06, and 2 and 4, respectively.

6. Comparison of Biological Properties among the Newer Macrolides

The antimicrobial activity and pharmacokinetics of newer macrolides are shown in Tables VII through X. The MIC_{50} values shown in Table VII are the values against respective EM-susceptible strains [55]. The MIC_{90} values of these newer macrolides show >128 against methicillin susceptible *to S. aureus*, >128 against *Staphylococcus epidermidis*, 0.015 to 0.12 against *S. pyogenes*, 0.015 to 0.12 against *S. pneumoniae*, >128 against *Enterococcus* sp., and 0.5 to 8 μg/ml against

TABLE VII
Comparative *In Vitro* Activity of Newer Macrolides (MIC_{50}: μg/ml)

Organism	No. of isolates	EM (**1**)	RXM (**18**)	CAM (**16a**)	AZM (**19**)
S. aureus	(174)[a]	0.39	0.78	0.20	0.78
S. epidermidis	(112)	0.39	0.39	0.20	0.78
S. pyogenes	(70)	≤0.025	0.10	≤0.025	0.10
S. pneumoniae	(53)	0.39	0.39	0.10	0.39
E. faecalis	(25)	3.13	6.25	1.56	6.25
H. influenzae	(81)	3.13	12.5	6.25	1.56

Source: Data from Matsunaga *et al.* [55].
[a]Parentheses indicate number of strains.

TABLE VIII
Comparative *In Vivo* Activity of EM, CAM, and AZM (ED_{50}: mg/kg)

	S. aureus	*S. pyogenes*	*S. pneumoniae*	*H. influenzae*
EM (**1**)	>156 (0.78)[a]	41.2 (0.20)	>172 (0.10)	189 (3.13)
CAM (**16a**)	19.0 (0.39)	6.78 (0.05)	69.5 (0.05)	113 (3.13)
AZM (**19**)	35.8 (1.56)	7.23 (0.39)	42.2 (0.20)	29.1 (0.78)

Source: Data from Matsunaga *et al.* [57].
Test organisms are MLS-susceptible strains.
[a]Parentheses indicate MIC (μg/ml).

TABLE IX
Influence of Medium pH on Antibacterial Activity
of Newer Macrolides

Compound	pH	S. aureus	E. faecalis	H. influenzae
EM (1)	6.0	1.87	5.66	7.46
	7.2	0.35	0.62	2.00
	8.0	0.18	0.35	0.81
RXM (18)	6.0	6.06	24.25	19.70
	7.2	0.81	3.48	5.30
	8.0	0.35	1.41	1.90
CAM (16a)	6.0	1.15	3.25	7.46
	7.2	0.25	0.61	2.30
	8.0	0.11	0.25	1.00
AZM (19)	6.0	64.00	111.43	6.50
	7.2	1.41	4.59	0.76
	8.0	0.31	0.76	0.23

Source: Data from Barry et al. [58].
MIC: (µg/ml).

TABLE X
Pharmacokinetics of Newer Macrolides in Healthy Adults[a]

Macrolide and dose		C_{max} (µg/ml)	T_{max} (hr)	$t_{1/2}$ (hr)	AUC (µg·hr/ml)
EM (1)	(500 mg)	(1.8)[c]	(1.7)	(1.7)	(14.2)
RXM (18)	(300 mg)	10.8	1.6	11.9	—
CAM (16a)	(400 mg)	2.1 (2.1)	1.7 (1.8)	4.7 (4.3)	17 (11.8)
AZM (19)	(500 mg)	0.4.(0.4)	2.0 (2.3)	41[b] (68)	4.5 (2.4)
DRM (20)	(500 mg)	(0.5)	(4)	20–50 (44)	(3.4)
FLM (21)	(500 mg)	1.2–2	1–2	8	16

Source: Data from Cunha [59] and Kirst and Sides [60].
[a]Route; single oral administration in human volunteers.
[b]i.v. administration.
[c]Parentheses indicate different experiment (500 mg).

H. influenzae [56]. Compound **19** exhibits remarkably lower MIC against *H. influenzae*. Comparative *in vivo* activity of **1**, **16a**, and **19** is shown in Table VIII [57].

The effect of medium pH on the antibacterial activity of these newer macrolides is shown in Table IX. The MIC of **19** is much affected by a change of pH, particularly with Gram-positive bacteria (150–200 times) compared with Gram-negative bacteria (30 times). The changes in MICs of **18** and **16a** were less than that of **19** (10–20 times) against three species at pH 6.0–8.0 [58].

The comparative pharmacokinetics of five newer macrolides are summarized in Table X [59, 60]. Compound **18** showed the highest C_{max} value, whereas **19** and DRM (**20**) showed longer half-life values than the others. The serum binding rate of **19** was found to be lower than those of **1** and **16a** [61]. Compound **19** has been used as oral and injectable formulations by its increased hydrophilicity.

Although most of the newer macrolides exhibited activity comparable to **1** against Gram-positive bacteria *in vitro*, **19** was found to have higher activity against Gram-negative bacteria such as *H. influenzae* than others [62]. It is noteworthy that all are remarkably improved in acid stability and bioavailability, which reflects on the *in vivo* activity. These facts have prompted further investigation and modification of macrolides. In addition, the newer macrolides have also changed the concept of serum drug concentration and *in vivo* activity relationships, which reflect more on the tissue drug concentration. Indeed, they gave much higher values of tissue concentration/serum concentration than those of other antibiotics, which seems to relate to the excellent *in vivo* activity.

C. Recent Modifications of Erythromycins

According to the accumulation of knowledge on functional group/activity relationships in EMs, the attention has been focused mainly to the functionalities at C-6, C-8, C-9, C-11, C-12, and C-4″ in EM (**1**) as the molecular parts to be modified. Based on the successful modifications resulting in the newer macrolides such as RXM (**18**), CAM (**16a**), and AZM (**19**), further trials to improve *in vitro* and *in vivo* activity have extended to combined modification of these functional groups. The derivatives thus obtained are classed into several groups, as follows.

1. 6-O-Substituted Erythromycin Derivatives

Although megalomicin (**48**) (Fig. 8) is a well known naturally occurring 14-membered macrolide bearing an amino sugar (megosamine) at C-6, chemical substitution of the tertiary 6-hydroxy group in EM has been considered to be limited because of steric hindrance [63]. Since the preparation of shorter chain alkyl ethers at C-6 was achieved in CAM (**16a**) [15], substitution of the 6-OH group by more bulky groups has been attempted [63]. As a new approach, 6-*O*-allyl EM

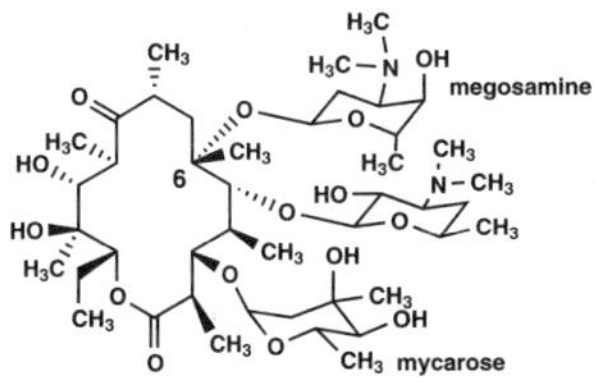

Megalomicin (**48**)

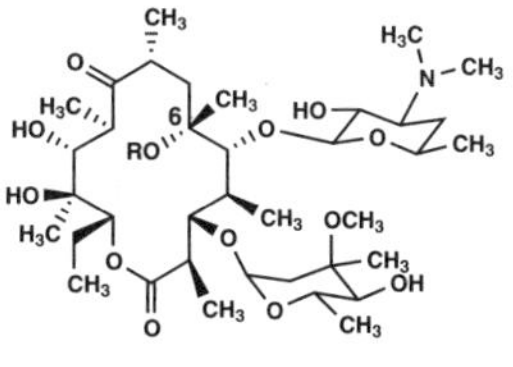

A-181785 (**50a**): R=-(CH$_2$)$_2$NH(CH$_2$)$_2$-Ph
A-181978 (**50b**): R=-CH$_2$CH=N-O-Ph

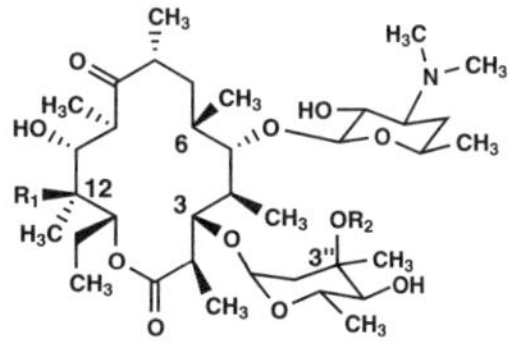

	R$_1$	R$_2$
A-77113 (**51a**):	-OH	-CH$_3$
A-77116 (**51b**):	-H	-CH$_3$
A-77112 (**51c**):	-OH	-H
A-77146 (**51d**):	-H	-H

Fig. 8.

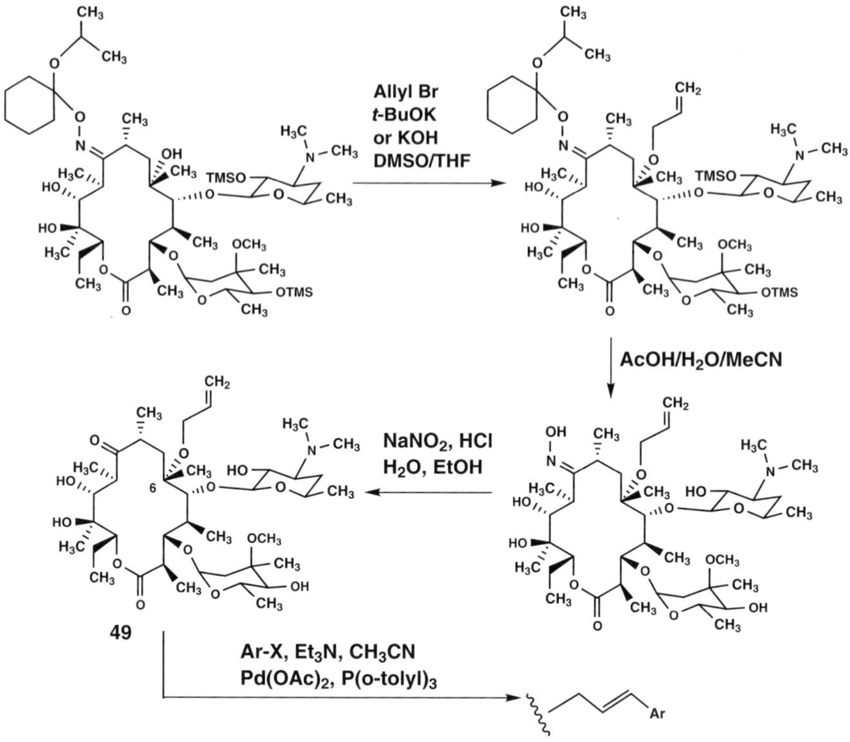

Scheme 7.

derivative (**49**) was synthesized as an intermediate to facilitate further modification (Scheme 7). Conversion of the allyl moiety to aldehyde and epoxide and subsequent reactions could lead to various novel derivatives [63].

6-O-Substituted derivatives thus prepared did not overcome CAM (**16a**) in activity *in vitro* except for the aromatic amine derivatives A-181785 (**50a**) and A-181978 (**50b**) (Fig. 8). Although none of them are effective against macrolide-resistant pathogens, this work was later extended to the synthesis of ABT-773 (**95**) (described in Section II.D), a ketolide. 6-Deoxy-EMs A (**51a**), B (**51b**), C (**51c**), and D (**51d**) (Fig. 8) were reported to show acid stability and reduced *in vitro* activity [64, 65].

2. 9-O-Substituted Erythromycin Derivatives

Following the preparation of 9-deoxo EM [66], 9-deoxo-12-deoxy-9,12-epoxy EM (**52**) derivatives, together with their respective 11-epi and 4″-epi isomers of hydroxy and/or amino, and 11-keto substituents (Fig. 9) were synthesized [67]. Among them, the 11-keto (A-63881) (**52a**), 11-epiamino (A-69334) (**52b**), 11-epiamino-4″-amino (A-71671) (**52c**), and 11-epiamino-4″-epiamino (A-73020)

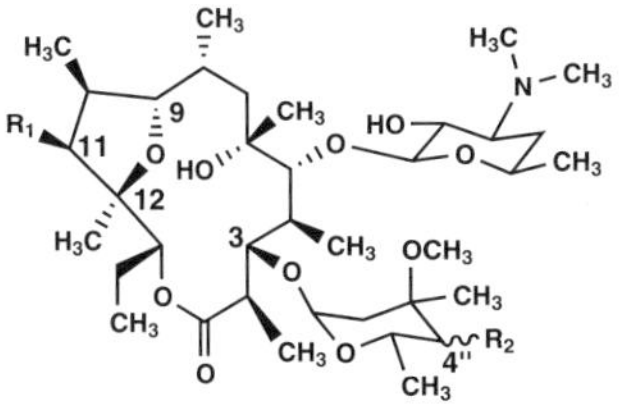

3. Chemical Modification of Macrolides   119

Fig. 9.

	R₁	R₂
A-63881 (**52a**):	=O	⋯OH
A-69334 (**52b**):	-NH₂	⋯OH
A-71671 (**52c**):	-NH₂	⋯NH₂
A-73020 (**52d**):	-NH₂	—NH₂

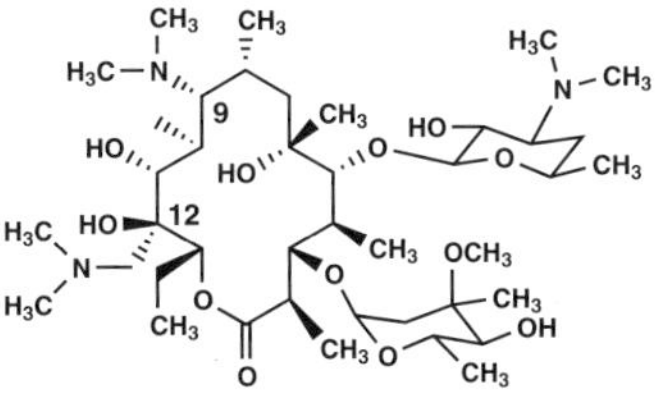

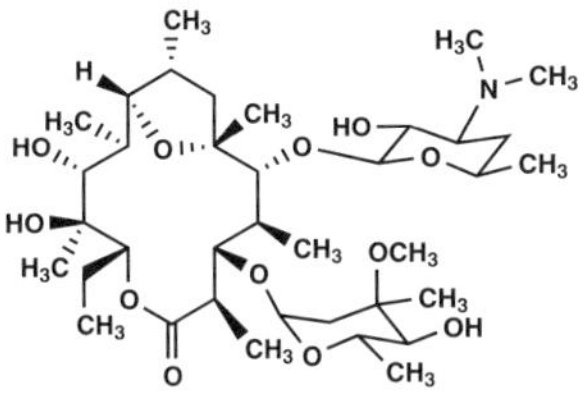

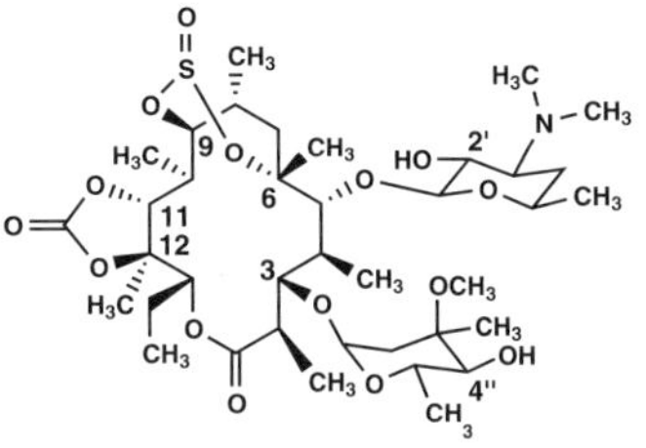

(**52d**) analogues increased the serum (1–4 times higher than **1**) and lung concentrations (3–14 times higher) and the serum half-life (**52c** and **52d** were 10 times longer) in mice. Thus, improved pharmacokinetics were reflected in their *in vivo* activities, although their *in vitro* activities were not so much improved.

(9*S*)- And (9*R*)-9-dihydro derivatives of EM and CAM (**16a**) were prepared by reduction of the 9-keto group, but they were less active than **16a** [68]. Novel analogues of (9*R*)-9-deoxo-9-(*N*-dimethylamino) EM, bearing an *N*-alkylamino group at C-21, were synthesized. C-21 Dimethyl amino derivative A-75729 (**53**) (Fig. 9) exhibited a two-fold lower potency against Gram-positive bacteria compared with the parent 9-dimethylamino compound, and **53** showed notably potent activity (2–8 times higher) against Gram-negative bacteria [69, 70]. (9*R*)-9-Dihydro-6,9-anhydro derivative of EM (**54**) and other 6,9-linked analogues such as 6,9-cyclic sulfite **55** (Fig. 9) are reported to show increased acid stability and moderate *in vitro* activity [71].

3. 9a- and 8a-Azalide Derivatives

Various 9a-*N*-acyl azalides and their 11,12-cyclic carbonates were synthesized starting from 9-deoxo-9-dihydro-9a-aza-homo EM (**37**) [26]. Their activities are generally lower than **1** against Gram-positive bacteria but slightly increased against Gram-negative bacteria. Instead of *N*-acyl substituents, a series of *N*-alkyl

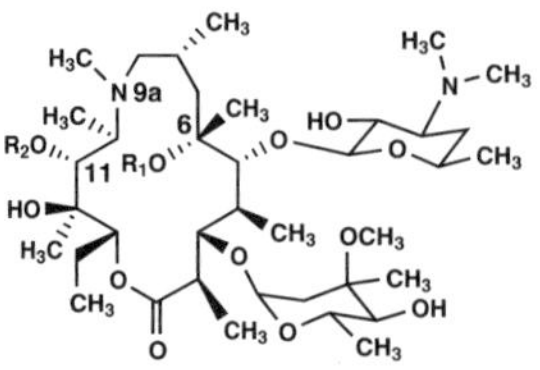

	R_1	R_2
56a:	$-CH_3$	-H
56b:	-H	$-CH_3$

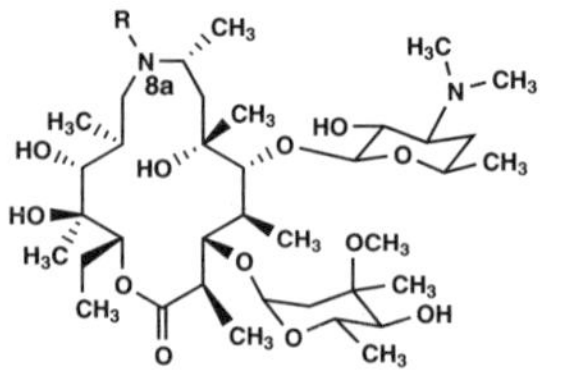

	R
58a:	-H
58b:	$-CH_3$
58c:	$-CH_2CH{=}CH_2$
58d:	$-CH_2CH_2F$

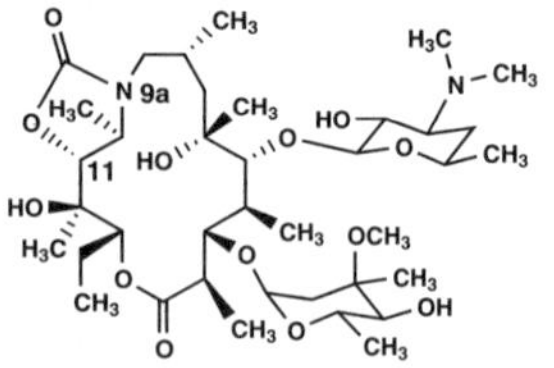

57

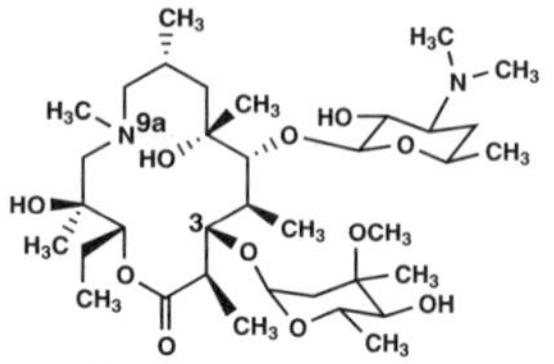

59

Fig. 10.

compounds were synthesized together with their derivatives having hydroxy, amino, keto, or oxime substituents (epimers included) at C-4″ [27]. These new azalides exhibited higher (8–64 times) activity than that of **1** against Gram-negative bacteria such as *E. coli* and *H. influenzae*. Their serum half-life and AUC were more than 10 times greater than that of **1** in monkey [26].

Various *N*-methyl and *O*-methyl derivatives (4″-, 6-, 11-, and 12-) of **37** were prepared and evaluated for their antibacterial activity. 6-*O*-Methyl AZM (**56a**) (Fig. 10) was slightly less active than AZM, while 11-*O*-methyl AZM (**56b**) exhibited remarkably increased antibacterial activity [72–74]. (Structure of 6-*O* methyl AZM was later revised to 12-*O*-methyl AZM [73, 74].) 9a-*N*, 11-Cyclic carbamate azalide (**57**), 8a-*N*-azalides (**58**), and 14-membered azalide (**59**) (Fig. 10) are interesting in view of their structure–activity relationship compared with **19**, although they generally exhibited the same or lower activity compared with AZM (**19**) [75–78].

4. 11-O- and/or 12-O-Substituted Erythromycin Derivatives

EM 11,12-methylene acetal (**60a**) and its 9-methoxime (**60b**), 9-dihydro (**60c**), and 8-hydroxy (**60d**) derivatives (Fig. 11) were also synthesized [79]. However, they were generally less active *in vitro* against Gram-positve bacteria, although **60a** exceeded EM by 2–8 times. 11,12-Cyclic carbamate A-62514 (**61**) was prepared by stereoselective intramolecular Michael addition of C-12-*O*-carbamate (**62**) to an α, β-unsaturated ketone moiety (Fig. 11) [80, 81]. Compound **61** was 8 times more potent *in vivo* against *S. pneumoniae* and was less active in gastrointestinal (G.I.) motility relative to **1**. The derivatives of (**63**) having 11-*N*-aryl substituents such as benzylaminoethyl (A-66173) (**63a**), benzylmethylaminoethyl (A-66005) (**63b**), benzylcyclopropylaminoethyl (A-64239) (**63c**), and phenyl butyl (A-66321) (**63d**) (Fig. 11) have been reported [82]. These four derivatives

3. Chemical Modification of Macrolides   121

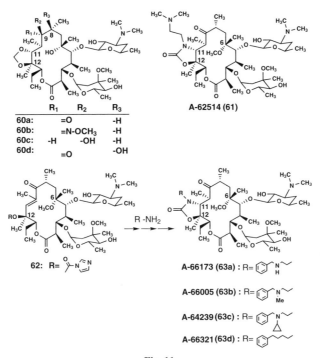

Fig. 11.

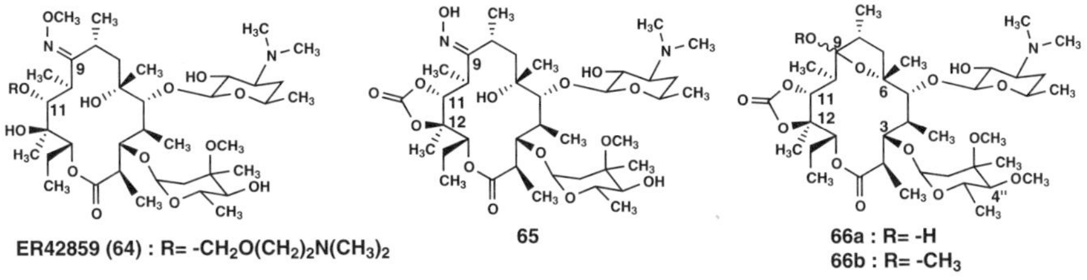

Fig. 12.

showed significant activity against macrolide-lincosamide-streptogramin B susceptible (MLS-s) and inducibly MLS-resistant (MLS-i) *S. pyogenes* (MIC: MLS-s, 0.03–0.12 µg/ml; MLS-i, 0.25–0.5 µg/ml). These improved activities correlate with the presence of phenyl group on the side chain but not amino group.

11-(2-Dimethylaminoethyl)-oxymethyl derivative (ER42859) (**64**) (Fig. 12) showed *in vivo* activity against *S. pneumoniae* in a mice infection model superior to **1** and similar to RXM (**18**). Its higher blood and prolonged tissue levels in rodents seemed to be promising for further study [83]. However, the blood levels were inferior to those of **1** in healthy volunteers.

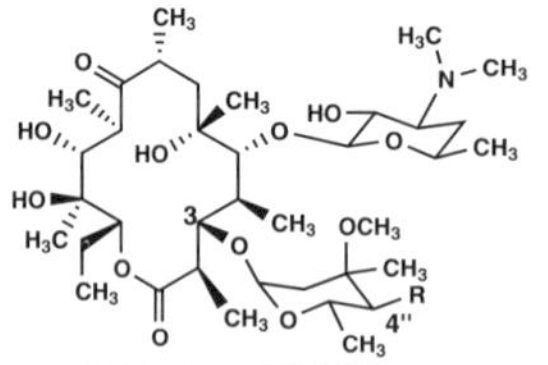

A-24108 (**67a**) : R= -OCOCH₃
A-63111 (**67b**) : R= -OCONH₂
A-62581 (**67c**) : R= -OCH₂-Ph
A-57752 (**67d**) : R= -H

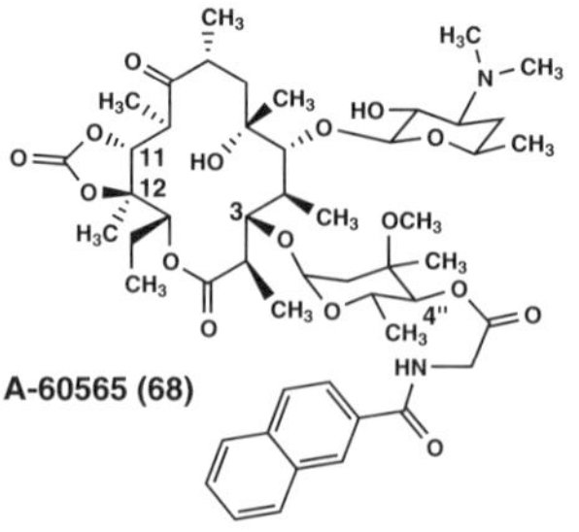

A-60565 (**68**)

Fig. 13.

9-Oxime EM 11,12-cyclic carbonate (**65**) (Fig. 12) and its oxime ethers have also been reported [84]. Their antibacterial activities are comparable to **1** and slightly more potent against *H. influenzae*. The derivatives of 4″-*O*-methyl EM A 11,12-cyclic carbonate (**15b**) were synthesized [85]. Among them, (9*S*)-hemiketal (**66a**) and its (9*S*)- and (9*R*)-*O*-methyl (**66b**) analogues (Fig. 12) exhibited activities comparable to **1** against EM-susceptible strains of *S. aureus* and were more potent against MLS-i and EM efflux-type resistant (*mef*) strains of staphylococci. However, 11,12-cyclic carbonate substitution seems to have been avoided thereafter because of its liver toxicity [81].

5. 4″-O-Substituted Erythromycin Derivatives

4″-*O*-Acetyl (A-24108) (**67a**), 4″-*O*-carbamyl (A-63111) (**67b**), 4″-*O*-benzyl (A-62581) (**67c**), and 4″-deoxo (A-57752) (**67d**) (Fig. 13) derivatives of EM and 4″-*O*-acyl EM 11,12-cyclic carbonate derivatives such as the 4″-*O*-naphthoyl-glycyl substituent of EM 11,12-cyclic carbonate (A-60565) (**68**) (Fig. 13) were synthesized and their antibacterial activities evaluated [82, 85]. Compound **68** showed significant activity against EM-susceptible (MIC: 0.12 µg/ml) and MLS-i (MIC: 0.12 µg/ml) and constitutively EM-resistant (MLS-c) (MIC: 16 µg/ml) strains of *S. pyogenes* (EM MICs: 0.06, 4, and >128 µg/ml). The 6-*O*-methyl derivative of **67b** (A-63075) exhibited significant *in vivo* activity and reduced GI motility compared with **16a** [86].

4″-*O*-Substituted derivatives CP-225,600 (**69a**) and CP-544,372 (**69f**) (Fig. 14) have been of much interest in the improvement process [87]. The first compound (**69a**) exhibited significant *in vitro* activity against Gram-positive bacteria but poor activity against *H. influenzae in vitro* and against Gram-positive pathogens *in vivo*. Derivative CP-279,107 (**69b**), carrying the same 4″-substituents as **69a**, were then prepared. Compound **69b** exhibited activity as strong as that of HMR-3004 (**86e**) (Fig. 22) against MLS-s, MLS-i, and *mef* type strains of Gram-positive pathogens. However, its *in vitro* activity against *H. influenzae* and *in vivo* activity against Gram-positive bacteria by oral administration were poor, although it was more effective by subcutaneous dose. To overcome these defects, CP-360,280 (**69c**) was synthesized to reduce molecular weight and lipophilicity.

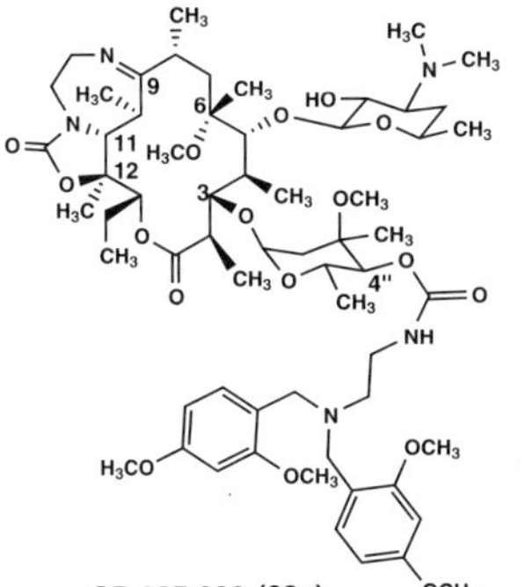

CP-225,600 (69a)

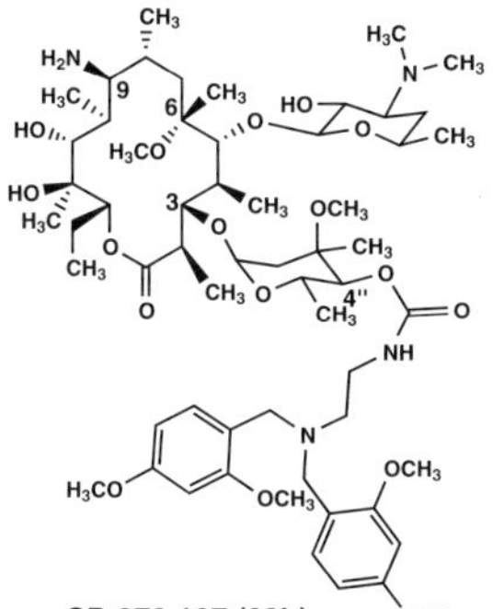

CP-279,107 (69b)

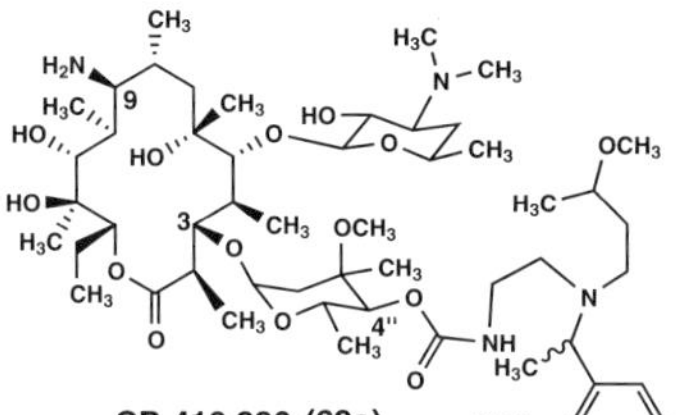

CP-416,890 (69e)

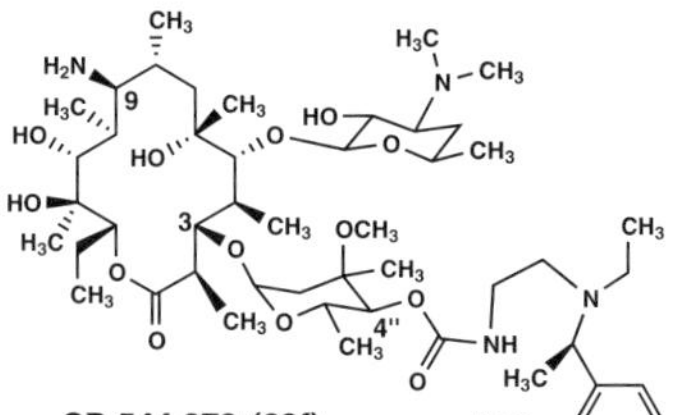

CP-544,372 (69f)

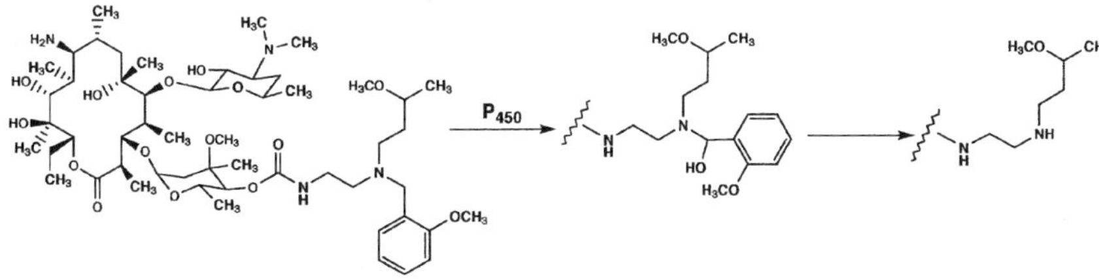

CP-360,280 (69c)

CP-418,001 (69d)

Fig. 14.

Compound **69c** was proved to maintain good activity *in vitro* but had a short half-life *in vivo* due to hydroxylation of the 4″-side chain by cytochome *P*-450 followed by conversion to CP-418,001 (**69d**), which was inactive. To avoid such metabolism, a methyl group was introduced to the α-side chain of **69c** to obtain CP-416,890 (**69e**). It exhibited increased stability to metabolism and good *in vivo* activity. Finally, synthesized **69f**, which bears a smaller side chain, demonstrated superior *in vitro* and *in vivo* activity, particularly against *S. pneumoniae*, including the MLS-i and *mef* type of resistant strains. Its pharmacokinetic profile is comparable to AZM (**19**) and telithromycin (TLM) (**91i**) [87–93]. However, the antibacterial activity of **69f** seems to be insufficient against MLS-c resistant strains of *S. pneumoniae* and *H. influenzae* compared with **91i** and ABT-773 (**95**) (Schemes 8 and 11).

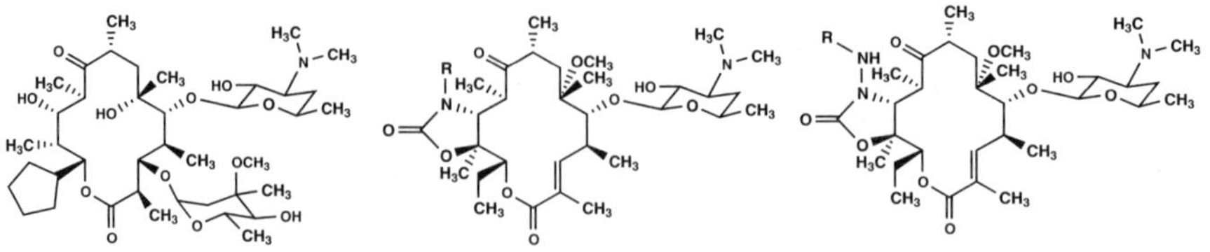

70

71

72a : R = 4-NO$_2$-PhCH=CHCH$_2$-
72b : R = 4-NH$_2$-PhCH=CHCH$_2$-

Fig. 15.

6. 13-Substituted Erythromycin Derivatives

Cycloalkane, cyclo alkene, or branched alkane bearing hetero atoms were introduced at C-13 instead of the ethyl group of EMs A and B by feeding short-chain fatty acid precursors into the fermentation broth of genetically modified *Saccharopolyspora erythraea*. The cyclopentyl EM B (**70**) (Fig. 15) obtained and the related derivatives exhibited activity comparable to that of EM B against *S. aureus* (MIC **70**: 1.56; **1**: µg/ml) and *E. coli* (MIC **70**: 0.1; **1**: 25 µg/ml) [94, 95].

7. 3-O-Decladinosyl Erythromycin Derivatives

a. *3-O-Decladinosyl 2,3-Anhydro Erythromycin Derivatives* 2,3-Anhydro-CAM 11, 12-cyclic carbamates (**71**) (Fig. 15) were found to have potent antibacterial activity against Gram-positive bacteria and to show improved activity against MLS-i resistant strains. This structure–activity study demonstrated that arylalkyl carbamates with two and four carbon atoms between the aromatic moiety and carbamate nitrogen have the best activity *in vitro*. However, they showed only moderate *in vivo* activity in mouse protection tests [96, 97], while 3-decladinosyl 2,3-anhydro CAM 11,12-carbazates (**72**) (Fig. 15) showed potent *in vitro* activity against EM-susceptible and EM-resistant Gram-positive pathogens except MLS-c strains [98]. Although these showed poor activity *in vivo*, two compounds having *p*-nitrophenylpropenyl (**72a**) and *p*-aminophenyl propenyl (**72b**) chains at the hydrazo terminal were more active than EM (**1**) against *S. pneumoniae* (MIC **72a**: MLS-s, 0.03; MLS-i, 1; MIC **72b**: MLS-s, 0.03; MLS-i, 1; MIC EM: MLS-s, 0.06; MLS-i, 16 µg/ml) *in vitro*. However, they were not so active against organisms with constitutive MLS resistance and *H. influenzae*. These 2,3-anhydro derivatives were shown not to induce EM resistance, the same as ketolides.

b. *3-O-Decladinosyl 3-O-Acylides* In modification of 3-decladinosyl EM and CAM, attention has been focused also on the introduction of 3-*O*-acyl (**73a, 73b**) (Fig. 16), 3-keto, 3-ether, 3-*O*-carbonate, and 3-*O*-carbamate. TEA-0769 (**74**) is one of the acyl derivatives that has *p*-nitrophenyl acetic acid ester at C-3 [99]. Its activity against MLS-s type of staphylococci was similar to that of **16a** and was higher against enterococci. In addition, it was more potent against MLS-i

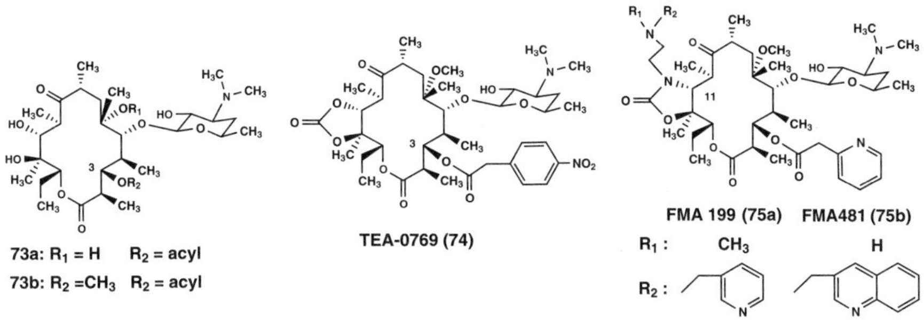

Fig. 16.

resistant strains of staphylococci and streptococci but less active compared with **16a** *in vivo* using staphylococcal and streptococcal infection models. By a similar procedure, 3-*O*-pyridyl acetic acid ester was introduced into CAM 11,12-cyclic carbamates. Among them, FMA199 (**75a**) and FMA481 (**75b**) (Fig. 16) showed significant activity against EM-resistant *S. pneumoniae* (MIC **75a**: MLS-s, 0.1 µg/ml; MLS-i, 0.39; mef, 0.39; MIC **75b**: MLS-s, 0.05; MLS-i, 0.1; mef, 0.1 µg/ml) compared with the **91** series of ketolides (Table XIII) (Scheme 8) [100]. Thus, some 3-*O*-decladinosyl-3-*O*-acylides were found to exhibit improved activity aginst EM-resistant Gram-positive bacteria.

c. Other 3-O-Decladinosyl Derivatives 3-*O*-Decladinosyl-3-deoxy CAM (**76**) (Fig. 17) has been reported but its activity seems to be reduced compared to **16a** [101]. Among several derivatives of **76**, 3-*O*-decladinosyl-3-deoxo-11-*N*-phenylbutyl-11,12-cyclic carbamate CAM (**77**) was the most potent, with MIC values of 1.56 µg/ml or lower against the majority of Gram-positive bacteria except MLS-c type strains of *S. aureus* (MIC, 50 µg/ml). It was particularly potent against *S. pyogenes*, with MIC values of 0.05–0.2 µg/ml. However, compound **77** was still less potent against macrolide-susceptible *S. aureus* and *S. pyogenes* compared with **1** and **16a**.

8. Miscellaneous Erythromycin Drivatives

The introduction of fluorine at C-3″ of EM 6,9-hemiketal-11,12-carbonate (**78**) (Fig. 17) improved its stability and efficacy compared to the parent compound [102]. 2′-*O*-(*N*-Alkyl succinamoyl)-EM derivatives were prepared as the water-soluble prodrug of **1** [103].

9. Considerations for Changing the Lactone Ring and the Sugar Moiety

The antimicrobial activity of **1** is extremely sensitive to small changes in the configuration and/or conformation of the lactone ring and in each functional group. This evidence was demonstrated by the behavior in acidic media [2]. Cleavage of the lactone ring by esterase, of course, destroys the activity [104]. Loss of activity has also been shown from the contracted 12-membered derivative

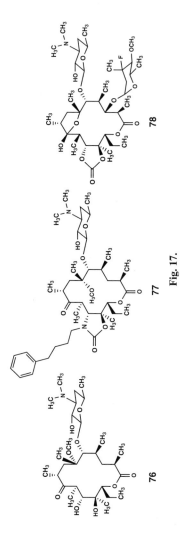

Fig. 17.

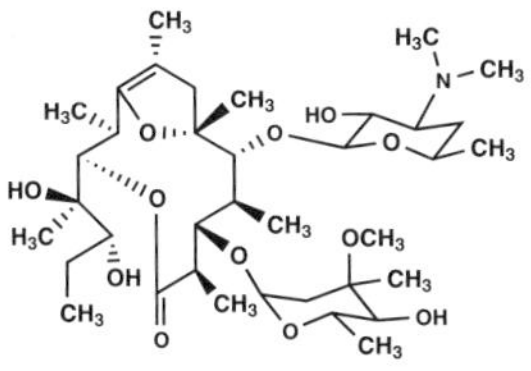

Fig. 18.

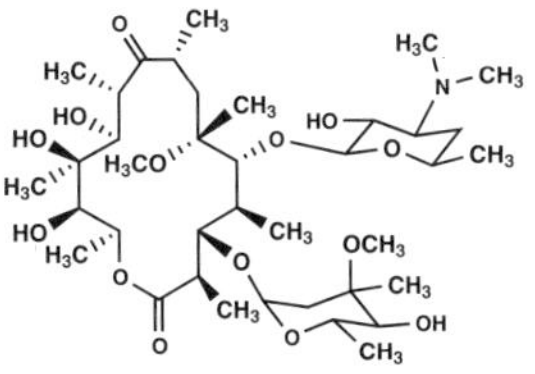

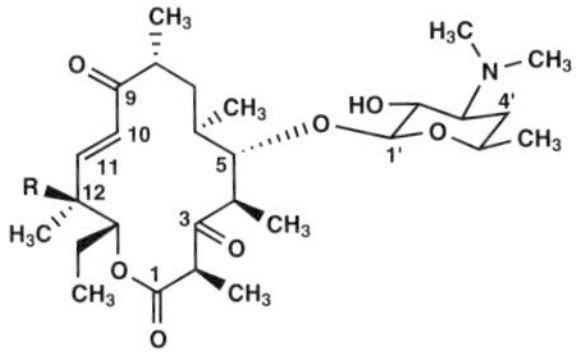

pikromycin (81a): R = OH
narbomycin (81b): R = H

Fig. 19.

(**79**) of 8,9-anhydro-6,9-hemiketal EM and the 15-membered derivative (**80**) of EM (Fig. 18) obtained by the translactonization of **1** and 14-hydroxy CAM (**35**), respectively [105–107]. In addition, 8a-, 9a-*N* azalides and 14-membered azalides are similarly referred for the conformational study of the lactone ring. Comparative study of the conformation of the lactone ring between **1** and its derivatives is required for the rational modification to select the compounds that excel in acid stability, *in vitro* and *in vivo* activity, and pharmacokinetics. For the purpose of conformational analysis of various derivatives of **1**, X-ray crystallography has often been used.

Besides, it is clearly confirmed from the substitution at C-3, such as with the acyl group in this section, that the 3-cladinose is not essential for antibacterial activity. This finding has led to the investigation and development of ketolides.

D. Ketolides

Allen reported in 1977 that naturally occurring 3-oxo (or 3-keto) macrolides such as pikromycin (**81a**) and narbomycin (**81b**) (Fig. 19) do not induce macrolide resistance in *S. aureus*, and they retained sufficient antibacterial activity [108]. On the other hand, derivatives of 3-decladinosyl-3-oxo-6-*O*-methyl EM (**82**) (Fig. 20) have been recently synthesized and designated as ketolides [109]. They generally exhibit significant activity against EM-resistant Gram-positive pathogens, and the features of their activity are fairly different from those of EM derivatives that bear cladinose at the C-3 position.

Originally, Agouridas *et al.* tried to prepare ketolide using EM 9-oxime, but 3-decladinosyl-3-oxo-EM 9-oxime (**83**) could not be obtained due to the formation

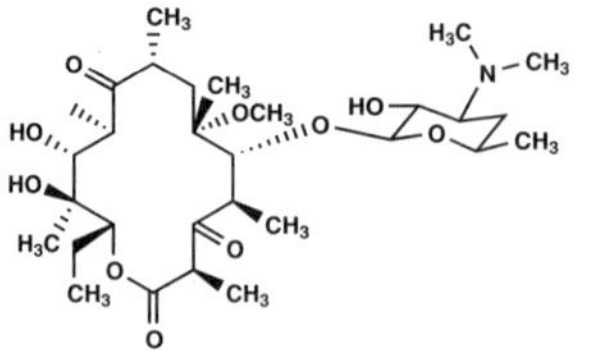

Fig. 20.

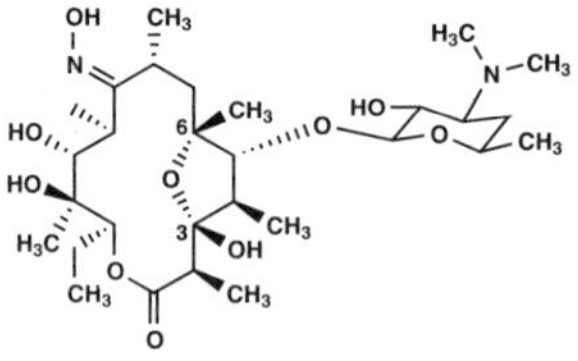

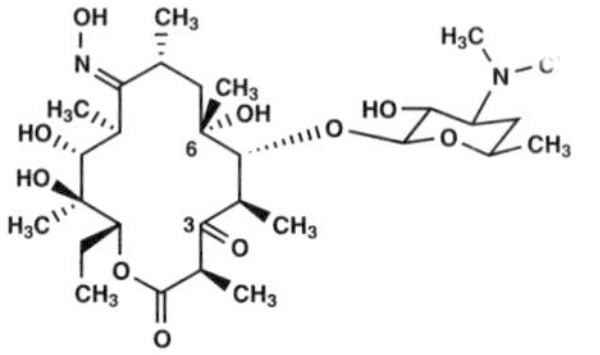

of C-3, 6-hemiacetal (**84**) (internal cyclization) (Fig. 20) [109]. However, it was possible to form ketolide from CAM (**16a**) by preventing cyclization [109–111]. At present, various 3-oxo-6-O-alkyl derivatives such as ABT-773 (**95**) are also classified as ketolides [112]. A number of ketolides were prepared by starting from **16a**. They include compounds that carry a variety of substituents such as 9-keto, 9-oxime, 6-O-alkyl, 11,12-cyclic carbonate, 11,12-cyclic carbamate, 11, 12-cyclic carbazate, and 9a-aza-lactone as well as their combinations. Such ketolides showed improved acid stability, antibacterial activity, and pharmacokinetics. Some of them exhibited excellent antibacterial activity, particularly against MLS-i and MLS-c resistant strains of *S. pneumoniae* and *H. influenzae*. For convenience, they are divided into several groups based on their substituents. Their chemical structures and biological properties are described here.

1. 9-Oxime Ketolides

A number of 9-oxime alkyl ether ketolides (**85**) (Fig. 21) were synthesized from **16a**. 9-Oxime (**85a**) and 9-O-(2-methoxyethoxy)methyloxime (**85b**) exhibited weak activity against MLS-i resistant *S. pneumoniae* (MIC **85b**, 10 μg/ml). 9-O-2-(Dimethylamino)ethyl derivative (**85c**), demonstrated remarkably improved activity against various EM-resistant strains of *S. aureus* and *S. pneumoniae*. The introduction of nitrogen into these substituents did not significantly change their activity against *H. influenzae*. Among the cyclic amine ethers, the (R)-3-piperidyl derivative (**85d**) was the first example to exhibit significant activity against both *H. influenzae* and EM-resistant Gram-positive pathogens (Table XI) [109].

TABLE XI
In Vitro Evaluation of 9-Oxime Ketolide Derivatives

	MIC (µg/ml)						
	S. aureus			S. pneumoniae			
Compound	MLS-s	MLS-i	MLS-c	MLS-s	MLS-i	MLS-c	H. influenzae
85a	0.3	40	40	0.15	40	40	20
85b	1.2	2.5	40	0.15	10	40	40
85c	0.08	1.2	40	0.15	0.15	2.5	20
85d	0.15	2.5	40	0.08	0.3	0.3	0.6
CAM (16a)	0.3	40	40	0.04	40	40	5
AZM (19)	0.3	40	40	0.15	40	40	1.2

Source: Data from Agouridas *et al.* [109].

85a, R = H; **85b**, R = 2-(2-Methoxyethoxy)methyl-; **85c**, 2-(Dimethylamino) ethyl-; **85d**, R = (*R*)-3-Piperidyl-.

MLS-s, EM sensitive; MLS-i, EM inducibly resistant; MLS-c, EM constitutively resistant.

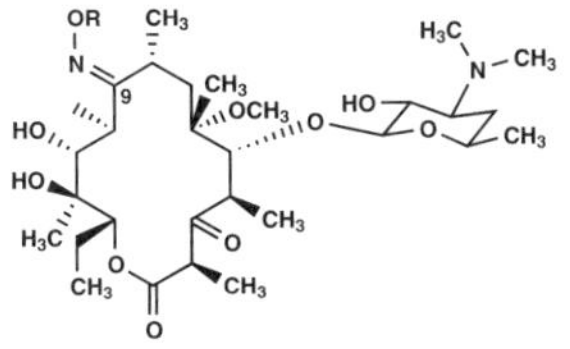

	R:
85a:	H
85b:	Methoxyethoxymethyl
85c:	Dimethylaminoethyl
85d:	3-(*R*)-Piperidinyl

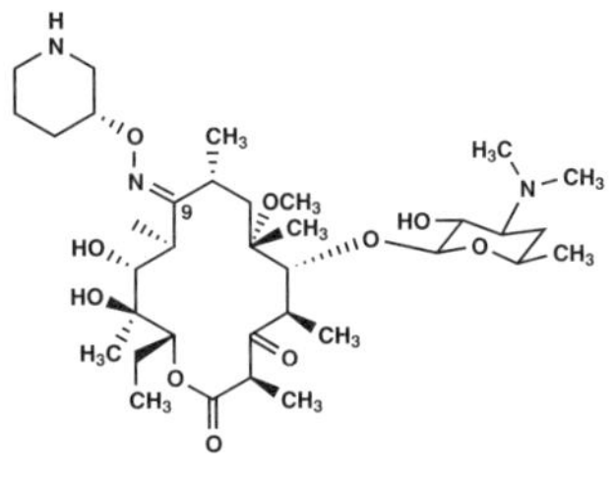

Fig. 21.

2. *11,12-Carbamate and 11,12-Hydrazone Carbamate (Carbazate) Ketolides*

Numerous derivatives of 11,12-carbamate and 11,12-carbazate ketolides were prepared by a synthesis procedure similar to those for CAM 11,12-carbamates and carbazates [109–112]. 11,12-Cyclic carbamate ketolide (**86a**) (Fig. 22) exhibited weak activity against EM-resistant Gram-positive pathogens, and the (10*S*)-methyl epimer of **86a** was almost inactive [109]. *N*-Arylalkyl substitution of carbamate moiety remarkably increased the activity against EM-resistant strains of *S. aureus*, *S. pneumoniae*, and *H. influenzae*. Thus, compounds **86b**, **86d**, and **86e** showed excellent activity against MLS-i and MLS-c resistant strains of *S. pneumoniae* and MLS-i resistant strains of *S. aureus* (Table XII). They also showed activity comparable to that of AZM (**19**) against *H. influenzae*. The alkyl chain lengths influence activity, and **86b** exhibited increased activity compared to **86c**. Introduction of heteroatoms into the aryl group demonstrated a more desirable feature of activity. The 11-hydrazino ketolide (carbazate) (**86d**) analogues with

TABLE XII
In Vitro Evaluation of 11,12-Carbamate and Carbazate Ketolides

Compound	MIC (µg/ml)						
	S. aureus			S. pneumoniae			
	MLS-s	MLS-i	MLS-c	MLS-s	MLS-i	MLS-c	H. influenzae
86b	0.08	0.15	40	0.02	0.3	0.15	2.5
86c	1.2	2.5	40	0.3	20	40	10
86d	0.08	0.15	40	0.02	0.15	5	1.2
86e	0.02	0.08	40	0.02	0.02	0.15	0.6
CAM (16a)	0.3	40	40	0.04	40	40	5
AZM (19)	0.3	40	40	0.15	40	40	1.2

Source: Data from Agouridas *et al.* [109].
86b, R = 4-Phenylbutyl; **86c**, R = 3-Phenylpropyl; **86d**, R = 3-Phenylpropylamino; **86e**, R = 4-Quinolylpropylamino-.
MLS-s, EM sensitive; MLS-i, EM inducibly resistant; MLS-c, EM constitutively resistant.

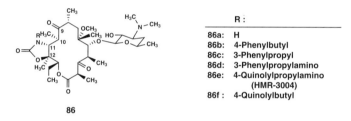

Fig. 22.

phenypropylamino chain at 11-*N* were less active than the analogues of **86b** against MLS-c resistant *S. pneumoniae*, but **86e** (HMR-3004) with quinolylpropylamino chain was significantly more potent against both EM-sensitive and EM-resistant Gram-positive pathogens except MLS-c resistant *S. aureus*, and it was more active than **86d** against *H. influenzae* [109]. It showed the importance of aryl moiety of 11,12-carbazate side chain. Compound **86e** was one of the candidates for development, together with **91i** (HMR 3647, TLM) (Scheme 8).

3. 9a-N-Ketolides

Numerous derivatives of 9a-aza-3-*O*-decladinosyl 3-keto azalide bearing 9a-*N*-methyl and other substituents are reported, but the antibacterial activity is not shown [113, 114].

4. Tricyclic and Tetracyclic Ketolides

TE-802 (**87a**) (Fig. 23) and its analogues, so-called tricyclic or tetracyclic ketolides, were prepared with the intention of increasing the conformational rigidity of ketolide, thereby improving the antimicrobial activity and bioavailability

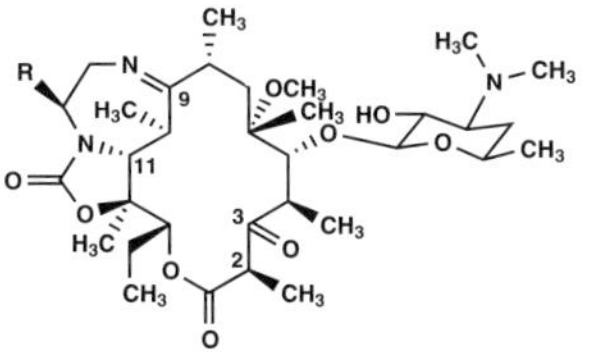

TE-802 (**87a**) : R = H
A-177511 (**87b**) : R = CH$_2$OCH$_2$-Ph
A-202094 (**87c**) : R = CH$_2$OC(O)-p-NO$_2$-Ph

Fig. 23.

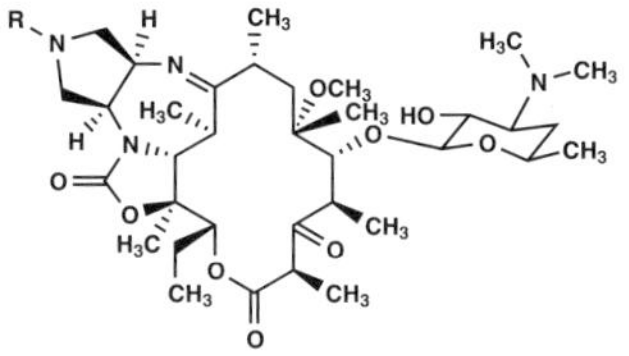

88

Fig. 24.

[115, 116]. The antibacterial activity of **87a**, a tricyclic ketolide, was comparable to that of CAM (**16a**) against EM-sensitive (MLS-s) strains of Gram-positive pathogens and superior against MLS-i and mef type pathogens (MIC *S. aureus* MLS-i: **87a**, 0.20–1.56; **16a**, >100; MIC *S. pneumoniae* intermediate EM-r: **87a**, 0.10; **16a**, 0.78–1.56 μg/ml). However, it was almost inactive against MLS-c resistant *S. aureus*.

Tricyclic ketolide analogues bearing a variety of substitutions on the 9-*N*,11-*N* cyclic imine ring were prepared. A-177511 (**87b**) and A-202094 (**87c**) (Fig. 23), having benzyloxymethyl and *p*-nitrobenzoyloxymethyl substituents, respectively, exhibited increased *in vitro* activity against mef type and MLS-i resistant Gram-positive pathogens (MIC: 0.05–2 μg/ml) and *H. influenzae* (MIC: 1–2 μg/ml) compared with **87a** [117, 118]. Tetracyclic ketolides (**88**) (Fig. 24) were generally less active than tricyclic ketolides *in vitro* [119].

5. C-2 Halogenated Ketolides

C-2 Halogenated derivatives of TE-802 (*X* = F, Cl, Br) have also been synthesized (Fig. 25) [118]. C-2-Fluorinated TE-802 (A-241550) (**89a**) exhibited more potent *in vitro* activity against MLS-s and MLS-i resistant Gram-positive pathogens (MIC MLS-i, *mef*: **89a**, 0.1–0.5; **87a**, 0.2–0.5; **1**, 3.1–32 μg/ml) and against *H. influenzae* (MIC: **89a**, 2; **87a**, 4; **1**, 4 μg/ml) compared with **1** and **87a**. Compound **89a** also showed higher *in vivo* activity against *S. aureus*, *S. pneumoniae*, and *H. influenzae* than CAM (**16a**), AZM (**19**), and TLM (**91i**) (ED$_{50}$: *S. aureus*,

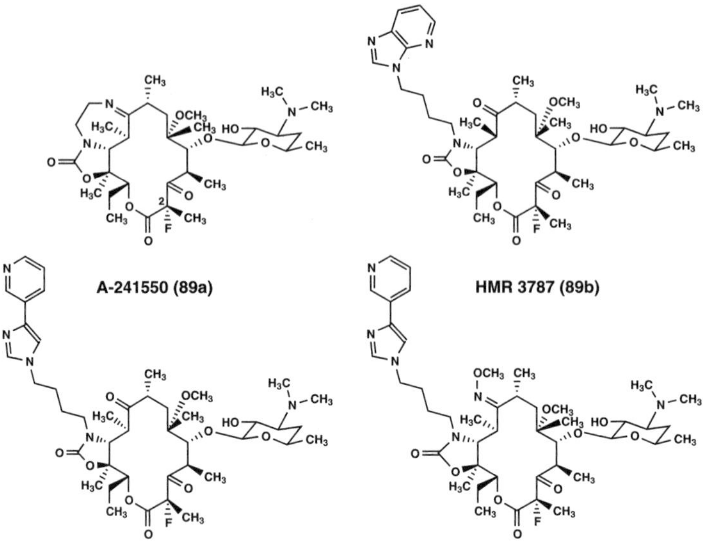

A-241550 (89a)   HMR 3787 (89b)

HMR 3562 (89c)   CP-654,743 (89d)

Fig. 25.

9.4, 30, 25, and 13 mg/kg; *S. pneumoniae*, 7.2, 37, 39, and 34 mg/kg; *H. influenzae*, 36, >100, 35, and 58 mg/kg in mice, respectively).

Recently, HMR 3787 (**89b**), HMR 3562 (**89c**), and CP-654,743 (**89d**) have been reported as C-2 fluorinated ketolides [120–122] (Fig. 25). Compound **89d** was synthesized from the corresponding ketolide by treatment with a base, potassium hexamethyldisilazide, followed by addition of an appropreate electrophile. Bulky substituents at C-2 such as methyl or chlor reduced activity, particularly against EM-resistant strains. However, a small substituent at C-2, such as fluorine, exhibited excellent activity against EM-resistant Gram-positive cocci and comparable activity with that of AZM (**19**) against *H. influenzae*. These facts suggested that, in ketolides, appropriate space is required near the C-3 keto group to retain activity. The MICs of **89b** and **89c** were commonly lower than 0.02 µg/ml against MLS-s, MLS-i, and mef type resistant strains of staphylococci, streptococci, or enterococci. Both compounds are inactive against MLS-c resistant staphylococci [121]. The *in vivo* antibacterial activities of **89b** and **89c** were higher than those of EM (**1**), CAM (**16a**), and AZM (**19**) in a murine septicaemia model by Gram-positive EM-resistant cocci or *H. influenzae* [122]. In addition, **89d**, which has 9-methyloxime and similar substituent with TLM (**91i**) at C-11 carbamate, were 2 to 4 times more potent in activity *in vitro* against Gram-positive pathogens, including EM-resistant strains and *H. influenzae* [123, 124]. However, *in vivo* activity in a mouse infection model was almost the same as that of **91i**. From *in vitro* and *in vivo* activity and pharmacokinetic data, **89d** seems to be practically equivalent with **91i**.

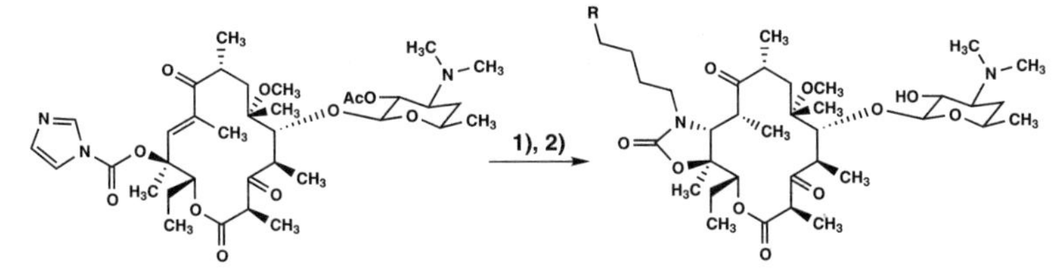

Scheme 8.

1) R(CH$_2$)$_4$NH$_2$, CH$_3$CN, H$_2$O, 60°C ; 2) MeOH, r.t.

Compd	91a	91b	91c	91d	91e
R	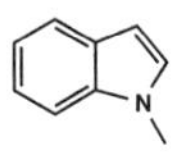 indole	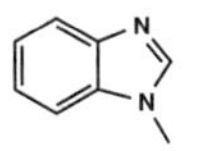 benzimidazole	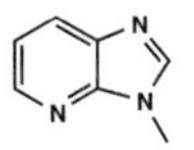 imidazopyridine	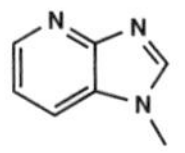 imidazopyridine	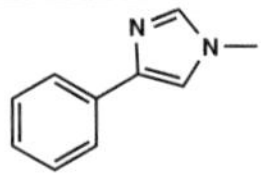 phenyl-imidazole

Compd	91f	91g	91h	91i (HMR 3647)
R	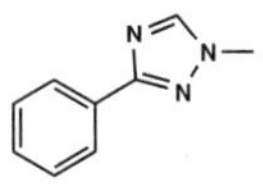 phenyl-triazole	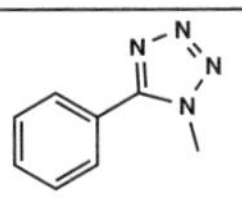 phenyl-tetrazole	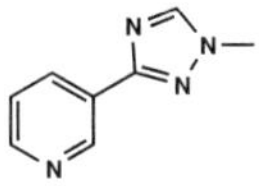 pyridyl-triazole	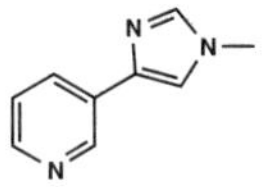 pyridyl-imidazole

6. Synthesis of Telithromycin (HMR-3647) and Its Biological Properties

HMR-3647 (telithromycin: TLM) (**91i**), one of the 11,12-cyclic carbamate ketolides bearing various heterocyclic butyl substituents at 11-*N* of carbamate (**91**), was synthesized from 10,11-anhydro-12-*O*-imidazolyl carbonyl ketolide (**90**) by the reaction with the corresponding amine (Scheme 8) [125]. The synthesis of the heterocyclicbutylamines (Scheme 8, a–i) was achieved by alkylation of the respective heterocyclic amine with 4-bromobutylphthalimide in the presence of NaH or K$_2$CO$_3$ in DMF, followed by hydrazinolysis of phthalimide intermediates. The overall yields of these steps were 24–64%.

The *in vitro* antimicrobial activities of these derivatives were compared with those of TE-802 (**87a**), CAM (**16a**), and AZM (**19**). All of these ketolides were more potent than control macrolides against EM-sensitive and EM-resistant strains of Gram-positive pathogens except MLS-c strains of *S. aureus*, and their activity against *H. influenzae* was comparable to that of AZM [125] (Table XIII). Their activity against various strains of EM- and methicillin-susceptible *S. aureus* was 4–16 times higher than that of EM, but they were inactive against MRSA [126]. They also exhibited excellent activity against macrolide- and penicillin-resistant pneumococci [127].

Among them, **91i** exhibited almost the same *in vitro* activity and antibacterial spectrum as those of HMR 3004 (**86e**), which was the first candidate for development in this series [126]. Both **91i** and **86e** showed considerable activity against EM-resistant strains of *E. faecalis* and *Enterococcus faecium*, whose resistance

TABLE XIII
In Vitro Evaluation of 11,12-Carbamate Ketolides with Heteroatom Aryl Chains

Compound	MIC (µg/ml)						
	S. aureus			S. pneumoniae			
	MLS-s	MLS-i	MLS-c	MLS-s	MLS-i	MLS-c	H. influenzae
86f	0.04	0.08	40	0.02	0.02	1.2	1.2
91a	0.08	0.15	40	0.02	0.08	10	2.5
91b	0.04	0.08	40	0.02	0.08	2.5	0.6
91c	0.04	0.15	40	0.08	0.04	1.2	0.6
91d	0.08	0.15	40	0.02	0.15	2.5	0.6
91e	0.04	0.08	40	0.02	0.04	5	1.2
91f	0.08	0.3	40	0.02	2.5	2.5	2.5
91g	0.02	0.15	40	0.04	0.3	0.6	1.2
91h	0.02	0.6	40	0.02	0.02	0.3	1.2
TLM (91i)	0.04	0.08	40	0.02	0.02	0.08	1.2
TE-802 (87a)	0.3	0.6	40	0.02	40	40	5
CAM (16a)	0.3	40	40	0.04	40	40	5
AZM (19)	0.3	40	40	0.15	40	40	1.2

Source: Data from Denis *et al.* [125].

86f, R = 4-quinolylbutyl; **91a–91i**; R of 11,12-carbamate are shown in Scheme 8.

MLS-s, EM sensitive; MLS-i, EM inducibly resistant; MLS-c, EM constitutively resistant.

has recently been recognized as a severe problem in clinical treatment. Their MIC values against EM-susceptible and EM-resistant enterococci were near 0.015 and 1–8 µg/ml, respectively. *E. faecalis* strains showing resistance to EM (MIC 128–>512 µg/ml) were more sensitive to **91i** (MIC 0.03–16 µg/ml) and **86e** (MIC 0.03–64 µg/ml). *E. faecalis* were more sensitive to both ketolides than *E. faecium*. However, vancomycin-resistant strains of enterococci showed poor susceptibility to both ketolides. These ketolides seem to be applicable to *H. influenzae, Neisseria gonorrhaeae, Neisseria meningitidis, M. catarrhalis, Legionella pneumophila, Mycoplasma*, and *Chlamydia*, judging from *in vitro* and *in vivo* activity of **91i** [127–129].

Consequently, **91i** exhibited the most potent antibacterial activity against Gram-positive pathogens with penicillin G and/or EM resistance. Furthermore, **91i** was shown not to induce MLS resistance from *in vitro* studies against *S. aureus, S. pneumoniae*, and *S. pyogenes*. These facts suggested that the activity may be related to the lack of the C-3 cladinose and the presence of the 3-keto group [130]. In addition, **91i** was reported to bind strongly to 23S rRNA at domain II and domain V, where there is a region of peptidyl transferase on the ribosome. That would be why **91i** shows increased activity against MLS-resistant strains [131–133]. In particular, the interaction of the 11,12-carbamate side chain

TABLE XIV
In Vivo Evaluation of TLM (**91i**) against Gram-Positive
Pathogens and H. influenzae

	ED_{50} in mice (mg/kg)[a]					
	S. aureus		S. pneumoniae			
Compound	MLS-s	MLS-i	MLS-s	MLS-i	MLS-c	H. influenzae
AMP[b]	ND	ND	ND	ND	ND	5.5
CAM (**16a**)	6	55	7.5	>50	>50	71
AZM (**19**)	30	>100	6	>50	>50	56
TLM (**91i**)	10	4.5	1	4	15	68

Source: Data from Denis *et al.* [125].
MLS-s, EM sensitive; MLS-i, EM inducibly resistant; MLS-c, EM constitutively resistant.
[a]Effective dosage that 50% of mice survive after oral administration.
[b]AMP, ampicillin.

in **91i** with domain II is known to correlate with the superior binding to the MLS-r ribosome [131]. 11,12-Carbamate side chain of **91i** was also shown to stimulate the ribosomal subunit assembly [132]. In addition, the presence of a 3-keto group demonstrated avoidance of MLS resistance induction and increased the acid stability.

In vivo activity of **91i** reflected its *in vitro* activity against EM-resistant Gram-positive pathogens and *H. influenzae* in mice (Table XIV). The results of pharmacokinetic study of the oral administration of **91i** (10 mg/kg) in mice were as follows: C_{max} = 2.9 µg/ml, bioavailability = 49%, AUC (0–8 hr) = 17.2 µg·hr/ml, $t_{1/2}$ = 1 hr. The pharmacokinetic data were also well correlated to *in vivo* activity [125]. Compound **91i** showed advantageous properties in increased ribosome binding, good balanced *in vitro* and *in vivo* activity, and improved acid stability and pharmacokinetics. Pharmacokinetic data on the oral administration of 800 mg of **91i** once daily to young healthy volunteers are summarized in Table XV. The results recommend that **91i** should be administrated once daily [134].

7. 6-O-Substituted Ketolides

3-*O*-Decladinosyl-3-oxo-EM having 6-*O*-alkyl, hetero alkyl, allyl, alkylamine, alkyloxime, arylalkyl, or heteroarylalkyl, alkenyl, alkynyl substituents were synthesized. The activities of ketolides bearing 6-*O*-arylalkyl substituents were superior to those having 6-*O*-aliphatic substituents. The former generally exhibited improved *in vitro* activity against MLS-i and *mef* type resistant strains of Gram-positive pathogens. 6-*O*-[3-(3-Quinolyl)-propenyl] ketolide (A-184656) (**92**) (Fig. 26) showed excellent activity against MLS-i and *mef* type of Gram-positive pathogens. The activity of these 6-*O*-substituted ketolides was almost the same as

TABLE XV
Pharmacokinetics of TLM (**91i**) 800 mg in Healthy Volunteers

Parameter (mean+SD)	Day 1 (single dose)	Day 10 (last dose)
C_{max} (μg/ml)	1.99 ± 0.84	1.84 ± 1.14
T_{max} median (range)	1.0 (1–2)	2.0 (0.5–3)
AUC (0–24 hr) (μg · hr/ml)	7.25 ± 2.33	8.4 ± 2.59
C_{24hr} (μg/ml)	0.025 ± 0.007	0.046 ± 0.016
$t_{1/2}$ Terminal (hr)	10.64 ± 2.53	13.4 ± 3.5

Source: Data from Lenfant *et al.* [134].
n = 12 (male).

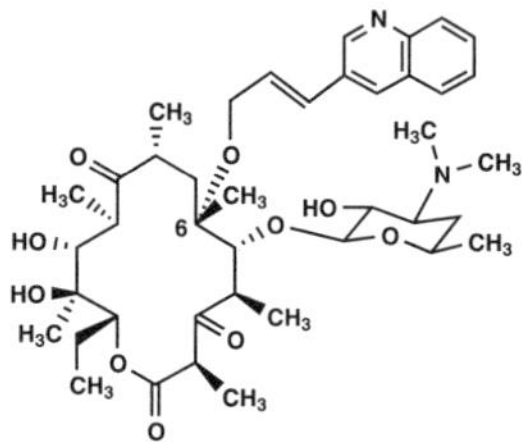

92

Fig. 26.

those of 6-*O*-substituted EM derivatives against EM-sensitive Gram-positive pathogens [135].

8. ABT-773 (A-195773)

ABT-773 (**95**) is one of the promising ketolides, like TLM (**91i**). Compound **95** has 3-keto-11,12-carbamate and a bulky 6-*O*-quinolyl-allyl group instead of the methyl group in CAM. The drug design of **95** was derived from evidence obtained from the structure–activity relationships of A-66321 (**63d**), TE-802 (**87a**), and TLM (**91i**). They all have 11,12-cyclic carbamate and were active against MLS-resistant pathogens [117, 136]. The main objectives of the research were to obtain compounds showing potent activity against MLS-resistant *S. pneumoniae* and *H. influenzae* and to improve drug tolerability and pharmacokinetic profiles. The synthetic process of **95** is divided into four steps: (1) introduction of the 6-*O* allyl group into EM; (2) formation of 11,12-carbamate; (3) deglycosidation of 3-*O*-cladinose, which leads to 3-keto as the key intermediate; and (4) a final reaction of the allyl key intermediate of 11,12-carbamate ketolide with bromoquinoline (Schemes 9 through 11). 2–8-Bromoquinoline carrying derivatives were prepared to compare the activity of each analogue. On the other hand, a series of 6-*O*-quinolylallyl-11,12-carbamates such as A-226234 (**93**) and A-222389 (**94**) (Scheme 9), 6-*O*-quinolylalkyl analogues such as A-201943 (**96**) and A-224281 (**97**) (Scheme 11), an 11-carbazate analogue (A-201316) (**98**),

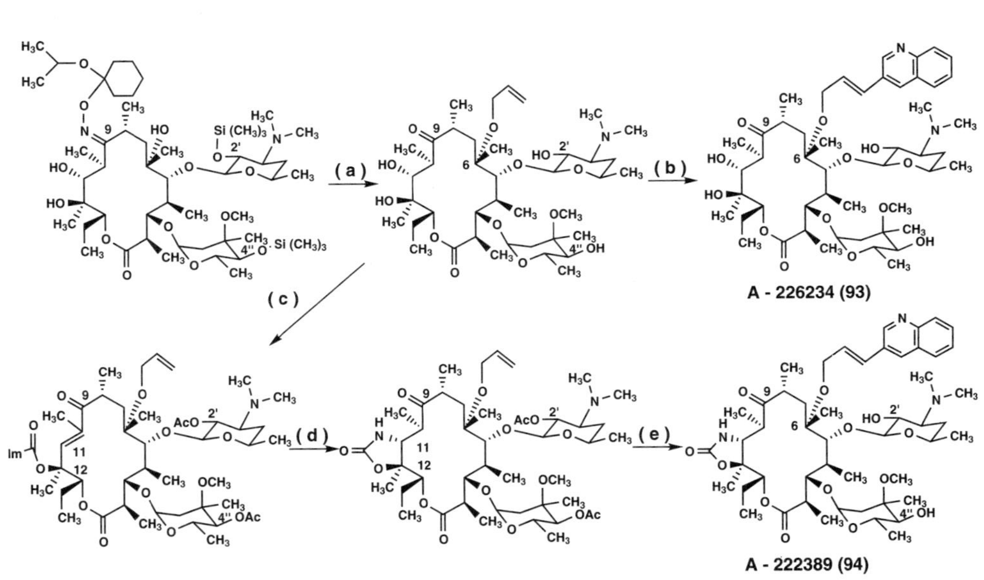

A - 226234 (93)

A - 222389 (94)

(a): (i) Allylbromide, KOt-Bu, DMSO-THF, (ii) HOAc, CH$_3$CN-H$_2$O
(iii) NaHSO$_3$, HCOOH, EtOH-H$_2$O
(b): 3-Bromoquinoline, Pd(OAc)$_2$, P(o-tolyl)$_3$, Et$_3$N, H$_3$CN

(c): (i) Ac$_2$O, DMAP, Et$_3$N, CH$_2$Cl$_2$, (ii) CDI, NaN(TMS)$_2$, THF-DMF
(d): NH$_4$OH, CH$_3$CN-THF
(e): (i) 3-Bromoquinoline, Pd(OAc)$_2$, P(o-tolyl)$_3$, Et$_3$N, CH$_3$CN
(ii) LiOH, MeOH

Scheme 9.

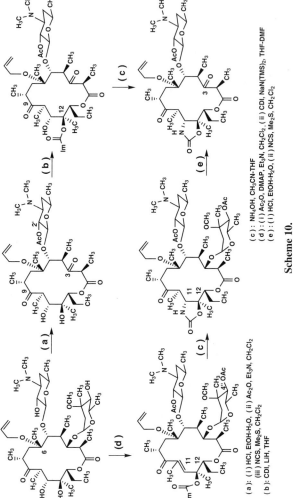

Scheme 10.

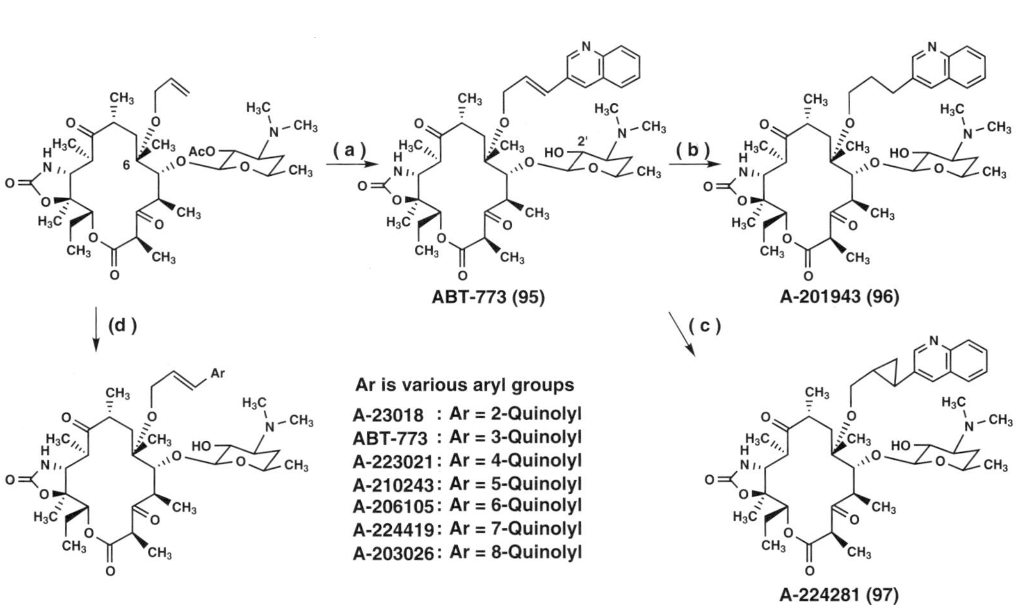
(a)
(b)
(c)
(d)
6
2'
ABT-773 (95)
A-201943 (96)
A-224281 (97)
Ar is various aryl groups
A-23018 : Ar = 2-Quinolyl
ABT-773 : Ar = 3-Quinolyl
A-223021 : Ar = 4-Quinolyl
A-210243 : Ar = 5-Quinolyl
A-206105 : Ar = 6-Quinolyl
A-224419 : Ar = 7-Quinolyl
A-203026 : Ar = 8-Quinolyl

(a): (i) 3-Bromoquinoline, $Pd(OAc)_2$, $P(o\text{-tolyl})_3$, Et_3N, CH_3CN
(ii) MeOH
(b): H_2, 10% Pd-C, MeOH

(c): CH_2N_2, $Pd(OAc)_2$, CH_2Cl_2
(d): (i) Ar-X (X = Br or OTf), $Pd(OAc)_2$, $P(o\text{-tolyl})_3$, Et_3N, CH_3CN
(ii) MeOH

Scheme 11.

TABLE XVI
In Vitro Antibacterial Activity of 6-O-Substituted Ketolides/Macrolides

Compound	MIC (µg/ml)							
	A-226234	A-222389	ABT-773	A-201943	A-224281	A-201316	A-197579	EM
S. aureus								
ATCC6538	0.39	0.2	0.05	0.01	0.05	0.02	0.05	0.39
A5177	3.1	3.1	0.05	0.01	0.05	0.02	0.05	6.2
A5278	>100	>100	>100	>100	>100	>100	>100	>100
S. pyogenes								
EES61	0.03	0.03	0.004	0.015	0.03	0.015	0.06	0.03
930	64	8	1	4	32	64	16	>128
PIU2548	4	1	0.125	0.5	0.5	0.5	2	32
S. pneumoniae								
ATCC6303	0.06	0.03	0.004	0.03	0.03	0.015	0.06	0.06
5737	128	8	0.25	64	4	64	32	>128
5649	4	1	0.25	0.5	0.25	0.25	1	16
H. influenzae								
DILL	8	2	2	2	4	4	4	8

Source: Data from Ma et al. [112].
S. aureus: ATCC6538 (MLS-s), A-5177 (MLS-i), A5278 (MLS-c).
S. pyogenes: EES61 (EM-s), 930 (MLS-r), PIU2548 (mef).
S. pneumoniae: ATCC6303 (EM-s), 5737 (MLS-r), 5649 (mef).
H. influenzae: DILL (ampicillin resistant).

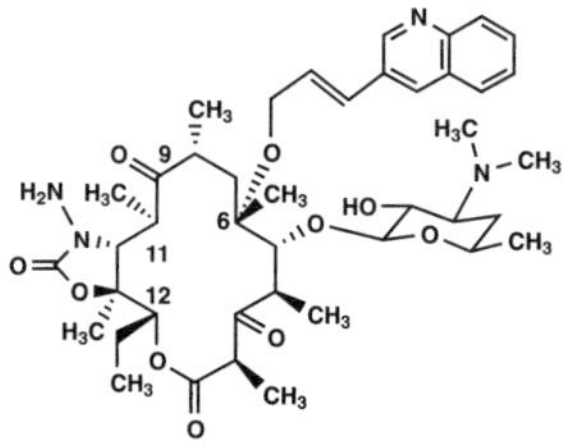

A - 201316 (98) 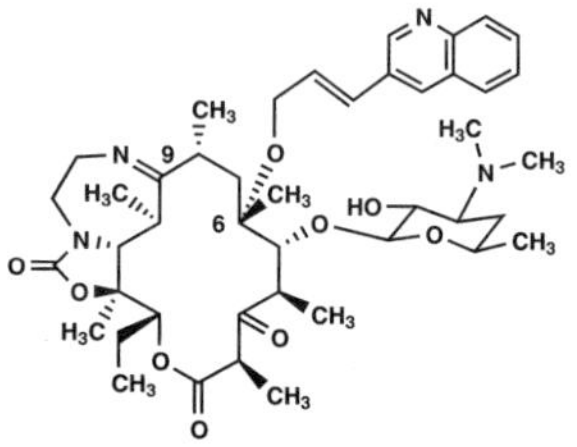   A - 197579 (99)

Fig. 27.

and a 9, 11-N,N-tricyclic analogue (A-197579) (**99**) (Fig. 27) were synthesized and their antibacterial activities compared (Table XVI). The structure–activity relationships of all these derivatives were evaluated and **95** was then selected as the most promising candidate for development [112].

Compound **95** showed excellent activity against MLS-i, MLS-c, and mef type resistant strains of S. aureus, S. pyogenes, and S. pneumoniae except the MLS-c type of S. aureus. The conformation of the 6-O-quinolylpropenyl side chain in ketolide was compared with that of the 11-N-quinolylpropylamino side chain in

3. Chemical Modification of Macrolides

TABLE XVII
In Vitro Antibacterial Activity of ABT-773 (**93**) and Reference Compounds

Organism	MIC (µg/ml)					
	ABT-773 (**95**)	TLM (**91i**)	TE-802 (**87a**)	CAM (**16a**)	AZM (**19**)	EM (**1**)
S. aureus MLS-i	0.05	0.1	0.1	3.1	12.5	6.2
S. aureus MLS-c	>100	>100	>100	>100	>100	>100
S. pyogenes EM-s	0.004	0.004	0.06	0.03	0.25	0.06
S. pyogenes MLS-r	1	8	>128	>128	>128	>128
S. pyogenes mef	0.125	2	1	32	64	32
S. pneumoniae EM-s	0.004	0.004	0.06	0.03	0.25	0.06
S. pneumoniae MLS-r	0.25	8	>128	>128	>128	>128
S. pneumoniae mef	0.25	0.5	1	16	32	16
H. influenzae Amp-r[a]	2	2	4	8	2	8

Source: Data from Ma et al. [112].
[a]Amp-r, ampicillin resistant.

HMR-3004 (**86e**). Then, the conformation analysis of **95** in solution indicated that 6-*O* position was placed at the center of hydrophilic face on macrolide ring and the quinolylpropenyl side chain in this position gave a desirable conformation for the activity, which almost coincided with the result given for **86e** [112, 136].

The *in vitro* antimicrobial activity of **95** was compared with that of other well known macrolides, as shown in Table XVII [112]. Compound **95** generally showed slightly more potent activity than TLM (**91i**) against MLS-s and MLS-r (MLS-resistant) strains of *S. aureus, S. pneumoniae*, and *S. pyogenes in vitro*. Compound **95** exhibited better MIC_{50} and MIC_{90} values against *S. pneumoniae* and *S. pyogenes* than those of EM (**1**) and TLM (**91i**), as shown in Table XVIII [137], and it showed almost the same MIC values as **91i** against *Moraxella, Haemophilus, Enterococcus,* and *H. pylori* [138]. The MIC_{50} values of both **95** and **91i** against these four pathogens were determined to be 0.12, 1, 0.12, and 0.25 µg/ml, respectively.

Compound **95** was shown to be more effective than **91i** against MLS-c resistant strains of *S. pneumoniae* (*erm*AM) and *mef* type strains of *S. pneumoniae* (*mef* E) (Table XIX) and *H. influenzae in vitro* and in rat plumonary infections [139], being 3–16 and 3 times more potent than **91i** against MLS-c and *mef* type strains of *S. pneumoniae* and 2–3 times more effective against *H. influenzae* (Table XX). In addition, *in vitro* and *in vivo* activity of **95** has been investigated against *Legionella, Chlamydia, M. avium, Toxoplasma gondii*, and other microorganisms. Similar investigations of newer macrolides and **91i** have also been reported [140–143].

Pharmacokinetics of **95** after a single oral dose in animals are shown in Table XXI [144]. Its concentration in the lung tissue of rats was markedly higher than the concentration in plasma. Then, the bioavailability and daily dosing frequency seem to be improved.

TABLE XVIII
Comparison of Activity against Clinically Isolated
EM-Susceptible *S. pneumoniae* and *S. pyogenes*

S. pneumoniae $n = 94$	ABT-773 (**95**)	TLM (**91i**)	EM (**1**)
MIC_{50}	0.001	0.004	≦0.03
MIC_{90}	0.002	0.008	≦0.06
MIC_{min}	≦0.0005	≦0.0005	≦0.03
MIC_{max}	0.004	0.015	0.5
S. pyogenes $n = 30$	ABT-773 (**95**)	TLM (**91i**)	EM (**1**)
MIC_{50}	≦0.002	0.008	0.06
MIC_{90}	≦0.002	0.015	0.06
MIC_{min}	≦0.002	0.004	≦0.03
MIC_{max}	≦0.002	0.015	0.12

Source: Data from Shortridge *et al.* [137].
MIC: μg/ml.

TABLE XIX
Comparison of Activity against Clinically Isolated
EM-Resistant *S. pneumoniae*

mef $n = 100$	ABT-773 (**95**)	TLM (**91i**)	EM (**1**)
MIC_{50}	0.004	0.06	2
MIC_{90}	0.12	1	8
MIC_{min}	≦0.002	0.015	0.25
MIC_{max}	0.5	2	>32
erm $n = 104$	ABT-773	TLM	EM
MIC_{50}	0.004	0.015	>32
MIC_{90}	0.015	0.12	>32
MIC_{min}	≦0.002	≦0.002	8
MIC_{max}	1	0.5	>32

Source: Data from Shortridge *et al.* [137].
MIC: μg/ml.

E. Discussion

Recent research on the modification of 14-membered macrolides and their structure–activity relationships has focused mainly on two subjects: Improvement of their antimicrobial activity and spectra, including those for macrolide-resistant and Gram-negative pathogens, and optimization of their pharmacokinetic

TABLE XX
Efficacy against MLS-r *S. pneumoniae* (*erm*AM, *mef*), and *H. influenzae*, Rat Pulmonary Infection

Compound	Dosage[a,b]	Mean log cfu	Percent responders[c]	ED$_{50}$ for 2 log reduction[d]	MIC[e] (µg/ml)
ABT-773 (**95**)	10	0.00	100	1.6	0.015
S. pneumoniae	2.5	2.38	100	(1.3–2.2)	
(*erm*AM)	0.63	5.64	20		
TLM (**91i**)	60	1.34	100	26.7	0.125
S. pneumoniae	20	4.38	60	(20.8–34.3)	
(*erm*AM)	5	4.40	0		
ABT-773 (**95**)	20	2.63	100	7.0	0.06
S. pneumoniae	10	3.46	60	(6.9–7.1)	
(*mef* E)	5	4.60	40		
TLM (**91i**)	30	3.98	60	25.8	0.25
S. pneumoniae	20	4.39	40	(9.3–72.1)	
(*mef* E)	10	5.23	20		
ABT-773 (**95**)	60	1.91	100	19.8	0.5
H. influenzae	20	3.19	60	(15.8–24.8)	
	5	6.38	0		
TLM (**91i**)	80	2.11	80	64.6	0.5
H. influenzae	25	6.13	0	(39.4–105.9)	
	5	5.23	0		

Source: Data from Mitten *et al.* [139].
[a]Drug administered orally, q.d., days 1–3, starting 18 hr postinfection.
[b]Treatment 1 ml vehicle (2% ethanol, 98% D5W) q.d., days 1–3 (cfu of each strain; 5.87, 6.15, 6.15).
[c]Responder is defined as an animal that had at least a 2 log reduction in plumonary bacterial burden as compared to the vehicle-treated control mean on day 4.
[d]ED$_{50}$ (mg/kg) for mean 2 log reduction compared to vehicle-treated control.
[e]Micro-broth dilution method.

TABLE XXI
Pharmacokinetics of ABT-773 (**95**) Following a Single Dose

Animal	Dose (mg/kg)	$t_{1/2}$ (hr)	C_{max} (µg/ml)	AUC (µg·hr/ml)	F (%)
Mouse	10	3.9	1.47	6.1	49.5
Rat	20	6.0	0.52	8.6	60
Monkey	10	2.3	0.56	3.2	35.8
Dog	5	4.2	0.84	8.4	44.1

Source: Data from Hernandez *et al.* [144].

profiles. The former has been successfully achieved by rational drug design to gain acid stability and ribosomal binding ability against various types of EM-resistant pathogens (Table XXII), such as EM-resistant staphylococci (without MLS-c), streptococci, and enterococci. Their antimicrobial spectra have been expanded to *H. influenzae* [145], pathogens of atypical pneumoniae (*Mycoplasma* [146], *Chlamydia* [147], *Legionella* [141]), *Moraxella* [145], *Neisseria* [148],

TABLE XXII
Recent Items for Evaluation of Newly Synthesized Macrolides and Ketolides

A. Physicochemical and biological properties of drugs
   1. Conformation and configuration
   2. Acid stability
   3. Lipophilicity [153]
   4. Water solubility
   5. Ionizable function group (pK value) [154]
   6. Mode of action
      a. Binding affinity to ribosome [153–156]
      b. Inhibition of protein synthesis [131–133] (b is not always parallel with a [157])
      c. Binding site (domain V, II, etc.) [131–133]
      d. Stimulation and inhibition of ribosomal subunit assembly [132, 158]
      e. Synergism by combination with other drugs [159, 160]
   7. Drug accumulation [156]
   8. Bactericidal or bacteriostatic [147, 161–164]
   9. Post antibiotic effect [162, 165–167]
   10. Resistance inducibility [168]
   11. Interaction with resistance peptide [169]
B. Response of microorganisms
   1. Antibiotic resistance
      a. Mutation of r-RNA
      b. Mutation of ribosomal protein [161, 167, 170]
      c. Efflux pump [33]
      d. Inactivation of antibiotics by enzymes
      e. Penetration
      f. Methylation level on ribosome [168]
      g. Multiresistance (heterogeneous) by these combinations
   2. Gram-positive or negative [33]
   3. Aerobic or anaerobic
   4. Intracellular or extracellular [23, 171]
C. Effect of host condition
   1. Pharmacokinetics (bioavailability)
   2. Metabolism [95]
   3. Tissue concentration [cells (PMN, macrophage), organs (lung), and fluid (urine)] [172, 173]
   4. pH circumstance [174]
   5. Response to cytokines or immunological factors [175–177]
   6. Acute or chronic [178]
   7. Adverse effect (drug interaction, *P*-450 inhibition) [179]

Helicobacter [24], *Mycobacterium* (*M. avium* [149a], *M. leprae* [149b]), *Toxoplasma* [143], *Plasmodium* [150], *Rickettsia* [151], etc. To improve the chemical and biological properties of macrolide derivatives, it has been necessary to clarify the drug–parasite interaction and mode of action in more detail. Such requirements stimulated the study of drug resistance on the molecular and gene levels.

In addition, progress in the research on drug–host and parasite–host interactions resulted in finding and application of unique pharmacokinetics and other biological properties of macrolides, such as extremely high tissue concentration (particularly in the lung, WBC [152]), acid stability in the stomach, long half-life (in **19, 20**), high urinary excretion of active metabolite (of **16a**), intracellular accumulation (*Chlamydia, Legionella*), immuno-modulating activity in DPB (diffuse panbronchitis), etc. [153–179] (Table XXII).

These improvements to 14- and 15-membered macrolides by structural modification has been achieved by developments of structural and conformational analysis techniques as well as by chemical synthesis methods. A number of derivatives were thus designed, synthesized, and evaluated by advanced methodologies. However, the problems of drug resistance in MLS-c resistant strains of *S. aureus* and of low potency against Gram-negative bacteria remain unsolved. Competition between parasites and macrolides will continue hereafter, and further efforts to overcome a number of infectious diseases, particularly those caused by newly occurring drug-resistant types of pathogens, will be required.

The reviews of macrolides and ketolides reported by H.A. Kirst, A.J. Bryskier, and D.T.W. Chu *et al.* are useful for references [180–186].

III. Sixteen-Membered Macrolide Antibiotics and the Avermectin Family

The 16-membered macrolide antibiotics are an important class of antimicrobial agents. Like 14-membered macrolides, they are active mainly against Gram-positive bacteria and mycoplasma. They show some advantages over 14-membered macrolides, having better gastrointestinal tolerance (gastrointestinal motor stimulating activity, GMSA) and activity against strains expressing resistance of the inducible type [1].

The 16-membered macrolide antibiotics are generally classified into two large groups, namely, the leucomycin-related family and the tylosin-related family, on the basis of the substitution patterns of their aglycons (Fig. 28) [187]. Interestingly, the leucomycin series, such as leucomycin, josamycin, midecamycin, and spiramycin, have been used clinically for humans, while the tylosin series has been utilized in veterinary medicine. In this section, we describe recent studies on the chemical modification of 16-membered macrolides and their structure–activity relationships.

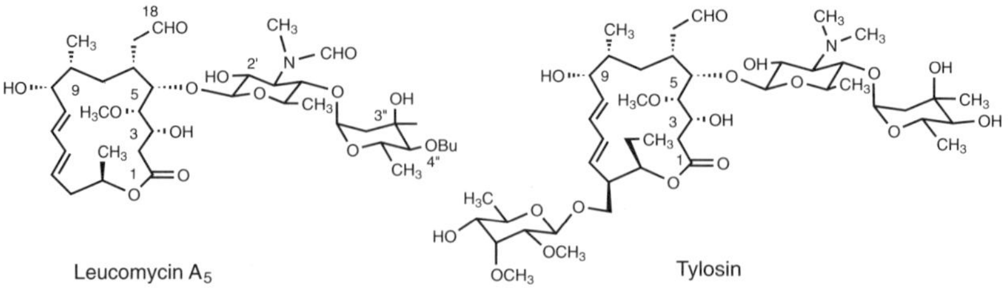

Fig. 28.

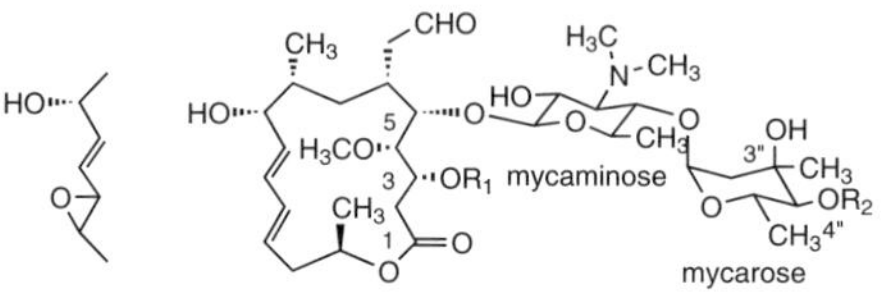

	chromophore	R₁	R₂
Leucomycin Fr group	I	H	H-COCH₂CH(CH₃)₂
Ac group	I	COCH₃	H-COCH₂CH(CH₃)₂
Midecamycin (SF837 A₁, A₂)	I	COCH₂CH₃	COCH₂CH₃-COCH₂CH₂CH
Platenomycin (YL-704 A₃, B₃, C₂, B₁, A₁, A₀, C₄, C₁, C₃)	I, II	COCH₃-COCH₂CH₂CH₃	COCH₃-COCH₂CH(CH₃)₂
Maridomycin	II	COCH₃, COCH₂CH₃	COCH₃-COCH₂CH(CH₃)₂
Josamycin	I	COCH₃	COCH₂CH(CH₃)₂

Fig. 29.

A. Chemical Modification of Sixteen-Membered Macrolides before 1984

A review of the research done before 1984 was reported by Sakakibara and Omura in the first edition of this book [1]. Here, we summarize the chemical modifications and structure–activity relationships of the 16-membered macrolides (mainly leucomycins) described in that edition (Fig. 29).

1. Modification of the 3-,9-,2´-, and 4´´-Hydroxyl Groups

Both *in vitro* and *in vivo* activity is reduced by an increase in the acyl group length in the 2´-*O*-acyl derivatives. The MIC values of the 2´-*O*-acetyl and 2´-*O*-propionyl derivatives are nearly the same as that of the parent antibiotic. Because the 2´-*O*-acyl is easily hydrolyzed in aqueous solution, the potent antibacterial activity has been explained by the hydrolysis of 2´-*O*-acyl group during bioassays. For this reason, the 2´-hydroxyl group has been considered to be one of the functional groups essential for the biological activity of the macrolide.

Among the 9-hydroxyl acylated derivatives, the acetyl and propionyl derivatives showed the same degree of antibacterial activity as the parent antibiotics. The *in vitro* antibacterial activity of the 9,2´-di-*O*-acyl derivatives from the leucomycin group are somewhat lower than those of the 9-*O*-acyl derivatives. The 9-hydroxyl compounds and the corresponding ketone derivatives showed essentially the same antimicrobial spectra. Modification of the 9-hydroxyl group has little influence on the MIC but does lead to other improvements, such as higher blood levels, enhancement of the chemotherapeutic effect, reduction of the bitterness, and better aqueous solubility.

It is generally known that the *in vitro* antibacterial activities of 3-*O*-alkanoyl derivatives are lower than those of the 3-hydroxyl compounds but the blood levels are increased. It is suggested that the hydroxyl group at C-3 is not directly related to the antibacterial activity.

On the other hand, the 3″-*O*-alkanoyl derivatives showed nearly the same degree of antibacterial activity *in vitro* as the parent antibiotics, and their *in vivo* activities were enhanced.

2. *Reduction of the Double Bond*

The respective tetrahydro derivatives showed slightly reduced activites both *in vitro* and *in vivo* compared with the parent antibiotics, but the difference was essentially negligible.

3. *Chemical Modification of the Aldehyde Group at C-18*

The 18-dihydro compounds showed significantly reduced antibacterial activity. It is suggested that the aldehyde group at C-18 is one of the functional groups essential for the biological activity of the macrolide.

4. *Chemical Modification of the 3′-Dimethylamino Group*

Demethylation of the 3′-dimethylamino group on the desosamine moiety reduced the *in vitro* antibacterial activity, as did the *N*-oxide compound. From these results, the 3′-dimethylamino group has been considered to be one of the functional groups essentials for the biological activity of the macrolide.

B. Newer Macrolides

Recently, 3″-*O*-acyl derivatives have been synthesized and were found to have a strong antibacterial activity *in vitro* and *in vivo*. Among them, miokamycin and rokitamycin have been launched (Fig. 30).

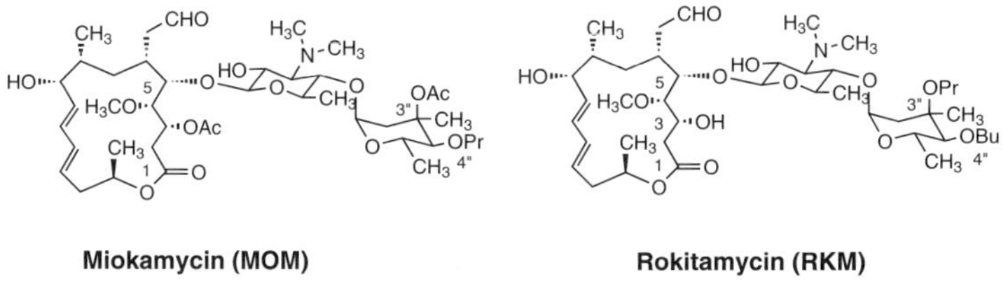

Miokamycin (MOM)   **Rokitamycin (RKM)**

Fig. 30.

1. Macrolides in the Leucomycin Family

a. Miokamycin Miokamycin is a semisynthetic derivative of midecamycin A_1 obtained by introducing two acetyl groups at C-3″ and C-9. Miokamycin (also spelled miocamycin and called midecamycin acetate, MOM, or ponsiomycin) is 9,3″-di-*O*-acetylmidecamycin A_1 [188]. It has been launched commercially in several countries.

Extensive early studies on the *in vitro* and *in vivo* structure–activity relationships of the leucomycin family demonstrated correlation between the numbers and types of *O*-acyl substituents and their antibacterial potency, efficacy in treating experimental infections, and serum antibiotic concentrations. Consequently, esterification of the hydroxyl groups in several leucomycin-related macrolides was conducted to find derivatives with better antibiotic activity and pharmaceutical properties (such as greater water solubility and a masking of their extremely bitter taste) [189]. After such an investigation with midecamycin, miokamycin was synthesized and characterized as a useful macrolide antibiotic.

Due to the facile intramolecular migration of the acyl groups between the 3″- and 4″-hydroxyl groups, miokamycin was originally synthesized by an indirect route (Scheme 12). Microbial hydrolysis of the 4″-*O*-propiony group of midecamycin A was followed by 9,2′,4″-tri-*O*-acetylation. Heating this intermediate with propionic anhydride in pyridine effected both the 4″- to 3″-transacetylation and then the return of the 4″-*O*-propionyl group [188]. A direct synthetic route for 3″-*O*-acylation without acyl migration was later developed [191]. Further investigations of this chemistry yielded additional *O*-acyl derivatives of midecamycin [192].

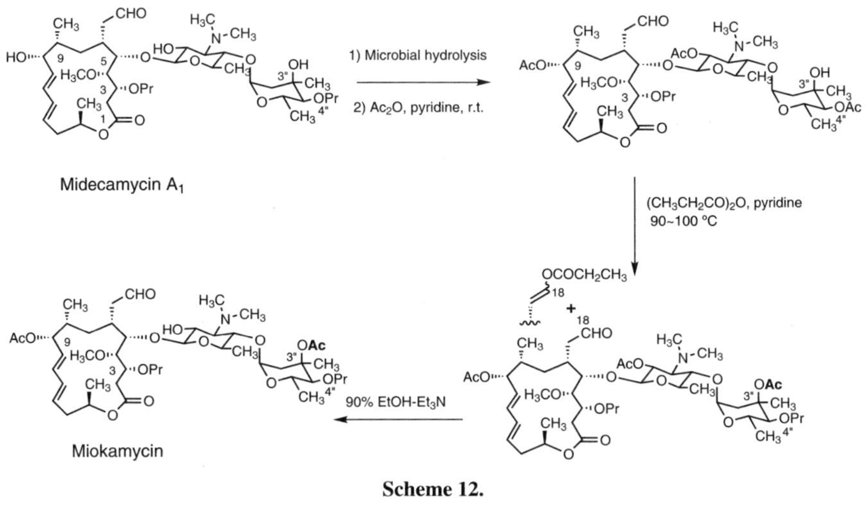

Scheme 12.

Miokamycin is 2 to 10 times more effective than midecamycin in the treatment of experimentally infected mice. Miokamycin is highly metabolized and 12 metabolites having been described. Two of them showed good antibacterial activity *in vitro* [190].

  b. Rokitamycin   Rokitamycin, a semisynthetic derivative of leucomycin A_5, introduced a propionate group at the C-3″ hydroxy group, another of the individual factors found in the leucomycin complex along with josamycin (leucomycin A_3). Rokitamycin, initially referred to as TMS-19-Q, is the 3″-*O*-propionyl derivative of leucomycin A_5 [193]. It has been launched commercially in several countries.

  The research lines that led to miokamycin were extended in another direction by other investigators, who synthesized additional acyl derivatives of the leucomycin factors, including monoesters of the hindered 3″-hydroxyl group. These later derivatives were prepared via a multistep protection including 2′-*O*-acetyl, 3,9-bis-*O*-trimethylsilyl, 3″-*O*-acylation (propinyl chloride in the presence of tribenzylamine), and a deprotection sequence (Scheme 13). 3″-*O*-Propionyl leucomycin A_5 is the most active derivative among the series of 3″-*O*-acyl leucomycin A_5 derivatives and shows the highest serum levels in dogs and monkeys. The 3″-*O*-acetyl and 3″-*O*-butyryl derivatives were less active than 3″-*O*-propionyl leucomycin A_5 [193].

  The *in vitro* activity of rokitamycin against Gram-negative bacteria is twice as high as that of leucomycin A_5, josamycin, and midecamycin. Rokitamycin is highly metabolized into four metabolites. 10″-Hydroxyrokitamycin, leucomycin A_7, leucomycin V, and 14-hydroxyleucomycin V (4″-debutyrylleucomycin A_5)

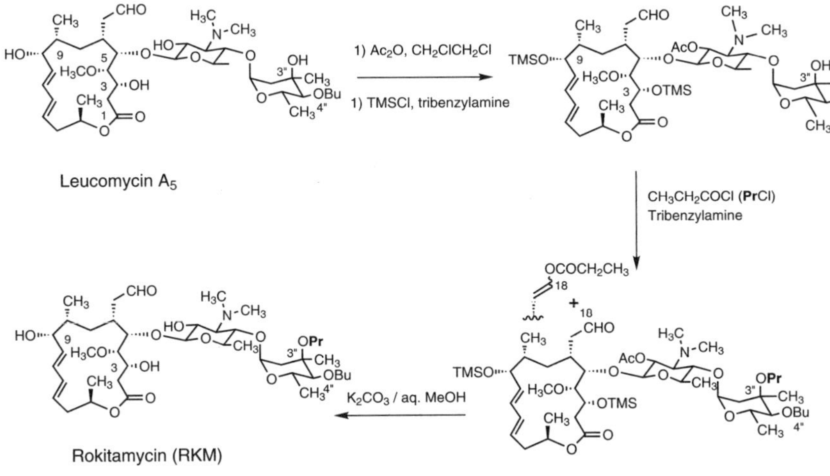

Scheme 13.

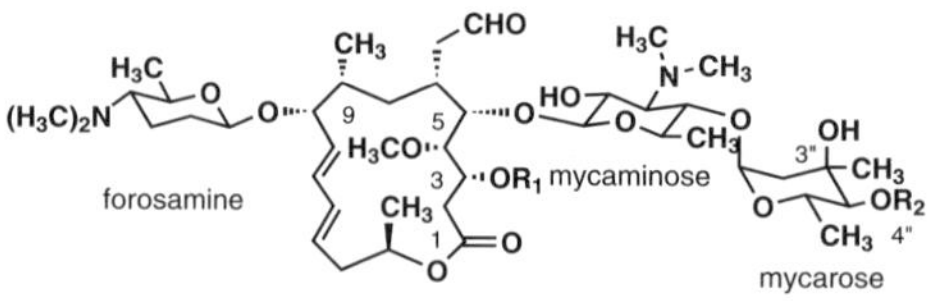

Spiramycin I $R_1=R_2=H$
Spiramycin II $R_1=H$, $R_2=COCH_3$
Spiramycin III $R_1=COCH_2CH_3$, $R_2=H$
Acetylspiramycin $R_1=R_2=COCH_3$

Fig. 31.

are the main metabolites. Their *in vitro* potencies are one-half and one-tenth, respectively, that of the parent compound [194].

2. Macrolides in the Spiramycin Family

The spiramycin family is often classified separately from the leucomycin family despite the close resemblance of their respective aglycones. The spiramycins are structurally distinct in that the 9-α-hydroxyl group is glycosylated by a β-D-forosaminyl moiety. Having this second amino sugar incorporated into the typical macrolide structure, the spiramycins are unique dibasic macrolides (Fig. 31).

Among the numerous spiramycin esters that were initially prepared, 3,4″-di-*O*-acetylspiramycin I (acetylspiramycin) exhibited greater *in vivo* activity than that of spiramycin and received clinical evaluation [195]. Chemical modification of the 4″-hydroxyl group to sulfonyl and alkyl groups retained the antibacterial activity [196]. Acylation of the 3″-hydroxyl group of spiramycin analogues showed better antibacterial activity.

Neospiramycins are demycarosyl derivatives of spiramycins and consist of a platenolide ring, D-mycaminose, and D-forosamine. Neospiramycin I shows antibacterial activity similar or superior to that of spiramycin *in vitro* but is practically inactive *in vivo*. The pharmacokinetic profiles of the two compounds appear to differ.

Sano *et al.* synthesized 3- and/or 4″-tetrahydrofuranyl (THF) and tetrahydropyranyl derivatives of neospiramycin I to improve its *in vivo* activity. Among the THF derivatives, those with the α-configuration are more active than those with the β-configuration, and 3,4′-di-*O*-THF derivatives are more active *in vivo* than the monosubstituted derivatives [197]. Among the 4′-deoxyneospiramycin I, the 12(Z) derivatives are more active *in vitro* than the corresponding 12(E) derivatives [198].

a. SN-41 SN-41, 3,3″,4″-tri-*O*-propionylspiramycin I, is a semisynthetic derivative of spiramycin I. Sano *et al.* prepared a series of 3″,4″-diacylates and 3,3″,4″-tri-*O*-acylates of spiramycin I (Scheme 14) [199]. The acyl derivatives and acetylspiramycin are more active than spiramycin I *in vivo*. 3,4″-*O*-Acetyl-3″-*O*-butyryl spiramycin I and 3,3″,4″-tri-*O*-propionyl spiramycin I (SN-41) are

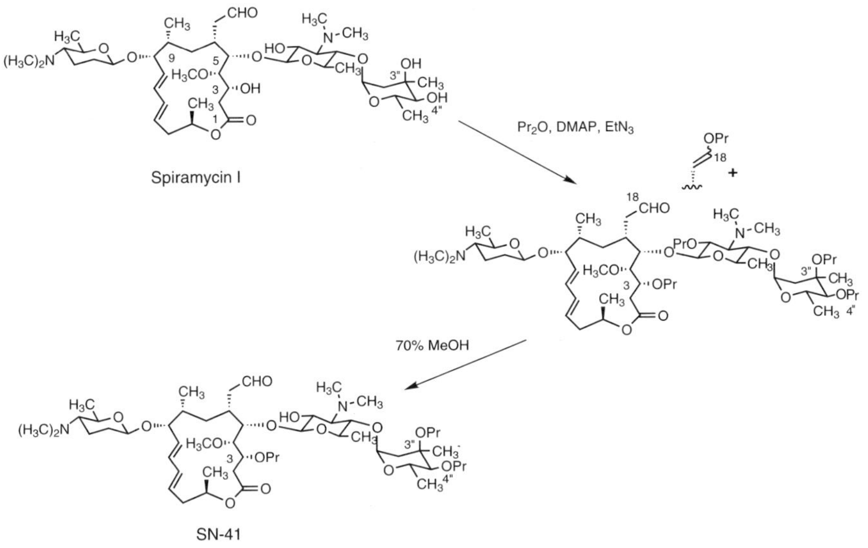

Scheme 14.

more effective therapeutically than acetylspiramycin. SN-41 has less bitter taste than spiramycin I and has been selected for further evaluation [200].

3. Macrolides in the Tylosin Family

Tylosin is the prototype of the second large division of 16-membered macrolides, which differ from the leucomycins in aglycone structure [201]. Tylosin is composed of the usual 16-membered macrocyclic lactone substituted with a disaccharide chain in position 3, composed of an aminosugar, D-mycaminose, and a neutral sugar, L-mycarose. In addition, it is substituted with another neutral sugar, L-mycinose, in position 23.

Tylosin is an important veterinary antibiotic produced by *Streptomyces fradiae*. A preliminary conformational study of tylosin has been recently published [201]. Minor compositions initially found in culture broths were 3'''-O-demethyltylosin (macrocin), demycarosyltylosin (desmycosin) [202], and 20-dihydrotylosin (relomycin) [203]. Other products differing in their degree of glycosylation and/or the oxidation state of their substituents were later isolated as biosynthetic intermediates of shunt metabolites. The abundance of new metabolites has been extensively utilized as starting material for diverse semisynthetic derivatives (Fig. 32)

Various modifications of tylosin have been reported [204]. The following chemical modifications have been made: 4''-Substitution, removal of the neutral sugars, and alteration of the 9-ketone and 20-aldehyde. 4''-O-Acylation of tylosin has been described, and all the derivatives obtained showed improved antimycoplasmal

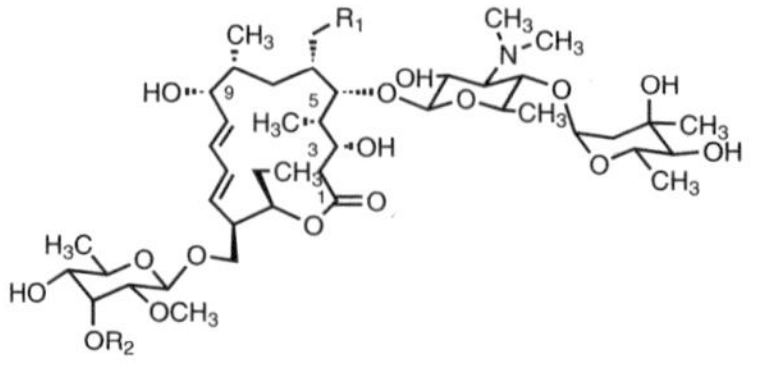

Tylosin R₁=CHO, R₂=CH₃
Relomycin R₁=CH₂OH, R₂=CH₃
Macrosin R₁=CHO, R₂=H

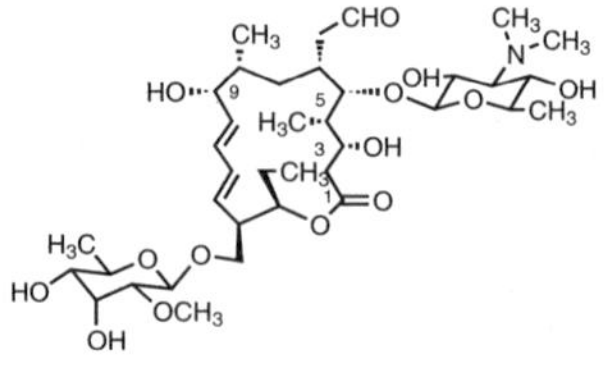

Desmycosin

Fig. 32.

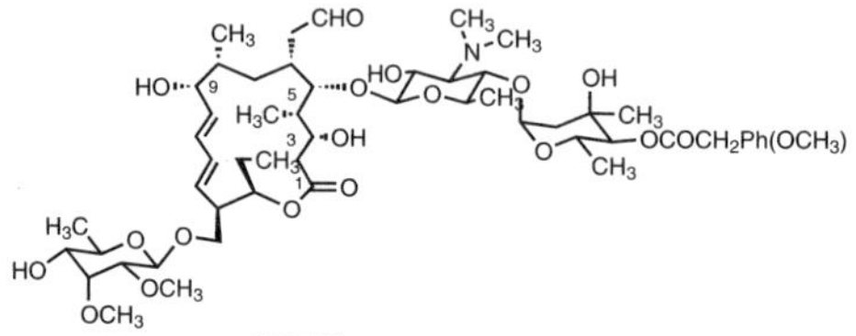

YM 133

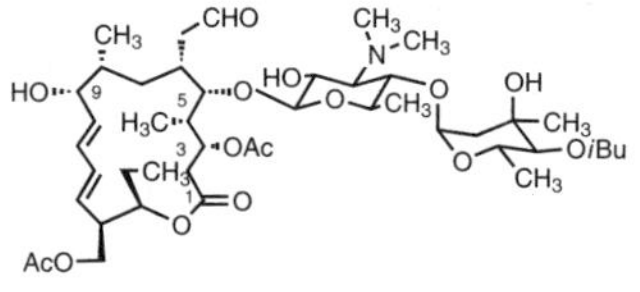

Sch 38646

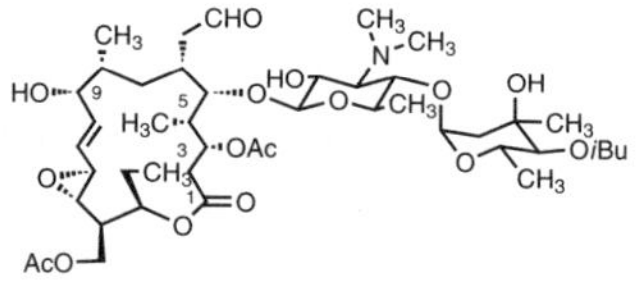

Sch 37644

Fig. 33.

activity *in vitro*. Among them, YM 133 (4″-*O*-(4-methoxyphenyl)acetyl tylosin) [204] was found to be active against inducible-type macrolide-resistant strains. Removal of the mycinose moiety made the derivatives active against macrolide-sensitive microorganisms but reduced their activity against macrolide-resistant strains.

Other closely related acyl demycinosyltylosins, such as 3,23,2′-tri-*O*-acetyl 23-*O*-demycinosyl 4″-*O*-isovaleryltylosin (Sch 37644) and its 12,13-epoxy derivative (Sch 38646) [205], showed improved pharmacokinetics but no advantages over erythromycin A against experimental Gram-positive infections in mice (Fig. 33) [206].

A variety of chemical modifications at the C-20 aldehyde group in tylosin and related macrolides have been reported [206]. The C-20 aldehyde can be radically modified or removed entirely without reducing the antibiotic activity. Reductive amination of C-20 aldehyde of tylosin and related macrolides has yielded a large group of derivatives with useful antibiotic properties. Omura *et al.* reported that 20-deoxy-20-aminotylosin and 20,20′-dideoxo-iminotylosin had lower antibacterial activity [207]. They also synthesized a series of tylosin derivatives carrying secondary and tertiary amino functions at C-20, but these did not show improved potency or a wide antibacterial spectrum relative to tylosin [208]. One of these derivatives, 20-deoxo-20-(3,5-dimethylpiperidin-1-yl) desmycosin (tilmicosin), was selected for further development as a therapeutic agent against pasteurellosis in animals [209]. (Methoxyethoxy)methyloximes of tylosin and demycarosyltylosin were prepared but showed no advantages over tylosin [210]. By analogy with the mycinamicins, a product of the *Micromonospora* sp. that has a methyl group instead of a formylmethyl group in position 6 and shows high antibacterial activity, the deformylation of desmycosin gave rise to better efficacy than that of desmycosin *in vitro* and *in vivo* [211]. The corresponding 4'-deoxy derivative, 19-deformyl-4'-deoxydesmycosin (TMC-016), showed increased *in vivo* activity and better bioavailability after oral administration [212]. Converting the 9-keto group of 16-membered macrolides to amino groups was first proposed for niddamycin [213].

Of the 9-(*N*,*N*-dimethylamino)tetrahydro derivatives, A-66208 showed improved antibacterial activity, and A-65352 was clearly more active than spiramycin and josamycin *in vitro*, except against *L. pneumophila* [214]. This type of modification has been extended to tylosin (A-74950) and rosaramicin (A-72299) [215] to give derivatives whose activities are comparable to that of the parent compound *in vitro*. Some derivatives improved activity significantly in experimental mouse infections.

a. Tilmicosin Modification of the aldehyde group in both tylosin and desmycosin was attempted extensively and was found to increase oral efficacy and bioavailability in experimental animals [206, 208]. Within this series of results, the optimum antimicrobial spectrum, especially including *Pasteurella* species responsible for pneumonia in cattle and pigs, was realized from certain 20-dialkylamino-20-deoxo derivatives synthesized by reductive aminations of desmycosin. Among these compounds, a derivative designated as tilmicosin was selected for commercial development [209]. It has been launched as a new veterinary antibiotic (Scheme 15).

Other reductive aminations of tylosin-related macrolides have been reported. They include the synthesis of dimeric compounds that retain antibacterial activity [207]. Other modifications of the aldehyde group include a variety of hydrozone derivatives, its reduction products, and its C-alkylated analogues [216].

C. The Avermectin Family

Avermectins, which were found by research groups at the Kitasato Institute, headed by Ōmura, and at Merck company, headed by Stapleg, have potent biocidal activities against a wide spectrum of nematodes, insects, and arachnids [217]. The avermectins, when administered orally to sheep, demonstrate activity against gastrointestinal parasites.

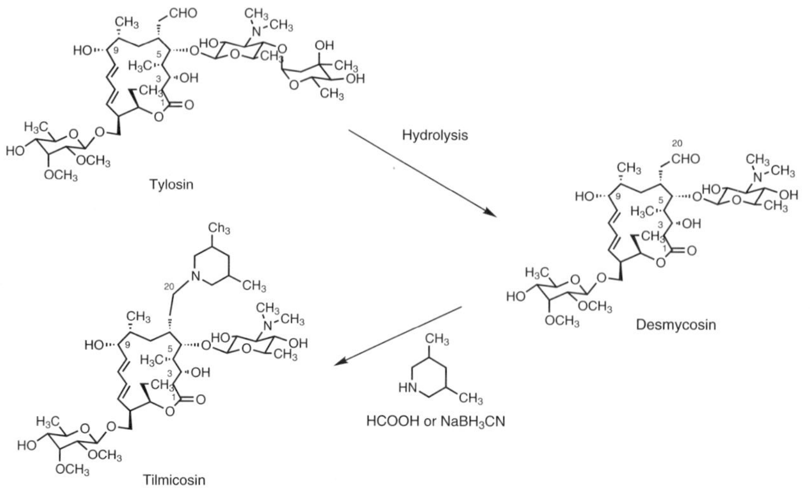

Scheme 15.

They are produced as a mixture of eight components by fermentation of the microbe *Streptomyces avermitilis* [218]. In general, compounds of the B-series, containing a 5′-hydroxy group, are more potent than those of the A-series, containing a 5-methoxy group. The 1-series and 2-series have similar potency against many parasites, but the 2-series compounds showed some important deficiencies that made them less interesting (Fig. 34).

A study was done to compare the anthelminitic activities of the principal naturally occurring avermectins, the monosaccharides and aglycones, and some reduced derivatives (Scheme 16). (Table XXIII).

1. Newer Avermectin Derivatives

Recently, chemical modifications have been carried out with the aim of increasing its insecticidal spectrum, residual activities, and chemical stability. Ivermectin, doramectin, emamectin, eprinomectin, and selamectin have been developed, and KAV-218 has been under development (Fig. 35).

a. Ivermectin Reduction of the 22,23-olefin with Wilkinson's catalyst conferred only a small effect on potency but an overall improvement in spectrum and safety sufficient that 22,23-dihydro avermectin B1 was selected for commercial development under the nonproprietary name ivermectin (Scheme 17) [219].

Ivermectin has found wide use as a systemic antiparasitic agent against endo and ectoparasites of animals. Ivermectin is also used as a treatment for human filarial worm infections (*Onchocerca volvulus*) [220]. The monosaccharides were 2 to 4 times less active than the disaccharide, and dihydroavermectin B1 aglycone was 30 times less potent [219].

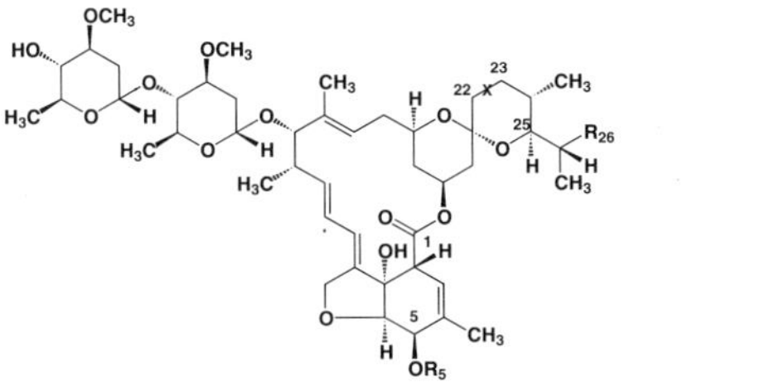

3. Chemical Modification of Macrolides 155

Avermectin	R_5	R_{26}	C_{22}-X-C_{23}
A1a	CH_3	C_2H_5	-CH=CH-
A1b	CH_3	CH_3	-CH=CH-
B1a	H	C_2H_5	-CH=CH-
B1b	H	CH_3	-CH=CH-
A2a	CH_3	C_2H_5	-CH$_2$-CH(OH)-
A2b	CH_3	CH_3	-CH$_2$-CH(OH)-
B2a	H	C_2H_5	-CH$_2$-CH(OH)-
B2b	H	CH_3	-CH$_2$-CH(OH)-
Ivermectin	H	>80% C_2H_5 <20% CH_3	-CH$_2$-CH$_2$-

Fig. 34.

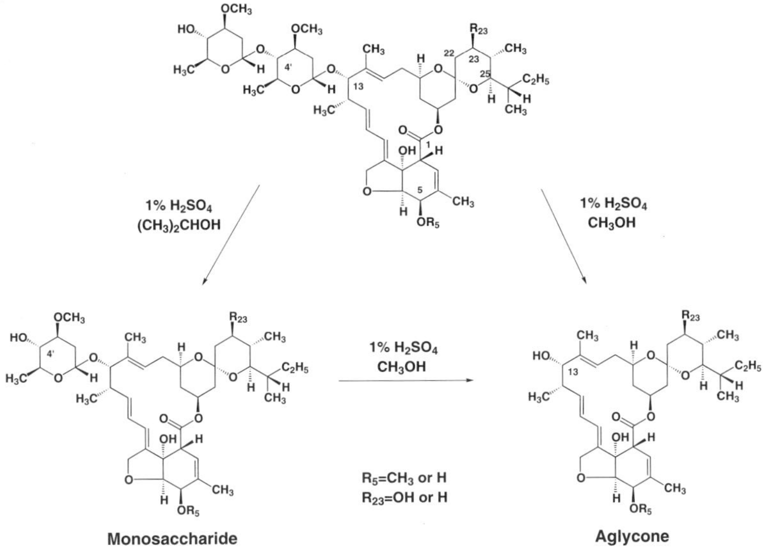

$R_5 = CH_3$ or H
$R_{23} = OH$ or H

Monosaccharide   Aglycone

Scheme 16.

TABLE XXIII
Biological Activities of Aminosubstituted Avermectin Derivatives

Avermectin Analogues	Brine shrimp IC_{100} (ng/ml)	Two-spotted spider mite assay EC_{90} (ppm)	Southern armyworm assay EC_{90} (ppm)
B1 (ABAMECTIN)	263	0.03	8.000
4″-amino-4″-deoxy B_1	2600	0.25	0.100
4″-epi-amino-4″-deoxy B_1	1730	0.25	0.020
4″-epi-amino-4″-deoxy-22,23-dihydro B_1		1.25	0.500
4′-epi-amino-4′-deoxy B_1 monosaccharide		>0.25	>0.500
4″-epi-amino-4″-deoxy-22,23-dihydro B_1 monosaccharide		>0.05	0.100
13-amino-13-deoxy-22,23-dihydro-B_1 aglycone		>0.10	>1.000
4″-epi-CH$_3$NH—	1730	0.25	0.004
4″-epi-(CH$_3$)$_2$N—	1300	>0.05	0.020
4″-epi-(CH$_3$)$_2$CHNH—	1730	0.25	0.020
4″-epi-C$_6$H$_5$CHNH—	>55500	0.25	0.020
4″-epi-H$_3$C(CH$_2$)$_7$NH—		0.25	0.100
4″-epi-CH$_3$CO—NH—	540	0.25	0.500
4″-epi-CH$_3$CO—MeN—	650	0.50	0.050
4″-epi-C$_6$H$_5$CO—NH—	28000		
4″-epi-(CH$_3$)$_2$NNH—	430	0.05	0.100
4″-epi-CH$_3$SO$_2$—NH—	430	0.05	0.100
4″-epi-(CH$_3$)$_2$NCH=NH—		0.25	0.100
4″-epi-HN—CN		>0.05	0.500
4″-epi-N$_3$		>0.05	2.000
4″-epi-CH$_3$O—CO—NH—	430	0.05	0.100
4″-epi-NHCONHCH$_3$		1.25	0.500
4′=N—NHCONH$_2$		0.05	0.500

Ivermectin has been on the market since 1980 as an endectocide for animals. Its special advantage over conventional anthelmintic agents is its wide spectrum against gastrointestinal and systemic parasites as well as against many ectoparasitic insects with a single application. Ivermectin is used at the unprecedented low dose of 200 mg/kg in cattle, sheep, and horses. The small dose allows for

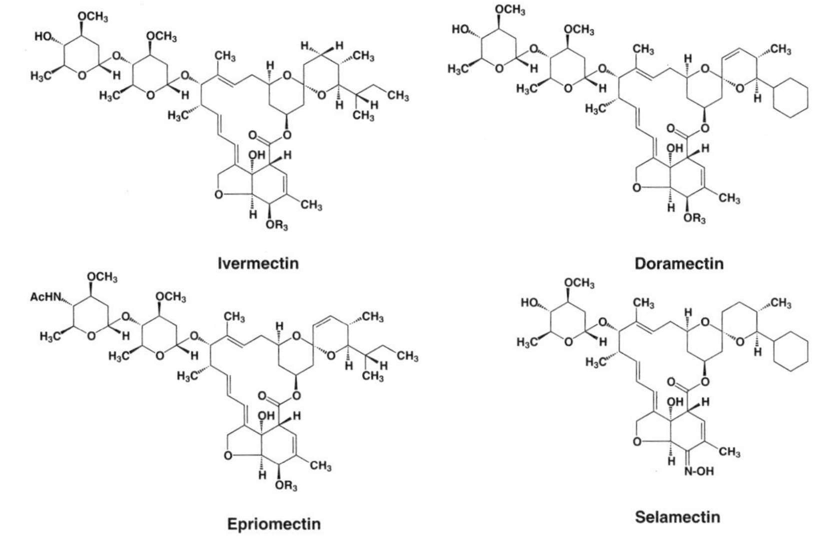

Fig. 35.

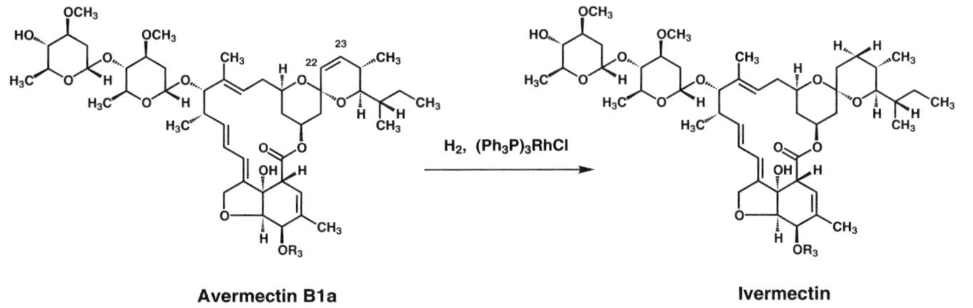

Scheme 17.

easy formulation in oral, parenteral, and topical applications. It is used in swine at 300 mg/kg, the higher dose being required due to higher metabolism in this species. In dogs it is used under the name Heartgard 30 for the prevention of heartworm infections at the low dose of 6 mg/kg and is effective with one single monthly application [221].

It has been used by several hundred thousand humans with a single dose every six months at 50 to 200 mg/kg to alleviate the most damaging symptoms of onchoceriasis, or river blindness. It is highly efficacious against a wide variety of mite, tick, and bot families. Representatives of these, for instance, are responsible for economic losses in cattle or sheep. No activity, however, has been observed against flatworms, protozoa, bacteria, or fungi [220].

b. Doramectin Additional C-25-substituted avermectins were obtained by directed biosynthesis. Incorporation studies revealed that natural avermectins are produced in the fermentation medium from 7-acetate and 5-propionate building blocks, which account for all the carbon atoms except C-25 and the attached C-25 substituent. These atoms are not labeled by acetate or propionate. Instead, it was found that isoleucine or 2-methylbutyrate are incorporated into the 2-butyl group of avermectin B1a [222].

Subsequently, it was shown that addition of 2-methylpentanoate and 2'-methylhexanoate to the avermectin fermentation with *S. avermitilis* gave additional products corresponding to mono and bis homologues of avermectin B1a [223]. A systematic approach by scientists at Pfizer using a mutant of *S. avermitilis* devoid of branched chain 2-oxo acid dehydrogenase, which blocks the fermentation of 2'-methylbutyrate, with the addition of a wide variety of carboxylic acids leads to avermectin derivatives with a modified C-25 side chain (Scheme 18) [224]. Among them, 25-cyclohexyl-25-de(1-methylpropyl)–avermectin B1a (doramectin) was selected. Doramectin possesses broad-spectrum antiparasitic activity *in vitro* and *in vivo* [225].

The activity of 2-epimer of avermectin B1 is approximately 1/100 as potent as avermectin B1, whereas the double bond isomer is 1/5 as potent as avermectin B1 [226]. It has been reported elsewhere that the aromatic derivative of avermectin B1 is devoid of biological activity [227].

A study of the structure–activity relationships of various acylated avemectins was done [227]. The 4"-*O*-acetates have the same potency as the unsubstituted compounds, whereas 5- and 23-acylation are generally detrimental to activity. In particular, the 5-hydroxy group is essential for good anthelmintic activity. The efficacy of acylated avermectins was also investigated against a broad spectrum of parasites in sheep.

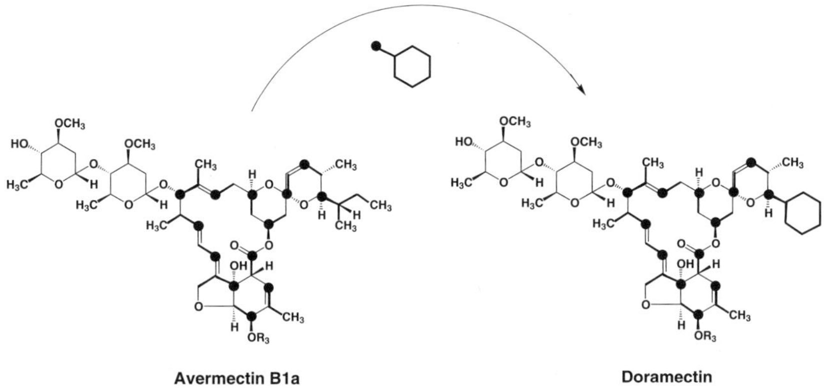

Scheme 18.

It was discovered that derivation of the 4″-hydroxy group retains the high anthelmintic activity against the sheep nematode *Trichstrongylus colubriformis* in a gerbil *in vivo* assay, but the free 5-hydroxy group was required for activity. A number of 4″-esters and carbamates were highly active anthelmintics. Selective protection of the 5- and 7- hydroxy groups is possible, and subsequent reaction of the 4″-hydroxy group with the fluorinating reagent diethylaminosulfur trifluoride (DAST) gave 4″-deoxy-4″-fluoro analogues plus a ring constructed isomer. The fluoro analogues, however, had somewhat reduced biological activity. On the other hand, 4″-epi-hydroxy, oxo, cyano, even deoxy, and in particular methyl analogues have good activity in twospotted spider mite and southern armyworm assays. One also often observed a shift in spectrum after a minor modification, as seen with the southern armyworm activities of the two epimeric 4″-methyl derivatives.

A Merck group was interested in the introduction of a basic amino group into the avermectin molecule [228]. This should change its physical properties, make it more polar, and result in a different tissue distribution. In addition, most of the antibacterially active macrolide antibiotics contain an aminosugar. After suitable protection and deprotection, reductive amination of a 4″-oxo intermediate gave the 4″-amino-4″-deoxy analogue as the major reaction product (Scheme 19). Since they also had observed better activity in the southern armyworm assay for certain monosaccharides, they prepared their 4′-amino derivatives as well as the 13-amino aglycone, but they did not observe further improvement of biological activity. They then looked at alkylamino analogues and found the monomethyl analogue to be the most promising of the group. Further modifications with acylamino, ureido, or hydrazono substituents resulted in mostly potent derivatives, and some of these are of considerable interest as endectocides for animal health, but none surpassed the epi-monomethyl derivatives in potency against lepidoptera species. Among them, emamectin and eprinomectin have been developed.

c. Emamectin 4″-Deoxy-4″-epi-*N*-methylaminoavermectin B1, or MK 244, with the generic name emamectin, is currently commercially available as an agricultural insecticide, particularly against lepidoptera species [229]. It is reserved initially for high value crops, since it will be a rather expensive compound. The foliar ingestion toxicities of MK 244 for a number of important insects are from 0.003 to 0.03 ppm (LD_{90}) for tobacco hornworm, cabbage looper, beet armyworm, fall armyworm, and Colorado potato beetle, 0.2 to 0.3 ppm for Mexican bean beetle and twospotted spider mite, and as high as 20 ppm for bean aphid. Emamectin showed topical potency superior to that of the amino analogue against the three lepidoptera species of southern armyworm, tobacco budworm, and corn earworm. Emamectin was considerably more potent in a foliar residue test against southern armyworm and tobacco budworm than thiodicarb or fenvalerate [230].

d. Eprinomectin Eprinomectin is a semisynthetic derivative of avermectin B1a. Eprinomectin (MK 397, or 4″-epi-acetylamino-4″-deoxy-avermectin B1a) is a novel avermectin selected for development as a topical endectocide for all cattle,

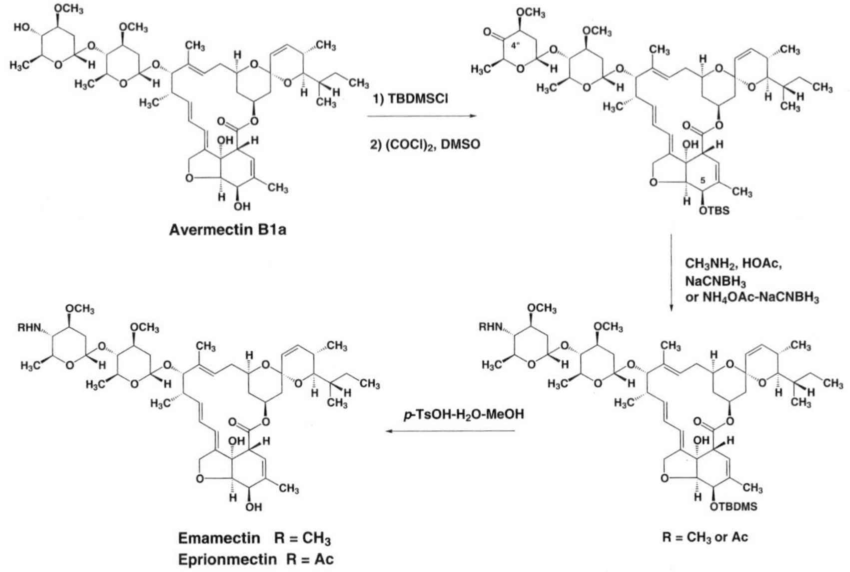

Scheme 19.

including lactating cows [229]. It has been launched commercially in several countries. The initial efficacy assessments were made in sheep to identify subclasses of the avermectin/milbemycins that possessed inherent activity against a spectrum of namatode parasites. Eprinomectin possessed potent activity against the range of nematodes when tested at 0.025 mg kg^{-1} per day. Milk and plasma concentration profiles were also made for these and other selected avermectin/milbemycins following topical admistration to lactating dairy cattle [231]. Therefore, not only is eprinomectin the most potent broad spectrum avermectin/milbemycin to date, but it also possesses one of the lowest milk partitioning coefficients in this class of antiparasitics [232].

  e. *Selamectin* Avermectin and milbemycin are well established for the treatment of a wide range of both endoparasite and ectoparasite infections of livestock and hence are referred to as endectocides. Although safe for use as an endectocide in livestock, ivermectin produces idiosyncratic toxicity in collie dogs and crossbred collies at doses required to give effective control of gastrointestinal nematodes. Its use in dogs, at the very low dose required for safety in all dog breeds, provides only prophylactic control of heartworm. Milbemycin oxime has also been commercialized. The compound's greater safety permits doses that provide additional control of gastrointestinal nematodes, but it is ineffective against fleas, the major ectoparasite of commercial importance for cats and dogs.

  A Pfizer group evaluated a wide range of avermectin derivatives for flea activity in an *in vitro* feeding screen using the cat flea *Ctenocephalides felis*, which

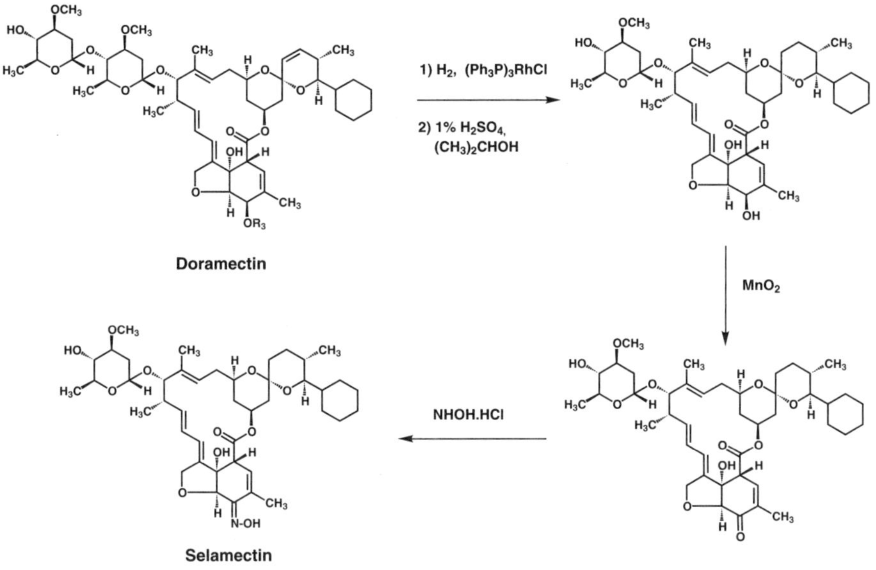

Scheme 20.

revealed a narrow structure–activity relationship, with activity surprisingly associated with monosaccharides and especially their C-5 oximes (Scheme 20) [233]. They discovered commercially exploitable flea activity in a single compound, selamectin, which also possessed the necessary antiparasitic spectrum and margin of safety for development as a broad-spectrum endectocide for companion animals (cats and dogs) [234].

f. KAV-218 Ōmura *et al.* developed a way to introduce the polar functional group (ex. cyano, amino, ester, carboxylic acid) at C-4″ on oleandorose by a Horner-Emons reaction, as shown in Scheme 21 [235]. Selective protection of the 5-hydroxy group was followed by Swern oxidation of 4″-hydroxy to obtain the 4″-oxo derivative, then Horner-Emons olefination with various phosphonoacetates to afford the Z-olefin as the major reaction product (Scheme 21). Among these derivatives, KAV-218 was the most active derivative, showing stronger activity than ivermectin and doramectin against a spectrum of ectoparasites and endoparasites *in vitro* and *in vivo* [235].

2. *Newer Milbemycin Derivatives*

Milbemycins, which were discovered by a Sankyo group, were isolated from *Streptomyces hygroscopicus* subspecies *aureolacrimosus*. Milbemycins exhibit exceptionally potent acaricidal and insecticidal activities. First, they were produced as a mixture of 13 components (α_1–α_{10} and β_1–β_3) by fermentation of strain SANK 60,576 [236]. Furthermore, the mutant strain SANK 60,576 produced milbemycin D, E, F, G, and H in high yield [237].

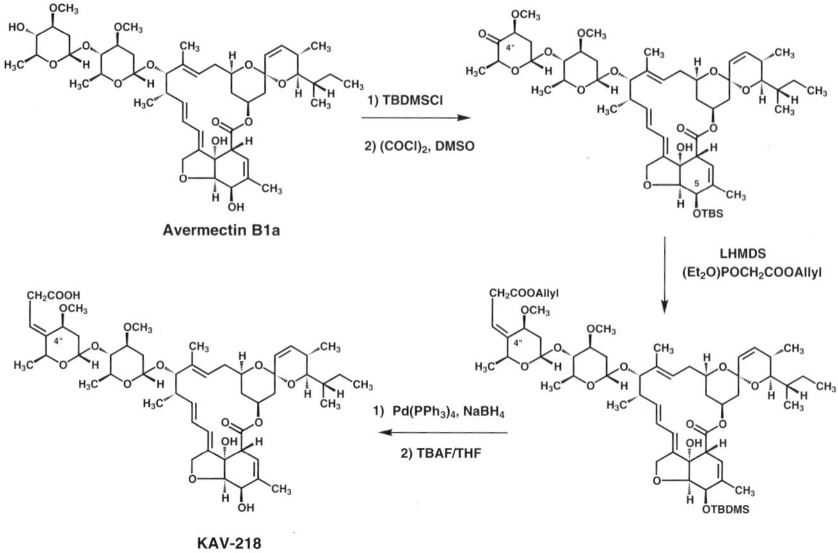

Scheme 21.

In 1990, milbemectin [a mixture of milbemycin A3 (α_1) and A4 (α_3)] was launched as an acaricide on tea and eggplant. In the animal health field, milbemycin D was developed for the control of fatal canine heartworm disease, caused by *Dirofilaria immitis*, and launched in Japan in 1986. A mixture of the oxime derivatives of milbemycin A3 and A4 (milbemycin oxime) is used as a canine parasiticide.

On the other hand, an American Cyanamid group isolated milbemycin analogues, nemadectins α, β, γ, and λ from *Streptomyces cyaneogriseus* sp. *noncyanogenus* in 1987 [238]. The main congener, nemadectin α, is currently used as starting material for the production of commercial moxidectin used in veterinary medicine (Fig. 36).

a. Milbemycin 5-Oxime The structure of milbemycin D includes hydroxyl groups at the C-5 and C-7 positions, isolated double bonds at the C-3 and C-14 positions, and a conjugated diene system from C-8 to C-11. These functional groups provide the clues for modifying these complex molecules. During a study on the chemistry of milbemycins, a Sankyo group has been especially interested in the chemical transformation of the allylic hydroxy group at the 5-position [239]. Starting from milbemycin D, A$_4$, and A$_3$, they synthesized a series of 5-keto-5-oxime derivatives. The C-5 hydroxyl group was oxidized to ketone by manganese oxide (Mn$_2$O) in nearly quantitative yield. Reaction of these α, β-unsaturated with ketones hydroxylamine hydrochloride in dioxane-methanol-water gave a series of 5-hydroxyiminomilbemycins in high yield (Scheme 22). The structure–activity

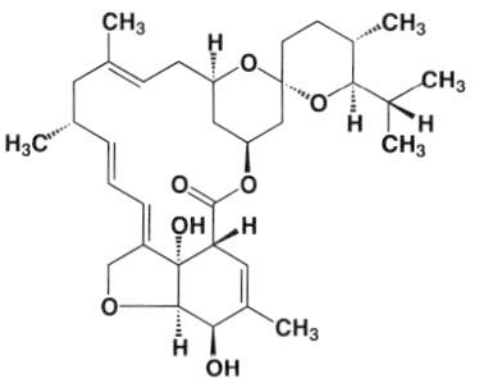

Milbemycin D

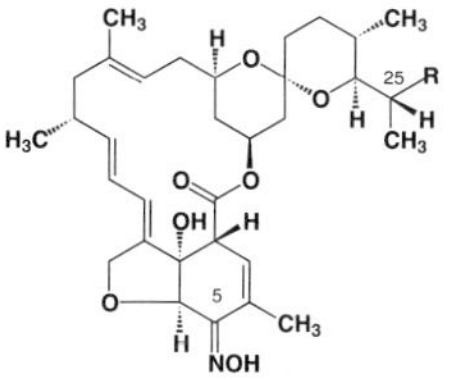

Milbemycin A₃/A₄ 5-oxime
R₂₅ = CH₃ and CH₄

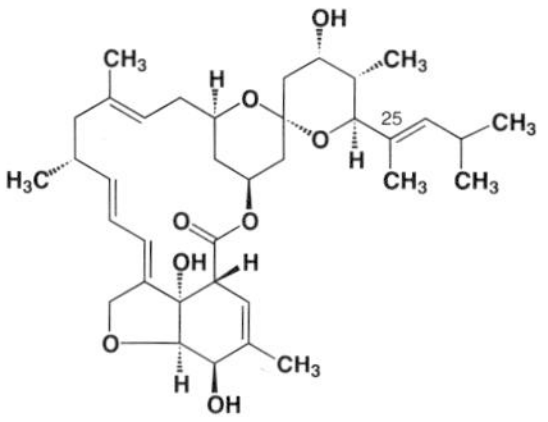

Nemadectin

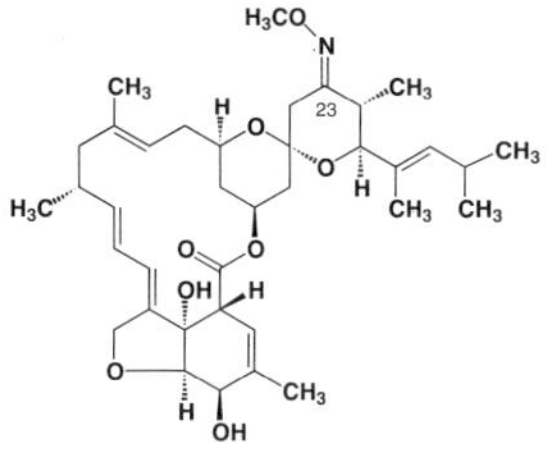

Moxidectin

Fig. 36.

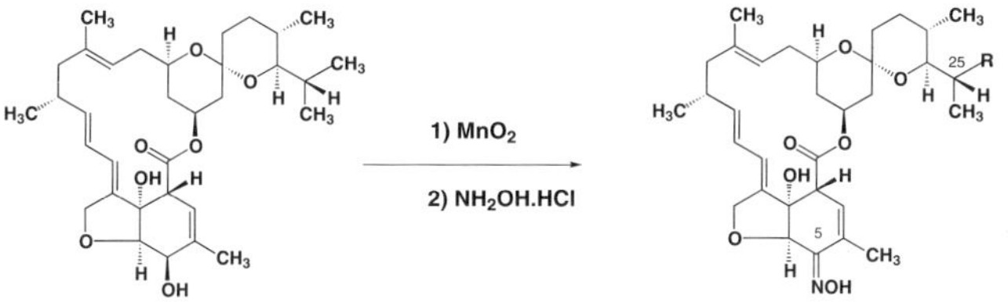

Milbemycin A₃/A₄
R₂₅ = CH₃ and C₂H₅

1) MnO₂
2) NH₂OH·HCl

Milbemycin A₃/A₄ 5-oxime
R₂₅ = CH₃ and C₂H₅

Scheme 22.

relationships at the 5-position of milbemycins showed that the hydroxyimino group in the 5-oxime turned out not only to act as a bioisoster of the hydroxy group but also to potentiate activity against microfilariae [240].

Consequently, in consideration of practical and economical aspects, a mixture of 5-hydroxyimino derivatives of milbemycin A_4 and A_3 ($A_4:A_3$ = 80:20) was judged to be the most promising candidate for development and was finally launched as a parasiticide for dogs.

b. Moxidectin A new family of antiparasitic macrolides has been isolated from *S. cyaneogriseus* sp. *noncyanogenus* by American Cyanamid [238]. The compounds, designated nemedectin (LL-F28249) α, β, γ, and λ, possess potent antiparasitic activity. The nemadectins are distinguished from the avermectins and the milbemycins by the presence of an unsaturated side chain at C-25 and by

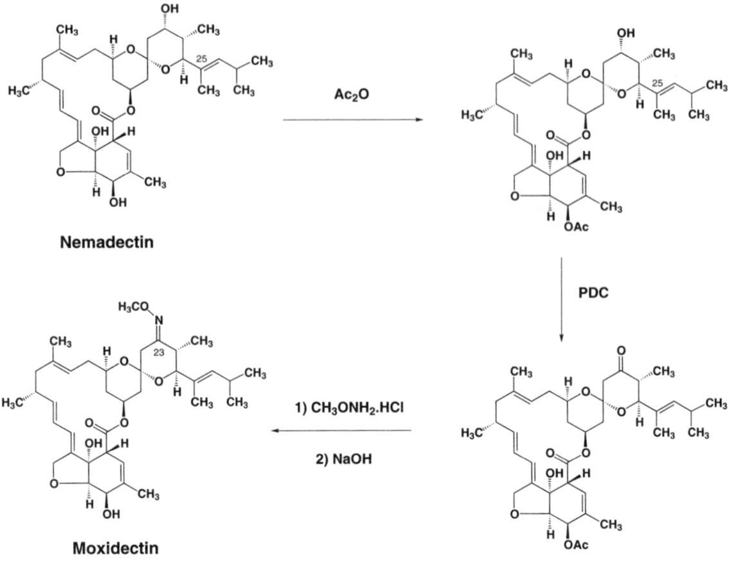

Scheme 23.

the hydroxy group at C-23. Nemadectin d, the main component, was launched as an agricultural insecticide, particularly against lepidoptera species.

Furthermore, while studing the chemistry of nemadectins, the American Cyanamide group became interested in the chemical modification of the C-5, C-7, and C-23 hydroxy groups. The secondary C5-OH is allylic and pseudo-equatorial and, therefore, can be functionalized more easily than the axial-OH at C-23 [241]. From such investigation with nemadectins, moxidectin was synthesized. Acetylation of the hydroxy group at C-5 was followed by pyridinium dichromate oxidation to afford C-23 ketone. Treatment with methoxyamine hydrochloride and sodium acetate in methanol-water followed by alkali hydrolysis gave moxidectin (Scheme 23) [242]. Moxidectin possessed the necessary antiparasitic spectrum and margin of safety for development as a broad-spectrum companion animal (cats and dogs) endectocide.

IV. Concluding Remarks

Macrolides with 14- and 16-membered rings have been productive sources of semisynthetic derivatives that have significantly extended the utility of the macrolide class as important antibiotics. New semisynthetic derivatives have exhibited a variety of improved features, such as an expanded antimicrobial spectrum, increased potency, greater efficacy, better oral bioavailability, extended chemical and metabolic stability, higher and more prolonged concentration in tissues and fluids, lower and less frequent dosing and diminished side effects, and antimicrobial activity against resistant strains. Similarly, new avemectin derivatives have been developed.

As new biotechnologies such as those promised by molecular biology and genetic engineering emerge, applications to the field of macrolide antibiotics are also expected. Newer semisynthetic macrolides that satisfy these important needs should be anticipated as the contributions from new fields, such as the genetic engineering of macrolide-producing organisms and more powerful computational chemistry, are combined with the more traditional disciplines of chemical synthesis, bioconversion, and screening fermentation broths.

Although the second activity, gastrointestinal prokinetic activity, is already being developed [243], other potentially promising areas of activity include anti-inflammatory agents and a variety of ways to influence and modulate the immune system. Consequently, it is a very interesting and potentially useful area of research for the discovery of new therapeutic agents.

In summary, research on structural modification of macrolide antibiotics appears to have a bright future.

References

1. Sakakibara, H., and Ōmura, S. (1984). Modification and structure–activity relationship. In "Macrolide Antibiotics: Chemistry, Biology, and Practice" (S. Omura, Ed.), pp. 85–126. Academic Press, New York.
2. Kurath, P., Jones, P. H., Egan, R. S., and Perun, T. J. (1971). Acid degradation of erythromycin A and erythromycin B. *Experientia* **27**, 362.
3. Chantot, J. F., Gasc, J. C., Gouin D'Ambrieres, S., and Lutz, A. (1983). New ether oxime derivatives of erythromycin A: Preparation and antibacterial activities. Presented at 23rd Intersci. Conf. Antimicrob. Agents Chemother. (Las Vegas, NV). Abstr. No. 447.
4. Kirst, H. A. (1990). Structure modification of macrolide antibiotics. In "Recent Progress in the Chemical Synthesis of Antibiotics" (G. Lukacs and M. Ohno, Eds.), pp. 40–63. Springer-Verlag, Heidelberg.
5. Chantot, J.-F., Bryskier, A., and Gasc, J.-C. (1986). Antibacterial activity of roxithromycin: A laboratory evaluation. *J. Antibiot.* **39**, 660–668.
6. Gasc, J.-C., D'Ambrieres, S. G., Lutz, A., and Chantot, J. F. (1991). New ether oxime derivatives of erythromycin A. A structure–activity relationship study. *J. Antibiot.* **44**, 313–330.
7. Morimoto, S., Takahashi, Y., Watanabe, Y., and Ōmura, S. (1984). Chemical modification of erythromycins. I. Synthesis and antibacterial activity of 6-O-methylerythromycins A. *J. Antibiot.* **37**, 187–189.
8. Bachet, B., Brassy, C., and Mornon, J. P. (1988). [O-(2,5-Dioxahexyl)oxime]-9 de l'erythromycine A hydratee. *Acta Cryst. C* **44**, 112–116.
9. Girault, J. P., Gharbi, J., Laoui, A., Artaud, I., and Delaforge, M. (1989). Conformational analysis in solution of roxithromycin, a new macrolide antibiotic. Presented at 16th Int. Congr. Chemother. (June 11–16, Jerusalem).

10. Delaforge, M., Sartori, E., and Mansuy, D. (1988). *In vivo* and *in vitro* effects of a new macrolide antibiotic roxithromycin on rat liver cytochrome P-450: Comparison with trolenadomycin and erythromycin. *Chem. Biol. Interact.* **68**, 179–188.
11. Manuel, C., Dellamonica, P., Rosset, M. J., Safran, C., Pirot, D., Audegond, L., and Pechere, J. C. (1988). Penetration of roxithromycin (ROX) into brain tissue (BT). Presented at 28th Intersci. Conf. Antimicrob. Agents Chemother. (Oct. 23–26, Los Angeles). Abstr. No. 1224.
12. Bergogne-Berezin, E. (1987). Tissue distribution of roxithromycin. *J. Antimicrob. Chemother.* **20** (Suppl. B), S113–S120.
13. Esumi, Y., Zin, Y., Ban, S., Ninomiya, S., Hayashi, K., Yokoshima, T., Yamamoto, T., Hirayama, M., Saitoh, K., and Okui, K. (1988). Absorption, distribution, metabolism and excretion of RU 28965 in animals. *Jpn. J. Chemother.* **36** (S-4), 148–163.
14. Koyama, M., and Tateno, M. (1988). Absorption, metabolism and excretion of RU 28,965 in humans. *Jpn. J. Chemother.* **36** (S-4), 164–183.
15. Morimoto, S., Misawa, Y., Adachi, T., Nagate, T., Watanabe, Y., and Omura, S. (1990). Chemical modification of erythromycins. II. Synthesis and antibacterial activity of *O*-alkyl derivatives of erythromycin A. *J. Antibiot.* **43**, 286–294.
16. Omura, S., Morimoto, S., Nagate, T., Adachi, T., and Kono, Y. (1992). Research and development of clarithromycin. *Yakugaku Zasshi.* **112**, 593–614.
17. Morimoto, S., Nagate, T., Sugita, K., Ono, T., Numata, K., Miyachi, J., Misawa, Y., Yamada, K., and Omura, S. (1990). Chemical modification of erythromycins. III. *In vitro* and *in vivo* antibacterial activities of new semisynthetic 6-*O*-methylerythromycin A, TE-031 (clarithromycin) and TE-032. *J. Antibiot.* **43**, 295–305.
18. Morimoto, S., Adachi, T., Misawa, Y., Nagate, T., Watanabe, Y., and Omura, S. (1990). Chemical modification of erythromycins. IV. Synthesis and biological properties of 6-*O*-methylerythromycin B. *J. Antibiot.* **43**, 544–549.
19. Morimoto, S., Adachi, T., Kashimura, M., Matsunaga, T., Asaka, T., Yoshida, T., Amano, T., Watanabe, Y., and Hatayama, K. (1992). Synthesis of erythromycin derivatives. IX. Novel procedure of clarithromycin (6-*O*-methylerythromycin A) synthesis. Presented at 112th Ann. Meet. Pharm. Soc. Jpn. (March 29–31, Fukuoka). Abstr. No. ZH 9-14.
20. Goto, H., Kawashima, Y., Kashimura, M., Morimoto, S., and Osawa, E. (1993). Origin of regioselectivity in the *O*-methylation of erythromycin as elucidated with the aid of computational conformational space search. *J. Chem. Soc. Perkin Trans. II* 1647–1654.
21. Adachi, T., Morimoto, S., Kondoh, H., Nagate, T., Watanabe, Y., and Sota, K. (1988). 14-Hydroxy-6-*O*-methyl erythromycins A, active metabolites of 6-*O*-methylerythromycin A in humans. *J. Antibiot.* **41**, 966–975.
22. Sasaki, J., Mizoue, K., Morimoto, S., Adachi, T., and Omura, S. (1988). Microbial transformation of 6-*O*-methylerythromycin derivatives. *J. Antibiot.* **41**, 908–915.
23. Suwa, T., Yoshida, H., Fukushima, K., and Nagate, T. (1988). Comparative pharmacokinetics of TE-031 and erythromycin stearate in rats and mice. *Jpn. J. Chemother.* **36** (S-3), 198–204.
24. Piccolomini, R., Di Bonaventura, G., Picciani, C., Laterza, F., Vecchiet, J., and Neri, M. (2001). *In vitro* activity of clarithromycin against intracellular *Helicobacter pylori*. *Antimicrob. Agents Chemother* **45**, 1568–1571.
25. Rastogi, N., Goh, K. S., Berchel, M., and Bryskier, A. (2000). *In vitro* activities of the ketolides telithromycin (HMR 3647) and HMR 3004 compared to those of clarithromycin against slowly growing mycobacteria at pHs 6.8 and 7.4. *Antimicrob. Agents Chemother.* **44**, 2848–2852.
26. Djokic, S., Kobrehel, G., and Lazarevski, G. (1987). Erythromycin series. XII. Antibacterial *in vitro* evaluation of 10-dihydro-10-deoxo-11-azaerythromycin A: Synthesis and structure–activity relationship of its acyl derivatives. *J. Antibiot.* **40**, 1006–1015.
27. Bright, G. M., Nagel, A. A., Bordner, J., Desai, K. A. Dibrino, J. N., Nawakowska, J., Vincent, L., Watrous, R. M., Sciavolino, F. C., English, A. R., Retsema, J. A., Anderson, M. R., Brennan,

L. A., Borovoy, R. J., Cimochowski, C. R., Faiella, J. A., Girard, A. E., Girard, D., Herbert, C., Manousos, M., and Mason, R. (1988). Synthesis, *in vitro* and *in vivo* activity of novel 9-deoxo-9a-aza-9a-homoerythromycin A. *J. Antibiot.* **41**, 1029–1047.

28. Retsema, J., Girard, A., Schelkly, W., Manousos, M., Anderson, M., Bright, G., Borovoy, R., Brennan, L., and Mason, R. (1987). Spectrum and mode of action of azithromycin (CP-62,993), a new 15-membered-ring macrolide with improved potency against gram-negative organisms. *Antimicrob. Agents Chemother.* **31**, 1939–1947.

29. Girard, A. E., Girard, D., English, A. R., Gootz, T. D., Cimochowski, C. R., Faiella, J. A., Haskell, S. L., and Retsema, J. A. (1987). Pharmacokinetic and *in vivo* studies with azithromycin (CP-62,993), a new macrolide with an extended half-life and excellent tissue distribution. *Antimicrob. Agents Chemother.* **31**, 1948–1954.

30. Matsunaga, T., Shimohira, H., Oishi, S., Ogawa, M., and Goto, S. (1995). *In vitro* and *in vivo* antimicrobial activity of azithromycin, a new macrolide antibiotic. *Jpn. J. Chemother.* **43** (S-6), 68–83.

31. Djokic, S., Kobrehel, G., Lopotar, N., Kamenar, B., Nagl, A., and Mrvos, D. (1988). Erythromycin series. Part 13. Synthesis and structure elucidation of 10-dihydro-10-deoxo-11-methyl-11-azaerythromycin A. *J. Chem. Res.* (S), 152–153.

32. Barber, J. (1991). Assignments of the ^{13}C and ^{1}H NMR spectra of azithromycin in CDCl$_3$. *Magn. Reson. Chem.* **29**, 740–743.

33. Goldman, R. C., Fesik, S. W., and Doran, C. C. (1990). Role of protonated and neutral forms of macrolides in binding to ribosomes from gram-positive and gram-negative bacteria. *Antimicrob. Agents Chemother.* **34**, 426–431.

34. Massey, E. H., Kitchell, B., Martin, L. D., Gerzon, K., and Murphy, H. W. (1970). Erythromycylamine. *Tetrahedron Lett.* 157–160.

35. Timms, G. H., and Wildsmith, E. (1971). The reduction of oximes with tervalent titanium, a mild deoxymation procedure, and the partial synthesis of erythromycylamine. *Tetrahedron Lett.* 195–198.

36. Wildsmith, E. (1972). The reaction of erythromycin hydrazone with nitrous acid, a new route to erythromycylamine. *Tetrahedron Lett.* 29–30.

37. Massey, E. H., Kitchell, B. S., Martin, L. D., and Gerzon, K. (1974). Antibacterial activity of 9(*S*)-erythromycylamine-aldehyde condensation products. *J. Med. Chem.* **17**, 105–107.

38. Ryden, R., Timms, G. H., Prime, D. M., and Wildsmith, E. (1973). *N*-Substituted derivatives of erythromycylamine. *J. Med. Chem.* **16**, 1059–1060.

39. Cockerill, A. F., Ellis, M. F., Rackham, D. M., and Wildsmith, E. (1973). Substituent effects on the pKa values and rates of hydrolysis of arylmethylene-erythromycylamines. *J. Chem. Soc. Perkin Trans II* 173–179.

40. Maier, R., Woitun, E., Wetzel, B., and Lechner, U. (1988). Synthesis, structure, and biological activity of tetrahydro-1,3-oxazines derived from 9(*S*)-erythromycylamine and substituted acetaldehydes. Presented at 28th Intersci. Conf. Antimicrob. Agents Chemother (Oct. 23–26, Los Angeles). Abstr. No. 917.

41. Kirst, H. A., Wind, J. A., Leeds, J. P., Willard, K. E., Debono, M., Bonjouklian, R., Greene, J. M., Sulliran, K. A., Paschal, J. W., Deeter, J. B., Jones, N. D., Ott, J. L., Felty-Duckworth, A. M., and Counter, F. T. (1990). Synthesis and structure–activity relationships of new 9-*N*-alkyl derivatives of 9(*S*)-erythromycylamine. *J. Med. Chem.* **33**, 3086–3094.

42. Kirst, H. A. (1995). Dirithromycin: Introduction and historical development. *Drugs Today* **31**, 89–97.

43. Colbert, W. E., Turk, J. A., Williams, P. D., and Buening, M. K. (1991). Cardiovascular and autonomic pharmacology of the macrolide antibiotic LY281389 in anesthetized beagles and in isolated smooth and cardiac muscles. *Antimicrob. Agents Chemother.* **35**, 1365–1369.

44. Counter, F. T., Ensminger, P. W., Preston, D. A., Wu, C.-Y. E., Greene, J. M., Felty-Duckworth, A. M., Paschal, J. W., and Kirst, H. A. (1991). Synthesis and antimicrobial evaluation of

dirithromycin (AS-E 136: LY237216), a new macrolide antibiotic derived from erythromycin. *Antimicrob. Agents Chemother.* **35**, 1116–1126.
45. McGill, J. M. (1993). Generation of stable synthetic equivalents of unstable α–α alkoxyacetaldehydes: An improved preparation of dirithromycin. *Synthesis* 1089–1091.
46. Davis, J. S., Everett, J. R., Hatton, I. K., Hunt, E., Tyler, J. W., Zomaya, I. I., Slawin, A. M. Z., and Williams, D. J. (1991). NMR-spectroscopic and X-ray crystallographic studies on the structure, stereochemistry and conformation of a series of 9,11-cyclic aminals of (9S)-9-N-methylerythromycylamine. *J. Chem. Soc. Perkin Trans. II* 201–214.
47. Luger, P., and Marier, R. (1979). Molecular structure of 9-deoxy-11-deoxy-9,11-(imino(2-(2-methoxyethoxy)ethylidene)oxy)-(9S)-erythromycin, a new erythromycin derivative. *J. Cryst. Mol. Struct.* **9**, 329–338.
48. Kirst, H. A., Greene, J. M., and Amos, J. G. (1992). Synthesis, characterization, epimerization and antimicrobial evaluation of epidirithromycin and related 9-N-11-O-oxazines of 9(S)-erythromycylamine. Presented at 32nd Intersci. Conf. Antimicrob. Agents Chemother (Oct. 11–14, Anaheim, CA). Abstr. No. 1365.
49. Toscano, L. and Seghetti, E. (1983). Transformation of 3-O-mycarosylerythronolide B, an intermediate of erythromycin biogenesis, into 3-O-mycarosyl-(8S)-8-fluoroerythronolide B using trifluoromethyl hypofluorite. *Tetrahydron Lett.* **24**, 5527–5530.
50. a) Toscano, L., Fioriello, G., Silingardi, S., and Inglesi, M. (1984). Preparation of (8S)-8-fluoroerythronolide A and (8S)-8-fluoroerythronolide B, potential substrates for the biological synthesisi of new macrolide antibiotics. *Tetrahedron.* **40**, 2177–2181.
    b) Toscano, L., Fioriello, G., Spagnoli, R., Cappelletti, L., and Zanuso, G. (1983). New fluorinated erythromycins obtained by mutasynthesis. *J. Antibiot.* **36**, 1439–1450.
51. Saverino, D., Debbia, E. A., Pesce, A., Lepore, A. M., and Schito, G. C. (1986). Antibacterial profile of flurithromycin, a new macrolide. *J. Antimicrob. Chemother.* **30**, 261–272.
52. Gialdroni Grassi, G., Alesina, R., Bersani, C., Ferrara, A., Fietta, A., and Peona, V. (1986). In vitro activity of flurithromycin, a novel macrolide antibiotic. *Chemotherapia* **5**, 177–184.
53. Benoni, G., Cuzzolin, L., Leone, R., Consolo, U., Ferronato, G., Bertrand, C., Puchetti, V., and Fraccaso, M. E. (1988). Pharmacokinetics and human tissue penetration of flurithromycin. *Antimicrob. Agents Chemother.* **32**, 1875–1878.
54. Benazzo, M., Giacopini, G., Oldini, C., Scheiber, E., Tombolini, A., and Mira, E. (1998). Flurithromycin versus clarithromycin in upper respiratory tract infections. *Curr. Ther. Res.* **59**, 28–38.
55. Matsunaga, T., and Ogawa, M. (1995). Antibacterial activity of azithromycin against fresh clinical isolates. *Jpn. J. Chemother.* **43** (S-6), 84–94.
56. Hardy, J. H., Hensey, D. M., Beyer, J. M., Vojtko, C., McDonald, E. J., and Fernandes, P. B. (1988). Comparative *in vitro* activities of new 14-, 15-, and 16-membered macrolides. *Antimicrob. Agents Chemother.* **32**, 1710–1719.
57. Matsunaga, T., Shimohira, H., Oishi, S., Ogawa, M., and Goto, S. (1995). *In vitro* and *in vivo* antimicrobial activity of azithromycin, a new macrolide antibiotic. *Jpn. J. Chemother.* **43** (S-6), 68–83.
58. Barry, A. L., Jones, R. N., and Thornsberry, C. (1988). *In vitro* activities of azithromycin (CP-62,993), clarithromycin (A-56268; TE-031), erythromycin, roxithromycin, and clindamycin. *Antimicrob. Agents Chemother.* **32**, 752–754.
59. Cunha, B. A. (1996). Newer macrolide antibiotics: Advantages and uses. *Adv. Therapy* **13**, 29–37.
60. Kirst, H. A., and Sides, G. D. (1989). New directions for macrolide antibiotics: Pharmacokinetics and clinical efficacy. *Antimicrob. Agents Chemother.* **33**, 1419–1422.
61. Muto, H., Kuboe, Y., Kimura, Y., Saito, S., Sekiguchi, K., Enogaki, K., and Shimooka, K. (1995). Pharmacokinetic study of azithromycin in experimental animals. *Jpn. J. Chemother.* **43** (S-6), 110–121.

62. Neu, H. C. (1993). Activity of macrolides against common pathogens *in vitro*. In "Macrolides—Chemistry, Pharmacology, and Clinical Uses" (A. J. Bryskier, J. P. Butzler, H. C. Neu, and P. M. Tulkens, Eds.), pp. 167–182. Arnette Blackwell, Paris.
63. Clark, R. F., Or, Y. S., Ma, Z., Wang. S., Tafano, M., Yong, H., Chu, D. T., Nilius, A., Bui, M.-H., and Raney, P. (1998). Novel 6-*O*-substituted erythromycin A derivatives: Synthesis and antibacterial activity. Presented at 38th Intersci. Conf. Antimicrob. Agents Chemother. (Sept. 24–27, San Diego). Abstr. No. F-125.
64. McAlpine, J. B., Swanson, S. J., Whittern, D. N., Buko, A. M., and Weber, J. M. (1990). 6-Deoxyerythromycins. I. Producing organisms, isolation, separation and structure elucidation. Presented at 30th Intersci. Conf. Antimicrob. Agents Chemother. (Oct. 21–24, Atlanta, GA). Abstr. No. 810.
65. Clement, J. J., Hanson, C., Shipkowitz, N., Hardy, D., and Swanson, R. (1990). 6-Deoxyerythromycins. II. Comparative *in vitro* activity and *in vivo* efficacy. Presented at 30th Intersci. Conf. Antimicrob. Agents Chemother (Oct. 21–24, Atlanta, GA). Abstr. No. 811.
66. Hauske, J. R., Kostek, G., and Guadliana, M. (1984). Regiospecific synthesis of 9-deoxoerythromycin A. *J. Org. Chem.* **49**, 712–714.
67. Hardy, D. J., Swanson, R. N., Shipkowitz, N. L., Freiberg, L. A., Larty, P. A., and Clement, J. J. (1991). *In vitro* activity and *in vivo* efficacy of a new series of 9-deoxo-12-deoxy-9,12-epoxyerythromycin A derivatives. *Antimicrob. Agents Chemother.* **35**, 922–928.
68. Faghih, R., Buytendorp, M., Stephens, R., Hardy, D., Plattner, J., and Lartey, P. (1990). Synthesis and antibacterial activity of (9*S*)-9-dihydroclarithromycin. *J. Antibiot.* **43**, 1334–1336.
69. Lartey, P. A., Deninno, S. L., Faghih, R., Clement, J. J., and Plattner, J. J. (1991). Effect of C-21 modifications on the antibacterial activity of (9*R*)–9-deoxo-9-dimethylamino erythromycin A. Presented at 31st Intersci. Conf. Antimicrob. Agents Chemother. (Sept. 29–Oct. 2, Chicago). Abstr. No. 382.
70. Lartey, P. A., DeNinno, S. L., Faghih, R., Hardy, D. J., Clement, J. J., and Plattner, J. J. (1992). Synthesis and activity of C-21 alkylamino derivatives of (9*R*)-erythromycylamine. *J. Antibiot.* **45**, 380–385.
71. Klein, L. L., Freiberg, L. A., Kurath, P., and Henry, R. F. (1993). Synthesis of (9*R*)-9-dihydro-6,9-anhydroerythromycin A. *J. Org. Chem.* **58**, 3209–3212.
72. Kobrehel, G., Lazarevski, G., Dokic, S., Kolacny-Babic, L., Kucisec-Tepes, N., and Cvrlje, M. (1992). Synthesis and antibacterial activity of *O*-methylazithromycin derivatives. *J. Antibiot.* **45**, 527–534.
73. Waddell, S. T., Santorelli, G. M., Blizzard, T. A., Graham, A., and Occi, J. (1998). Synthesis and antibacterial activity of *O*-methyl derivatives of azalide antibiotics: I. 4″,11 and 12-*O*-Me via direct methylation. *Bioorg. Med. Chem. Lett.* **8**, 549–554.
74. Waddell, S. T., Santorelli, G. M., Blizzard, T. A., Graham, A., and Occi, J. (1998). Synthesis and antibacterial activity of *O*-methyl derivatives of azalide antibiotics: II. 6-*O*-Me derivatives via clarithromycin. *Bioorg. Med. Chem. Lett.* **8**, 1321–1326.
75. Wilkening, R. R., Ratcliffe, R. W., Szymonifka, M. J., Shankaran, K., May, A. M., Blizzard, T. A., Heck, J. V., Herbert, C. M., Graham, A. C., and Bartizal, K. (1993). Synthesis and *in vitro* activities of 9-deoxo-8a-aza-8a-homoerythromycin derivatives, a new series of azalide antibiotics. Presented at 33rd Intersci. Conf. Antimicrob. Agents Chemother. (Oct. 17–20, New Orleans, LA). Abstr. No. 426.
76. Kujundzic, N., Kobrehel, G., and Kelneric, Z. (1995). 9a-*N*-(*N*′-Carbamoyl) and 9a-*N*-(*N*′-thiocarbamoyl) derivatives of 9-deoxo-9a-aza-9a-homoerythromycin A. Patent EP657464-A.
77. Kobrehel, G., Lazarevski, G., and Djokic, S. (1994). 9-Deoxo-9a-aza-11-deoxy-9a-homoerythromycin A 9a, 11-cyclic carbamates. Patent EP606062-A1.
78. Jones, A. B. (1992). New macrolide antibiotics: Synthesis of a 14-membered azalide. *J. Org. Chem.* **57**, 4361–4367.

79. Hunt, E., Knowles, D. J. C., Shillingford, C., and Zomaya, I. I. (1988). Erythromycin A 11,12-methylene acetal. *J. Antibiot.* **41**, 1644–1648.
80. Baker, W. R., Fernandes, P. B., Bopp, B., Marsh, K., Nellans, H., Clark, J., Herrin, T., and Hannick, S. (1987). Synthesis and biological activity of erythromycin A 11,12-cyclic carbamates. Presented at 27th Intersci. Conf. Antimicrob. Agents Chemother (Oct. 4–7, New York). Abstr. No. 221.
81. Baker, W. R., Clark, J. D., Stephens, R. L., and Kim, K. H. (1988). Modification of macrolide antibiotics. Synthesis of 11-deoxy-11(carboxyamino)-6-*O*-methylerythromycin A 11,12-(cyclic esters) via an intramolecular Michael reaction of *O*-carbamates with an α,β-unsaturated ketone. *J. Org. Chem.* **53**, 2340–2345.
82. Fernandes, P. B., Baker, W. R., Freiberg, L. A., Hardy, D. J., and McDonald, E. J. (1989). New macrolides active against *Streptococcus pyogenes* with inducible or constitutive type of macrolide-lincosamide-streptogramin B resistance. *Antimicrob. Agents Chemother.* **33**, 78–81.
83. Wilson, J. M., Hannan, P. C. T., Shillingford, C., and Knowles, D. J. C. (1989). Biological properties of ER 42859, a novel erythromycin derivative. *J. Antibiot.* **42**, 454–462.
84. Brain, E. G., Forrest, A. K., Hunt, E., Shillingford, C., and Wilson, J. M. (1989). Erythromycin A oxime 11,12-carbonate and its oxime ethers. *J. Antibiot.* **42**, 1817–1822.
85. Morimoto, S., Misawa, Y., Kondoh, H., Watanabe, Y., and Omura, S. (1990). Chemical modification of erythromycins. V. Synthesis and antibacterial activity of 4″-*O*-methyl derivatives of erythromycin A 11,12-cyclic carbonate. *J. Antibiot.* **43**, 566–569.
86. Freiberg, L. A., Klein, L., Hannick, S., Nellans, H. N., Fernandes, P. B., and Pernet, A. G. (1987). A-63075, a potent 14-membered macrolide with minimal gastrointestinal side-effects. Presented at 27th Intersci. Conf. Antimicrob. Agents Chemother (Oct. 4–7, New York). Abstr. No. 224.
87. Bryskier, A. (1999). New research in macrolides and ketolides since 1997. *Exp. Opin. Invest. Drugs* **8**, 1171–1194.
88. Su, W. G. (1999). Novel macrolide C-4″ carbamate derivative: Chemistry and structure activity relationship. Presented at 3rd International Antibacterial Drug Discovery Development (March 8–9, Princeton).
89. Su, W. G., Smyth, K. T., Rainville, J. P., Kaneko, T., Sutcliffe, J. A., Brennan, L. A., Duignan, J. M., Girard, A. E., Finegan, S. M., and Cimochowski, C. R. (1998). Novel erythromycylamine 4″-carbamate antibacterial agents: Synthesis and biological evaluation resulting in discovery of CP-554,372. Presented at 38th Intersci. Conf. Antimicrob. Agents Chemother. (Sept. 24–27, San Diego). Abstr. No. F-122.
90. Wu, Y. J., Wons, R., Durkin, Jr., D., Goldsmith, M., Su, W.G., Rainville, J., Smyth, K., Yang, B. V., Massa, M., Kane, J., Brighty, K., Blair, K., Monahan, R., Sutcliffe, J., Brennan, L., Duiagnan, J., Pettipas, J., Tait-Kamratd, A., Linde, R., Miller, S., Kaneko, T., Keaney, M., Birsner, N., Brickner, S., Roache, J. Masamune, H., Tickner, J., and Vazquez, E. (1998). Synthesis and *in vitro* activity of novel C-4″ carbamate derivatives of 14- and 15-membered macrolides. Presented at 38th Intersci. Conf. Antimicrob. Agents Chemother. (Sept. 24–27, San Diego). Abstr. No. F-123.
91. Girard, D., Mathieu, H. W., Girard, A. E., Finegan, S. M., Menard, C. A., Su, W. G., Smith, K. T., and Rainville, J. P. (1998). *In vivo* oral efficacy of CP-554,372, a new erythromycylamine C-4″ carbamate, against macrolide resistant pneumococci and *H. influenzae*. Presented at 38th Intersci. Conf. Antimicrob. Agents Chemother. (Sept. 24–27, San Diego). Abstr. No. F-121
92. Yee, S., Finegan, S. M., Girard, A. E., Su, W. G., Smyth, K. T., Rainville, J. P., Gernert, L. M., and Regan, C. M., (1998). Pharmacokinetics of CP-544,372 in preclinical species. Presented at 38th Intersci. Conf. Antimicrob. Agents Chemother. (Sept. 24–27, San Diego). Abstr. No. F-119.
93. Girard, D., Mathieu, H. W., Girard, A. E., Menard, C. A., Su, W. G., Smyth, K. T., and Rainville, J. P. (1998). Pharmacokinetics and pharmacodynamics of CP-554,372, a new erythromycylamine C-4″ carbamate, in murine pneumococcal pulmonary infections. Presented at 38th Intersci. Conf. Antimicrob. Agents Chemother. (Sept. 24–27, San Diego). Abstr. No. F-120.

94. Pacey, M. S., Dirlam, J. P., Geldart, R. W., Leadley, P. F., McArthur, H. A., McCormick, E. L., Monday, R. A., O'connell, T. N., Staunton, J., and Winchester, T. J. (1998). Novel erythromycins from a recombinant *Saccharopolyspora erythraea* strain NRRL 2338 pIG1. I. Fermentation, isolation, and biological activity. *J. Antibiot.* **51**, 1029–1034.
95. Parsons, I. C., Everett, J. R., Pacey, M. S., Ruddock, J. C., Swanson, A. G., and Thompson, C. M. (1999). Structural elucidation of a novel erythromycin, 13-cyclopentyl-13-desethyl erythromycin B, from a recombinant *Saccharopolyspora erythreaea* strain, NRRL 2338pIG/1. *J. Antibiot.* **52**, 190–192.
96. Asaka, T., Tanikawa, T., Ishii, T., and Kashimura, M. (1998). Preparation of erythromycin A derivatives as bactericides. Patent WO9813373.
97. Elliott, R. L., Pireh, D., Griesgraber, G., Nilius, A. M., Ewing, P. J., Bui, M. H., Raney, P. M., Flamm, R. K., Kim, K., Henry, R. F., Chu, D. T. W., Plattner, J. J., and Or, Y. S. (1998). Anhydrolide macrolides. 1. Synthesis and antibacterial activity of 2,3-anhydro-6-*O*-methyl 11,12 carbamate erythromycin A analogues. *J. Med. Chem.* **41**, 1651–1659.
98. Griesgraber, G., Kramer, M. J., Elliott, R. L., Nilius, A. M., Ewing, P. J., Raney, P. M., Bui, M. H., Flamm, R. K., Chu, D. T. W., Plattner, J. J., and Or, Y. S. (1998). Anhydrolide macrolides. 2. Synthesis and antibacterial activity of 2,3-anhydro-6-*O*-methyl 11,12- carbazate erythromycin A analogues. *J. Med. Chem.* **41**, 1660–1670.
99. Asaka, T., Kashimura, M., Ishii, T., Matsuura, A., Suzuki, K., Ohyauchi, R., Matsumoto, K., Numata, K., Akashi, T., Adachi, T., and Morimoto, S. (1997). New macrolide antibiotics, acylides (3-*O*-acyl-5-*O*-dedosaminyl-6-*O*-methylerythronolides); Synthesis and biological properties. Presented at 37th Intersci. Conf. Antimicrob. Agents Chemother. (Sept. 28–Oct. 1, Toronto). Abstr. No. F-262.
100. Asaka, T., Kashimura, M., Manaka, A., Tanikawa, T., Ishii, T., Sugimoto, T., Suzuki, K., Sugiyama, H., Akashi, T., Saito, H., Adachi, T., and Morimoto, S. (1999). Structure activity studies leading to potent acylides: 3-*O*-Acyl-5-*O*-desosaminylerythronolide 11,12-carbamates. Presented at 39th Intersci. Conf. Antimicrob. Agents Chemother. (Sept. 26–29, San Francisco). Abstr. No. 2159.
101. Elliott, R. L., Pireh, D., Nilius, A. M., Johnson, P. M., Flamm, R. K., Chu, D. T. W., Plattner, J. J., and Or, Y. S. (1997). Novel 3-deoxy-3-descladinosyl-6-*O*-methyl erythromycin A analogues. Synthesis and *in vitro* activity. *Bioorg. Med. Chem. Lett.* **7**, 641–646.
102. Edwards, C., and Lartey, P. A. (1995). Antibacterial compounds comprising C-3″ derivatives of erythromycin A. Patent WO9501794.
103. Ackland, M. J., Atkins, P. J., and Jones, N. B. (1991). Approaches to novel water-soluble prodrugs of erythromycin A. Synthesis of 2′-*O*-(*N*-alkylsuccinamoyl)-erythromycin derivatives incorporating anionic and cationic groups. *J. Chem. Res.* (M), 1265–1278.
104. Barthelemy, P., Autissier, D., Gerbaud, G., and Courvalin, P. (1984). Enzymatic hydrolysis of erythromycin by a strain of *Escherichia coli*. A new mechanism of resistance. *J. Antibiot.* **37**, 1692–1696.
105. Kirst, H. A., Wind, J. A., and Paschal, J. W. (1987). Synthesis of ring-contracted derivatives of erythromycin. *J. Org. Chem.* **52**, 4359–4362.
106. Kibwage, I. O., Busson, R., Jannsen, G., Hoogmartens, J., Vanderhaeghe, H., and Bracke, J. (1987). Translactonization in erythromycins. *J. Org. Chem.* **52**, 990–996.
107. Adachi, T. (1989). 15-Membered macrolides via translactonization in 14-hydroxy-6-*O*-methylerythromycin A. *J. Org. Chem.* **54**, 3507–3510.
108. Allen, N. E. (1977). Macrolide resistance in *Staphylococcus aureus*: Inducers of macrolide resistance. *Antimicrob. Agents Chemother.* **11**, 669–674.
109. Agouridas, C., Denis, A., Auger, J.-M., Benedetti, Y., Bonnefoy, A., Bretin, F., Chantot, J.-F., Dussarat, A., Fromentin, C., D'Ambrieres, S. G., Lachaud, S., Laurin, P., Martret, O. L., Loyau, V., and Tessot, N. (1998). Synthesis and antibacterial activity of ketolides (6-*O*-methyl-3-oxo-erythromycin derivatives): A new class of antibacterials highly potent against macrolide-resistant and susceptible respiratory pathogens. *J. Med. Chem.* **41**, 4080–4100.

110. Agouridas, C., Benedetti, Y., Denis, A., LeMartret, O., and Chantot, J. F. (1995). Ketolides: A new distinct class of macrolide antibacterials—Synthesis and structural characteristics of RU 004. Presented at 35th Intersci. Conf. Antimicrob. Agents Chemother. (Sept. 17–20, San Francisco). Abstr. No. F-157.
111. Asaka, T., Kashimura, M., Misawa, Y., Ono, T., Suzuki, K., Yoshida, H., Matsumoto, K., Akashi, T., Nagate, T., and Morimoto, S. (1995). A new macrolide antibiotic, TE-810: Synthesis and biological properties. Presented at 35th Intersci. Conf. Antimicrob. Agents Chemother. (Sept. 17–20, San Francisco). Abstr. No. F-177.
112. Ma, Z., Clark, R. F., Wang, S., Nilius, A. M., Flamm, R. K., and Or, Y. S. (2000). Design, synthesis and characterization of ABT-773. Presented at 5th ICMAS-KO (Jan. 26–28, Seville). Poster No. 02.03.
113. Blizzard, T. A., Waddell, S. T., Santorelli, G. M., and Morgan, J. D. (1999). Preparation of 9a-aza-3-ketolide erythromycin analogs as bactericides. Patent WO9900125.
114. Blizzard, T. A., Morgan, J. D., Santorelli, G. M., and Waddell, S. T., (1999). Preparation of 9a-aza-3-ketolide erythromycins as antibacterial agents. Patent GB2327084.
115. Asaka, T., Kashimura, M., Misawa, Y., Ono, T., Suzuki, K., Yoshida, H., Yoshida, T., Akashi, T., Yokoo, C., Nagate, T., and Morimoto, S. (1995). A new macrolide antibiotic, TE-802: Synthesis and biological properties. Presented at 35th Intersci. Conf. Antimicrob. Agents Chemother. (Sept. 17–20, San Francisco) Abstr. No. F-176.
116 Kashimura, M., Asaka, T., Misawa, Y., Matsumoto, K., and Morimoto, S. (2001). Synthesis and antibacterial activity of the tricyclic ketolides TE-802 and its analogs. *J. Antibiot.* **54**, 664–678.
117. Phan, T., Or, Y. S., Spina, K. P., Chen, Y., Tufano, M., Chu, D. T. W., Nilius, A. M., Bui, M.-H., and Plattner, J. J. (1997). Tricyclic ketolides, mono-substitution on the imine ring. Synthesis and *in vitro* antibacterial activity. Presented at 37th Intersci. Conf. Antimicrob. Agents Chemother. (Sept. 28–Oct. 1, Toronto). Abstr. No. F-263.
118. Phan, T., Or, Y. S., Chen, Y., Chu, D. T. W., Ewing, P., Nilius, A. M., Bui, M.-H., Raney, P. M., Hensey-Rudloff, D., Mitten, M., Henry, R. F., and Plattner, J. J. (1998). 2-Substituted tricyclic ketolides: New antibacterial macrolides—Synthesis and biological activity. Presented at 38th Intersci. Conf. Antimicrob. Agents Chemother. (Sept. 24–27, San Diego). Poster No. F-127.
119. Phan, L. T., Or, Y. S., Chen, Y., Chu, D. T. W., Nilius, A. M., Bui, M.-H., and Plattner, J. J. (1997). Tetracyclic ketolides, a new antibacterial macrolide. Synthesis and *in vitro* antibacterial activity. Presented at 37th Intersci. Conf. Antimicrob. Agents Chemother. (Sept. 28–Oct. 1, Toronto). Abstr. No. F-264.
120. Denis, A., Agouridas, C., Bonnefoy, A., Bretin, F., Fromentin, C., and Bonnet, A. (1999). Synthesis and antibacterial activity of 2-halogeno, 2-methyl, and 2,3 enol-ether ketolides using β-keto-ester chemistry. Presented at 39th Intersci. Conf. Antimicrob. Agents Chemother. (Sept. 26–29, San Francisco). Abstr. No. 2152.
121. Bonnefoy, A., Denis, A., Bretin, F., Fromentin, C., and Agouridas, C. (1999). *In vitro* antibacterial activity of novel 2-fluoro ketolides. Presented at 39th Intersci. Conf. Antimicrob. Agents Chemother. (Sept. 26–29, San Francisco). Abstr. No. F-2153.
122. Bonnefoy, A., Denis, A., Bretin, F., Fromentin, C., and Agouridas, C. (1999). *In vivo* antibacterial activity of two ketolides HMR 3562 and HMR 3787 highly active against respiratory pathogens. Presented at 39th Intersci. Conf. Antimicrob. Agents Chemother. (Sept. 26–29, San Francisco). Abstr. No. F-2156.
123. Kaneko, T., McMillen, W., Sutcliffe, J., Duignan, J., and Petitpas, J. (2000). Synthesis and *in vitro* activity of C2-substituted C9-oxime ketolide. Presented at 40th Intersci. Conf. Antimicrob. Agents Chemother. (Sept. 17–20, Toronto). Abstr. No. 1815.
124. Girard, D., Mathieu, H. W., Finegan, S. M., Cimochowski, C. R., Kaneko, T., and McMillen, W. (2000). *In vitro* antibacterial activity of CP-654,743, a new C2-fluoro ketolide, against

macrolide-resistant pneumococci and *Haemophilus influenzae*. Presented at 40th Intersci. Conf. Antimicrob. Agents Chemother. (Sept. 17–20, Toronto). Abstr. No. F-1816.

125. Denis, A., Agouridas, C., Auger, J.-M., Benedetti, Y., Bonnefoy, A., Bretin, F., Chantot, J.-F., Dussarat, A., Fromentin, C., D'Ambrieres, S. G., Lachaud, S., Laurin, P., Martret, O.L., Loyau, V., Tessot, N., Pejac, J.-M., and Perron, S. (1999). Synthesis and antibacterial activity of HMR 3647, a new ketolide highly potent against erythromycin-resistant and susceptible pathogens. *Bioorg. Med. Chem. Lett.* **9**, 3075–3080.

126. Malathum, K., Coque, T. M., Singh, K. V., and Murray, B. E. (1999). *In vitro* activities of two ketolides, HMR 3647 and HMR 3004, against gram-positive bacteria. *Antimicrob. Agents Chemother.* **43**, 930–936.

127. Pankuch, G. A., Jacobs, M. R., and Applebaum, P. C. (1997). Antipneumococcal activity (MIC) of HMR 3647 (RU 66647), a new ketolide, compared to 16 other agents. Presented at 37th Intersci. Conf. Antimicrob. Agents Chemother. (Sept. 28–Oct. 1, Toronto). Abstr. No. F-108.

128. Bebear, C. M., Renaudin, H., Bryskier, A., and Bebear, C. (2000). Activity of telithromycin (HMR 3647), a new ketolide antimicrobial, against human mycoplasma. Presented at 5th ICMAS-KO (Jan. 26–28, Seville). Poster No. 02.35.

129. Roblin, P. M., and Hammerschlag, M. R. (1998). *In vitro* activity of a new ketolide antibiotic, HMR 3647, against *Chlamydia pneumoniae*. *Antimicrob. Agents Chemother.* **42**, 1515–1516.

130. Mauvaris, P., and Bonnefoy, A. (2000). Lack of *in vitro* MLS$_B$ resistance induction by the ketolide telithromycin (HMR 3647): Role of the 3-keto group. Presented at 5th ICMAS-KO (Jan. 26–28, Seville). Poster No. 02.10.

131. Douthwaite, S., Mauvaris, P., and Haastrup Hansen, L. (2000). The C 11,12-carbamate side chain of the ketolide antimicrobial telithromycin (HMR 3647) enhances binding to both MLS-sensitive and -resistant ribosome. Presented at 5th ICMAS-KO (Jan. 26–28, Seville). Poster No. 02.01.

132. Mankin, A., Xiong, L., and Khaitovich, P. (2000). Interaction of macrolides and ketolides with the ribosome. Presented at 5th ICMAS-KO (Jan. 26–28, Seville). Poster No. 01.02.

133. Douthwaite, S., Mauvaris, P., Champney, S., and Bryskier, A. (2000). Structure–activity relationship of the ketolide telithromycin (HMR 3647). Presented at 5th ICMAS-KO (Jan. 26–28, Seville). Poster No. 02.02.

134. Lenfant, B., Sultan, E., Wable, C., Pascual, M. H., and Meyer, B. H. (2000). Pharmacokinetics of 800 mg once-daily oral dosing of the ketolide antibacterial telithromycin (HMR 3647) in healthy young subjects. Presented at 5th ICMAS-KO (Jan. 26–28, Seville). Poster No. 09.21.

135. Ma, Z., Or, Y. S., Clark, R. F., Wang, S., Brazzale, A., Yong, H., Tufano, M., Nilius, A. M., Bui, M.-H., Raney, P., Flamm, R. K., Chu, D. T. W., and Plattner, J. J. (1998). Synthesis and antibacterial activity of 6-*O*-substituted ketolides. Presented at 38th Intersci. Conf. Antimicrob. Agents Chemother. (Sept. 28–Oct. 1, San Diego). Abstr. No. F-126.

136. Ma, Z., Clark, R. F., and Or, Y. (1999). Design, synthesis, and characterization of ABT-773: A novel ketolide highly active against multidrug-resistant pathogens. Presented at 39th Intersci. Conf. Antimicrob. Agents Chemother. (Sept. 26–29, San Francisco). Abstr. No. 2133.

137. Shortridge, D., Ramer, N. C., Beyer, J., Ma, Z., Or, Y., and Flamm, R.K. (2000). The *in vitro* activity of ABT-773 against gram-positive respiratory pathogens. Presented at 5th ICMAS-KO (Jan. 26–28, Seville). Poster No. 02.17.

138. Shortridge, D., Ramer, N. C., Beyer, J., Ma, Z., Or, Y., and Flamm, R. K. (2000). The *in vitro* activity of ABT-773 against gram-negative respiratory pathogens. Presented at 5th ICMAS-KO (Jan. 26–28, Seville). Poster No. 02.28.

139. Mitten, M., Meulbroek, J., Paige, L., Nukkala, M., Mollison, K. W., Nilius, A. M., Flamm, R. K., Ma, Z., and Djuric, S. (2000). *In vivo* evaluations of ABT-773 and HMR-3647 against respiratory pathogens in murine and rat models of infection. Presented at 5th ICMAS-KO (Jan. 26–28, Seville). Poster No. 03.10.

140. Strigl, S., Roblin, P. M., Reznik, T., and Hammerschlag, M. R. (2000). *In vitro* activity of ABT-773, a new ketolide antibiotic, against *Chlamydia pneumoniae*. Presented at 5th ICMAS-KO (Jan. 26–28, Seville). Poster No. 02.32.
141. Dubois, J., and St. Pierre, C. (2000). Intracellular activity and post-antibiotic effect of ABT-773 against *Legionella*. Presented at 5th ICMAS-KO (Jan. 26–28, Seville). Poster No. 02.33.
142. Cynamon, M. H., Carter, J. L., and Shoen, C. M. (2000). Activity of ABT-773 against *Mycobacterium avium* complex in the Beige mouse model. Presented at 5th ICMAS-KO (Jan. 26–28, Seville). Poster No. 03.14.
143. Khan, A. A., Araujo, F. G., and Craft, J. C. (2000). The ketolide ABT-773 protects against lethal *Toxoplasma gondii*. Presented at 5th ICMAS-KO (Jan. 26–28, Seville). Poster No. 03.07.
144. Hernandez, L., Krill, S., Ma, Z., and Marsh. K. (2000). Pharmacokinetics of ABT-773 in mouse, rat, monkey and dog. Presented at 5th ICMAS-KO (Jan. 26–28, Seville). Poster No. 09.33.
145. Brueggemann, A. B., Doern, G. V., Huynh, H. K., Wingert, E. M., and Rhomberg, P. R. (2000). *In vitro* activity of ABT-773, a new ketolide, against recent clinical isolaters of *Streptococcus pneumoniae, Haemophilus influenzae*, and *Moraxella catarrhalis*. *Antimicrob. Agents Chemother.* **44**, 447–449.
146. Yamaguchi, T., Hirakata, Y., Izumikawa, K., Miyazaki, Y., Maesaki, S., Tomono, K., Yamada, Y., Kamihira, S., and Kohno, S. (2000). *In vitro* activity of telithromycin (HMR 3647), a new ketolide, against clinical isolates of *Mycoplasma pneumoniae* in Japan. *Antimicrob. Agents Chemother.* **44**, 1381–1382.
147. Strigl, S., Robin, P. M., Renznik, T., and Hammerschlag, M. R. (2000). *In vitro* activity of ABT-773, a new ketolide antibiotic, against *Chlamydia pneumoniae*. *Antimicrob. Agents Chemother.* **44**, 1112–1113.
148. Saez-Nieto, J. A., and Vazquez, J. A. (1999). *In vitro* activities of ketolide HMR 3647 and HMR 3004, Levofloxacin, and other quinolones and macrolides against *Neisseria* spp. and *Moraxella catarrhalis*. *Antimicrob. Agents Chemother.* **43**, 983–984.
149. a) Cynamon, M. H., Carter, J. L., and Shoen, C. M. (2000). Activity of ABT-773 against *Mycobacterium avium* complex in the Beige mouse model. *Antimicrob. Agents Chemother.* **44**, 2895–2896.
    b) Consigny, S., Bentoucha, A., Bonnafous, P., Grosset, J., and Ji, B. (2000). Bactericidal activities of HMR 3647, moxifloxacin, and rifapentine against *Mycobacterium leprae* in mice. *Antimicrob. Agents Chemother.* **44**, 2919–2921.
150. Pechere, J. (2001). New perspectives on macrolide antibiotics. *Int. J. Antimicrob. Agents* **18**, (Suppl. 1), 93–97.
151. Rolain, J.-M., Maurin, M., Bryskier, A., and Raoult, D. (2000). *In vitro* activities of telithromycin (HMR3647) against *Rickettsia rickettsii, Rickettsia conorii, Rickettsia africae, Rickettsia typhi, Rickettsia prowazekii, Coxiella burnetii, Bartonella henselae, Bartonella quintana, Bartonella bacilliformis*, and *Ehrlichia chaffeensis*. *Antimicrob. Agents Chemother.* **44**, 1391–1393.
152. Gia, H. P., Roeder, V., Namour, F., Sultan, E., and Lenfant, B. (2000). Telithromycin (HMR 3647) achieves high and sustained concentrations in white blood cells in man. Presented at 5th ICMAS-KO (Jan. 26–28, Seville). Poster No. 09.27.
153. Goldman, R. C., Zakula, D., Flamm, R., Beyer, J., and Capobianco, J. (1994). Tight binding of clarithromycin, its 14-(R)-hydroxy metabolite, and erythromycin to *Helicobacter pylori* ribosomes. *Antimicrob. Agents Chemother.* **38**, 1496–1500.
154. Goldman, R. C., Feski, S. W., and Doran, C. (1990). Role of protonated and neutral forms of macrolides in binding to ribosomes from gram-positive and gram-negative bacteria. *Antimicrob. Agents Chemother.* **34**, 426–431.
155. Cao, Z., Hammond, R., Pratt, S., Saiki, A., Lerner, C., and Zhong, P. (1999). Mechanism of action for novel ketolide ABT-773. Presented at 39th Intersci. Conf. Antimicrob. Agents Chemother. (Sept. 26–29, San Francisco). Abstr. No. F-2135.

156. Capobianco, J. O., Shortridge, V., Ma, Z., Phan, L., Or, Y., and Zhong, P. (1999). Rapid drug accumulation of ABT-773 in *Streptococcus pneumoniae*. Presented at 39th Intersci. Conf. Antimicrob. Agents Chemother. (Sept. 26–29, San Francisco). Abstr. No. 2137.
157. Cao, Z., Hammond, R., Pratt, S., Saiki, A., Lerner, C., Flamm, R., and Zhong, P. (2000). Mechanism of action for novel ketolide-ABT-773. Presented at 5th ICMAS-KO (Jan. 26–28, Seville). Poster No. 02.04.
158. Champney, W. S., and Tober, C. L. (2000). Evermicin (SCH27899) inhibits both translation and 50S ribosomal subunit formation in *Staphylococcus aureus* cells. *Antimicrob. Agents Chemother.* **44**, 1413–1417.
159. Bin, X. X., Wolf, K., Schaffner, T., and Malinverni, R. (2000). Effect of azithromycin plus rifampin versus amoxicillin alone on eradication and inflammation in the chronic course of *Chlamydia pneumoniae* pneumonitis in mice. *Antimicrob. Agents Chemother.* **44**, 1761–1764.
160. Pendland, S. L., Prause, J. L., Neuhauser, M. M., Boyea, N., Hackleman, J. M., and Danziger, L. H. (1999). *In vitro* activity of ABT-773, a new ketolide antibiotic, alone and in combination with metronidazole, amoxicillin, or tetracycline against *Helicobacter pylori*. Presented at 39th Intesci. Conf. Antimicrob. Agents Chemother. (Sept. 26–29, San Francisco). Abstr. No. F-2145.
161. Tait-Kamrard, A., Davies, T., Cronan, M., Jacobs, M. R., Applebaum, P. C., and Sutcliffe, J. (2000). Mutations in 23S rRNA and ribosomal protein L4 account for resistance in pneumococcal strains selected *in vitro* by macrolide passage. *Antimicrob. Agents Chemother.* **44**, 2118–2125.
162. Jung, R., Messick, C. R., Pendland, S. L., Tesoro, E. P., Losendahl, K. J., Schriever, C. A., and Danziger, L. H. (2000). Postantibiotic effects and bactericidal activities of clarithromycin-14-hydroxyl-clarithromycin, versus those of amoxicillin-clavulanate, against anaerobes. *Antimicrob. Agents Chemother.* **44**, 778–779.
163. Bui, M. H., Almer, L. S., Hensey, D. M., Ma, Z., Or, Y., Nilius, A. M., and Flamm, R. K. (1999). Antibacterial effects of ABT-773 against respiratory tract pathogens. Presented at 39th Intersci. Conf. Antimicrob. Agents Chemother. (Sept. 26–29, San Francisco). Abstr. No. F-2138.
164. Neuhauser, M. M., Prause, J. L., Jung, R., Boyea, N., Hackleman J. M., Danziger, L. H., and Pendland, S. L. (1999). *In vitro* bactericidal activity of ABT-773, a new ketolide, versus clarithromycin (CL) and azithromycin (AZ) against penicillin/erythromycin-sensitive and -resistant *Streptococcus pneumoniae* (SP). Presented at 39th Intersci. Conf. Antimicrob. Agents Chemother. (Sept. 26–29, San Francisco). Abstr. No. 2139.
165. Munckhof, W. J., Borlace, G., and Turnidge, J. D. (2000). Postantibiotic suppression of growth of erythromycin A-susceptible and -resistant gram-positive bacteria by the ketolides telithromycin (HMR 3647) and HMR 3004. *Antimicrob. Agents Chemother.* **44**, 1749–1753.
166. Gustafsson, I., Hjelm, E., and Cars, O. (2000). *In vitro* pharmacodynamics of the new ketolides HMR 3004 and HMR 3647 (telithromycin) against *Chlamydia pneumoniae*. *Antimicrob. Agents Chemother.* **44**, 1846–1849.
167. Davis, T. A., Ednie, L. M., Hoellman, D. M., Pankuch, G. A., Jacobs, M. R., and Applebaum, P.C. (2000). Antipneumococcal activity of ABT-773 compared to those of 10 other agents. *Antimicrob. Agents Chemother.* **44**, 1894–1899.
168. Zhong, P., Hammond, R., Cao, Z., Chen, Y., Shortridge, A., Nilius, A., Flamm, R. K., and Or, Y. (1999). Molecular basis of ABT-773 active against *erm*-containing macrolide-resistant *S. pneumoniae*. Presented at 39th Intersci. Conf. Antimicrob. Agents Chemother. (Sept. 26–29, San Francisco). Abstr. No. 2134.
169. Mankin, A. (2000). Antibiotics and ribosomes: Interaction of macrolide with ribosome and resistance. Presented at 40th Intersci. Conf. Antimicrob. Agents Chemother. (Sept. 17–20, Toronto). Abstr. No. G-1132.
170. Tait-Kamradt, A., Davies, T., Applebaum, P. C., Depardieu, F., Courvalin, P., Petitpas, J., Wondrack, L., Walker, A., Jacobs, M. R., and Sutcliffe, J. (2000). Two new mechanisms of macrolide resistance

in clinical strains of *Streptococcus pneumoniae* from Eastern Europe and North America. *Antimicrob. Agents Chemother.* **44**, 3395–3401.
171. Jung, R., Li, D. H., and Pendland, S. L. (2000). Intracellular activity of ABT-773 and other antimicrobial agents against *Legionella pneumoniae* isolates. Presented at 5th ICMAS-KO (Jan. 26–28, Seville). Poster No. 02.34.
172. Labro, M. T., Vazifeh, D., and Bryskier, A. (2000). Uptake of two new fluoroketolides, HMR 3562 and HMR 3787, by human neutrophils (PMN) *in vitro*. Presented at 40th Intersci. Conf. Antimicrob. Agents Chemother. (Sept. 17–20, Toronto). Abstr. No. F-1819.
173. Kadota, J., Ishimatsu, Y., Iwashita, T., Matsubara, Y., Kohno, S., Tateno, M., and Ishihara, R. (2000). The ketolide antimicrobial, telithromycin (HMR 3647), achieves high and sustained concentration in alveolar macrophages and bronchoalveolar epithelial lining fluid in healthy Japanese volunteers. Presented at 40th Intersci. Conf. Antimicrob. Agents Chemother. (Sept. 17–20, Toronto). Abstr. No. 2143.
174. Ramer, N. C., McDaniel, D. F., Johnson, P. M., Shortridge, D., Ma, Z., Or, Y., and Flamm, R. K. (1999). Effect of pH and media on the *in vitro* minimal inhibitory concentration of ABT-773. Presented at 39th Intersci. Conf. Antimicrob. Agents Chemother. (Sept. 26–29, San Francisco). Abstr. No. F-2143.
175. Vazifeh, D., Bryskier, A., and Labro, M. T. (2000). Effect of proinflammatory cytokines on the interplay between roxithromycin, HMR 3647, or HMR 3004 and human polymorphonuclear neutrophils. *Antimicrob. Agents Chemother.* **44**, 511–521.
176. Sato, E., Nelson, D. K., Koyama, S., Hoyt, J., and Robbins, R. A. (2000). Erythromycin modulates eosinophil chemotactic cytokine production by human lung fibroblasts *in vitro*. *Antimicrob. Agents Chemother.* **44**, 401–406.
177. Jung, R., Bearden, D. T., and Danziger, L. H. (2000). Effect of ABT-773 and other antimicrobial agents on the morphology and the release of interleukin-1β and interleukin-8 against *Haemophilus influenzae* and *Streptococcus pneumoniae* in whole blood. Presented at 5th ICMAS-KO (Jan. 26–28, Seville). Poster No. 02.25.
178. Deabate, C. A., Heyder, A., Leroy, B., Sidadous, E., and Backstrom, J. (2000). Oral telithromycin (HMR 3647; 800mg od) for 5 days is well tolerated and as effective as cefuroxime axetil (500 mg bid) for 10 days in adults with acute exacerbations of chronic bronchitis (AECB). Presented at 40th Intersci. Conf. Antimicrob. Agents Chemother. (Sept. 17–20, Toronto). Abstr. No. L-2228.
179. Labbe, G., Flor, M., and Lenfant, B. (2000). Cytochrome P-450 (CYP-450) activity is not inhibited *in vitro* by telithromycin (HMR 3647), a new ketolide. Presented at 5th ICMAS-KO (Jan. 26–28, Seville). Poster No. 09.28.
180. Kirst, H. A. (1992). Antibiotics (macrolides). *In* "Kirk–Othmer Encyclopedia of Chemical Technology," 4th Ed., Vol. 3, pp. 169–213. John Wiley & Sons, Inc., New York.
181. Bryskier, A. J., Butzler, J. P., Neu, H. C., and Tulkens, P. M. (Eds.) (1993). "Macrolides: Chemistry, Pharmacology and Clinical Uses." Arnette Blackwell, Paris.
182. Kirst, H. A. (1998). Recent developments with macrolide antibiotics. *Exp. Opin. Ther. Patients* **8**, 111–120.
183. Nakajima, Y. (1999). Mechanisms of bacterial resistance to macrolide antibiotics. *J. Infect. Chemother.* **5**, 61–74.
184. Chu, D. T. W. (2000). Progress in macrolide and ketolide antibacterials. *In* "Annual Reports in Medicinal Chemistry. Section III: Cancer and Infectious Diseases" (J. J. Plattner, Ed.). Vol. 35, Chap. 13, pp. 145–155. Academic Press, New York.
185. O'hara, K. (2000). Up-to-date report on the resistance of macrolide antibiotics and future macrolides. *Jpn. J. Chemother.* **48**, 169–190.
186. Bryskier, A., Agouridas, C., and Chantot, J. F. (2000). Ketolide: Novel antibacterial agent designed to overcome erythromycin A resistance. *In* "New Considerations for Macrolides,

Azalides, Streptogramins, and Ketolides" (S. H. Zinner, L. S. Young, J. F. Acar, and C. Ortiz-Neu, Eds.), pp. 79–102. Marcel Dekker, New York.
187. Ōmura, S., Thishler, M., Nakagawa, A., Yamada, H., Umezasa, I., Komiyama, K., and Hata, T. (1972). Relationship of structures and microbiological activities of the 16-membered macrolides. *J. Med. Chem.* **15**, 1011–1015.
188. Omoto, S., Iwamatsu, K., Inoue, S., and Niida, T. (1976). Modification of a macrolide antibiotic midecamycin (SF837). I. Synthesis and structure of 9,3″-diacetyl midecamycin. *J. Antibiot.* **39**, 536–548.
189. Kawaharajo, K., Sekizawa, Y., and Inoue, M. (1981). *In vitro* and *in vivo* antibacterial activity of 9,3″-di-*O*-acetyl midecamycin (MOM), a new macrolide antibiotic. *J. Antibiot.* **34**, 436–442.
190. Shomura, T., Someya, S., Murata, S., Umemara, K., and Nishio, M. (1981). Metabolism of 9,3″-diacetyl midecamycin. II. The structures of several metabolites of 9,3″-diacetyl midecamycin. *Chem. Pharm. Bull.* **29**, 2413–2419.
191. Nakamura, T., Fukatsu, S., Seki, S., and Niida, T. (1978). A convenient method for the preparation of the acylated macrolide antibiotic midecamycin using molecular sieves and acylchloride. *Chem. Lett.* 1293–1296.
192. Inoue, S., Omoto, S., Iwamatsu, K., and Niida, T. (1980). Modification of macrolide antibiotic midecamycin. II. Reaction of midecamycin and 9-acetyl midecamycin with dimethylsulfoxide and acetic anhydride. *J. Anibiot.* **33**, 61–71.
193. Sakakibara, H., Okekawa, O., Fujiwara, T., Otani, M., and Ōmura, S. (1981). Acyl derivatives of 16-membered macrolides. I. Synthesis and biological properties of 3″-*O*-propionyl leucomycin A$_5$. *J. Antibiot.* **34**, 1001–1010.
194. Sakakibara, H., Okekawa, O., Fujiwara, T., Aizawa, M., and Ōmura, S. (1981). Acyl derivatives of 16-membered macrolides. II. Antibacterial activities and serum levels of 3″-*O*-acyl deviratives of leucomycin. *J. Antibiot.* **34**, 1011–1018.
195. Harada, S., Muroi, M., Kondo, M., Tsuchiya, K., Matsuzawa, T., Fugono, T., Kishi, T., and Ueyanagi, J. (1973). Chemical modification of maridomycin, a new macrolide antibiotic. *Antimicrob. Agents Chemother.* **4**, 140–148.
196. Sano, H., Sunazuka, T., Tanaka, H., Yamashita, K., Okachi, R., and Ōmura, S. (1984). Chemical modification of spiramycins. III. Synthesis and antimicrobial activity of 4″-sulfonates and 4″-alkylethers of spiramycin I. *J. Antibiot.* **37**, 750–759.
197. Sano, H., Inoue, M., Yamashita, K., Okachi, R., and Ōmura, S. (1983). Chemical modification of spiramycin. I. Synthesis of the acetal derivatives of neospiramycin I. *J. Antibiot.* **36**, 1336–1344.
198. Sano, H., Tanaka, H., Yamashita, K., Okachi, R., and Ōmura, S. (1984). Chemical modification of spiramycin, V. Synthesis and antimicrobial activity of 3′ or 4‴-de-*N*-methylspiramycin I and 12(Z)-isomers. *J. Antibiot.* **38**, 738–749.
199. Sano, H., Sunazuka, T., Tanaka, H., Yamashita, K., Okachi, R., and Ōmura, S. (1984). Chemical modification of spiramycin. IV. Synthesis and *in vitro* and *in vivo* activities of 3″, 4″-diacylates and 3,3″,4″-triacylates of spiramycin I. *J. Antibiot.* **37**, 766–772.
200. Ōmura, S., Sano, H., and Sunazuka, T. (1984). Structure–activity relationships of spiramycins. *J. Antimicrob. Chemother.* **16** (Suppl. A), 1–11.
201. Ōmura, S., Nakagawa, A., Machida, M., and Imai, H. (1977). Evidence for configurational identity between leucomycin and tylosin. *Tethahedron Lett.* 1045–1048.
202. Bhuwapathanapun, S., and Gray, P. (1977). High pressure liquid chromatography of the macrolide antibiotic tylosin. *J. Antibiot.* **30**, 673–674.
203. Whaley, H. A., Patterson, E. L., Dornbush, A. C., Backus, E. J., and Ohonos, N. (1963). Isolation and characterization of relomycin, a new antibiotic. *Antimicrob. Agents Chemother.* 45–48.
204. Yoshioka, J., Kiyoshima, K., Maeda, M., Sakamoto, M., Ishikura, T., Fukagawa, Y., Sawa, T., Hamada, M., Naganawa, H., and Takeuchi, T. (1988). Synthesis and structure–activity studies of new 4″-*O*-acyltylosin derivatives of therapeutic interest. *J. Antibiot.* **41**, 1617–1628.

205. Cacciapuoti, A. F., Loebenberg, D., Moss, Jr., E. L., Menzel, Jr., F. W., Rudeen, J. A., Naples, L. R., Cramer, C. L., Hare, R. S., Mallams, A. K., and Miller, G. H. (1990). Microbiological and pharmacokinetic studies of acyl desmycarosyl tylosin and related tylosin derivatives. *J. Antibiot.* **43**, 1131–1136.
206. Matsubara, H., Inokoshi, J., Nakagawa, A., Tanaka, H., and Ōmura, S. (1983). Chemical modification of tylosin: Synthesis of amino derivatives at C-20 position of tylosin and demycarosyl tylosin. *J. Antibiot.* **36**, 1713–1721.
207. Ōmura, S., Miyano, K., Matsubara, H., and Nakagawa, A. (1982). Novel dimeric derivatives of leucomycin and tylosin, sixteen-membered macrolides. *J. Med. Chem.* **25**, 271–275.
208. Kirst, H. A., Willard, K. E., Debono, M., Toth, J. E., Truedell, B. A., Leeds, J. P., Ott, J. L., Felty-Duckworth, A. M., and Counter, F. T. (1989). Structure–activity studies of 20-deoxo 20-amino derivatives of tylosin related macrolides. *J. Antibiot.* **42**, 1673–1683.
209. Debono, M., Willard, K. E., Kirst, H. A., Wind, J. A., Crouse, G. D., Tao, E. V., Viceuzi, J. T., Counter, F. T., Ott, J. L., Ose, E. E., and Ōmura, S. (1989). Synthesis and antimicrobial evaluation of 20-deoxo 20-(3,5-dimethylpiperidin 1-yl) desmycosin (tilmicosin, EL-870) and related cyclic amino derivatives. *J. Antibiot.* **42**, 1253–1267.
210. Ruggeri, C., Laborde, M. L. A., Dessinges, A., Ming, L., Olesker, A., and Lukacs, G. (1989). Synthesis and antibacterial activity of 9-*O*-((2-methoxyethoxy)methyl)-oxime of tylosin and desmycarosyl tylosin. *J. Antibiot.* **42**, 1442–1445.
211. Satoi, S., Muto, N., Hayashi, M., Fujii, T., and Otani, M. (1980). Mycamicins, new macrolide antibiotics. I. Taxonomy, production, isolation, characterization and properties. *J. Antibiot.* **33**, 364–376.
212. Fujiwara, T., Watanabe, H., Kogami, Y., Shiritani, Y., and Sakakibara, H. (1989). 19-Deformyl-4′-deoxydesmycosin (TMC-016): Synthesis and biological properties of a unique 16-membered macrolide antibiotic. *J. Antibiot.* **42**, 903–912.
213. Lartey, P. A., Hannich, S., Maring, C. J., Freiberg, L. A., Grampovnick, D., Swanson, R., Hardy, D., and Fernandes, P. B. (1989). Synthesis and evaluation of niddamycin analogs with bacterial activity. Presented at 29th Intersci. Conf. Antimicrob. Agents Chemother. (Sept. 17–20, Houston). Abstr. No. 1033.
214. Freiberg, L. A., Maring, C. J., Lartey, P. A., Edwards, D., Grampovnick, D., Hardy, D., and Fernandes, P. B. (1989). Synthesis and antimicrobial activities of 9-*N,N*-dialkylamino-9-deoxo-10,11,12,13-tetrahydroniddamycins. Presented at 29th Intersci. Conf. Antimicrob. Agents Chemother. (Sept. 17–20, Houston). Abstr. No. 1032.
215. Maring, C. J., Grampovnick, D., Bacino, D., Lartey, P. A., Hardy, D., and Clement, J. (1990). Synthesis and SAR of 9(*R*)-*N,N*-dialkylamino-9-deoxoderivatives of 16-membered macrolides. Presented at 30th Intersci. Conf. Antimicrob. Agents Chemother. (Oct. 21–24, Atlanta, GA). Abstr. No. 812.
216. Mallams, A. K., and Rossman, R. R. (1989). Semisynthetic macrolide antibacterials derived tylosin. Synthesis of 23-*O*-demycinosyl tylosin and related compounds. *J. Chem. Soc. Perkin Trans. I*, 775–785.
217. Burg, R. W., Miller, B. M., Baker, E. E., Birnbaum, J., Currie, S. A., Hartman, R., Kong, Y.-L., Monaghan, R. L., Olsen, G., Putter, I., Yunac, J., Wallich, H., Stapley, E. O., Oiwa, R., and Omura, S. (1979). Avermectins, a new family of potent anthelmintic agents: Producing organism and fermentation. *Antimicrob. Agents Chemother.* **15**, 361–367.
218. Miller, T. W., Chaiet, L., Cole, D. J., Cole, L. J., Flor, J. E., Goegelman, R. T., Gullo, V. P., Joshue, H., Kempf, A. J., Krellwits, W. R., Monaghan, R. L., Ormond, R. E., Wilson, K. E., Albers-Schonberg, G., and Puter, I. (1979). Avermectins, a new family of potent anthelmintic agents: Isolation and chromatographic properties. *Antimicrob. Agents Chemother.* **15**, 368–371.
219. Chabala, J. C., Mrozik, H., Tolman, R. L., Eskola, P., Lusi, A., Peterson, L. H., Wood, M. F., and Fisher, M. H. (1980). Ivermectin, a new broad-spectrum antiparasitic agent. *J. Med. Chem.* **23**, 1134–1136.

220. Taylor, H. R., and Greene, B. M. (1989). The status of ivermectin in the treatment of onchocerciasis. *Am. J. Trop. Hyg.* **41**, 460–466.
221. Cambell, W. C., Fisher, M. H., Stapley, E. O., Albers-Schonberg, G., and Jacob, T. A. (1983). A potent new antiparasitic agent. *Science* **221**, 823–828.
222. Cane, D. E., Liang, T. C., Kaplan, L., Nallin, M. K., Schulman, M. D., Hensens, O. D., Douglas, A. W., and Albers-Schonberg, G. L. (1983). Biosynthetic origin of the carbon skeleton and oxygen atoms of the avermectins. *J. Am. Chem. Soc.* **105**, 4110–4112.
223. Chen, T. S., Inamine, E. S., Hensens, O. D., Zink, D., and Ostlind, D. A. (1989). Directed biosynthesis of avermectins. *Arch. Biochem. Biophys.* **269**, 544–547.
224. Hafner, E. W., Holley, B. W., Holdom, K. S., Lee, S. E., Wox, R. G., Bech, D., McArthur, H. A. I., and Wernau, W. C. (1991). Branched chain acid requirement for avermectin produced by a mutant of *Streptomyces avermitilis* lacking branched-chain 2-oxo acid dehydrogenase activity. *J. Antibiot.* **44**, 349–356.
225. Dutton, C. J., Gibson, S. P., Goudie, A. C., Holdom, K. S., Pacey, M. S., Ruddock, J. C., Bu'Lock, J. P., and Richards, M. K. J. (1991). Novel avermectins produced by mutational biosynthesis. *J. Antibiot.* **44**, 357–365.
226. Pivnichny, J. V., Arison, B. H., Preiser, F. A., Shim, J. S. K., and Mrozik, H. (1988). Base-catalyzed isomerization of avermectins. *J. Agric. Food Chem.* **36**, 826–828.
227. Mrozik, H., Eskola, P., Fisher, M. H., Egerton, J. R., Cifelli, S., and Ostlind, D. A. (1982). Avermctin acyl derivatives with antihelminthic activity. *J. Med. Chem.* **25**, 658–663.
228. Cvetovick, R. J., Kelly, D. H., DiMichele, L. M., Shuman, R. F., and Grabowski, E. J. J. (1994). Syntheses of 4″-*epi*-amino-4″-deoxyavermectin B1. *J. Org. Chem.* **59**, 7704–7708.
229. Mrozik, H., Eskola, P., Arison, B. H., Linn, B. O., Lusi, A., Matzuk, A., Shih, T. L., Tischler, M., Waksmunski, F. S., Wyvratt, M. J., Blizzard, T. A., Margiatto, G. M., Fisher, M. H., Shoop, W. L., and Egerton, J. R. (1995). 2-Deoxy-4″-aminoavermectins with potent broad spectrum antiparasitic activities. *Bioorg. Med. Chem. Lett.* **5**, 2435–2440.
230. Mrozik, H., Eskola, P., Linn, B. O., Lusi, A., Shih, T. L., Tischler, M., Waksmunski, F. S., Wyvratt, M. J., Hilton, N. J., Anderson, T. E., Babu, J. H., Dybas, R. A., Preiser, F. A., and Fisher, M. H. (1989). Discovery of novel avermectins with unprecedented insecticidal activity. *Experientia* **45**, 315–316.
231. Shoop, W. L., DeMontigny, P., Fink, D. W., Williams, J. B., Egerton, J. R., Mrozik, H., Fisher, M. H., Skelly, B. J., and Turner, M. J. (1996). Efficacy in sheep and pharmacokinetics in cattle that led to the selection of eprinomectin as a topical endectocide for cattle. *Int. J. Parasitol.* **26**, 1227–1235.
232. Shoop, W. L., Egerton, J. R., Eart, C. H., Haines, H. W., Michael, B. F., Mrozik, H., Eskola, P., Fisher, M. H., Slayton, L., Ostlind, D. A., Skelly, B. J., Fulton, R. K., Barth, D., Costa, S., Gregory, L. M., Campbell, W. C., Seward, R. L., and Turner, M. J. (1996). Eprinomectin: A novel avermectin for use as a topical endectocide for cattle. *Int. J. Parasitol.* **26**, 1237–1242.
233. Banks, B. J., Bishop, B. F., Evans, N. A., Gibson, S. P., Goudie, A. C., Gration, K. A. F., Pacey, M. S., Perry, D. A., and Witty, M. J. (2000). Avermectin and flea control: Structure–activity relationships and the selection of selamectin for development as an endectocide for companion animals. *Bioorg. Med. Chem. Lett.* **8**, 2027–2025.
234. Bishop, B. F., Bruce, C. J., Evans, N. A., Goudie, A. C., Gration, K. A. F., Grison, S. P., Pacey, M. S., Perry, D. A., Walshe, N. D. A., and Witty, M. J. (2000). Selamectin: A novel broad-spectrum Endectocide for dogs and cats. *Vet. Parasitol.* **91**, 163–176.
235. Ōmura, S., Sunazuka, T., Turberg, A., VonSamson-Himmelstjerma, G., Hansen, O., and Harder, A. (2000). Preparation of avermectin derivatives as antiparasitic agents. PCT/JP00/0069, The Kitasato Institute.

236. Takiguchi, Y., Mishima, H., Okuda, M., Terao, M., Aoki, A., and Fukada, R. (1980). A new family of macrolide antibiotics: Fermentation, isolation and physico-chemical properties. *J. Antibiot.* **33**, 1120–1127.
237. Mishima, H., Ide, J., Muramatsu, S, and Ono, M. (1983). Milbemycins, a new family of macrolide antibiotics: Structure determination of milbemycin D, E, F, G, H, I, J and K. *J. Antibiot.* **36**, 980–990.
238. Carter, G. T., Nietsche, J. A., Hertz, M. R., Williams, D. R., Siegel, M. M., Morton, G. O., James, J. C., and Border, D. B. (1988). LL-F28249 antibiotics complex: A new family of antiparasitic macrocyclic lactones. Isolation, characterization, and structures of LL-F28249 α, β, γ, λ. *J. Antibiot.* **41**, 519–529.
239. Tsukamoto, Y., Sato, K., Mio, S., Sugai, S., Yamai, T., Kitano, N., Muramatsu, S., Nakadi, Y., and Ide, J. (1991). Synthesis of 5-keto-5-oxime derivatives of milbemycins and their activities against microfilariae. *Agric. Biol. Chem.* **55**, 2615–2621.
240. Ide, J., Okazaki, T., Ono, A., Nakagawa, K., Naito, S., Sato, K., Tanaka, K., Yoshikawa, H., Ando, M., Katsumi, S., Matsumoto, K., Toyawa, T., Shibano, M., and Abe, M. (1993). Milbemycin: Discovery and development. *Annu. Rep. Sankyo Res. Lab.* **45**, 1–98.
241. Ramsay, M. V. J., Robert, S. M., Russell, J. C., Shingeler, A. H., Slawin, A. M. Z., Sutherland, D. R., Tiley, E. P., and Williams, D. J. (1987). Novel antiparasitic agents derived by modification of a new natural product series. *Tetrahedron Lett.* **28**, 5353–5356.
242. Beddall, N. E., Howes, P. D., Ramsay, M. V. J., Roberts, S. M., Slawin, A. M. E., Sutherland, D. R., Tiley, E. P., and Williams, D. J. (1988). Chemical transformation of S541 factors (A)-(D): Preparation and reactions of the 23-ketones. *Tetrahedron Lett.* **29**, 2595–2598.
243. Ōmura, S., Tsuzuki, K., Sunazuka, T., Toyoda, H., Takahashi, I., and Itoh, Z. (1987). Macrolides with gastrointestinal motor stimulating activity. *J. Med. Chem.* **30**, 1941–1943.

Chapter 4

Total Synthesis of Macrolides

TADASHI NAKATA

RIKEN, Synthetic Organic Chemistry Laboratory
Saitama, Japan

I.	Introduction	181
II.	Synthetic Strategy for Macrolide Synthesis	182
	A. Asymmetric Synthesis of 1,3-Diol	182
	B. Synthetic Methodology for Macrolactone	190
	C. Glycosidation	202
III.	Total Synthesis of Selected Macrolides	210
	A. FK506 (Tacrolimus)	210
	B. Rapamycin (Sirolimus)	220
	C. Avermectin	232
	D. Altohyrtin (Spongistatin)	243
	E. Epothilone	257
	F. Others	268
IV.	Concluding Remarks	271
	Appendix: Abbreviations	272
	References	275

I. Introduction

Macrolide antibiotics have attracted the attention of numerous synthetic chemists due to their remarkable biological activity and their synthetically challenging unique structures. The structural complexity, the large number of chiral centers, the characteristic macrolactone rings, and the various types of attached sugars have presented a formidable challenge to chemists. Since the first total synthesis of methymycin as a polyoxomacrolide by Masamune *et al.* in 1975 [1], progress in macrolide synthesis has been dramatic because of the recent great advances in organic synthesis: developments in strategy, methodology, useful reactions, stereoselective reactions, effective protective groups, and so on. The most important tasks for the total synthesis of macrolide antibiotics involve (1) the stereoselective construction of chiral centers on the macrolide ring, (2) construction of the macrocyclic ring, and (3) attachment of a sugar or sugars to the aglycon.

At the present time, the synthesis of the chiral centers has been accomplished mainly via acyclic stereoselection, in which great progress has been made since the 1980s. A new methodology developed for the construction of the macrolide ring and glycosidation has also contributed to complex macrolide synthesis. This chapter introduces the progress in the total synthesis of macrolides from the mid-1980s to the present.

In Section II, the synthetic strategies for macrolide synthesis are introduced and focus in particular on asymmetric synthesis of 1,3-diol, synthetic methodology for macrolactone, and glycosidation. In Section III, the total synthesis of selected macrolide antibiotics is introduced: FK506 (tacrolimus; **1**), rapamycin (sirolimus; **2**), avermectins (**3**), altohyrtins (spongistatins; **4**), and epothilones (**5**) (Fig. 1). Several other synthesized macrolides are also illustrated.

II. Synthetic Strategy for Macrolide Synthesis

A. Asymmetric Synthesis of 1,3-Diol

The 1,3-diol systems are often found in the polyoxomacrolide ring. Thus, two major synthetic transformations for the construction of 1,3-diols are described in this section: (1) asymmetric aldol reaction and (2) asymmetric epoxidation and epoxide ring-opening.

1. Asymmetric Aldol Reaction

The aldol reaction is a versatile method for the construction of new carbon–carbon bonds in a regio-, diastereo-, and enantioselective manner. During the last two decades, major progress toward the total synthesis of macrolide antibiotics was made as a result of the development of the stereoselective aldol reaction in acyclic systems. This section is concerned mainly with the boron-mediated aldol reaction, which is particularly effective for the efficient synthesis of β-hydroxy carbonyl compounds [2].

In most boron-mediated aldol reactions, (Z)-boron enolates **7** and (E)-boron enolates **8** stereoselectively afford *syn*-aldols **9** and *anti*-aldols **10**, respectively, via a chelation-controlled transition state (Scheme 1). The (Z)-boron enolates **7** can be prepared from **6** by a combination of n-Bu$_2$BOTf and i-Pr$_2$NEt, while a combination of (c-hex)$_2$BCl and Et$_3$N gives the (E)-boron enolates **8**. (Abbreviations are defined in the Appendix at the end of this chapter.)

The simplest asymmetric induction involves a reaction of an achiral enolate with a chiral aldehyde. In this case, if the boron enolate geometry and facial selectivity to the aldehyde are well controlled, the stereoselective aldol reaction will proceed. For example, treatment of (Z)-boron enolate **11** with chiral aldehyde **12** effected stereoselective aldol reaction to give *syn*-aldol adduct **13** as a single product [3].

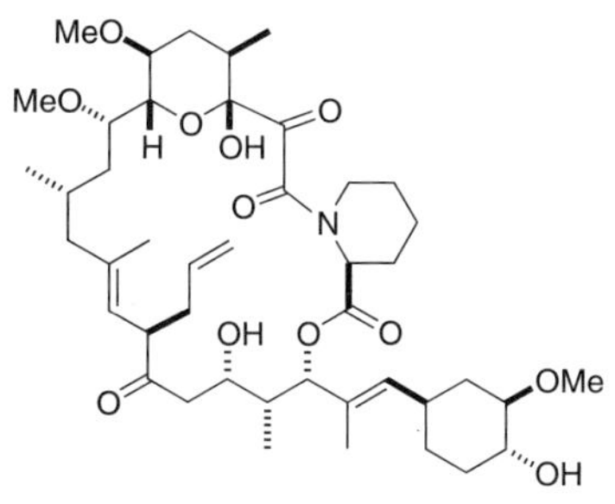

4. Total Synthesis of Macrolides

FK-506 (Tacrolimus) (**1**)

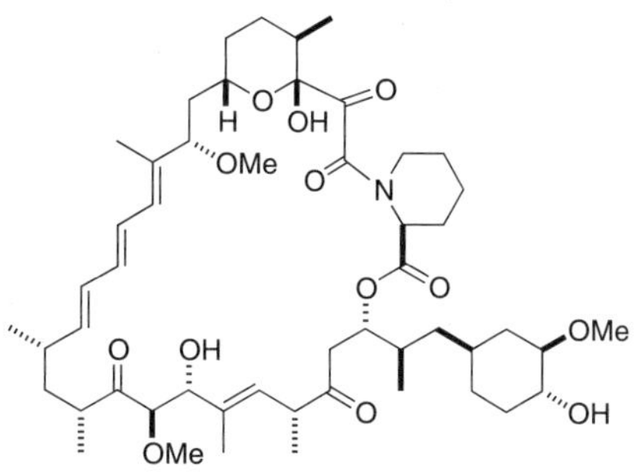

Rapamycin (Sirolimus) (**2**)

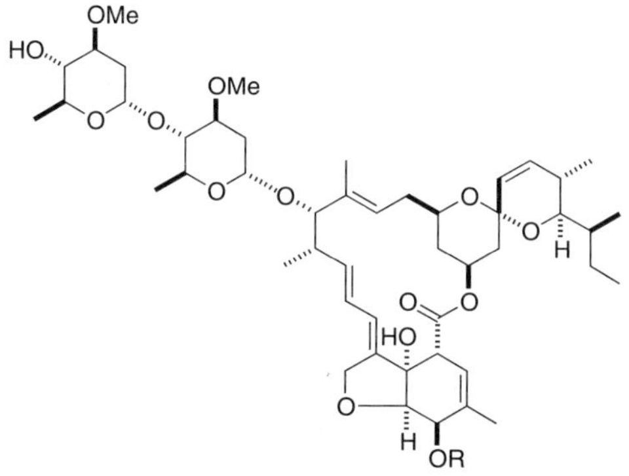

Avermectin A$_{1a}$ (**3a**): R = Me
Avermectin B$_{1a}$ (**3b**): R = H

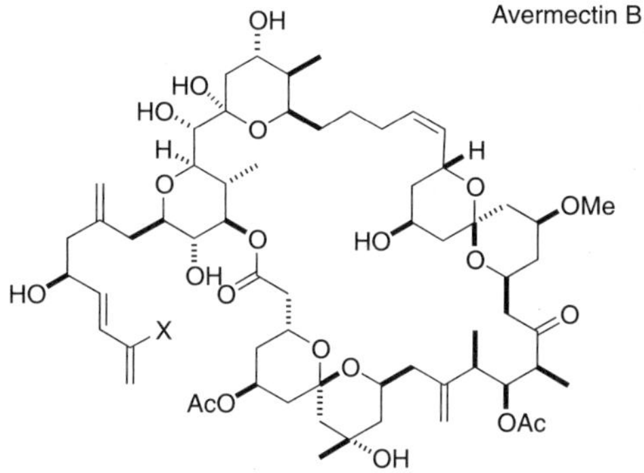

Altohyrtin A (Spongistatin 1) (**4a**): X = Cl
Altohyrtin C (Spongistatin 2) (**4b**): X = H

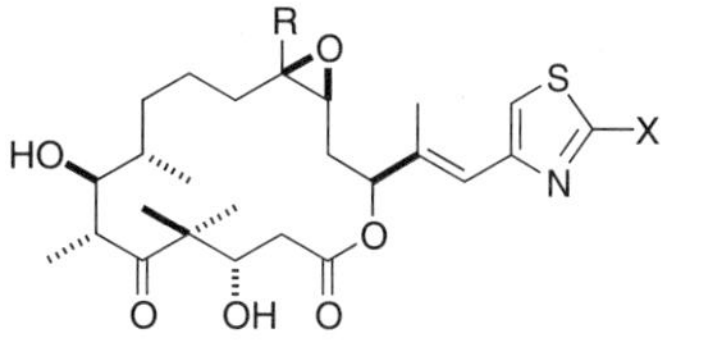

Epothilone A (**5a**): R = H, X = Me
Epothilone B (**5b**): R = Me, X = Me
Epothilone E (**5c**): R = H, X = CH$_2$OH

Fig. 1.

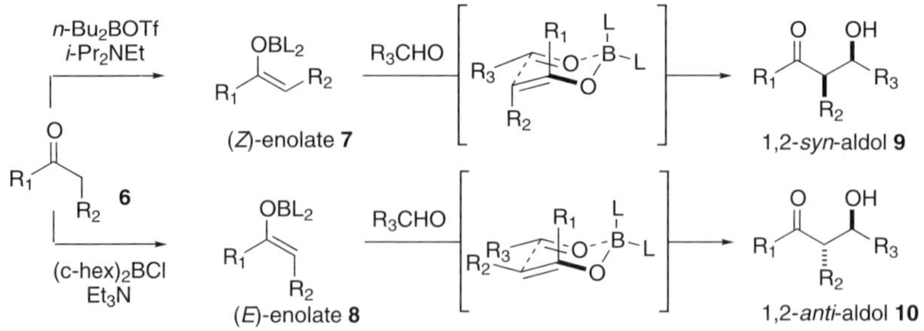

Scheme 1.

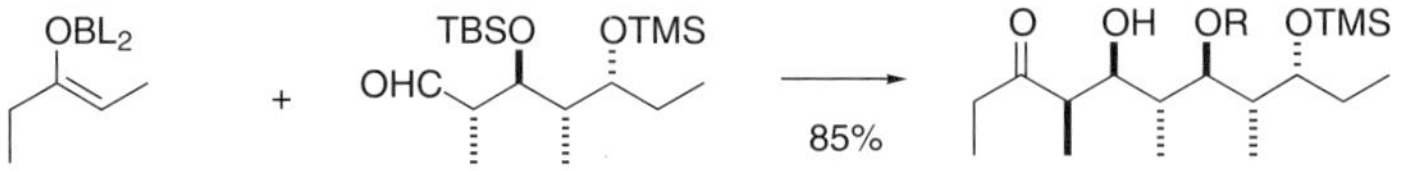

A highly efficient and widely used method for diastereoselective aldol reactions is the Evans aldol reaction using boron enolate derived from a chiral imide (Scheme 2) [4]. Upon treatment of imide **14** with n-Bu$_2$BOTf and i-Pr$_2$NEt in CH$_2$Cl$_2$ followed by addition of aldehyde, stereoselective aldol reaction proceeds smoothly through the chelation transition state to afford 1,2-*syn*-aldol adduct **16** in high yield and with excellent diastereoselectivity. After the aldol reaction, the chiral auxiliary is cleaved by hydrolysis to carboxylic acid, reduction to aldehyde or alcohol, conversion to Weinreb amide, etc. Various chiral imides, prepared from the corresponding amino alcohols, are applicable to the Evans asymmetric aldol reaction with high diastereoselectivity. The use of TiCl$_4$, instead of n-Bu$_2$BOTf, in the presence of amine (i-Pr$_2$NEt, Et$_3$N, or TMEDA) also afforded *syn*-aldol in high yield [5]. In contrast, addition of a Lewis acid (TiCl$_4$, SnCl$_4$, or Et$_2$AlCl) to the boron enolate **15** provides either *anti*-diol **17** or "non-Evans" 1,2-*syn*-aldol **18** with 80–95% diastereoselectivity [6]. Reaction using acyloxazolidine-thione **19** with 1 equiv. of TiCl$_4$ and 2.5 equiv. of TMEDA or sparteine yielded the normal 1,2-*syn*-aldol **21** with excellent diastereoselectivity, while the use of 2 equiv. of TiCl$_4$ and 1 equiv. of i-Pr$_2$NEt led to dramatic reversal to give the "non-Evans" 1,2-*syn*-aldol **22** with complete diastereoselectivity [7]. These results considerably expand the scope of the asymmetric aldol reaction.

The Evans aldol reaction using chiral β-keto imide **23** as a dipropionate building block is also very effective for the construction of polypropionate segments in polyoxomacrolides (Scheme 2) [8]. The diastereoselective aldol reaction of **23** via different metal enolates (Ti, Sn, and B enolates) afforded three kind of aldols, *syn-syn*-**24**, *anti-syn*-**25**, and *anti-anti*-**26**, with high diastereoselectivity, respectively. The subsequent stereoselective reduction of the resulting β-hydroxy ketones **24**–**26** provides various types of dipropionate units. Based on this strategy, the

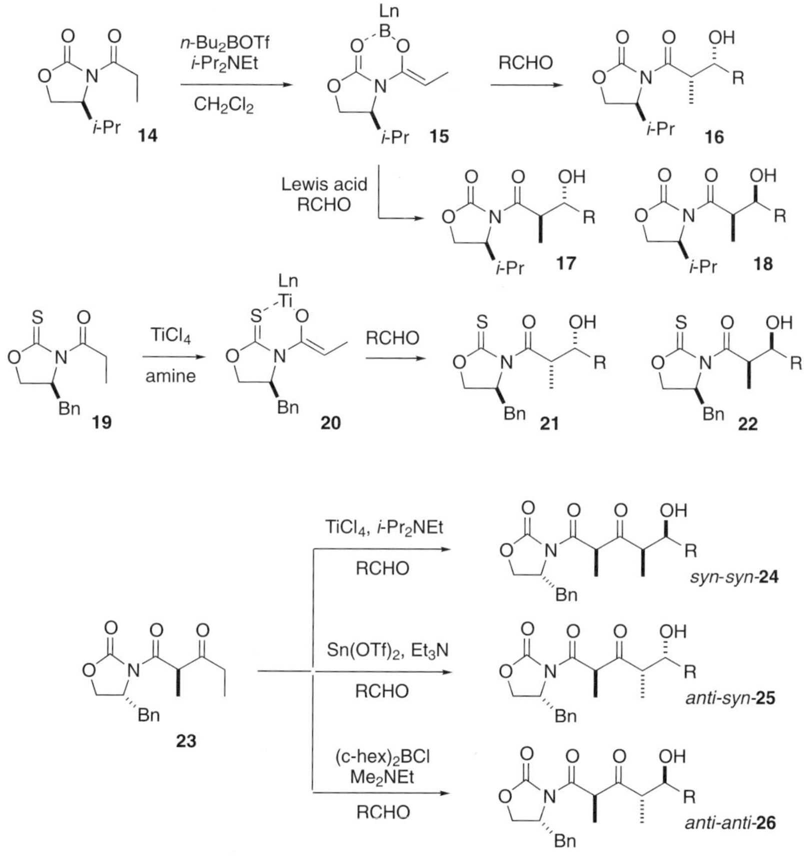

Scheme 2.

total syntheses of 6-deoxyerythronolide B and oleandolide have been achieved very efficiently, each in 18 linear steps (Section III.F) [171c].

The aldol reaction using chiral α-oxygenated ketones proceeds in a diastereoselective manner (Scheme 3); (Z)-boron enolate **28** derived from benzyloxy ketone **27** afforded *syn*-aldol **29** via chelation control, while benzoyloxy ketone **30** afforded *anti*-aldol **32** via (E)-enolate **31** through a nonchelation transition state [9].

The aldol reaction using chiral α-methyl ketone **33** having a β-benzyloxy group, corresponding to dipropionate, is often more direct and effective for the synthesis of polypropionate natural products (Scheme 4). Reaction of (E)-boron enolate **34**, prepared from **33** by treatment with (c-hex)$_2$BCl, and aldehyde gave *anti-anti*-aldol adduct **35** with high diastereoselectivity [10]. On the other hand, treatment of the same ketone **33** with Sn(OTf)$_2$ afforded *syn-syn*-aldol **37** [11]. Reaction of α-methyl-β-silyloxy ketones **38** gave *anti-syn*-aldol adduct **39**

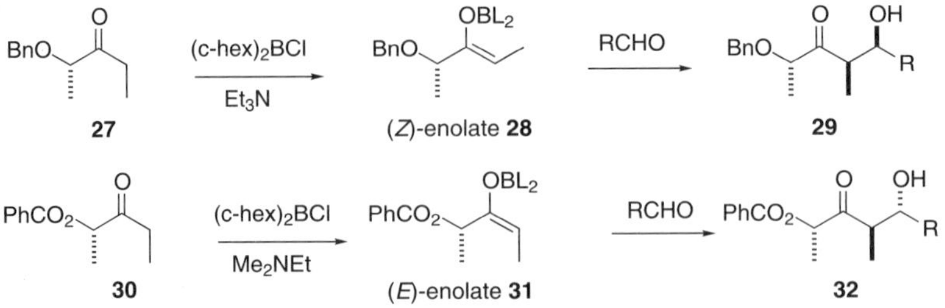

Scheme 3.

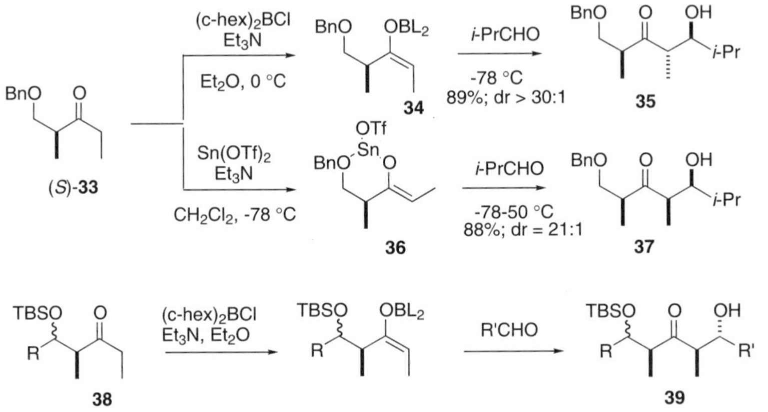

Scheme 4.

regardless of the stereochemistry of the silyloxy group [12]. The present strategy using chiral α-methyl ketones is also effectively applied to the total synthesis of oleandolide in 20 steps from (S)-ethyl ketone **33**, achieving excellent stereochemical control (Section III.F) [171a].

Asymmetric aldol reactions using chiral boron reagents (for example, **40**, **41**, and **42**) have been developed (Scheme 5). Reaction of diethyl ketone and aldehyde with (−)-Ipc$_2$BOTf (**40**) and i-Pr$_2$NEt proceeded via the (Z)-enolate of the ketone to afford syn-aldol **43** in good yield with high enantioselectivity [13]. Reaction of thioester **44** and aldehyde with (S,S)-boron triflate **41** produced anti-aldol **45** with high enantioselectivity via the (E)-enolate [14]. The chiral diazaborolidine reagent **42** effected the asymmetric anti-aldol reaction of t-butyl ester **46a** and the syn-aldol reaction of phenyl thioester **46b** with high enantioselectivity, respectively [15]. Reaction of aldehydes with (Z)-enolate **48** of thioester in the presence of chiral diamine **49**, Sn(OTf)$_2$, and dibutyltin diacetate produced syn-aldol **50** in high yields with complete syn-selectivity and enantioselectivity (>98% ee) [16].

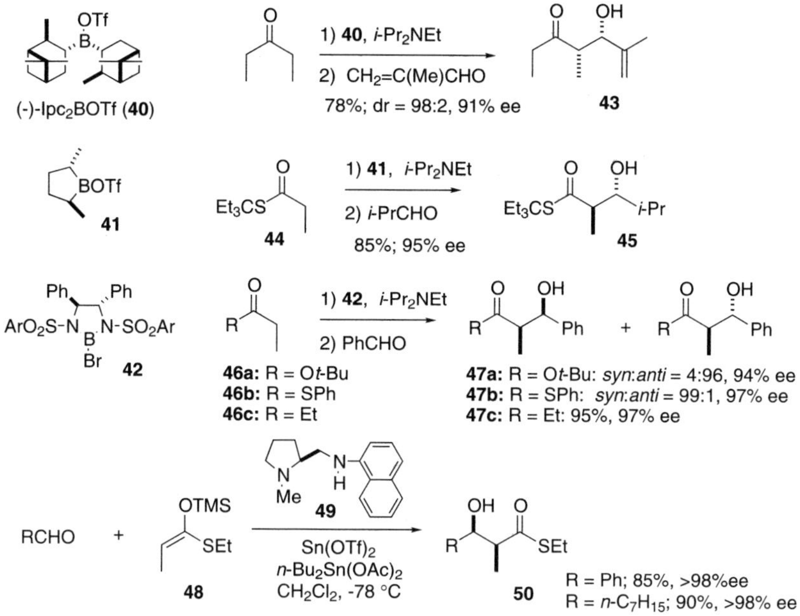

Scheme 5.

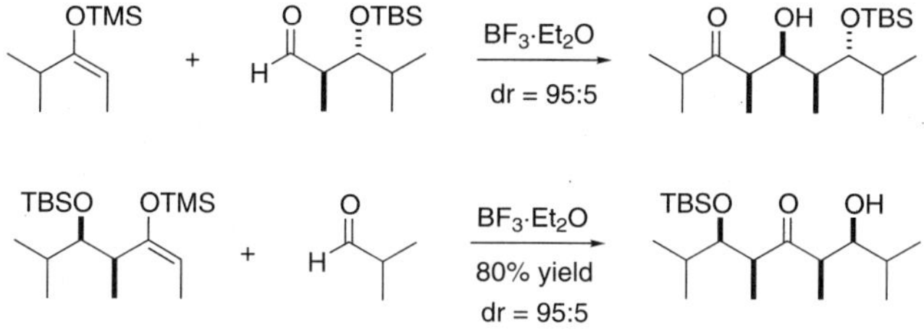

Scheme 6.

In addition to boron-mediated aldol reactions, the Lewis acid-catalyzed reactions of silyl enol ethers with aldehydes are also useful as shown in Scheme 6 [17].

Asymmetric allylation and crotylation, synthetically equivalent to the aldol reaction, have been extensively studied and have become a very useful procedure for preparation of propionate units. Among various chiral ligands on boron-developed, isopinocampheyl- and tartrate-derived reagents, **51** and **52**, which were developed by Brown et al. [18] and Roush et al. [19], respectively, are the most commonly used (Scheme 7). Reaction of aldehyde with (E)-**51a** or **52a** gave anti-adduct **54**, while that using (Z)-**51b** or **52b** afforded syn-adduct **53** with high asymmetric selectivity.

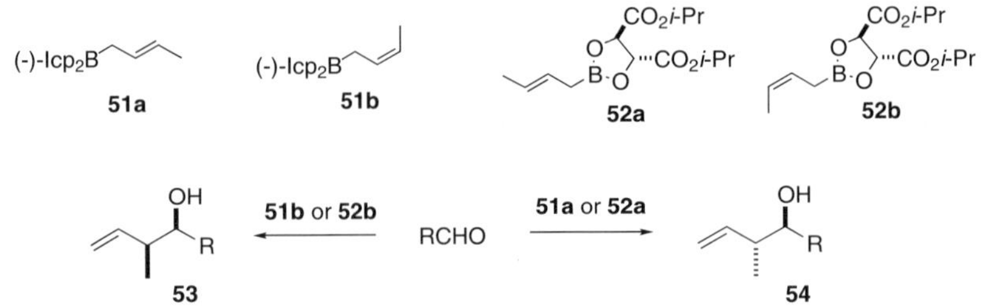

Scheme 7.

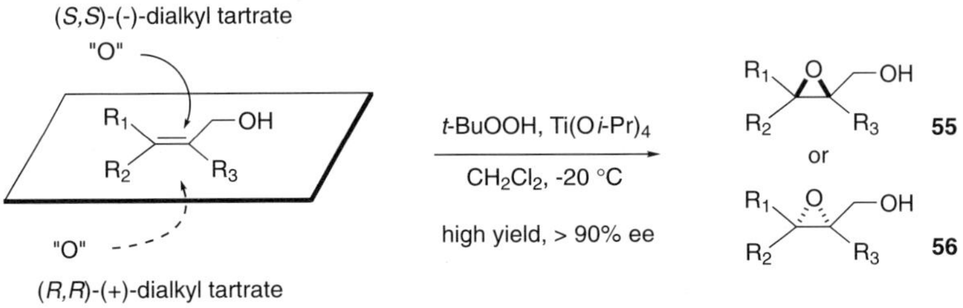

Scheme 8.

The development of catalytic asymmetric aldol reactions is being extensively studied [20]. Further progress in this field will expand the impact of these reactions on the total synthesis of natural products.

2. Asymmetric Epoxidation and Epoxide Ring-Opening

Since the first report by Katsuki and Sharpless in 1980 [21], the asymmetric epoxidation of allylic alcohols has been widely applied to the synthesis of various compounds [22]. Among several asymmetric epoxidations developed, the Sharpless asymmetric epoxidation (AE) is undoubtedly the most popular and versatile method, which has greatly contributed to the progress of natural product synthesis.

Epoxidation of allylic alcohols with t-BuOOH in the presence of Ti(Oi-Pr)$_4$ and a dialkyl tartrate as a chiral ligand affords the corresponding epoxides in good yield with high regio- and enantioselectivity (Scheme 8). The enantioselectivity can be anticipated by empirical rule as shown in Scheme 8; when (S,S)-(−)-dialkyl tartrate is used as the chiral auxiliary, β-epoxide **55** is obtained, while (R,R)-(+)-dialkyl tartrate gives α-epoxide **56**.

Representative examples are shown in Scheme 9. The Sharpless AE of geraniol (**57**) with (+)-diethyl tartrate (DET) gave α-epoxide **58** with 95% ee. In a double asymmetric induction, epoxidation of allylic alcohol **59** with (−)- and (+)-DET provided α- and β-epoxides, **60** and **61**, in ratios of 40:1 and 1:14, respectively [23]. It is noteworthy that high asymmetric selectivity was induced even in the mismatched case. The Sharpless AE is also effective for the kinetic resolution of racemic allylic alcohols. In the reaction of **62** with 0.6 equiv. of t-BuOOH and

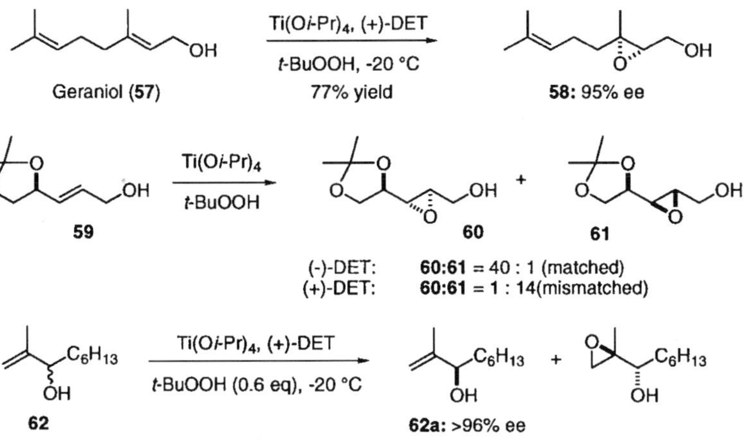

Scheme 9.

(+)-DET, the (S)-enantiomer reacted faster than the (R)-enantiomer to give (R)-alcohol **62a** with high enantiomeric purity [24]. Addition of catalytic amounts of CaH_2 and silica gel to the reaction system enhances the epoxidation rate [25]. Furthermore, the asymmetric epoxidation and kinetic resolution with a catalytic amount of Ti(Oi-Pr)$_4$ and a chiral tartrate can be achieved in the presence of MS 4A, without impairing the enantioselectivity [26].

Stereoselective ring-opening of 2,3-epoxy alcohols is highly valuable for the synthesis of diversely functionalized compounds (Scheme 10). In the presence of Ti(Oi-Pr)$_4$, various nucleophiles such as secondary amine, alcohol, thiol, azide, and carboxylic acid attack predominantly at the C3 position of **63** with inversion of the configuration to give 1,2-diol **64** [27]. Regioselective reductive ring-opening of 2,3-epoxy alcohols was accomplished; reduction of epoxide **65** with Red-Al® led to the almost exclusive formation of 1,3-diol **66** [28, 29], while DIBAH or LiBH$_4$-Ti(Oi-Pr)$_4$ reduction of **65** provided 1,2-diol **67** as the major product [28, 30]. In reduction of 2,4-dimethyl-2,3-epoxy alcohol **68** with Red-Al®, the regio- and stereoselective ring-opening took place to give 2,3-*syn*-1,3-diol **69** [31].

Nucleophilic addition of a methyl group at the C2 position of 2,3-epoxy alcohols is useful for the synthesis of propionate segments. Treatment of epoxy alcohols **70** and **72** with organocuprates gave 2-methyl-1,3-diols **71** and **73**, respectively, with high regio- and stereoselectivity (Scheme 11) [32]. Recently, nucleophilic addition using alkyl- and alkynylaluminum ate complexes has been reported; reaction of **74** with Me$_3$Al alone occurred selectively at the C3 position, while pretreatment with *n*-BuLi followed by Me$_3$Al effected a dramatic change in the regioselectivity to give 2-methyl-1,3-diol **75** as a major product [33].

In addition to the Sharpless AE, the Sharpless asymmetric dihydroxylation [34] is also important in the total synthesis of natural products including macrolide antibiotics.

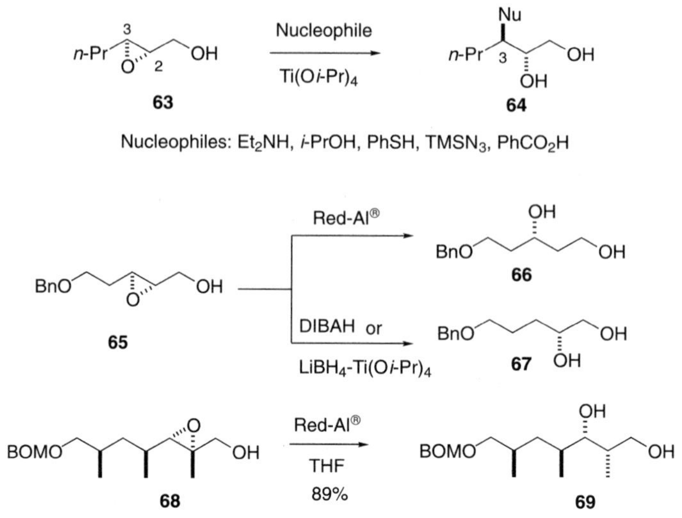

Scheme 10.

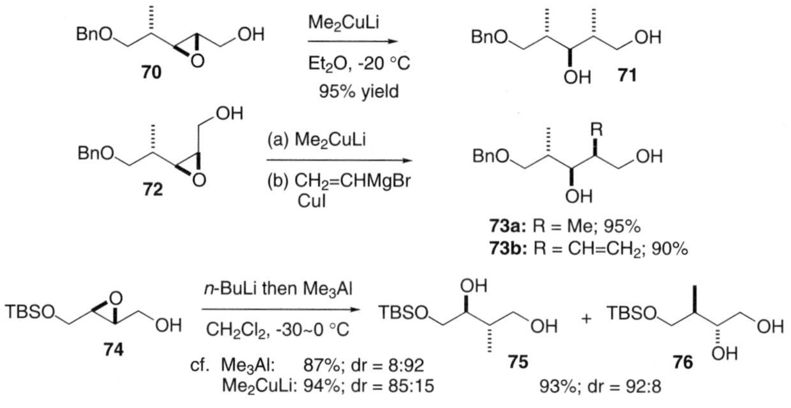

Scheme 11.

B. Synthetic Methodology for Macrolactone

Efficient construction of the macrocyclic ring is essential for achievement of the total synthesis of macrolide antibiotics. Macrolactonization and intramolecular Wittig-type reaction have been frequently used so far. In recent years, olefin metathesis and Stille-type reaction for formation of C–C bonds have been efficiently applied to the construction of macrocyclic rings.

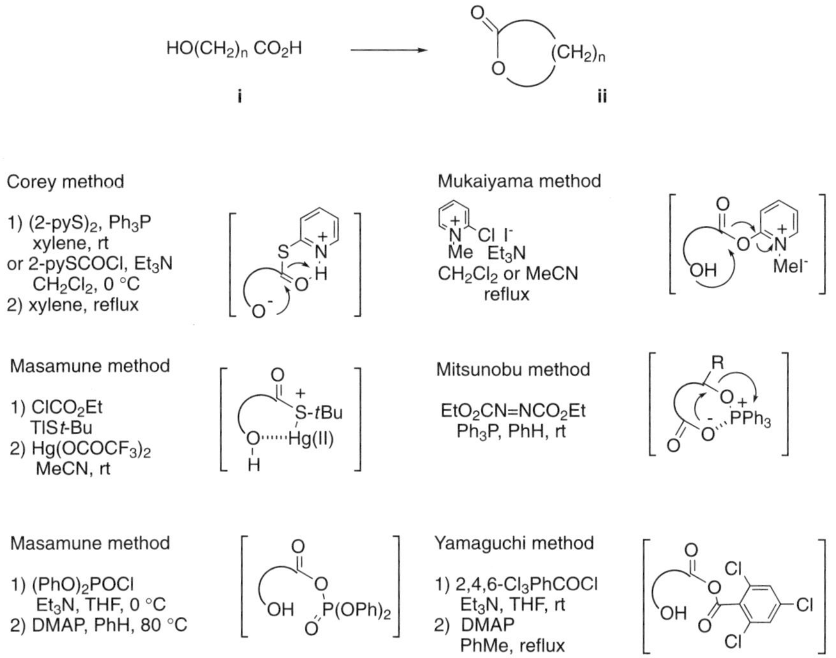

Fig. 2.

HO(CH$_2$)$_n$ CO$_2$H ⟶ [macrolactone ii with (CH$_2$)$_n$]

i → **ii**

Corey method
1) (2-pyS)$_2$, Ph$_3$P
   xylene, rt
   or 2-pySCOCl, Et$_3$N
   CH$_2$Cl$_2$, 0 °C
2) xylene, reflux

Mukaiyama method
[pyridinium intermediate]
Me Et$_3$N
CH$_2$Cl$_2$ or MeCN
reflux

Masamune method
1) ClCO$_2$Et
   TlS-tBu
2) Hg(OCOCF$_3$)$_2$
   MeCN, rt

Mitsunobu method
EtO$_2$CN=NCO$_2$Et
Ph$_3$P, PhH, rt

Masamune method
1) (PhO)$_2$POCl
   Et$_3$N, THF, 0 °C
2) DMAP, PhH, 80 °C

Yamaguchi method
1) 2,4,6-Cl$_3$PhCOCl
   Et$_3$N, THF, rt
2) DMAP
   PhMe, reflux

1. Macrolactonization

A highly effective method for the construction of macrolactone ring **ii** is the intramolecular lactonization of the corresponding seco-acid **i** (Fig. 2). Thus, various effective methods for the synthesis of macrolactones were developed during the 1970s and 1980s and have been successfully applied to the total synthesis of macrolide antibiotics. In the 1990s, new methods for macrolactonization were reported [35]; however, they have not yet been applied frequently to macrolide synthesis. Thus, in this section, the established methods applied to macrolide synthesis are introduced, and selected recent applications are shown in Scheme 12 and Section III.F.

Corey et al. [36] developed an efficient and mild lactonization method using 2-pyridinethiol ester. Slow addition of 2-pyridinethiol esters, prepared from ω-hydroxy acids by reaction with 2,2′-dipyridyl disulfide and Ph$_3$P or 2-thiopyridyl chloroformate and Et$_3$N, to refluxing xylene under dilution conditions yielded 7- to 16-membered macrolactones. A similar method catalyzed by AgBF$_4$ or AgClO$_4$ was reported by Gerlach [37]. The Corey method was successfully applied to macrolactonization of the seco-acid in Woodward's total synthesis of erythromycin A [38], total synthesis of tylosin [39], and so on. In the total synthesis of tuckolide (**78**), lactonization of **77** was performed under the conditions with

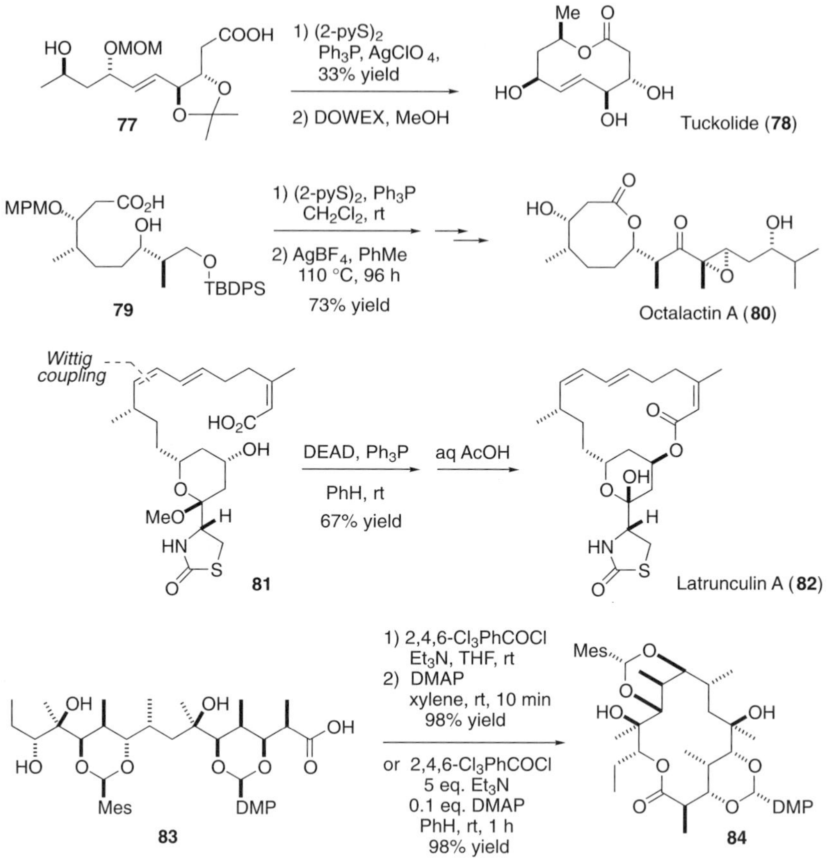

Scheme 12.

added AgClO$_4$; this lactonization did not proceed by other methods. The first total syntheses of octalactin A (**80**) and B were also achieved under the conditions with AgBF$_4$ via an unprecedented intramolecular esterification of a saturated acid **79** to form 8-membered lactone [40]. Corey *et al.* [41] further developed an alternative method using 2,2′-bis-(4-*t*-butyl-*N*-isopropyl)imidazoyl disulfide, which was applied to the synthesis of their erythronolide B.

Masamune *et al.* [42] developed the macrolactonization of ω-hydroxy *t*-butyl thioester with Hg(OCOCF$_3$)$_2$ in MeCN at room temperature, and this method accomplished the first total synthesis of methynolide [1]. They further developed an alternative method using a phosphoric acid mixed anhydride, which was applied to the synthesis of narbonolide [43] and tylonolide [44].

Mukaiyama *et al.* [45] developed a direct lactonization of ω-hydroxy acids by treatment with 1-methyl-2-chloropyridinium iodide. The Mukaiyama method is frequently applied to the construction of macrolactam or macrolactone for FK506 (**1**), rapamycin (**2**), avermectins (**3**), and others (Sections III.A, B, C, and F).

Mitsunobu *et al.* [46] reported an efficient macrolactonization using diethyl azodicarboxylate (DEAD) and Ph$_3$P. In the case of ω-hydroxy acids having a secondary alcohol, this cyclization takes place with inversion of the configuration of the alcohol. In the total synthesis of latrunculin A (**82**) and B, the Mitsunobu reaction was used for the macrolactonization of the seco-acid **81** with inversion of the secondary alcohol [47].

Yamaguchi *et al.* reported a very efficient esterification by means of mixed anhydride and its application to large-ring lactonization [48]: treatment of ω-hydroxy acids with 2,4,6-trichlorobenzoyl chloride and Et$_3$N in THF followed by addition of the resulting mixed anhydrides to a refluxing solution of 4-dimethylaminopyridine (DMAP) in toluene under dilution conditions. Since the application to the total synthesis of methynolide [49], the Yamaguchi method has been most frequently and successfully applied to macrolide synthesis. The applications to macrolactonization of altohyrtins (**4**), epothilones (**5**), and others are described in Sections III.D, E, and F. In the synthesis of erythronolide A, Yonemitsu *et al.* [50] reported an efficient macrolactonization of the seco-acid **83** by modification of the Yamaguchi method with a high concentration of DMAP. They further found that macrolactonization of **83**, which has a favorable conformation for cyclization, proceeded even without high dilution technique and prepreparation of the mixed anhydride; on treatment of **83** with excess Et$_3$N, cat. DMAP, and 2,4,6-trichlorobenzoyl chloride in benzene at room temperature, the cyclization occurred very rapidly to give **84** in almost quantitative yield [51].

In the synthetic studies of verrucarins, Roush *et al.* [52] developed a mixed anhydride method by successive treatment with pivaloyl chloride and Et$_3$N, and then *in situ* treatment with 4-pyrrolidinopyridine.

Boden and Keck [53] developed a practical method for macrolactonization by treatment with DCC and DMAP in the presence of DMAP·HCl; the original esterification procedures using DCC and DMAP or 4-pyrrolidinopyridine were reported by Neises and Steglich [54] and Hassner and Alexanian [55] independently. This method worked very efficiently for the macrolactonization (92% yield) in the total synthesis of cytovaricin (**196**) (see Scheme 29 later in this chapter).

2. *Ring-Closing Olefin Metathasis*

Recently, ring-closing olefin metathesis (RCM) has been extensively studied, and this new strategy has had great impact on the total synthesis of natural products [56]. Among several reagents developed, two main types of catalyst are often used as effective reagents at the present time (Fig. 3): the molybdenum-based catalyst **85**, developed by Schrock *et al.* [57] and the ruthenium-based catalysts **86a** and **86b**, developed by Grubbs *et al.* [58]. Both reagents are now commercially available.

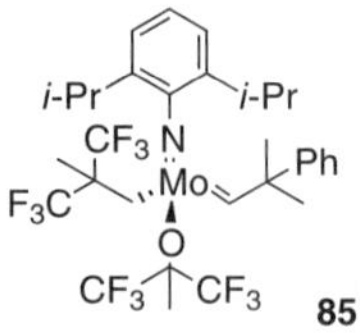

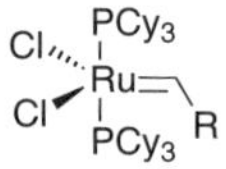

86a: R = CH=CPh$_2$
86b: R = Ph

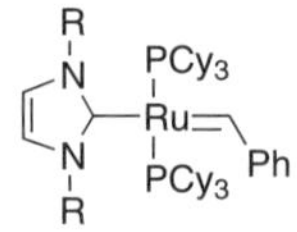

87: R = 2,4,6-Me$_3$Ph-

Fig. 3.

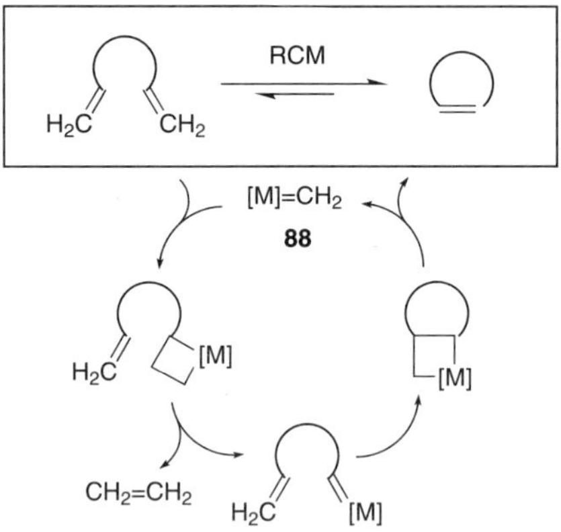

Scheme 13.

The reaction proceeds via metallacyclobutanes as shown in Scheme 13. A [2+2] cycloaddition occurs between the olefin substrate and the metal alkylidene catalyst to produce a metallacyclobutane. Retrocycloaddition then occurs to afford an olefin metathesis product and a new metal alkylidene **88**, which works as a further catalyst.

In general, ruthenium catalysts **86** are less active than **85** with respect to the formation of tri- and tetra-substituted alkenes. Although molybdenum catalyst **85** is appreciably sensitive to air and moisture, ruthenium catalysts **86** are not significantly affected. Both catalysts are tolerant of functionality in the substrate; for example, ketones, esters, amides, epoxides, acetals, silyl ethers, amines, sulfides, and alcohols.

Many applications of these catalysts to the formation of 5-, 6-, and 7-membered carbo- and heterocycles have been reported. Furthermore, RCM is found to be very efficient for the construction of macrolides. Reaction of dienes **89** with the

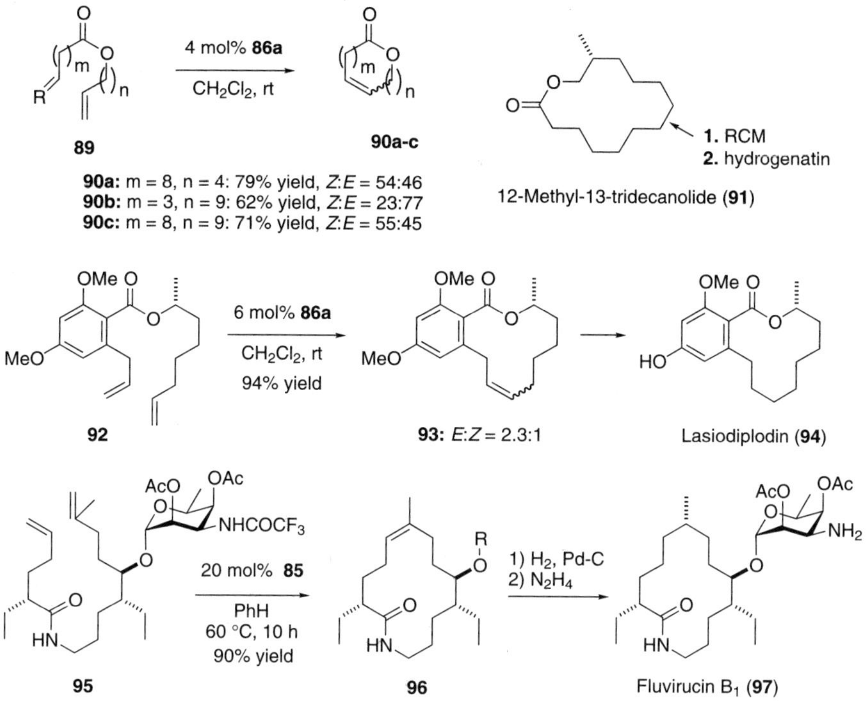

Scheme 14.

Grubbs catalyst **86a** efficiently produced 16- or 21-membered lactone **90a–c** in good yields (Scheme 14) [59]. The method was further applied to the synthesis of (+)-12-methyl-13-tridecanolide (**91**), a minor musk-odor component. Lasiodiplodin (**94**) was efficiently synthesized via the formation of 12-membered lactone **93** by RCM of **92** with **86a** as the key step [60]. Schrock catalyst **85** was successfully applied to the synthesis of fluvirucin B$_1$ (Sch 38516) (**97**). Reaction of the fully functionalized diene **95** with **85** afforded macrolactam **96** as a single (Z)-olefin [61]. RCM has been also utilized in the total synthesis of epothilones (**5**) (Section III.E). In all of these cases mentioned above, the double bond produced by RCM was altered later in the synthesis, either by hydrogenation or, in the case of epothilones (**5**), by epoxidation.

In the total synthesis of (−)-zearalenone (**99**), the catalyst **86b** was ineffective for cyclization of the styrenyl precursor **98** (Scheme 15). However, the cyclization of **98** was performed by using the "second-generation" ruthenium complex **87** bearing an N-heterocyclic ligand for the crucial ring closure [62]. The total syntheses of (+)-ricinelaidic acid lactone (**101**) and (−)-gloeosporone (**104**) have been accomplished based on RCM [63]. Although diene **102** was not cyclized by **86a**, the cyclization proceeded smoothly by treatment with **86b** in the presence of

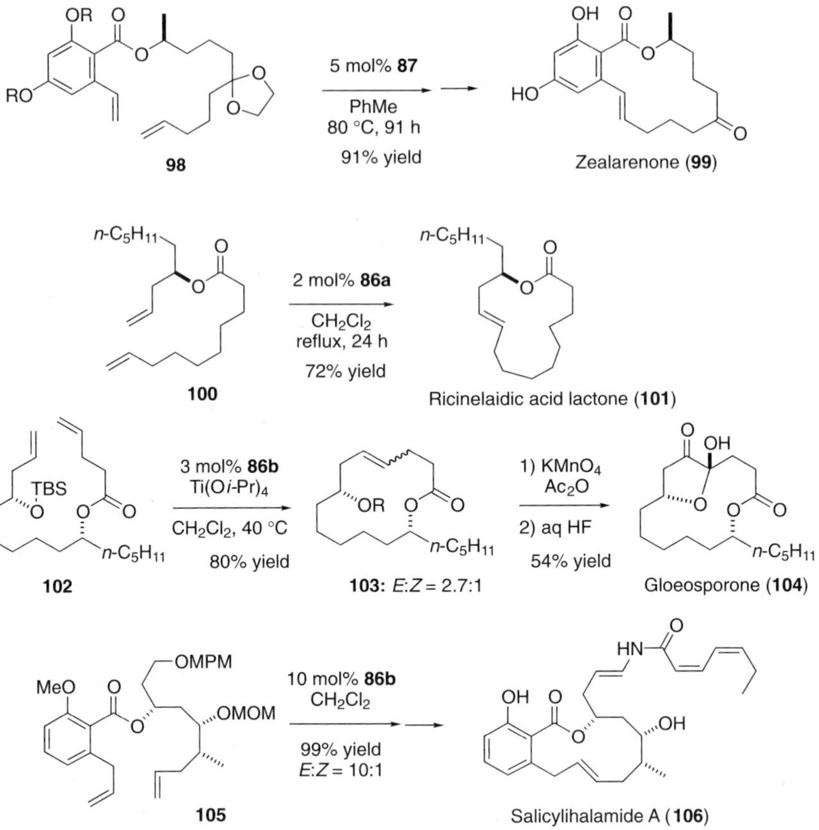

Scheme 15.

catalytic amounts of Ti(O*i*-Pr)$_4$, which clearly features the performance of the new binary catalyst system in RCM. Reaction of diene **105** with **86b** preferentially produced the desired (*E*)-isomer, which led to the total synthesis of salicylihalamide A (**106**) [64].

The ring-closing *alkyne* metathesis of **iii** using a tungsten complex, (*t*-BuO)$_3$W ≡ CCMe$_3$, or Mo(CO)$_6$ and *p*-chlorophenol, has been developed (Scheme 16) [65], which provides an efficient synthesis of macrocyclic (*Z*)-alkenes **v** by the subsequent Lindlar reduction of the resulting cycloalkynes **iv**. This method was applied to the synthesis of ambrettolide (**108**) and yuzu lactone (**109**). The total synthesis of sophorolipid lactone (**111**) containing a (*Z*)-olefin was also accomplished based on the ring-closing alkyne metathesis [66]. Reaction of diyne **110** with a new catalyst Mo[N(*t*-Bu)(3,5-Me$_2$Ph)]$_3$ activated *in situ* by CH$_2$Cl$_2$ produced a cyclized acetylene product, which was subjected to hydrogenation to (*Z*)-olefin and removal of the protective group.

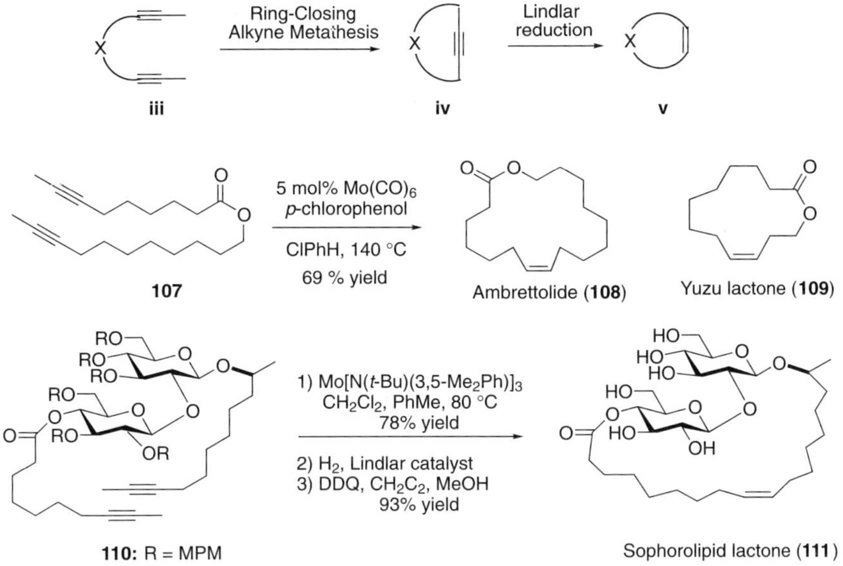

Scheme 16.

The RCM strategy represents an effective, reliable, and powerful approach for the construction of complex and highly functionalized medium and large rings.

3. Stille Coupling Reaction

In recent years, great advances in organic chemistry have been made in organotransition metal chemistry represented by palladium-catalyzed reactions. There are several useful palladium-mediated coupling reactions, for example, the Heck reaction using olefin [67], the Suzuki reaction involving organoborane [68], the Negishi reaction with organozinc [69], the Stille reaction with organostannane [70], the Sonogashira reaction using alkyne [71], and the Tsuji–Trost reactions involving a π-allylpalladium intermediate [72]. This section mainly introduces the Stille coupling reaction, which is often applied to not only coupling of segments but also intramolecular coupling for the construction of macrocyclic rings.

Stille *et al.* developed a very effective and versatile palladium-mediated direct coupling reaction of organic electrophiles with organostannanes and a carbonylative coupling reaction as shown in Scheme 17, respectively. R_1 is typically an unsaturated moiety (for example, vinyl, aryl, heteroaryl, alkynyl, allyl) and R, the nontransferable ligand, is almost always butyl or methyl [70]. Electrophiles participating in the coupling include halides or a triflate. The catalytic cycles for both reactions are shown in Scheme 17. In general, the present reaction proceeds in high yields under mild reaction conditions and tolerates a variety of functional groups such as esters, ketones, nitriles, alcohols and so on.

198   Tadashi Nakata

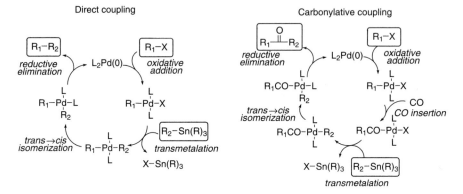

Scheme 17.

Intramolecular versions of this coupling reaction yield a variety of ring sizes from four and five to medium-size rings, and even macrocycles. The first example of the Stille reaction for macrolide construction is the cyclization of **112** containing the vinylstannane and vinyl triflate groups at the termini of the ester chain under high dilution to afford 12- to 15-membered lactones **113** (Scheme 18) [73]. The present reaction under an atmosphere of CO effected an insertion of CO to give keto lactone **114** [74]. The cyclization of acid chloride **115** to give **116** was more efficient in an atmosphere of CO [75]. A series of γ-oxo-α,β-unsaturated 10- to 20-membered macrolides was prepared by this method, which was successfully applied to the total synthesis of C_2-symmetric 16-membered macrolide (–)-pyrenophorin (**117**) by macrodimerization [75c]. The cyclization of organostannanes **118** involving an allylic halide has also been reported, in which large rings **119** were constructed in fair yields [76].

Intramolecular coupling of **120** having an aryl iodide group and a *trans*-vinylstannane group accomplished the total synthesis of (–)-zealarenone (**99**) (Scheme 19) [77]. The first total synthesis of macrolactin A (**124**) was efficiently accomplished based on the Stille reaction of **122** for both stereospecific construction of the diene moieties and closure of the 24-membered macrocyclic ring [78]. The key precursor **122** was synthesized via two Stille couplings and Mitsunobu esterification. An alternative route to the dimethyl ether **125** was reported by cyclization of **123**, which was prepared by the Stille and Suzuki couplings followed by DCC-DMAP esterification

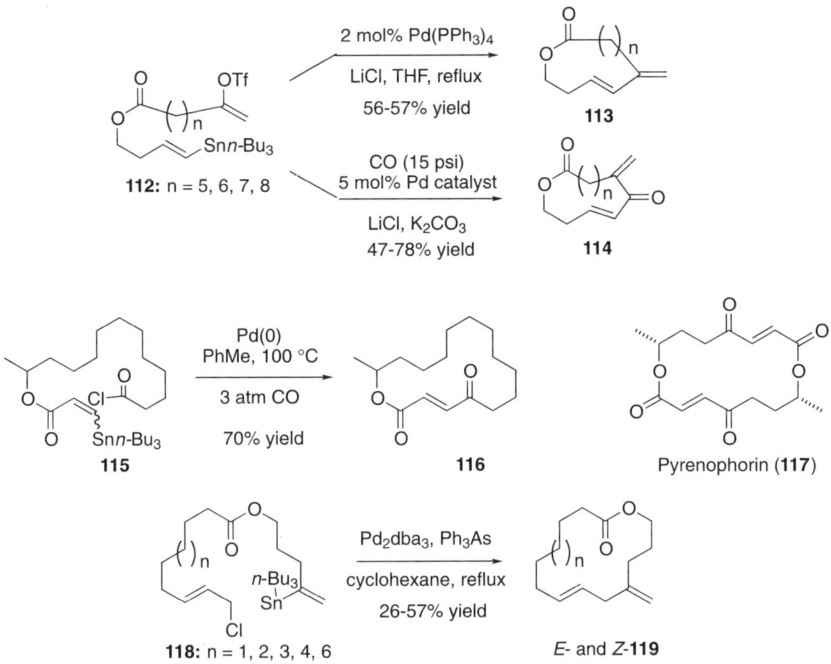

Scheme 18.

[79]. A concise and convergent synthesis of the polyene thiazole-containing macrolide (−)-pateamine A (**127**) has been accomplished via intra- and intermolecular Stille reactions for the construction of 19-membered dilactone and installation of the side chain, respectively [80]. The power of the Stille coupling reaction was demonstrated in the synthesis of a C_2-symmetric 16-membered macrolide **130**, corresponding to the macrocyclic core of elaiophylin (**181**), by an efficient cyclodimerization of **128** [81]. Treatment of **128** with copper(I) thiophene-2-carboxylate **129**, a new Cu(I) reagent [82], promoted the double Stille coupling reactions under mild conditions in the absence of Pd catalyst to give dilactone **130**.

The most spectacular application of the Stille reaction is represented by the final step of Nicolaou's elegant total synthesis of rapamycin (**2**) (see Section III.B), in which a tandem Stille coupling is carried out on the fully functionalized skeleton.

As shown above, the mildness and generality of the palladium-mediated coupling reaction are demonstrated by its frequent application to the late stages of the synthesis of complex natural product, including macrolide antibiotics.

4. Wittig Coupling Reaction

The Wittig coupling reaction is an important reaction for effective olefination of carbonyl groups. The Wittig and related coupling reactions have frequently been used for coupling of two segments and also formation of macrocyclic rings [83].

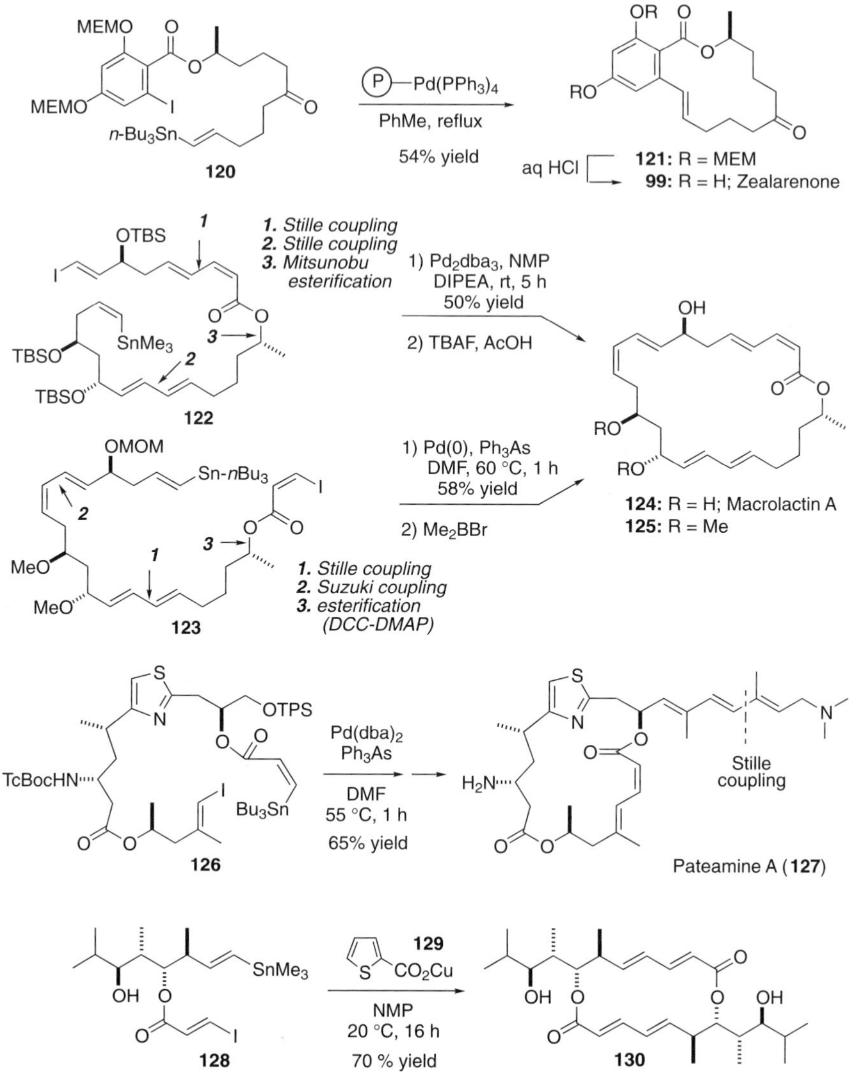

Scheme 19.

Since the first report of the Wittig reaction with phosphonium ylide **vi** [84], several variants have been reported so far (Scheme 20). The reaction of carbonyl compounds with diphenylphosphine oxide **x** [Horner–Wittig (HW) reaction] gives alkenes **xi** [85]. The Horner–Wadsworth–Emmons (HWE) reaction using dialkylphosphonate **xii** has the advantage of increasing reactivity over the original Wittig ylides and is particularly useful for the preparation of (*E*)-α,β-unsaturated esters

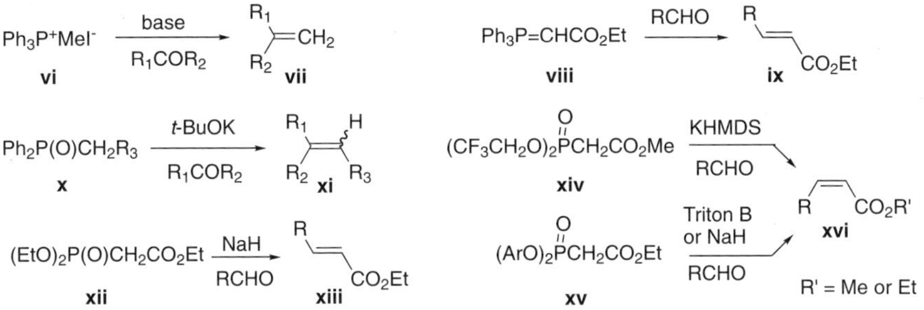

Scheme 20.

xiii and ketones [86]. Still and Gennari [87] and Ando [88] independently reported the highly (Z)-selective HWE reaction using their reagents **xiv** and **xv**.

The intramolecular HWE cyclizations using phosphonoacetate (*i*-PrOLi or LHMDS, HMPA, THF-PhH) or ketophosphonate (NaH, DME or Na, PhMe) under high dilution conditions were developed as variable methods for the construction of macrolactones by Stork and Nakamura [89] and Nicolaou *et al.* [90], respectively. The very mild and efficient conditions for the intramolecular HWE reaction of β-ketophosphonate or phosphonoacetate were reported: (1) treatment with K_2CO_3, in the presence of 18-crown-6 [91], or (2) treatment with tertiary amines such as Et_3N or Hunig's base in the presence of LiBr or LiCl [92]. These methods are applied to the total synthesis of polyene macrolides [93a–95] such as amphotericin B (**170**) [96] and polyoxomacrolides [97–100].

In the total synthesis of amphotericin B (**170**), the Wittig-type reactions were effectively used five times in the construction of the basic skeleton of aglycon **139** (Scheme 21). The HWE reaction of **131** with **132** and the HW reaction of **134** with **135** provided **136**, which was converted into ketophosphonate **137** via DCC-DMAP esterification followed by construction of the polyene part by two consecutive HWE reactions. The final HWE reaction of **137** under K_2CO_3 and 18-crown-6 or DBU and LiCl conditions efficiently constructed the 38-membered ring to give ketone **138** in 70% yield, which was converted into aglycon **139**. The completion of the total synthesis of amphotericin B (**170**) by glycosidation is described in Section II.C.

In the total synthesis of filipin III (**141**) [95], macrocyclization under K_2CO_3/ 18-crown-6 conditions proceeded (Fig. 4), although the cyclization was unsuccessful under LiCl/DBU conditions, which were useful for roflamycoin (**140**) [94] and roxaticin [93]. The total synthesis of ulapualide A (**142**) was accomplished via three Wittig-type couplings [101].

There are several attractive methods for the construction of macrolides (Scheme 22). An intramolecular alkenylzinc/aldehyde addition from ω-alkynal ester **144** in the presence of (−)-DAIB furnished 18-membered lactone **145**, which led to (+)-aspicilin (**146**) [102]. In the total synthesis of 10-deoxymethynolide

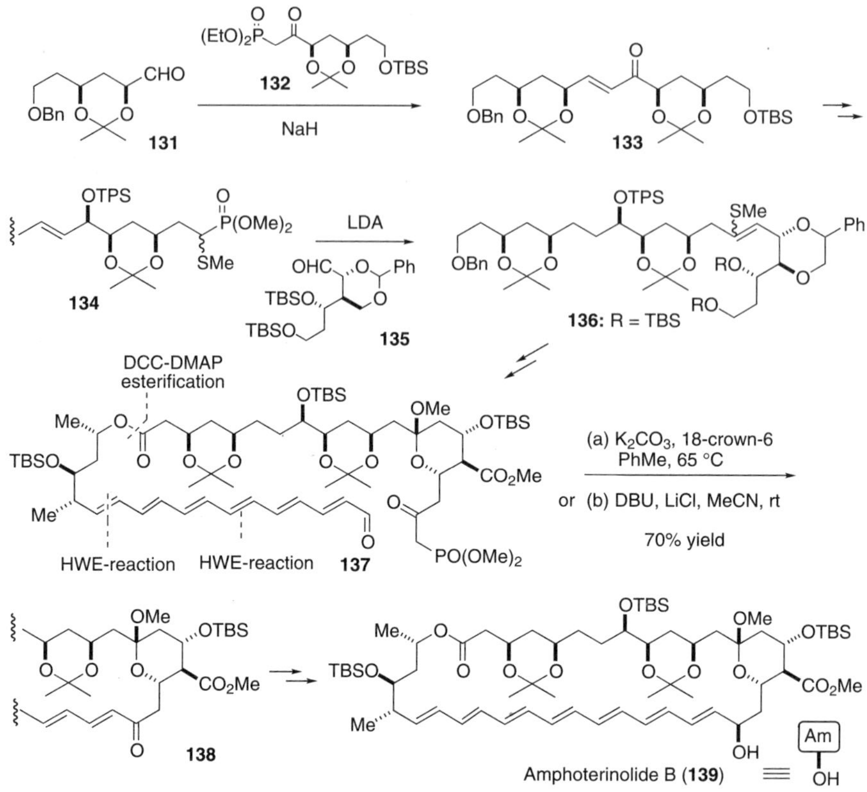

Scheme 21.

(**149**), an intramolecular Nozaki–Hiyama–Kishi coupling reaction of **147** was very effective for construction of the macrocyclic ring **148** [103].

C. Glycosidation

The total synthesis of macrolide antibiotics is not complete without glycosidation to the aglycon. Thus, a variety of methods for glycosidation have been extensively studied by coupling of a glycosyl donor and an acceptor in the presence of an activator [104]. The glycosyl donors involve (1) glycosyl halide, (2) thioglycoside, (3) glycal, (4) 1-*O*-acyl glycoside, (5) 1-*O*- and S-carbonate, (6) 1-*O*-sulfonyl glycoside, (7) 1-*O*-silylated glycoside, (8) 1-*O*-aryl glycoside, (9) 1-hydroxyl sugar, (10) trichloroimidate, (11) 4-pentenyl glycoside, (12) phosphate derivative, (13) ortho ester, (14) 1,2-anhydrosugar, and so on. Among them, only the glycosidations used in macrolide synthesis are described in this section.

Several types of glycosidation were investigated in the synthetic studies on avermectins (**3**) (Section III.C). Nicolaou *et al.* [105] developed a practical synthesis of

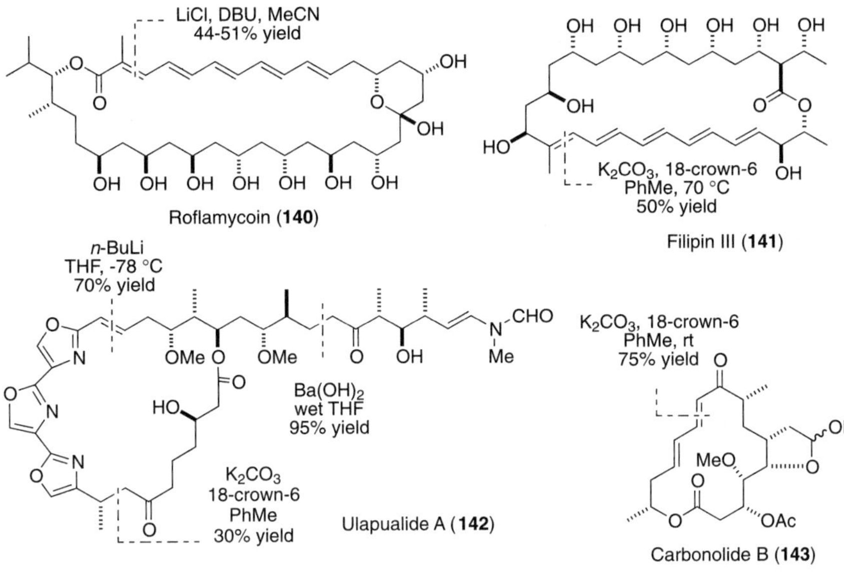

4. Total Synthesis of Macrolides

Fig. 4.

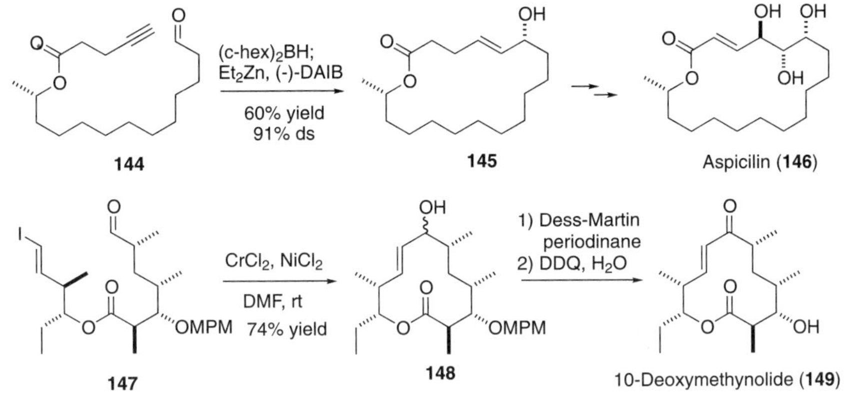

Scheme 22.

oligosaccharides from phenylthio sugars via glycosyl fluorides that was applied to a partial synthesis of avermectin B_{1a} (**3b**) (Scheme 23A). Coupling of fluoride **150** and thioglycoside **151** with $AgClO_4$-$SnCl_2$ exclusively afforded disaccharide **152** as an α-anomer. Activation of **152** as fluoride **153** followed by coupling to aglycon **154a** prepared from natural avermectin B_{1a} (**3b**), furnished the di-TBS ether of **3b**. Deprotection of the silyl groups completed the partial synthesis of avermectin B_{1a} (**3b**).

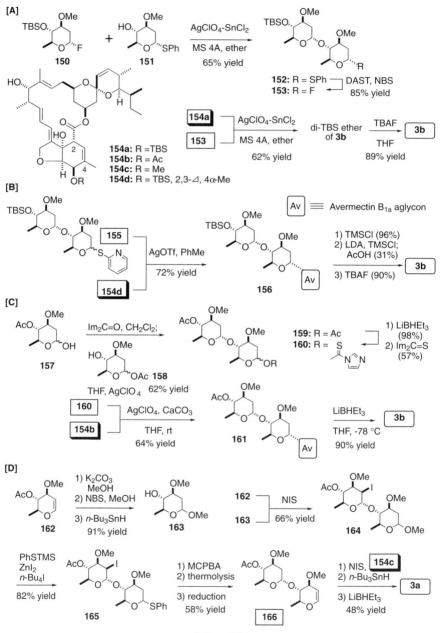

Scheme 23.

In Hanessian's total synthesis of **3b** (Scheme 23B), silver-induced glycosidation [106] of synthetic aglycon **154d** (Section III.C.1) with 2-pyridylthio glycoside **155**, prepared from natural **3b**, stereoselectively yielded **156**. After TMS-silylation of the angular hydroxyl group, deconjugation of the C2–C3 double bond to the C3,4 position followed by deprotection of all silyl groups completed the total synthesis of avermectin B_{1a} (**3b**).

In Ley's total synthesis of **3b** (Scheme 23C), the synthesis of the disaccharide **159** was accomplished via silver-mediated coupling [107] of imidazolylcarbonyl derivative of **157** with **158**. The selective deacetylation with $LiBHEt_3$ followed by treatment with thiocarbonyldiimidazole gave **160** as the active coupling partner. Glycosidation of synthetic 5-acetyl aglycon **154b** (Section III.C.2) with **160** in the presence of $AgClO_4$ and $CaCO_3$ gave avermectin B_{1a} diacetate (**161**). Finally, the selective cleavage of the diacetate with $LiBHEt_3$ afforded avermectin B_{1a} (**3b**).

In Danishefsky's total synthesis of avermectin A_{1a} (**3a**) (Scheme 23D), L-2,6-dideoxyglycal derivative **162** was synthesized in optically active form through their developed Lewis acid-catalyzed diene-aldehyde cyclocondensation (LACDAC) [108]. The coupling partner **163** was prepared from **162** by methanolysis, methoxybromination, and debromination. Coupling between **162** and **163** with NIS [109] produced the required axial α-glycoside **164** with high stereoselectivity. The Hanessian reaction of **164** with PhSTMS regioselectively afforded **165**, which was converted into the unsaturated disaccharide **166**. NIS treatment of **166** with synthetic avermectin A_{1a} aglycon (**154c**) effected the coupling, and deiodination followed by reduction of the acetate provided avermectin A_{1a} (**3a**).

In the glycosidation of amphoterinolide B (**139**) toward amphotericin B (**170**), the glycosyl donor **167** with the C2 acetoxy group was employed so that formation of the requisite β-anomer could be secured through neighboring group participation (Scheme 24). Glycosidation of **139** with trichloroacetimidate **167** could be

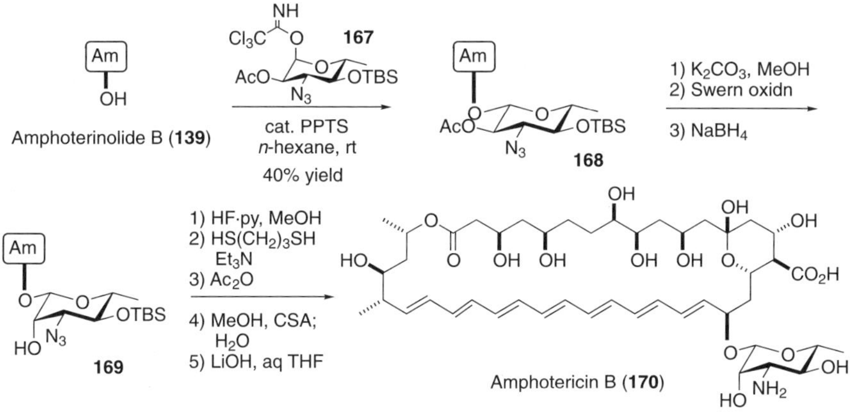

Scheme 24.

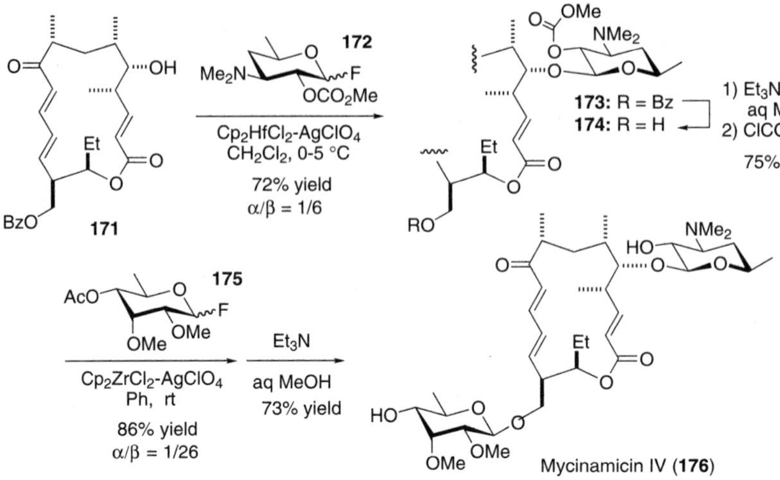

Scheme 25.

achieved in the presence of a catalytic amount of PPTS. The requisite inversion of the C2 configuration was accomplished via reduction of ketone, derived from **168**, to give **169**. The final functional group manipulation achieved the total synthesis of amphotericin B (**170**).

Suzuki et al. [110] developed a selective glycosidation method for the introduction of β-D-mycinosyl and β-D-desosaminyl linkages. This method was successfully applied to the first total synthesis of mycinamicin IV (**176**) (Scheme 25) [110c]. Glycosidation of synthetic aglycon **171** with fluoride **172** in the presence of Cp_2HfCl_2-$AgClO_4$ proceeded smoothly to afford **173** with 1:6 ratio of the α- and β-anomers. The second glycosidation of **174** with fluoride **175** in the presence of Cp_2ZrCl_2-$AgClO_4$ took place efficiently to afford the desired β-anomer.

In the first total synthesis of elaiophylin (azalomycin B; **181**), glycosidation of β-hydroxy ketone **178** and glycal **177** was examined (Scheme 26) [111]. NBS-promoted glycosidation [112] followed by debromination with n-Bu_3SnH-AIBN was first applied to give **179** in only a 30% yield. Then, the Wakamatsu procedure using CSA-MS 4A [113] was found to afford the desired **179** in 80% yield as the sole anomer, which led to **181** via aldol reaction of the (Z)-boron enolate of the ketone **179** with dialdehyde **180**.

The glycosidation of oleandolide (**182**) with **183**, prepared from rhamnosides, was first accomplished by a modified Woodward procedure (Scheme 27). Reaction of **182** with thioglycoside **183** in the presence of AgOTf afforded the desired β-glycoside **184**, regio- and stereoselectively. The second glycosidation with glycal **185** was widely investigated by a variety of methods; treatment with NBS, NIS, etc. The best result was again realized by using CSA-MS 4A to afford **186**, which completed the first total synthesis of oleandomycin (**187**) [114].

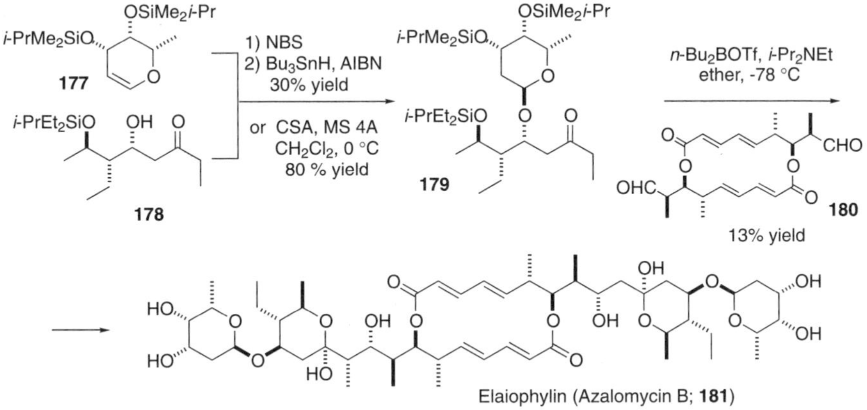

4. Total Synthesis of Macrolides

Scheme 26.

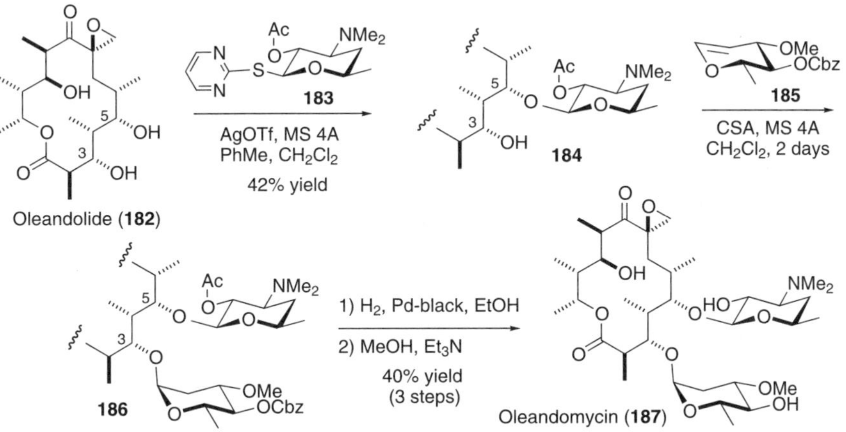

Scheme 27.

Toshima, Tatsuta *et al.* [115] developed highly stereoselective and powerful glycosidation using 2,6-anhydro-2-thio sugar, which was applied to glycosidation in their total synthesis of erythromycin A (**193**) (Scheme 28) [115e]. The first glycosidation of **188** with **189** was performed by the modified Woodward procedure in the presence of AgOTf to give **190**. Next, their glycosidation method of **190** with activated 2,6-anhydro-2-thio sugar **191** in the presence of NIS-TfOH afforded the desired α-glycoside **192** in 90% yield as the only isolated product. In contrast to this result, glycosidation of **190** with glycal **194**, fluoride or thioglycoside **195** under some appropriate conditions was not effective. The total synthesis of erythromycin A (**193**) was accomplished via desulfurization of **192** and hydrogenolysis with Raney-Ni. The efficiency of this novel

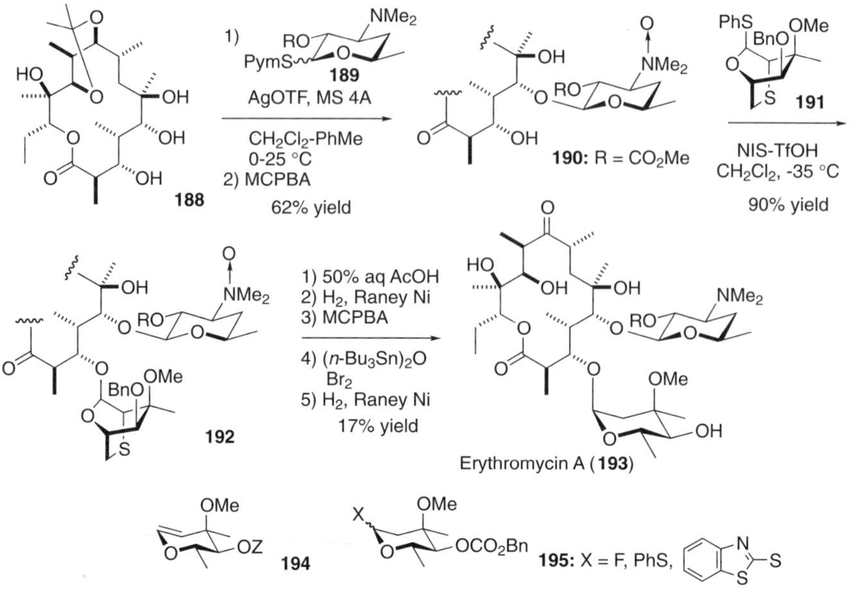

Scheme 28.

method was also demonstrated by the synthesis of disaccharide, corresponding to the glycon of avermectins (**3**) [115d].

During the course of synthesis of cytovaricin (**196**) [116], a number of glycosidation methods were surveyed for the preparation of the required polyol glycoside subunit **199** (Scheme 29). After numerous attempts, the most successful protocol was the Mukaiyama method using Ph_3CClO_4 as an activator [117]. Treatment of **197** and **198** with a catalytic amount of Ph_3CClO_4 in toluene effected glycosidation to afford a 1:3 mixture in favor of the undesired α-anomer **200**. However, on warming the reaction mixture to −3°C, anomeric equilibration took place to give a 4:1 mixture in favor of the desired β-anomer **199**. After the success of the glycosidation, the total synthesis of cytovaricin (**196**) was accomplished via Julia coupling between the spiroketal subunit and the polyol glycoside subunit followed by Keck macrolactonization (92%).

In the total synthesis of (+)-A83543A [(+)-lepicidin A; **208**] (Scheme 30) [118], the aglycon **201** was synthesized via Yamaguchi lactonization, Stille coupling, intramolecular Diels–Alder reaction, and intramolecular aldol reaction as the key reactions. Glycosidation of **201** with D-rhamnose derivative **202** was again realized by the Mukaiyama method using Ph_3CClO_4 to afford the desired α-anomer **203** in 87% yield. The second glycosidation of pseudoaglycon **204** with L-furosamine derivative **205** was performed by treatment of silver zeolite. However, the desired β-anomer **207** was a minor product in 10% yield (α-anomer

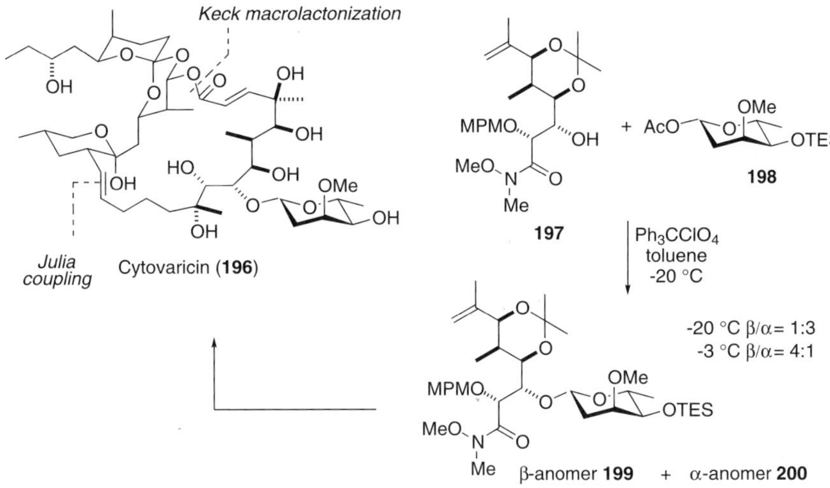

4. Total Synthesis of Macrolides

Scheme 29.

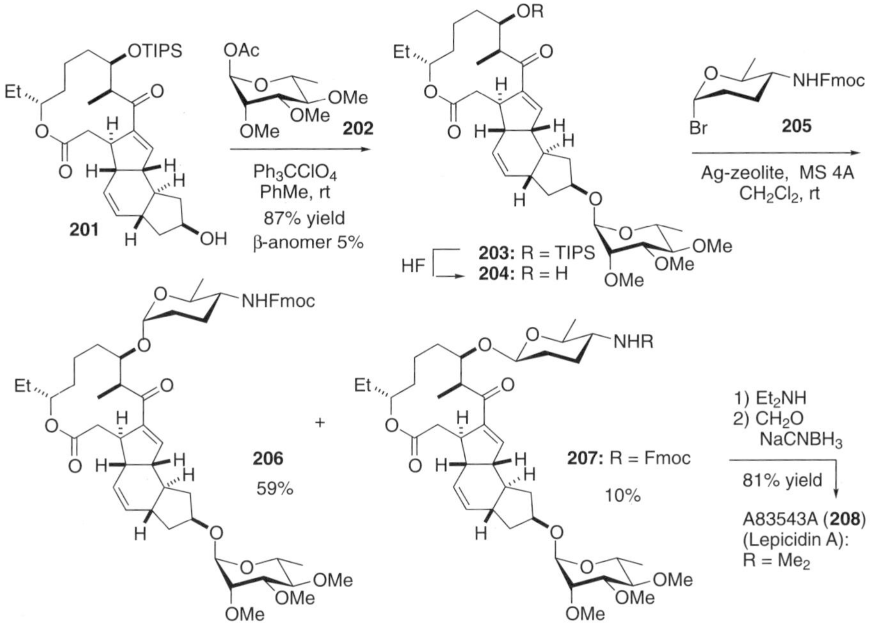

Scheme 30.

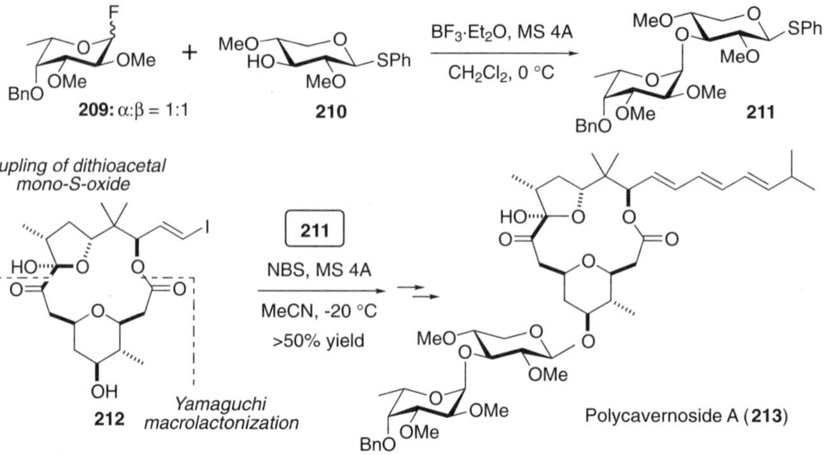

Scheme 31.

206: 59%, recovered **204**: 20%). Removal of the Fmoc protective group in **207** followed by methylation provided (+)-A83543A (**208**).

In the first total synthesis of (−)-polycavernoside A (**213**) by Murai et al. (Scheme 31) [119], fucosyl-xylose disaccharide **211** was prepared by coupling of fluoride **209** and phenylthioglycoside **210** by treatment of $BF_3 \cdot Et_2O$ or $SnCl_2$-$AgClO_4$. NBS-promoted glycosidation [120] of **212** with the activated disaccharide **211** delivered exclusively the β-glycoside in >50% yield, which led to **213**.

III. Total Synthesis of Selected Macrolides

A. FK506 (Tacrolimus)

FK506 (tacrolimus; **1**), isolated from the fermentation broth of *Streptomyces tsykubaensis* in 1987, exhibits strong immunosuppressive activity that is approximately 100 times greater than that of cyclosporin A. The structural feature of FK506 (**1**) (Fig. 1), a unique 23-membered macrolide containing both lactam and lactone linkages, involves an unusual α,β-diketoamide partially masked via C10 hemiketal formation, δ-pipecolinyl-α′-allyl-β-hydroxy ketone, two trisubstituted olefins, 1,2,4-trisubstituted cyclohexane and highly substituted tetrahydropyran rings.

The total synthesis of FK506 has been completed by the Merck group [121] in 1989, by Schreiber et al. [122] in 1990, and by Ireland et al. [123] in 1996. In addition, Danishefsky et al. [124], Gu and Sih [125], and Smith et al. [126] have each completed formal total synthesis. The synthetic strategy of each group is outlined in Fig. 5. The order of bond connection of significant segments for the synthesis of the target molecule is described by number and reaction names. The several

4. Total Synthesis of Macrolides 211

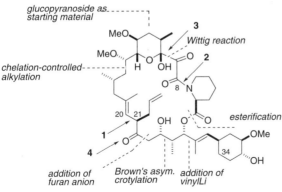

Fig. 5.

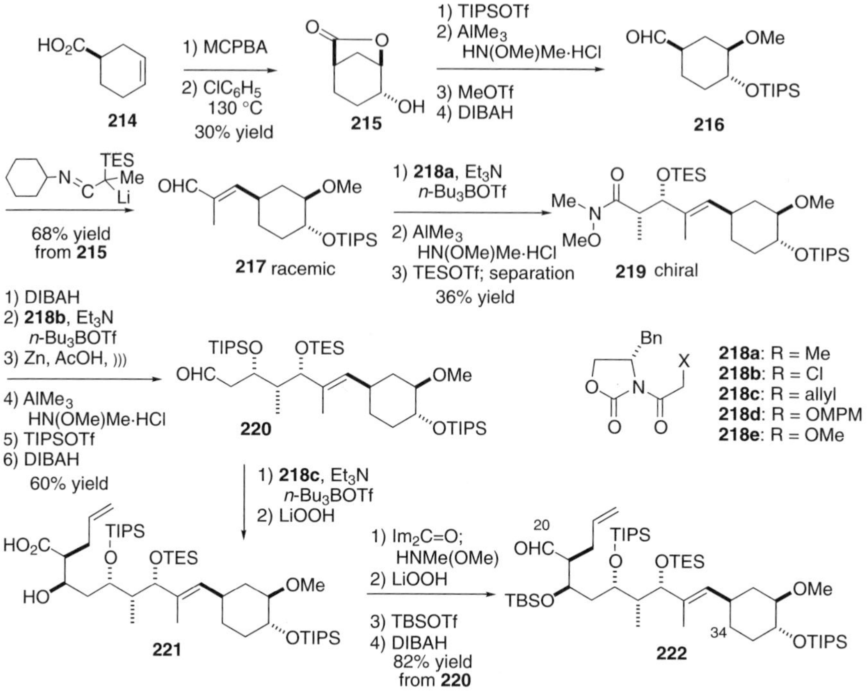

Scheme 32.

coupling positions of small fragments for the synthesis of the segments are also indicated by dotted lines and reaction names in italic.

1. Merck's Total Synthesis [121]

The first total synthesis by the Merck group was accomplished via coupling of the C10–C19 and C20–C34 segments, attachment of (S)-pipecolic acid to the C26 hydroxyl group, aldol addition of glycolate derivatives to the C10 aldehyde, and final macrolactamization (Fig. 5).

The synthesis of the C20–C34 bottom segment **222** is shown in Scheme 32. The synthesis was accomplished by effective repetition of the Evans asymmetric aldol reactions. Epoxidation of racemic **214** followed by thermolysis effected lactonization to give **215**, which was transformed to aldehyde **216** through functional group manipulation, including protection of hydroxyl groups, and conversion of lactone to Weinreb amide and reduction. α,β-Unsaturated aldehyde unit in **217** was introduced in one step by addition of 2-lithio-2-TES-propanal cyclohexylimine. The Evans aldol reaction using the boron enolate of chiral imide **218a** provided 2,3-*syn*-adduct, which was transformed to Weinreb amide **219**. Separation of the diastereomers at this stage gave optically pure **219**. After DIBAH reduction, the resulting

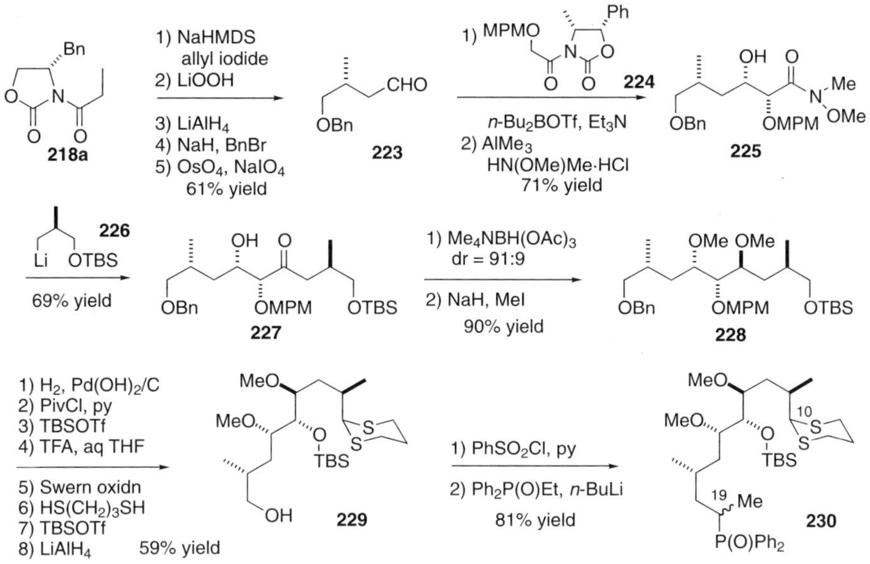

4. Total Synthesis of Macrolides   213

Scheme 33.

aldehyde was again subjected to the Evans aldol reaction using **218b** and further functional group manipulation to afford aldehyde **220**. The third Evans aldol reaction using **218c** followed by hydrolysis afforded carboxylic acid **221**, which was converted into aldehyde **222** via a standard procedure.

The Evans asymmetric alkylation [127] and aldol reactions were also effectively applied to the synthesis of the C10–C19 top segment **230** (Scheme 33). The starting chiral unit **223** was synthesized via the Evans asymmetric alkylation of **218a**. The subsequent Evans aldol reaction of **223** with **224** followed by transamidation yielded 2,3-*syn*-diol derivative **225** with complete stereoselectivity. Addition of alkyl lithium **226** to the Weinreb amide **225** produced ketone **227**, which was stereoselectively reduced and methylated to give dimethyl ether **228**. The standard functional group manipulation afforded thioacetal **229**, which was converted into phosphine oxide **230**.

The completion of the total synthesis of FK506 (**1**) is described in Scheme 34. The coupling of the two segments **222** and **230** was accomplished by phosphine oxide-mediated HW olefination. The addition of **222** to **230** afforded a separable 1:1 mixture, and the less polar diastereomer yielded (*E*)-olefin **231**. After selective removal of the TES group, esterification of **231** with (*S*)-*N*-Boc-pipecolic acid (**232**) under DCC-DMAP conditions followed by dethioacetalization afforded aldehyde **233**. The Evans aldol reaction of **233** with **218d** installed a glycolate unit, and hydrolysis of the chiral auxiliary followed by TES protection provided **234** ready for macrocyclization. The macrolactamization of **234** was effectively

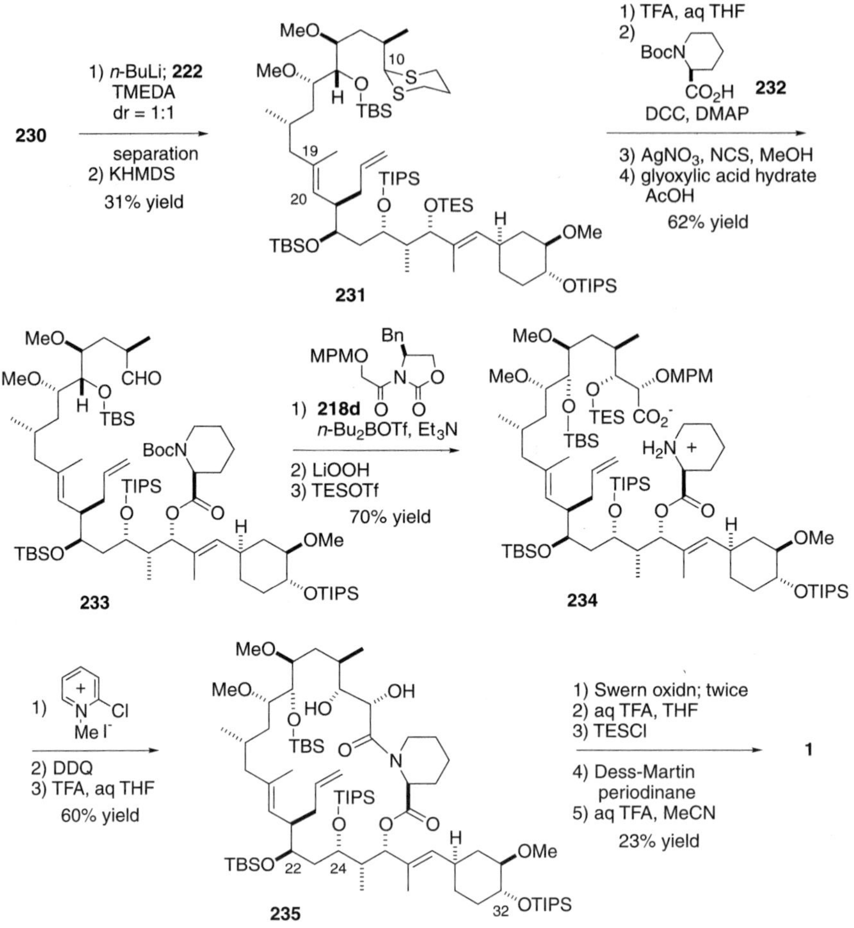

Scheme 34.

accomplished with Mukaiyama's reagent, 2-chloro-1-methlpyridinium iodide, under high-dilution conditions in high yield. Subsequent deprotection of the MPM and TES groups afforded diol **235**. Two successive Swern oxidations gave α,β-diketoamide, which was subjected to deprotection of all silyl groups followed by regioselective TES protection at C24 and C32, and Dess–Martin oxidation at C22. The final treatment with aqueous HF completed the first total synthesis of FK506 (**1**).

2. Schreiber's Total Synthesis [122]

The total synthesis by Schreiber et al. was accomplished via coupling of the C10–C19 and C20–C34 segments having a pipecolic acid moiety (the C1–C7 part), aldol coupling of a glycolate derivative to the C10-aldehyde, and final macrolactonization (Fig. 5).

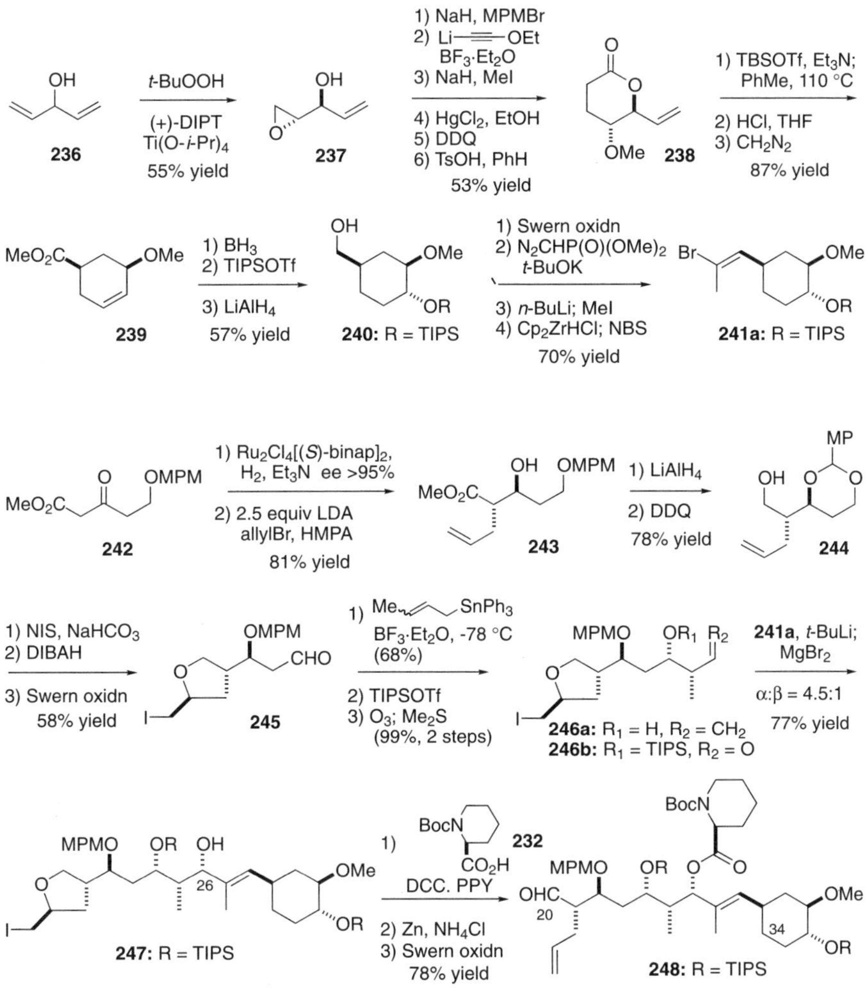

4. Total Synthesis of Macrolides 215

Scheme 35.

The synthesis of the C20–C34 bottom segment **248** is outlined in Scheme 35. The synthesis of cyclohexanyl fragment **241a** started with the catalytic Sharpless AE of divinylcarbinol **236** to give **237**. The epoxide **237** was converted into δ-lactone **238** by a sequence of reactions, including addition of lithium ethoxyacetylide, methylation, hydrolysis, and lactonization. The Ireland–Claisen rearrangement of the ketene acetal derived from **238**, hydrolysis and esterification afforded ester **239**. Regio- and stereoselective hydroboration of the double bond, TIPS protection, and ester reduction afforded alcohol **240**, which was then converted into the desired vinylbromide **241a**, via Swern oxidation, propyne formation, and hydrozirconation/bromination.

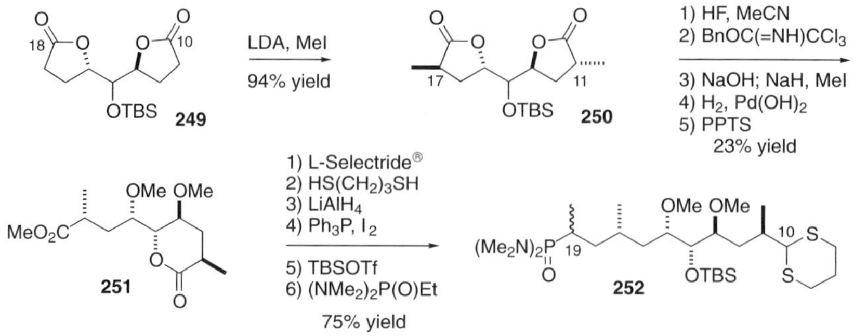

Scheme 36.

The synthesis of the coupling partner **246b** started with β-keto ester **242**. Catalytic asymmetric hydrogenation of **242** under Noyori's conditions afforded β-hydroxy ester with excellent enantioselectivity. The subsequent Frater–Seebach alkylation provided 2,3-*anti*-ester **243**. Reduction followed by acetalization afforded **244**. The allyl group of **244** was masked as the iodo ether, and reductive cleavage of acetal followed by oxidation gave aldehyde **245**. Treatment of **245** with (E/Z)-crotylstannane in the presence of $BF_3 \cdot Et_2O$ gave the desired α-alcohol **246a** as the major product, which was transformed to aldehyde **246b** by TIPS protection and oxidative cleavage of the olefin. With the desired **241a** and **246b**, the coupling was investigated. The reaction of the aldehyde **246b** with vinyl lithium prepared from **241a** in the presence of $MgBr_2$ afforded 26α-alcohol **247** as the major isomer. Esterification of **247** with (S)-N-Boc-pipecolic acid (**232**) was realized under DCC and 4-pyrrolidinopyridine conditions. Subsequent treatment with zinc dust revealed an allyl side chain, and the Swern oxidation afforded aldehyde **248**.

The C10–C19 top segment **252** was constructed based on a unique two-directional chain synthesis strategy (Scheme 36). The synthesis started with C_2-symmetric bis(lactone) **249**, prepared from commercially available arabitol. Methylation of the dianion of **249** effectively took place on both lactone rings to afford 11,17-dimethyl bis(lactone) **250** with high stereoselectivity. The C_2-symmetric bis(lactone) **250** was converted into **252** through unique differentiation of the left and right sides. After exchange of TBS to a benzyl group, **250** was subjected to hydrolysis followed by methylation to provide the C_2-symmetric acyclic bis(methyl ester). Hydrogenolysis of benzyl ether followed by PPTS treatment effected differentiated cyclization to afford δ-lactone **251**, which led to the desired **252** in a sequence of reactions: (1) selective reduction of the lactone to lactol, (2) thioacetalization with concomitant lactonization, (3) lactone reduction to diol, (4) selective iodination of the primary alcohol, (5) TBS protection of the secondary alcohol, and (6) introduction of phosphonamide.

The completion of the total synthesis is described in Scheme 37. The reaction of a lithio derivative of **252** with **248** gave two separable pairs of diastereomeric

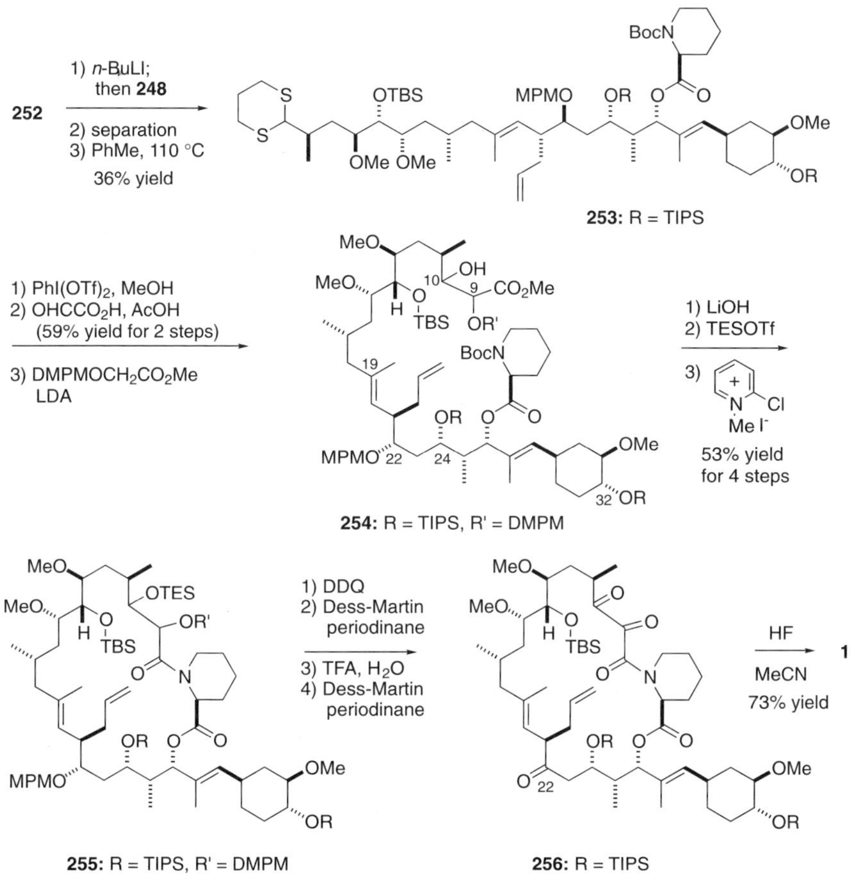

4. Total Synthesis of Macrolides

Scheme 37.

adducts. The major pair of diastereomers underwent stereospecific elimination by heating to give the desired (*E*)-olefin **253**. Dethioacetalization followed by aldol reaction with protected methyl glycolate produced a mixture of hydroxy esters **254**. Hydrolysis of the methyl ester and TES protection of the C10 hydroxyl group afforded a diastereomeric mixture of amino acids, which was treated with Mukaiyama's reagent to give macrolactam **255**. Deprotection of the MPM and DMPM groups followed by oxidation and subsequent deprotection of the TES group followed by oxidation gave 22-oxo-α,β-diketoamide **256**. Final deprotection of TBS and TIPS ethers furnished FK506 (**1**).

The Schreiber group has further carried out structural and mechanistic studies combined with chemistry and biology [128]. The solution structure of FK506 and rapamycin binding protein (FKBP) has been determined by NMR and the structure of FKBP, complexed with FK506, has also been determined by X-ray crystallography.

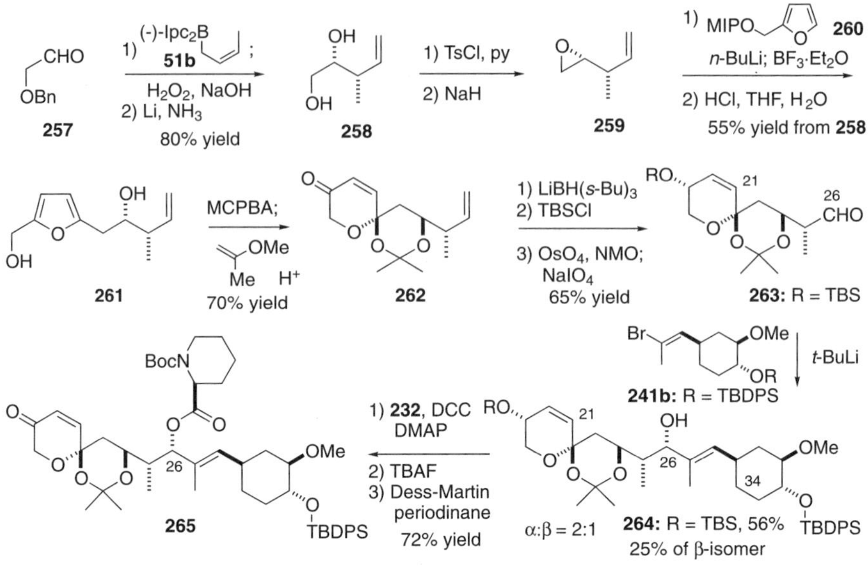

218   Tadashi Nakata

Scheme 38.

They have also investigated the design, synthesis, and analysis of 506BD, a high-affinity ligand for FKBP, which inhibits the rotamase activity of FKBP. Further pioneering study in this field is in progress.

3. Ireland's Total Synthesis [123]

The total synthesis by the Ireland group was accomplished via Michael addition of the C8–C20 segment to the C21–C34 segment having a pipecolic ester moiety, and macrolactonization (Fig. 5). Their synthetic feature is that the C21–C24 α'-allyl-β-hydroxy ketone system is masked as 6,6-spiroketal during the synthesis and the labile α,β-diketoamide hemiketal system is prepared by oxidation of a masked enediol at a late stage.

The synthesis of the C21–C34 bottom segment **264** is illustrated in Scheme 38. The C21–C26 fragment was constructed as spiroketal **263**, which has a masked α'-allyl-β-hydroxy ketone unit. The Brown asymmetric crotylation of **257** with (−)-Ipc$_2$B(crotyl) (**51b**) followed by deprotection of the benzyl group afforded *syn*-diol **258** with high enantio- and stereoselectivity. Lewis acid-catalyzed addition of lithiated **260** to epoxide **259**, prepared from **258**, followed by deprotection afforded **261**. Oxidation with MCPBA and acetonization produced the spiroketal **262** as a masked α'-allyl-β-hydroxy ketone, which was transformed to coupling partner **263**, by reduction of the ketone followed by silyl protection and oxidative cleavage of the terminal double bond. The coupling of vinyl lithium, derived from **241b**, with the aldehyde **263** afforded the desired 26α-alcohol **264** and its β-isomer in a 2:1 ratio. The cyclohexyl unit **241b** was synthesized starting from optically active

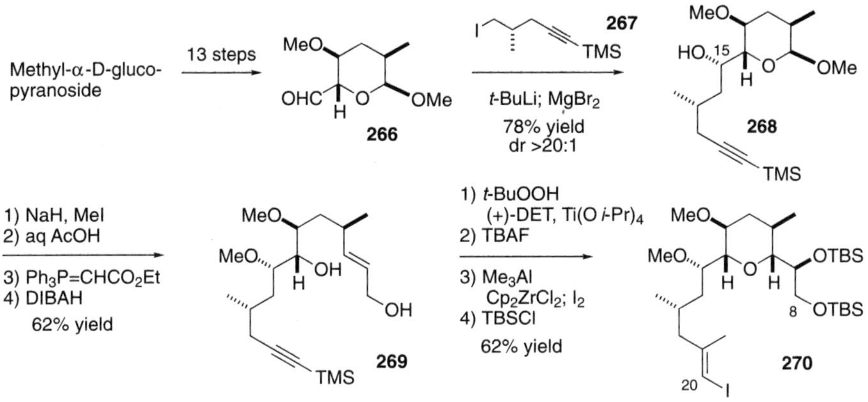

4. Total Synthesis of Macrolides

Scheme 39.

cyclohexenyl carboxylic acid **214** (98% ee) through similar procedures to those of the Merck and Schreiber groups. Treatment of the desired **264** with N-Boc-L-pipecolic acid in the presence of DCC and DMAP afforded the ester of the C26 alcohol. Deprotection of the TBS group followed by oxidation gave enone **265**.

The synthesis of the C8–C20 top segment **270** started with **266**, prepared from methyl-α-D-glucopyranoside (Scheme 39). Reaction of aldehyde **266** with alkyllithium of **267** in the presence of MgBr$_2$ afforded the desired 15α-alcohol **268** with high stereoselectivity. Methylation, hydrolysis, Wittig reaction, and DIBAH reduction afforded allylic alcohol **269**. The Sharpless AE stereoselectively furnished β-epoxide, which was converted *in situ* to the desired tetrahydropyran diol. After removal of the TMS group, the resulting acetylene was subjected to zirconium-catalyzed carboalumination followed by iodide quenching to provide vinyl iodide, which was then protected to afford TBS ether **270**.

The coupling of the top and bottom segments and completion of the total synthesis are summarized in Scheme 40. Successive treatment of the iodide **270** with *t*-BuLi, hexynylcopper·2HMPT, and TMSCl effected stereoselective conjugate addition to the spiroenone **265** to give **271** after TBS deprotection. Selective oxidation of the primary alcohol of **271** followed by esterification gave methyl ester **272**. After functional group manipulation of **272**, the resulting 9-keto amino acid was treated with Mukaiyama's reagent to give macrolactam **273**. The remaining task was to reveal the α′-allyl-β-hydroxy ketone system from the spiroketal and oxidation of the C10 position. Stereoselective reduction of the ketone on the spiroketal followed by iodination afforded iodide **274**. After conversion of C9 ketone to the TES-enolate, treatment with active Zn/Ag-graphite effected reductive fragmentation of the spiroketal to afford the α′-allyl-β-hydroxy ketone **275**. Oxidation of the TES-enolate with dimethyldioxirane (DMDO) generated the α,β-diketoamide hemiketal system. Final desilylation completed the total synthesis of FK506 (**1**).

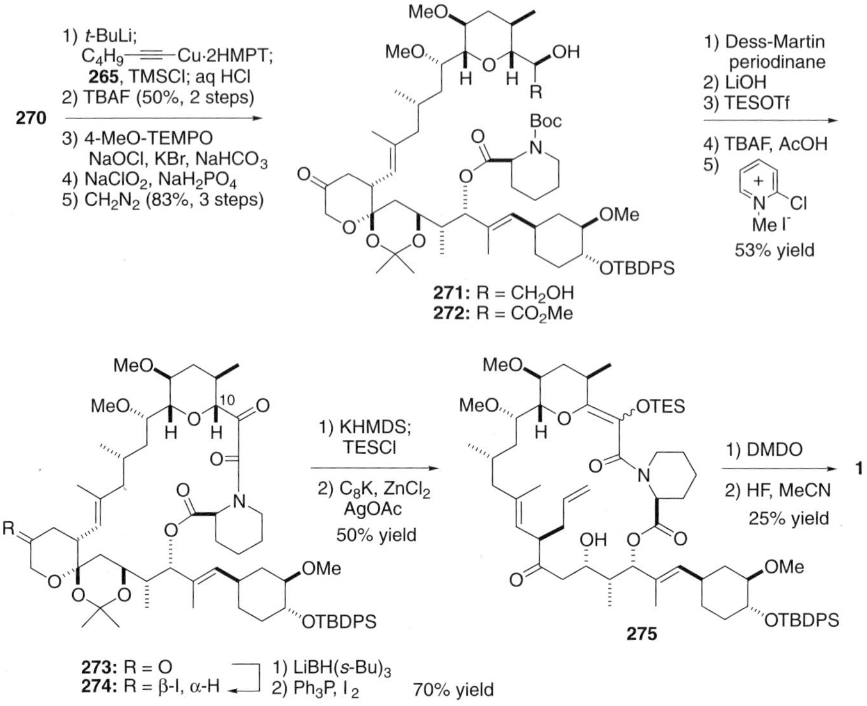

Scheme 40.

B. Rapamycin (Sirolimus)

Rapamycin (sirolimus; **2**), isolated from *Streptomyces hygroscopicus*, is a highly functionalized 31-membered macrolide that exhibits potent antibiotic, cytotoxic, and immunosuppressive activity. FK506 (**1**) and rapamycin (**2**) are the structurally related macrolides (Fig. 1); thus, rapamycin possesses an α,β-diketoamide hemiketal system, a pipecolic acid moiety, 1,2,4-trisubstituted cyclohexane, and trisubstituted tetrahydropyran rings, which are similar to those of FK506. In addition to these units, rapamycin (**2**) includes an (*E,E,E*)-triene moiety, two stereochemically complex aldol units, and 15 chiral centers beyond those found in FK506.

The total synthesis of rapamycin (**2**) has been accomplished by the Nicolaou [129], Schreiber [130], Danishefsky [131], and Smith [132] groups.

1. Nicolaou's Total Synthesis [129]

The first total synthesis of rapamycin (**2**) was accomplished by Nicolaou *et al.* in 1993. The synthetic feature includes esterification of the C34 hydroxyl group of the C21–C42 segment with pipecolic acid, amide formation with the C8–C18 segment, and extremely novel macrocyclization by double Stille coupling using vinyl stannane at the C18 and C21 positions (Fig. 6).

4. Total Synthesis of Macrolides 221

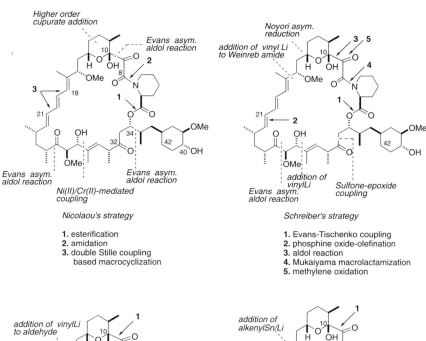
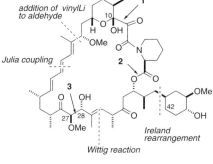
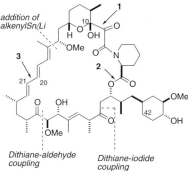

Fig. 6.

The synthesis of three fragments **278**, **282**, and **285** for the C21–C42 bottom segment is summarized in Scheme 41. The Eschenmoser–Claisen rearrangement of amido acetal of **276**, which was prepared via 2-bromocyclohexenone by Corey's asymmetric reduction, afforded amide **277**. Functional group manipulation including chain elongation provided Evans-type amide **278**. The Evans aldol reaction of boron enolate of **279** with aldehyde **280** stereoselectively afforded **281**, which was converted into aldehyde **282** through a sequence of seven steps

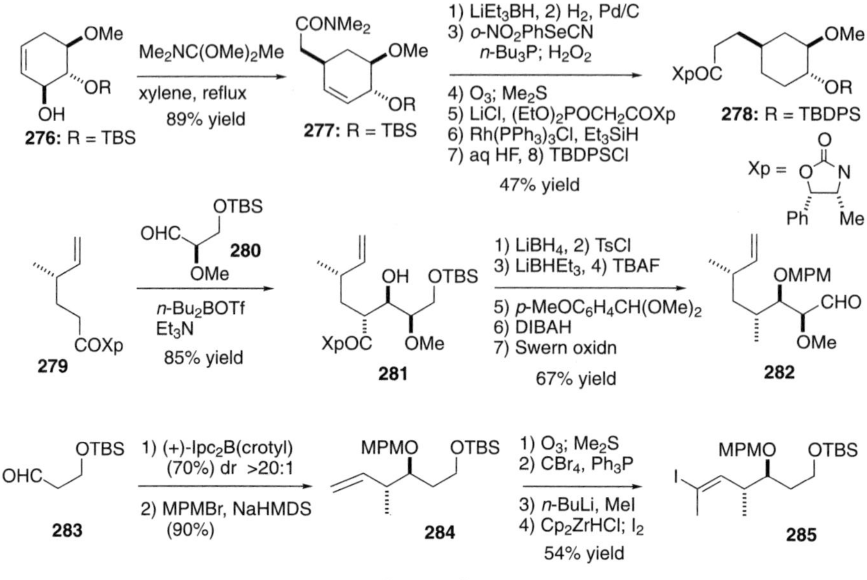

Scheme 41.

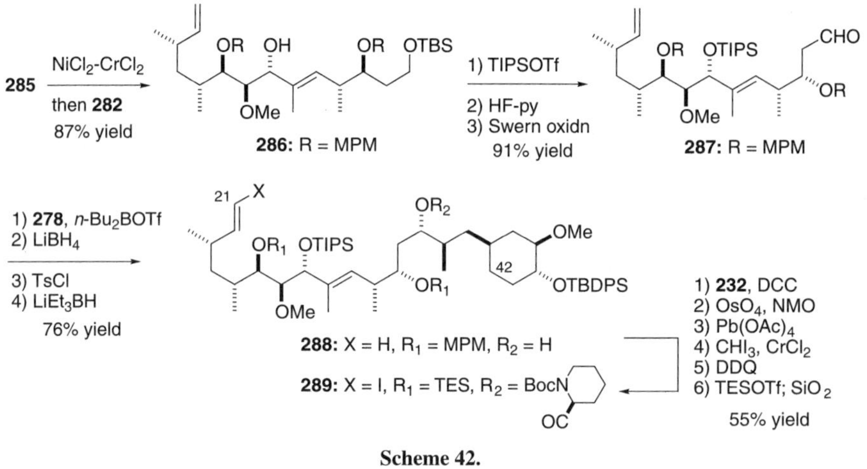

Scheme 42.

involving conversion of the amide to a methyl group. The Brown asymmetric crotylboration of **283** followed by MPM protection provided **284** with high stereo- and enantioselectivity. Oxidative cleavage of the double bond, propargylation, and hydrozirconation/iodination afforded vinyl iodide **285**.

The synthesis of the C21–C42 bottom segment **289** was achieved via coupling of the three subunits **285**, **282**, and **278** (Scheme 42). The key coupling

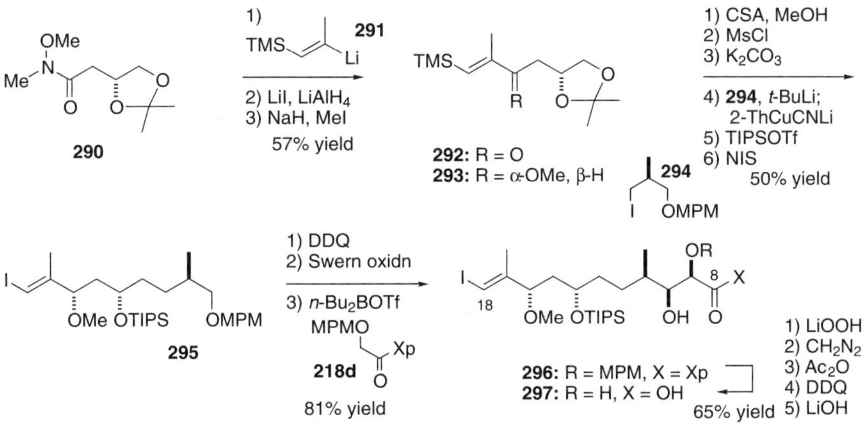

4. Total Synthesis of Macrolides

Scheme 43.

reaction of **285** and **282** was realized by a Ni(II)–Cr(II)-mediated addition to give **286**. TIPS protection, TBS deprotection, and Swern oxidation afforded aldehyde **287**, which was then subjected to the Evans aldol reaction using the boron enolate of **278** followed by conversion of amide to a methyl group to give **288** stereoselectively. Finally, the intermediate **288** was joined with (S)-N-Boc-pipecolic acid (**232**) and the terminal olefin was converted into vinyl iodide **289** via cleavage of olefin, chromium-mediated iodo olefination and protective group exchange.

The synthesis of the C8–C18 top segment **297** started with Weinreb amide **290** (Scheme 43). Addition of vinyl lithium **291** afforded ketone **292**, which was subjected to stereoselective reduction with LiI-LiAlH$_4$ followed by methylation to give **293**. Epoxide formation from **293** followed by addition of a higher order cuprate of **294**, TIPS protection, and exchange of TMS to iodide gave vinyl iodide **295**. Deprotection of the MPM group, Swern oxidation, and the Evans aldol reaction with **218d** gave **296**, which was converted into the desired carboxylic acid **297** by functional group manipulation.

The coupling of **289** and **297** was carried out by amidation to give **298** (Scheme 44). After oxidation of this diol **298** to diketone, selective removal of the TES groups followed by Swern oxidation furnished 26,32-diketone, whose remaining TIPS and TBDPS protective groups were removed to give fully deprotected intermediate **299**, in which all carbons were in their proper oxidation state. The crucial macrocyclization was achieved through a tandem intra- and intermolecular palladium-mediated Stille coupling reaction between the acyclic bis(vinyl iodide) **299** and distannylethene **300**. This unique final process accomplished in one step the installation of the remaining two carbons of rapamycin (**2**) and the completion of the first total synthesis.

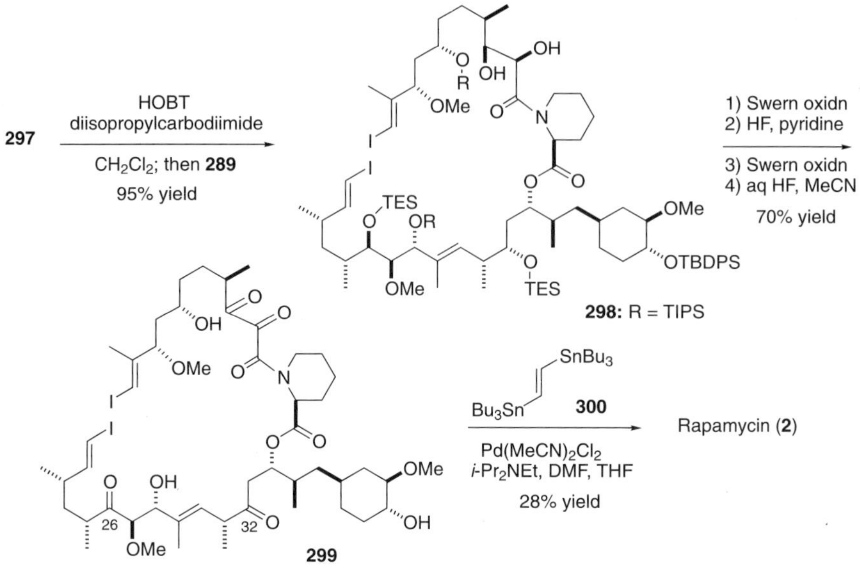

Scheme 44.

2. Schreiber's Total Synthesis [130]

The total synthesis of rapamycin (**2**) by Schreiber *et al.* was accomplished via unique SmI$_2$-mediated attachment of pipecolic acid to the C22–C42 segment, subsequent coupling with the C10–C21 segment, addition of a glycolate derivative at C10, and macrolactamization. The C9 methylene was finally oxidized to ketone (Fig. 6).

The synthesis of the C22–C42 bottom segment **314** is summarized in Scheme 45. The synthesis of the C33–C42 fragment **303** started with the known alcohol **240**. After iodination followed by addition of lithiated allylic sulfide **301**, sigmatropic rearrangement took place via sulfoxide to afford an allylic alcohol, which was subjected to the Sharpless AE to give **302** with high diastereoselectivity. Regioselective introduction of the C35 methyl group to the epoxide **302** with Me$_3$Al followed by epoxide formation gave **303**. The C22–C28 Weinreb amide **307** as a coupling partner was then synthesized via hydroxy-directed hydrogenation of **304** with rhodium catalyst, the Evans aldol reaction with chiral imide **218e**, transamidation of **306**, and MPM protection.

With coupling partners **303** and **307**, these segments were coupled with vinyl lithium **308**. Reaction of **307** with **308** gave an enone, which was subjected to stereoselective Zn(BH$_4$)$_2$ reduction followed by DEIPS protection and oxidation to give sulfone **309**. Coupling of lithiated **309** with epoxide **303** in the presence of BF$_3$·Et$_2$O gave alcohol **310**. The Julia olefination of **310** followed by oxidative cleavage of the newly formed exo olefin provided ketone **311**. The coupling of

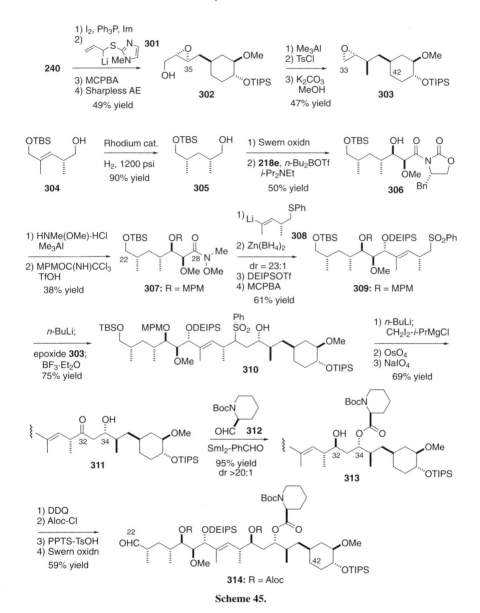

Scheme 45.

311 and the pipecolic acid moiety was regio- and stereoselectively realized by a unique method. The Evans–Tischenko coupling of alcohol **311** with (S)-N-Boc-pipecolinal (**312**) in the presence of SmI_2 and benzaldehyde effected stereoselective reduction of C32 ketone and esterification at C34 to give **313**. Functional group manipulation of **313** furnished the bottom segment **314**.

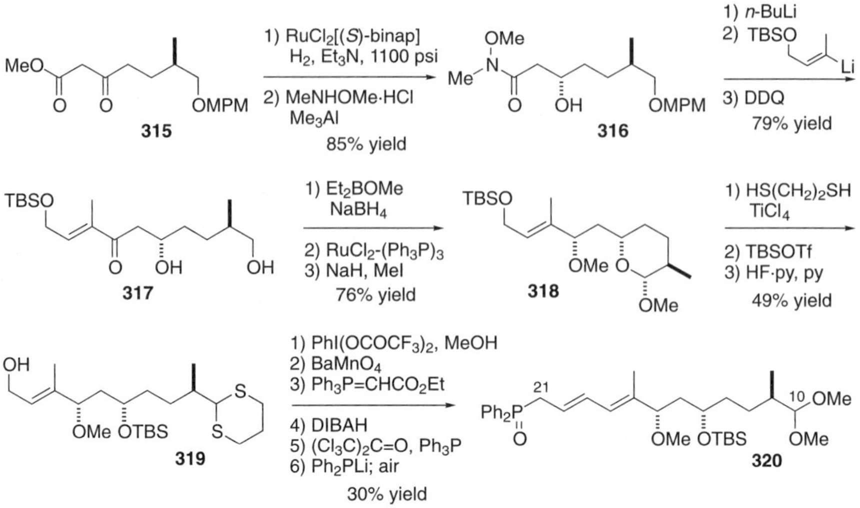

Scheme 46.

The synthesis of the C10–C21 top segment **320** started with β-keto ester **315** (Scheme 46). Catalytic reduction of **315** under Noyori's conditions followed by transamidation provided the Weinreb amide **316**, which was converted into ketone **317** via addition of vinyl lithium. The 1,3-*syn*-selective reduction of **317** with NaBH$_4$-Et$_2$BOMe and selective oxidation of the primary alcohol provided lactol, which was doubly methylated to give **318**. The acetal **318** was converted into diphenylphosphine oxide **320** via thioactal **319** by a series of transformations.

The completion of the total synthesis is described in Scheme 47. The HW reaction of phosphine oxide **320** and **314** efficiently afforded (*E*)-triene **321**. After hydrolysis of dimethyl acetal in **321** followed by aldol reaction, macrolactamization was carried out under Mukaiyama macrocyclization conditions to give an ~1:1 mixture of diastereomers **322**. Removal of three allylcarbonates followed by treatment with the Dess–Martin reagent resulted in oxidation of the three alcohols and subsequent oxidation of the C9 methylene. Final deprotection of the resulting tetraketone with HF·py completed the total synthesis of rapamycin (**2**).

3. *Danishefsky's Total Synthesis* [131]

Danishefsky's total synthesis features aldol coupling of the C10–C27 segment and *N*-[2-(phenylsulfoxy)acetyl] pipecolate, subsequent esterification with the C28–C42 segment, and highly novel intramolecular macroaldolization at the C27 and C28 positions (Fig. 6).

The synthesis of the C28–C42 bottom segment **333** is outlined in Scheme 48. The carboxylic acid **326** having four chiral centers at C31, 32, 34, and 35 was

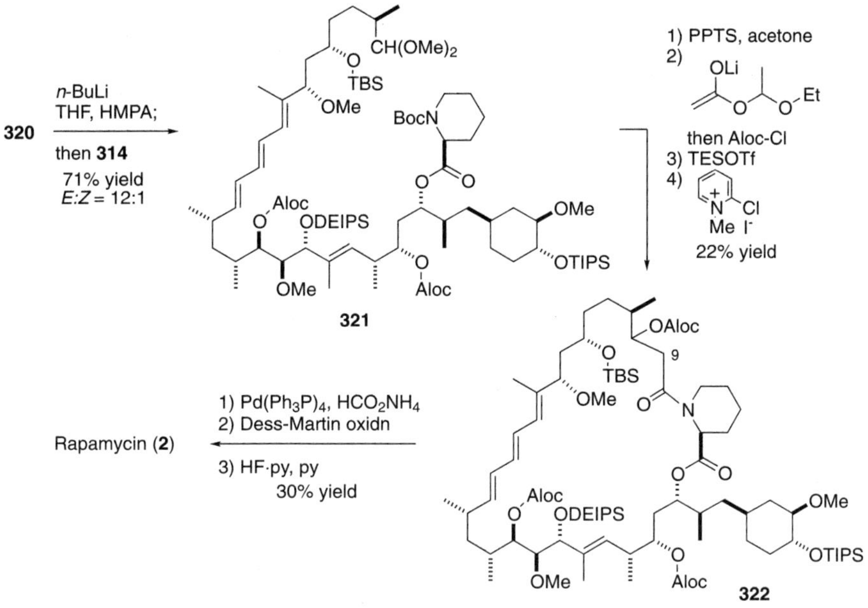

Scheme 47.

synthesized starting from aldehyde **323**. Reaction of **323** with allyltrimethylsilane in the presence of TiCl$_4$ gave a 7:1 mixture of the desired β-alcohol **324** and its isomer. After TBS protection of **324** and ozonolysis, the resulting aldehyde was subjected to the Roush asymmetric crotylboronation to afford a 3.5:1 mixture of α- and β-alcohols, which were separated after deprotection of the silyl ether. TBS protection of **325** followed by hydroboration and oxidation gave carboxylic acid **326**. The synthesis of the C30–C42 segment **329** was accomplished via Ireland–Claisen rearrangement after esterification of **326** and **327**, which was prepared from 2-deoxy-D-glucose via the Ferrier rearrangement and Luche reduction. Thermolysis of the ketene acetal **328** in toluene under reflux effected the rearrangement, and successive hydrolysis gave carboxylic acid **329**. The remaining problems include differentiation of the C32 and C34 alcohols and removal of the C36 carboxyl group. Treatment of **329** with (COCl)$_2$ provided γ-lactone **330**, differentiating the two alcohols at C32 and C34. Diimide reduction of the double bond followed by DIBAH reduction and Suarez oxidation provided iodoformate **331**. Deiodination followed by functional group manipulation gave ester **332**, which was transformed to the desired aldehyde **333** in standard fashion.

The synthesis of the C10–C27 top segment **341** started with aldehyde **334** (Scheme 49). Ni(II)–Cr(II)-mediated coupling of vinyl iodide **335** with **334** afforded the adduct, which was oxidized and debenzylated to give ketone **336**.

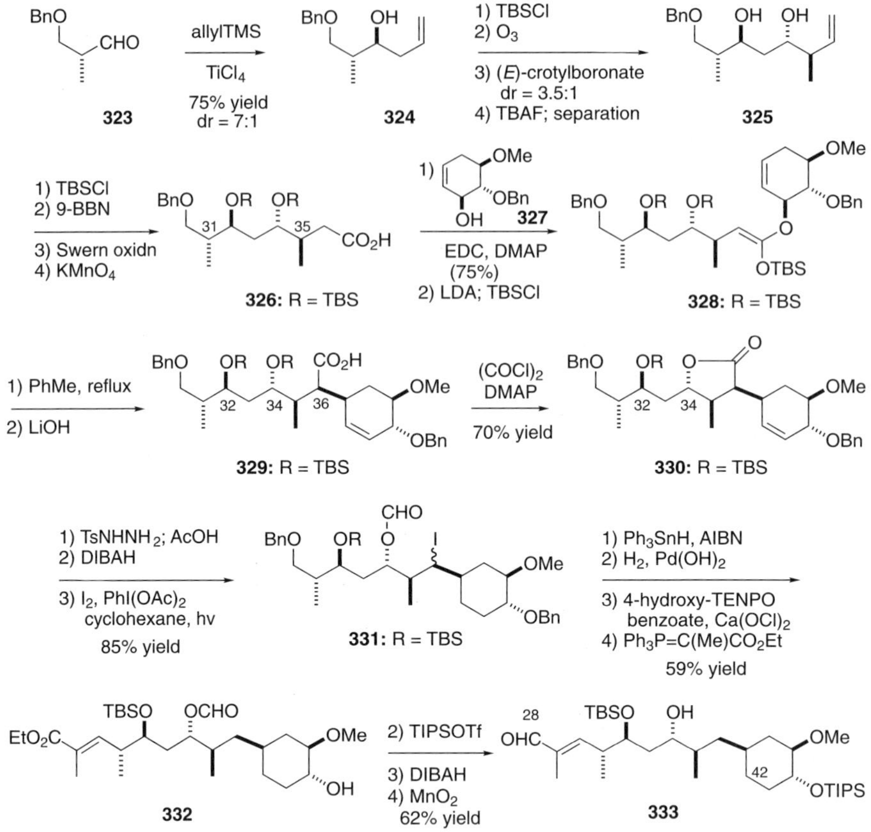

Scheme 48.

After 1,3-*syn*-selective reduction with NaBH$_4$-Et$_2$BOMe, the ester **336** was converted into **338** via lactone **337** through further functional group manipulation. The Julia coupling of **338** and aldehyde **339** followed by β elimination gave **340**, which was converted into methoxymethyl ketone **341** in straightforward steps.

The completion of the total synthesis was achieved by successive coupling of **341**, **342**, and **333** and final intramolecular macroaldolization (Scheme 50). Coupling of **341** with the lithium enolate of **342** afforded **343**. The Dess–Martin oxidation of **343** followed by TBS deprotection afforded hemiketal of α,β-diketoamide, which was converted into allyl ester **344**. After TMS protection of the hemiketal and cleavage of the allyl ester, the resulting acid was subjected to esterification with **333** using DCC-DMAP followed by TMS deprotection to afford **345**. The Dess–Martin oxidation of **345** gave the C32 ketone, corresponding to 40-TIPS seco-rapamycin. After several attempts for macroaldolization, it

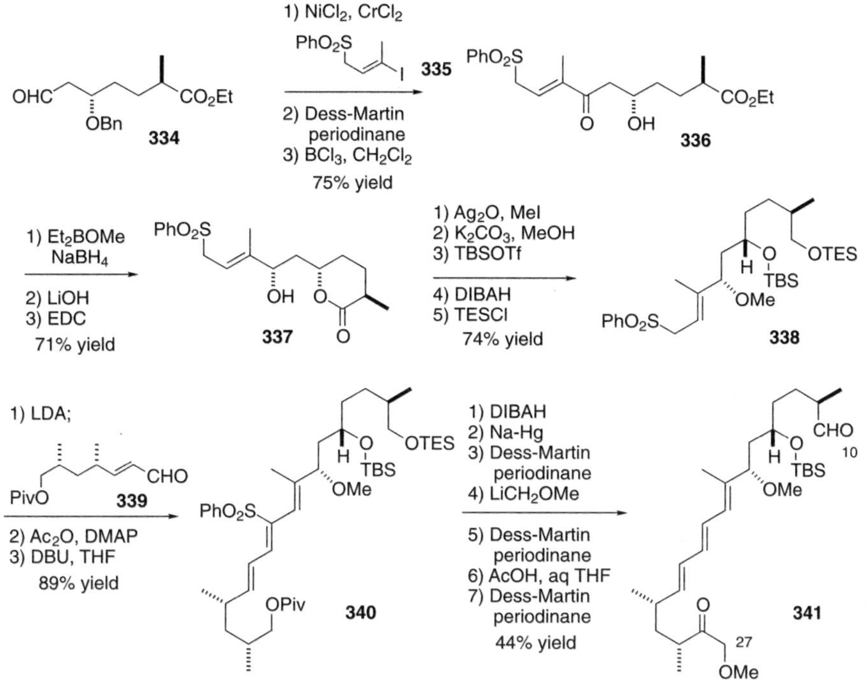

Scheme 49.

was found that the use of TiCl$_3$(Oi-Pr) afforded 10% of the desired 40-TIPS rapamycin **346** and 23% of its stereoisomer. Final desilylation of **346** afforded rapamycin (**2**).

4. Smith's Total Synthesis [132]

The Smith group has also achieved the total synthesis of rapamycin (**2**) and demethoxyrapamycin (Fig. 6). Their total synthesis of **2** features aldol reaction of acetylpipecolic acid and the C10–C20 segment, esterification with the C21–C42 segment, and final Stille macrocyclization at the C20 and C21 positions.

The union of several fragments for the C21–C42 bottom segment **360** was based on efficient sulfone- and dithiane-mediated couplings (Scheme 51). The synthesis of the C33–C42 fragment **350** having a cyclohexyl unit started with sulfone **347**. Regioselective addition of lithiated **347** to epoxide **348** in the presence of BF$_3$·Et$_2$O followed by sodium-amalgam reduction gave **349**, which was converted into the desired iodide **350** via addition of iodide to epoxide. The coupling partner **354** was synthesized from sulfone **351**. The coupling of **351** and aldehyde **352**, Swern oxidation and aluminum-amalgam reduction gave ketone **353**. The C29 methyl group was introduced by addition of dimethylcuprate to the enol

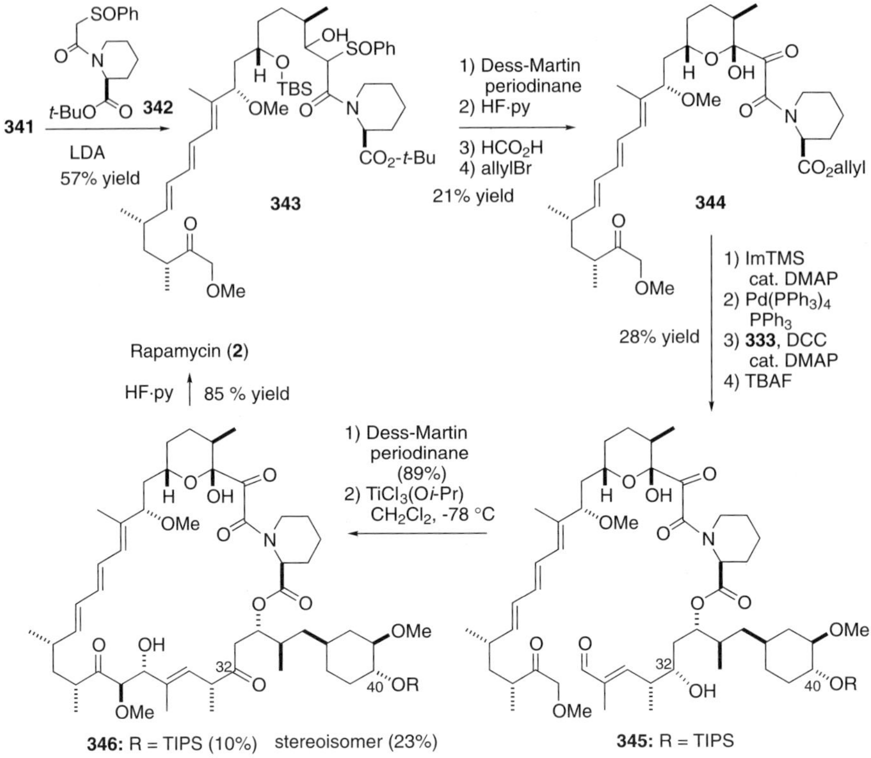

Scheme 50.

triflate derived from **353**. Functional group manipulation in six steps afforded acetal **355**. The union of both segments was accomplished by treatment of the lithium anion of **355** with iodide **350**. The subsequent deacetalization gave aldehyde **356**. Addition of thioacetal **357** to this aldehyde **356** afforded a separable 1.2:1 mixture of the α- and β-alcohols. The desired β-alcohol was methylated to give **358**, which was converted into vinylstannane **360** via acetylene **359** by acetylene formation, deprotection of MPM, dethioketalization, and palladium-mediated hydrostannylation.

The synthesis of the C1–C20 top segment **367** started with aldehyde **361** (Scheme 52). Addition of vinylstannane **362** to **361** gave a 1.3:1 mixture of the desired α-alcohol **363** and its β-isomer. The β-alcohol was transformed to the α-alcohol **363** by PDC oxidation followed by Corey asymmetric reduction using chiral oxaborolidine catalyst. After methylation of the alcohol, **363** led to aldehyde **364** through a standard procedure. Condensation of the dianion of *N*-acetylpipecolic acid (**365**) with **364** followed by methylation and the Dess–Martin oxidation

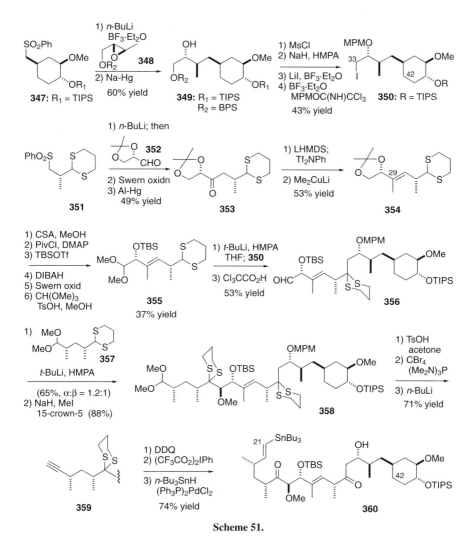

Scheme 51.

yielded the desired α,β-diketoamide, which, after removal of the O- and C-silyl groups, produced **366**. The hemiketal **366** was converted into the requisite (E,E)-dienyl iodide **367** by TES protection, radical hydrostannylation, and tin–iodide exchange.

Coupling of two segments **360** and **367** by EDC-DMAP-induced esterification gave **368** (Scheme 53). The macrocyclization of **368** was effectively achieved by intramolecular Pd-mediated Stille coupling in 74% yield. Final desilylation completed the total synthesis of rapamycin (**2**).

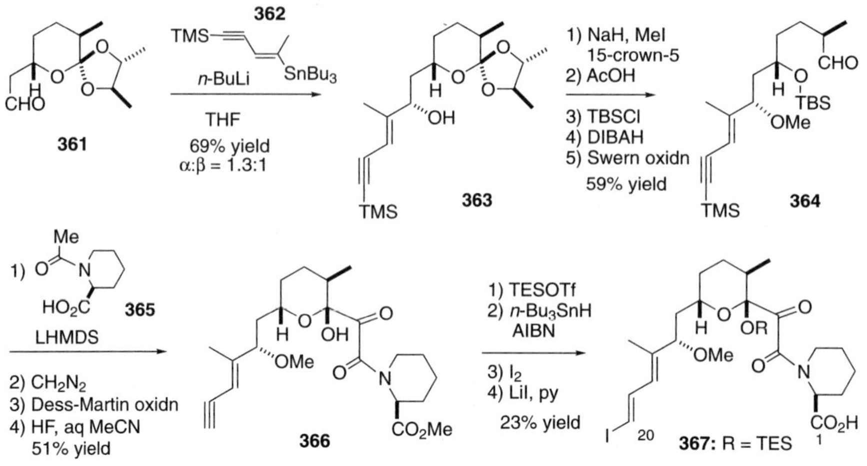

Scheme 52.

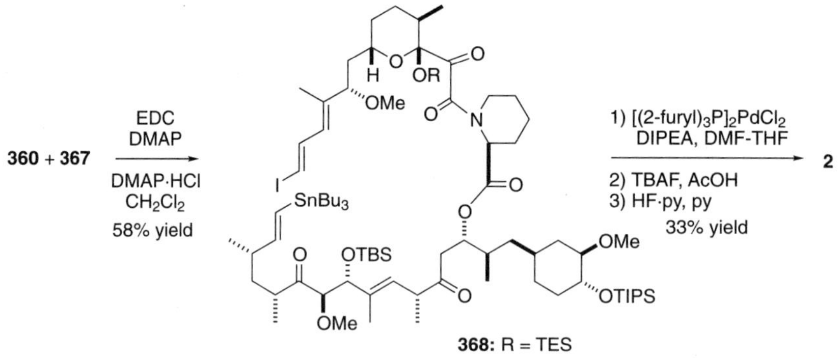

Scheme 53.

C. Avermectin

Avermectins and milbemycins are naturally occurring 16-membered macrolactones that show a broad spectrum of anthelmintic and pesticidal activity with relatively low toxicity to both humans and animals. The milbemycins differ from the avermectins primarily in the lack of a disaccharide moiety and oxygen functionality at C13.

The structural feature of avermectins (**3**) (Fig. 1) includes 6,6-spiroketal and a disaccharide unit in the northern part and an oxahydrindene unit in the southern part. Because of the sensitive hexahydrobenzofuran subunit, the C3–C4 double bond was incorporated at a late stage during the course of total synthesis.

Total syntheses of avermectin B_{1a} (**3b**) have been achieved by Hanessian et al. [133], Ley et al. [134], and White et al. [135], and a total synthesis of the closely related avermectin A_{1a} (**3a**) has been reported by Danishefsky *et al.* [136]. In addition, syntheses of the structurally less complex milbemycins have been reported [137]. In this section, the total syntheses of avermectins are described. Attachment of a disaccharide moiety to aglycon for completion of the total syntheses is introduced in Section II.C.

1. Hanessian's Total Synthesis [133]

The first synthesis of avermectin B_{1a} (**3b**), the most active avermectin component, was accomplished by the Hanessian group in 1986. The key steps of the synthesis involve coupling of the C11–C28 northern and C1–C10 southern segments under Julia conditions, macrolactonization, stereocontrolled glycosidation, and deconjugation to the C3–C4 double bond (Fig. 7).

The synthesis of three fragments, **371, 374**, and **377**, for the C11–C28 northern segment is summarized in Scheme 54. The acetylene **371** having three consecutive chiral centers was synthesized from **369**, prepared from L-isoleucine. The Sharpless AE of **369** followed by regio- and stereoselective addition of a methyl group afforded diol **370**, which was converted into **371** by functional group manipulation. Frater's *anti*-selective methylation of ethyl (*S*)-malate (**372**), hydrolysis, and ketalization gave **373**, which was converted into methyl ketone **374** by a standard procedure. Addition of a Grignard reagent to aldehyde **375**, prepared from (*S*)-malic acid, and benzylation stereoselectively afforded **376**, which was converted into δ-lactone **377** by functional group manipulation, including homologation.

The northern segment **383** was synthesized via successive coupling of **371, 377**, and **374** (Scheme 55). Coupling of the lithium acetylide of **371** with lactone **377** gave hemiketal **378**. Hydrogenation with Lindlar catalyst followed by $BF_3 \cdot Et_2O$ treatment afforded spiroketal **379**. Deprotection of the TBDPS ether, phenylsulfination, and oxidation gave sulfone **380**. The Julia coupling of lithiated **380** with ketone **374** gave **381**. Sodium-amalgam reduction followed by deprotection of TBDPS and benzyl ethers provided triol **382**. Protective group manipulation and phenylsulfonylation afforded the northern segment **383**.

The synthesis of the aglycon **154d** of avermectin B_{1a} (**3b**) is summarized in Scheme 56. The Julia coupling of sulfone **383** and southern segment **384**, which was prepared by degradation of natural avermectin B_{1a}, gave (*E*)-olefin **385**. After removal of the TBS group and hydrolysis, the resulting carboxylic acid **386** was subjected to macrocyclization under DCC-DMAP conditions to afford lactone **387**. After regioselective TBS protection at C5, the resulting alcohol **154d** was subjected to silver-induced glycosidation (Section II.C), which led to the total synthesis of avermectin B_{1a} (**3b**).

2. Ley's Total Synthesis [134]

The total synthesis of avermectin B_{1a} (**3b**) by Ley *et al.* was accomplished via Julia coupling of the C11–C28 northern segment with the southern C1–C10

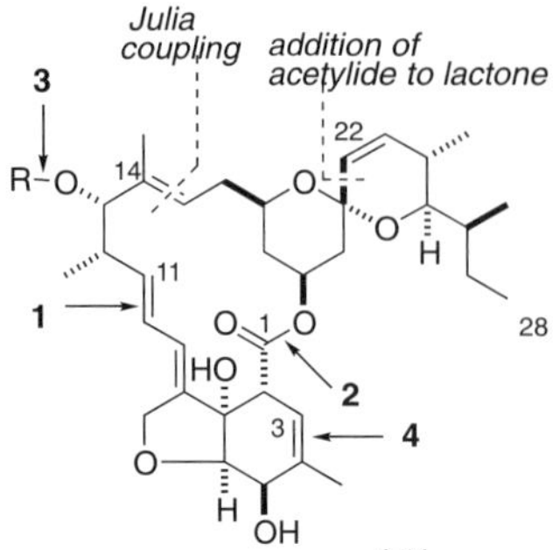

Hanessian's strategy

1. Julia coupling
2. macrolactonization (DCC-DMAP)
3. glycosidation (pyS-glycoside-AgOTf)
4. deconjugation of double bond

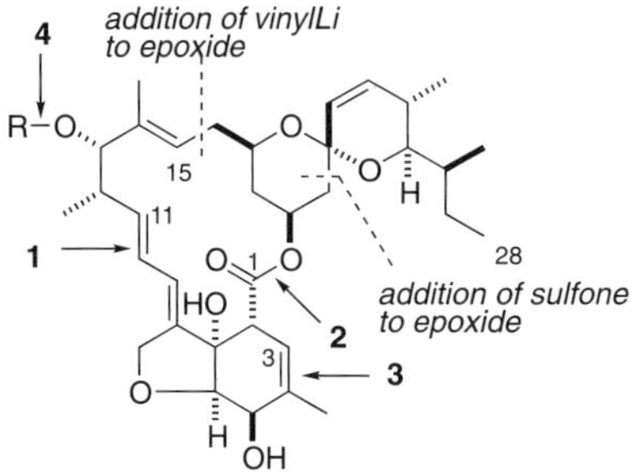

Ley's strategy

1. Julia coupling
2. Mukaiyama macrolactonization
3. introduction of double bond
4. glycosidation (ImC(S)-glycoside-AgClO$_4$)

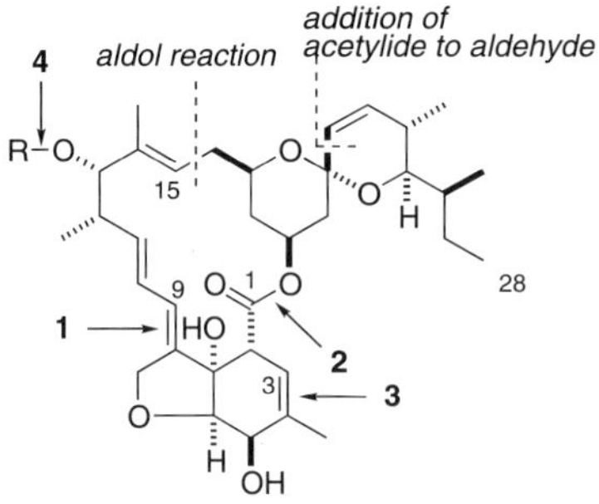

White's strategy

1. Julia coupling
2. Mukaiyama macrolactonization
3. deconjugation of double bond
4. glycosidation (pyS-glycoside-AgOTf)

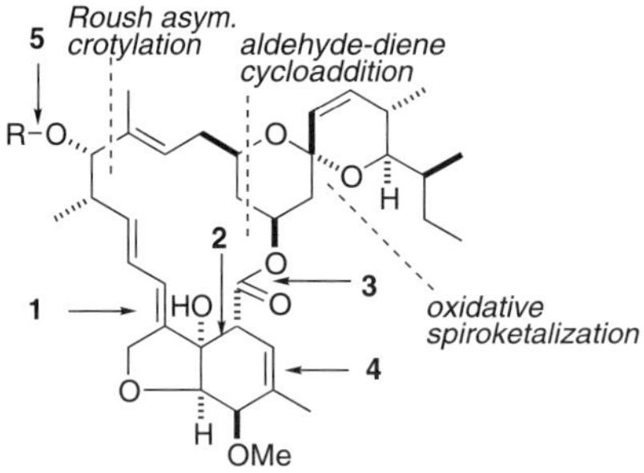

Danishefsky's strategy

1. aldol condensation
2. Niozaki aldol cyclization
3. Mukaiyama macrolactonization
4. deconjugation of double bond
5. glycosidation (NIS, deiodination)

Fig. 7.

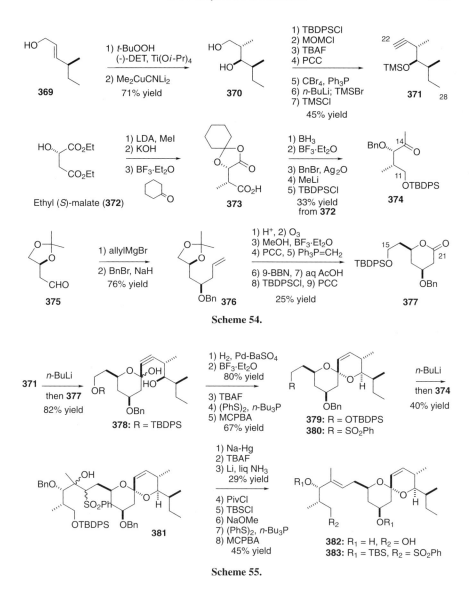

Scheme 54.

Scheme 55.

segment, macrolactonization, introduction of the C3–C4 double bond, and glycosidation (Fig. 7).

The synthesis of the northern spiroacetal segment **394** features the selective ring-opening of a symmetric 1,4-bis-epoxide **389** by successive introduction of the C11–C15 unit **388** and C21–C28 phenylsulfonylpyran **391** (Scheme 57). Addition of vinyl lithium of **388**, prepared via Brown asymmetric crotylboration, to a symmetric

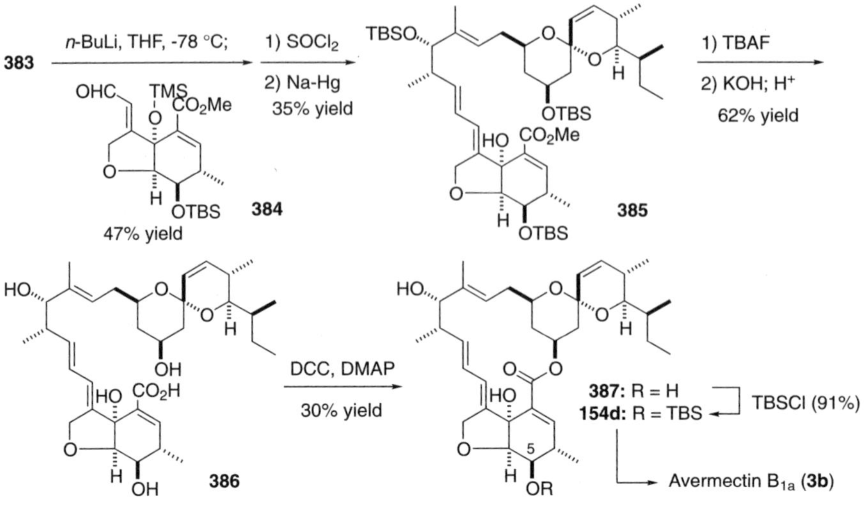

Scheme 56.

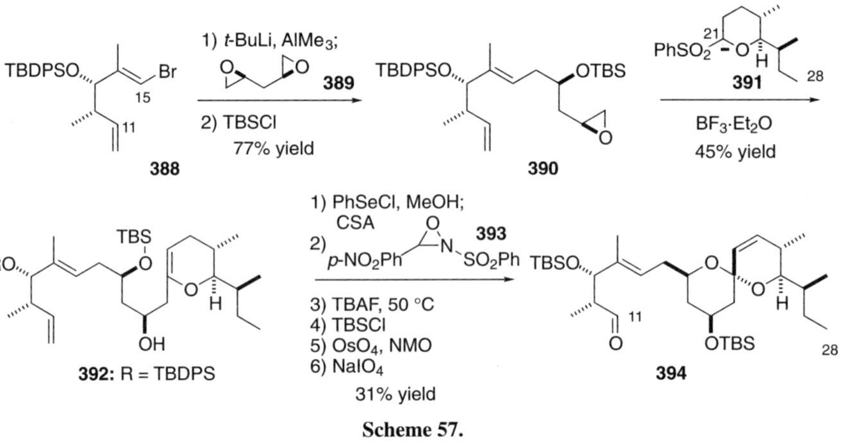

Scheme 57.

1,4-bis-epoxide **389** followed by TBS protection afforded **390**. Treatment of **390** with the anion derived from phenylsulfonylpyran **391** afforded alcohol **392**. Phenylselenation-methyl hemiketalization of glycal **392** and successive oxidation with the Davis reagent **393**, *p*-nitrophenyl-*N*-sulphonyloxaziridine, gave phenylselenoxide, which underwent elimination to give an olefin. TBAF treatment effected removal of the silyl groups and cyclization to give spiroketal, which was transformed to aldehyde **394** by TBS protection and selective cleavage of the terminal olefin.

The C1–C10 hydrobenzufuran segment **400** was synthesized from cyclohexanone derivative **395** (Scheme 58). The C3–C4 double bond of the cyclohexane

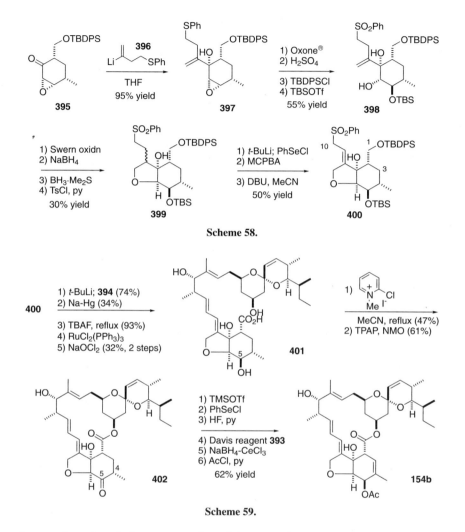

Scheme 58.

Scheme 59.

ring was introduced after construction of the macrolactone ring. Addition of vinyl lithium **396** to ketone **395** stereoselectively afforded **397**. Oxidation of sulfide to sulfone, stereoselective epoxide opening, and regioselective silyl protection afforded **398**. Inversion of the secondary alcohol, hydroboration of *exo*-methylene followed by tosylation effected cyclization to give tetrahydrofuran **399**. Phenylselenation of sulfone followed by oxidative elimination gave vinylsulfone, which was deconjugated by treatment with DBU to give **400**.

The coupling of the northern and southern segments, **394** and **400**, was accomplished by Julia olefination (Scheme 59). After coupling of **394** and **400**, desilylation followed by oxidation afforded carboxylic acid **401**. Macrolactonization was accomplished under Mukaiyama's conditions and selective oxidation of the C5 hydroxyl group gave ketone **402**. Phenylselenation of **402** at the C4 position via the TMS enol ether and conversion to selenoxide effected *syn*-elimination to give an enone, which was stereoselectively reduced and acetylated to give 5-acetyl aglycon **154b**. Total synthesis of avermection B_{1a} (**3b**) by glycosidation of **154b** is described in Section II.C.

3. White's Total Synthesis [135]

The total synthesis of avermectin B_{1a} (**3b**) by White *et al.* was accomplished via Julia coupling of the C9–C28 northern and C1–C8 southern segments, macrolactonization, deconjugation of the C2–C3 double bond to the C3–C4 position, and glycosidation via the pyridylthio glycoside (Fig. 7).

Similar fragments **405**, **408**, and **409** to those in Hanessian's synthesis were used for the construction of the northern segment (Scheme 60). Fragment **405** was synthesized via stereoselective crotylation of **403** under Hiyama–Nozaki conditions, (*E*)-crotyl bromide in the presence of chromium(III) chloride and $LiAlH_4$. Ketone **408** was prepared from **406** by Sharpless AE, regioselective addition of lithium dimethylcuprate and protective group manipulation. The first coupling was undertaken by addition of a Grignard reagent prepared from **405** to aldehyde **409**, derived from laevoglucosan. The adduct was then subjected to oxidation, removal of acetonide, and TBS deprotection to give hemiketal **410**. Lindlar-catalyzed hydrogenation followed by CSA treatment induced spiroketalization, and Swern oxidation afforded aldehyde **411**. Aldol reaction of the lithium enolate of ketone **408** with **411** afforded aldol **412**. Elimination of the alcohol followed by methylation gave **413**, which underwent a series of transformations, including 1,3-migration of the double bond via sulfenate to sulfoxide, to provide **414**. The alcohol **414** was converted into sulfone **415** in standard fashion.

The synthesis of the southern segment **421** started with the racemic anhydride **416** (Scheme 61), which was prepared by the Diels–Alder reaction of maleic anhydride and 3-methyl-1-[(trimethylsilyl)oxy]-1,3-butadiene. Cleavage of the TMS ether with TBAF induced spontaneous intramolecular acylation to give γ-lactone **417**, which was obtained as an optically pure form by optical resolution at this stage. Homologation to diazo ketone followed by acidic conversion gave *cis*-fused furan-3-one, which was esterified to **418**. Formation of bromohydrin followed by DBU treatment afforded 4,5-β-epoxide, which was successively treated with sodium methoxide and then TESOTf to induce γ-lactonization and epoxide opening, respectively, yielding **419**. Conversion of TES to the SEM group and basic methanolysis gave α,β-unsaturated ester **420**, resulting in double-bond migration. Finally, MCPBA oxidation of the TES enol ether, derived from ketone **420**, followed by TES protection, furnished **421**.

The completion of the total synthesis of **3b** is summarized in Scheme 62. The Julia coupling of **415** and **421** gave **422**. Sodium methoxide treatment of **422** afforded γ-lactone accompanying double-bond migration to the 3,4 position.

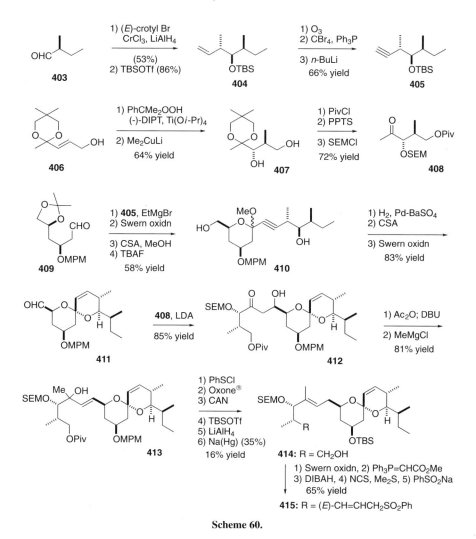

Scheme 60.

Subsequent sodium-amalgam reduction yielded (*E,E*)-dienoic 2β-carboxylic acid **423**. After TBS deprotection, macrolactonization was accomplished under Mukaiyama's conditions to give 2β-lactone **424**. Epimerization at the C2 position under Hanessian's conditions [133b], with imidazole in benzene under reflux, afforded the desired 2α-isomer and 2β-**424** in a ratio of 3:2. Cleavage of the SEM group followed by selective TBS protection at C5 afforded **154a**. Glycosidation was accomplished by coupling with 2-pyridylthio glycoside **155**, prepared from natural avermectin B_{1a} (**3b**), in the presence of AgOTf. Deprotection of the TBS group completed the total synthesis of **3b**.

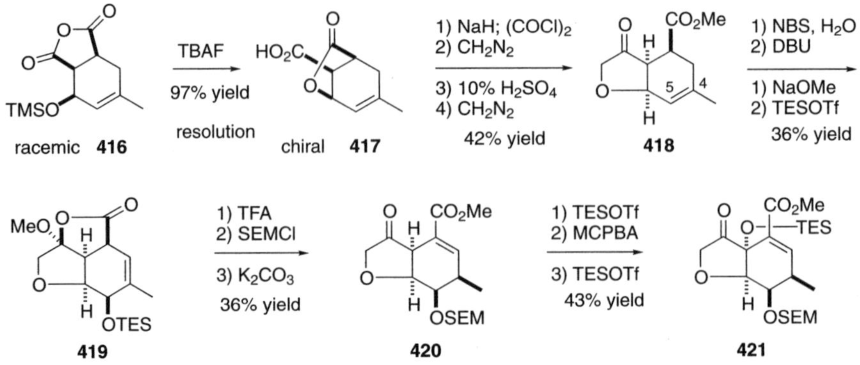

Scheme 61.

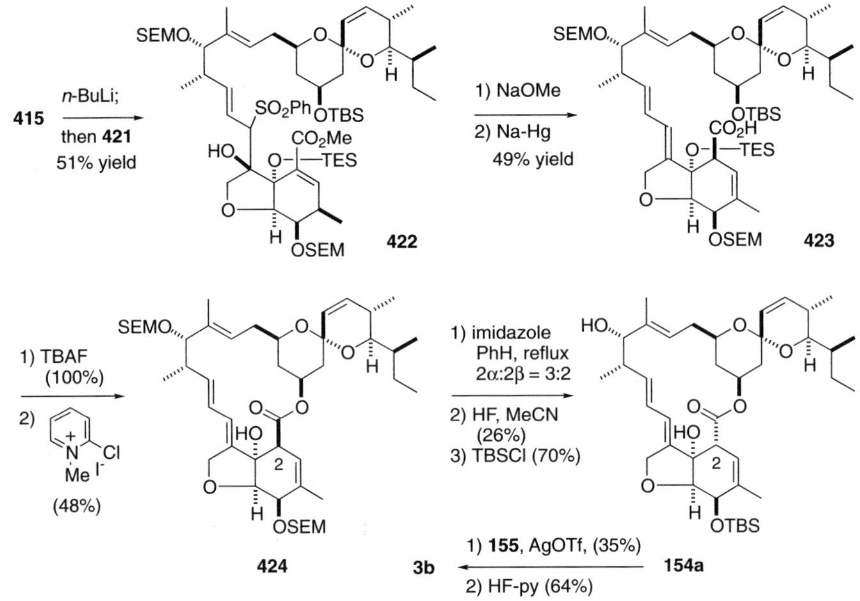

Scheme 62.

4. Danishefsky's Total Synthesis [136]

Danishefsky's total synthesis of avermectin A_{1a} (**3a**) was accomplished via an aldol reaction to join the northern and southern segments, intramolecular Nozaki aldol cyclization [138] to construct the southern ring system, Mukaiyama macrolactonization, deconjugation to the C3–C4 double bond, and NIS-mediated glycosidation (Fig. 7).

The 6,6-spiroketal system of the northern segment was constructed via their developed LACDAC reaction and oxidative cyclization at C21 (Scheme 63).

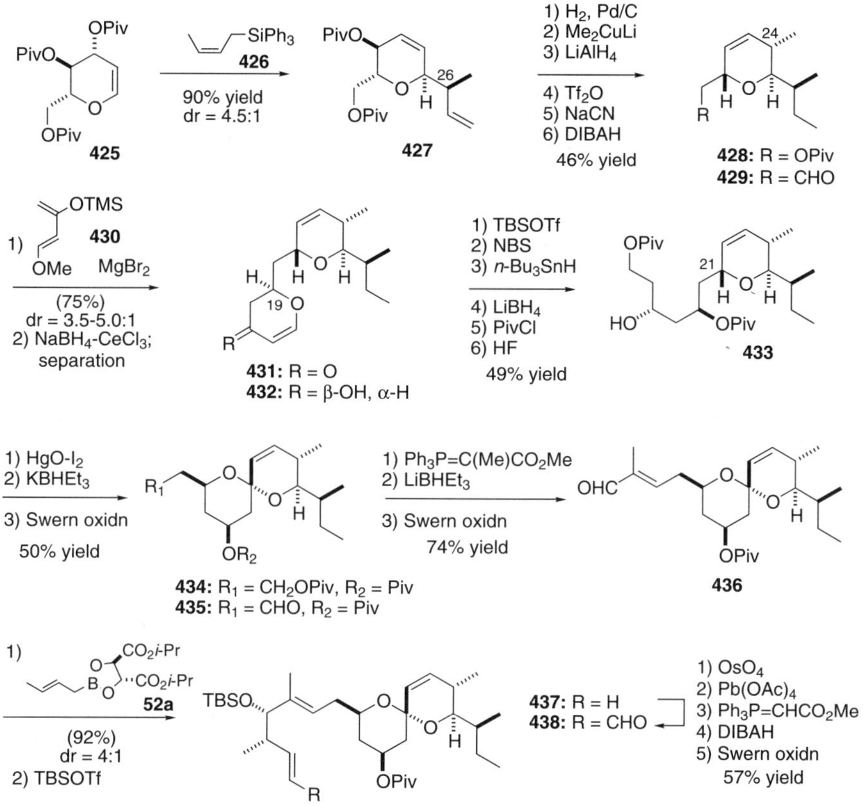

4. Total Synthesis of Macrolides   241

Scheme 63.

Reaction of D-glucal derivative **425** with (Z)-crotyltriphenylsilane (**426**) afforded a separable 4.5:1 mixture of **427** and its C26 epimer. Selective hydrogenation of the terminal double bond and stereoselective introduction of the C24 methyl group effectively afforded **428**, which was homologated to aldehyde **429**. The LACDAC reaction of **429** with diene **430** in the presence of $MgBr_2$ afforded a 3.5–5:1 mixture of pyrone-pyran **431** and its C19 isomer. Luche reduction followed by purification gave the pure **432**. The conversion of **432** into **433** was carried out by standard procedure. The critical oxidative cyclization of **433** was accomplished by treatment with $HgO-I_2$ in carbon tetrachloride in the presence of light to give spiroketal **434**, which was further converted into **436** via **435** in standard fashion. Roush asymmetric crotylboration using **52a** followed by silylation afforded a 4:1 mixture of **437** and its diastereomer. The homologation of **437** afforded the northern aldehyde **438** by straightforward steps.

The synthesis of aglycon **154c** is summarized in Scheme 64. The synthesis of **442**, corresponding to the southern segment, started with aldehyde **439** derived

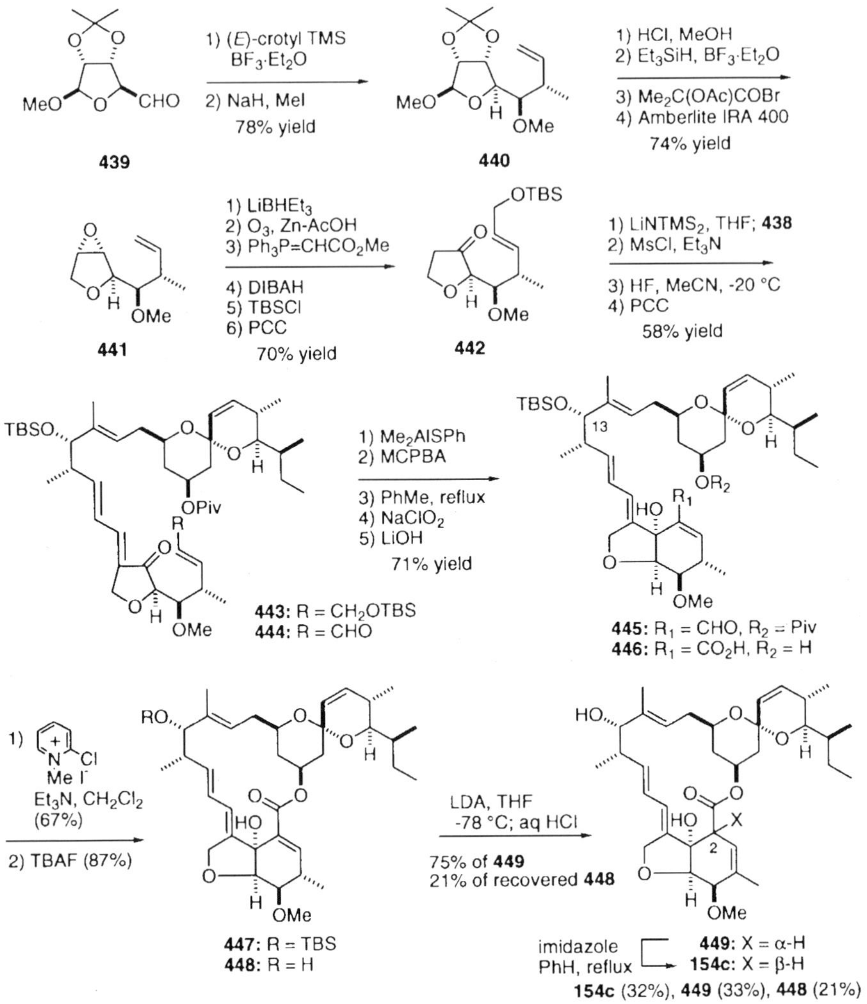

Scheme 64.

from D-ribose. Reaction of **439** with (*E*)-crotyltrimethylsilane followed by methylation stereoselectively afforded **440**. Methanolysis of acetonide, reduction of acetal, and oxirane formation afforded epoxide **441**. After regioselective reductive opening of the epoxide, **441** was converted into ketone **442** in standard fashion. The stage was now set for coupling of the northern and southern segments. Coupling of the lithium enolate of **442** with aldehyde **438** followed by dehydration afforded (*E,E*)-diene **443**. The selective desilylation of the primary TBS group followed by PCC oxidation gave aldehyde **444**. The southern oxahydrindene system was constructed by intramolecular Nozaki aldol cyclization. Reaction of **444** with dimethylaluminum benzenethiolate followed by MCPBA treatment and thermolysis afforded seco-aldehyde **445**. Oxidation and depivaloylation provided C13-TBS seco acid **446**. Macrolactonization of **446** under Mukaiyama's conditions followed by desilylation afforded **448**. Deconjugation of the double bond of **448** leading to 2β-H epimer **154c** was required for the completion of the total synthesis. The deconjugation was accomplished by LDA treatment followed by aqueous HCl quenching. However, the major product was α-H isomer **449**, and **448** was recovered. Hanessian's conditions [133b] were then applied to this epimerization; reaction of **449** with imidazole in benzene under reflux provided the desired avermectin A_{1a} aglycon (**154c**) (32%) along with **449** (33%) and **448** (21%). The total synthesis of avermectin A_{1a} (**3a**) by glycosidation is described in Section II.C.

D. Altohyrtin (Spongistatin)

In 1993, the Pettit, Fusetani, and Kitagawa groups independently reported the isolation and structural elucidation of spongistatins, cinachyrolide, and altohyrtins. The new spongipyran macrolides, isolated from marine sponges, exhibit extraordinarily potent antitumor activities against a wide variety of human cancer cell lines. While all three families have been assigned the same carbon skeleton, the proposed structures differed in the relative stereochemical relationships. The Kitagawa group has proposed the complete structure of altohyrtins to be that shown in structure **4** (Fig. 1). These assignments were subsequently confirmed by the total synthesis of (+)-altohyrtin C (spongistatin 2; **4b**) by Evans *et al.* [139] and (+)-altohyrtin A (spongistatin 1; **4a**) by the Kishi group [140]. In addition to these total syntheses, Smith *et al.* [141] have recently accomplished the formal and total synthesis of **4a** and **4b**, respectively.

The structural feature of the members in this family involves a 51-carbon chain having 24 stereogenic centers, two 6,6-spiroketal moieties, two highly substituted tetrahydropyran rings, and a 42-membered lactone ring. In addition, spongistatins 1 (altohyrtin A), 4, 5, and 9 contain a novel unprecedented chlorodiene functionality.

1. Evans's Total Synthesis [139]

The first total synthesis of altohyrtin C (spongistatin 2; **4b**) was accomplished via diastereoselective aldol coupling of the C1–C15 AB and C16–C28 CD segments, Wittig coupling of the ABCD and the C29–C43 EF segments, addition of the C44–C51 side chain to the fully elaborated ABCDEF system, and the regioselective macrolactonization (Fig. 8).

The synthesis of the C1–C15 AB spiroketal **457** is described in Scheme 65. The synthesis of the C8–C15 fragment **453** started with allyl TMS **450**, prepared by Evans asymmetric alkylation. Addition of **450** to aldehyde **451** with $SnCl_4$ stereoselectively afforded α-alcohol **452**, which was converted into methyl ketone **453** through a sequence of four steps involving Mitsunobu inversion. The synthesis of

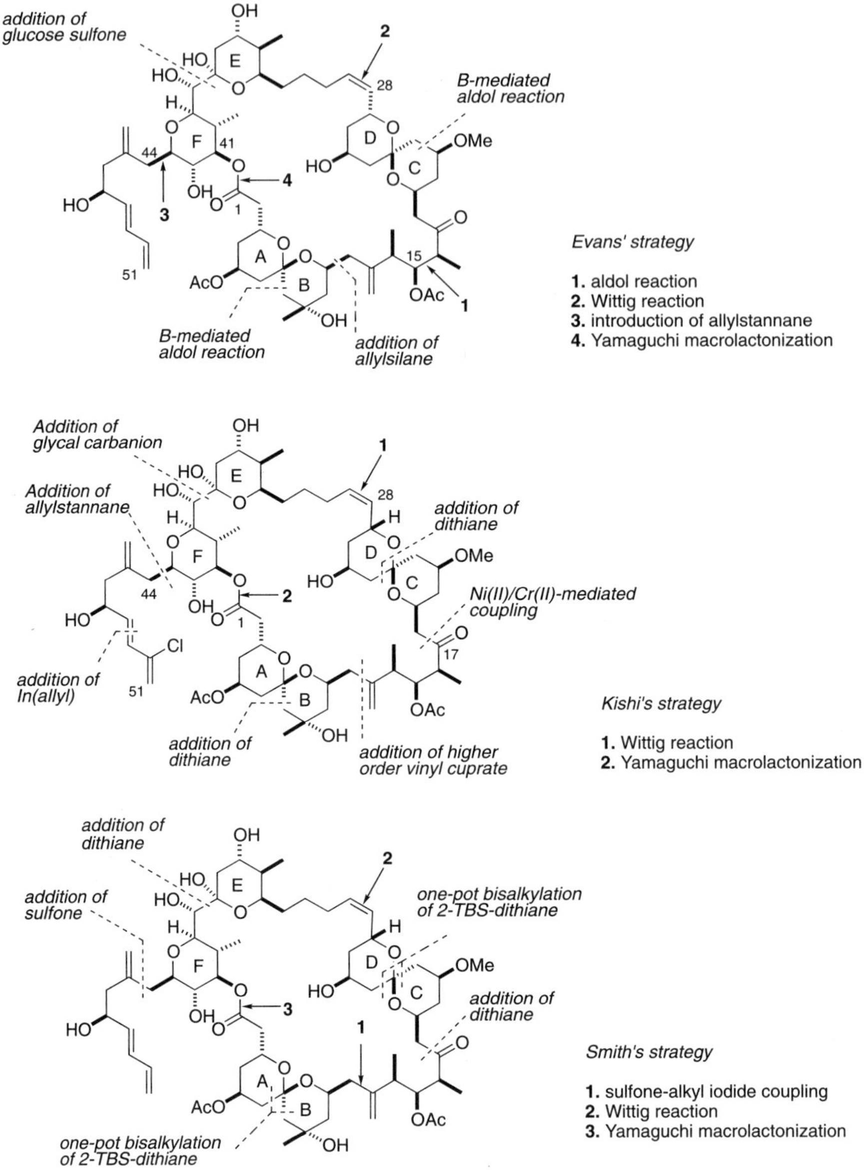

Evans' strategy

1. aldol reaction
2. Wittig reaction
3. introduction of allylstannane
4. Yamaguchi macrolactonization

Kishi's strategy

1. Wittig reaction
2. Yamaguchi macrolactonization

Smith's strategy

1. sulfone-alkyl iodide coupling
2. Wittig reaction
3. Yamaguchi macrolactonization

Fig. 8.

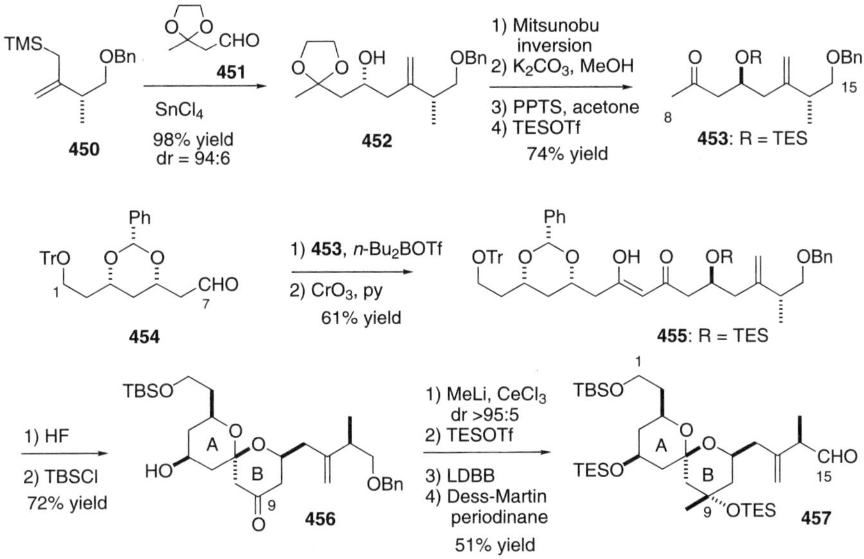

Scheme 65.

the C1–C7 fragment **454** was accomplished via intramolecular Michael addition of the benzaldehyde hemiacetal to afford 1,3-*syn*-acetal. Aldol reaction of the boron enolate of **453** with aldehyde **454** gave a 1:1 mixture of aldols, which was oxidized to diketone **455**. Upon treatment with HF, multiple deprotection and simultaneous spiroketalization took place to give 6,6-spiroketal as a single product, which was protected to give TBS ether **456**. The stereoselective introduction of the C9 methyl group was achieved by treatment with MeLi/CeCl$_3$ to give an *axial* alcohol (dr > 95:5), which led to aldehyde **457** in standard fashion.

The synthesis of the C16–C28 CD spiroketal **464** started with δ-hydroxy-α,β-unsaturated Weinreb amide **458** (Scheme 66). Treatment of **458** with benzaldehyde in the presence of *t*-BuOK induced intramolecular hetero Michael addition of the resulting hemiacetal to afford 1,3-*syn* acetal, which was treated with MeLi to give ketone **459**. Aldol reaction of the boron enolate of **459** with **460** stereoselectively produced the desired aldol **461**. Methylation followed by removal of the TBS and benzylidene groups effected spiroketalization to give a separable 6:1 mixture of 6,6-spiroketals **462** and **463**. The major compound **462** has the undesired configuration of the spiroketal. Equilibration of the mixture with Mg(O$_2$CCF$_3$)$_2$ in CF$_3$CO$_2$H-CH$_2$Cl$_2$ afforded a 1:2.2 mixture of **462** and **463**. The desired **463** was converted into ethyl ketone **464** via protection of the hydroxyl groups and addition of EtMgBr to the Weinreb amide.

As shown in Scheme 67, aldol coupling of the boron enolate of **464** with aldehyde **457** stereoselectively provided aldol **465**, which was converted into **466** through standard functional group manipulation.

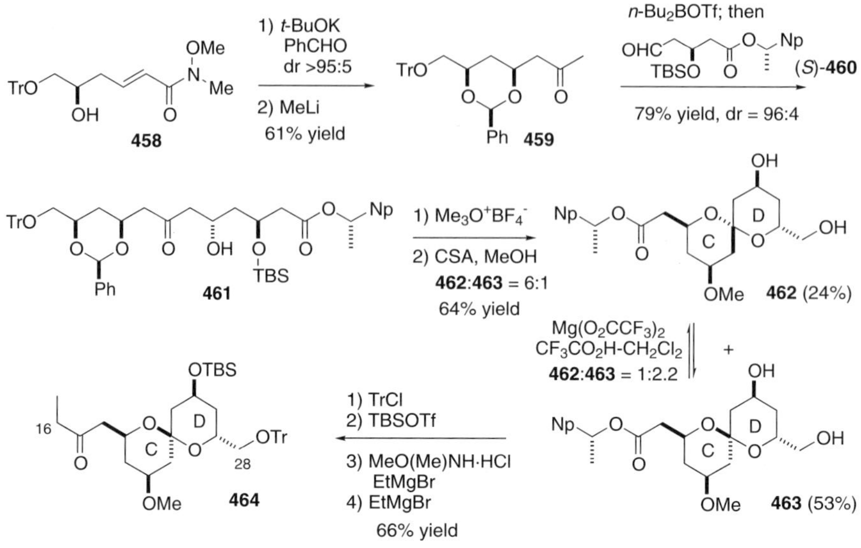

Scheme 66.

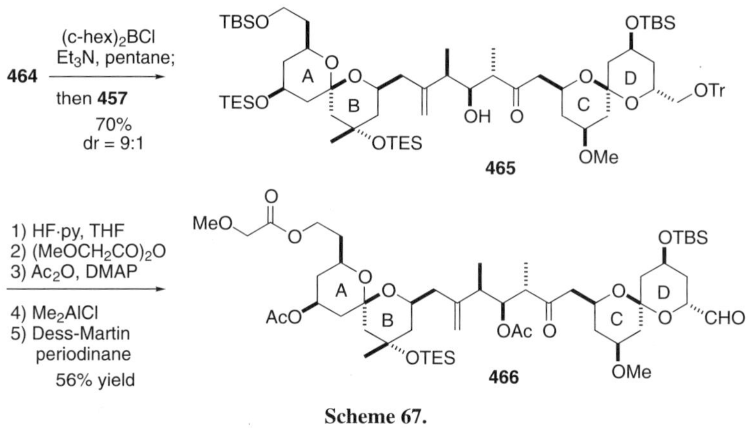

Scheme 67.

The synthesis of the C29–C43 EF segment **479** is summarized in Scheme 68. Methylation of **467** under Frater's conditions stereoselectively afforded 2,3-*anti* compound **468** (dr = 5–8:1). After TES protection followed by thioester reduction, aldol reaction of the resulting aldehyde with thioketene acetal afforded α-alcohol **469** under Felkin–Anh selectivity. TES deprotection and silver-mediated lactonization followed by TES protection and lactone reduction gave lactol **470**. Dehydration, debenzylation, and oxidation furnished aldehyde **471**, which was transformed to benzotriazolyl amide **472**.

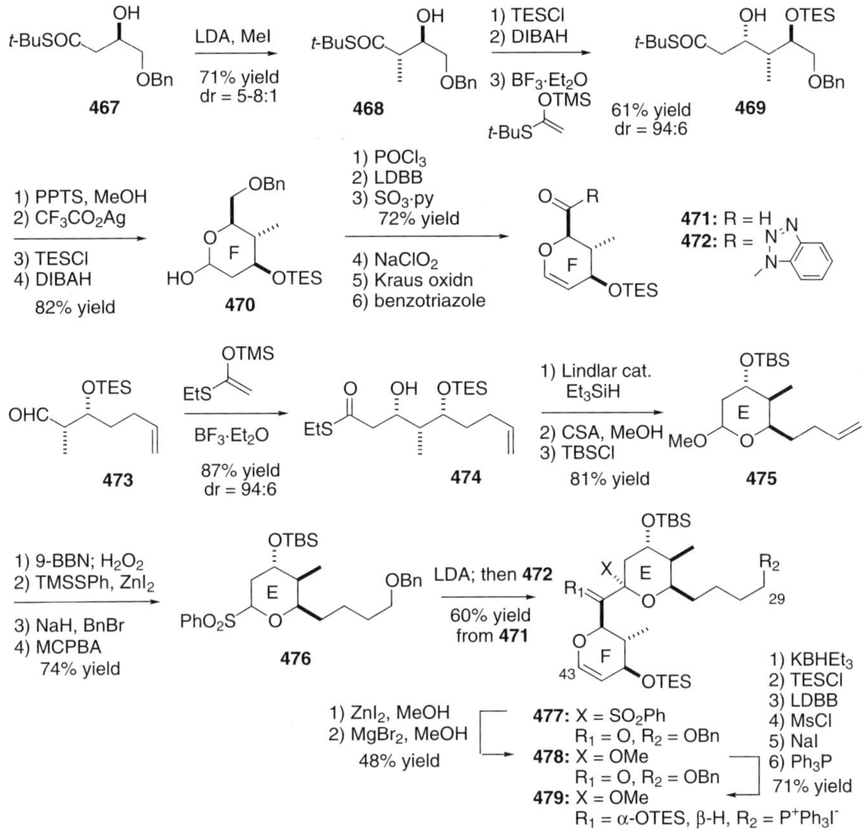

Scheme 68.

The synthesis of the E-ring **476** started with aldehyde **473**, prepared via boron-mediated aldol reaction. Felkin-selective Lewis acid-catalyzed aldol reaction of **473** with thioketene acetal afforded α-alcohol **474** (dr = 94:6). Fukuyama reduction of thioester to aldehyde followed by hemiacetalization and TBS protection furnished the E-ring **475**, which was converted into phenylsulfone **476** in standard fashion.

The union of the E and F rings was accomplished by coupling of the lithium anion derived from **476** with **472** to give EF bis(pyran) **477**. Methanolysis of **477** provided methyl ketal **478**. Reduction of ketone with KBHEt$_3$ afforded the desired α-alcohol as a single product, which was converted into the phosphonium salt **479** by functional group manipulation.

The completion of the total synthesis is described in Scheme 69. The Wittig reaction of **479** with LHMDS followed by addition of **466** afforded the desired (Z)-olefin **480** exclusively. Selective removal of methoxyacetate of **480** followed by oxidation and protection afforded TIPS ester **481**. Introduction of the diene

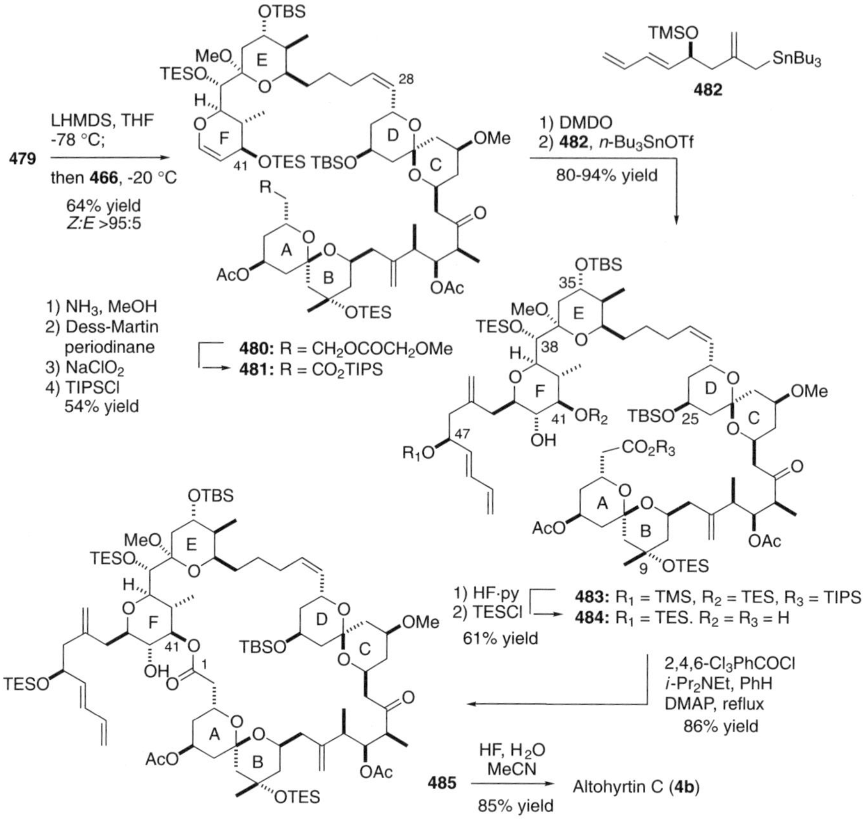

Scheme 69.

side chain was then investigated. Treatment of the dihydropyran F ring of **481** with DMDO gave α-epoxide with complete chemo- and stereoselectivity. The introduction of the side chain was realized by addition of allylstannane **482** with n-Bu$_3$SnOTf to afford the desired adduct **483** as a single diastereomer. HF·py treatment of **483** effected selective deprotection of the TIPS ester, the C47 TMS ether, and the C41 TES ether, with four silyl groups remaining at C9, C25, C35, and C38. After TES protection at C47, treatment of the resulting **484** under Yamaguchi's conditions effected macrolactonization to give a single regioisomeric lactone **485** that exhibited discrimination between the C41 and C42 diol functionalities of the F ring. Finally, complete deprotection of the silyl ethers with HF furnished the desired (+)-altohyrtin C (**4b**).

2. Kishi's Total Synthesis [140]

The first total synthesis of (+)-altohyrtin A (spongistatin 1; **4a**) by Kishi *et al.* was accomplished via Wittig coupling of the C1–C28 ABCD and C29–C51 EF

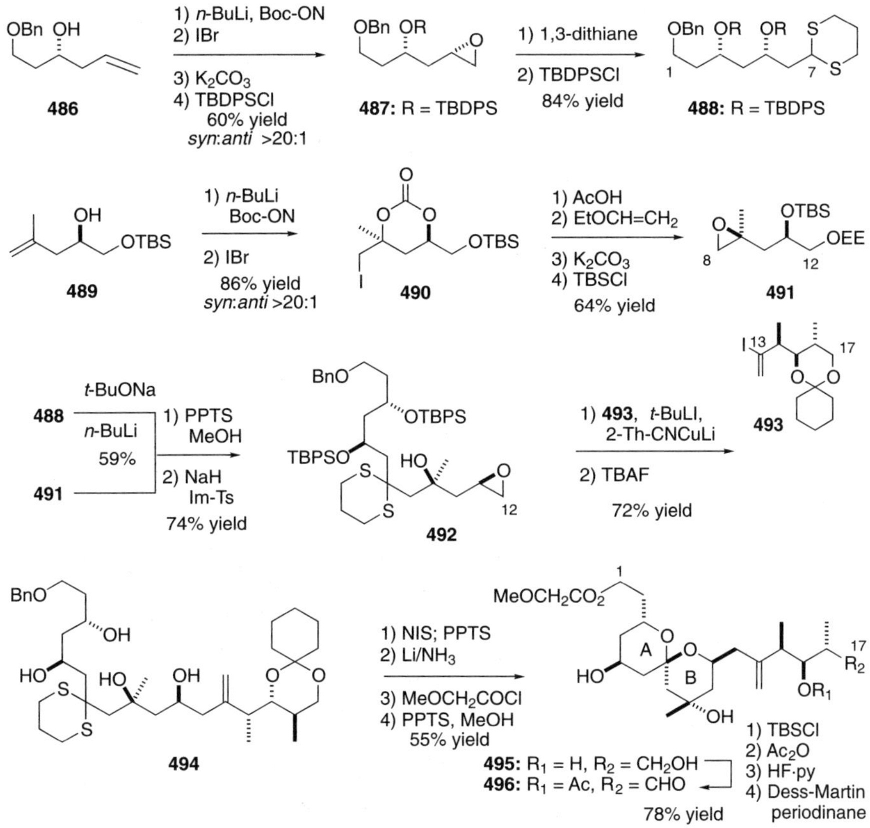

4. Total Synthesis of Macrolides

Scheme 70.

segments, and the regioselective macrolactonization (Fig. 8). The C44–C51 side chain was thus brought in at an early stage.

The C1–C17 AB segment **496** was synthesized as shown in Scheme 70, based on stereoselective halocarbonylation and ring-opening of the terminal epoxides with vinyl cuprates or 1,3-dithiane. Iodocarbonylation of **486** afforded 1,3-*syn*-carbonate, which was successively treated with a base and TBDPSCl to give epoxide **487**. Addition of the anion of 1,3-dithiane followed by TBDPS protection yielded the C1–C7 fragment **488**. The C8–C12 segment **491** was also synthesized from **489** via **490** based on similar iodocarbonylation. The coupling between the anion of **488** and epoxide **491** followed by formation of the terminal epoxide gave **492**. The C12–C13 bond formation was realized by ring-opening of epoxide **492** with a higher order 2-thienyl cyanocuprate derived from **493** to give **494**. Deprotection of thioketal afforded 6,6-spiroketal, which was converted into **496** via **495** by protective group manipulation.

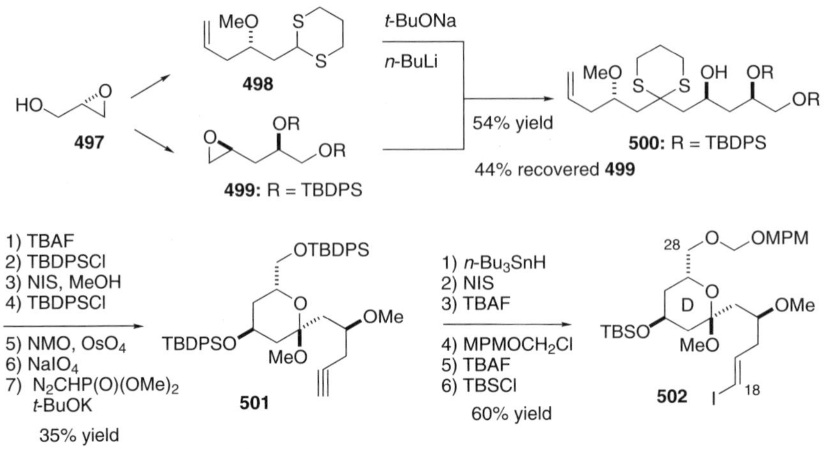

Scheme 71.

The C18–C28 segment **502** was synthesized starting from epoxide **497** as shown in Scheme 71. The thioketal **500** was synthesized by coupling of **498** and **499**, which were derived from epoxide **497**. The thioketal **500** was converted into hemiketal **501** via adjustment of the silyl protective group, deprotection of thioketal, oxidative cleavage of the double bond, and acetylene formation. Hydrostannylation/iodination of acetylene **501** gave vinyliodide, and protective group manipulation afforded **502**.

Construction of the C1–C28 ABCD segment **507** is described in Scheme 72. Enone **503** was synthesized from Ni(II)–Cr(II)-mediated coupling of **502** with **496** followed by Dess–Martin oxidation. After hydrolysis of hemiketal, the crucial intramolecular hetero Michael cyclization was effected with Triton-B to give **504** as the single isomer. The stereochemistry of the CD spirocenter in **504** was unsatisfactory. Epimerization was expected by deprotection of the TBS group of **504**. Indeed, desilylation of **504** with HF·py provided a separable 1:1 mixture of the desired diastereomer **506** and undesired **505**, which could be recycled under acidic conditions. Functional group manipulation converted **506** into aldehyde **507**, corresponding to the ABCD segment.

The synthesis of the C29–C51 EF segment **518** started with olefin **508** having three chiral centers, prepared by utilizing crotyl- and allyl-boronate chemistry (Scheme 73). Oxidative cleavage of olefin **508**, debenzylation, and dehydration afforded glycol **509**, which was converted into iodoglycal **510** by the Freisen method.

The olefinic side chain was introduced to F ring glycal **511** at an earlier stage. Epoxidation of **511** with DMDO gave the single product **512**. The coupling of **512** and a higher order cuprate prepared from allylstannane **513** gave **514**, which was converted into aldehyde **515** having identical protective groups by

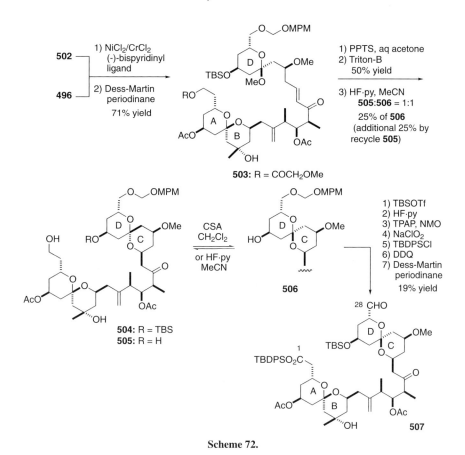

Scheme 72.

straightforward reactions. The next crucial step was an unprecedented introduction of the chlorodiene. Coupling of the allylindium reagent with aldehyde **515** afforded two homoallylic alcohols, which were dehydrated with Martin's sulfurane to afford exclusively the *trans*-chlorodiene. TBS deprotection followed by Swern oxidation gave aldehyde **516**.

The crucial coupling of the E and F rings was accomplished by the addition of a novel Grignard reagent prepared from **510** to **516**, giving **517** with high stereoselectivity. The glycal **517** was converted into phosphonium salt **518** via iodomethanolysis followed by reductive dehalogenation, switching the TIPS to the more labile TBS group.

The completion of the total synthesis is described in Scheme 74. The Wittig reaction of **507** with **518** afforded the (Z)-olefin **519**. After removal of the MPM and TBDPS protective groups, macrolactonization of the resulting 41,42-diol was carried out under Yamaguchi's conditions to give the desired macrolactone **520**.

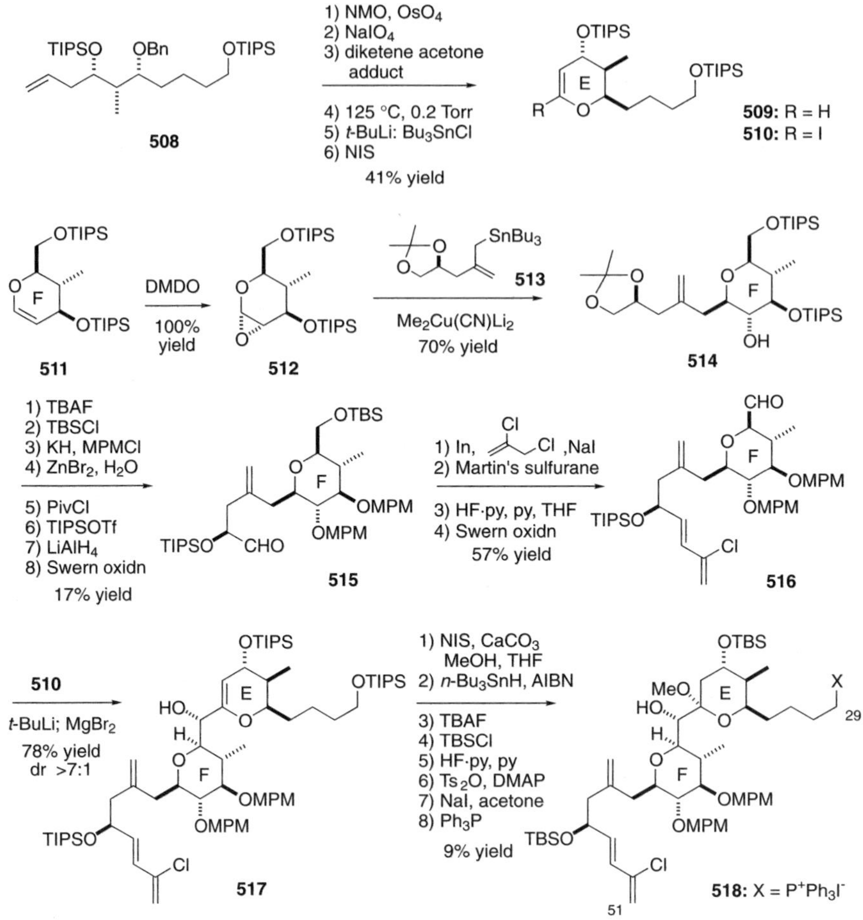

Scheme 73.

As anticipated, the macrolactonization regioselectively occurred with C41 alcohol. Finally, cleavage of the three TBS groups in **520** furnished (+)-altohyrtin A (**4a**).

3. Smith's Total Synthesis [141]

The total synthesis of altohyrtin C (spongistatin 2; **4b**) by the Smith group was accomplished via the sulfone-mediated coupling of the AB and CD subunits, subsequent Wittig coupling with the EF fragment having the side chain, and regioselective macrolactonization (Fig. 8).

The syntheses of the C1–C12 AB and C13–C28 CD segments, **528** and **535**, are summarized in Scheme 75. The syntheses of the two segments are based on a one-pot unsymmetric bisalkylation of 2-TBS-1,3-dithiane (**525**) with Brook rearrangement. Epoxides **521**, **522**, **523**, and **524** were synthesized as coupling partners.

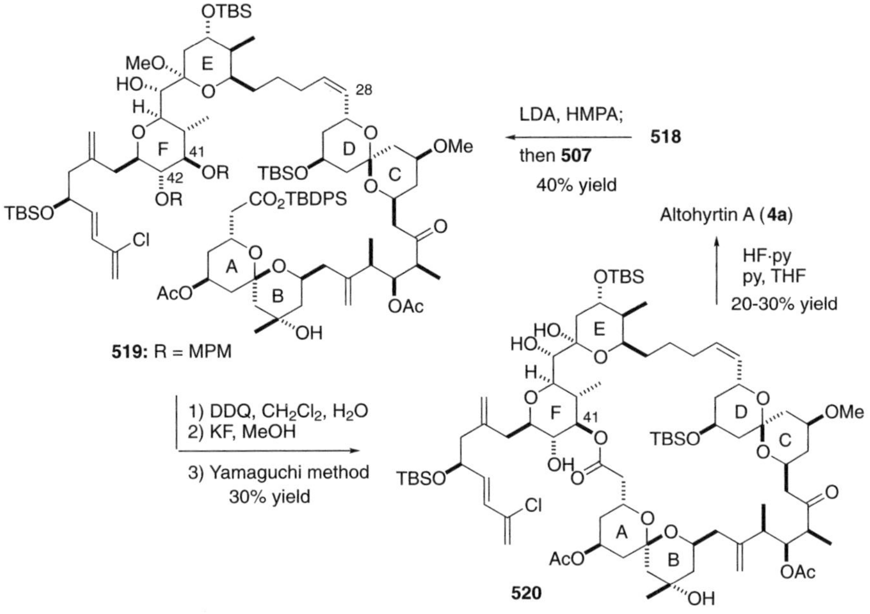

Scheme 74.

The unsymmetric bisalkylation of the dithiane **525** with epoxides **522** and **521** was efficiently carried out in one pot to give **526**. Deprotection of TES and acetonide followed by tosylation gave triol **527**, which was subjected to acetonization, dethioacetalization, intramolecular hemiketalization, and iodination to give B ring **528**.

The synthesis of the CD segment **532** also started with dithiane **525**. One-pot unsymmetric bisalkylation of **525** with **523** and **524** also effectively afforded coupling product **529**, after methylation. Deprotection of TBS and acetonide followed by dethioketalization afforded a 2:1 mixture of **530** and **531**. Treatment of **530** with $HClO_4$ effected epimerization to **531**. Functional group manipulation converted **531** to iodide **532**, which was coupled with thioacetal **533**, prepared by Roush asymmetric crotylboration, to give adduct **534**. The thioketal **534** was converted into **535** via reduction of ketone (dr = 1:3.5) and introduction of a phenylsulfone group.

The ABCD segment **538** was then constructed (Scheme 76). The Julia coupling of **528** and **535** followed by methylenation afforded adduct **536**. The AB spiroketal system was constructed by functional group manipulation to give **537**, which was converted into ketoaldehyde **538** via protective group manipulation and oxidation. The spiroketal **539**, which is a key intermediate for Kishi's total synthesis of altohyrtin A (**4a**), was also synthesized from **537**.

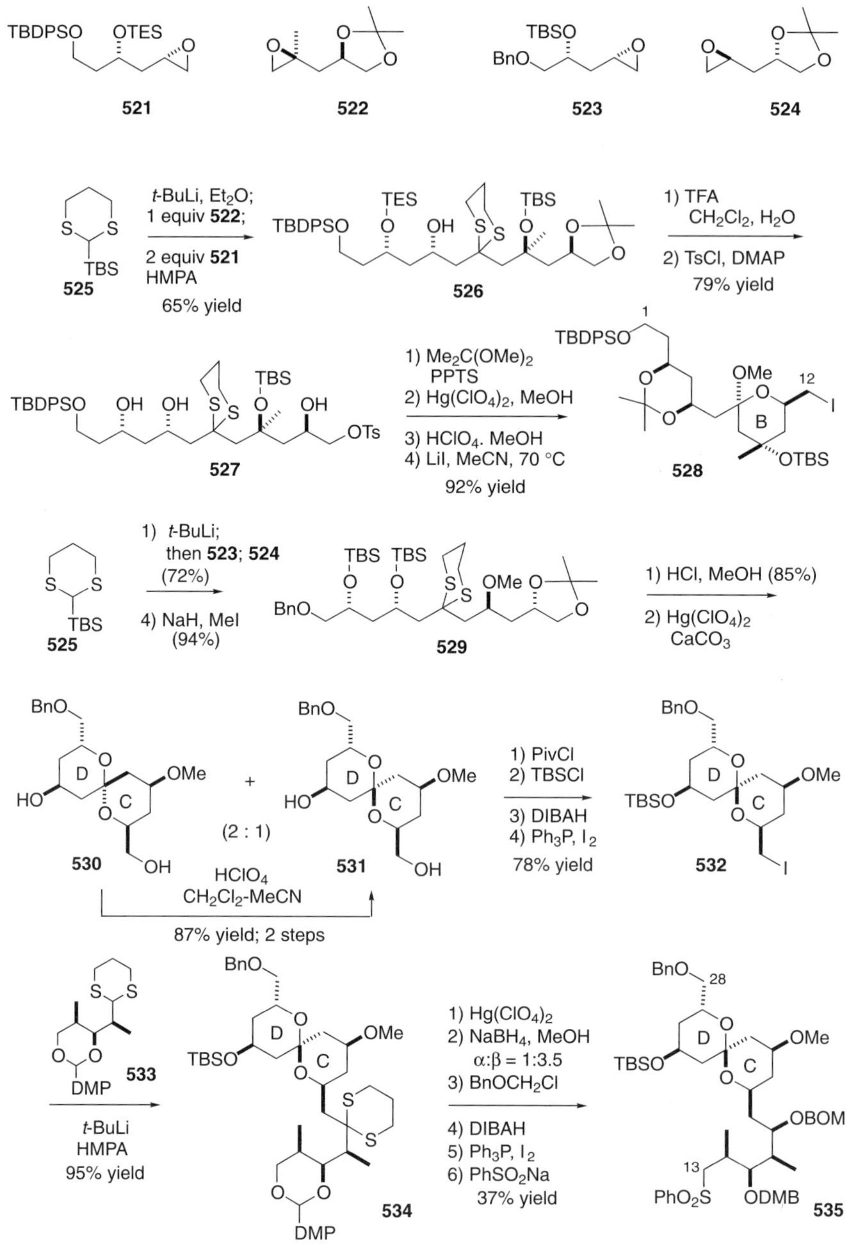

Scheme 75.

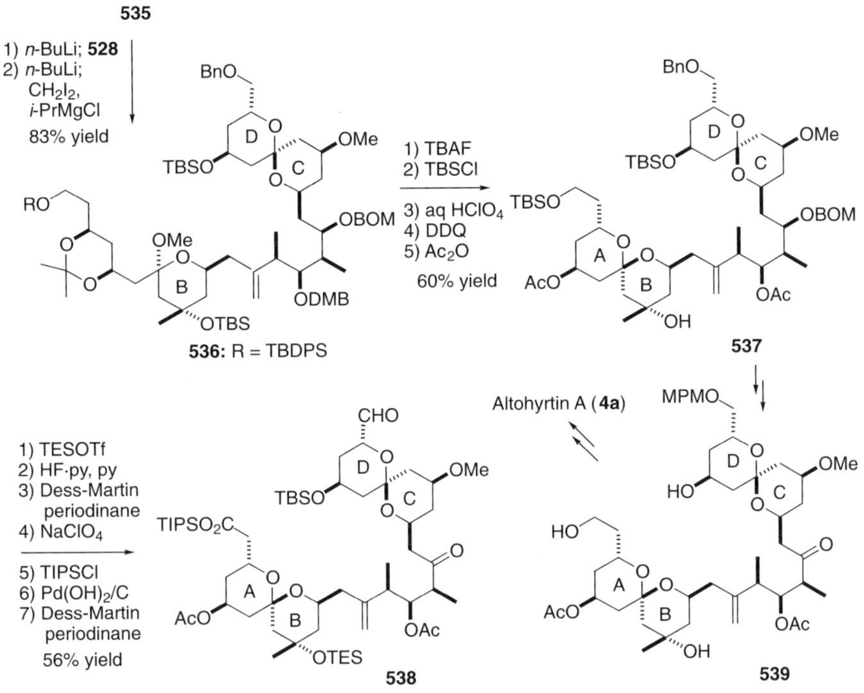

Scheme 76.

The synthesis of the C29–C51 EF segment **552** is summarized in Scheme 77. The C29–C37 fragment **542** was synthesized via iodo-carbonation of **540** and addition of 1,3-dithiane to epoxide. The F ring was constructed by Sharpless AE of **543** followed by regioselective epoxide opening, and Hg-mediated cyclization to give a 1:1 mixture of tetrahydropyrans **545** having α- and β-hydroxymethyl groups. The mixture was converted to a single aldehyde **546** via acetonization and oxidation. Addition of the anion of **542** to the aldehyde **546** stereoselectively afforded α-alcohol **547**. Deprotection of acetonide and thioketal afforded hemiketal **548**, which was converted into iodide **549** by functional group manipulation. Introduction of the side chain was carried out at this stage. Addition of the anion of phenylsulfone **550** to **549** afforded **551**. The Julia methylenation of **551**, deprotection of the benzyl ether, oxidation, and Wittig reaction afforded a diene, which was transformed to phosphonate salt **552** to be coupled with the ABCD segment.

The completion of the total synthesis is described in Scheme 78. The Wittig reaction of **552** and **538** effected the coupling, and subsequent deprotection of TMS with KF afforded 41,42-dihydroxy carboxylic acid **553**. Yamaguchi conditions effected macrolactonization of **553** in high yield and complete deprotection of the silyl groups afforded **4b** and **554** in a ratio of 1:3. Treatment of (−)-23-epi-altohyrtin

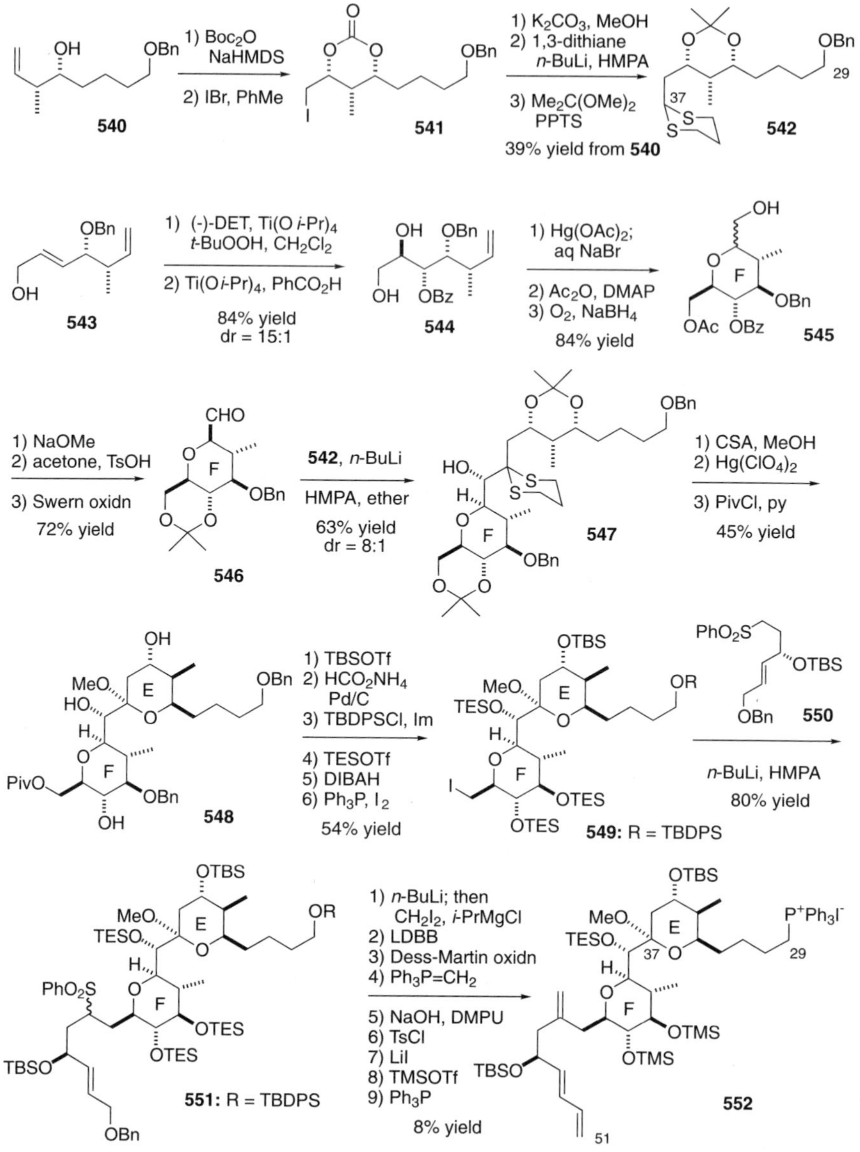

Scheme 77.

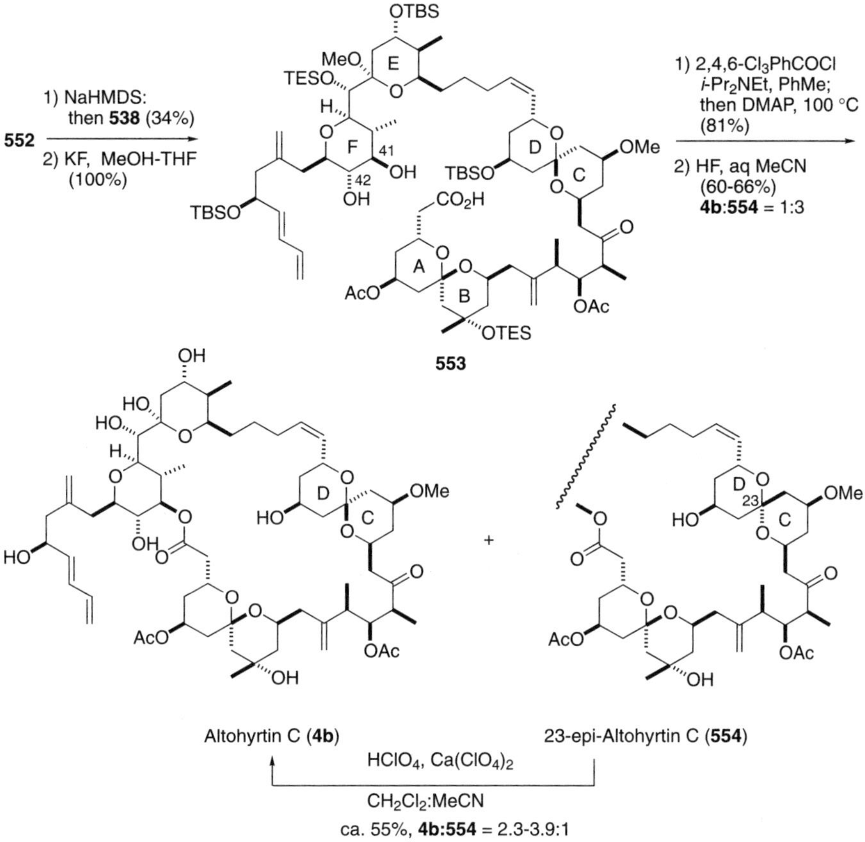

Scheme 78.

C (**554**) with HClO$_4$ and Ca(ClO$_4$)$_2$ in CH$_2$Cl$_2$-MeCN effected epimerization to (+)-altohyrtin C (**4b**) in a ratio of 2.3–3.9:1 of **4b** and **554**.

E. Epothilone

Epothilones, isolated from the mycobacterium *Sorangium cellulosum* in 1996, are a new class of cytotoxic macrolides that are especially effective against drug-resistant tumor cell lines by the same mechanism as the taxoids. The structural feature of epothilones (Fig. 1) includes a 16-membered lactone ring, β-hydroxyl carbonyl units and *cis*-epoxide on the macrocyclic ring, and a thiazole unit as the side chain.

The syntheses of epothilones (**5**) have been extensively studied by numerous research groups [142–153], notably those of Danishefsky [142] and Nicolaou [143].

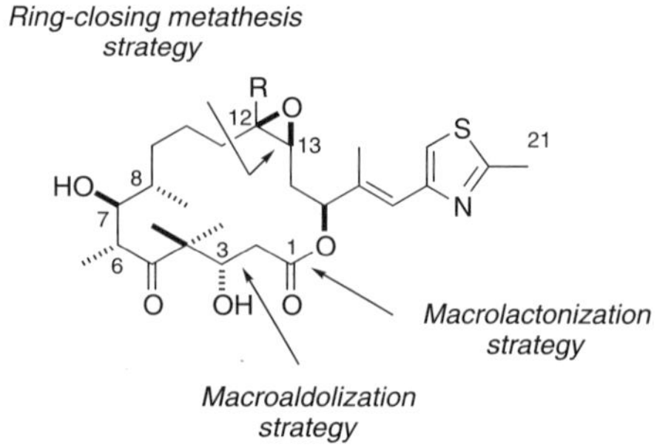

Fig. 9.

These studies have culminated in the total synthesis of epothilones and a large number of related analogues [143e–m, 154].

The problems hindering the total synthesis of epothilones involve (1) stereoselective construction of C3, C6, C7, and C8 chiral centers, (2) synthesis of the thiazole unit having a 15S-hydroxyl group, (3) construction of (12Z)-olefin, (4) macrocyclization, and (5) stereoselective epoxidation of (12Z)-olefin. The total syntheses of epothilones have been achieved via three strategies for the macrocyclization (Fig. 9): (1) macroaldolization, (2) ring-closing olefin metathesis, and (3) macrolactonization.

1. Macroaldolization Strategy

The first total synthesis of epothilone A (**5a**) was accomplished by Danishefsky *et al.* [142a,c,d] based on macroaldol reaction for the construction of the macrocyclic ring (Fig. 9). The synthetic feature includes stereoselective synthesis of the C3–C11 segment based on their LACDAC reaction, Suzuki coupling with the C12–C21 segment, and unique intramolecular macroaldolization at the C2 and C3 positons.

The synthesis of the C3–C11 chiral segment **561** started with $TiCl_4$-induced cycloaddition of aldehyde **555** and diene **556** (Scheme 79). The LACDAC reaction stereoselectively afforded enone **557**, which was subjected to $LiAlH_4$ reduction and Simmons–Smith reaction to give cyclopropane **558** stereoselectively. The geminal methyl groups were constructed by oxidative solvolytic fragmentation of the cyclopropane and reductive deiodination to afford **559**. The standard protective group manipulation led to **560**, which was converted into acetal **561**, as a coupling partner, by functional group manipulation involving homologation. The thioacetal **560** is the key intermediate in their total synthesis of epothilone A (**5a**) through ring-closing olefin metathesis strategy [142b,d]. The acetal **561** was converted into ester **562** as a substrate for their other strategy, macrolactonization [142d].

The (Z)-vinyl segment **565** was first synthesized from commercially available (R)-glycidol [142a] and later by an alternative route [142c] using Brown asymmetric allylation (Scheme 80). The asymmetric allylation of aldehyde **563** under

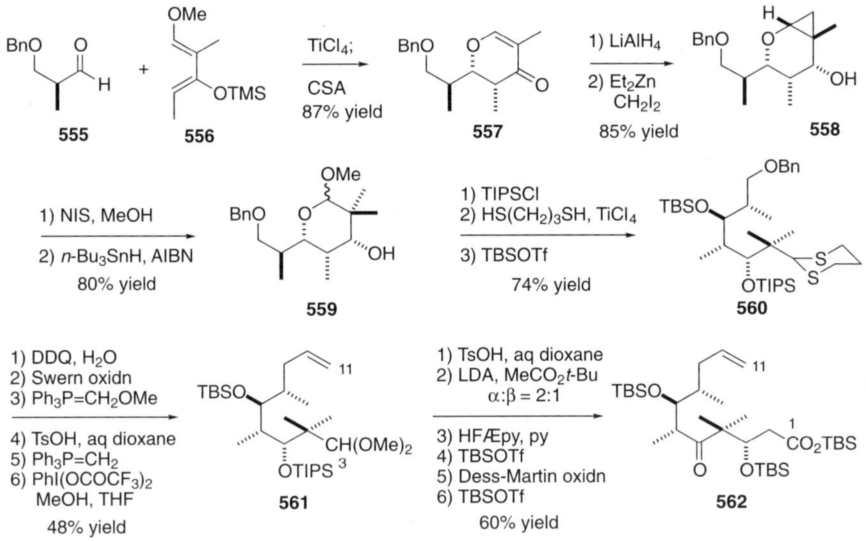

Scheme 79.

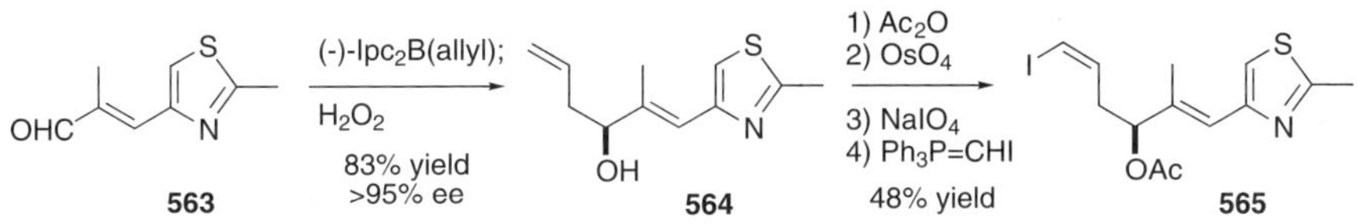

Scheme 80.

Brown's conditions afforded **564** with high ee. Acylation, oxidative cleavage of double bond, and Wittig olefination gave (Z)-vinyl iodide **565**.

As shown in Scheme 81, the coupling of two segments, **561** and **565**, was effectively achieved by a B-alkyl Suzuki coupling to give the desired (12Z)-olefin, which led to aldehyde **566** by hydrolysis. The stage was then set for unique intramolecular macroaldolization. Upon treatment of **566** with KHMDS, an intramolecular aldol reaction took place to give a 6:1 mixture of 3α- and 3β-hydroxy macrolactone. The desired 3α-hydroxy lactone was then converted into 5-keto lactone **567** by adjustment of the silyl protective groups, and oxidation. After removal of the TBS groups, epoxidation of the (12Z)-olefin was examined. It was found that treatment with DMDO induced highly stereoselective epoxidation to give epothilone A (**5a**).

2. Ring-Closing Olefin Metathesis Strategy

The syntheses of epothilones based on the RCM strategy were extensively investigated by the Danishefsky [142b,d], Nicolaou [143a,c], Schinzer [144a,c], and Grieco [145] groups (Fig. 9).

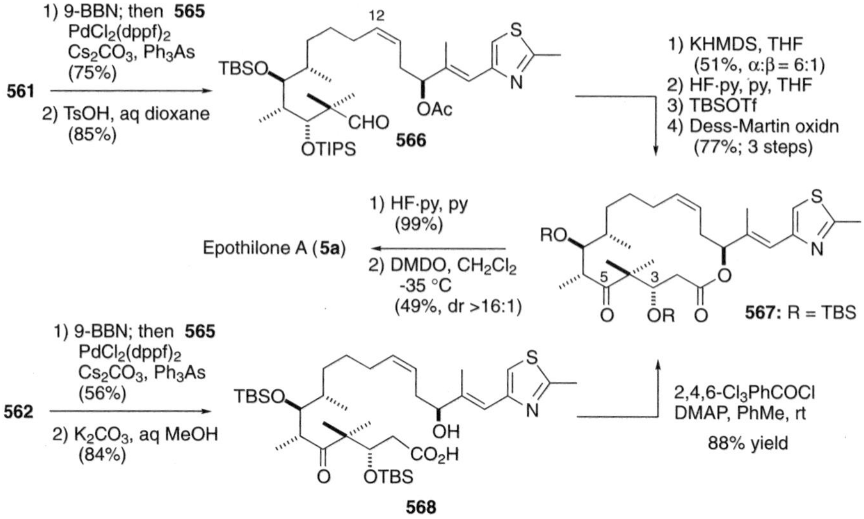

Scheme 81.

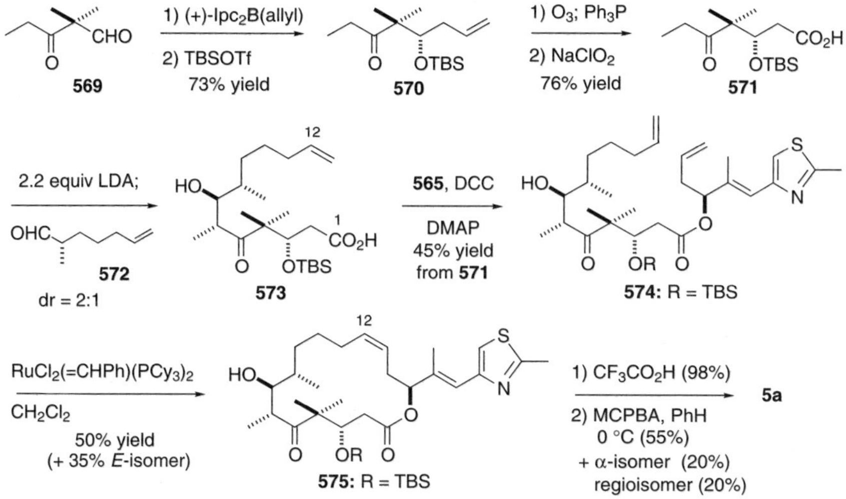

Scheme 82.

Only Nicolaou's synthesis [143a,c] is described in this section (Scheme 82), because the results of olefin metathesis at the C12 and C13 positions, the key reaction of this strategy, are almost the same. The C1–C12 acid **573** was synthesized as the coupling partner. The Brown asymmetric allylboration of β-keto aldehyde **569** followed by TBS protection furnished **570**, which was oxidized

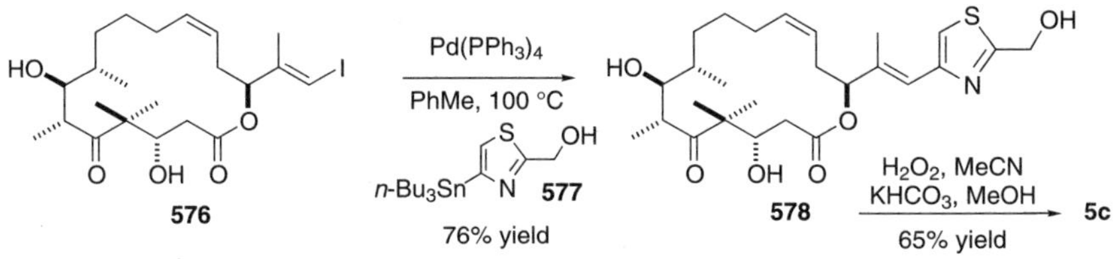

4. Total Synthesis of Macrolides   261

Scheme 83.

to carboxylic acid **571**. Aldol reaction of the lithium enolate of **571** with chiral aldehyde **572** afforded **573** and its diastereomer in a 2:1 ratio. The other coupling partner **565** was also prepared by the Brown asymmetric allylation (see Scheme 80). Esterification of **565** and **573** with DCC-DMAP afforded **574**, a substrate for olefin metathesis. Olefin metathesis using Grubbs reagent effected ring-closing to give (12Z)-olefin **575** and (12E) isomer in 50% and 35% yield, respectively. Deprotection of **575** followed by epoxidation yielded epothilone A (**5a**).

The RCM strategy was applied to the synthesis of epothilone E (**5c**) [143i] (Scheme 83). The vinyl iodide **576**, which was synthesized through C12–C13 ring-closing olefin metathesis, was subjected to the Stille coupling with **577** to afford **578**. The epoxidation of the C12–C13 double bond in **578** completed the total synthesis of epothilone E (**5c**).

Olefin metathesis is very effective for the construction of the macrocyclic ring; however, in the synthesis of epothilones, this approach has been reported thus far to result in low (E)/(Z) selectivity in construction of the C12–C13 double bond.

The RCM strategy was also successfully applied to solid-phase synthesis of epothilones using polystyrene resin, which permits simultaneous formation of the macrocyclic ring and its release from the resin. The present method provides a large library of epothilone analogues for biological screening [143c,e,f,i,l].

3. Macrolactonization Strategy

The syntheses based on the macrolactonization strategy were extensively studied by numerous synthetic groups. Most of these groups applied Yamaguchi's conditions to the macrolactonization. Thus, the feature of each group in this strategy is how to construct seco-acid derivatives, which should be constructed by coupling of several segments.

Danishefsky's first macrolactonization [142d] to **5a** was carried out using **568** (Scheme 81), which was prepared starting from **562**. The C1–C10 ester **562** was synthesized from acetal **561** (Scheme 79). After hydrolysis of the acetal **561**, the resulting aldehyde was subjected to aldol reaction with *t*-butyl acetate to give α- and β-alcohols in a 2:1 ratio. The desired α-alcohol was further converted to TBS ester **562** in standard fashion. As shown in Scheme 81, Suzuki coupling of **562** and **565** followed by hydrolysis of the ester afforded hydroxy carboxylic

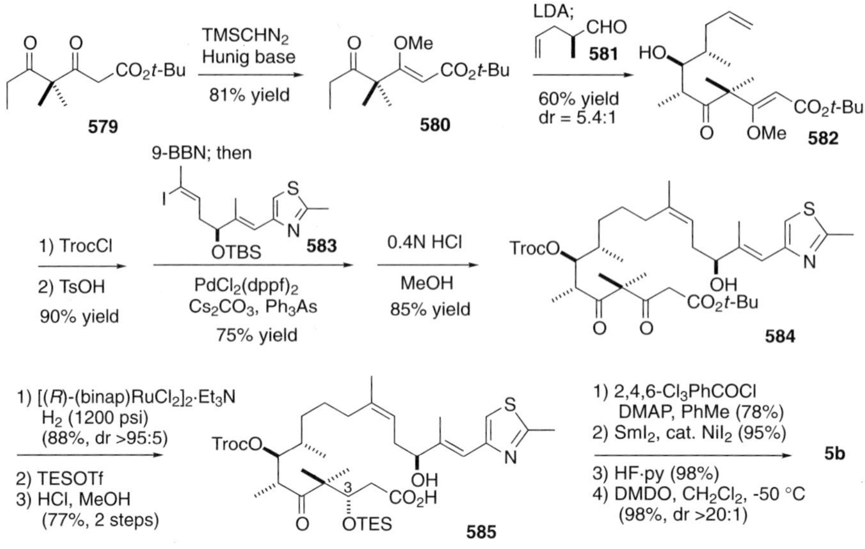

Scheme 84.

acid **568**. Macrolactonization was carried out under Yamaguchi's conditions to give macrolactone **567** in 88% yield. The lactone **567** was already converted into epothilone A (**5a**) in the first synthesis through the macroaldolization strategy [142a].

An alternative route by the Danishefsky group was developed [142e–g] (Scheme 84). The aldol reaction of ethyl ketone **580**, prepared from β-keto ester **579**, with aldehyde **581** stereoselectively afforded **582** (dr = 5.4:1). After Troc protection followed by hydrolysis of the enol ether, Suzuki coupling with **583** followed by TBS deprotection gave the desired (12Z)-olefin **584**. The Noyori reduction of the β-keto ester **584** gave 3α-alcohol with high stereoselectivity, which was converted into hydroxy carboxylic acid **585**. Macrolactonization of **585** was accomplished by the Yamaguchi method, and subsequent deprotection and DMDO oxidation efficiently afforded epothilone B (**5b**).

In Nicolaou's synthesis through the macrolactonization strategy [143b,d], three segments **586**, **587**, and **571** were used for the synthesis of seco-acid (Scheme 85). The C7–C12 phosphonium salt **586** was prepared via Enders SAMP asymmetric alkylation. The Wittig olefination of **586** with aldehyde **587**, prepared from **564**, predominantly afforded (12Z)-olefin, which underwent selective deprotection of the primary TBS group and oxidation to give aldehyde **588**. Aldol reaction of **588** and ketone **571** afforded a 1:1 mixture of hydroxy esters, which was, after silylation and methanolysis, purified to give **589**. After selective deprotection of the C15 TBS group, macrolactonization under Yamaguchi's conditions afforded lactone **567** in

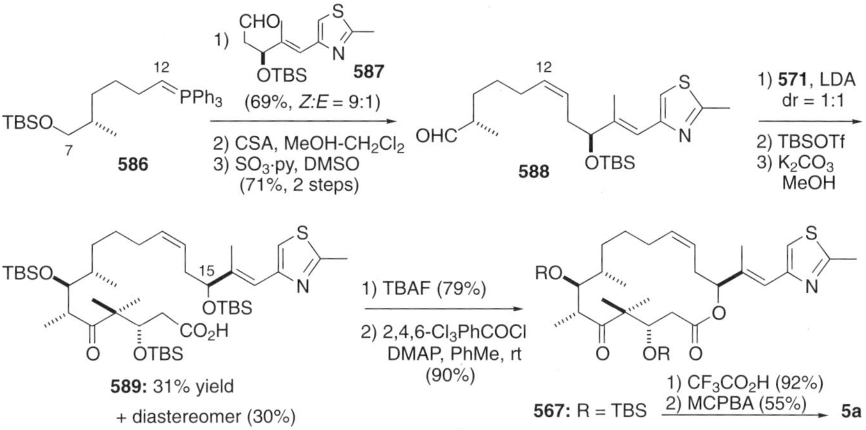

4. Total Synthesis of Macrolides 263

Scheme 85.

90% yield. Deprotection of the TBS group followed by epoxidation completed the total synthesis of epothilone A [143b] (**5a**).

Nicolaou et al. [143d] also reported the total synthesis of epothilone B (**5b**) via macrolactonization (Scheme 86). Aldol reaction of the lithium enolate of ketone **591** with aldehyde **590**, prepared via Wittig reaction, yielded a 3:1 mixture of aldols. The desired isomer **592** was converted into carboxylic acid **593**, which led to **5b**. Because of the moderate and variable selectivities in the aldol coupling using ketone **591**, they further investigated the reaction conditions. After extensive studies, the present reaction led to significant improvement of stereoselectivity (dr > 10:1) [143m].

The total syntheses of **5b** were independently accomplished via the same intermediates **592** and **593** via **590** by the Schinzer [144b,d] and Mulzer groups [146] (Scheme 86). In the Schinzer synthesis [144b,d], palladium-mediated coupling of vinyl iodide **583** with an organozinc species, derived from **594**, efficiently produced (12Z)-olefin, which, on desilylation and oxidation, led to **590**. Aldol reaction of **590** with acetonide ketone **595** induced high selectivity to give the desired adduct (dr = 9:1), which was converted into **592**, identical with Nicolaou's intermediate **592**. The chiral centers of three segments **583**, **594**, and **595** were derived from (S)-malic acid, by Evans asymmetric alkylation, and by asymmetric aldol reaction using (S)-2-hydroxy-1,2,2-triphenyl acetate, respectively.

The key reactions of Mulzer's synthesis [146a,b] involved coupling of sulfone **596** and allyl iodide **597** for the synthesis of (12Z)-olefin, and aldol reaction with ketone **591**. The chiral sources of **596**, **597**, and **591** were derived from (R)-3-hydroxy-2-methylpropionate, (S)-malic acid, and by Kiyooka asymmetric aldol reaction, respectively. The aldehyde **590** was synthesized through addition of sulfone **596** to allyl iodide **597**. Aldol reaction of **590** with **591** resulted in a 4:1 ratio

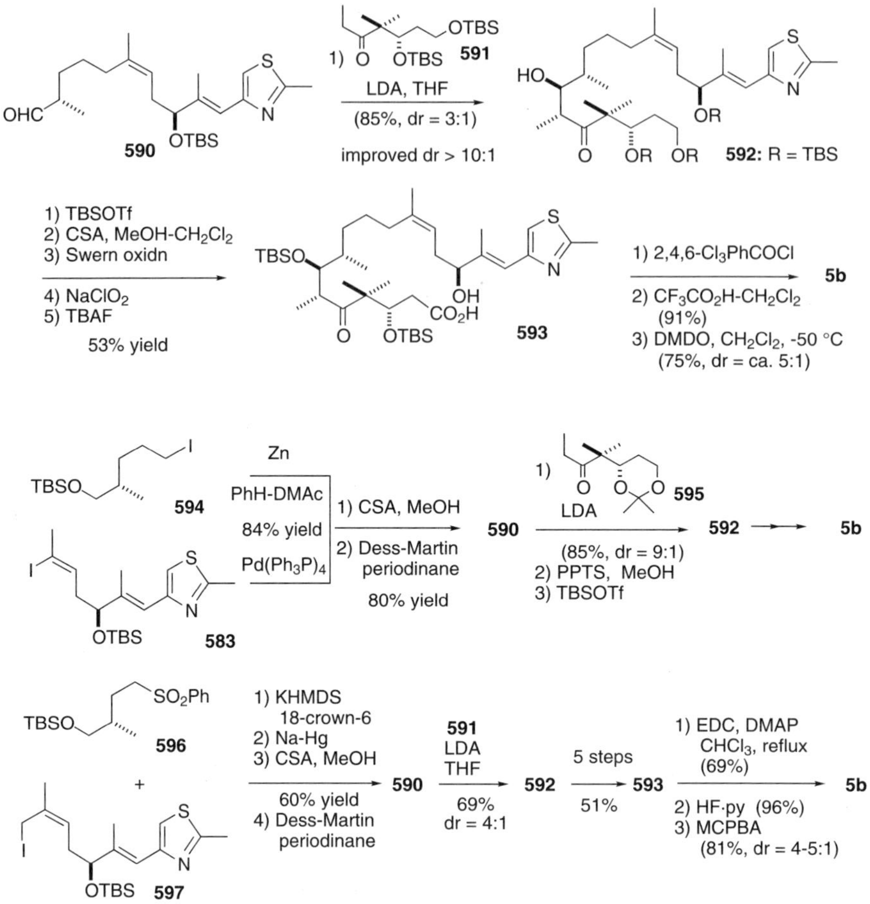

Scheme 86.

of stereoselectivity. The macrolactonization of **593** under the modified Keck (EDC-DMAP) conditions afforded macrolactone in 69% yield, which led to **5b**.

White's strategy involves acetylide coupling at C10 and C11 [147b,c] (or Wittig coupling at C9 and C10 [147a,c]) of two principal segments, and reduction of the resulting alkyne (or olefin) (Scheme 87). The C1–C10 acetylene segment **601** was synthesized from ketone **598**. Aldol reaction of **598** and aldehyde **599** yielded **600** with complete stereoselectivity. The standard functional group manipulation led to acetylene **601**. Coupling of the copper(I) derivative of **601** with allylic bromide **602** afforded **603**, which was subjected to semihydrogenation with Lindlar catalyst, hydrolysis, and desilylation to give seco acid **604**. After macrolactonization under Yamaguchi's conditions and TBS deprotection,

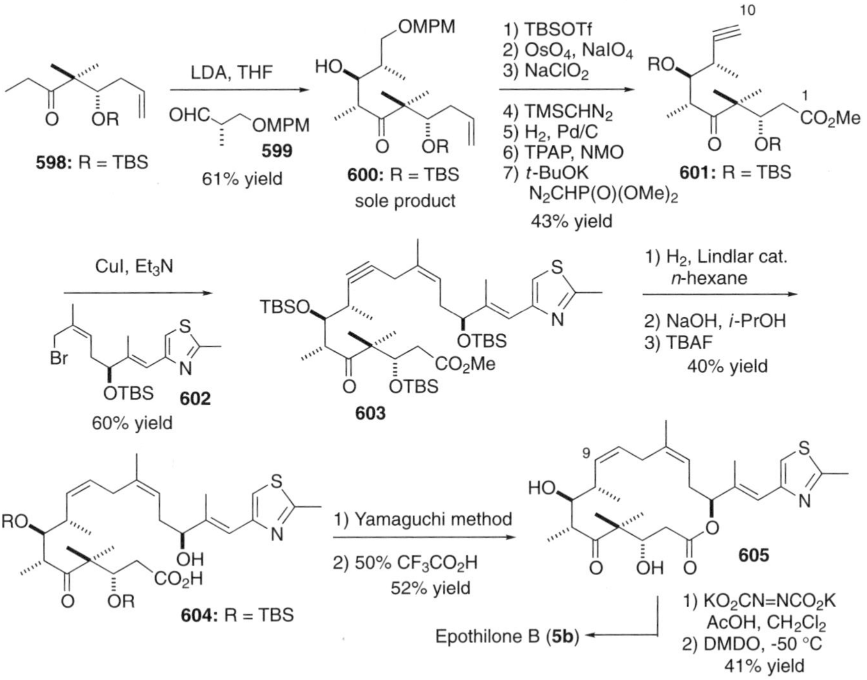

Scheme 87.

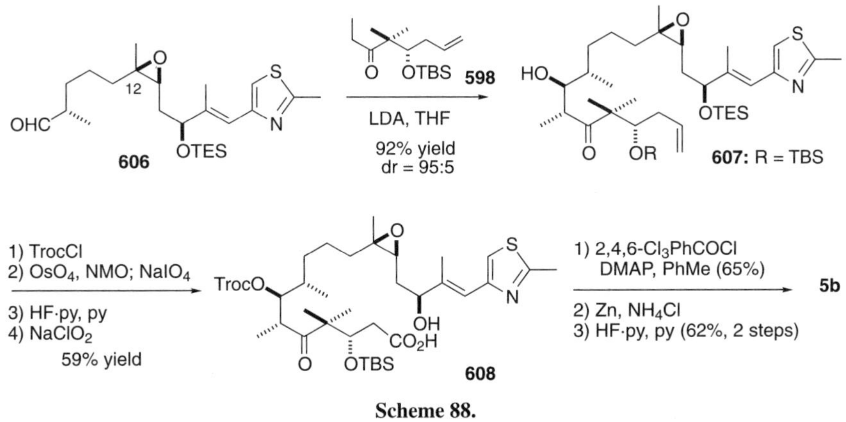

Scheme 88.

the resulting lactone **605** was subjected to reduction of the C9–C10 double bond with diimide and epoxidation to afford epothilone B (**5b**).

Mulzer *et al.* reported a unique total synthesis using aldehyde **606** that involves 12,13β-epoxide [146d] (Scheme 88). Aldol reaction of aldehyde **606** with ethyl

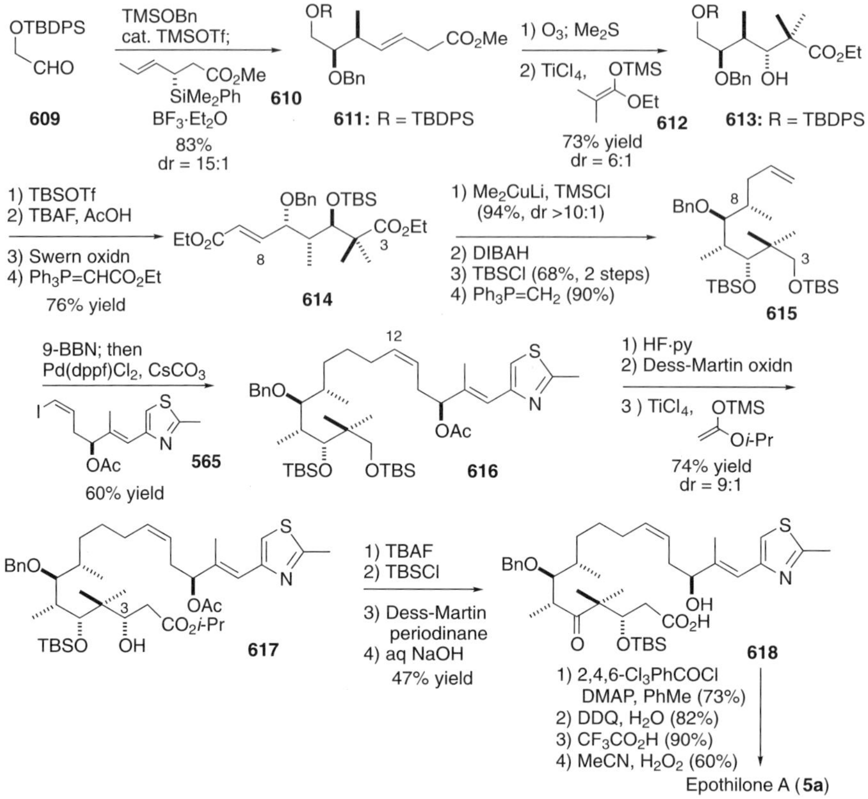

Scheme 89.

ketone **598** afforded **607** with high stereoselectivity. Protective group manipulation and oxidation provided carboxylic acid **608**. The Yamaguchi macrolactonization of **608** afforded lactone in 65% yield. Removal of the Troc and TBS groups completed the total synthesis of epothilone B (**5b**). During the course of the synthesis, the epoxide remains stable.

Zhu and Panek's total synthesis [148] is described in Scheme 89. After conversion of aldehyde **609** to di-benzyl acetal, treatment with chiral crotylsilane **610** afforded 1,2-*syn*-**611** with high stereo- and enantioselectivity. The oxidative cleavage of the double bond and subsequent aldol reaction with silyl ketene acetal **612** provided **613**, which was converted into α,β-unsaturated ester **614** via Wittig olefination. The C8 methyl group was stereoselectively introduced by treatment with dimethylcuprate in the presence of TMSCl. DIBAH treatment differentially reduced the C3 and C10 esters to alcohol and aldehyde, respectively. Protection of the alcohol as silyl ether followed by the Wittig reaction afforded **615**. In a manner similar to Danishefsky's synthesis [142d], an intermolecular Suzuki

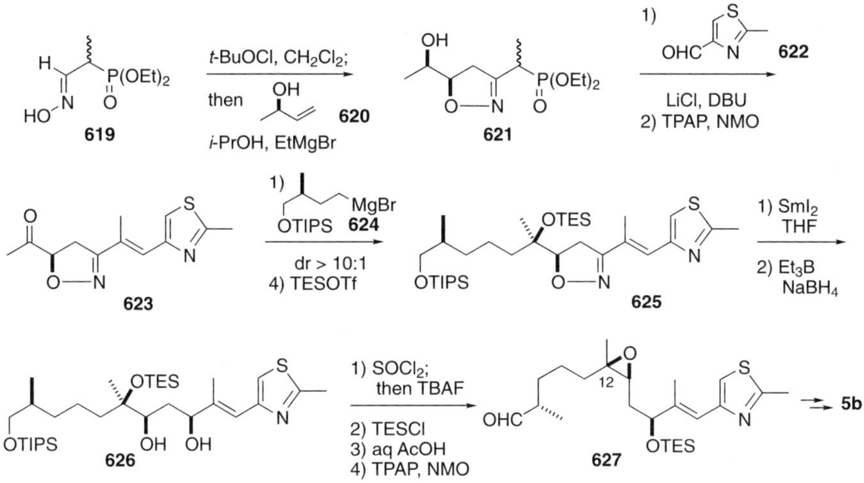

Scheme 90.

coupling of **615** and **565** gave (12Z)-olefin **616**. After selective TBS deprotection and oxidation, aldol reaction of the resulting aldehyde with silyl ketene acetal gave 3α-hydroxy ester **617** (dr = 9:1). Functional group manipulation afforded hydroxy acid **618**, which was subjected to Yamaguchi macrolactonization to give lactone. Deprotection followed by epoxidation completed the total synthesis of epothilone A (**5a**).

The Shibasaki group has reported an enantioselective total synthesis of epothilones A (**5a**) and B (**5b**) using their developed multifunctional asymmetric catalysts for the synthesis of the key segments [149]. The enantioselective total synthesis of epothilone A has also been accomplished using an antibody catalyst or an enzyme by the Sinha–Lerner [150] and Wong [151] groups, respectively.

Very recently, Bode and Carreira [152] have reported total synthesis of epothilones A and B via directed nitrile oxide cycloaddition (Scheme 90). Oxidation of **619** to the nitrile oxide followed by treatment with (R)-3-butene-2-ol (**620**) effected highly diastereoselective cycloaddition to furnish a single isoxazoline diastereomer **621**. Subsequent Wittig-type reaction with **622** followed by oxidation afforded ketone **623**, which was subjected to chelation-controlled Grignard reaction with **624** to give **625** after silylation. Reduction of the isoxazole **625** to β-hydroxy ketone was realized by treatment with SmI$_2$. The reduction of the resulting ketone with NaBH$_4$-Et$_3$B afforded 1,3-syn-diol **626**. 1,3-Cyclic sulfite formation followed by TBAF treatment afforded 12β-epoxy alcohol, which was converted into the known **627**, the key intermediate in Mulzer's total synthesis [146d] of epothilone B (**5b**), in the standard manner. Under a similar procedure, the total synthesis of epothilone A (**5a**) was also achieved.

268   Tadashi Nakata

F. Others

In addition to the total syntheses mentioned in Sections II.B and III.A–E, several selected complex macrolide antibiotics are illustrated in Fig. 10, suggesting the strategy of bond connections and key reactions in the total synthesis [155–174].

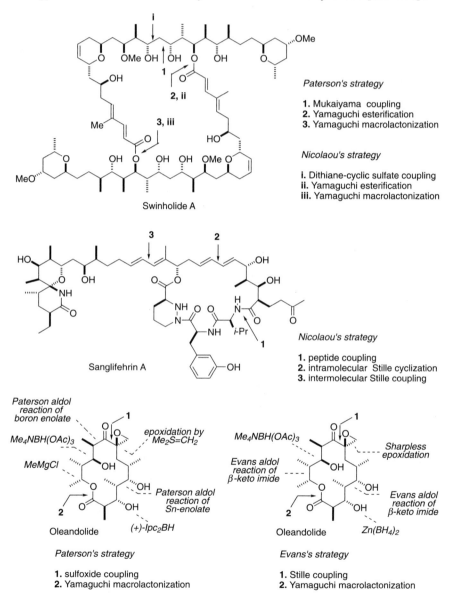

Fig. 10.

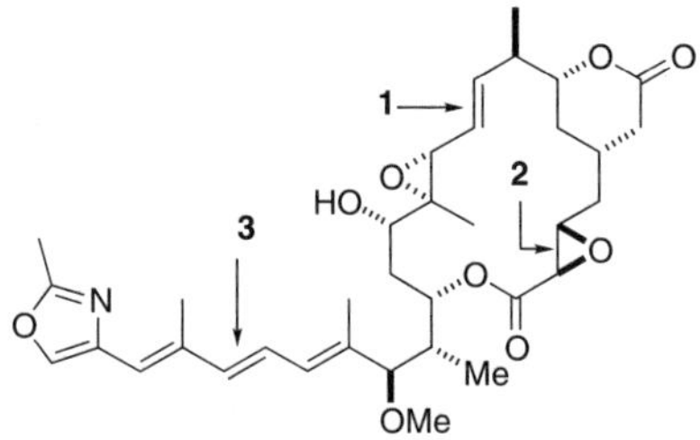

4. Total Synthesis of Macrolides

Rhizoxin
Ohno's strategy

1. Julia coupling
2. HWE macrocyclization
3. Horner-Wittig reaction

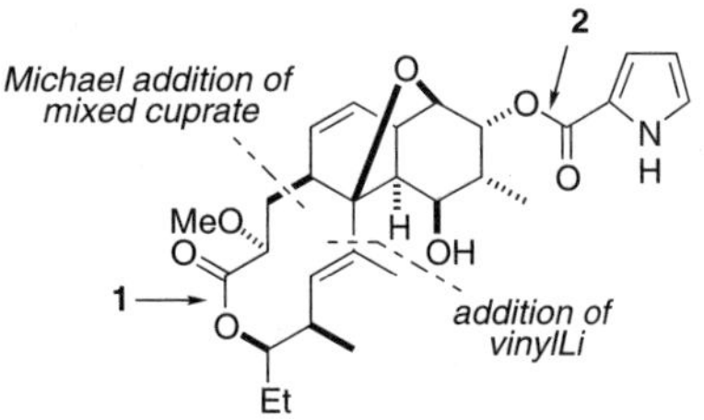

18-Deoxynargenicin A₁
Kallmerten's strategy

1. Corey macrolactonization
2. DCC-DMAP esterification

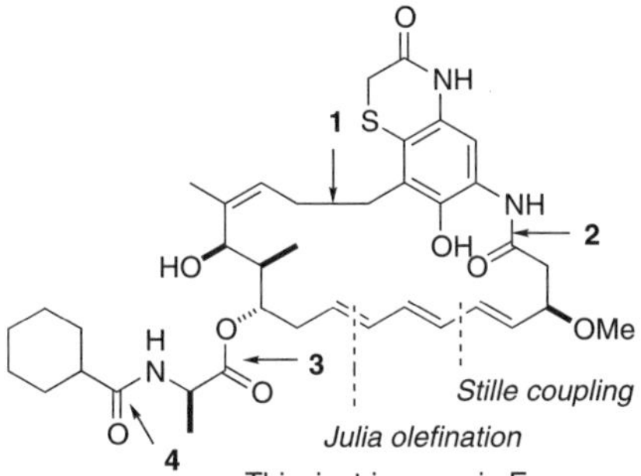

Thiazinotrienomycin E
Smith's strategy

1. sulfone alkylation
2. Mukaiyama macrolactamization
3. acylation
4. BOP-mediated coupling

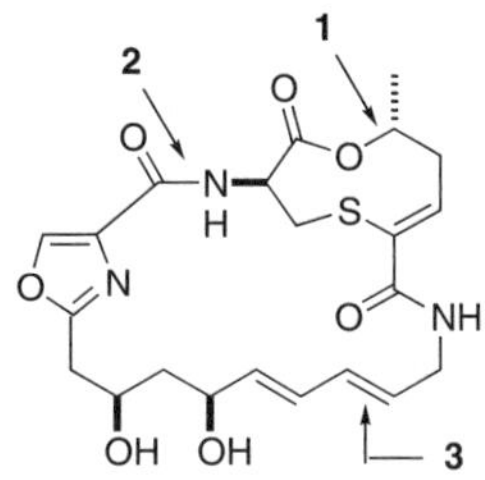

Griseoviridin
Meyer's strategy

1. Mitsunobu lactonization
2. amidation (EDC-HOBT)
3. Grubbs RCM

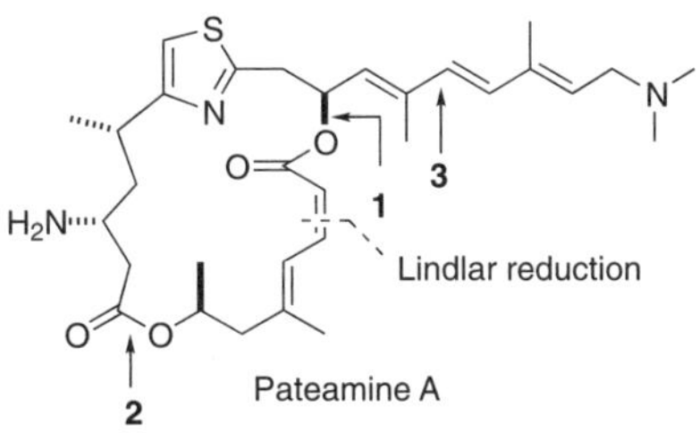

Pateamine A
Romo-Liu's strategy

1. Mitsunobu esterification
2. β-lactam based macrocyclization
3. Stille coupling

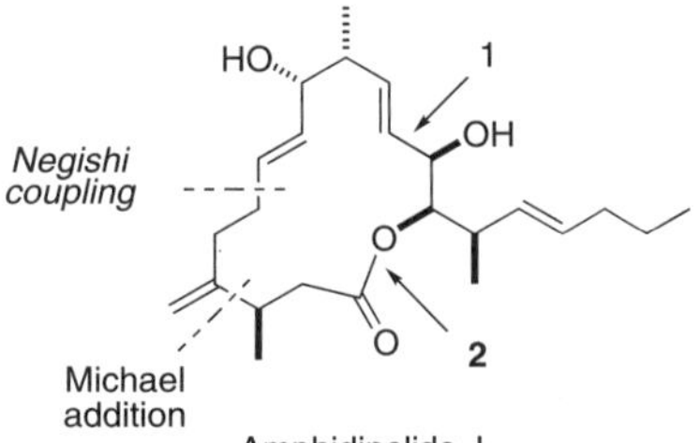

Amphidinolide J
Williams' strategy

1. addition of zincate
2. Yamaguchi macrolactonization

Fig. 10. *Continued*.

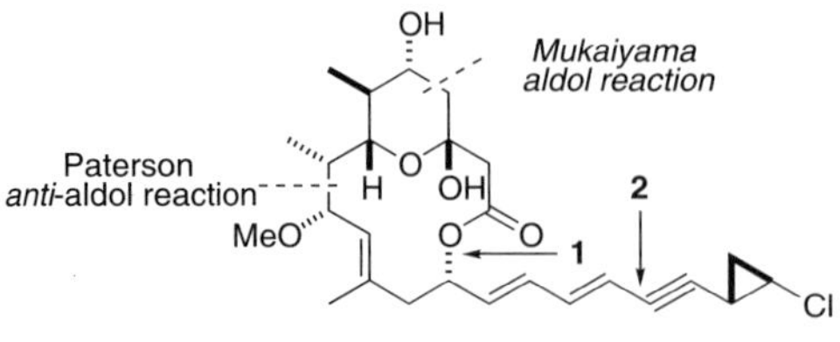

Callipeltoside aglycone
Paterson's strategy

1. Yamaguchi macrolactonization
2. Sonogashira coupling

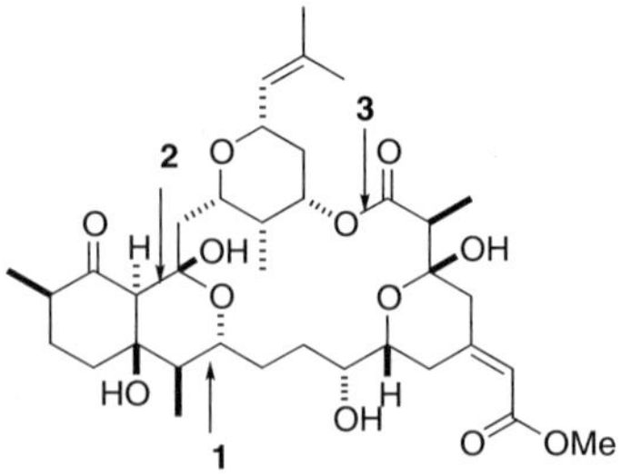

Miyakolide
Evans' strategy

1. aldol reaction
2. [3+2] dipolar cycloaddition
3. Yamaguchi macrolactonization

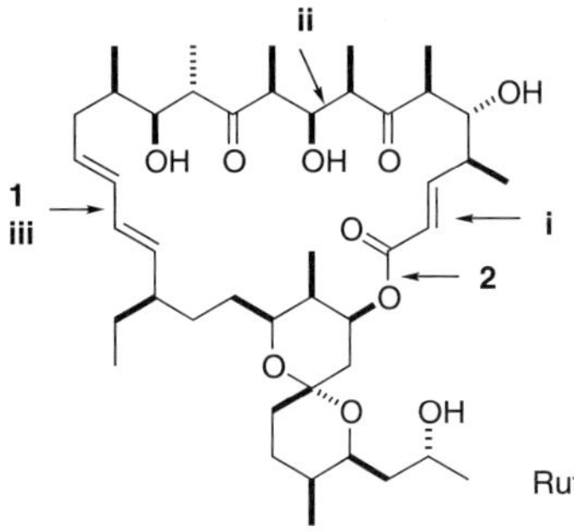

Rutamycin B
Evans's strategy

1. Suzuki coupling
2. Yamaguchi macrolactonization

White's strategy

i. HWE coupling
ii. aldol reaction
iii. Suzuki macrocyclization

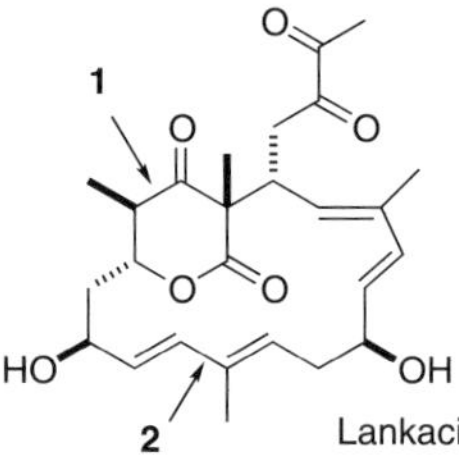

Lankacidin C
Kende's strategy

1. acylation of β-lactam
2. Stille coupling
3. Stork cyanohydrin macrocyclization

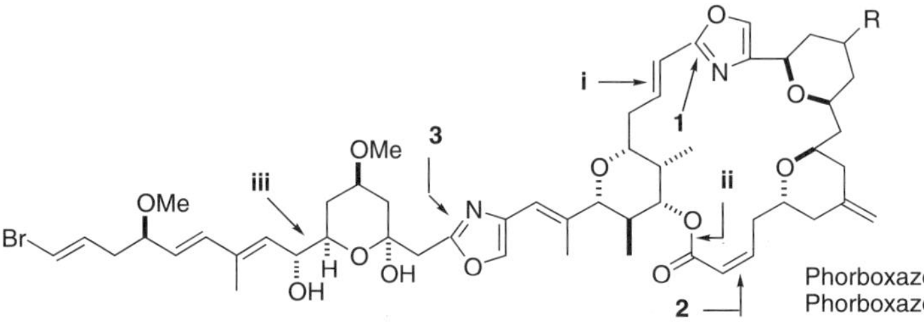

Phorboxazole A : R = α-OH
Phorboxazole B : R = β-OH

Forsyth's strategy

1. amidation (EDCI, HOBT)
2. HWE-coupling under Still's conditions
3. amidation (EDCI, DMAP)

Evans' strategy

i. Wittig coupling
ii. Yamaguchi macrolactonization
iii. addition of vinyl MgBr

Fig. 10. *Continued.*

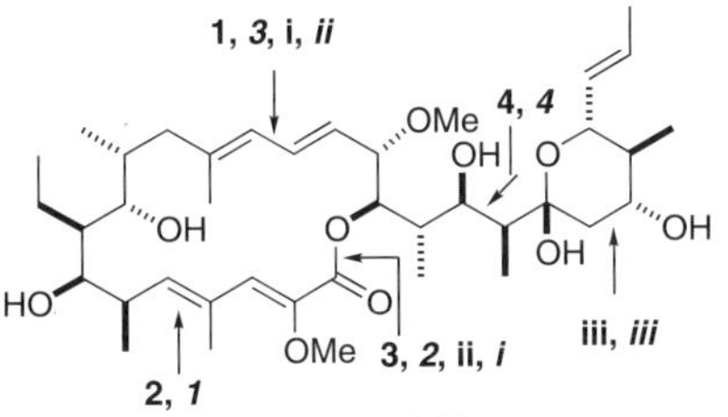

4. Total Synthesis of Macrolides

Concanamycin F

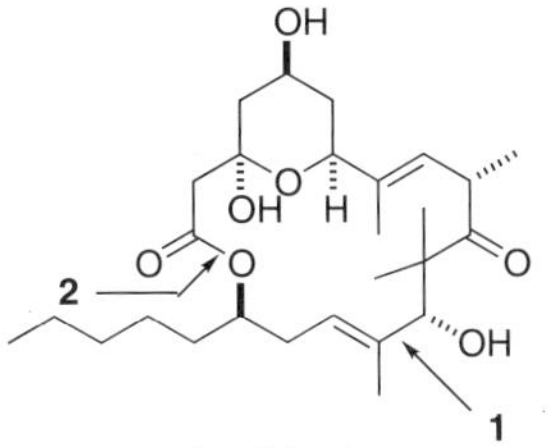

Toshima's strategy A
1. Stille coupling
2. Wittig reaction
3. Yamaguchi macrolactonization
4. aldol reaction

Toshima's strategy B
1. Wittig reaction
2. Yamaguchi esterification
3. Stille macrocyclization
4. aldol reaction

Paterson's strategy A
i. Stille type reaction
ii. Yamaguchi macrolactonization
iii. Mukaiyama aldol reaction

Paterson's strategy B
i. Yamaguchi esterification
ii. Stille macrocyclization
iii. Mukaiyama aldol reaction

Acutiphycin

Smith's strategy
1. vinyl Grignard addition
2. Yamaguhi macrolactonization

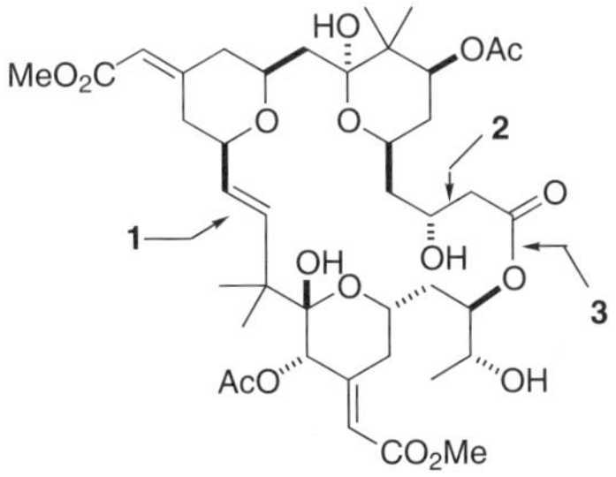

Bryostatin 7

Masamune's strategy
1. Julia coupling
2. aldol reaction using thioester
3. DCC-py-PPTS macrolactonization

Fig. 10. *Continued.*

After completion of the manuscript, several total syntheses of epothilones have been reported [175].

IV. Concluding Remarks

A number of new macrolide antibiotics with interesting bioactivity have been isolated, and the unique and complex structures have been determined. Toward the total synthesis of such attractive macrolide antibiotics, very efficient synthetic strategies and useful methodologies have been developed. Recent advances in macrolide synthesis based on newly developed strategies and methodologies are remarkable; various complex macrolides having many chiral centers have been efficiently synthesized with excellent stereoselectivity. Recently, combinatorial

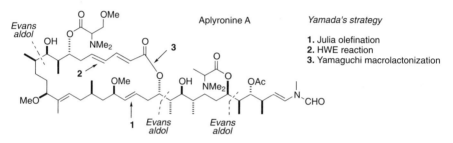

Fig. 10. *Continued.*

chemistry for preparation of a large number of derivatives has been extensively investigated based on these synthetic studies, and mechanistic studies combined with synthetic chemistry and biology are also in progress. Further synthetic studies on macrolide antibiotics will make a great contribution to progress in organic chemistry, preparation of various kinds of derivatives, design of new bioactive compounds, and biological mechanistic studies.

Appendix: Abbreviations

AE	asymmetric epoxidation
AIBN	2,2′-azobisisobutyronitrile
Aloc	allyloxycarbonyl
aq	aqueous
9-BBN	9-borabicyclo[3.3.1]nonane
binap	2,2′bis(diphenylphosphino)-1,1′-binaphthyl
Bn	benzyl
Boc	*t*-butoxycarbonyl

BOC-ON	2-(*t*-butoxycarbonyloxyimino)-2-phenylacetonitrile
BOM	benzyloxymethyl
Bz	benzoyl
c-hex	cyclohexyl
CAN	cerium ammonium nitrate
cat.	catalytic or catalyst
Cbz	benzyloxycarbonyl
Cp	cyclopentadienyl
CSA	10-camphorsulfonic acid
DAIB	3-*exo*-(dimethylamino)isoborneol
DAST	(diethylamino)sulfur trifluoride
DBU	1,8-diazabicyclo[5.4.0]undec-7-ene
DCC	dicyclohexylcarbodiimide
DDQ	2,3-dichloro-5,6-dicyano-1,4-benzoquinone
DEAD	diethyl azodicarboxylate
DEIPS	diethylisopropylsilyl
DET	diethyl tartrate
DIBAH	diisobutylaluminum hydride
DIPT	diisopropyl tartrate
DMAP	4-dimethylaminopyridine
DMDO	dimethyldioxirane
DME	ethylene glycol dimethyl ether
DMF	N,N-dimethylformamide
DMB	3,4-dimethoxybenzyl
DMP	3,4-dimethoxyphenyl
DMPM	3,4-dimethoxyphenylmethyl
DMPU	1,3-dimethyl-3,4,5,6-tetrahydro-2(1H)-pyrimidinone
DMSO	dimethylsulfoxide
dppf	1,3-bis(diphenylphosphino)ferrocene
dr	diastereomer ratio
EDC	N-ethyl-N'-3-dimethylaminopropyl carbodiimide
EE	1-ethoxyethyl
ee	enantiomeric excess
Fmoc	9-fluorenylmethoxycarbonyl
HMPA	hexamethylphosphoramide
HMPT	hexamethylphosphorous triamide
HOBT	1-hydroxybenzotriazole
i	iso
Im	imidazol-1-yl or imidazole
Ipc	isopinocampheyl
KHMDS	potassium hexamethyldisilazide
LDA	lithium diisopropylamide

LDBB	di-*t*-butylbiphenyllithium
LHMDS	lithium hexamethyldisilazide
L-Selectride®	lithium tri-*s*-butylborohydride
MCPBA	*m*-chloroperoxybenzoic acid
MEM	methoxyethoxymethyl
Mes	2,4,6-trimethylphenyl
MIP	1-methoxyisopropyl
MOM	methoxymethyl
MPM	4-methoxyphenylmethyl
MS	molecular sieves
Ms	methanesulfonyl
n	normal
NaHMDS	sodium hexamethyldisilazide
NBS	*N*-bromosuccinimide
NIS	*N*-iodosuccinimide
NMO	*N*-methylmorpholine *N*-oxide
NMP	*N*-methylpyrrolidinone
Np	2-naphtyl
o	orto
Oxone®	potassium peroxomonosulfate
p	para
PCC	pyridinium *p*-toluenesulfonate
PDC	pyridinium dichromate
Piv	pivaloyl (trimethylacetyl)
PPTS	pyridinium *p*-toluenesulfonate
PPY	4-pyrolidinopyridine
py	pyridine
Red-Al®	sodium bis(2-methoxyethoxy)aluminum hydride
SEM	2-(trimethylsilyl)ethoxymethyl
s	secondary
t	tertiary
TBAF	tetrabutylammonium fluoride
TBDPS	*tert*-butyldiphenylsilyl
TBS	*t*-butyldimethylsilyl
TEMPO	2,2,6,6-tetramethyl-1-piperidinyloxy radical
TES	triethylsilyl
Tf	trifluoromethanesulfonyl
TFA	trifluoroacetic acid
Tf_2O	trifluoromethanesulfonic anhydride
TfOH	trifluoromethanesulfonic acid
Th	thienyl
THF	tetrahydrofuran

TIPS	triisopropylsilyl
TMEDA	N,N,N',N'-tetramethylethylenediamine
TMS	trimethylsilyl
TPAP	tetrapropylammonium perruthenate
Tr	triphenylmethyl (trityl)
Triton-B	N-benzyltrimethylammonium hydroxide
Troc	2,2,2-trichloroethoxycarbonyl
Ts	p-toluenesulfonyl
TsOH	p-toluenesulfonic acid.

References

1. Masamune, S., Yamamoto, H., Kamata, S., and Fukuzawa, A. (1975). *J. Am. Chem. Soc.* **97**, 3513–3515.
2. For a review, see Cowden, C. J., and Paterson, I. (1997). In "*Org. React.*" (L. A. Paquette, Ed.), Vol. 51, pp. 1–200. John Wiley & Sons, Inc., New York.
3. Hoffmann, R. W., and Dahmann, G. (1994). *Chem. Ber.* **127**, 1317–1322.
4. (a) Evans, D. A., Bartroli, J., and Shih, T. L. (1981). *J. Am. Chem. Soc.* **103**, 2127–2129. (b) Gage, J. R., and Evans, D. A. (1990). *Org. Synth.* **68**, 83–91.
5. Evans, D. A., Rieger, D. L., Bilodeau, M. T., and Urpi, F. (1991). *J. Am. Chem. Soc.* **113**, 1047–1049.
6. Walker, M. A., and Heathcock, C. H. (1991). *J. Org. Chem.* **56**, 5747–5750.
7. Crimmins, M. T., King, B. W., and Tabet, E. A. (1997). *J. Am. Chem. Soc.* **119**, 7883–7884.
8. Evans, D. A., Clark, J. S., Metternich, R., Novack, V. J., and Sheppard, G. S. (1990). *J. Am. Chem. Soc.* **112**, 866–868.
9. (a) Paterson, I., Wallace, D. J., and Velazquez, S. M. (1994). *Tetrahedron Lett.* **35**, 9083–9086. (b) Paterson, I., Wallace, D. J., and Cowden, C. J. (1998). *Synthesis* 639–652.
10. (a) Paterson, I., Goodman, J. M., and Isaka, M. (1989). *Tetrahedron Lett.* **30**, 7121–7124. (b) Paterson, I., Norcross, R. D., Ward, R. A., Romea, P., and Lister, M. A. (1994). *J. Am. Chem. Soc.* **116**, 11,287–11,314.
11. Paterson, I., and Tillyer, R. D. (1992). *Tetrahedron Lett.* **33**, 4233–4236.
12. Evans, D. A., Ng, H., Clark, J. S., and Rieger, D. L. (1992). *Tetrahedron* **48**, 2127–2142.
13. Paterson, I. (1992). *Pure Appl. Chem.* **64**, 1821–1830.
14. Masamune, S., Sato, T., Kim, B., and Wollman, T. A. (1986). *J. Am. Chem. Soc.* **108**, 8279–8281.
15. (a) Corey, E. J., and Lee, D.-H. (1993). *Tetrahedron Lett.* **34**, 1737–1740. (b) Corey, E. J. (1990). *Pure Appl. Chem.* **62**, 1209–1216.
16. Mukaiyama, T. (1996). *Aldrichim. Acta,* **29**, 59–76.
17. Evans, D. A., Yang, M. G., Dart, M. J., Duffy, J. L., and Kim, A. S. (1995). *J. Am. Chem. Soc.* **117**, 9598–9599.
18. (a) Brown, H. C., and Jadhav, P. K. (1983). *J. Am. Chem. Soc.* **105**, 2092–2093. (b) Brown, H. C., and Bhat, K. S. (1986). *J. Am. Chem. Soc.* **108**, 293–294.
19. (a) Roush, W. R., Walts, A. E., and Hoong, L. K. (1985). *J. Am. Chem. Soc.* **107**, 8186–8190. (b) Roush, W. R., Ando, K., Powers, D. B., Palkowitz, A. D., and Halterman, R. L. (1990). *J. Am. Chem. Soc.* **112**, 6339–6348.
20. For a review, see Nelson, S. G. (1998). *Tetrahedron Asymmetry* **9**, 357–389.
21. Katsuki, T., and Sharpless, K. B. (1980). *J. Am. Chem. Soc.* **102**, 5974–5976.

22. For reviews see: (a) Katsuki, T., and Martin, V. S. (1996). In "*Org. React.*" (L. A. Paquette, Ed.), Vol. 48, pp. 1–300. John Wiley & Sons, Inc., New York. (b) Lin, G.-Q., Li, Y.-M., and Chan, A. S. C. (2001). In "*Principles and Applications of Asymmetric Synthesis,*" pp. 195–266. John Wiley & Sons, Inc., Chichester, England.
23. (a) Minami, N., Ko, S. S., and Kishi, Y. (1982). *J. Am. Chem. Soc.* **104**, 1109–1111. (b) Katsuki, T., Lee, A. W. M., Ma, P., Martin, V. S., Masamune, S., Sharpless, K. B., Tuddenham, D., and Walker, F. J. (1982). *J. Org. Chem.* **47**, 1373–1378.
24. Martin, V. S., Woodward, S. S., Katsuki, T., Yamada, Y., Ikeda, M., and Sharpless, K. B. (1981). *J. Am. Chem. Soc.* **103**, 6237–6240.
25. Wang, Z. M., Zhou, W. S., and Lin, G. Q., (1985). *Tetrahedron Lett.* **26**, 6221–6244.
26. Gao, Y., Hanson, R. M., Klunder, J. M., Ko, S. Y., Masamune, H., and Sharpless, K. B. (1987). *J. Am. Chem. Soc.* **109**, 5765–5780.
27. Caron, M., and Sharpless, K. B. (1985). *J. Org. Chem.* **50**, 1557–1560.
28. Finan, J. M., and Kishi, Y. (1982). *Tetrahedron Lett.* **23**, 2719–2722.
29. (a) Ma, P., Martin, V. S., Masamune, S., Sharpless, K. B., and Viti, S. M. (1982). *J. Org. Chem.* **47**, 1378–1380. (b) Viti, S. M. (1982). *Tetrahedron Lett.* **23**, 4541–4544. (c) reference 23a.
30. Dai, L., Lou, B., Zhang, Y., and Guo, G. (1986). *Tetrahedron Lett.* **27**, 4343–4346.
31. Honda, M., Katsuki, T., and Yamaguchi, M. (1984). *Tetrahedron Lett.* **25**, 3857–3860.
32. (a) Nagaoka, H., and Kishi, Y. (1981). *Tetrahedron* **37**, 3873–3888. (b) Johnson, M. R., Nakata, T., and Kishi, Y. (1979). *Tetrahedron Lett.* **20**, 4343–4346. (c) Wood, R. D., and Ganem, B. (1982). *Tetrahedron Lett.* **23**, 707–710.
33. Sasaki, M., Tanino, K., and Miyashita, M. (2001). *Org. Lett.* **3**, 1765–1767.
34. Hentges, S. G., and Sharpless, K. B. (1980). *J. Am. Chem. Soc.* **102**, 4263–4265.
35. (a) Mukaiyama, T., Izumi, J., Miyashita, M., and Shiina, I. (1993). *Chem. Lett.* 907–910. (b) Shiina, I., and Mukaiyama, T. (1994). *Chem. Lett.* 677–680. (c) Ishihara, K., Kubota, M., Kurihara, H., and Yamamoto, H. (1996). *J. Org. Chem.* **61**, 4560–4567. (d) Mukaiyama, T., Izumi, J., and Shiina, I. (1997). *Chem. Lett.* 187–188. (e) Nagarajan, M., Kumar, V. S., and Rao, B. V. (1999). *Tetrahedron* **55**, 12,349–12,360.
36. (a) Corey, E. J., and Nicolaou, K. C. (1974). *J. Am. Chem. Soc.* **96**, 5614–5616. (b) Corey, E. J., and Brunelle, D. J. (1976). *Tetrahedron Lett.* **17**, 3409–3412. (c) Corey E. J., and Clark, D. A. (1979). *Tetrahedron Lett.* **20**, 2875–2878.
37. Gerlach, H., and Thalmann, A. (1974). *Helv. Chim. Acta* **57**, 2661–2663.
38. Woodward, R. B., *et al.* (1981). *J. Am. Chem. Soc.* **103**, 3210–3213, 3213–3215, 3215–3217.
39. (a) Tatsuta, K., Amemiya, Y., Kanemura, Y., and Kinoshita, M. (1981). *Tetrahedron Lett.* **22**, 3997–4000. (b) Tatsuta, K., Amemiya, Y., Kanemura, H., Takahashi, H., and Kinoshita, M. (1982). *Tetrahedron Lett.* **23**, 3375–3378.
40. Buszek, K. R., Sato, N., and Jeong, Y. (1994). *J. Am. Chem. Soc.* **116**, 5511–5512.
41. (a) Corey, E. J., Hopkins, P. B., Kim, S., Yoo, S., Nambier, K. P., and Falck, J. R. (1979). *J. Am. Chem. Soc.* **101**, 7131–7134. (b) Corey, E. J., Nicolaou, K. C., and Melvin, Jr., L. S. (1975). *J. Am. Chem. Soc.* **97**, 653–654, 654–655.
42. Masamune, S., Kamata, S., and Schilling, W. (1975). *J. Am. Chem. Soc.* **97**, 3515–3516.
43. Kaiho, T., Masamune, S., and Toyoda, T. (1982). *J. Org. Chem.* **47**, 1612–1614.
44. (a) Masamune, S., Lu, L. D.-L., Jackson, W. P., Kaiho, T., and Toyoda, T. (1982). *J. Am. Chem. Soc.* **104**, 5523–5526. (b) Masamune, S., Kaiho, T., and Garvey, D. S. (1982). *J. Am. Chem. Soc.* **104**, 5521–5523.
45. Mukaiyama, T., Usui, M., and Saigo, K. (1976). *Chem. Lett.* 49–50.
46. Kurihara, T., Nakajima, Y., and Mitsunobu, O. (1976). *Tetrahedron Lett.* **17**, 2455–2458.
47. (a) Zibuck, R., Liverton, N. J., and Smith III, A. B. (1986). *J. Am. Chem. Soc.* **108**, 2451–2453. (b) White, J. D., and Kawasaki, M. (1990). *J. Am. Chem. Soc.* **112**, 4991–4993. (c) Smith III,

A. B., Leahy, J. W., Noda, I., Remiszewski, S. W., Liverton, N. J., and Zibuck, R. (1992). *J. Am. Chem. Soc.* **114**, 2995–3007.
48. Inanaga, J., Hirata, K., Saeki, H., Katsuki, T., and Yamaguchi, M. (1979). *Bull. Chem. Soc. Jpn.* **52**, 1989–1993.
49. Inanaga, J., Katsuki, T., Takimoto, S., Ouchida, S., Inoue, K., Nakano, A., Okuda, N., and Yamaguchi, M. (1979). *Chem. Lett.* 1021–1024.
50. (a) Hikota, M., Tone, H., Horita, K., and Yonemitsu, O. (1990). *J. Org. Chem.* **55**, 7–9. (b) Hikota, M., Tone, H., Horita, K., and Yonemitsu, O. (1990). *Tetrahedron* **46**, 4613–4628.
51. Hikota, M., Sakurai, Y., Horita, K., and Yonemitsu, O. (1990). *Tetrahedron Lett.* **31**, 6367–6370.
52. (a) Roush, W. R., and Blizzard, T. A. (1983). *J. Org. Chem.* **48**, 759–761. (b) Roush, W. R., and Blizzard, T. A. (1984). *J. Org. Chem.* **49**, 4332–4339.
53. Boden, E. P., and Keck, G. E. (1985). *J. Org. Chem.* **50**, 2394–2395.
54. Neises, B., and Steglich, W. (1978). *Angew. Chem. Int. Ed. Engl.* **17**, 522–523.
55. Hassner, A., and Alexanian, V. (1978). *Tetrahedron Lett.* **19**, 4475–4478.
56. For reviews, see (a) Schuster, M., and Blechert, S. (1997). *Angew. Chem. Int. Ed. Engl.* **36**, 2036–2057. (b) Armstrong, S. K. (1998). *J. Chem. Soc., Perkin Trans. I* 371–388. (c) Grubbs, R. H., and Chang, S. (1998). *Tetrahedron* **54**, 4413–4450. (d) Grubbs, R. H., Miller, S. J., and Fu, G. C. (1995). *Acc. Chem. Res.* **28**, 446–452. (d) Phillips, A. J., and Abell, A. D. (1999). *Aldrichim. Acta* **32**, 75–89.
57. (a) Schrock, R. R., Murdzek, J. S., Bazan, G. C., Robbins, J., DiMare, M., and O'Regan, M. (1990). *J. Am. Chem. Soc.* **112**, 3875–3888. (b) Bazan, G. C., Khosravi, E., Schrock, R. R., Feast, W. J., Gibson, V. C., O'Regan, M. B., Thomas, J. K., and Davis, W. M. (1990). *J. Am. Chem. Soc.* **112**, 8378–8387. (c) Bazan, G. C., Oskam, J. H., Cho, H.-N., Park, L. Y., and Schrock, R. R. (1991). *J. Am. Chem. Soc.* **113**, 6899–6907.
58. (a) Nguyen, S. T., Johnson, L. K., Grubbs, R. H., and Ziller, J. W. (1992). *J. Am. Chem. Soc.* **114**, 3974–3975. (b) Nguyen, S. T., Grubbs, R. H., and Ziller, J. W. (1993). *J. Am. Chem. Soc.* **115**, 9858–9859. (c) Schwab, P., France, M. B., Ziller, J. W., and Grubbs, R. H. (1995). *Angew. Chem. Int. Ed. Engl.* **34**, 2039–2041. (d) Schwab, P., Grubbs, R. H., and Ziller, J. W. (1996). *J. Am. Chem. Soc.* **118**, 100–110.
59. Fürstner, A., and Langemann, K. (1996). *J. Org. Chem.* **61**, 3942–3943.
60. Fürstner, A., and Kindler, N. (1996). *Tetrahedron Lett.* **37**, 7005–7008.
61. (a) Xu, Z., Johannes, C. W., Salman, S. S., and Hoveyda, A. H. (1996). *J. Am. Chem. Soc.* **118**, 10,926–10,927. (b) Houri, A. F., Xu, Z., Cogan, D. A., and Hoveyda, A. H. (1995). *J. Am. Chem. Soc.* **117**, 2943–2944.
62. Fürstner, A., Thiel, O. R., Kindler, N., and Bartkowska, B. (2000). *J. Org. Chem.* **65**, 7990–7995.
63. Fürstner, A., and Langemann, K. (1997). *J. Am. Chem. Soc.* **119**, 9130–9136.
64. Wu, Y., Esser, L., and De Brabander, J. K. (2000). *Angew. Chem. Int. Ed. Engl.* **39**, 4308–4310.
65. Fürstner, A., Guth, O., Rumbo, A., and Seidel, G. (1999). *J. Am. Chem. Soc.* **121**, 11,108–11,113.
66. Fürstner, A., Radkowski, K., Grabowski, J., Wirtz, C., and Mynott, R. (2000). *J. Org. Chem.* **65**, 8758–8762. Cf. Tricolorin A and G: (b) Fürstner, A., and Müller, T. (1999). *J. Am. Chem. Soc.* **121**, 7814–7821.
67. (a) Heck, R. F. (1979). *Acc. Chem. Res.* **12**, 146–151. (b) Heck, R. F. (1982). in *"Org. React."* (W. G. Dauben, Ed.) Vol. 27, pp. 345–390, John Wiley & Sons, Inc., New York. (d) Heck, R. F. (1985). In *"Palladium Reagents in Organic Syntheses,"* p. 179. Academic Press, New York. (e) Heck, R. F. (1991). In *"Comprehensive Organic Synthesis"* (B. M. Trost and I. Freming, Eds.), Vol. 4, pp. 833–863. Pergamon Press, New York.
68. Suzuki, A. (1982). *Acc. Chem. Res.* **15**, 178–184. (b) Suzuki, A. (1991). *Pure Appl. Chem.* **63**, 419–422. (c) Suzuki, A. (1992). In *"Organic Synthesis in Japan, Past, Present, and Future,"* pp. 411–423. Tokyo Kagaku Dozin, Tokyo.

69. (a) Negishi, E., Valente, L. F., and Kobayashi, M. (1980). *J. Am. Chem. Soc.* **102**, 3298–3299. (b) Negishi, E. (1982). *Acc. Chem. Res.* **15**, 340–348. (c) Negishi, E., Ay, M., Gulevich, Y. V., and Noda, Y. (1993). *Tetrahedron Lett.* **34**, 1437–1440.
70. (a) Milstein, D., and Stille, J. K. (1978). *J. Am. Chem. Soc.* **100**, 3636–3638. (b) Stille, J. K. (1986) *Angew. Chem. Int. Ed. Engl.* **25**, 508–524. (c) Farina, V., and Krishnamurthy, V. (1997). In *"Org. React."* (L. A. Paquette, Ed.), Vol. 50, pp.1–652. John Wiley & Sons, Inc., New York.
71. (a) Sonogashira, K., Tohda, Y., and Hagihara, N. (1975). *Tetrahedron Lett.* **16**, 4467–4470. (b) Sonogashira, K. (1991). In *"Comprhensive Organic Synthesis"* (B. M. Trost and I. Freming, Eds.), Vol. 3, pp. 521–549. Pergamon Press, New York.
72. (a) Tsuji, J. (1995). *"Palladium Reagents and Catalysts,"* Wiley, New York. (b) Trost, B. M. (1989). *Angew. Chem. Int. Ed. Engl.* **28**, 1173–1192.
73. Stille, J. K., and Tanaka, M. (1987). *J. Am. Chem. Soc.* **109**, 3785–3786.
74. Stille, J. K., Su, H., Hill, D. H., Schneider, P., Tanaka, M., Morrison, D. L., and Hegedus, L. S. (1991). *Organometallics* **10**, 1993–2000.
75. (a) Baldwin, J. E., Adlington, R. M., and Ramcharitar, S. H. (1991). *J. Chem. Soc. Chem. Commun.* 940–942. (b) Baldwin, J. E., Adlington, R. M., and Ramcharitar, S. H. (1992). *Tetrahedron* **48**, 2957–2976. (c) Baldwin, J. E., Adlington, R. M., and Ramcharitar, S. H. (1992). *Synlett.* 875–877.
76. Boden, C., and Pattenden, G. (1994). *Synlett* 181–182.
77. (a) Gyorkos, A. C., Stille, J. K., and Hegedus, L. S. (1990). *J. Am. Chem. Soc.* **112**, 8465–8472. (b) Kalivreyenos, A., Stille, J. K., and Hegedus, L. S. (1991). *J. Org. Chem.* **56**, 2883–2894.
78. Smith III, A. B., and Ott, G. R. (1996). *J. Am. Chem. Soc.* **118**, 13,095–13,096.
79. Boyce, R. J., and Pattenden, G. (1996). *Tetrahedron Lett.* **37**, 3501–3504.
80. Remuinan, M. J., and Pattenden, G. (2000). *Tetrahedron Lett.* **41**, 7367–7371.
81. Paterson, I., and Man, J. (1997). *Tetrahedron Lett.* **38**, 695–698.
82. Allred, G. D., and Liebeskind, L. S. (1996). *J. Am. Chem. Soc.* **118**, 2748–2749.
83. For reviews, see (a) Maryanoff, B. E., and Reitz, A. B. (1989). *Chem. Rev.* **89**, 863–927. (b) Nicolaou, K. C., Haerter, M. W., Gunzner, J. L., and Nadin, A. (1997). *Liebigs Ann. Recl.* **7**, 1283–1302.
84. Wittig, G., and Geissler, G. (1953). *Liebigs Ann. Chem.* **580**, 44–57.
85. (a) Horner, L., Hoffmann, H., and Wippel, H. G. (1958). *Chem. Ber.* **91**, 61–63. (b) Horner, L. (1964). *Pure Appl. Chem.* **9**, 225–244.
86. (a) Horner, L., Hoffmann, H., Wippel, H. G., and Klaahre, G. (1959). *Chem. Ber.* **92**, 2499–2505. (b) Horner, L., Hoffmann, H., Klink, W., Ertel, H., and Toscano, V. G. (1962). *Chem. Ber.* **95**, 581–601. (c) Wadsworth, W. S., and Emmons, W. D. (1961). *J. Am. Chem. Soc.* **83**, 1733–1738.
87. Still, W. C., and Gennari, C. (1983). *Tetrahedron Lett.* **41**, 4405–4408.
88. Ando, K. (1997). *J. Org. Chem.* **62**, 1934–1939.
89. Stork, G., and Nakamura, E. (1979). *J. Org. Chem.* **44**, 4010–4011.
90. (a) Nicolaou, K. C., Seitz, S. P., Pavia, M. R., and Petasis, N. A. (1979). *J. Org. Chem.* **44**, 4011–4013. (b) Nicolaou, K. C., Pavia, M. R., and Seitz, S. P. (1981). *J. Am. Chem. Soc.* **103**, 1224–1226.
91. (a) Nicolaou, K. C., Seitz, S. P., and Pavia, M. R. (1982). *J. Am. Chem. Soc.* **104**, 2030–2031. (b) Aristoff, P. A. (1981). *J. Org. Chem.* **46**, 1954–1957.
92. Blanchette, M. A., Choy, W., Davis, J. T., Essenfeld, A. P., Masamune, S., Roush, W. R., and Sakai, T. (1984). *Tetrahedron Lett.* **25**, 2183–2186.
93. Roxaticin: (a) Rychnovsky, S. D., and Hoye, R. C., (1994). *J. Am. Chem. Soc.* **116**, 1753–1765. (b) Mori, Y., Asai, M., Kawade, J., and Furukawa, H. (1995). *Tetrahedron* **51**, 5315–5330.
94. Roflamycoin: Rychnovsky, S. D., Khire, U. R., and Yang, G. (1997). *J. Am. Chem. Soc.* **119**, 2058–2059.

95. Filipin III: Richardson, T. I., and Rychnovsky, S. D. (1997). *J. Am. Chem. Soc.* **119**, 12,360–12,361. Idem. (1999). *Tetrahedron* **55**, 8977–8996.
96. Amphotericin B: (a) Nicolaou, K. C., Daines, R. A., Uenishi, J., Li, W. S., Papahatjis, D. P., and Chakraborty, T. K. (1988). *J. Am. Chem. Soc.* **110**, 4672–4685. (b) Nicolaou, K. C., Daines, R. A., Chakraborty, T. K., and Ogawa, Y. (1988). *J. Am. Chem. Soc.* **110**, 4685–4696. (c) Nicolaou, K. C., Daines, R. A., Ogawa, Y., and Chakraborty, T. K. (1988). *J. Am. Chem. Soc.* **110**, 4696–4705.
97. Carbonolide B: (a) Nakajima, N., Uoto, K., and Yonemitsu, O. (1990). *Heterocycles* **31**, 5–8. (b) Keck, G. E., Palani, A., and McHardy, S. F. (1994). *J. Org. Chem.* **59**, 3113–3122.
98. Methynolide: Tanaka, T., Oikawa, Y., Nakajima, N., Hamada, T., and Yonemitsu, O. (1987). *Chem. Pharm. Bull.* **35**, 2203–2208.
99. Tylonolide: Tanaka, T., Oikawa, Y., Hamada, T., and Yonemitsu, O. (1987). *Chem. Pharm. Bull.* **35**, 2219–2227.
100. Pikronolide: Nakajima, N., Tanaka, T., Hamada, T., Oikawa, Y., and Yonemitsu, O. (1987). *Chem. Pharm. Bull.* **35**, 2228–2237.
101. Chattopadhyay, S. K., and Pattenden, G. (2000). *J. Chem. Soc. Perkin Trans. I* 2429–2454.
102. (a) Oppolzer, W., Radinov, R. N., and De Brabander, J. (1995). *Tetrahedron Lett.* **36**, 2607–2610. (b) Oppolzer, W., and Radinov, R. N. (1993). *J. Am. Chem. Soc.* **115**, 1593–1594. Other total syntheses: (c) Nishioka, T., Iwabuchi, Y., Irie, H., and Hatakeyama, S. (1998). *Tetrahedron Lett.* **39**, 5597–5600. (d) Sinha, S. C., and Keinan, E. (1997). *J. Org. Chem.* **62**, 377–386 and references cited therein.
103. Pilli, R. A., De Andrade, C. K. Z., Souto, C. R. O., and De Meijere, A. (1998). *J. Org. Chem.* **63**, 7811–7819.
104. For reviews, see (a) Toshima, K., and Tatsuta, K. (1993). *Chem. Rev.* **93**, 1503–1531. (b) Boons, G.-J. (1996). *Tetrahedron* **52**, 1095–1121. (c) Danishefsky, S. J., and Bilodeau, M. T. (1996). *Angew. Chem. Int. Ed. Engl.* **35**, 1380–1419.
105. Nicolaou, K. C., Dolle, R. E., Papahatjis, D. P., and Randall, J. L. (1984). *J. Am. Chem. Soc.* **106**, 4189–4192.
106. Hanessian, S., Bacquet, C., and LeHong, N. (1980). *Carbohydr. Res.* **80**, C17–C22.
107. Ford, M. J., and Ley, S. V. (1990). *Synlett* 255–256.
108. (a) Danishefsky, S. J., Larson, E., Atkins, D., and Kato, N. (1985). *J. Am. Chem. Soc.* **107**, 1246–1255. (b) Danishefsky, S. J., Pearson, W. H., Harvey, D. F., Maring, C. J., and Springer, J. P. (1985). *J. Am. Chem. Soc.* **107**, 1256–1268. (c) Danishefsky, S. J., Myles, D. C., and Harvey, D. F. (1987). *J. Am. Chem. Soc.* **109**, 862–867.
109. Thiem, J., Karl, H., and Schwentner, J. (1978). *Synthesis* 696–698.
110. (a) Matsumoto, T., Maeta, H., Suzuki, K., and Tsuchihashi, G. (1988). *Tetrahedron Lett.* **29**, 3567–3570. (b) Suzuki, K., Maeta, H., Matsumoto, T. and Tsuchihashi, G. (1988). *Tetrahedron Lett.* **29**, 3571–3574. (c) Matsumoto, T., Maeta, H., Suzuki, K., and Tsuchihashi, G. (1988). *Tetrahedron Lett.* **29**, 3575–3578.
111. Toshima, K., Tatsuta, K., and Kinoshita, M. (1988). *Bull. Chem. Soc. Jpn.* **61**, 2369–2381.
112. Tatsuta, K., Fujimoto, K., Kinoshita, M., and Umezawa, S. (1977). *Carbohydr. Res.* **54**, 85–104.
113. Wakamatsu, T., Nakamura, H., Nara, E., and Ban, Y. (1986). *Tetrahedron Lett.* **27**, 3895–3898.
114. (a) Tatsuta, K., Ishiyama, T., Tajima, S., Koguchi, Y., and Gunji, H. (1990). *Tetrahedron Lett.* **31**, 709–712. (b) Tatsuta, K., Kobayashi, Y., Gunji, H., and Masuda, H. (1988). *Tetrahedron Lett.* **29**, 3975–3978.
115. (a) Toshima, K., Mukaiyama, S., Ishiyama, T., and Tatsuta, K. (1990). *Tetrahedron Lett.* **31**, 3339–3342. (b) Toshima, K., Mukaiyama, S., Ishiyama, T., and Tatsuta, K. (1990). *Tetrahedron Lett.* **31**, 6361–6362. (c) Toshima, K., Nozaki, Y., Mukaiyama, S., and Tatsuta, K. (1992). *Tetrahedron Lett.* **33**, 1491–1494. (d) Toshima, K., Nozaki, Y., Inokuchi, H., Nakata, M., Tatsuta, K.,

and Kinoshita, M. (1993). *Tetrahedron Lett.* **34**, 1611–1614. (e) Toshima, K., Mukaiyama, S., Yoshida, T., Tamai, T., and Tatsuta, K. (1991). *Tetrahedron Lett.* **32**, 6155–6158.
116. Evans, D. A., Kaldor, S. W., Jones, T. K., Clardy, J., and Stout, T. J. (1990). *J. Am. Chem. Soc.* **112**, 7001–7031.
117. Mukaiyama, T., Kobayashi, S., and Shoda, S. (1984). *Chem. Lett.* 907–910.
118. Evans, D. A., and Black, W. C. (1993). *J. Am. Chem. Soc.* **115**, 4497–4513.
119. (a) Murai, A., Fujiwara, K., Yotsu-Yamashita, M., and Yasumoto, T. (1998). *J. Am. Chem. Soc.* **120**, 10,770–10,771. (b) Fujiwara, K., Amano, S., and Murai, A. (1995). *Chem. Lett.* 855–856. (c) Fujiwara, K., Amano, S., and Murai, A. (1995). *Chem. Lett.* 191–192. Other total syntheses: (d) Paquette, L. A., Barriault, L., Pissarnitski, D., and Johnston, J. N. (2000). *J. Am. Chem. Soc.* **122**, 619–631. (e) White, J. D., Blakemore, P. R., Browder, C. C., Hong, J., Lincoln, C. M., Nagornyy, P. A., Robarge, L. A., and Wardrop, D. J. (2001). *J. Am. Chem. Soc.* **123**, 8593–8595.
120. Nicolaou, K. C., Seitz, S. P., and Papahatjis, D. P. (1983). *J. Am. Chem. Soc.* **105**, 2430–2434.
121. (a) Jones, T. K., Mills, S. G., Reamer, R. A., Askin, D., Desmond, R., Volante, R. P., and Shinkai, I. (1989). *J. Am. Chem. Soc.* **111**, 1157–1159. (b) Jones, T. K., Reamer, R. A., Desmond, R., and Mills, S. G. (1990). *J. Am. Chem. Soc.* **112**, 2998–3017.
122. (a) Nakatsuka, M., Ragan, J. A., Sammakia, T., Smith, D. B., Uehling, D. E., and Schreiber, S. L. (1990). *J. Am. Chem. Soc.* **112**, 5583–5601. (b) Andrus, M. B., and Schreiber, S. L. (1993). *J. Am. Chem. Soc.* **115**, 10,420–10,421.
123. (a) Ireland, R. E., Gleason, J. L., Gegnas, L. D., and Highsmith, T. K. (1996). *J. Org. Chem.* **61**, 6856–6872. (b) Ireland, R. E., Highsmith, T. K., Gegnas, L. D., and Gleason, J. L. (1992). *J. Org. Chem.* **57**, 5071–5073. (c) Ireland, R. E., Liu, L., and Roper, T. D. (1997). *Tetrahedron* **53**, 13,221–13,256. (d) Ireland, R. E., Liu, L., Roper, T. D., and Gleason, J. L. (1997). *Tetrahedron* **53**, 13,257–13,284.
124. Jones, A. B., Villalobos, A., Linde III, R. G., and Danishefsky, S. J. (1990). *J. Org. Chem.* **55**, 2786–2797.
125. Gu, R. L., and Sih, C. J. (1990). *Tetrahedron Lett.* **31**, 3287–3290.
126. (a) Smith III, A. B., Chen, K., Robinson, D. J., Laakso, L. M., and Hale, K. J. (1994). *Tetrahedron Lett.* **35**, 4271–4274. (b) Smith III, A. B., and Hale, K. J. (1989). *Tetrahedron Lett.* **30**, 1037–1040. (c) Smith III, A. B., Hale, K. J., Laakso, L. M., Chen, K., and Riera, A. (1989). *Tetrahedron Lett.* **30**, 6963–6966.
127. Evans, D. A., Ennis, M. D., and Mathre, D. J. (1982). *J. Am. Chem. Soc.* **104**, 1737–1739.
128. (a) Wandless, T. J., Michnick, S. W., Rosen, M. K., Karplus, M., and Schreiber, S. L. (1991). *J. Am. Chem. Soc.* **113**, 2339–2341. (b) Van Duyne, G. D., Standaert, R. F., Karplus, P. A., Schreiber, S. L., and Clardy, J. (1991). *Science* **252**, 839–842. (c) Somer, P. K., Wandless, T. J., and Schreiber, S. L. (1991). *J. Am. Chem. Soc.* **113**, 8045–8056.
129. (a) Nicolaou, K. C., Chakraborty, T. K., Piscopio, A. D., Minowa, N., and Bertinato, P. (1993). *J. Am. Chem. Soc.* **115**, 4419–4420. (b) Piscopio, A. D., Minowa, N., Chakraborty, T. K., Bertinato, P., and Nicolaou, K. C. (1993). *J. Chem. Soc. Chem. Commun.* 617–618. (c) Nicolaou, K. C., Bertinato, P., Piscopio, A. D., Chakraborty, T. K., and Minowa, N. (1993). *J. Chem. Soc. Chem. Commun.* 619–622.
130. (a) Romo, D., Meyer, S. D., Johnson, D. D., and Schreiber, S. L. (1993). *J. Am. Chem. Soc.* **115**, 7906–7907. (b) Meyers, S. D., Miwa, T., Nakatsuka, M., and Schreiber, S. L. (1992). *J. Org. Chem.* **57**, 5058–5060. (c) Romo, D., Johnson, D. D., Plamondon, L., Miwa, T., and Schreiber, S. L. (1992). *J. Org. Chem.* **57**, 5060–5063.
131. (a) Hayward, C. M., Yohannes, D., and Danishefsky, S. J. (1993). *J. Am. Chem. Soc.* **115**, 9345–9346. (b) Hayward, C. M., Fisher, M. J., Yohannes, D., and Danishefsky, S. J. (1993). *Tetrahedron Lett.* **34**, 3989–3992. (c) Horvath, R. F., Linde II, R. G., Hayward, C. M., Joglar, J., Yohannes, D., and Danishefsky, S. J. (1993). *Tetrahedron Lett.* **34**, 3993–3996.

132. (a) Smith III, A. B., Condon, S. M., McCauley, J. A., Leazer, Jr., J. L., Leahy, J. W., and Maleczka, Jr., R. E. (1995). *J. Am. Chem. Soc.* **117**, 5407–5408. (b) Smith III, A. B., Condon, S. M., McCauley, J. A., Leazer, Jr., J. L., Leahy, J. W., and Maleczka, Jr., R. E. (1997) *J. Am. Chem. Soc.* **119**, 947–961, 962–973. (c) Smith III, A. B., Condon, S. M., McCauley, J. A., Leahy, J. W., Leazer, Jr., J. L., and Maleczka, Jr., R. E. (1994). *Tetrahedron Lett.* **35**, 4907–4910. (d) Smith III, A. B., Maleczka, Jr., R. E., Leazer, Jr., J. L., Leahy, J. W., McCauley, J. A., and Condon, S. M. (1994). *Tetrahedron Lett.* **35**, 4911–4914.
133. (a) Hanessian, S., Ugolini, A., Dube, D., and Andre, C. (1986). *J. Am. Chem. Soc.* **108**, 2776–2778. (b) Hanessian, S., Dube, D., and Hodges, P. J. (1987). *J. Am. Chem. Soc.* **109**, 7063–7067. (c) Hanessian, S., Ugolini, A., Hodges, P. J., Beaulieu, P., Dube, D., and Andre, C. (1987). *Pure Appl. Chem.* **59**, 299–316.
134. (a) Ley, S. V., Armstrong, A., Diez-Martin, D., Ford, M. J., Grice, P., Knight, J. G., Kolb, H. C., Madin, A., Marby, C. A., Mukherjee, S., Shaw, A. N., Slawin, A. M. Z., Vile, S., White, A. D., Williams, D. J., and Woods, M. (1991). *J. Chem. Soc. Perkin Trans. I* 667–692. (b) Armstrong, A., and Ley, S. V. (1990). *Synlett.* 323–325. (c) Diez-Martin, D., Brice, P., Kolb, H. C., Ley, S. V., and Madin, A. (1990). *Synlett.* 326–328. (d) Armstrong, A., Ley, S. V. Madin, A., and Mukherjee, S. (1990). *Synlett.* 328–330. (e) Ford, M. J., Knight, J. G., Ley, S. V., and Vile, S. (1990). *Synlett.* 331–332.
135. (a) White, J. D., and Bolton, G. L. (1990). *J. Am. Chem. Soc.* **111**, 1626–1628. (b) White, J. D., Bolton, G. L., Dantanarayana, A. P., Fox, C. M. J., Hiner, R. N., Jackson, R. W., Sakuma, K., and Warrier, U. S. (1995). *J. Am. Chem. Soc.* **117**, 1908–1939.
136. (a) Danishefsky, S. J., Armistead, D. M., Wincott, F. E., Selnick, and H. G., Hungate, R. (1987). *J. Am. Chem. Soc.* **109**, 8117–8119. (b) Danishefsky, S. J., Selnick, H. G., Armistead, D. M., and Wincott, F. E. (1987). *J. Am. Chem. Soc.* **109**, 8119–8120. (c) Danishefsky, S. J., Armistead, D. M., Wincott, F. E., Selnick, H. G., and Hungate, R. (1989). *J. Am. Chem. Soc.* **111**, 2967–2980.
137. (a) Ferezou, J. P., Julia, M., Liu, L. W., and Pancrazi, A. (1991). *Synlett.* 614–617. (b) Ferezou, J. P., Julia, M., Liu, L. W., and Pancrazi, A. (1991). *Synlett.* 618–620. (c) Ferezou, J. P., Julia, M., Li, Y., Liu, L. W., Pancrazi, A., and Porteu, F. (1994). *Bull. Soc. Chim. Fr.* **131**, 865–894. (d) Ferezou, J. P., Julia, M., Li, Y., Liu, L. W., and Pancrazi, A. (1995). *Bull. Soc. Chim. Fr.* **132**, 428–452. (e) Williams, D. R., Barner, B. A., Nishitani, K., and Phillips, J. G. (1982). *J. Am. Chem. Soc.* **104**, 4708–4710. (f) Show, S. R., Bloom, J. D., Thompson, A. S., Winzenberg, K. N., and Smith III, A. B. (1986). *J. Am. Chem. Soc.* **108**, 2662–2674. (g) Barrett, A. G., Carr, R. A. E., Attwood, S. V., Richardson, G., and Walshe, N. D. A. (1986). *J. Org. Chem.* **51**, 4840–4856. (h) Baker, R., O'Mahony, M. J., and Swain, C. J. (1987). *J. Chem. Soc. Perkin Trans. I* 1623–1633. (i) Kocienski, P. J., Street, S. D. A., Yeates, C., and Campbell, S. F. (1987). *J. Chem. Soc. Perkin Trans. I* 2171–2182. (j) Crimmins, M. T., Bankaitis-Davis, D. M., and Hollis, Jr., W. G. (1988) *J. Org. Chem.* **53**, 652–657. (k) Anthony, N. J., Armstrong, A., Ley, S. V., and Madin, A. (1989). *Tetrahedron Lett.* **30**, 3209–3212. (l) Ley, S. V., Anthony, N. J., Armstrong, A., Brasca, M. G., Clarke, T., Culshaw, D., Greck, C., Grice, P., Jones, A. B., Lygo, B., Madin, A., Sheppard, R. N., Slawin, A. M. Z., and Williams, D. J. (1989). *Tetrahedron* **45**, 7161–7194. (m) Parmee, E. F., Steel, P. J., and Thomas, E. J. (1989) *J. Chem. Soc. Chem. Commun.* 1250–1253. (n) Hirama, M., Noda, T., Yasuda, S., and Ito, S. (1991). *J. Am. Chem. Soc.* **113**, 1380–1382.
138. Itoh, A., Ozawa, S., Oshima, K., and Nozaki, H. (1981). *Bull. Chem. Soc. Jpn.* **54**, 274–278.
139. (a) Evans, D. A., Trotter, B. W., Cote, B., Coleman, P. J., Dias, L. C., and Tyler, A. N. (1997). *Angew. Chem. Int. Ed. Engl.* **36**, 2744–2747. (b) Evans, D. A., Coleman, P. J., and Dias, L. C. (1997). *Angew. Chem. Int. Ed. Engl.* **36**, 2738–2741. (c) Evans, D. A., Trotter, B. W., Cote, B., and Coleman, P. J. (1997). *Angew. Chem. Int. Ed. Engl.* **36**, 2741–2744. (d) Evans, D. A., Trotter,

B. W., Coleman, P. J., Cote, B., Dias, L. C., Rajapakse, H. A., and Tyler, A. N. (1999). *Tetrahedron* **55**, 8671–8726.
140. (a) Guo, J., Duffy, K. J., Stevens, K. L., Dalko, P. I., Roth, R. M., Hayward, M. M., and Kishi, Y. (1998). *Angew. Chem. Int. Ed.* **37**, 187–192. (b) Hayward, M. M., Roth, R. M., Duffy, M. J., Dalko, P. I., Stevens, K. L., Guo, J., and Kishi, Y. (1998). *Angew. Chem. Int. Ed.* **37**, 192–196.
141. (a) Smith III, A. B., Doughty, V. A., Lin, Q., Zhuang, L., McBriar, M. D., Boldi, A. M., Moser, W. H., Murase, N., Nakayama, K., and Sobukawa, M. (2001). *Angew. Chem. Int. Ed.* **40**, 191–195. (b) Smith III, A. B., Lin, Q., Doughty, V. A., Zhuang, L., McBriar, M. D., Kerns, J. K., Brook, C. S., Murase, N., and Nakayama, K. (2001). *Angew. Chem. Int. Ed.* **40**, 196–199.
142. (a) Balog, A., Meng, D., Kamenecka, T., Bertinato, P., Su, D.-S., Sorensen, E. J., and Danishefsky, S. J. (1996). *Angew. Chem. Int. Ed. Engl.* **35**, 2801–2803. (b) Meng, D., Su, D.-S., Balog, A., Bertinato, P., Sorensen, E. J., Danishefsky, S. J., Zheng, Y.-H., Chou, T.-C., He, L., and Horwitz, S. B. (1997). *J. Am. Chem. Soc.* **119**, 2733–2734. (c) Su, D.-S., Meng, D., Bertinato, P., Balog, A., Sorensen, E. J., Danishefsky, S. J., Zheng, Y.-H., Chou, T.-C., He, L., and Horwitz, S. B. (1997). *Angew. Chem. Int. Ed. Engl.* **36**, 757–759. (d) Meng, D., Bertinato, P., Balog, A., Su, D.-S., Kamenecka, T., Sorensen, E. J., and Danishefsky, S. J. (1997). *J. Am. Chem. Soc.* **119**, 10,073–10,092. (e) Balog, A., Harris, C., Savin, K., Zhang, X.-G., Chou, T. C., and Danishefsky, S. J. (1998). *Angew. Chem. Int. Ed.* **37**, 2675–2678. (f) Harris, C. R., Kuduk, S. D., Savin, K., Balog, A., and Danishefsky, S. J. (1999). *Tetrahedron Lett.* **40**, 2263–2266. (g) Harris, C. R., Kuduk, S. D., Balog, A., Savin, K., Glunz, P. W., and Danishefsky, S. J. (1999). *J. Am. Chem. Soc.* **121**, 7050–7062.
143. (a) Yang, Z., He, Y., Vourloumis, D., Vallberg, H., and Nicolaou, K. C. (1997). *Angew. Chem. Int. Ed. Engl.* **36**, 166–168. (b) Nicolaou, K. C., Sarabia, F., Ninkovic, S., and Yang, Z. (1997). *Angew. Chem. Int. Ed. Engl.* **36**, 525–527. (c) Nicolaou, K. C., He, Y., Vourloumis, D., Vallberg, H., Roschangar, F., Sarabia, F., Ninkovic, S., Yang, Z., and Trujillo, J. I. (1997). *J. Am. Chem. Soc.* **119**, 7960–7973. (d) Nicolaou, K. C., Ninkovic, S., Sarabia, F., Vourloumis, D., He, Y., Vallberg, H., Finlay, M. R. V., and Yang, Z. (1997). *J. Am. Chem. Soc.* **119**, 7974–7991. (e) Nicolaou, K. C., Winssinger, N., Pastor, J., Ninkovic, S., Sarabia, F., He, Y., Vourloumis, D., Yang, Z., Li, T., Giannakakou, P., and Hamel, E. (1997). *Nature* **387**, 268–272. (f) Nicolaou, K. C., Vallberg, H., King, N. P., Roschanbar, F., He, Y., Vourloumis, D., and Nicolaou, C. G. (1997). *Chem. Eur. J.* **3**, 1957–1970. (g) Nicolaou, K. C., Sarabia, F., Finlay, M. R. V., Ninkovic, S., King, N. P., Vourloumis, D., and He, Y. (1997). *Chem. Eur. J.* **3**, 1971–1986. (h) Nicolaou, K. C., Sarabia, F., Ninkovic, S., Finlay, M. R. V., and Boddy, C. N. C. (1998). *Angew. Chem. Int. Ed.* **37**, 81–84. (i) Nicolaou, K. C., He, Y., Roschangar, F., King, N. P., Vourloumis, D., and Li, T. (1998). *Angew. Chem. Int. Ed.* **37**, 84–87. (j) Nicolaou, K. C., Finlay, M. R. V., Ninkovic, S., and Sarabia, F. (1998). *Tetrahedron* **54**, 7127–7166. (k) Nicolaou, K. C., Finlay, M. R. V., Ninkovic, S., King, N. P., He, Y., Li, T., Sarabia, F., and Vourloumis, D. (1998). *Chem. Biol.* **5**, 365–372. (l) Nicolaou, K. C., King, N. P., Finlay, M. R. V., He, Y., Roschangar, F., Vourloumis, D., Vallberg, H., Sarabia, F., Ninkovic, S., and Hepworth, D. (1999). *Bioorg. Med. Chem.* **7**, 665–697. (m) Nicolaou, K. C., Hepworth, D., Finley, M. R. V., King, N. P., Werschkun, B., and Bigot, A. (1999). *Chem. Commun.* 519–520. (n) Nicolaou, K. C., Hepworth, D., King, N. P., Finlay, M. R. V., Scarpelli, R., Pereira, M. M. A., Bollbuck, B., Bigot, A., Werschkun, B., and Winssinger, N. (2000). *Chem. Eur. J.* **6**, 2783–2800. (o) Nicolaou, K. C., He, Y., Vourloumis, D., Vallberg, H., and Yang, Z. (1996). *Angew. Chem. Int. Ed. Engl.* **35**, 2399–2401.
144. (a) Schinzer, D., Limberg, A., Bauer, A., Bohm, O. M., and Cordes, M. (1997). *Angew. Chem. Int. Ed. Engl.* **36**, 523–524. (b) Schinzer, D., Bauer, A. and Schieber, J. (1998). *Synlett.* 861–864. (c) Schinzer, D., Bauer, A., Bohm, O. M., Limberg, A., and Cordes, M. (1999). *Chem. Eur. J.* **5**, 2483–2491. (d) Schinzer, D., Bauer, A., and Schieber, J. (1999). *Chem. Eur. J.* **5**, 2492–2500. (e) Schinzer, D., Limberg, A., and Bohm, O. M. (1996). *Chem. Eur. J.* **2**, 1477–1482.
145. May, S. A., and Grieco, P. (1998). *Chem. Commun.* 1597–1598.

146. (a) Mulzer, J., Mantoulidis, A., and Ohler, E. (1998). *Tetrahedron Lett.* **39**, 8633–8636. (b) Mulzer, J., Mantoulidis, A., and Ohler, A. (2000). *J. Org. Chem.* **65**, 7456–7467. (c) Mulzer, J., Karig, G., and Pojarliev, P. (2000). *Tetrahedron Lett.* **41**, 7635–7638. (d) Martin, H. J., Drescher, M., and Mulzer, J. (2000). *Angew. Chem. Int. Ed.* **39**, 581–583.
147. (a) White, J. D., Carter, R. G., and Sundermann, K. F. (1999). *J. Org. Chem.* **64**, 684–685. (b) White, J. D., Sundermann, K. F., and Carter, R. G. (1999). *Org. Lett.* **1**, 1431–1434. (c) White, J. D., Carter, R. G., Sundermann, K. F., and Wartmann, M. (2001). *J. Am. Chem. Soc.* **123**, 5407–5413.
148. Zhu, B., and Panek, J. S. (2000). *Org. Lett.* **2**, 2575–2578.
149. (a) Sawada, D., and Shibasaki, M. (2000). *Angew. Chem. Int. Ed.* **39**, 209–213. (b) Sawada, D., Kanai, M., and Shibasaki, M. (2000). *J. Am. Chem. Soc.* **122**, 10,521–10,532.
150. Sinha, S. C., Barbas III, C. F., and Lerner, R. A. (1998). *Proc. Natl. Acad. Sci. USA* **95**, 14,603–14,608. (b) Sinha, S. C., Sun, J., Miller, G., Barbas III, C. F., and Lerner, R. A. (1999). *Org. Lett.* **1**, 1623–1626.
151. Machajewski, T. D., and Wong, C.-H. (1999). *Synthesis* 1469–1472.
152. Bode, J. W., and Carreira, E. M. (2001). *J. Am. Chem. Soc.* **123**, 3611–3612.
153. Tayler, R. E., Galvin, G. M., Hilfiker, K. A., and Chen, Y. (1998). *J. Org. Chem.* **63**, 9580–9583.
154. For reviews of chemical biology of epothilones, see Nicolaou, K. C., Roschangar, F., and Vourloumis, D. (1998). *Angew. Chem. Int. Ed.* **37**, 2014–2045. (b) Nicolaou, K. C., Hepworth, D., King, N. P., and Finlay, M. R. V. (1999). *Pure Appl. Chem.* **71**, 989–997. (c) Nicolaou, K. C., Ritzen, A., and Namoto, K. (2001). *Chem. Commun.* 1523–1535.
155. Swinholide A: (a) Paterson, I., Watson, C., Yeung, K.-S., Ward, R. A., and Wallace, P. A. (1998). *Tetrahedron* **54**, 11,955–11,970. (b) Paterson, I., Watson, C., Yeung, K.-S., Wallace, P. A., and Ward, R. A. (1997). *J. Org. Chem.* **62**, 452–453. (c) Nicolaou, K. C., Patron, A. P., Ajito, K., Richter, P. K., Khatuya, H., Bertinato, P., Miller, R. A., and Tomaszewski, M. J. (1996). *Chem. Eur. J.* **2**, 847–868.
156. Sanglifehrin A: (a) Nicolaou, K. C., Murphy, F., Barluenga, S., Ohshima, T., Wei, H., Xu, J., Gray, D. L. F., and Baudoin, O. (2000). *J. Am. Chem. Soc.* **122**, 3830–3838. (b) Duan, M., and Paquette, L. A. (2001). *Angew. Chem. Int. Ed.* **40**, 3632–3636.
157. Rhizoxin: (a) Nakada, M., Kobayashi, S., Shibasaki, S., Iwasaki, S., and Ohno, M. (1993). *Tetrahedron Lett.* **34**, 1039–1042. Rhizoxin D: (b) Kende, A. S., Blass, B. E., and Henry, J. R. (1995). *Tetrahedron Lett.* **36**, 4741–4744. (c) Williams, D. R., Werner, K. M., and Feng, B. (1997). *Tetrahedron Lett.* **38**, 6825–6828. (d) Lafontaine, J. A., Provencal, D. P., Gardelli, C., and Leahy, J. W. (1999). *Tetrahedron Lett.* **40**, 4145–4148. (e) Keck, G. E., Wager, C. A., Wager, T. T., Savin, K. A., Covel, J. A., Mclaws, M. D., Krishnamurthy, D., and Cee, V. J. (2001). *Angew. Chem. Int. Ed.* **40**, 231–234.
158. 18-Deoxynargenicin A_1: (a) Plata, D. J., and Kallmerten, J. (1988). *J. Am. Chem. Soc.* **110**, 4041–4042. Nargenicin A_1: (b) Roush, W. R., Koyama, K., Curtin, M. L., and Moriarty, K. J. (1996). *J. Am. Chem. Soc.* **118**, 7502–7512.
159. Amphidinolide J: (a) Williams, D. R., and Kissel, W. S. (1998). *J. Am. Chem. Soc.* **120**, 11,198–11,199. Amphidinolide P: (b) Williams, D. R., Myers, B. J., and Mi, L. (2000). *Org. Lett.* **2**, 945–948. Amphidinolide K: (c) Williams, D. R., and Meyer, K. G. (2001). *J. Am. Chem. Soc.* **123**, 765–766.
160. Thiazinotrienomycin E: Smith III, A. B., and Wan, Z. (2000). *J. Org. Chem.* **65**, 3738–3753.
161. Callipeltoside aglycone: (a) Paterson, I., Davies, R. D. M., and Marquez, R. (2001). *Angew. Chem. Int. Ed.* **40**, 603–607. Deschlorocallipeltoside A: (b) Trost, B. M., and Gunzner, J. L. (2001). *J. Am. Chem. Soc.* **123**, 9449–9450.
162. Pateamine A: (a) Romo, D., Rzasa, R. M., Shea, H. A., Park, K., Langenhan, J. M., Sun, L., Akhiezer, A., and Liu, J. O. (1998). *J. Am. Chem. Soc.* **120**, 12,237–12,254. (b) Remuinan, M. J., and Pattenden, G. (2000). *Tetrahedron Lett.* **41**, 7367–7371.

163. Griseoviridin: Dvorak, C. A., Schmitz, W. D., Poon, D. J., Pryde, D. C., Lawson, J. P., Amos, R. A., and Meyers, A. I. (2000). *Angew. Chem. Int. Ed.* **39**, 1664–1666.
164. Miyakolide: Evans, D. A., Ripin, D. H. B., Halstead, D. P., and Campos, K. R. (1999). *J. Am. Chem. Soc.* **121**, 6816–6826.
165. Rutamycin B: (a) Evans, D. A., Ng, H. P., and Rieger, D. L. (1993). *J. Am. Chem. Soc.* **115**, 11,446–11,459. (b) White, J. D., Hanselmann, R., Jackson, R. W., Porter, W. J., Ohba, Y., Tiller, T., and Wang, S. (2001). *J. Org. Chem.* **66**, 5217–5231. (c) Panek, J. S., and Jain, N. F. (2001). *J. Org. Chem.* **66**, 2747–2756.
166. Lankacidin C: Kende, A. S., Liu, K., Kaldor, I., Dorey, G., and Koch, K. (1995). *J. Am. Chem. Soc.* **117**, 8258–8270.
167. Phorboxazole A: (a) Forsyth, C. J., Ahmed, F., Cink, R. D., and Lee, C. S. (1998). *J. Am. Chem. Soc.* **120**, 5597–5598. (b) Smith III, A. B., Minbiole, K. P., Verhoest, P. R., and Schelhaas, M. (2001). *J. Am. Chem. Soc.* **123**, 10,942–10,953. Phorboxazole B: (c) Evans, D. A., Cee, V. J., Smith, T. E., Fitch, D. M., and Cho, P. S. (2000). *Angew. Chem. Int. Ed.* **39**, 2533–2536. (d) Evans, D. A., and Fitch, D. M. (2000). *Angew. Chem. Int. Ed.* **39**, 2536–2540.
168. Concanamycin F: (a) Toshima, K., Jyojima, T., Miyamoto, N., Katohno, M., Nakata, M., and Matsumura, S. (2001). *J. Org. Chem.* **66**, 1708–1715. (b) Paterson, I., Doughty, V. A., McLeod, M. D., and Trieselmann, T. (2000). *Angew. Chem. Int. Ed.* **39**, 1308–1312. Bafilomycin A_1: (c) Evans, D. A., and Calter, M. A. (1993). *Tetrahedron Lett.* **34**, 6871–6847. (d) Toshima, K., Jyojima, T., Yamaguchi, H., Noguchi, Y., Yoshida, T., Murase, H., Nakata, M., and Matsumura, S. (1997). *J. Org. Chem.* **62**, 3271–3284. (e) Scheidt, K. A., Tasaka, A., Bannister, T. D., Wendt, M. D., and Roush, W. R. (1999). *Angew. Chem. Int. Ed.* **38**, 1652–1655. Hygrolidin: (f) Makino, K., Nakajima, N., Hashimoto, S., and Yonemitsu, O. (1996). *Tetrahedron Lett.* **37**, 9077–9080.
169. Bryostatin 7: (a) Kageyama, M., Tamura, T., Nantz, M. H., Roberts, J. C., Somfai, P., Whritenour, D. C., and Masamune, S. (1990). *J. Am. Chem. Soc.* **112**, 7407–7408. Bryostatin 2: (b) Evans, D. A., Carter, P. H., Carreira, E. M., Charette, A. B., Prunet, J. A., and Lautens, M. (1999). *J. Am. Chem. Soc.* **121**, 7540–7552. Bryostatin 3: (c) Ohmori, K., Ogawa, Y., Obitsu, T., Ishikawa, Y., Nishiyama, S., and Yamamura, S. (2000). *Angew. Chem. Int. Ed.* **39**, 2290–2294.
170. Aplyronine A: Kigoshi, H., Suenaga, K., Mutou, T., Ishigaki, T., Atsumi, T., Ishiwata, H., Sakakura, A., Ogawa, T., Ojika, M., and Yamada, K. (1996). *J. Org. Chem.* **61**, 5326–5351.
171. Oleandolide: (a) Paterson, I., Norcross, R. D., Ward, R. A., Romea, P., and Lister, M. A. (1994). *J. Am. Chem. Soc.* **116**, 11,287–11,314. (b) Hu, T., Takenaka, N., and Panek, J. S. (1999). *J. Am. Chem. Soc.* **121**, 9229–9230. Oleandolide and 6-Deoxyerythronolide B: (c) Evans, D. A., Kim, A. S., Metternich, R., and Novack, V. J. (1998). *J. Am. Chem. Soc.* **120**, 5921–5942.
172. Mycalolide A: Panek, J. S., and Liu, P. (2000). *J. Am. Chem. Soc.* **122**, 11,090–11,097.
173. Acutiphycin: Smith III, A. B., Chen, S. S.-Y., Nelson, F. C., Reichert, J. M., and Salvatore, B. A. (1997). *J. Am. Chem. Soc.* **119**, 10,935–10,946.
174. Halichondrin B: Aicher, T. D., Buszek, K. R., Fang, F. G., Forsyth, C. J., Jung, S. H., Kishi, Y., Matelich, M. C., Scola, P. M., Spero, D. M., and Yoon, S. K., (1992). *J. Am. Chem. Soc.* **114**, 3162–3164.
175. Epothilones: (a) Martin, N., and Thomas, E. J. (2001). *Tetrahedron Lett.* **42**, 8373–8377. (b) Valluri, M., Hindupur, R. M., Nijoy, P., Labadie, G., Jung, J.-C., and Avery, M. A. (2001). *Org. Lett.* **3**, 3607–3609. (c) Hindupur, R. M., Panicker, B., Valluri, M., and Avery, M. A. (2001). *Tetrahedron Lett.* **42**, 7341–7344. (d) Tayler, R. E., and Chen, Y. (2001). *Org. Lett.* **3**, 2221–2224. (e) Frustner, A., Mathes, C., and Grela, K. (2001). *Chem. Commun.* 1057–1059. (f) Martin, H. J., Pojarliev, P., Kahling, H., and Mulzer, J. (2001). *Chem. Eur. J.* **7**, 2261–2271. (g) Zhu, B., and Panek, J. S. (2001). *Eur. J. Org. Chem.* 1701–1714. (h) Lee, C. B., Wu, Z., Zhang, F., Chappell, M. D., Stachel, S. J., Chou, T.-C., Guan, Y., and Danishefsky, S. J. (2001). *J. Am. Chem. Soc.* **123**, 5249–5259. (i) Sinha, S. C., Sun, J., Miller, G. P., Wartmann, M., and Lerner, R. A. (2001). *Chem. Eur. J.* **7**, 1691–1702.

Chapter 5

Biosynthesis, Regulation, and Genetics of Macrolide Production

HARUO IKEDA
SATOSHI ŌMURA

Kitasato Institute for Life Sciences
Kitasato University, Tokyo, Japan

I. Introduction 286
II. Reaction Mechanism of Polyketide Biosynthesis 287
III. Polyketide Synthase 289
    A. β-Ketoacyl Synthase Domain 291
    B. Acyltransferase Domain 291
    C. Acyl Carrier Protein Domain 292
    D. Dehydratase, Enoyl Reductase, β-Ketoreductase,
        and Thioesterase Domains 292
IV. Genes Encoding Modular Polyketide Synthase 295
    A. Erythromycin 295
    B. Tylosin 295
    C. Spiramycin and Niddamycin 296
    D. Pikromycin 300
    E. Avermectin 301
    F. Nystatin and Pimaricin 301
    G. Oligomycin 301
    H. Rifamycin 309
    I. Rapamycin and FK520 309
    J. Other Polyketide Compounds 314
    K. Post-Polyketide Modification 314
V. Sugar Biosynthesis 314
    A. 2-Deoxysugar 314
    B. Deoxyaminosugar 317
VI. Genetic Manipulation of PKS Genes 319
    A. Loss of Function 319
    B. Gain of Function 320
    C. Alternation of Function 320
    References 320

I. Introduction

Macrolide compounds are a class of natural products synthesized by bacteria, fungi, and plants through the successive condensation of simple short-chain fatty acids [1]. They are attracting increasing attention as useful agents for medicine, veterinary medicine, and agriculture and as unique biochemical tools. They have antibacterial, antifungal, antiparasite, insecticidal, or antitumor activity. Some antifungal macrolide compounds also have immunosuppressive activity. They consist of polyketide-derived aglycons linked to saccharide(s), and some macrolides contain aglycon alone.

The polyketides can be classified into two major classes: the aromatic and the complex polyketides including macrolide compounds. Aromatic polyketides are produced through the condensation of acetates, which generate β-keto groups that remain largely unreduced during and after initiation of the acyl chain. The acyl chains, either during or after completion of their synthesis, undergo enzymatic aldol condensations resulting in the formation of six-membered rings. The aromatic rings are subsequently reduced through enzymatic dehydrations.

Complex polyketides are structurally more diverse than the aromatic polyketides and are composed of acetates, propionates, and butyrates in varying ratios. The biosynthetic mechanism for the formation of complex polyketides, represented by the aglycons of the various classes of macrolides, is quite different from that for the aromatic polyketides. In complex polyketides, because of processive β-carbonyl reduction and because of the structural strains imposed by the presence of methyl side chains, molecules do not undergo folding leading to aromatization. Polyketide chains cyclize through lactonization to form aglycon rings.

Several of the macrolide aglycons are highly modified after polyketide chain synthesis and cyclization, and most of them are linked to sugars. In general, the polyketide chain is composed of a single carbon chain with a very complex structure. Among them the best studied example is the macrolide antibiotic erythromycin, which is the first example for which we understood the genetics and biochemistry of polyketide biosynthesis [2, 3]. Erythromycin, produced by the gram-positive filamentous bacterium *Saccharopolyspora erythraea*, is composed of a 14-membered macrocyclic lactone ring to which is attached the 6-deoxysugars; D-desosamine at the C5 and L-cladinose at the C3 position [4]. The macrocyclic lactone ring, 6-deoxyerythronolide, which is synthesized by polyketide synthase, undergoes cytochrome *P*-450-mediated hydroxylation at C6 [5, 6] and C12 [7, 8]. In the final step, modified lactone is attached above two sugars. Although the synthesis of the deoxysugars and modification of the polyketide backbone are highly controlled, some steps of the pathway leading to the final product might be bypassed.

II. Reaction Mechanism of Polyketide Biosynthesis

The polyketide synthesis chemically and biochemically resembles that of fatty acids. The reaction of fatty acid synthesis is inhibited by the fungal product cerulenin [9]. It inhibits all known types of fatty acid synthases, both multifunctional enzyme complex and unassociated enzyme from different sources like that of some bacteria, yeast, plants, and mammalians [10]. Cerulenin also blocks synthesis of polyketides in a wide variety of organisms, including actinomycetes, fungi, and plants [11, 12]. The inhibition of fatty acid synthesis by cerulenin is based on binding to the cysteine residue in the condensation reaction domain [13]. Synthesis of both polyketide and fatty acids is initiated by a Claisen condensation reaction between a starter carboxylic acid and a dicarboxylic acid such as malonic or methylmalonic acid. An example of this type of synthesis is shown in Fig. 1. An acetate and malonate as enzyme-linked thioesters are used as starter and extender, respectively. The starter unit is linked through a thioester linkage to the cysteine residue in the active site of the enzymatic unit, β-ketoacyl ACP synthase (KS), which catalyzes the condensation reaction. On the other hand, the extender

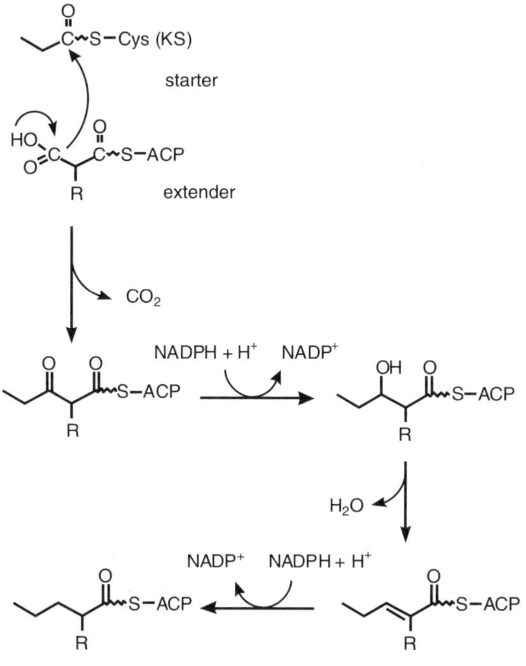

Fig. 1. Reduction cycle of β-carbonyl formed from the Claisen condensation of carboxylic acids. ACP, acyl carrier protein domain of PKS; KS, β-ketoacyl ACP synthase domain of PKS.

unit is linked as a thioester to the panthotheine residue, which is attached as a phosphoester to a serine residue in an acyl carrier protein (ACP) unit. The KS and ACP units are the common components of polyketide synthases or vertebrate fatty acid synthases which are multifunctional polypeptides. The resulting β-ketoacyl residue (diketide) is two carbons longer than the starter. If the extender is replaced by methylmalonate, the resulting β-ketoacyl residue carries an α-methyl side chain. The energy supplied by the exergonic decarboxylation of the extender permits breakage of the thioester bond in the starter and formation of β-ketoacyl chain. In fatty acid synthesis, the next step is reduction of the β-carbonyl that takes place after every condensation step but the reduction is not always involved in polyketide synthesis.

Formation of a fully reduced saturated carbon chain is a three-step process requiring three distinct enzymatic functions. The first step is ketoreduction (β-ketoacyl ACP reductase; KR) to produce the secondary alcohol residue in that an electron is supplied by NADPH to the carbonyl group followed by protonation. The second step is dehydration (dehydratase; DH) to lead to the α,β unsaturated acyl group. The final step is enoyl reduction (enoyl reductase; ER), which employs NADPH as an electron donor and proton to result in the formation of a methylene function at the β-carbon. After the reduction steps are completed, the generated acyl chain enters the KS domain and is equivalent to the starter for the next cycle of the reaction to condense with the next extender unit.

Two differences exist between fatty acid and complex polyketide syntheses (Fig. 2). First, in fatty acid synthesis, synthase uses only malonyl moieties as extender units to build an acyl chain. In general, acetate is used as the starter unit in vertebrate fatty acid synthase, but bacterial fatty acid synthase may use a branched-chain carboxylic acid as the starter unit because bacterial fatty acids sometimes contain branched-chain fatty acids. In contrast, polyketide synthesis in bacteria uses malonyl, methylmalonyl, and ethylmalonyl units as extenders. In the polyketide synthase, respective extender units are used at every step of the condensation. The polyketide synthase in fungi uses malonyl units as extenders and methyl groups at α positions are added by C-methylation using S-adenosyl-L-methionine.

Second, in polyketide synthesis, the full steps of β-carbonyl reduction (ketoreduction, dehydration, and enoylreduction) are not always involved after every condensation as in fatty acid synthesis. The keto group in the polyketide backbone results from the lack of ketoreduction after the condensation step. Similarly, the hydroxyl group at the β position results from ketoreduction and failure to undergo the subsequent dehydration step, the α,β unsaturated group results from ketoreduction and dehydration steps, and the methylene group results from the full steps of reduction after the condensation. After all steps of condensation and β-carbonyl reduction are completed, the resulting acyl residue is released from polyketide synthases by thioesterase (TE) and spontaneous lactonization proceeds.

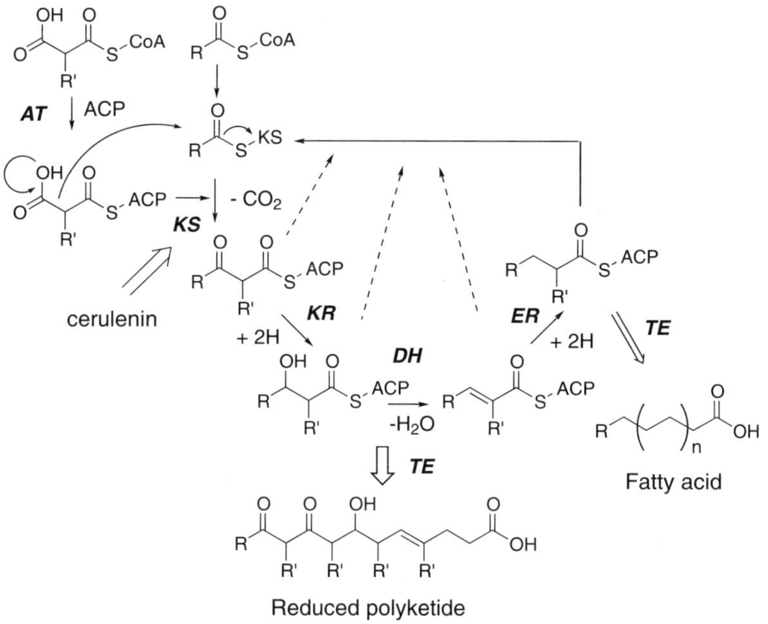

Fig. 2. The basic pathway of fatty acid and polyketide biosynthesis, showing the roles of the various activities carried out by the subunits or domains of the fatty acid or polyketide synthase. A fungal product, cerulenin, inhibits the KS reaction.

III. Polyketide Synthase

Many attempts were made to isolate active cell-free extracts of the polyketide synthase from the macrolide-producing organisms without success. It was generally recognized, however, that a parallel pathway exists with the fatty acid biosynthetic pathway that has been investigated at the enzyme level [14, 15]. A fatty acid synthase forms its acyl chain through a series of reactions that is initiated by condensation of a starter unit (commonly acetyl) with an extender unit (malonyl). The resulting β-keto acyl ester is then processed by reduction, dehydration, and reduction again to give an elongated saturated acyl chain and the cycle is resumed by the condensations of extender units.

To carry out these processes, the fatty acid synthase required a set of catalytic activities that are responsible for the respective step of a set of the cycle reactions. Acyltransferase (AT) transfers both the acetyl starter unit and malonyl chain-extender units from the respective coenzyme A esters to the appropriate thiol group of the synthase. β-Ketoacyl synthase and acyl carrier protein are responsible for chain elongation. β-Ketoreductase, dehydratase, and enoyl reductase are for reductive processing of the β-keto group, and a thioesterase catalyzes release of the full-length chain.

The constituent proteins of a fatty acid synthase can be organized in markedly different ways depending on the organism. There are two extremes. At one extreme found in bacteria, the individual activities (domains) are freely dissociable and can be isolated separately [16]. At the other extreme, in animals or yeast, the domains are bound together by peptide links to form large nondissociable multidomain proteins [14, 15]. The dissociable synthases are designated "type II" and the multidomain proteins are designated "type I." The same nomenclature has been adapted for the polyketide synthase. The sequence of domains from the amino to the carboxyl terminus of the type I fatty acid synthase is KS, AT, DH, ER, KR, ACP, and TE. A long stretch of amino acids is sited between the DH and ER domains that appears to have no specific catalytic function; therefore, the part is considered to be responsible for specific overall structure [14, 15]. In its spatial organization and structure, it shows strong similarities with the macrolide polyketide synthases.

Before the discovery of macrolide polyketide synthases themselves, information was obtained regarding the biosynthetic properties of modular polyketide synthases through incorporation experiments with labeled precursors and their analogues [17, 18]. Isotope-labeling experiments, especially those using ^{13}C-labeled substrates, demonstrated that the carbon chain backbones of macrolide compounds are derived through C—C bond formation between acetate, propionate, and butyrate building block units [17, 18]. Furthermore, the incorporation of exogenously added analogues of putative biosynthetic intermediates, such as N-acylcysteamine thioesters, into the corresponding macrolide compounds has unequivocally proven that modular polyketide synthases act via a processive mechanism in which the stereochemistry of the growing polyketide chain is adjusted immediately after each step of polyketide chain elongation [19, 20]. However, the biochemical basis for these highly controlled synthetic processes remained virtually unknown until the discovery of erythromycin polyketide synthase [2, 3, 21, 22].

The polyketide synthesis underscores the complex programming required for the enzymes involved to form the correct structure. To form the macrolide lactone ring, the polyketide synthase must select the correct units for the condensation, determine the specific stereochemistry for the side chains, determine the specific β-carbonyl reduction at each cycle, and permit specific condensation cycles before terminating the elongation of acyl chain. In 1990, for the first time, a complex polyketide synthase was sequenced—the 6-deoxyerythronolide B synthase involved in the synthesis of the polyketide backbone of erythromycin [2, 3]. The sequence analysis revealed that the structural genes responsible for the formation of 6-deoxyerythronolide B consist of three contiguous open reading frames of about 10 kb each, encoding three giant multifunctional proteins,

6-deoxyerythronolide B synthase 1, 2, and 3 [22]. Furthermore, sequence comparisons also showed that each of these proteins contains 8 to 10 functional domains considered to carry each of the individual steps of fatty acid biosynthesis. After this finding, more than 10 genes encoding complex modular polyketide synthases in macrolide compound-producing *Streptomyces* strains were cloned and sequenced. In each case, polyketide synthase is organized in modules similar to that of 6-deoxyerythronolide B synthase. On the basis of comparison of putative amino acid sequences with known sequences of fatty acid synthases, the modular polyketide synthase was composed of enzymatic domains resembling the KS, AT, KR, DH, ER, and ACP domains as well as of a single domain resembling TE near to the carboxy terminus of synthase, which carries the final step of polyketide synthesis.

A. β-Ketoacyl Synthase Domain

The KS domains of type I fatty acid and polyketide synthases are the most highly conserved among all the constituent domains in these multidomain proteins. About a 40% identity of the amino acid sequence was found for all of the domains. The putative active site Cys^{173} residue and two other conserved $His^{308/346}$ residues were present in KS domains [23, 24]. It is tempting to speculate that one of these His residues might increase the nucleophilicity of the active site Cys by acting as a general base. In contrast to the erythromycin and avermectin polyketide synthases in which the N-terminal enzymatic motifs are a loading AT and ACP, the loading AT and ACP of the other polyketide synthases (tylosin, pikromycin, nystatin, oligomycin, and so on) follow a KS domain that is not predicted to be necessary for the initiation of the polyketide synthesis.

The homology of these KS domains with the other functional KS domains in the cluster ranges is high. Sequence analysis of these domains indicates that KS domains in the loading modules are enzymatically nonfunctional, because a critical Cys residue in the consensus motif TVDTGCSSSLV, which is highly conserved among KS domains [23, 24], is replaced by a Gln or Ser residue at position 173. The Cys-to-Gln replacement in the active sites of KS domains might be related to the decarboxylase activity shown for the KS^Q domains to be largely dependent on the presence of a Gln residue in the active site [25]. It is tempting to speculate that Cys-to-Ser replacement in the active site of the KS domain might also have some significance for the above reaction.

B. Acyltransferase Domain

AT domains show more sequence variability than the KS domains. In all of the AT domains from type I fatty acid and polyketide synthases (PKS), apart from the active site sequence GHSxG where Ser is involved in the formation of the acyl enzyme intermediate, there is a second His residue at about the 100-amino-acid C terminal of the active site in the consensus sequence AxHs, which is invariant in all AT domains [23, 24]. Polyketide-derived compounds are composed of a variety of acyl building blocks. Sequence alignment generated by a number of programs to perform pairwise comparisons showed that AT domains of macrolide polyketide synthases are clustered as at least three groups [26–29]. The first group contains only the proposed malonate leading functions, the second group contains methylmalonate leading functions, and the third group contains acyl residues derived from monocarboxylic acid leading functions (Fig. 3). As shown in Fig. 3 the second group contains ethylmalonyl (to extend butyrate) and hydroxymalonyl (to extend glycolate) loading functions in tylosin and niddamycin PKS. Furthermore, the signature sequences for malonyl and methylmalonyl, and other acyl loading domains, were identified and are shown in Fig. 4. The alignment of the domain sequence of the third group is similar to that of methylmalonyl loading domains, but the consensus sequence was not found around the active serine residue. The actual assignment of the substrate specificity for each domain will have to await identification of the polyketide encoded by this pathway.

C. Acyl Carrier Protein Domain

The ACP domains, the smallest among all of the domains, are identified in each of the modules and after the loading AT domain and are also highly conserved. The pantotheine-binding Ser residue in the GFDSL motif is present in all of the ACP domains [23].

D. Dehydratase, Enoyl Reductase, β-Ketoreductase, and Thioesterase Domains

In polyketide synthases, all modules encoding putative KS, AT, and ACP domains are responsible for all of the acyl condensation processes and for a variable number of functions involved in the processing of β-carbon. The sequences of the DH domains of all of the polyketide synthases fall into two clearly distinguishable groups. The DH domains in one group are in good alignment with the DH domains from the fatty acid synthase, and active site motif HxxxGxxxxP is present [21, 23]. In contrast, the DH domains of the other group contain a few amino acid deletions or His-to-Tyr replacement in the active site, which readily accounts for their inactivity [28]. But some DH domains are intact, even though these cycles do not apparently require this activity [24]. In some DH domains in niddamycin and oligomycin polyketide synthases, an Asp occurs in place of the Gly residue (HxxxDxxxxP) [27]. The DH domain of module 2 of avermectin polyketide synthase contains Pro-to-Ser replacement in the active site [28]. In view of the niddamycin, oligomycin, and avermectin structures, this DH domain

5. Macrolide Production

Fig. 3. Phylogenetic analysis of acyltransferases. Phylogenetic tree of amino acid sequences of acyltransferase domains from actinomycete type I PKSs. Multiple alignment and phylogenetic analysis using the bootstrapping method were performed by using CLUSTALW. AVE, avermectin PKS module; ERY, erythromycin PKS module; FKB, FK520 PKS module; MEG, megalomicin PKS module; NID, niddamycin PKS module; NYS, nystatin PKS module; OLM, oligomycin PKS module; PIK, pikromycin PKS module; RAP, rapamycin PKS module; RIF, rifamycin PKS module; TYL, tylosin PKS module.

Fig. 4. Putative consensus sequences of malonyl, methylmalonyl and other loading domains. All letters shown represent invariant amino acids for the malonyl, methylmalonyl, or other loading domain.

should be active; hence, this divergence from consensus may not adversely affect the activity of the enzyme.

A few ER domains are present in polyketide synthases and no ER domains are present in avermectin [28] and pimaricin [30] polyketide synthases. ER domains contain the putative NADP(H)-binding motif LxHxg(a)xGGVG, which was originally proposed for the rat ER [31, 32], or GxGxxAxxxA [33].

KR domains are found in most modules. KR domains contain a potential motif for NADP(H)-binding (GxGxxAxxxA) [33]. The first invariant Ala in this motif had been found to be substituted by Gly in many polyketide synthases [21, 23, 34]. In avermectin polyketide synthase, the last invariant Ala in the motif sequence is substituted to Thr [28]. KR domain activity is predicted to be required for the formation of avermectin aglycons. In another group of KR domains, the motif sequence corresponding region is not found, which readily accounts for their inactivity.

TE domains are found at the C terminus of the module to terminate the chain elongation. Alignment of the sequences of TE domains revealed the invariant GxSxg and GdH motifs, which are considered to be essential for TE activity [23]. Although TE is involved in releasing the polyketide chain from the polyketide synthase, rapamycin, FK520 and rifamycin polyketide synthases lack TE domains at the C terminus. In polyketide synthases of rapamycin and its related compound, FK520, chain termination is not carried out by a TE as in the macrolide polyketide synthases, but is thought to be effected by a specialized multidomain protein coded by the gene *rapP* [35] or *fkbP* [29]. These genes are quite similar to the gene involved in nonribosomal peptide biosynthesis and corresponding protein is believed to catalyze the formation of the ester and amide bonds to pipecolic acid. In the gene cluster for rifamycin biosynthesis, peptide synthetase is not found; however, an amide synthase gene is located downstream of the genes encoding polyketide synthases [36, 37]. It is proposed that the gene product might catalyze the release of the completed polyketide chain from the polyketide synthase by intramolecular amide formation to generate the macro-cyclic lactam structure.

At the end of the polyketide chain extension, there is usually a covalently linked TE domain, which is defined as type I TE (TE-I). Some gene clusters for macrolide biosyntheses contain a second TE gene, designated TE-II, outside of the cluster. Disruption of TE-II genes from several polyketide synthase clusters has shown that TE-II plays an important role in maintaining normal levels of macrolide production. The detailed experiments in which TE-II protein was overexpressed in *Escherichia coli* were done in the tylosin producer. The experiments show that the TE-II of the tylosin cluster has an activity for releasing the polyketide chain from polyketide synthase. It is significant that the best substrate analogues for TE-II are the short acyl chains acetate, propionate, and butyrate of *N*-acetylcysteamine derivatives, which would be formed by aberrant decarboxylation of chain extenders in the various chain extension modules of the tylosin polyketide synthase. Representative polyketide chains are also cleaved at a significant rate, indicating that tylosin TE-II could also remove aberrant intermediates arising from incorrect operation of chain extension processes [38].

IV. Genes Encoding Modular Polyketide Synthase

A. Erythromycin

Most of our knowledge about modular polyketide synthases emanates from studies on the erythromycin polyketide synthases [2, 3, 5, 6, 21, 39–44] (Fig. 5). Erythromycin-resistant gene, *ermE* [39], was used as a hybridization probe to clone genes for erythromycin biosynthesis from the genome of *S. erythraea*. The DNA fragments isolated by genomic walking from *ermE* were sequenced and used in gene disruption and complementation experiments. The sequence revealed that three multidomain proteins were involved in the formation of 6-deoxyerythronolide B. Recently, the gene for megalomicin biosynthesis in *Micromonospora megalomicea* has been cloned and sequenced [45]. The organization and composition of its biosynthetic genes is quite similar to that of erythromycin.

B. Tylosin

Tylosin [46–49] (Fig. 6), produced by *Streptomyces fradiae*, was one of the first antibiotics for which a comprehensive set of blocked mutants was isolated. They were used to help define the biosynthetic pathway. Whereas erythromycin is a 14-membered macrolide, tylosin has a 16-membered lactone structure. The gene for the final step of biosynthesis was cloned by reverse genetics from the protein sequence of the enzyme [46], and specific segments of surrounding DNA were found to complement other classes of blocked mutants. The formation of tylosin aglycon, protylonolide, is involved in five polyketide synthases.

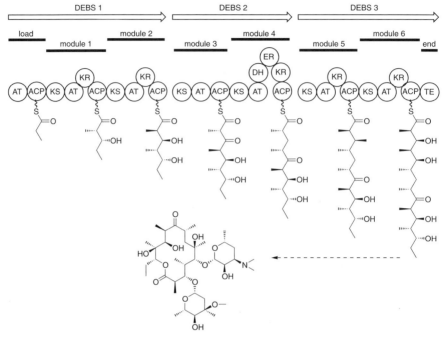

Fig. 5. Predicted domain organization and biosynthetic intermediates of the erythromycin synthase. Each circle represents an enzymatic domain as follows: ACP, acyl carrier protein; AT, acyltransferase; DH, dehydratase; ER, β-ketoacyl-ACP enoyl reductase; KR, β-ketoacyl-ACP reductase; KS, β-ketoacyl-ACP synthase; TE, thioesterase. Zero indicates dysfunctional domain.

C. Spiramycin and Niddamycin

Spiramycin [50] (Fig. 7) and niddamycin [27] (Fig. 8), produced by *S. ambofaciens* and *S. caelestis*, respectively, are two further 16-membered macrolide antibiotics both derived from the same primary polyketide intermediate, platenolide. In the case of spiramycin biosynthetic genes, the cloning of one or more resistance genes in a heterologous *Streptomyces* host led to identification of linked biosynthetic genes that were revealed by complementation of blocked mutants with segments of the cloned DNA. On the other hand, a strategy was devised to isolate the polyketide synthase genes of the niddamycin biosynthetic pathway by using sequence conservation in the KS and AT regions encoded by erythromycin polyketide synthase genes. A pair of degenerate primers spanning conserved regions was designed from sequence in database with the expectation of amplifying from *S. caelestis* chromosomal DNA. The amplified DNA segments were used as a hybridization probe to clone genes for niddamycin biosynthesis from the genome of *S. caelestis*. Both polyketide synthases of spiramycin

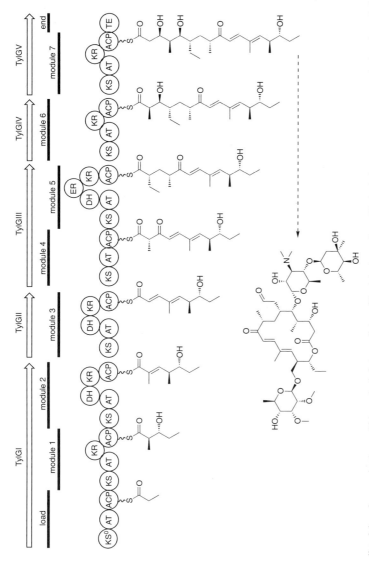

Fig. 6. Predicted domain organization and biosynthetic intermediates of the tylosin synthase. Each circle represents an enzymatic domain as defined in Fig. 5.

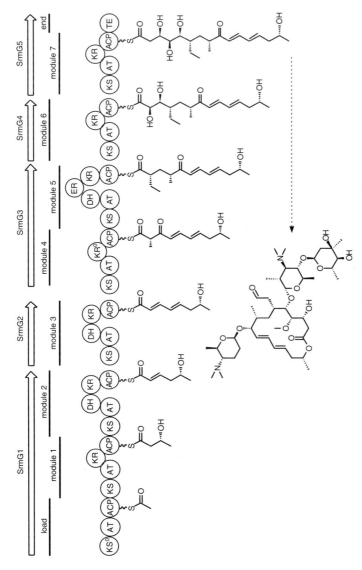

Fig. 7. Predicted domain organization and biosynthetic intermediates of the spiramycin synthase. Each circle represents an enzymatic domain as defined in Fig. 5.

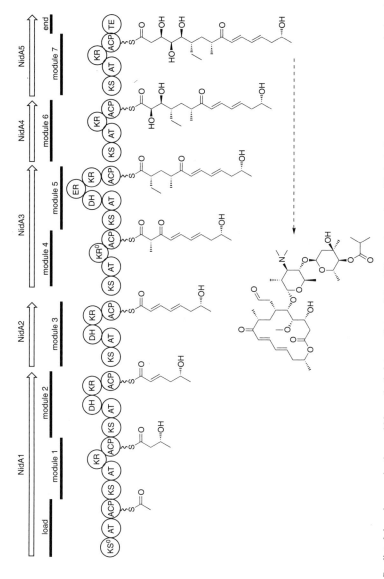

Fig. 8. Predicted domain organization and biosynthetic intermediates of the niddamycin synthase. Each circle represents an enzymatic domain as defined in Fig. 5.

and niddamycin are the same composition, and five polyketide synthases are involved in the formation of these 16-membered lactone rings.

D. Pikromycin

S. venezuelae produces a 14-membered macrolide, pikromycin [51, 52] (Fig. 9), and a 12-membered macrolide, methymycin, simultaneously. Heterologous hybridization was used to identify genes for pikromycin biosynthesis in *S. venezuelae*. Initial hybridization analysis with DNA segments that encode a heterologous polyketide synthase as probe revealed two type I polyketide synthase clusters of uncharacterized function in the genome. A second hybridization analysis with α-glucose-1-phosphate thymidylyltransferase DNA probe, which is involved in the biosynthesis of deoxysugar linked to macrolide aglycons, established that only one cluster was concerned with pikromycin biosynthesis. Although three polyketide synthases are involved in 14-membered macrolide erythromycin aglycon, 6-deoxyerythronolide B, the formation of the 14-membered aglycon of pikromycin is performed by four kinds of polyketide synthases. Interestingly, both 14-membered macrolide pikromycin and 12-membered macrolide methymycin are formed by the same polyketide synthases. The formation of the smaller ring aglycons might be concerned with the alternative transcription of later steps of polyketide extension.

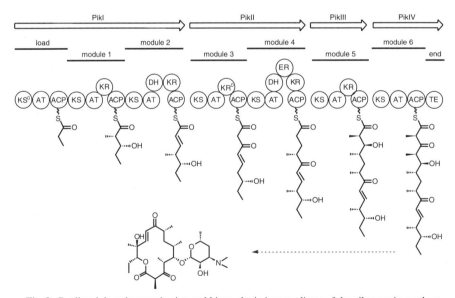

Fig. 9. Predicted domain organization and biosynthetic intermediates of the pikromycin synthase. Each circle represents an enzymatic domain as defined in Fig. 5.

E. Avermectin

A biosynthetic gene cluster for the avermectins [28, 53–59] (Fig. 10), important antiparasitic polyketide macrocyclic lactones produced by *S. avermitilis*, was isolated by complementing mutants blocked in a post-polyketide modification step [58] and glycosylation steps [55]. The discovery of the entire gene cluster for avermectin biosynthesis was made by chromosomal walking between the two complementing regions and by gene disruptions. Centrally located in a cluster, two converging regions encode four polyketide synthases containing 12 modules [28]. The starter unit of the polyketide extension is 2-methylbutyryl or isobutyryl residue, which is different from that of other macrolide biosyntheses. Use of avermectin polyketide synthase sequences as probes against DNA of *S. cyaneogriseus*, the producer of nemadectin, with an aglycon structure similar to that of avermectin led to the isolation of the genes encoding nemadectin polyketide synthases [57]. The organization of polyketide synthases of nemadectin in the chromosome is extremely similar to that of avermectin [60].

F. Nystatin and Pimaricin

Nystatin [61] (Fig. 11) and pimaricin [30, 62] (Fig. 12) are members of the polyene macrolide class of complex antifungal polyketides. The nystatin molecule contains a polyketide moiety represented by a 38-membered macrocyclic lactone ring to which the deoxysugar is attached. The formation of nystatin agylcon requires 18 cycles of condensations, in which six polyketide synthases are involved in the polyketide extension. One of the polyketide synthase-encoding genes, *nysC*, was found to encode the largest (11,096 amino acids long and six modules) modular polyketide synthase described to date [61].

Pimaricin, produced by *S. natalensis*, is also a 26-membered antifungal polyene macrolide antibiotic. The formation of pimaricin backbone requires 12 condensation cycles in which five polyketide synthases are involved. The domain composition in the loading module of the pimaricin polyketide synthase is different from other polyketide synthases. The formation of pimaricin is predicted to start on PIMS0, a single module polyketide synthase with an N-terminal ATP-dependent carboxylic acid:CoA ligase domain (CoL), and continue to the next polyketide synthase with the formation of aglycon.

G. Oligomycin

Oligomycin [63] (Fig. 13) is a 26-membered macrocyclic lactone produced by avermectin-producing *Streptomyces avermitilis*. The formation of oligomycin backbone requires 16 cycles of condensations. A region encoding polyketide synthase of oligomycin biosynthesis was cloned by transposon-induced mutagenesis.

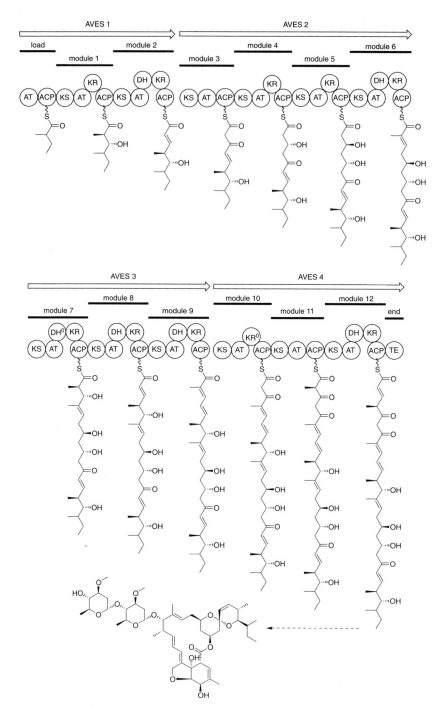

Fig. 10. Predicted domain organization and biosynthetic intermediates of the avermectin synthase. Each circle represents an enzymatic domain as defined in Fig. 5.

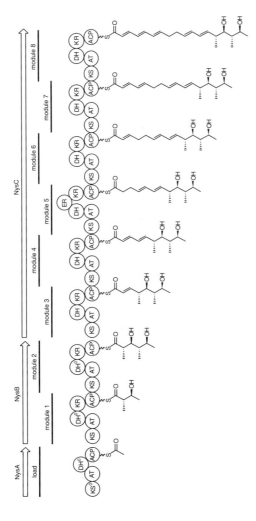

Fig. 11. Predicted domain organization and biosynthetic intermediates of the nystatin synthase. Each circle represents an enzymatic domain as defined in Fig. 5. (*Continued*)

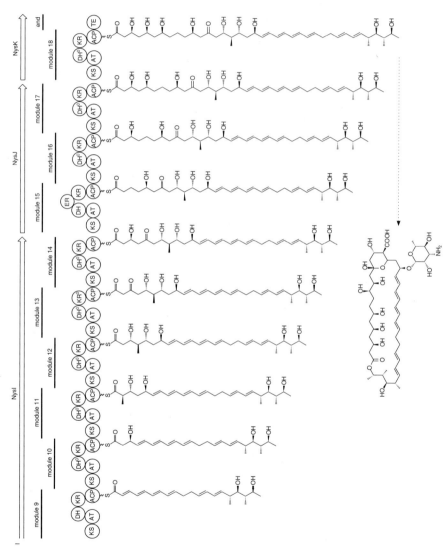

Fig. 11. *Continued.*

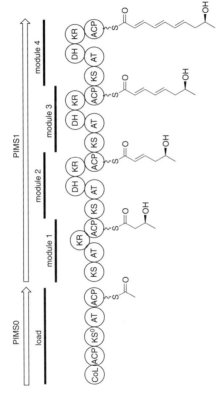

Fig. 12. Predicted domain organization and biosynthetic intermediates of the pimaricin synthase. Each circle represents an enzymatic domain as defined in Fig. 5. CoL, carboxylic acid:CoA ligase. (*Continued*)

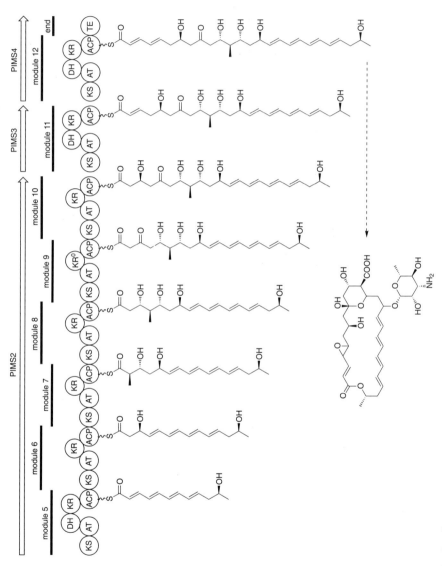

Fig. 12. *Continued.*

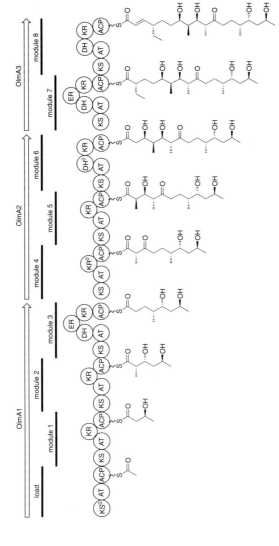

Fig. 13. Predicted domain organization and biosynthetic intermediates of the oligomycin synthase. Each circle represents an enzymatic domain as defined in Fig. 5. (*Continued*)

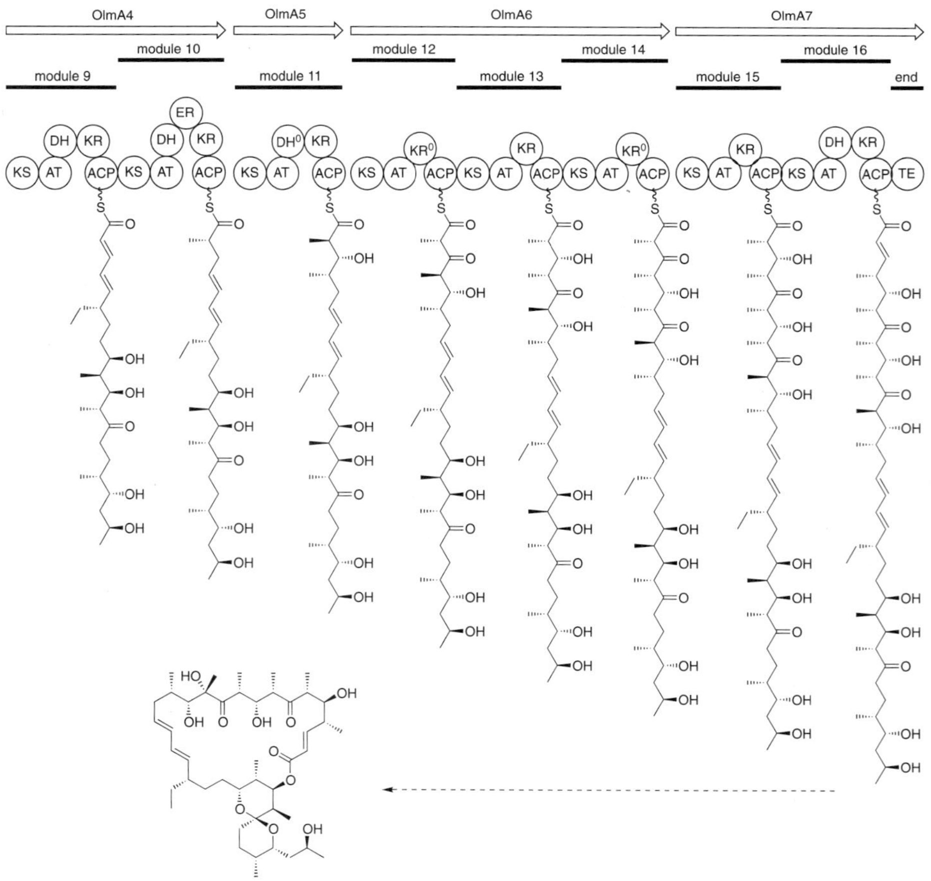

Fig. 13. *Continued.*

The discovery of the entire gene cluster for oligomycin biosynthesis was accomplished by means of chromosomal walking. Seven polyketide synthases are involved in the formation of oligomycin backbone.

H. Rifamycin

Rifamycin B, produced by *Amycolatopsis mediterranei*, is one of the most notable members of the ansamycin family [36, 37, 64, 65] (Fig. 14). It has been used clinically in a synthetically modified form called rifampicin and it is still one of the first-line therapies effective in the treatment of tuberculosis and other mycobacterial infections. The starter unit for rifamycin polyketide assembly is part of the chromophore and is derived from 3-amino-5-hydroxybenzoic acid. Five polyketide synthases are involved in the formation of rifamycin chromophore and the first polyketide synthase contains at the N terminus the loading domain for 3-amino-5-hydroxybenzoic acid, which consists of an acyl-CoA ligase linked to ACP, and module 1–3. The rifamycin polyketide synthase lacks a TE domain at the C terminus. The release of polyketide chain from polyketide synthase and the formation of amide to generate the macrocyclic lactam will be catalyzed by RifF, which is very similar to arylamine N-acetyltransferase.

I. Rapamycin and FK520

Rapamycin [24, 35, 66] (Fig. 15) and FK520 [29] (Fig. 16) are related macrolide compounds produced by *Streptomyces hygroscopicus*, respectively. They are important immunosuppressants. In this group of macrolides, FK506 is used to prevent rejection of transplanted organs in humans. The formation of rapamycin backbone requires 14 condensation cycles, which are involved in three polyketide synthases. The three polyketide synthases for FK520 biosynthesis are concerned with the formation of FK520 backbone, which requires 10 condensation cycles. In contrast to the other macrolide cases, typical starter and TE domains are not found in rapamyin and FK520 polyketide synthases. Both compounds start with incorporation of a cyclohexane carboxylic acid unit and end with incorporation of a pipecolic acid unit. Chain initiation on the rapamycin and FK520 polyketide synthases will require an ATP-dependent carboxylic acid:CoA ligase, which catalyzes the formation of CoA ester of dihydroxycyclohexene carboxylic acid. Such a region was identified at the N terminus of the first polyketide synthase, adjacent to ER, which is involved in the reduction of cyclohexene residue to generate cyclohexane residue. The rapamycin and FK520 polyketide synthases also have an unusual mechanism for chain termination and cyclization. A gene in both macrolide biosynthetic genes is strikingly similar to activation domains of nonribosomal peptide synthetases. The polyketide chain of both macrolides will be transferred from thioester linkage on polyketide synthase directly to the amino

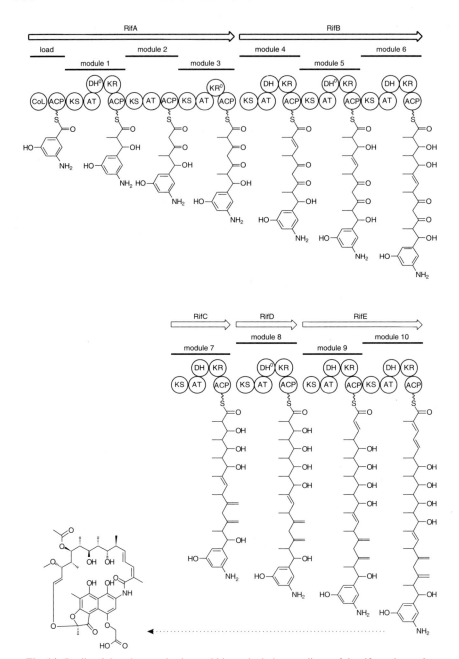

Fig. 14. Predicted domain organization and biosynthetic intermediates of the rifamycin synthase. Each circle represents an enzymatic domain as defined in Figs. 5 and 12.

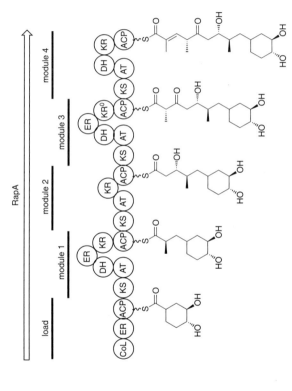

Fig. 15. Predicted domain organization and biosynthetic intermediates of the rapamycin synthase. Each circle represents an enzymatic domain as defined in Figs. 5 and 12. PS, nonribosomal peptide synthetase. (*Continued*)

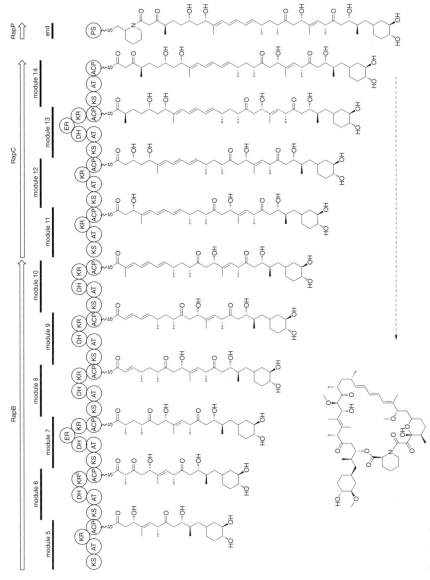

Fig. 15. *Continued.*

5. Macrolide Production

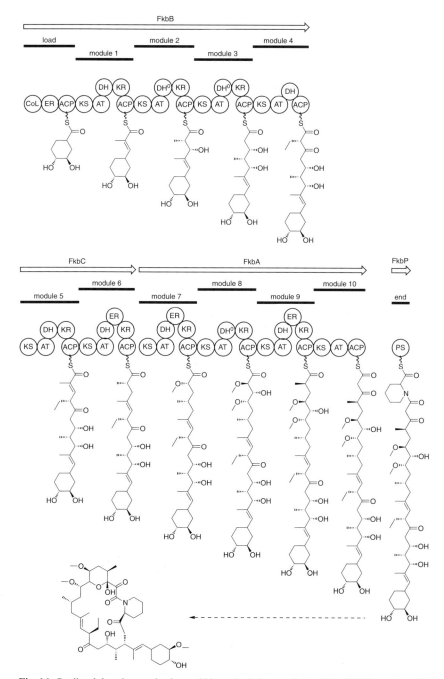

Fig. 16. Predicted domain organization and biosynthetic intermediates of the FK520 synthase. Each circle represents an enzymatic domain as defined in Figs. 5 and 12. PS, nonribosomal peptide synthetase.

group of an enzyme-bound pipecolyl moiety, which in turn is attacked by the hydroxyl residue to form the macrolactam ring.

J. Other Polyketide Compounds

Several other polyketide synthase genes have been cloned and characterized. In their PKSs, some variations occur in the content and organization of different systems but the key features of the modular hypothesis remain unchanged [67–81].

K. Post-Polyketide Modification

After synthesis of the lactone ring by polyketide synthases, some biosynthetic steps are involved in the modification of the lactone. In several macrolide biosyntheses, modifying by means of hydroxylation occurs more frequently. The enzyme involving hydroxylation of the polyketide-derived lactone ring is mainly cytochrome *P*-450 hydroxylase. The other modification is *O*-methylation or reduction of the lactone ring.

V. Sugar Biosynthesis

Many macrolide compounds contain deoxysugar(s) and/or deoxyaminosugar(s) by glycoside linkage and these sugars are often important for the their bioactivity. From the experimental feeding of the labeled precursors, these deoxysugars are derived from glucose and belong to the 6-deoxyhexoses group. The genes involved in the biosynthesis of deoxysugar are located in the polyketide synthase cluster. These sugars are activated by nucleotide sugar derivatives in which dTDP sugar is general.

A. 2-Deoxysugar

The biosyntheses of 2-deoxysugars, L-mycarose, L-cladinose (methylated L-mycarose), and L-oleandrose, are well charaterized (Fig. 17). These 2-deoxysugars are found in antimicrobial 14- and 16-membered macrolides, erythromycin, oleandomycin, tylosin, spiramycin, niddamycin, and anthelmintic macrolide avermectin. The genes involved in the biosynthesis of L-mycarose are designated *eryBI–eryBVII* in erythromycin producer [4]. It is generally accepted that early steps in the biosynthesis of dTDP-mycarose require the activation of D-glucose-1-phosphate into dTDP-D-glucose by dTDP-D-glucose synthase, followed by dehydration at C4 and C6 by a TDP-D-glucose 4,6-dehydratase [82]. The formation of NDP-4-keto-6-deoxyhexoses from NDP-hexose is catalyzed by NDP-hexose-4,6- dehydratases. The stereochemical course of the intramolecular H-shift was

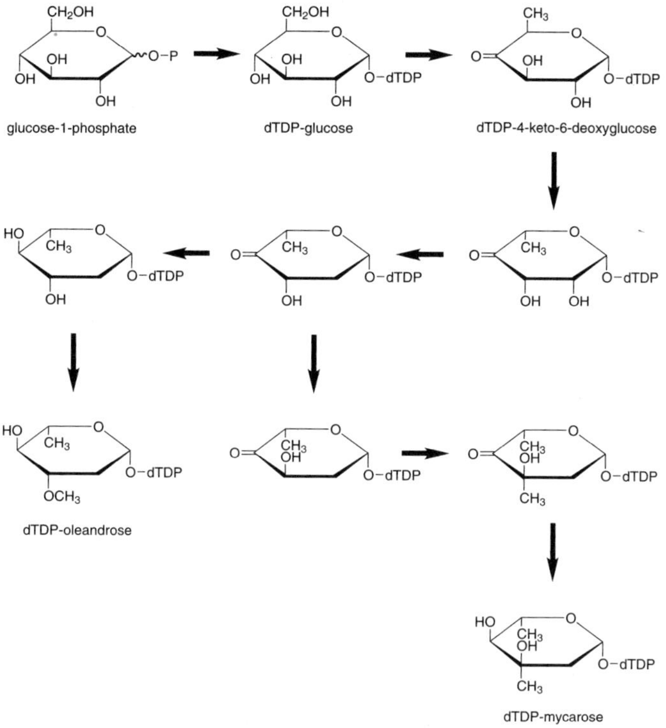

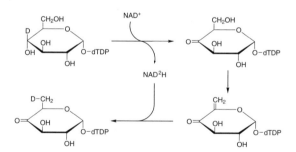

Fig. 17. Possible pathways for the formation of 2,6-dideoxysugars.

Fig. 18. Proposed mechanism for the C6 deoxygenation reaction.

determined by using (6R)- and (6S)-[4-2H, 6-3H]-D-glucose as substrates on enzymes of various organisms [83–87]. The analysis of chiral acetic acid, which was obtained by Kuhn–Roth oxidation of the resulting 6-deoxysugar nucleotide, confirmed that the hydrogen shift from C4 to C6 occurs intramolecularly and showed (Fig. 18) that the migrating hydrogen replaces the 6-OH group with inversion of configuration [88].

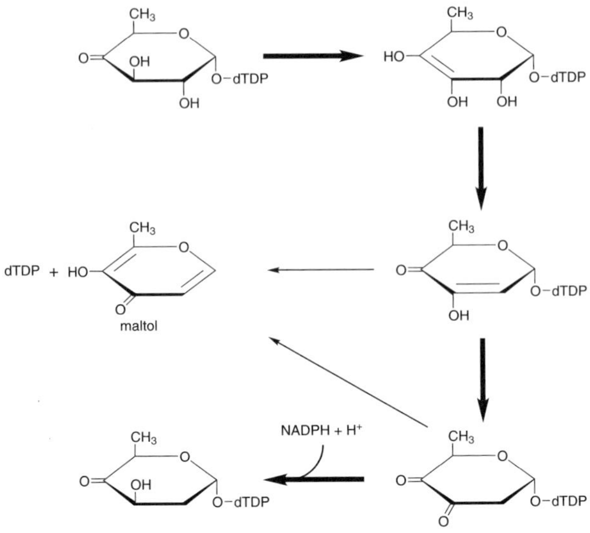

Fig. 19. Proposed mechanism for the C2 deoxygenation reaction.

The involvement in the dehydroxylation at C2 of two genes (*gra orf27* and *gra orf26*) from an aromatic polyketide (granatisin) cluster has been demonstrated [89]. These genes were overexpressed in *E. coli* as soluble proteins. Incubation of Gra Orf27 with dTDP-4-keto-6-deoxy-D-glucose gave dTDP and compound, which was identified as maltol. Evidently the enzymatic reaction produces an unstable compound, presumably dTDP-3,4-diketo-2,6-dideoxy-D-glucose or its 2,3-enol, which undergoes facile elimination of dTDP. Incubation of dTDP-4-keto-6-deoxy-D-glucose with Gra Orf26 and NAD(P)H alone gave no reaction, but addition of Gra Orf27 resulted in the almost quantitative conversion to dTDP-4-keto-2,6-dideoxy-D-glucose. These results established the first mechanism of removal of the 2-OH group in the biosynthesis of 2,6-dideoxyhexose (Fig. 19).

From the formation of dTDP-4-keto-6-deoxy-D-glucose to dTDP-L-mycarose, six proteins are involved, five of which have potential functions assigned: 3,5-epimerization (EryVII), 2,3-dehydration and reduction (EryBVI and EryBII), 3-C-methylation (EryBIII), and 4-ketoreduction (EryBIV) [4].

The seven genes are involved in the biosynthesis of dTDP-L-oleandrose from D-glucose-1-phosphate. Since the early stage of the biosynthetic pathway of L-oleandrose is extremely similar to that of erythromycin, some of the genes are very similar to EryB genes. From the formation of dTDP-4-keto-6-deoxy-D-glucose to dTDP-L-oleandrose in avermectin biosynthesis, five proteins are involved, in which these proteins have potential functions assigned: 3,5-epimerization (AveV), 2,3-dehydration and reduction (AveBVI and AveBVII), 4-ketoreduction (AveBIV), and 3-*O*-methylation (AveBVII) [28, 59].

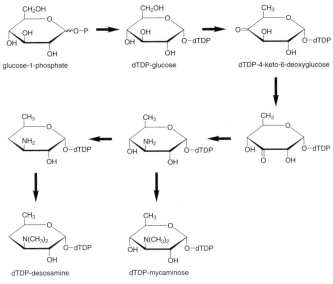

Fig. 20. Possible pathway for the formation of amino sugars.

B. Deoxyaminosugar

The amino sugars D-desosamine and D-mycaminose are found in 12-, 14- and 16-membered (D-desosamine) and 16-membered macrolide antibiotics (D-mycaminose). In polyene macrolide compounds, a different type of amino sugar, mycosamin, is linked to the lactone ring. dTDP-D-Desosamine and dTDP-D-mycaminose also require the activation of D-glucose-1-phosphate into dTDP-D-glucose, followed by dehydration at C4 and C6 (Fig. 20).

In the erythromycin biosynthesis, five genes have been involved in the formation of dTDP-D-desosamine. Starting from dTDP-4-keto-6-deoxy-D-glucose, EryCII could be a dTDP-4-keto-6-deoxy-D-glucose isomerase, EryCI would be a 3-aminotransferase, EryCIV a 3,4-dehydratase, EryCV a 3,4-reductase, and EryCVI a 3-N-methyltransferase [4]. A similar five genes are also involved in the pikromycin biosynthesis from dTDP-4-keto-6-deoxy-D-glucose: DesVIII could be a dTDP-4-keto-6-deoxy-D-glucose isomerase, DesV would be a 3-aminotransferase, DesI a 3,4-dehydratase, DesII a 3,4-reductase, and DesVI a 3-N-methyltransferase [90, 91]. The D-desosamine is a 4,6-dideoxy sugar and the replacement of the 4-oxygen by hydrogen could be catalyzed by two enzymes. For the formation of 4-deoxysugar, cofactor is involved in the catalytic reaction. First, a pyridoxamine-phosphate (PMP)-linked dehydratase attacks the 4-keto group to yield the imine followed by a 1,4-elimination of water. The resulting dTDP-6-deoxy-D-hexosene intermediate is then reduced by a reductase. Cleavage of the PMP-enzyme yields a dTDP-4,6-dideoxyhexose intermediate (Fig. 21).

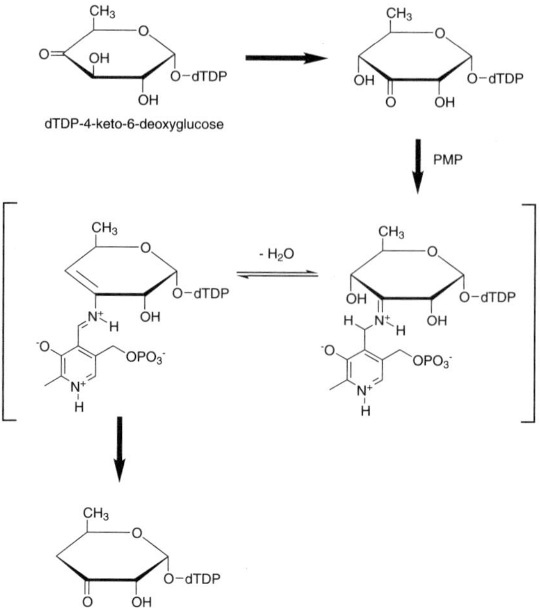

Fig. 21. Proposed mechainsm for the C4 deoxygenation reaction.

Tylosin contains a 16-membered lactone to which three deoxysugar moieties are attached. A disaccharide D-mycaminosyl-D-mycarose is linked at C5 and a D-mycinose at C23. Several genes have been cloned that encode deoxysugar biosynthetic enzymes; most of them are involved in dTDP-D-mycaminose biosynthesis [48, 49]. TylA1 and TylA2 correspond to the dTDP-D-glucose synthase and dTDP-4,6-dehydratase, respectively. A putative isomerase not yet identified acts on dTDP-4-keto-6-deoxy-D-glucose, and then aminotransferase (TylB) and C-methyltransferase (TylM1) to generate the final dTDP-D-mycaminose.

The biosynthetic genes involving amino sugar in polyene macrolide compounds, amphotericin B, nystatin, and pimaricin, are somewhat different from that of D-desosamine and D-mycaminose [30, 61, 69]. The early steps in the biosynthesis of NDP-D-mycosamine require the activation of D-mannose-1-phosphate into NDP-mannose by NDP-D-mannose synthetase, followed by dehydration at C4 and C6 by an NDP-D-mannose 4,6-dehydratase. NysDIII and PimJ in nystatin and pimaricin biosynthetic genes, respectively, more closely resemble the GDP-D-mannose-4,6-dehydratases than the dTDP-D-glucose-4,6-dehydratases, suggesting that the D-mycosamine biosynthetic pathway in nystatin and pimaricin producers differs from those of other macrolide deoxysugar moieties. A putative isomerase not yet found in the cluster acts on GDP-4-keto-6-deoxy-D-mannose to convert GDP-3-keto-6-deoxy-D-mannose and then aminotransferase (NysDII that resembles aminotransferase and PimC?) to generate the final GDP-D-mycosamine.

The transfer of dTDP sugars into the lactone moiety to form glycoside linkage is mediated by individual glycosyltransferase as follows: EryBV [4] and TylCV [48] for L-mycarose, EryCIII [4] and DesVII [51] for D-desosamine, TylM2 [49] for D-mycaminose, and AveBI [28, 59] for L-oleandrose, respectively. The avermectins are pentacyclic polyketide-derived compounds linked to a disaccharide of the methylated deoxy sugar (L-oleandrosyl-L-oleandrose). Only one glycosyltransferase, AveBI, is involved in the transfer of dTDP-L-oleandrose, and the disruption of its gene causes the production of avermectin aglycons. This suggests that AveBI has relatively wide substrate specificity in which the glycosyltransferase catalyzes to transfer L-oleandrose into the aglycons and their monosaccharides [59]. The transfer of GDP-D-mycosamine into the pimaricin or nystatin macrolactone is also mediated by specific glycosyltransferases, PimK or NysDI, respectively [30, 61].

VI. Genetic Manipulation of PKS Genes

Current genetic manipulation must be available to introduce a wide variety of specified genetic changes (mutagenesis) into PKS genes. In principle, the domain mutagenesis of a modular PKS can be used to achieve loss of function, alteration of substrate specificity, or even gain of function. The theoretical number of ways in which a functionally unique sequence of domains and/or modules can be constructed is astronomical. However, at least two potential barriers have to be surmounted before the biosynthetic relevance of this genetic capability can be realized. First, strategies have to be developed for engineering hybrid PKSs without deleteriously affecting the protein–protein interactions required for intra- and intermodule chain processing. Second, the intrinsic tolerance of individual active sites in each domain must be determined in order to design modules with functionally compatible modules. The experimental observations in the erythromycin PKS are good references [3, 92–103].

A. Loss of Function

The three domains involved in β-carbonyl processing function in the hierarchical order KR-DH-ER. Thus, if all three are present in a given domain, loss of KR function obviates the requirement for the DH and ER functions, and loss of the DH domain renders the otherwise functional ER domain unnecessary. Loss of the KR domain results in the appearance of a keto group in place of the original hydroxyl. Loss of the DH domain leaves the hydroxyl side chain in place of the methylene function, if a functional ER domain is present within the affected module; if an ER domain is not present, the olefin formerly present at the site in the polyketide is replaced by a hydroxyl residue.

Because the number of condensations, each of which is determined by a single module within a PKS, determines the length of the acyl chain, removing or adding modules can change the chain length. Only removal of modules, resulting in the shortening of the polyketide chain, has been demonstrated to accumulate 12-membered macrolide aglycones using erythromycin PKSs [95].

The formation of avermectin aglycons involves KRs that resemble type I and II FAS. The C5-keto group is reduced by a monofunctional KR (AveF) that is not within module 11 of the AVES 4 (Fig. 10). The mutants defective in C5 ketoreductase gene, *aveF*, which were obtained by introducing an insertion mutation in the promoter region of *aveD-aveF* operon produced 5-oxoavermectins [104].

B. Gain of Function

Structural changes resulting from gain of β-carbonyl-processing functions in PKS modules are the reverse of those mentioned above for corresponding losses of function. Replacement of a hydroxyl side chain with a methylene group in the polyketide chain through gain of DH function cannot occur without the presence of the ER function within the affected module.

C. Alternation of Function

Since the structure of the side chain at a given site in the polyketide chain is determined by the structure of the extender unit incorporated at the corresponding site in the nascent chain, changes in the side chains that appear in the final product will result from replacement of one extender unit for another. Because the AT domains determine the nature of the cognate extender unit incorporated (described in Section III), AT swapping results in corresponding side-chain changes.

The starter unit, acetyl-CoA, propionyl-CoA, or the other acyl-CoA, is also determined by an AT (or ligase in rapamycin and FK520 PKSs) function present at the first functional domain of the N terminus of the PKS polypeptide. Domain swaps at these sites can result in changes in the starting side chain.

References

1. O'Hagan, D. (1991). "The Polyketide Metabolites." Ellis Horwood, Chichester, U.K.
2. Cortes, J., Haydock, S. F., Robert, G. A., Bevitt, D. J., and Leadlay, P. F. (1990). An unusually large multifunctional polypeptide in the erythromycin-producing polyketide synthase of *Saccharopolyspora erythraea*. *Nature* **348**, 176–178.
3. Donadio, S., Staver, M. J., McAlpine, J. B., Swanson, S. J., and Katz, L. (1991). Modular organization of genes required for complex polyketide biosynthesis. *Science* **252**, 675–679.
4. Summers, R. G., Donadio, S., Staver, M. J., Wendt-Pienkowski, E., Hutchinson, C. R., and Katz, L. (1997). Sequencing and mutagenesis of genes from the erythromycin biosynthetic gene

cluster of *Saccharopolyspora erythraea* that are involved in L-mycarose and D-desosamine production. *Microbiology* **143**, 3251–3262.
5. Weber, J. M., Leung, J. O., Swanson, S. J., Idler, K. B., and McAlpine, J. B. (1991). An erythromycin derivative produced by targeted gene disruption in *Saccharopolyspora erythraea*. *Science* **252**, 114–117.
6. Haydock, S. F., Dowson, J. A., Dhillon, N., Roberts, G. A., Cortes, J., and Leadlay, P. F. (1991). Cloning and sequence analysis of genes involved in erythromycin biosynthesis in *Saccharopolyspora erythraea*: Sequence similarities between EryG and a family of S-adenosylmethionine-dependent methyltransferases *Mol. Gen. Genet.* **230**, 120–128.
7. Stassi, D. L., Donadio, S., Staver, M. J., and Katz, L. (1993). Identification of a *Saccharopolyspora erythraea* gene required for the final hydroxylation step in erythromycin biosynthesis *J. Bacteriol.* **175**, 182–189.
8. Lambalot, R. H., Cane, D. E., Aparicio, J. J., and Katz, L. (1995). Overproduction and characterization of the erythromycin C-12 hydroxylase, EryK. *Biochemistry* **34**, 1858–1866.
9. Omura, S. (1976). The antibiotic cerulenin, a novel tool for biochemistry as an inhibitor of fatty acid synthesis. *Bacteriol. Rev.* **40**, 681–697.
10. Vance, D., Goldberg, I., Mitsuhashi, O., Bloch, K., Omura, S., and Nomura, S. (1972). Inhibition of fatty acid synthetases by the antibiotic cerulenin. *Biochem. Biophys. Res. Commun.* **48**, 649–656.
11. Omura, S., and Takeshima, H. (1975). Inhibition of the biosynthesis of leucomycin, a macrolide antibiotic, by cerulenin. *J. Biochem.* **75**, 193–195.
12. Ohno, H., Ohno, T., Awaya, J., and Omura, S. (1975). Inhibition of 6-methylsalicylic acid synthesis by the antibiotic cerulenin. *J. Biochem.* **78**, 1149–1152.
13. Funabashi, H., Kawaguchi, A., Tomoda, H., Omura, S., Okuda, S., and Iwasaki, S. (1989). Binding site of cerulenin in fatty acid synthetase *J. Biochem.* **105**, 751–755.
14. Wakil, S. J. (1989). Fatty acid synthase, a proficient multifunctional enzyme. *Biochemistry* **28**, 4523–4530.
15. Smith, S. (1994). The animal fatty acid synthase: One gene, one polypeptide, seven enzymes. *FASEB* **8**, 1248–1259.
16. Fulco, A. J. (1983). Fatty acid metabolism in bacteria. *Prog. Lipid. Res.* **22**, 133–160.
17. Omura, S., Nakagawa, A., Neszmelyi, A., Gero, S. D., Sepulchre, A. M., Piriou, F., and Lukacs, G. (1975). Carbon-13 nuclear magnetic resonance spectral analysis of 16-membered macrolide antibiotics. *J. Am. Chem. Soc.* **97**, 4001–4009.
18. Omura, S., Nakagawa, A., Takeshima, H., Atusmi, K., Miyazawa, J., Piriou, F., and Lukacs, G. (1975). Biosynthetic studies using ^{13}C enriched precursors on the 16-membered macrolide antibiotic leucomycin A3. *J. Am. Chem. Soc.* **97**, 6600–6602.
19. Yue, D., Duncan, J. S., Yamamoto, Y., and Hutchinson, C. R. (1987). Macrolide biosynthesis. Tylactone formation involves the processive addition of three carbon units. *J. Am. Chem. Soc.* **109**, 1253–1255.
20. Cane, D. E., and Yang, C.-C. (1987). Macrolide biosynthesis. 4. Intact incorporation of a chain-elongation intermediate into erythromycin. *J. Am. Chem. Soc.* **109**, 1255–1257.
21. Bevitt, D. J., Cortes, J., Haydock, S. F., and Leadlay, P. F. (1992). 6-Deoxyerythronolide-B synthase 2 from *Saccharopolyspora erythraea*. Cloning of the structural gene, sequence analysis and inferred domain structure of the multifunctional enzyme. *Eur. J. Biochem.* **204**, 39–49.
22. Caffrey, P., Bevitt, D. J., Staunton, J., and Leadlay, P. F. (1992). Identification of DEBS 1, DEBS 2 and DEBS 3, the multienzyme polypeptides of the erythromycin-producing polyketide synthase from *Saccharopolyspora erythraea*. *FEBS Lett.* **304**, 225–228.
23. Donadio, S., and Katz, L. (1992). Identification of a *Saccharopolyspora erythraea* gene required for the final hydroxylation step in erythromycin biosynthesis. *Gene* **111**, 51–60.

24. Aparicio, J. F., Molar, I., Schwecke, T., König, A., Haydock, S. F., Khaw, L. E., Staunton, J., and Leadlay, P. F. (1996). Organization of the biosynthetic gene cluster for rapamycin in *Streptomyces hygroscopicus*: Analysis of the enzymatic domains in the modular polyketide synthase. *Gene* **169**, 9–16.
25. Bisang, C., Long, P. F., Cortes, J., Westcott, J., Crosby, J., Matharu, A. L., Cox, R. J., Simpson, T. J., Staunton, J., and Leadlay, P. F. (1999). A chain initiation factor common to both modular and aromatic polyketide synthases. *Nature* **401**, 502–505.
26. Ruan, X., Stassi, D., Lax, S. A., and Katz, L. (1997). A second type-I PKS gene cluster isolated from *Streptomyces hygroscopicus* ATCC 29253, a rapamycin-producing strain. *Gene* **203**, 1–9.
27. Kakavas, S. J., Katz, L., and Stassi, D. (1997). Identification and characterization of the niddamycin polyketide synthase genes from *Streptomyces caelestis*. *J. Bacteriol.* **179**, 7515–7522.
28. Ikeda, H., Nonomiya, T., Usami, M., Ohta, T., and Omura, S. (1999). Organization of the biosynthetic gene cluster for the polyketide anthelmintic macrolide avermectin in *Streptomyces avermitilis*. *Proc. Natl. Acad. Sci. USA* **96**, 9509–9514.
29. Wu, K., Chung, L., Revill, W. P., Katz, L., and Reeves, C. D. (2000). The FK520 gene cluster of *Streptomyces hygroscopicus* var. *ascomyceticus* (ATCC 14891) contains genes for biosynthesis of unusual polyketide extender units. *Gene* **251**, 81–90.
30. Aparicio, J. F., Fouces, R., Mendes, M. V., Olivera, N., and Martin, J. F. (2000). A complex multienzyme system encoded by five polyketide synthase genes is involved in the biosynthesis of the 26-membered polyene macrolide pimaricin in *Streptomyces natalensis*. *Chem. Biol.* **7**, 895–905.
31. Amy, C. M., Witkowski, A., Naggert, J., Williams, B., Randhawa, Z., and Smith, S. (1989). Molecular cloning and sequencing of cDNAs encoding the entire rat fatty acid synthase. *Proc. Natl. Acad. Sci. USA* **86**, 3114–3118.
32. Witkowski, J., Rangan, V. S., Randhawa, Z. I., Amy, C. M., and Smith, S. (1991). Structural organization of the multifunctional animal fatty-acid synthase. *Eur. J. Biochem.* **198**, 571–579.
33. Scrutton, N. S., Berry, A., and Perham, R. N. (1990). Redesign of the coenzyme specificity of a dehydrogenase by protein engineering. *Nature* **343**, 38–43.
34. Swan, D. G., Rodriguez, A. M., Vilches, C., Mendes, C., and Salas, J. A. (1994). Characterisation of a *Streptomyces antibioticus* gene encoding a type I polyketide synthase which has an unusual coding sequence. *Mol. Gen. Genet.* **242**, 358–362.
35. Schwecke, T., Aparicio, J. F., Molnar, I., König, A., Khaw, L. E., Haydock, S. F., Oliynyk, M., Cafferey, P., Cortes, J., Lester, J. B., Bohn, G. A., Staunton, J., and Leadlay, P. F. (1995). The biosynthetic gene cluster for the polyketide immunosuppressant rapamycin. *Proc. Natl. Acad. Sci. USA* **92**, 7839–7843.
36. Tang, L., Yoon, Y. J., Choi, C.-Y., and Hutchinson, C. R. (1998). Characterization of the enzymatic domains in the modular polyketide synthase involved in rifamycin B biosynthesis by *Amycolatopsis mediterranei*. *Gene* **216**, 255–165.
37. Yu, T. W., Shen, Y., Doi-Katayama, Y, Tang, L., Park, C., and Hutchinson, C. R. (1999). Direct evidence that the rifamycin polyketide synthase assembles polyketide chains processively. *Proc. Natl. Acad, Sci. USA* **96**, 9051–9056.
38. Heathcote, M. L., Staunton, J., and Leadlay, P. F. (2001). Role of type II thioesterases: Evidence for removal of short acyl chains produced by aberrant decarboxylation of chain extender units. *Chem. Biol.* **8**, 207–220.
39. Thompson, C. J., Ward, J. M., and Hopwood, D. A. (1982). Biochemical characterization of resistance determinants cloned from antibiotic-producing streptomycetes. *J. Bacteriol.* **151**, 668–677.
40. Stanzak, R., Matsushima, P., Baltz, R. H., and Rao, R. N. (1986). Cloning and expression in *Streptomyces lividans* of clustered erythromycin biosynthesis genes from *Streptomyces erythreus*. *Bio/Technology* **4**, 229–232.

41. Dhillon, N., Hale, R. S., Cortes, J., and Leadlay, P. F. (1989). Molecular characterization of a gene from *Saccharopolyspora erythraea* (*Streptomyces erythraeus*) which is involved in erythromycin biosynthesis. *Mol. Microbiol.* **3**, 1405–1414.
42. Vara, J., Lewandowska-Skarbek, M., Wang, Y.-G., Donadio, S., and Hutchinson, C. R. (1989). Cloning of genes governing the deoxysugar portion of the erythromycin biosynthesis pathway in *Saccharopolyspora erythraea* (*Streptomyces erythreus*). *J. Bacteriol.* **171**, 5872–5881.
43. Paulus, T. J., Tuan, J. S., Luebke, V. E., Main, G. T., SeWitt, J. P., and Katz, L. (1990). Mutation and cloning of *eryG*, the structural gene for erythromycin *O*-methyltransferase from *Saccharopolyspora erythraea*, and expression of *eryG* in *Escherichia coli*. *J. Bacteriol.* **172**, 2541–2546.
44. Tuan, J. S., Weber, J. M., Staver, M. J., Leung, J. O., Donadio, S., and Katz, L. (1990). Cloning of genes involved in erythromycin biosynthesis from *Saccharopolyspora erythraea* using a novel actinomycete–*Escherichia coli* cosmid. *Gene* **90**, 21–29.
45. Volchegursky, Y., Hu, Z., Katz, L., and McDaniel, R. (2000). Biosynthesis of the antiparasitic agent megalomicin: Transformation of erythromycin to megalomicin in *Saccharopolyspora erythraea*. *Mol. Microbiol.* **37**, 752–762.
46. Fishman, S. E., Cox, K., Larson, J. L., Reynolds, P. A., Seno, E. T., Yeh, W.-K., van Frank, R., and Hershberger, C. L. (1987). Cloning genes for the biosynthesis of a macrolide antibiotic. *Proc. Natl. Acad. Sci. USA* **84**, 8248–8252.
47. Kuhstoss, S., Huber, M., Turner, J. R., Paschal, J. W., and Rao, R. N. (1996). Production of a novel polyketide through the construction of a hybrid polyketide synthase. *Gene* **183**, 231–236.
48. Merson-Davies, L. A., and Cundliffe, E. (1994). Analysis of five tylosin biosynthetic genes from the *tylIBA* region of the *Streptomyces fradiae* genome. *Mol. Microbiol.* **13**, 349–355.
49. Gandecha, A. R., Large, S. L., and Cundliffe, E. (1997). Analysis of four tylosin biosynthetic genes from the *tylLM* region of the *Streptomyces fradiae* genome. *Gene* **184**, 197–203.
50. Geistlich, M., Losick, R., Turner, J. R., and Rao, R. N. (1992). Characterization of a novel regulatory gene governing the expression of a polyketide synthase gene in *Streptomyces ambofaciens*. *Mol. Microbiol.* **6**, 2019–2019.
51. Xue, Y., Zhao, L., Liu, H.-W., and Sherman, D. (1998). A gene cluster for macrolide antibiotic biosynthesis in *Streptomyces venezuelae*: Architecture of metabolic diversity. *Proc. Natl. Acad. Sci. USA* **95**, 12,111–12,116.
52. Xue, Y., and Sherman, D. (2000). Alternative modular polyketide synthase expression controls macrolactone structure. *Nature* **403**, 571–575.
53. Ikeda, H., Kotaki, H., and Omura, S. (1987). Genetic studies of avermectin biosynthesis in *Streptomyces avermitilis*. *J. Bacteriol.* **169**, 5615–5621.
54. MacNeil, D. J., Occi, J. L., Gewain, K. M., and MacNeil, T. (1992). Complex organization of the *Streptomyces avermitilis* genes encoding the avermectin polyketide synthase. *Gene* **115**, 119–125.
55. MacNeil, T., Gewain, K. M., and MacNeil, D. J. (1993). Deletion analysis of the avermectin biosynthetic genes of *Streptomyces avermitilis* by gene cluster displacement. *J. Bacteriol.* **175**, 2552–2563.
56. Ikeda, H., and Omura, S. (1995). Control of avermectin biosynthesis in *Streptomyces avermitilis* for the selective production of a useful component *J. Antibiot.* **48**, 669–682.
57. Ikeda, H., and Omura, S. (1997). Avermectin biosynthesis. *Chem. Rev.* **97**, 2591–2609.
58. Ikeda, H., Wang, R.-L., Ohta, T., and Omura, S. (1998). Cloning of the gene encoding avermectin B 5-*O*-methyltransferase in avermectin-producing *Streptomyces avermitilis*. *Gene* **206**, 175–180.
59. Ikeda, H. (1999). Genetic analysis of biosynthesis of polyketide anthelmintic macrolide avermectin in *Streptomyces avermitilis*. *Actinomycetology* **13**, 94–112.

60. MacNeil, D. J., Occi, J. L., Gewain, K. M., MacNeil, T., Gibbons, P. H., Foor, F., and Morin, N. (1993). In "Industrial Microorganisms: Basic and Applied Molecular Genetics" (R. H. Baltz, G. D. Hegeman, and P. L. Skatrud, eds.), pp. 245–256. American Society for Microbiology, Washington, D.C.
61. Brautaset, T., Sekurova, O. N., Sletta, H., Ellingsen, T. E., Strøm, A. R., Valla, S., and Zotchev, S. B. (2000). Biosynthesis of the polyene antifungal antibiotic nystatin in *Streptomyces noursei* ATCC 11455: Analysis of the gene cluster and deduction of the biosynthetic pathway. *Chem. Biol.* **7**, 395–403.
62. Aparicio, J. F., Colina, A. J., Ceballos, E., and Martin, J. F. (1999). The biosynthetic gene cluster for the 26-membered ring polyene macrolide pimaricin. A new polyketide synthase organization encoded by two subclusters separated by functionalization genes. *J. Biol. Chem.* **274**, 10,133–10,139.
63. Omura, S., Ikeda, H., Ishikawa, J., Hanamoto, A., Takahashi, C., Shinose, M., Takahashi, Y., Horikawa, H., Nakazawa, H., Osonoe, T., Kikuchi, H., Shiba, T., Sakaki, Y., and Hattori, M. (2001). Genome sequence of an industrial microorganism. *Streptomyces avermitilis*: Deducing the ability of producing secondary metabolites. *Proc. Natl. Acad. Sci. USA* **98**, 12,215–12,220.
64. Schupp, T., Toupet, C., Engel, N., and Goff, S. (1998). Cloning and sequence analysis of the putative rifamycin polyketide synthase gene cluster from *Amycolatopsis mediterranei*. *FEMS Microbiol. Lett.* **159**, 201–207.
65. August, P. R., Tang, L., Yoon, Y. J., Ning, S., Müller, R., Yu, T.-W., Taylor, M., Hoffmann, D., Kim, C.-G., Zhang, X., Hutchinson, C. R., and Floss, H. G. (1998). Biosynthesis of the ansamycin antibiotic rifamycin: Deductions from the molecular analysis of the *rif* biosynthetic gene cluster of *Amycolatopsis mediterranei* S699. *Chem. Biol.* **5**, 69–79.
66. Molnar, I., Aparicio, J. F., Haydock, S. F., Khaw, L. E., Schwecke, T., König, A., Staunton, J., and Leadlay, P. F. (1996). Organisation of the biosynthetic gene cluster for rapamycin in *Streptomyces hygroscopicus*: Analysis of genes flanking the polyketide synthase. *Gene* **169**, 1–7.
67. Motamedi, H., and Shafiee, A. (1998). The biosynthetic gene cluster for the macrolactone ring of the immunosuppressant FK506. *Eur. J. Biochem.* **256**, 528–534.
68. Hu, Z., Bao, K., Zhou, X., Zhou, Q., Hopwood, D. A., Kieser, T., and Deng, Z. (1994). Repeated polyketide synthase modules involved in the biosynthesis of a heptaene macrolide by *Streptomyces* sp. FR-008. *Mol. Microbiol.* **14**, 163–172.
69. Caffrey, P., Lynch, S., Flood, E., Finnan, S., and Oliynyk, M. (2001). Amphotericin biosynthesis in *Streptomyces nodosus*: Deductions from analysis of polyketide synthase and late genes. *Chem. Biol.* **8**, 713–723.
70. Molnar, I., Schupp, T., Ono, M., Zinkle, R. E., Milamow, M., Nowak-Thompson, B., Engle, N., Toupet, C., Stratmann, A., Cyr, D. D., Gorlach, J., Mayo, J. M., Hu, A., Goff, S., Schmid, J., and Ligon, J. M. (2000). The biosynthetic gene cluster for the microtubule-stabilizing agents epothilones A and B from *Sorangium cellulosum* So ce90. *Chem. Biol.* **7**, 97–109.
71. Tang, L., Shah, S., Chung, L., Carney, J., Katz, L., Khosla, C., and Julien, B. (2000). Cloning and heterologous expression of the epothilone gene cluster. *Science* **287**, 640–642.
72. Beck, J., Ripka, S., Signer, A., Schiltz, E., and Schweizer, E. (1990). The multifunctional 6-methylsalicylic acid synthase gene of *Penicillium patulum*. Its gene structure relative to that of other polyketide synthases. *Eur. J. Biochem.* **192**, 487–498.
73. Hendrickson, L., Davis, C. R., Roache, C., Nguyen, D. K., Aldrich, T., McAda, P. C., and Reevers, C. D. (1999). Lovastatin biosynthesis in *Aspergillus terreus*: Characterization of blocked mutants, enzyme activities and a multifunctional polyketide synthase gene. *Chem. Biol.* **6**, 429–439.
74. Mayorga, M. E., and Timberlake, W. E. (1992). The developmentally regulated *Aspergillus nidulans* wA gene encodes a polypeptide homologous to polyketide and fatty acid synthases. *Mol. Gen. Genet.* **235**, 205–212.

75. Takano, Y., Kubo, Y., Shimizu, K., Mise, K., Okuno, T., and Furusawa, I. (1995). Structural analysis of PKS1, a polyketide synthase gene involved in melanin biosynthesis in *Colletotrichum lagenarium*. *Mol. Gen. Genet.* **249**, 162–167.
76. Yu, J.-H., and Leonard, T. J. (1995). Sterigmatocystin biosynthesis in *Aspergillus nidulans* requires a novel type I polyketide synthase. *J. Bacteriol.* **177**, 4792–4800.
77. Feng, G. H., and Leonard, T. J. (1995). Characterization of the polyketide synthase gene (*pksL1*) required for aflatoxin biosynthesis in *Aspergillus parasiticus*. *J. Bacteriol.* **177**, 6246–6254.
78. Proctor, R. H., Desjardins, A. E., Plattner, R. D., and Hohn, T. M. (1999). A polyketide synthase gene required for biosynthesis of fumonisin mycotoxins in *Gibberella fujikuroi* mating population A. *Fungal Genet. Biol.* **27**, 100–112.
79. Silakowski, B., Nordsiek, G., Kunze, B., Blöcker, H., and Müller, R. (2001). Novel features in a combined polyketide synthase/non-ribosomal peptide synthetase: The myxalamid biosynthetic gene cluster of the myxobacterium *Stigmatella aurantiaca* Sga15. *Chem. Biol.* **8**, 59–69.
80. Nishizawa, T., Ueda, A., Asayama, M., Fujii, K., Harada, K., Ochi, K., and Shirai, M. (2000). Polyketide synthase gene coupled to the peptide synthetase module involved in the biosynthesis of the cyclic heptapeptide microcystin. *J. Biochem.* **127**, 779–789.
81. Silakowski, B., Schairer, H. U., Ehret, H., Kunze, B., Weinig, S., Nordsiek, G., Brandt, P., Blocker, H., Hofle, G., Beyer, S., and Müller, R. (1999). New lessons for combinatorial biosynthesis from myxobacteria. The myxothiazol biosynthetic gene cluster of *Stigmatella aurantiaca* DW4/3-1. *J. Biol. Chem.* **274**, 37,391–37,399.
82. Vara, J. A., and Hutchinson, C. R. (1988). Purification of an inducible L-valine dehydrogenase of *Streptomyces coelicolor* A3(2). *J. Biol. Chem.* **263**, 14992–14995.
83. Brahmbhatt, H. N., Quigley, N. B., and Reeves, P. R. (1986). Cloning part of the region encoding biosynthetic enzymes for surface antigen (*O*-antigen) of *Salmonella typhimurium*. *Mol. Gen. Genet.* **203**, 172–176.
84. Brahmbhatt, H. N., Wyk, P., Quigley, N. B., and Reeves, P. R. (1988). Complete physical map of the *rfb* gene cluster encoding biosynthetic enzymes for the *O* antigen of *Salmonella typhimurium* LT2. *J. Bacteriol.* **170**, 98–102.
85. Jiang, X. M., Neal, B., Santiago, F., Lee, S. J., Romana, L. K., and Reeves, P. R. (1991). Structure and sequence of the *rfb* (*O* antigen) gene cluster of *Salmonella* serovar *typhimurium* (strain LT2). *Mol. Microbiol.* **5**, 695–713.
86. Wyk, P., and Reeves, P. R. (1989). Identification and sequence of the gene for abequose synthase, which confers antigenic specificity on group B salmonellae: Homology with galactose epimerase. *J. Bacteriol.* **171**, 5687–5693.
87. Wang, L., Romana, L. K., and Reeves, P. R. (1992). Molecular analysis of a *Salmonella enterica* group E1 *rfb* gene cluster: *O* antigen and the genetic basis of the major polymorphism. *Genetics* **130**, 429–443.
88. Snipes, C. E., Brillinger, G. U., Sellers, L., Mascaro, L., and Floss, H. G. (1977). Stereochemistry of the dTDP-glucose oxidoreductase reaction. *J. Biol. Chem.* **252**, 8113–8117.
89. Lee, J. J., Lee, J. P., Keller, P. J., Cottrell, C. E., Chang, C., Zähner, H., and Floss, H. G. (1986). Further studies on the biosynthesis of chlorothricin. *J. Antibiot.* **39**, 1123–1134.
90. Zhao, L., Sherman, D. H., and Liu, H.-W. (1998). *J. Am. Chem. Soc.* **120**, 9375.
91. Zhao, L., Que, N. L. S., Xue, Y., Sherman, D. H., and Liu, H.-W. (1998). *J. Am. Chem. Soc.* **120**, 12,159.
92. Donadio, S., McAlpine, J. B., Sheldon, P. J., Jackson, M., and Katz, L. (1993). An erythromycin analog produced by reprogramming of polyketide synthesis. *Proc. Natl. Acad. Sci. USA* **90**, 7119–7123.
93. Donadio, S., Staver, M. J., McAlpine, J. B., Swanson, S. J., and Katz, L. (1992). Biosynthesis of the erythromycin macrolactone and a rational approach for producing hybrid macrolides. *Gene* **115**, 97–103.

94. Ruan, X., Pereda, A., Stassi, D. L., Zeidner, D., Summers, R. G., Jackson, M., Shivakumar, A., Kakavas, S., Staver, M. J., Donadio, S., and Katz, L. (1997). Acyltransferase domain substitutions in erythromycin polyketide synthase yield novel erythromycin derivatives. *J. Bacteriol.* **179**, 6416–6425.
95. Kao, C. M., Luo, G., Katz, L., Cane, D. E., and Khosla, C. (1995). Manipulation of macrolide ring size by directed mutagenesis of a modular polyketide synthase. *J. Am. Chem. Soc.* **117**, 9105–9106.
96. Kao, C. M., Luo, G., Katz, L., Cane, D. E., and Khosla, C. (1996). Engineered biosynthesis of structurally diverse tetraketides by a trimodular peptide synthase. *J. Am. Chem. Soc.* **118**, 9184–9185.
97. Stassi, D. L., Kakavas, S. J., Reynolds, K. A., Gunawardana, G., Swanson, S., Zeidner, D, Jackson, M., Liu, H., Bunko, A., and Katz, L. (1998). Ethyl-substituted erythromycin derivatives produced by directed metabolic engineering. *Proc. Natl. Acad. Sci. USA* **95**, 7305–7309.
98. Bohm, I., Holzbaur, I. E., Hanefeld, U., Cortes, J., Staunton, J., and Leadlay, P. F. (1998). Engineering of a minimal modular polyketide synthase, and targeted alteration of the stereospecificity of polyketide chain extension. *Chem. Biol.* **5**, 407–412.
99. McDaniel, R., Thamchaipenet, A., Gustafsson, C., Fu, H., Betlach, M., and Ashley, G. (1999). Multiple genetic modifications of the erythromycin polyketide synthase to produce a library of novel "unnatural" natural products. *Proc. Natl. Acad. Sci. USA* **96**, 1846–1851.
100. McDaniel, R., Kao, C. M., Hwang, S. J., and Khosla, C. (1997). Engineered intermodular and intramodular polyketide synthase fusions. *Chem. Biol.* **4**, 667–674.
101. McDaniel, R., Kao, C. M., Fu, H., Hevezi, P., Gustafsson, C., Betlach, M., Ashley, G., Cane, D. E., and Khosla, C. (1997). Gain-of-function mutagenesis of a modular polyketide synthase. *J. Am. Chem. Soc.* **119**, 4309–4310.
102. Kao, C. M., McPherson, M., McDaniel, R., Fu, H., Cane, D. E., and Khosla, C. (1997). Gain of function mutagenesis of the erythromycin polyketide synthase. 2. Engineered biosynthesis of an eight-membered ring tetraketide lactone. *J. Am. Chem. Soc.* **119**, 11,339–11,340.
103. Salah-Bey, K., Doumith, M., Michel, J. M., Haydock, S., Cortes, J., Leadlay, P. F., and Raynal, M. C. (1998). Targeted gene inactivation for the elucidation of deoxysugar biosynthesis in the erythromycin producer *Saccharopolyspora erythraea*. *Mol. Gen. Genet.* **257**, 542–553.
104. Ikeda, H., Takada, Y., Pang, C.-H., Matsuzaki, K., Tanaka, H., and Omura, S. (1995). Direct production of 5-oxo derivatives of avermectins by a recombinant strain of *Streptomyces avermitilis*. *J. Antibiot.* **48**, 95–97.

Chapter 6

Pharmacokinetics and Metabolism of Macrolides

YOSHIRO KOHNO
Research Center, Taisho Pharmaceutical Co., Ltd.
Saitama, Japan

I.	Introduction	327
II.	Pharmacokinetics and Metabolism	328
	A. Erythromycin	329
	B. Roxithromycin	330
	C. Clarithromycin	334
	D. Azithromycin	343
	E. Telithromycin (HMR 3647)	346
	F. ABT-773	347
III.	Drug Interaction	350
	A. Roxithromycin	352
	B. Clarithromycin	352
	C. Azithromycin	353
	D. Telithromycin	353
IV.	Concluding Remarks	354
	References	354

I. Introduction

Since the development of erythromycin in the 1950s, various other macrolides, most with improved pharmacokinetics, have been marketed throughout the world. The main rationale for the research into newer macrolides was to remedy some of the adverse features of erythromycin: its instability at acidic pH, which is principally responsible for its poor intestinal absorption; its short elimination half-life; its low degree of tissue and cellular penetration; its poor digestive tolerance; and its interactions with a number of other drugs [1]. New macrolides in wide use with clinically significant advances include clarithromycin and azithromycin. Clarithromycin has a 14-membered ring structure that is an erythromycin analogue; azithromycin has a 15-membered ring structure analogue with a nitrogen inserted into the lactone ring. These chemical modifications conferred quite dramatic new properties. They have been developed to the point at which they now display acid stability, improved absorption, tissue and cellular penetration, and longer persistence in the body. The efforts expended on chemical and biochemical modifications of 16-membered ring macrolides have been less successful. Only a few molecules, for example, rokitamycin and miocamycin, have improved bioavailability [2].

Most infections occur at sites outside the vascular compartment, and consequently extravascular drug distribution is one obvious factor of importance for therapeutic success. A significant difference is evident in the ability of various antimicrobial agents to penetrate into tissues: For instance, the β-lactam antibiotics do not penetrate particularly well into tissues and this is reflected in a relatively low volume of distribution (V_d). On the other hand, the macrolides penetrate well into tissues as shown by their much larger V_d [3]. The antimicrobial action of macrolides occurs at two sites. The first action is extracellular. Then, as the drug concentrates intracellularly, the organism is barraged by both innate phagocytic killing mechanisms and the concentrated amount of drug within the cell. Consequently, traditional *in vitro* measures of antibiotic activity (MIC) may not optimally be applied to the macrolide antibiotics.

New macrolides share the high tissue affinity of erythromycin, but marked variations are seen in the pharmacokinetic parameters such as C_{max}, T_{max}, and the half-lives of compounds in blood and in deep compartments. Variation is also seen in intracellular penetration and apparent bioavailability.

One of the important limitations to the widespread use of the macrolides has been the propensity to interact with other commonly administered medications. Serious, sometimes life-threatening, consequences have resulted from the administration of macrolides to patients receiving routine medications including theophylline, carbamazepine, terfenadine, and other frequently prescribed medications [4–6]. Most of these interactions involve inhibition of drug metabolism via cytochrome *P*-450 microsomal enzyme. However, not all macrolides have been associated with such drug interactions.

This chapter reviews the current pharmacokinetic information available on the new macrolides, including the pharmacokinetics, metabolism, and drug–drug interaction potency of 14- and 15-membered ring macrolides (the second generation of semisynthetic macrolides). In addition, ketolides currently under development (the third generation of semisynthetic macrolides) are included here as well.

II. Pharmacokinetics and Metabolism

As a general consideration of the pharmacokinetic profile of macrolides, Williams and Sefton [7] have described the following uptake process (Fig. 1). Briefly, after oral administration of macrolides, some of the dose is inactivated in the liver, but it is predominantly excreted into the bile and then enters the enterohepatic circulation. Only after biliary excretion is saturated does it flow into the hepatic vein. All macrolides flowing from the liver are carried through the right heart and

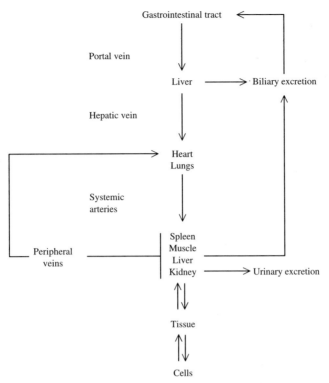

Fig. 1. Absorption of macrolide antibiotics following oral administration. (From Williams and Sefton [7], Fig. 1, p. 15.)

pulmonary artery to the lung. The heart then distributes the remaining portion to the other organs in the body and only that which is not retained eventually reaches the systemic veins where peripheral blood levels can be measured. In such a complicated situation, it is difficult to piece together the results in such a way as to compare the disposition of macrolides precisely. Therefore, the pharmacokinetic profiles of each new macrolide are described separately in this section. Five new macrolides have been selected for review; two are 14-membered ring macrolides (roxithromycin and clarithromycin), one is a 15-membered ring macrolide (azithromycin), and two are ketolides currently under development [telithromycin (HMR 3647) and ABT-773]. An outline of erythromycin pharmacokinetics has also been included at the beginning as a reference.

A. Erythromycin

This is an old drug that has been available for four decades. Under the acidic conditions present in the stomach, erythromycin essentially decomposes to two antibacterially weak compounds, the 8,9-anhydro-6,9-hemiketal, which is further degraded to 6,9;9,12-spiroketal [8] (for structures see Chapter 3). Because of the instability of erythromycin in an acidic environment, absorption is erratic and incomplete, leading to unpredictable serum and tissue concentrations [9]. In addition to decreased bioavailability, the anhydrohemiketal metabolite has been implicated as a primary agent responsible for some of the erythromycin-associated gastrointestinal side effects [10, 11]. To enhance bioavailability, minimize unpredictability in absorption, and reduce the incidence of adverse effects, particle-coating, enteric-tablet coating, and chemical modifications of erythromycin have been performed. Despite numerous attempts to improve the predictability and extent of erythromycin absorption, erythromycin bioavailability is formulation dependent and remains highly variable and erratic. Current estimates suggest that it averages ~25% [12, 13].

Enteric-coated formulations dissolve in the duodenum; a 250-mg dose produces a peak serum level of about 1 µg/ml [14]. Food in the stomach decreases absorption of the drug, except for the estolate form [15, 16]. The normal serum half-life of erythromycin is approximately 2 hr [17], and serum levels are maintained for 6 hr. The recommended orally administered dosage of erythromycin is 250 mg to 1 g every 6 hr.

Erythromycin diffuses readily into most tissues except the brain and cerebrospinal fluid. Erythromycin crosses the placental barrier and is present in maternal milk [18]. In many animal species and humans, erythromycin is metabolized by the hepatic microsomal cytochrome P-450 mixed function oxidase system. More specifically erythromycin appears to act as a substrate for the cytochrome P-450 3A4 (CYP 3A4) isoform [13]. Erythromycin is excreted primarily in the bile; only 2–5% is excreted in the urine. Concentrations in the bile may exceed 10 times those in plasma [19].

Renal impairment has only a minor impact on the pharmacokinetics of erythromycin [20], while alcoholic liver disease increases the elimination half-life from 2.0 to 3.2 hr [21]. No adjustment of the dose of erythromycin in these categories of patients is necessary.

Despite decades of clinical availability and coadministration of erythromycin with theophylline-containing compounds, the nature, extent, and severity of this potentially serious metabolic drug–drug interaction was not immediately recognized. Recently, however, erythromycin drug–drug interactions have been extensively studied and are now well recognized (see Section III).

B. Roxithromycin

1. Absorption and Elimination

Roxithromycin (RXM) is acid-stable macrolide that is similar to erythromycin but yields high plasma, tissue, and body fluid concentrations and has a long half-life [22]. A 150-mg roxithromycin tablet produces a maximum serum concentration that is more than twice that of the erythromycin base pellet (500 mg), and an area under the serum concentration time curve (AUC) eight to ten times greater, indicating an improved bioavailability, and/or a lower clearance of roxithromycin [23]. Birkett et al. [24] reported that the dose-normalized mean maximal plasma concentration for roxithromycin in 12 healthy subjects was approximately 5-fold greater than that for erythromycin and, as a result, the normalized AUC was 27-fold greater for roxithromycin than for erythromycin. The elimination half-lives ($T_{1/2}$) of roxithromycin and erythromycin are different, ranging from 10 to 13 hr for roxithromycin and from 2 to 3 hr for erythromycin after single oral doses (roxithromycin: 150, 300, 450 mg; erythromycin: 250, 500, 1000 mg) [23].

In a single dose rising study (150, 300, and 450 mg), plasma levels of roxithromycin increased in a dose-dependent manner (Table I). After reaching peak levels between 1.6 and 1.9 hr, the drug was eliminated with a $T_{1/2}$ of about 10 hr. Neither time to reach the peak level (T_{max}) nor $T_{1/2}$ was affected by an increase in dose. The AUCs and the amount of drug excreted in the urine increased as a function of dose, but dose proportionality could be demonstrated only with urine data. Renal clearance was increased as the dose was increased [25]. During multiple dosing, a steady state was reached by day 4 after which there was no further accumulation.

TABLE I
Pharmacokinetic Estimates of Roxithromycin after Three Single Doses to Healthy Man

Parameters	Mean values ± S.D.		
	150 mg	300 mg	450 mg
C_{max} (mg/l)	6.8 ± 1.2	9.1 ± 1.7	10.8 ± 1.4
	[13.6]——— c ———[9.1]——— c ———[7.2]		
T_{max} (hr)	1.9 ± 1.2	1.9 ± 1.3	1.6 ± 1.0
$AUC_{0-48\,hr}$ (mg/l)	70.8 ± 20.0	113.6 ± 23.8	148.7 ± 31.8
	[141.6]——— c ———[113.6]——— a ———[99.1]		
$AUC_{0-\infty\,hr}$ (mg/l·hr)	72.6 ± 23.2	116.5 ± 28.6	150.6 ± 39.5
	[145.2]——— b ———[116.5]——— c ———[102.9]		
$T_{1/2}$	8.4 ± 4.2	10.5 ± 5.2	10.3 ± 2.6
CL_r (l/hr/1.73 m^2)	0.23 ± 0.08	0.25 ± 0.07 —— b —— 0.34 ± 0.13	
	——— c ———		
$U_{0-\infty\,hr}$ (mg)	17.2 ± 6.4	31.3 ± 9.9	53.1 ± 20.3
	[34.4]	[31.3]	[35.4]

Source: From Puri and Lassman [25], Table III, p. 93.
Normalized values given in brackets.
[a] $P < 0.05$.
[b] $P < 0.01$.
[c] $P < 0.001$.

The minimum plasma concentrations of roxithromycin at steady state (days 4–11) ranged from 3.22 to 3.69 mg/l with 150 mg b.i.d. and from 2.02 to 2.22 mg/l with 300 mg q.d. These results indicate that roxithromycin exhibited nonlinear kinetics and may involve a saturable process as evidenced by increased elimination rates and lack of dose proportionality between doses [25]. The nonlinear kinetics is most probably explained by saturation of roxithromycin serum protein binding [26], as discussed later.

In a study of four healthy subjects administered 150 mg of ^{14}C labeled roxithromycin, 74% of the administered dose was accounted for, with fecal excretion being predominant (53.4%), followed by pulmonary excretion (13.4%) and urinary excretion (7.4%) [25].

2. Distribution

Of the new macrolides, roxithromycin appears to achieve the highest serum concentrations while maintaining high concurrent tissue concentrations [27]. Both roxithromycin and erythromycin produce tonsillar tissue levels similar to their serum levels. The decline of drug concentration in tonsillar tissue is similar to that found in serum. However, the serum and tonsillar tissue levels of roxithromycin are still above 1 mg/l (mg/kg) 12 hr after the last dose, while this level is reached for erythromycin after about 5 hr, indicating that the serum elimination half-lives of both compounds are consistent with their rate of elimination from the tonsillar tissue [23, 28]. After repeated administration of roxithromycin (150 mg twice daily) and erythromycin (ethyl succinate 1 g twice daily), lung tissue concentrations are higher for roxithromycin (3.7 ± 0.5 mg/kg after 12 hr) than for erythromycin (2.2 ± 1.1 mg/kg after 4 hr), although the lung/plasma concentration ratio was higher for erythromycin than for roxithromycin [29].

In bronchial secretions, roxithromycin reached high concentrations in pooled sputum samples collected in patients with superinfected chronic bronchitis at consecutive intervals until 24 hr after a single dose of 150 mg; stable levels ranging from 4.52 ± 0.59 to 5.27 ± 0.62 mg/l from 2 to 8 hr were followed by a slow decrease (2.90 ± 0.38 mg/l at 24 hr). These values were equal to simultaneous serum concentrations [30]. These results suggest that bronchial levels equilibrate rapidly with serum concentrations as a result of transport of roxithromycin by passive diffusion; this has been shown with other macrolides [31]. In prostatic tissue samples, collected in surgical conditions after multiple doses of roxithromycin (300 mg, followed by 150-mg doses), the mean concentrations ranged from 2.16 ± 0.21 to 2.81 ± 0.68 mg/kg; they were still 2.35 ± 0.57 mg/kg at 12 hr, equaling those in plasma (2.79 ± 0.41 mg/l) [30].

For bacterial infections located at sites where the antibiotic is distributed by passive diffusion, the most relevant concentrations relating to *in vitro* antimicrobial activity and other pharmacodynamic parameters depend on the unbound levels of the drug in serum [32]. Roxithromycin is bound extensively and saturably to α_1-acid glycoprotein with a significant increase in serum-free roxithromycin concentration when the roxithromycin total concentration in serum exceeds about 4 mg/l. The free concentration of roxithromycin increases from 4.3 to 13.4% (212%) when the total serum level rises from 3.3 to 8.4 mg/l [26]. This is in agreement with the observed increase in roxithromycin renal clearance during the first 8 hr after increasing single oral doses from 150 to 450 mg [33]. Less than 0.05% of the dose is transferred into the breast milk of lactating women [25].

3. Metabolism

Roxithromycin has been shown to be metabolized *in vivo* to form several metabolites, including a decladinosyl derivative (M1), *O*-dealkyl derivative (M2), *N*-monodemethyl derivative (M3), and didemethyl derivative (M4) in rats, dogs, and humans (Fig. 2) [34, 35]. In human plasma, only roxithromycin was detected [25].

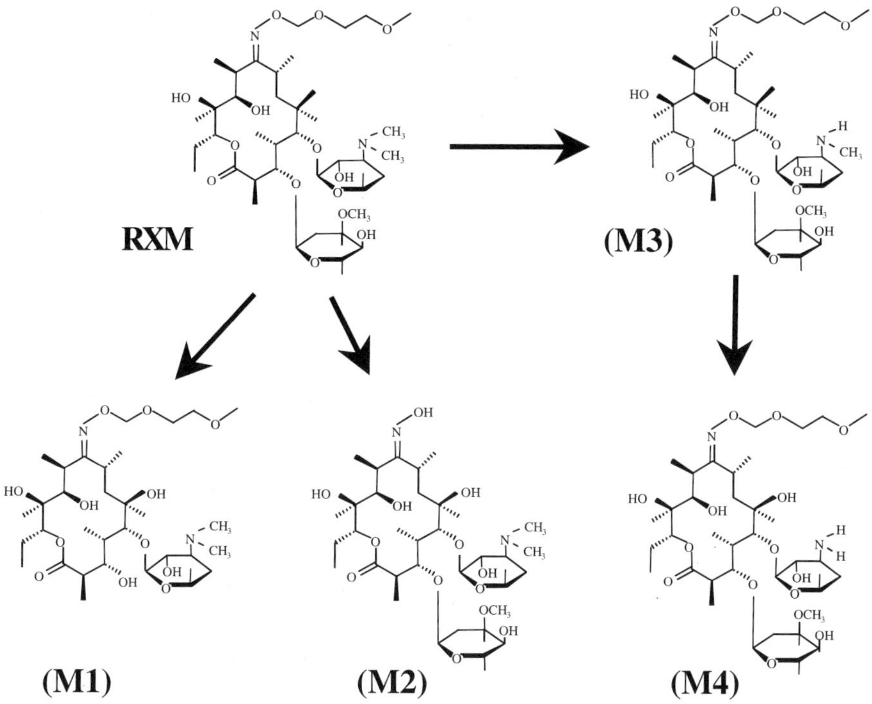

Fig. 2. Chemical structures of roxithromycin (RXM) and its metabolites (M1, M2, M3, and M4). (Reprinted with permission from Yamazaki and Shimada [35], Fig. 1, p. 1054.)

In urine collected between 0 and 48 hr after oral dosing with roxithromycin, M1, M2, and M3 have been shown to be expected at levels of 1.2, 0.9, and 0.14% of the total dose, respectively. Small amounts (0.06% of the administered dose) of didemethyl roxithromycin (M4), the oxidation product of M3, have also been detected [34]. In feces, the profile was almost similar to urine [25]. The formation of these mono- and didemethylated metabolites of roxithromycin has been suggested to be mediated by *P*-450 enzymes, particularly by CYP3A4/5 in humans [36, 37]. *In vitro* experiments have suggested that CYP3A4 is a major enzyme involved in the *N*-demethylation of roxithromycin in human liver microsomes [38].

4. Special Population

Although the maximum concentration (C_{max}), AUC, and $T_{1/2}$ of roxithromycin are increased by 98, 160, and 70%, respectively, in the presence of renal insufficiency, no dosing modification has been suggested [39, 40].

The pharmacokinetics of roxithromycin have been assessed in patients with histologically and biologically documented cirrhosis [41]. Although $T_{1/2}$ increased to 25.5 hr (vs. 7.9 hr in healthy subjects), C_{max} (5.8 mg/l), T_{max} (2.1 hr), and AUCs (118 mg/l·hr) in the patients were similar to those of healthy subjects after a single dose of 150 mg. The amount of drug excreted in the urine (28.0 mg) and renal clearance (0.42 l/hr) were increased, partially compensating for the reduction in hepatic clearance. Therefore, a dosage adjustment may not be necessary in these patients, especially if short-term therapy is anticipated.

The mean AUCs and $T_{1/2}$ of roxithromycin were significantly increased in the elderly group compared to the young group. The increase in AUC and $T_{1/2}$ in the elderly may be due to decreased renal and/or hepatic function [25].

C. Clarithromycin

1. Absorption and Elimination

Clarithromycin (6-*O*-methylerythromycin) (see Fig. 7 in a later section) is synthesized by substituting a methoxy group for the C-6 hydroxy group of erythromycin [42]. This substitution creates a more acid-stable antimicrobial and prevents the degradation of the erythromycin base to the hemiketal intermediate. The increased acid stability of clarithromycin results in improved oral bioavailability and reduced gastrointestinal intolerance [43]. A 250-mg oral dose of clarithromycin resulted in a C_{max} of 0.7–0.8 mg/l, achieved at 2 hr in healthy subjects [44]. Double this parameter would be expected for a 500-mg dose. The bioavailability of clarithromycin, as indicated by the appearance of the parent compound in plasma, is 52–55% after administration of the drug in tablet form [44]. In other cases, clarithromycin has oral bioavailability varying between 55% and 68% [3]. Concomitant ingestion of food increases its oral bioavailability by 25%

based on AUC data [45], but does not affect the extent of drug bioavailability. Consequently, clarithromycin may be given with or without food [46]. The serum peak of clarithromycin is about eightfold higher than that achieved with a comparable dose of erythromycin, and its half-life exceeds 5 hr (4–7 hr), allowing once- or twice-daily dosing [47, 48]. Clarithromycin is metabolized by hepatic microsomal enzymes and the peak serum concentration of the major metabolite, 14-(R)-hydroxy-clarithromycin, in humans is microbiologically active and reaches 30–40% of the parent compound level [3]. A steady state is usually achieved after five doses [44, 49]. Mean peak serum concentrations in the steady state with orally administered dosages of 250 and 500 mg every 12 hr are 1 µg/ml and 2–3 µg/ml, respectively [50].

When single doses of 100, 200, 400, 600, 800, and 1200 mg of clarithromycin were compared in healthy subjects, the pharmacokinetics of the parent drug and metabolite were nonlinear [51], with apparent capacity-limited formation of the 14-(R)-hydroxy metabolite at doses of ≥600 mg. Nonlinear kinetics were also seen in studies of single and multiple doses of clarithromycin, where increases in C_{max} and AUC of the parent drug were more than proportionate with the dosages [52]. In another study, the AUC for clarithromycin increased 13-fold, with a 4.8-fold increase in dose. Pharmacokinetic data suggest that nonlinearity was due predominantly to a decrease in the apparent metabolic clearance, which fell from 913 to 289 ml/min (Table II) [50].

Clarithromycin undergoes both hepatic and renal elimination. After a single 250-mg dose of radioactive clarithromycin in healthy subjects, 28% of the radioactivity was detected in the urine within 12 hr, and 38% was detected within 5 days. Fecal excretion averaged 40% of a dose. Urinary excretion was proportionately

TABLE II
Pharmacokinetic Parameters of Clarithromycin in Subjects Given a 250- or 1200-mg Oral Dose of Clarithromycin-^{14}C

Subject	Dose (mg)	C_{max} (µg/ml)	T_{max} (hr)	$T_{1/2}$ (hr)	AUC$_{0-120\,hr}$ (mcg · hr/ml)	CL total (ml/min)	CL renal (ml/min)
1	250	0.46	4.0	5.30	3.08	1353	254
2		0.53	2.0	3.87	3.16	1319	222
3		0.74	2.0	4.23	6.16	676	133
	Mean	0.58		4.39^a	4.13	1116	203
4	1200	3.42	3.0	11.18	76.72	261	92.3
5		1.88	4.0	11.76	46.49	430	119
6		2.68	4.0	10.90	38.46	520	131
	Mean	2.66		11.27^a	53.89	403	114

Source: Reprinted with permission from Ferrero *et al.* [50], Table 1, p. 443.
^{a}Harmonic mean.

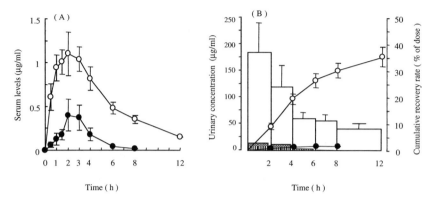

Fig. 3. (A) Serum levels and (B) urinary excretion after oral administration of clarithromycin and erythromycin to healthy volunteers. ○, clarithromycin; ●▩, erythromycin stearate. Each value represents the mean ±S.E. of six subjects. Dose: 200 mg/man. (From Saito *et al.* [53], Fig. 3, p. 527.)

slower but greater (46–53% of the total) after a 1200-mg rather than a 250-mg dose [50]. Both unchanged clarithromycin and the active 14-(R)-hydroxy metabolite are the main components in the urine [45]. Saito *et al.* [53] have demonstrated a remarkably higher level of clarithromycin in both serum and urine as compared with the same oral dose of erythromycin stearate (200-mg base equivalent) in healthy subjects (Fig. 3). Urinary recovery rates of clarithromycin were >20-fold those of erythromycin stearate. This favorable urinary excretion is clinically useful for urinary tract infections, for which most previously known macrolides have failed to have significant impact. Transintestinal elimination accounts for excretion of ~10% of the total dose [46]. The C_{max} of clarithromycin and 14-(R)-hydroxy-clarithromycin in breast milk were found to be 25% and 75% of the corresponding values in plasma of nursing mothers treated with the drug for puerperal infections [54].

2. Distribution

Clarithromycin is not extensively bound to plasma proteins. Over the serum concentration range achieved with recommended doses, its protein binding varies from 42% to 70% with the greatest affinity for α_1-acid glycoprotein [55]. The apparent volume of distribution (V_d) for clarithromycin following oral administration of a single 250- or 500-mg dose to healthy subjects is 226–266 liters [56]. This enables clarithromycin to concentrate at the site of infection, where its *in vivo* efficacy is enhanced. Because macrolide antibiotics are often used in the treatment of community-acquired respiratory tract infections due to pathogens such as *Streptococcus pneumoniae, Moraxella catarrhalis, Mycoplasma pneumoniae, Haemophilus influenzae,* and *Legionella pneumophila*, adequate penetration of clarithromycin into respiratory tract cells is of importance [57]. In the respiratory

TABLE III
Concentrations of Clarithromycin and 14-OH-Clarithromycin in Plasma and Lung Tissues

Drug	Concentration ± S.D. in:		Lung tissue (µg/g)	Ratio ± S.D. (lung/ intraoperative plasma drug concentration)
	Plasma (µg/ml)			
	Minimum	Intraoperative		
Clarithromycin	1.38 ± 0.99	1.89 ± 0.66	54.32 ± 33.82	29.6 ± 15.3
14-OH-clarithromycin	0.67 ± 0.41	0.80 ± 0.29	5.12 ± 3.18	6.5 ± 3.4
Ratio (clarithromycin/ 14-OH-clarithromycin)	1.98 ± 0.96	2.42 ± 0.74	11.25 ± 3.80	—[a]

Source: From Fish *et al.* [57], Table 1, p. 887.
[a] Not applicable.

tract of patients, clarithromycin rapidly achieves concentrations that exceed those in plasma by ratios of up to 3.1:1 in bronchial secretions, 8.82:1 in middle ear fluid, 5.17:1 in epithelial lining fluid (ELF), 94.1:1 in alveolar macrophages (AM), up to 28.7:1 in lung tissue and up to 27.5:1 in nasal mucosa [58]. Of particular note are the very high ratios of concentrations within ELF, AM, and lung tissue, which show high penetration of the drug into respiratory tract cells, a feature that should enhance efficacy against typical and atypical pathogens.

The concentrations of clarithromycin and its active principal metabolite, 14-(*R*)-hydroxy-clarithromycin, were determined in lung tissue obtained during lung resection and compared with concomitant concentrations in plasma. Fifteen patients studied were given 500 mg orally every 12 hr for a minimum of five doses to achieve steady-state concentrations. The concentrations of clarithromycin and 14-(*R*)-hydroxy-clarithromycin (14-OH-clarithromycin) obtained by HPLC are shown in Table III. Clarithromycin was highly concentrated in lung tissues. The concentrations of the parent and metabolite in lung tissue averaged 54.3 and 5.12 µg/g, respectively, with a mean calculated ratio of concentrations of the parent to metabolite of 11.3. In plasma, the mean concentrations of the parent and metabolite 4 hr after the final dose (intraoperative) were 1.89 and 0.80 µg/ml, respectively, with the parent to metabolite ratio being 2.4. The ratio of lung to intraoperative plasma clarithromycin concentrations ranged from 8.3 to 60.4 (mean, 29.6 ± 15.3) [57]. This penetration of the active metabolite, 14-(*R*)-hydroxy-clarithromycin, into intrapulmonary regions is important, since *in vitro* activity against *H. influenzae* is enhanced by the combination of clarithromycin and this metabolite compared to that of either compound alone [59].

The high concentrations in lung tissue can probably be explained by the intracellular accumulation of clarithromycin. Actually, extensive tissue and intracellular uptake of clarithromycin, especially in the lung, has been demonstrated in rats

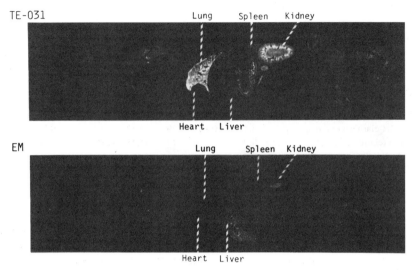

Fig. 4. Whole-body autoradiograms showing the distribution of radioactivity 5 min after intravenous administration of ^{14}C-clarithromycin (TE-031) and ^{14}C-erythromycin (EM) (5 mg/kg) to rats. (From Kohno et al. [60], Fig. 2, p. 754.)

using autoradiography and biochemical techniques [60, 61]. Comparative whole-body autoradiograms of ^{14}C-clarithromycin and ^{14}C-erythromycin clearly indicated that clarithromycin itself has a significant affinity to lung tissue (Fig. 4). Further, light microscope autoradiography revealed that radioactivity in the lung was confined almost entirely to the alveolar wall, but was rather low in the vascular system (Fig. 5). In isolated lung cells, clarithromycin was also found in greater concentrations than erythromycin. The amount of clarithromycin was ten times that of erythromycin after 5 min of incubation. This uptake profile was quite different from that observed in isolated liver cells (Fig. 6). Uptake by lung cells for both antibiotics was shown to be an active process, as revealed by the need for cell viability, a suitable environmental temperature, and ATP. Clarithromycin uptake proved to be dependent in part on mitochondrial oxidative respiration (Table IV).

Fietta et al. [62] found that energy, driven from cell metabolism, was not necessary for clarithromycin uptake by human neutrophils. They suggested a role for extracellular calcium in the incorporation of clarithromycin, while a decreased transport ratio by the inhibitors of phosphorylation metabolism has been reported in another study [63]. At least two mechanisms may contribute to the accumulation of macrolides in phagocytes. One could be an energy-independent system, leading to diffusion through cell membranes, followed by binding to intracellular sites or trapping by protonation into cellular acidic compartments [64–67]. The other could be an active transport mechanism that is dependent on the supply of

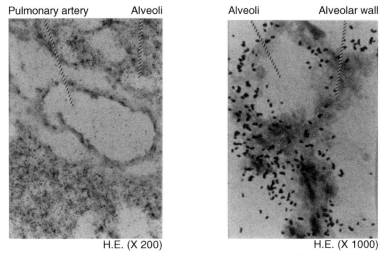

Fig. 5. Microautoradiograms of lung tissue after oral administration of ^{14}C-clarithromycin to rat.

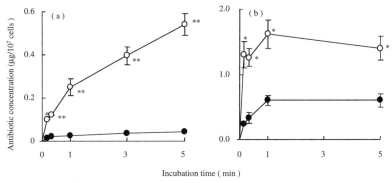

Fig. 6. Uptake of ^{14}C-clarithromycin (○) and ^{14}C-erythromycin (●) by isolated lung (a) or liver (b) cells. Each value represents the mean ± S.E. of six experiments. *$P < 0.01$ and **$P < 0.001$ versus erythromycin. (From Kohno et al. [61], Fig. 2, p. 507.)

metabolic energy [68–70]. Nevertheless, several points remain controversial and the respective contribution of the two mechanisms described above in macrolide uptake differs from one member to another, despite the similarity of the chemical structure.

The time above the MIC ($T >$ MIC) has recently been suggested to be a predictive pharmacodynamic parameter for the *in vivo* bacteriological response to clarithromycin. In plasma, ELF, and AM, the $T >$ MIC are maintained at steady-state concentrations and throughout the 12-hr dosing interval. The steady-state levels of clarithromycin in the ELF and AM exceed those observed in plasma,

TABLE IV
Influence of Cell Viability, Temperature and Metabolic
Inhibitors on the Uptake of [^{14}C]Clarithromycin and
[^{14}C]Erythromycin by Isolated Rat Lung Cells

Incubation condition	Uptake (% of control)	
	Clarithromycin	Erythromycin
Dead cells	9.1 ± 3.1[b]	23.9 ± 4.5[b]
Viable cells, 4	10.6 ± 3.5[b]	17.4 ± 2.9[b]
2,4-Dinitrophenol[a]	34.4 ± 3.1[b]	52.9 ± 2.7[b]
Sodium cyanide[a]	41.2 ± 2.9[b]	97.4 ± 10.4
Iodoacetic acid[a]	103.9 ± 4.6	111.9 ± 7.9
Ouabain[a]	114.2 ± 7.3	105.9 ± 6.0

Source: From Kohno et al. [61], Table I, p. 508.
Data are expressed as the mean ($n = 6$) ± S.E.
[a]Concentration of each metabolic inhibitor was 1 mM.
[b]$P < 0.001$.

achieving concentrations in all three matrixes equal to or higher than those of azithromycin described subsequently [59]. Consequently, the AUC/MIC ratio for clarithromycin appears to be an important pharmacodynamic parameter together with the T > MIC [3].

3. Metabolism

Clarithromycin is metabolized in humans by both oxidative and hydrolytic mechanisms. A number of metabolites have been detected in the urine and feces of subjects receiving low (250-mg) and high (1200-mg) doses of clarithromycin [71]. The nature of these metabolites revealed the involvement of three metabolic pathways: (1) hydroxylation at the 14-position to form the *R* and *S* epimers, (2) *N*-demethylation, and (3) hydrolysis of the cladinose sugar. Secondary metabolism via these pathways was also evident [50]. These have included *N*-desmethyl-clarithromycin, 14-(*R*)-hydroxy-clarithromycin, 14-(*S*)-hydroxy-clarithromycin, 14-(*R*)-hydroxy-*N*-desmethyl-clarithromycin, 14-(*S*)-hydroxy-*N*-desmethyl-clarithromycin, descladinosyl-clarithromycin, 14-(*R*)-hydroxy-descladinosyl-clarithromycin, and *N,N'*-didesmethyl-clarithromycin. However, *N*-desmethyl-clarithromycin and 14-(*R*)-hydroxy-clarithromycin are considered to be the major metabolites, and cladinose ring hydrolysis occurs nonenzymatically (Fig. 7) [50]. The metabolic profile of clarithromycin is unique, because no other 14-membered macrolide has been shown to undergo 14-hydroxylation in humans. In addition, the formation of 14-(*R*)-hydroxy-clarithromycin is clinically relevant, because this metabolite is pharmacologically active [72].

Clarithromycin undergoes extensive first-pass metabolism, with the C_{max} of the active 14-(*R*)-hydroxy metabolite in plasma being higher after oral rather than

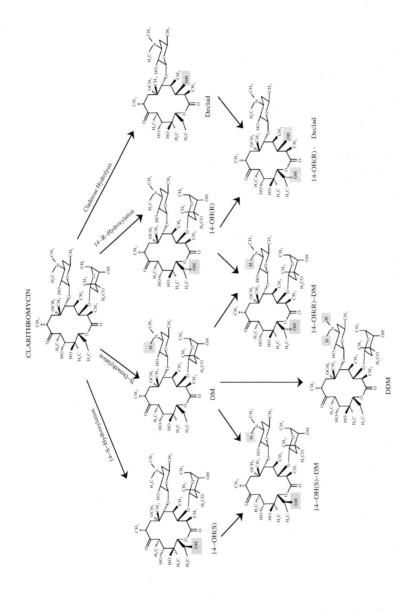

Fig. 7. Metabolic pathway of clarithromycin in man. DM, N-desmethylclarithromycin; DDM, N,N'-didesmethylclarithromycin; 14-OH(R), 14-hydroxy(R)-clarithromycin; 14-OH(R)-DM, 14-hydroxy-(R)-N-desmethylclarithromycin; 14-OH(S), 14-hydroxy(S)-clarithromycin; 14-OH(R)-Declad, 14-hydroxy-(R)-descladinosylclarithromycin; Declad, descladinosylclarithromycin; 14-OH(S)-DM, 14-hydroxy(S)-N-demethylclarithromycin. (Reprinted with permission from Ferrero et al. [50], Scheme 1, p. 446.)

341

TABLE V
Total Recovery of N-Demethylation and 14-Hydroxylation Products Following Oral Administration of a 250- or 1200-mg Dose of Clarithromycin-^{14}C

Products	Percent of dose		Percent change
	250 mg	1200 mg	
N-Demethylation (DM + DDM + 14-OH[R]-DM)	20.6	11.4	−45
14-Hydroxylation (14-OH[R] + 14-OH[S] + 14-OH[R]-DM + 14-OH[R]-declad)	40.5	24.1	−40
Clarithromycin	22.8	40.0	+75

Source: Reprinted with permission from Ferrero *et al.* [50], Table 3, p. 446.

intravenous administration and occurring earlier after administration [73]. The formation of this metabolite is capacity limited due to saturation of the 14-hydroxylation and N-demethylation pathways, partially accounting for its nonlinear pharmacokinetics [45, 51]. Products of N-demethylation and 14-hydroxylation decreased 40–45% at the 1200-mg dose (vs. 250-mg dose) while the amount of clarithromycin recovered unchanged increased 75% (Table V) [50]. In urine, clarithromycin and 14-(R)-hydroxy-clarithromycin account for the majority of radioactivity following radioactive oral dose. Metabolites accounted for most of the radioactivity in feces and included N,N′-didesmethyl-clarithromycin, and both R and S forms of 14-hydroxy-clarithromycin [58]. The predominance of N-demethylated derivatives in the feces may result from biliary secretion. Of all metabolites, only 14-(R)-hydroxy-clarithromycin was found in substantial concentrations in plasma [50]. Several lines of evidence in the *in vitro* study using human liver microsomes have demonstrated that member(s) of the CYP3A subfamily, most likely CYP3A4, are the principal human liver microsomal enzymes involved in the metabolism of clarithromycin [71].

4. Special Population

Both C_{max} and AUC values for clarithromycin and 14-(R)-hydroxy-clarithromycin increased markedly in patients with renal impairment after five consecutive 500-mg doses of the parent drug [73]. Creatinine clearance (CL_{Cr}) was significantly ($p < 0.03$) correlated with AUC, $T_{1/2}$, C_{max}, C_{min}, and the elimination rate constant. In patients with $CL_{Cr} > 80$ ml/min, clarithromycin C_{max} was 2.5 mg/l, AUC was 21.1 mg/l·hr, and $T_{1/2}$ was 6.7 hr. However, in patients with CL_{Cr} of 10–29 ml/min, C_{max} increased to 8.3 mg/l, AUC to 88.4 mg/l·hr, and $T_{1/2}$ to 21.6 hr. Thus, in patients with severe renal impairment (i.e., creatinine clearance <30 ml/min), the dose should be reduced [45]. In patients with moderate-to-severe hepatic impairment and normal renal function, there is less metabolism of clarithromycin to the 14-hydroxy form resulting in decreased peak plasma concentrations of the metabolite and increased renal excretion of unchanged clarithromycin. Dosing modifications do not appear to be necessary for these patients [74].

D. Azithromycin

1. Absorption and Elimination

Azithromycin (9-deoxo-9a-aza-9a-methyl-9a-homoerythromycin) is formed by inserting a methyl-substituted nitrogen at the 9a position of the aglycone ring (see Chapter 3). This structural change makes the compound more stable in acid, and significantly increases the $T_{1/2}$ and tissue penetration compared with erythromycin [75]. The oral bioavailability of a single 500-mg capsule of azithromycin in the fasting state in healthy subjects was 37% [76]. A more recent report described the absolute bioavailability of azithromycin from a sachet formulation in 12 healthy subjects as 44% and 50% when the sachet was administered after an overnight fast or immediately after a high fat breakfast, respectively [77]. The peak serum concentration of azithromycin following a 500-mg dose is approximately 0.4 µg/ml, fivefold lower than that achieved with a comparable dose of clarithromycin [43]. The $T_{1/2}$ of azithromycin exceeds 40 hr allowing once-daily dosing compared to clarithromycin, for which the $T_{1/2}$ of 4–5 hr necessitated twice-daily dosing [48]. Azithromycin, in contrast to clarithromycin and erythromycin, was initially said to be poorly absorbed in the presence of food, but recent studies show no reduction of absorption with food, at least in adults [77].

Azithromycin is eliminated primarily in the bile and through transintestinal excretion [78]; the urine is a minor elimination route. After a single 500-mg dose, 12.2% and 4.5% of the administered dose after i.v. and oral administration, respectively, were recovered as unchanged drug in the urine over a 72-hr collection period [79]. The transintestinal route is believed to account for elimination of 30–35% of the total administered dose [46]. No dosage modification of azithromycin is necessary for patients with class A or B liver cirrhosis [80].

2. Distribution

The extent to which azithromycin is bound to serum proteins varies with the serum drug concentration. As serum concentrations increase, the extent to which the drug is bound decreases. At serum concentrations of 0.02–0.05 mg/l binding approximated 50%; when concentrations in serum were increased to 0.1, 0.3, and 1 mg/l, the extent of azithromycin binding to serum protein averaged 23, 18 and 7.1%, respectively [79]. Based on this saturation characteristic it is presumed that, like erythromycin, azithromycin is bound primarily to α_1-acid glycoprotein [81, 82].

Following oral administration, azithromycin is distributed extensively into both animal tissues [83, 84] and those of man [76]. This pharmacokinetic property of azithromycin results in high tissue concentrations that are sustained long after serum concentrations have declined to very low levels (Fig. 8). A study of the localization of azithromycin in specific potential sites of infection within human pulmonary tissues showed peak concentrations, following a single 500-mg oral dose,

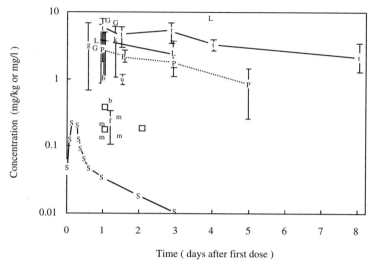

Fig. 8. Tissue concentrations of azithromycin in man after administration of 500 mg (mean ± S.D.). S, serum; P, prostate; t, tonsil; L, lung; k, kidney; g, gynecological; u, urological; G, gastric; m, muscle; f, fat; b, bone; ☐, gastric mucosa. (From Foulds et al. [76], Fig. 4, p. 78.)

of 1.6 mg/l in sputum, 3.9 mg/l in bronchial mucosa, and 2.2 mg/l in epithelial lining fluid [85]. These concentrations are generally similar to the gross pulmonary tissue concentrations (3.1 and 2.56 mg/kg at 24 and 72 hr, respectively) following a single 500-mg dose [86]. In a multiple-dose study, azithromycin (500-mg capsules) has been shown to penetrate tissues rapidly and extensively; steady-state serum levels were 0.64 µg/ml at 2–4 hr, 0.1 µg/ml at 10–12 hr, and 0.012 µg/ml at 72–96 hr. Steady-state tissue or fluid concentration levels are substantially increased in relationship to serum levels: lung, 4.0 µg/ml at 72–96 hr [tissue to serum ratio (TSR) = more than 100]; sputum, 2.9 µg/ml at 10–12 hr (TSR = 30); cervix, 2.8 µg/ml at 19 hr (TSR = 70); and skin, 0.4 µg/ml at 72–96 hr (TSR = 35) [19]. Azithromycin is widely distributed in brain tissue, whereas very little or none of the drug is detectable in aqueous humor or cerebrospinal fluid [87]. As would be expected from the distribution characteristics of the drug described above, azithromycin is concentrated in breast milk [88].

Incubating polymorphonuclear cells in the presence of azithromycin results in a ratio of intracellular to extracellular drug concentration of approximately 80, compared with a ratio of approximately 15 for erythromycin [89]. Azithromycin contains two amino groups, which may allow for greater protonation and trapping in the acidic intracellular compartment than other macrolides [62]. In human, the high concentrations of azithromycin (23 mg/l) in alveolar macrophages following a single 500-mg oral dose [85] and in leucocytes (45 mg/l) following

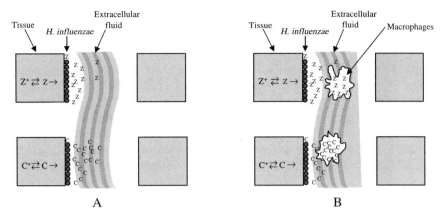

Fig. 9. Pharmacodynamic model for exposure of *H. influenzae* to macrolides. (A) *H. influenzae* adjacent or attached to the exterior cell surface is exposed to azithromycin (Z) or clarithromycin (C) released from cells into the extracellular fluid. (B) Macrophages transport and release azithromycin or clarithromycin at the site of infection. (From Nightingale [46], Fig. 1, p. 441.)

doses of 500 mg/day for 3 days [90] are also consistent with the extensive *in vitro* uptake of azithromycin.

The failure of serum concentrations of azithromycin, clarithromycin, and its active metabolite, 14-(*R*)-hydroxy-clarithromycin, to reach MIC_{90} levels suggests a lack of effectiveness against *H. influenzae* that is, however, contradicted by clinical data. To explain this apparent paradox, Nightingale [46] has proposed the pharmacodynamic model for the cellular location of *H. influenzae* organisms in relation to the tissue pharmacokinetics of the macrolides (Fig. 9). *H. influenzae* may reside adjacent to or attached to the exterior cell surface. Microorganisms at that site will be exposed to high antibiotic concentrations caused by efflux of drug out of the tissues. Additionally drug may be delivered to the infection site by macrophages and released there, thus further increasing the drug concentration to which the pathogen is exposed. This mechanism is probably applicable to both azithromycin and clarithromycin.

3. Metabolism

Azithromycin is essentially unchanged within the body and has no known active metabolites [78]. Data derived from the rat and dog suggest that only a small amount of systemically absorbed drug is metabolized, primarily via demethylation at 3′-*N*-dimethylamino group [91]. It has been suggested that as many as 10 metabolites of azithromycin appear in the bile, all of which possess insignificant antimicrobial activity [78]. To date, published descriptions of azithromycin disposition characteristics fail to delineate its metabolic fate and the activity and fate of any metabolites in humans [13]. Azithromycin does not induce cytochrome *P*-450 enzymes [46].

E. Telithromycin (HMR 3647)

During the last decade, antimicrobial resistance has spread at an alarming rate among respiratory tract pathogens and represents a significant public health threat [92–94]. New antimicrobial agents are, therefore, urgently needed to treat respiratory tract infections. Telithromycin is the first of a new family of antimicrobial agents, the ketolides, and has potent *in vitro* activity against both common and atypical respiratory pathogens, including those exhibiting β-lactam and/or macrolide resistance [95–97]. Telithromycin was derived from erythromycin A, with a 3-keto group on the 14-membered macrolactone ring instead of a cladinose sugar and an *N*-substituted cyclic carbamate moiety at positions 11 and 12.

1. Absorption and Elimination

In a randomized, three-period crossover study, healthy subjects received a single oral dose of 400, 800, or 1600 mg of telithromycin followed 4 days later by the same dose once daily for 7 days. Following oral administration, telithromycin was rapidly absorbed, reaching C_{max} within 1 hr of dosing. Telithromycin achieved steady state within 2–3 days of once-daily dosing. The pharmacokinetic parameters deviated moderately from dose proportionality. At a dose of 800 mg/kg, telithromycin attained mean maximal and trough plasma concentrations of 2.27 and 0.070 mg/l, respectively. Elimination was biphasic; initial and terminal half-lives were 2.87 and 9.81 hr. After 7 days of dosing, moderate accumulation was seen of telithromycin with AUC values approximately 1.5-fold higher than those attained following a single dose. This moderate accumulation might be explained by a slight decrease in nonrenal clearance with multiple dosing, since the main elimination half-life increased by 20–30% while $CL_{r(0-24)}$ remained unchanged [98]. In another study, data from a 10-day study in healthy subjects suggest that a once-daily 800-mg oral dose of telithromycin is tolerated well and provides adequate plasma concentrations to maintain activity against the major respiratory pathogens, including resistant strains [99]. The rate and extent of absorption is unaffected by food; thus telithromycin can be administered without regard for meals [100].

Telithromycin and its metabolites were rapidly eliminated, primarily by the fecal route. Ninety-three percent of the radioactive dose was recovered in 3 days [101]. Following administration of a single oral dose of telithromycin of 400, 800 or 1600 mg, 7.64, 13.0, and 19.0%, respectively, of the dose was eliminated unchanged in the urine in 3 days. Corresponding values following the final dose of the 7-day treatment period were 9.93, 18.4, and 25.8% [98].

Telithromycin was rapidly absorbed in elderly subjects, reaching a C_{max} of 3.0 mg/l within 30 min of an 800-mg dosing. Steady-state trough plasma concentrations were reached after 2 days of once-daily 800-mg dosing for 10 consecutive days, with a mean trough concentration of 0.14 mg/l [102].

2. Distribution

In vitro studies have shown that radiolabeled telithromycin is avidly concentrated by neutrophils, reaching intracellular levels 348-fold higher than extracellular levels [103]. Telithromycin is rapidly concentrated by white blood cells (WBCs), reaching levels 44-fold greater than in plasma 1 hr after 600-mg dosing to healthy subjects. Telithromycin is retained in WBCs so that the WBC concentration exceeds the MIC of respiratory pathogens 48 hr after multiple dosing. WBC concentrations were 20.9 and 8.9 mg/l 24 and 48 hr after the last dose following dosing for 10 days. Concentration of telithromycin in WBCs may facilitate transport of this agent to the site of infection [104].

Telithromycin rapidly penetrates bronchopulmonary tissue, achieving high and sustained concentrations, particularly in epithelial lining fluid and alveolar macrophages following 5 days of treatment with 800 mg once daily to healthy subjects. ELF concentrations far exceed the MIC of the common etiologic agents of pneumonia (maximum: 5.4 mg/l). The concentrations in AM peaked at 100 mg/l, 8 hr after drug administration. At the last time point (48 hr), the levels in AM were 2.15 mg/l. The high concentrations of telithromycin reached in AM are suggestive of good efficacy against facultative or obligate intracellular pathogens such as *Legionella* and *Chlamydia* spp. [105].

3. Metabolism

After oral administration of 800 mg of telithromycin labeled with ^{14}C to healthy subjects, two-thirds of the administered dose was eliminated as metabolites and one-third was unchanged. Telithromycin is the main circulating plasma species and represents 57% of plasma radioactivity. RU 76363, the main metabolite found in plasma and urine, is 4- to 16-fold less active than telithromycin *in vitro* (Fig. 10) [101]. The AUC_{0-24} of this metabolite was 10–12% that of the parent compound, a figure that was constant across the dose range [98]. RU 78849 was the main metabolite in feces [101].

F. ABT-773

ABT-773, 11-amino-3-*O*-descladinosyl-11-deoxy-3-oxo-6-*O*-(3″-quinolyl-2′-propenyl) erythromycin A 11,12-cyclic carbamate, is a novel ketolide antimicrobial being developed for clinical use. ABT-773 has demonstrated *in vitro* activity against community-acquired respiratory pathogens including penicillin/erythromycin sensitive and resistant strains of *S. pneumoniae*. Due to its *in vitro* activity against resistant pathogens, ABT-773 may represent another therapeutic option for community-acquired pneumonia and other respiratory infections [106]. To date, published pharmacokinetic information about clinical trials is not available. The following preclinical results could be used to assist in the prediction of potential *in vivo* human pharmacokinetic profiles with ABT-773.

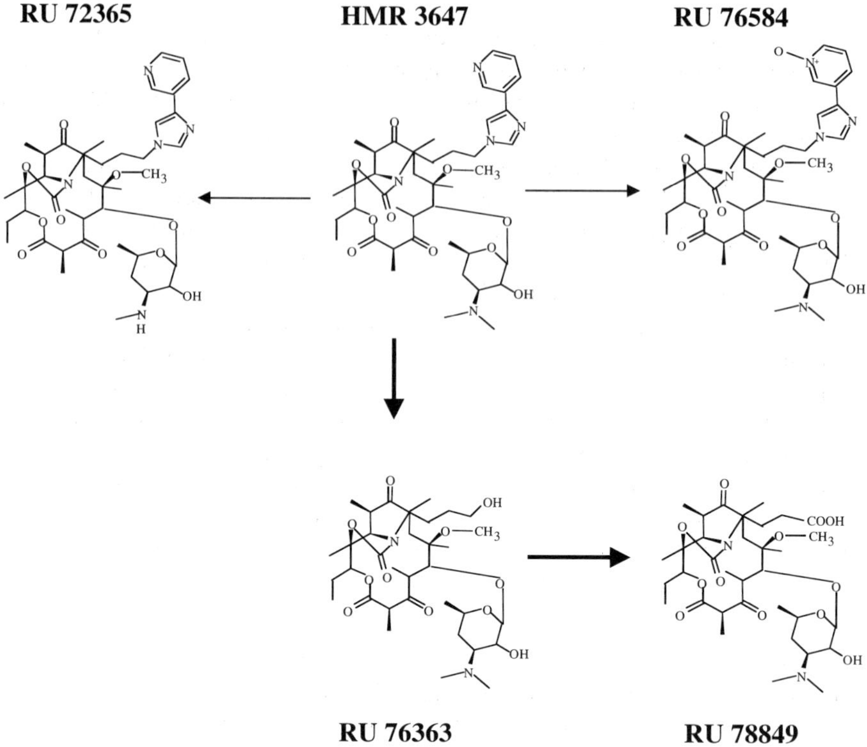

Fig. 10. Metabolic pathways of telithromycin (HMR 3647). (From Sultan *et al.* [102], Fig. 3, p. 54.)

1. Absorption and Elimination

The pharmacokinetic profile of ABT-773 was evaluated in monkeys (10 mg/kg), dogs (5 mg/kg), rats (20 mg/kg), and mice (10 mg/kg), using the HPLC method. Parent compound was distributed rapidly after intravenous dosing, with terminal elimination half-lives averaging 1.6, 4.5, 3.0, and 5.9 hr in mice, rats, monkeys, and dogs, respectively. Volume of distribution (V_β) values ranged from 1.5 l/kg in dog to 9.2 l/kg in the rats. After oral dosing, peak plasma concentrations averaged 1.47, 0.52, 0.56, and 0.84 μg/ml with bioavailability of 49.5, 60.0, 35.8, and 44.1% in mice, rats, monkeys, and dogs, respectively [107].

^{14}C-Labeled ABT-773 was rapidly cleared in rats, dogs, and monkeys with the excretion of parent drug and several metabolites almost quantitatively 3–5 days after dosing. Excretion was primarily via hepatic metabolism with small fractions in urine in all three species [108].

2. Distribution

Rat tissue distribution studies indicated that ^{14}C-ABT-773 was widely distributed throughout the body after a single oral dose. The concentrations in most tissues, except the brain, were several times higher than the circulating drug levels.

One of the highest concentrations was found in the lungs (tissue to plasma ratio >35:1) [108]. Lung concentrations of the parent compound measured by HPLC were >25-fold higher than plasma concentrations [107]. The distribution into lungs is clinically relevant since it is the primary site of antibacterial action of this drug. The disappearance of radioactivity from tissues was relatively rapid with most cleared 24 hr after a single oral dose [108].

3. Metabolism

^{14}C-ABT-773 was metabolized by liver microsomes and hepatocytes from mouse, rat, dog, monkey, and human. The metabolic pathway was oxidation, and its profile was similar in all species to give N-desmethyl-ABT-773 (M-1) as the major metabolite (Fig. 11). The kinetics of ABT-773 metabolism were examined in four human liver microsomes over a drug concentration range of 1.5–50 µM. The metabolism of ABT-773 followed monophasic Michaelis–Menten kinetics with K_m = 22.3 µM and V_{max} = 5.2 nmol/mg protein/min [108].

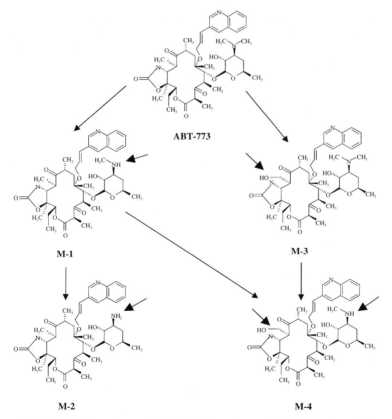

Fig. 11. Metabolic pathways of ABT-773. (From Guan *et al.* [108], Fig. 1, p. 350.)

The major metabolite and several other oxidative metabolites of ^{14}C-ABT-773 were isolated and purified from rat and monkey bile. As noted in the *in vitro* studies, the metabolic pathway was primarily oxidation, giving *N*-desmethyl-ABT-773 as the major metabolite and several other oxidative products as minor metabolites. In rat, dog, or monkey plasma, parent compound was the only or major component in circulating radioactivity, with M-1 as the major metabolite. The metabolic profile was similar in bile and feces with M-1, M-4 (10-hydroxy-*N*-desmethyl ABT-773), or M-6 (oxidation of ABT-773) as major metabolites in all species. In urine, the parent compound was the major component with M-1 as the primary metabolite [108].

III. Drug Interaction

Macrolide antibiotics are known to cause a variety of drug interactions. This may become clinically significant when erythromycin is administered concurrently with medications such as theophylline [109], warfarin [110], carbamazepine [111], digitalis, and cyclosporine [112]. The primary mechanism by which macrolide antibiotics interact with other drugs is inhibition of hepatic microsomal metabolism, but the magnitude of this effect varies among the macrolides (Tables VI and VII) [113]. The mechanism by which the inhibition of metabolism occurs is somewhat complex. Erythromycin drug–drug interactions have been extensively studied and have been the subject of numerous reviews [114]. Erythromycin and also troleandomycin actually induce some cytochrome *P*-450 drug metabolizing enzymes in the liver. The induced enzymes then metabolize the macrolide, which contains an aminosugar with a tertiary amine function,

TABLE VI
Inhibition of Hepatic Drug Metabolism by Macrolide Antibiotics

Macrolide	Comments
Azithromycin	Early evidence suggests that azithromycin does not inhibit hepatic drug metabolism. It does not appear to interact with carbamazepine, theophylline, or terfenadine.
Clarithromycin	Appears to act as an inhibitor of hepatic metabolism, but relative potency compared with erythromycin is not clear; it appears to inhibit the metabolism of theophylline, carbamazepine, and terfenadine.
Erythromycin	Inhibits the metabolism of various drugs; is known to inhibit cytochrome CYP3A4, but may affect other isozymes of cytochrome *P*-450 as well.
Troleandomycin	Inhibits hepatic drug metabolism; examples include carbamazepine, corticosteroids, and theophylline.

Source: From Horn and Hansten [113], Table 2, p. 83.

TABLE VII
Macrolide Drug Interactions of Potential Clinical Importance

Interacting drug	Comments
Astemizole	See terfenadine below.
Benzodiazepines	Erythromycin can considerably increase the serum concentrations of oral midazolam and triazolam (both of these benzodiazepines are metabolized by CYP3A4).
Bromocriptine	Preliminary study suggests that erythromycin markedly increases serum concentrations of bromocriptine.
Carbamazepine	Marked increase in plasma carbamazepine concentration with toxicity noted in several cases; confirmed in healthy subjects. Probably due to inhibition of carbamazepine metabolism.
Cyclosporine	Marked increases in plasma cyclosporine following erythromycin in several cases and healthy subjects; probably due to inhibition of cyclosporine metabolism.
Digoxin	Increased serum digoxin following erythromycin in selected patients (only 10% of population appears to be at risk); mechanism appears to be erythromycin-induced reduction in the bacterial degradation of digoxin in the intestine.
Felodipine	Isolated case reports suggest that erythromycin increases felodipine adverse effects (felodipine is metabolized by CYP3A4).
Lovastatin	Isolated cases of myopathy (muscle pain, weakness) with the combination of lovastatin and erythromycin but causal relationship not established.
Terfenadine	Terfenadine undergoes extensive first-pass metabolism, so almost no parent drug reaches the systemic circulation. Erythromycin and propably clarithromycin inhibit this metabolism, and the increased terfenadine serum concentration can result in cardiac arrhythmias (e.g., *torsades de pointes*). Early evidence suggests that astemizole can produce the same result if combined with erythromycin. Erythromycin also inhibits the metabolism of loratadine; however, loratadine does not appear to be cardiotoxic.
Theophylline	Increased serum theophylline, usually beginning only after several days of erythromycin; effect is modest in many cases, but severe toxicity has occurred. Clarithromycin and troleandomycin also inhibit theophylline metabolism.
Warfarin	Markedly enhanced hypoprothrombinemic response to warfarin following erythromycin noted in several cases; probably due to inhibition of warfarin metabolism. The interaction was less marked in controlled studies in healthy subjects.

Source: From Horn and Hansten [113], Table 3, p. 84.

R—N(CH$_3$)$_2$. This amine is demethylated and oxidized by cytochrome P-450 into a metabolite, probably the nitroso derivative (R—N=O), which forms a stable, inactive complex with the iron (Fe^{2+}) of cytochrome P-450 [115]. This complex prevents the enzyme from metabolizing other drugs any further [116]. These inactive complexes have been found in patients taking erythromycin. However, not all macrolides have been associated with the formation of these complexes [117]. The enzyme CYP3A4 appears to be most susceptible to inhibition by the macrolide antibiotics. This binding to and reduction in CYP3A4 activity is the basis for macrolide-associated metabolic drug–drug interactions. In the case of digoxin, the interaction with macrolide is thought to occur via the enhancement of digoxin bioavailability, secondary to inhibition of digoxin-metabolizing bacteria in the large bowel [118].

A. Roxithromycin

Roxithromycin has an effect on theophylline kinetics that is similar to that shown by erythromycin, but of lesser magnitude, with an increased elimination half-life of theophylline of about 8%, and decreased theophylline oral clearance of 16%. The dose of theophylline may not need to be reduced in the case of roxithromycin [119]. However, drug monitoring of theophylline blood levels during concomitant therapy is recommended [23]. It was shown previously that roxithromycin inhibits the CYP3A4-dependent oxidation of testosterone and nifedipine to a lesser extent than do erythromycin and troleandomycin in human liver microsomal and recombinant human *P*-450 systems *in vitro* [37]. This difference may be the result of roxithromycin being metabolized to M1, M2, and M3 relatively slowly by *P*-450 enzymes in human liver microsomes [38]. Roxithromycin itself is not as potent an inhibitor of CYP3A4 activities as are erythromycin and troleandomycin. The *N*-demethylamine derivative (M3) of roxithromycin is a more potent inhibitor of CYP3A4, after activation by CYP3A4 itself, than is the parent drug roxithromycin [35]. Roxithromycin appears to produce fewer clinically significant drug interactions [120].

Another potential for drug interaction of roxithromycin may involve changes in binding of drugs to plasma proteins [26, 39]. In *in vitro* binding experiments, Zini *et al.* [26] found that disopyramide displaced roxithromycin and increased the free concentration of roxithromycin. The magnitude of the increase was not reported. Similar results were not seen with lidocaine or prednisolone [26]. Thus, the clinical significance of this interaction is unknown.

B. Clarithromycin

As with erythromycin, drug interactions are extremely important with clarithromycin. Because clarithromycin inhibits the hepatic cytochrome *P*-450 system, it may result in increased levels of multiple medications metabolized by the liver [19]. Clarithromycin appeared to increase the mean steady-state plasma theophylline concentration and AUC from 15.6 µg/ml and 249 µg hr/ml, respectively, in the absence of clarithromycin, to 18.4 µg/ml and 291 µg hr/ml in the presence of clarithromycin ($p < 0.001$ for both concentration and AUC). Although a modest increase was seen in the plasma theophylline concentration, the concentration remained within the therapeutic range, and concurrent administration of clarithromycin and theophylline was safe and well tolerated [121]. Although the magnitude of elevation in theophylline was small, caution should be used when adding clarithromycin to existing theophylline regimens. A significant ($p < 0.05$) reduction in the systemic clearance of midazolam from 28 ± 9 to 10 ± 3 l/hr occurred after clarithromycin was administered in healthy subjects. Oral midazolam availability was significantly increased from 0.31 ± 0.1 to 0.75 ± 0.2 after clarithromycin dosing [122]. On the other hand, interactions of clarithromycin with zidovudine resulting in lower zidovudine levels have been reported [123, 124], and similar interactions have been reported with other antiretroviral agents [125].

Potent inhibitors ($K_i \leq 1.0$ μM) of CYP3A are likely to have a pronounced effect on the metabolism of clarithromycin. This is clinically relevant because 14-(R)-hydroxy-clarithromycin is a pharmacologically active metabolite [72, 126]. In agreement, in patients receiving ritonavir, increased plasma concentrations of both ritonavir and clarithromycin and decreased 14-(R)-hydroxy-clarithromycin (AUC: 15.7 μg·hr/ml vs. 0.04 μg·hr/ml) have been reported [127]. Interactions via the P-450 system also occur between clarithromycin and both rifabutin and rifampicin. Both of the latter agents induce CYP3A and the metabolism of clarithromycin. The administration of rifampicin has been shown to decrease the mean serum level of clarithromycin (5.4 to 0.7 μg/ml) in subjects with lung disease due to MAC [128]. In addition, because of the presence of CYP3A in the gut, it is probable that the first-pass metabolism of clarithromycin is mediated by both the intestine and liver. In accordance, the gut is thought to play a role in the observed interaction between cyclosporine and clarithromycin [129]. In this instance, clarithromycin enhances the oral bioavailability of cyclosporine by inhibiting CYP3A in the gut and liver. The potential exists for an interaction between commonly administered antihistamine terfenadine and clarithromycin, such that clarithromycin should not be administered to terfenadine recipients with cardiac or electrolyte abnormalities [130].

C. Azithromycin

Studies of the concomitant use of azithromycin with carbamazepine, terfenadine, and zidovudine have not reported drug interactions [131–133]. With the potential exception of antacids, no drug interactions have been reported with azithromycin, which does not appear to be metabolized by the cytochrome P-450 system [4, 134]. Both azithromycin and clarithromycin have been associated with digoxin toxicity. The postulated mechanism is by eradication of *Eubacterium lentum*, an anaerobic gram-positive bacteria responsible for the metabolism of digoxin in some patients [43].

D. Telithromycin

Telithromycin does not form nitrosoalkane inhibitor complexes with cytochrome P-450 (especially CYP3A) in rat microsomes *in vitro* [135]. These data suggest that common drug–drug interactions with some macrolides described above are unlikely to occur with telithromycin in man.

IV. Concluding Remarks

The pharmacokinetics of macrolides is complicated and factors such as the half-life of the compound in blood and deep compartments, its tissue affinity including its relative serum/tissue concentrations, and its intracellular penetration all need to be taken into account when choosing a macrolide. On the basis of pharmacokinetic profiles, new macrolides offer significant improvements over erythromycin and a choice between (1) roxithromycin, which offers much improved serum concentrations, an improved half-life, and greater cellular penetration, but retains high protein binding; (2) azithromycin, which has lower serum levels but a very prolonged half-life, low protein binding and high tissue and cellular concentrations; and (3) clarithromycin, which offers a combination of the beneficial attributes of the first two agents. Debate is ongoing regarding the pharmacokinetics/pharmacodynamics of the new macrolides and whether they are concentration dependent or independent. The parameter most closely associated with efficacy for erythromycin seems to be the time above MIC ($T >$ MIC), whereas for azithromycin, the AUC correlates best with outcome, because this agent concentrates highly in tissues and WBCs, which then act as a delivery system to the site of infection. On the other hand, both the $T >$ MIC and AUC values are linked to outcome for clarithromycin, and so it seems to be a combination of a time- and a concentration-dependent macrolide antibiotic [3].

Review of *in vitro* evaluations and clinical studies suggests that 14-membered ring macrolides appear to be associated with the propensity to interact with other CYP 3A4-metabolized drugs, whereas metabolic drug–drug interactions are very unlikely with 15- or 16-membered ring compounds. New macrolides, especially clarithromycin, should be used cautiously in patients receiving concomitant medications that are metabolized by the cytochrome *P*-450 system and are known to interact with erythromycin.

References

1. Bryskier, A. (1992). Newer macrolides and their potential target organisms. *Curr. Opin. Infect. Dis.* **5**, 764–772.
2. Mazzei, T., Mini, E., Novelli, A., and Periti, P. (1993). Chemistry and mode of action of macrolides. *J. Antimicrob. Chemother.* **31** (Suppl. C), 1–9.
3. Periti, P., and Mazzei, T. (1999). Clarithromycin: Pharmacokinetic and pharmacodynamic interrelationships and dosage regimen. *J. Chemother.* **11**, 11–27.
4. Periti, P., Mazzei, T., Mini, E., and Novelli, A. (1992). Pharmacokinetic interactions of macrolides. *Clin. Pharmacokinet.* **23**, 106–131.
5. Von Rosenstiel, N. A., and Adam, D. (1995). Macrolide antibacterials: Drug interactions of clinical significance. *Drug Safety* **13**, 105–122.
6. Amsden, G. W. (1995). Macrolides versus azalides: A drug interaction update. *Ann. Pharmacother.* **29**, 906–917.

7. Williams, J. D., and Sefton, A. M. (1993). Comparison of macrolide antibiotics. *J. Antimicrob. Chemother.* **31** (Suppl. C), 11–26.
8. Lazarevski, T., Radobolja, G., and Djokic, S. (1978). Erythromycin VI: Kinetics of acid-catalyzed hydrolysis of erythromycin oxime and erythromycylamine. *J. Pharm. Sci.* **67**, 1031–1033.
9. Wilson, J. T., and vanBoxtel, C. J. (1978). Pharmacokinetics of erythromycin in man. *Antibiot. Chemother.* **25**, 181–203.
10. Omura, S., Tsuzuki, K., and Sunazuka, T. (1987). Macrolides with gastrointestinal motor stimulating activity. *J. Med. Chem.* **30**, 1941–1943.
11. Kirst, H. A., and Sides, G. D. (1989). New directions for macrolide antibiotics: Structural modifications and *in vitro* activity. *Antimicrob. Agents Chemother.* **33**, 1413–1418.
12. Ginsburg, C. M. (1986). Pharmacology of erythromycin in infants and children. *Pediatr. Infect. Dis. J.* **5**, 124–129.
13. Reed, M. D., and Blumer, J. L. (1997). Azithromycin: A critical review of the first azilide antibiotic and its role in pediatric practice. *Pediatr. Infect. Dis. J.* **16**, 1069–1083.
14. Stein, G. E., and Havlichek, D. H. (1992). The new macrolide antibiotics. Azithromycin and clarithromycin. *Postgrad. Med.* **92**, 269–281.
15. Griffith, R. S., and Black, H. R. (1964). Comparison of the blood levels obtained after single and multiple doses of erythromycin estolate and erythromycin stearate. *Am. J. Med. Sci.* **247**, 69–74.
16. Osono, T., and Umezawa, H. (1985). Pharmacokinetics of macrolides, lincosamides and streptogramins. *J. Antimicrob. Chemother.* **16** (Suppl. A), 151–166.
17. Chun, A. H., and Seitz, J. A. (1977). Pharmacokinetics and biological availability of erythromycin. *Infection* **5** (Suppl. 1), 14–22.
18. Philipson, A., Sabath, L. D., and Charles, D. (1973). Transplacental passage of erythromycin and clindamycin. *N. Engl. J. Med.* **288**, 1219–1221.
19. Alvarez-Elcoro, S., and Enzler, M. J. (1999). The macrolides: Erythromycin, clarithromycin, and azithromycin. *Mayo Clin. Proc.* **74**, 613–634.
20. Disse, B., Gundert-Remy, U., Weber, E., Andrassy, K., Sietzen, W., and Lang, A. (1986). Pharmacokinetics of erythromycin in patients with different degrees of renal impairment. *Int. J. Clin. Pharmacol. Ther. Toxicol.* **24**, 460–464.
21. Hall, K. W., Nightingale, C. M., Gibaldi, M., Nelson, E., Bates, T. R., and Di Santo, A. R. (1982). Pharmacokinetics of erythromycin in normal and alcoholic liver disease subjects. *J. Clin. Pharmacol.* **25**, 321–325.
22. Markham, A., and Faulds, D. (1994). Roxithromycin: An update of its antimicrobial activity, pharmacokinetic properties and therapeutic use. *Drugs* **48**, 297–326.
23. Nilsen, O. G. (1987). Comparative pharmacokinetics of macrolides. *J. Antimicrob. Chemother.* **20** (Suppl. B), 81–88.
24. Birkett, D. J., Robson, R. A., Grgurinovich, N., and Tonkin, A. (1990). Single oral dose pharmacokinetics of erythromycin and roxithromycin and the effects of chronic dosing. *Ther. Drug Monit.* **12**, 65–71.
25. Puri, S. K., and Lassman, H. B. (1987). Roxithromycin: A pharmacokinetic review of a macrolide. *J. Antimicrob. Chemother.* **20** (Suppl. B), 89–100.
26. Zini, R., Fournet, M. P., Barre, J., Tremblay, D., and Tillement, J. P. (1988). *In vitro* study of roxithromycin binding to serum proteins and erythrocytes in humans. *Br. J. Clin. Pract.* **42** (Suppl. 55), 54.
27. Bahal, N., and Nahata, M. C. (1992). The new macrolide antibiotics: Azithromycin, clarithromycin, dirithromycin, and roxithromycin. *Ann. Pharmacother.* **26**, 46–55.
28. Falchi, M., Teodori, F., Carraro, A., Cioce, C., Scaglione, F., and Braga, P. C. (1985). Penetration of erythromycin into tonsillar tissue. *Curr. Med. Res. Opin.* **9**, 611–615.

29. Tremblay, D., Mignot, A., Courand, L., and Manuel, C. (1986). Concentrations of roxithromycin (RU28965), a new macrolide, in lung tissue after repeated dosing. Presented at the Third World Conference on Clinical Pharmacology and Therapeutics (Stockholm). Abstr. No. 1205.
30. Bergogne-Bérézin, E. (1987). Tissue distribution of roxithromycin. *J. Antimicrob. Chemother.* **20** (Suppl. B), 113–120.
31. Brun, Y., Forey, F., and Gammondes, J. P. (1981). Levels of erythromycin in pulmonary tissue and bronchial mucus compared to those of amoxycillin. *J. Antimicrob. Chemother.* **8**, 459–466.
32. Cars, O. (1991). Pharmacokinetics of antibiotics in tissue and tissue fluids: A review. *Scand. J. Infect. Dis.* Suppl. **74**, 23–33.
33. Tremblay, D., Jaeger, H., Fourtillan, J. B., and Manual, C. (1988). Pharmacokinetics of three single doses (150, 300, 450 mg) of roxithromycin in young volunteers. *Br. J. Clin. Pract.* **42** (Suppl. 55), 51.
34. Koyama, M., Tateno, M., Shirotsuka, M., Yamamoto, M., Hirayama, M., Saitoh, K., and Okui, K. (1988). Absorption, metabolism and excretion of RU 28965 in humans. *Chemotherapy* **36**, 164–183.
35. Yamazaki, H., and Shimada, T. (1998). Comparative studies of *in vitro* inhibition of cytochrome P-450 3A4-dependent testosterone 6β-hydroxylation by roxithromycin and its metabolites, troleandomycin, and erythromycin. *Drug Metab. Dispos.* **26**, 1053–1057.
36. Delaforge, M., Sartori, E., and Mansuy, D. (1988). *In vivo* and *in vitro* effects of a new macrolide antibiotic roxithromycin on rat liver cytochrome P-450: Comparison with troleandomycin and erythromycin. *Chem. Biol. Interact.* **68**, 179–188.
37. Yamazaki, H., Urano, T., Hiroki, S., and Shimada, T. (1996). Effects of erythromycin and roxithromycin on oxidation of testosterone and nifedipine catalyzed by CYP3A4 in human liver microsomes. *J. Toxicol. Sci.* **21**, 215–226.
38. Yamazaki, H., Hiroki, S., Urano, T., Inoue, K., and Shimada, T. (1996). Effects of roxithromycin, erythromycin and troleandomycin on their *N*-demethylation by rat and human cytochrome P-450 enzymes. *Xenobiotica* **26**, 1143–1153.
39. Petri, P., and Mazzei, T. (1987). Pharmacokinetics of roxithromycin in renal and hepatic failure and drug interactions. *J. Antimicrob. Chemother.* **20** (Suppl. B), 107–112.
40. Tremblay, D., Verger, C., Saint-Salvi, B., Robinet, D., and Manuel, C. (1988). Pharmacokinetics of roxithromycin in chronic renal insufficiency. *Br. J. Clin. Pract.* **42** (Suppl. 55), 61–62.
41. Lebrec, D., Benhamou, J. P., Fourtillan, J. B., Tremblay, D., Saint-Salvi, B., and Manuel, C. (1988). Roxithromycin: Pharmacokinetics in patients suffering from alcoholic cirrhosis. *Br. J. Clin. Pract.* **42** (Suppl. 55), 63.
42. Morimoto, S., Takahashi, Y., Watanabe, Y., and Ōmura, S. (1984). Chemical modification of erythromycin. I. Synthesis and antibacterial activity of 6-*O*-methylerythromycins A. *J. Antibiot.* **37**, 187–189.
43. Zuckerman, J. M. (2000). The newer macrolides: Azithromycin and clarithromycin. *Infect. Dis. Clin. North Am.* **14**, 449–462.
44. Chu, S. Y., Deaton, R., and Cavanaugh, J. (1992). Absolute bioavailability of clarithromycin after oral administration in humans. *Antimicrob. Agents Chemother.* **36**, 1147–1150.
45. Hardy, D. J., Guay, D. R. P., and Jones, R. N. (1992). Clarithromycin, a unique macrolide: A pharmacokinetic, microbiological, and clinical overview. *Diagn. Microbiol. Infect. Dis.* **15**, 39–53.
46. Nightingale, C. H. (1997). Pharmacokinetics and pharmacodynamics of newer macrolides. *Pediatr. Infect. Dis. J.* **16**, 438–443.
47. Wise, R. (1989). The development of macrolides and related compounds—Leading article. *J. Antimicrob. Chemother.* **23**, 299–300.
48. Piscitelli, S. C., Danziger, L. H., and Rodvold, K. A. (1992). Clarithromycin and azithromycin: New macrolide antibiotics. *Clin. Pharm.* **11**, 137–152.
49. LeBel, M. (1993). Comparative pharmacokinetics of new macrolides. *Can. J. Infect. Dis.* **4**, 149–151.

50. Ferrero, J. L., Bopp, B. A., Marsh, K. C., Quigley, S. C., Johnson, M. J., Anderson, D. J., Lamm, J. E., Tolman, K. G., Sanders, S. W., Cavanaugh, J. H., and Sonders, R. C. (1990). Metabolism and disposition of clarithromycin in man. *Drug Metab. Dispos.* **18**, 441–446.
51. Chu, S. Y., Sennello, L. T., Bunnell, S. T., Varga, L. L., Wilson, D. S., and Sonders, R. C. (1992). Pharmacokinetics of clarithromycin, a new macrolide, after single ascending oral doses. *Antimicrob. Agents Chemother.* **36**, 2447–2453.
52. Chu, S. Y., Wilson, D. S., Deaton, R. L., Mackenthum, A. V., Eason, C. N., and Cavanaugh, J. H. (1993). Single- and multiple-dose pharmacokinetics of clarithromycin, a new macrolide antimicrobial. *J. Clin. Pharmacol.* **33**, 719–726.
53. Saito, A., Ishikawa, Y., Shinohara, M., Fukuhara, I., Nakayama, I., Tomizawa, M., and Sato, K. (1988). Preclinical and clinical studies on TE-031 (A-56268). *Chemotherapy (Tokyo)* **36** (Suppl. 3), 521–537.
54. Sedlmayr, T., Peters, F., Raasch, W., and Kees, F. (1993). Clarithromycin, a new macrolide antibiotic. Effectiveness in puerperal infections and pharmacokinetics in breast milk [in German]. *Geburtshilfe Frauenheilkd.* **53**, 488–491.
55. Carbon, C. (1995). Clinical relevance of intracellular and extracellular concentrations of macrolides. *Infection* **23**, 10–14.
56. Peters, D. H., and Clissold, S. P. (1992). Clarithromycin; A review of its antimicrobial activity, pharmacokinetic properties and therapeutic potential. *Drugs* **44**, 117–164.
57. Fish, D. N., Gotfried, M. H., Danziger, L. H., and Rodvold, K. A. (1994). Penetration of clarithromycin into lung tissues from patients undergoing lung resection. *Antimicrob. Agents Chemother.* **38**, 876–878.
58. Langtry, H. D., and Brogden, R. N. (1997). Clarithromycin. A review of its efficacy in the treatment of respiratory tract infections in immunocompetent patients. *Drugs* **53**, 973–1004.
59. Rodvold, K. A., Gotfried, M. H., Danziger, L. H., and Servi, R. J. (1997). Intrapulmonary steady-state concentrations of clarithromycin and azithromycin in healthy adult volunteers. *Antimicrob. Agents Chemother.* **41**, 1399–1402.
60. Kohno, Y., Yoshida, H., Suwa, T., and Suga, T. (1989). Comparative pharmacokinetics of clarithromycin (TE-031), a new macrolide antibiotic, and erythromycin in rats. *Antimicrob. Agents Chemother.* **33**, 751–756.
61. Kohno, Y., Yoshida, H., Suwa, T., and Suga, T. (1990). Uptake of clarithromycin by rat lung cells. *J. Antimicrob. Chemother.* **26**, 503–513.
62. Fietta, A., Merlini, C., and Gialdroni-Grassi, G. (1997). Requirements for intracellular accumulation and release of clarithromycin and azithromycin by human phagocytes. *J. Chemother.* **9**, 23–31.
63. Ishiguro, M., Koga, H., Kohno, S., Hayashi, T., Yamaguchi, K., and Hirota, M. (1989). Penetration of macrolides into human polymorphonuclear leukocytes. *J. Antimicrob. Chemother.* **24**, 719–729.
64. Raghoebar, M., Lindeyer, E., Van den Berg, W. B., and Van Ginneken, C. A. M. (1988). On the mechanisms of association of the macrolide antibiotic erythromycin with isolated human polymorphonuclear leucocytes. *Biochem. Pharmacol.* **38**, 3221–3227.
65. Laufen, H., Wildfeuer, A., and Lach, P. (1990). Mechanism of azithromycin uptake in human polymorphonuclear leucocytes. *Arzneim-Forsch./Drug Res.* **40**, 686–689.
66. Mtairag, E. M., Abdelghaffar, H., and Labro, M. T. (1994). Investigation of dirithromycin and erythromycylamine uptake by human neutrophils *in vitro*. *J. Antimicrob. Chemother.* **33**, 523–536.
67. Hand, W. L., and Hand, D. L. (1993). Interactions of dirithromycin with human polymorphonuclear leukocytes. *Antimicrob. Agents Chemother.* **37**, 2557–2562.
68. Prokesch, R. C., and Hand, W. L. (1982). Antibiotic entry into human polymorphonuclear leukocytes. *Antimicrob. Agents Chemother.* **21**, 373–380.

69. Hand, W. L., King-Thompson, N., and Holman, J. W. (1987). Entry of roxithromycin (RU965), imipenem, cefotaxime, trimethoprim and metronidazole into human polymorphonuclear leukocytes. *Antimicrob. Agents Chemother.* **31**, 1553–1557.
70. Laufen, H., and Wildfeuer, A. (1989). Kinetics of the uptake of antimicrobial agents by human polymorphonuclear leucocytes. *Arzneim-Forsch./Drug Res.* **39**, 233–235.
71. Rodrigues, A. D., Roberts, E. M., Mulford, D. J., Yao, Y., and Quellet, D. (1997). Oxidative metabolism of clarithromycin in the presence of human liver microsomes. Major role for the cytochrome P-450 3A (CYP3A) subfamily. *Drug Metab. Dispos.* **25**, 623–630.
72. Sturgill, M. G., and Rapp, R. P. (1992). Clarithromycin: Review of a new macrolide antibiotic with improved microbiological spectrum and favorable pharmacokinetic and adverse effect profiles. *Ann. Pharmacother.* **26**, 1099–1108.
73. Davey, P. G. (1991). The pharmacokinetics of clarithromycin and its 14-OH metabolite. *J. Hosp. Infect.* **19** (Suppl. A), 29–37.
74. Chu, S. Y., Granneman, G. R., Pichotta, P. J., Decourt, J. P., Girault, J., and Fourtillar, J. B. (1993). Effect of moderate or severe hepatic impairment on clarithromycin pharmacokinetics. *J. Clin. Pharmacol.* **33**, 480–485.
75. Zuckerman, J. M., and Kaye, K. M. (1995). The new macrolides. Azithromycin and clarithromycin. *Inf. Dis. Clin. North Am.* **9**, 731–745.
76. Foulds, G., Shepard, R. M., and Johnson, R. B. (1990). The pharmacokinetics of azithromycin in human serum and tissues. *J. Antimicrob. Chemother.* **25** (Suppl. A), 73–82.
77. Foulds, G., Luke, D. R., Teng, R., Willavize, S. A., Freidman, H., and Curatolo, W. J. (1996). The absence of an effect of food on the bioavailability of azithromycin administered as tablets, sachet or suspension. *J. Antimicrob. Chemother.* **37** (Suppl. C), 37–44.
78. Schentag, J. J., and Ballow, C. H. (1991). Tissue-directed pharmacokinetics. *Am. J. Med.* **91** (Suppl. 3A), 5S–11S.
79. Nahata, M. C., Koranyi, K. I., Luke, D. R., and Foulds, G. (1995). Pharmacokinetics of azithromycin in pediatric patients with acute otitis media. *Antimicrob. Agents Chemother.* **39**, 1875–1877.
80. Mazzei, T., Surrenti, C., Novelli, A., Crispo, A., Fallani, S., Carla, V., Surrenti, E., and Periti, P. (1993). Pharmacokinetics of azithromycin in patients with impaired hepatic function. *J. Antimicrob. Chemother.* **31** (Suppl. E), 57–63.
81. Dette, G. A., and Knothe, H. (1986). The binding of erythromycin in human serum. *Biochem. Pharmacol.* **35**, 959–966.
82. Bothe, R., Mattie, H., and van den Broek, P. J. (1995). Levels of azithromycin and alpha-1 acid glycoprotein in serum in patients with community-acquired pneumonia. *Antimicrob. Agents Chemother.* **39**, 2801–2802.
83. Girard, A. E., Girard, D., English, A. R., Gootz, T. D., Cimochowski, C. R., Faiella, J. A., Haskell, S. L., and Retsema, J. A. (1987). Pharmacokinetics and *in vivo* studies with azithromycin (CP-62,993), a new macrolide with an extended half-life and excellent tissue distribution. *Antimicrob. Agents Chemother.* **31**, 1948–1954.
84. Davila, D., Kolacny-Babic, L., and Plavsic, F. (1991). Pharmacokinetics of azithromycin after single oral dosing of experimental animals. *Biopharm. Drug Dispos.* **12**, 505–514.
85. Baldwin, D. R., Wise, R., Andrews, J. M., Ashby, J. P., and Honeybourne, D. (1990). Azithromycin concentrations at the sites of pulmonary infection. *Eur. Respir. J.* **3**, 886–890.
86. Morris, D. L., De Souza, A., Jones, J. A., and Morgan, W. E. (1991). High and prolonged pulmonary tissue concentrations of azithromycin following a single oral dose. *Eur. J. Clin. Microbiol. Infect. Dis.* **10**, 859–861.
87. Jaruratanasirikul, S., Hortiwakul, R., Tantisarasart, T., Phuenpathom, N., and Tussanasunthornwong, S. (1996). Distribution of azithromycin into brain tissue, cerebrospinal fluid, and aqueous humor of the eye. *Antimicrob. Agents Chemother.* **40**, 825–826.

88. Kelsey, J. J., Moser, L. R., Jennings, J. C., and Munger, M. A. (1994). Presence of azithromycin breast milk concentrations: A case report. *Am. J. Obstet. Gynecol.* **170**, 1375–1376.
89. Gladue, R. P., Bright, G. M., Isaacson, R. E., and Newborg, M. F. (1989). In vitro and in vivo uptake of azithromycin (CP-62,993) by phagocytic cells: Possible mechanism of delivery and release at sites of infection. *Antimicrob. Agents Chemother.* **33**, 277–282.
90. Bonnet, M., and Van der Auwera, P. (1992). *In vitro* and *in vivo* intraleukocytic accumulation of azithromycin (CP-62,993) and its influence on *ex vivo* leukocyte chemiluminescence. *Antimicrob. Agents Chemother.* **36**, 1302–1309.
91. Shepard, R. M., and Falkner, F. C. (1990). Pharmacokinetics of azithromycin in rats and dogs. *J. Antimicrob. Chemother.* **25** (Suppl. A), 49–60.
92. Doern, G. V., Brueggemann, A., Holley, H. P., and Rauch, A. M. (1996). Antimicrobial resistance of *Streptococcus pneumoniae* recovered from outpatients in the United States during the winter months of 1994 to 1995: Results of a 30-center national surveillance study. *Antimicrob. Agents Chemother.* **40**, 1208–1213.
93. Doern, G. V., Brueggemann, A., Pierce, G., Hogan, T., Holley, H. P., and Rauch, A. (1996). Prevalence of antimicrobial resistance among 723 outpatient clinical isolates of *Moraxella catarrhalis* in the United States in 1994 and 1995: Results of a 30-center national surveillance study. *Antimicrob. Agents Chemother.* **40**, 2884–2886.
94. Doern, G. V., Brueggemann, A., Pierce, G., Holley, H. P., and Rauch, A. (1997). Antibiotic resistance among clinical isolates of *Haemophilus influenzae* in the United States in 1994 to 1995 and detection positive strains resistant to amoxicillin-clavulanate: Results of national multicenter surveillance study. *Antimicrob. Agents Chemother.* **41**, 292–297.
95. Biedenbach, D. J., Barrett, M. S., and Jones, R. N. (1998). Comparative antimicrobial activity and kill-curve investigations of novel ketolide antimicrobial agents (HMR 3004 and HMR 3647) tested against *Haemophilus influenzae* and *Moraxella catarrhalis* strains. *Diagn. Microbiol. Infect. Dis.* **31**, 349–353.
96. Roblin, P. M., and Hammerschlag, M. R. (1998). *In vitro* activity of a new ketolide antibiotic, HMR 3647, against *Chlamydia pneumoniae*. *Antimicrob. Agents Chemother.* **42**, 1515–1516.
97. Pankuch, G. A., Visalli, M. A., Jacobs, M. R., and Appelbaum, P. C. (1998). Susceptibilities of penicillin- and erythromycin-susceptible and –resistant pneumococci to HMR 3647 (RU 66647), a new ketolide, compared with susceptibilities to 17 other agents. *Antimicrob. Agents Chemother.* **42**, 624–630.
98. Namour, F., Wessels, D. H., Pascual, M. H., Reynolds, D., Sultan, E., and Lenfant, B. (2001). Pharmacokinetics of the new ketolide telithromycin (HMR 3647) administered in ascending single and multiple doses. *Antimicrob. Agents Chemother.* **45**, 170–175.
99. Lenfant, B., Sultan, E., Wable, C., Pascual, M. H., and Meyer, B. H. (1998). Pharmacokinetics of 800-mg once-daily oral dosing of the ketolide HMR 3647 in healthy young volunteers. *In* "Program of the 38th Interscience Conference of Antimicrobial Agents and Chemotherapy," Abstr. No. A-49, p. 16. American Society of Microbiology, Washington, DC.
100. Lenfant, B., Perret, C., and Pascual, M. H. (1999). The bioavailability of HMR 3647, a new once-daily ketolide antimicrobial, is unaffected by food. *In* "Proceedings of the 21st International Congress on Chemotherapy," Vol. 44, p. 69. Oxford University Press, Oxford, UK.
101. Sultan, E., Namour, F., Mauriac, C., Lenfant, B., and Scholtz, H. (1999). HMR 3647, a ketolide antimicrobial, is metabolised and excreted mainly in faeces in man. *In* "Proceedings of the 21st International Congress on Chemotherapy," Vol. 44, p. 63. Oxford University Press, Oxford, UK.
102. Sultan, E., Lenfant, B., Wable, C., Pascual, M. H., and Meyer, B. (1999). Pharmacokinetic profile of HMR 3647 800 mg once-daily in elderly volunteers. *In* "Proceedings of the 21st International Congress on Chemotherapy," Vol. 44, p. 66. Oxford University Press, Oxford, UK.

103. Vazifeh, D., Preira, A., Bryskier, A., and Labro, M. T. (1998). Interactions between HMR 3647, a new ketolide, and human polymorphonuclear neutrophils. *Antimicrob. Agents Chemother.* **42**, 1944–1951.
104. Gia, H. P., Roeder, V., Namour, F., Sultan, E., and Lenfant, B. (1999). HMR 3647 achieves high and sustained concentrations in white blood cells in man. *In* "Proceedings of the 21st International Congress on Chemotherapy," Vol. 44, p. 79. Oxford University Press, Oxford, UK.
105. Serieys, C. M., Cantalloube, C., Soler, P., Lemaitre, F., Gia, H. P., Brunner, F., and Andremont, A. (1999). HMR 3647 achieves high and sustained concentrations in broncho-pulmonary tissues. *In* "Proceedings of the 21st International Congress on Chemotherapy," Vol. 44, p. 78. Oxford University Press, Oxford, UK.
106. Neuhauser, M. M., Prause, J. L., Li, D. H., Jung, R., Boyea, N., Hackleman, J. M., Danziger, L. H., and Pendland, S. L. (1999). *In vitro* bactericidal activity of ABT-773, a new ketolide, versus clarithromycin, azithromycin, ciprofloxacin, amoxicillin/clavulanate against penicillin/erythromycin sensitive and resistant *Streptococcus pneumoniae* and *Haemophilus influenzae*. *In* "Program of the 39th Interscience Conference of Antimicrobial Agents and Chemotherapy," Abstr. No. 2139, p. 347. American Society of Microbiology, Washington, DC.
107. Hernandez, L., Sadrzadeh, N., Krill, S., Ma, Z., and Marsh, K. (1999). Preclinical pharmacokinetic profile of ABT-773 in mouse, rat, monkey, and dog. *In* "Program of the 39th Interscience Conference of Antimicrobial Agents and Chemotherapy," Abstr. No. 2148, p. 350. American Society of Microbiology, Washington, DC.
108. Guan, Z., Jayanti, V., Johnson, M., Nequist, G., Reisch, T., Roberts, E., Schmidt, J., Rotert, G., Surber, B., Thomas, S., Rodriguez, C., Lee, R., Kumar, G., Roberts, S., and Lin, J. (1999). *In vitro* and *in vivo* metabolism of [^{14}C]ABT-773. *In* "Program of the 39th Interscience Conference of Antimicrobial Agents and Chemotherapy," Abstr. No. 2149, p. 350. American Society of Microbiology, Washington, DC.
109. Reisz, G., Pingletone, S. K., Melethil, S., and Ryan, P. B. (1983). The effect of erythromycin on theophylline pharmacokinetics in chronic bronchitis. *Am. Rev. Respir. Dis.* **127**, 581–584.
110. Backmann, K., Schwartz, J. I., Forney, R., Frogameni, A., and Jauregui, L. E. (1984). The effect of erythromycin on the disposition kinetics of warfarin. *Pharmacology* **28**, 171–176.
111. Wong, Y. Y., Ludden, T. M., and Bell, R. D. (1983). Effect of erythromycin on carbamazepine kinetics. *Clin. Pharmacol. Ther.* **33**, 460–464.
112. Martell, R., Heinrichs, D., Stiller, C. R., Jenner, M., Keown, P. A., and Dupre, J. (1986). The effects of erythromycin in patients treated with cyclosporine. *Ann. Intern. Med.* **104**, 660–661.
113. Horn, J. R., and Hansten, P. D. (1995). Drug interactions with antibacterial agents. *J. Fam. Pract.* **41**, 81–90.
114. Guay, D. R. P. (1995). Formulary management of macrolide antibiotics. *Pharmacoeconomics* **8**, 491–512.
115. Babany, G., Larrey, D., and Pessayre, D. (1988). Macrolide antibiotics as inducers and inhibitors of cytochrome P-450 in experimental animals and man. *In* "Progress in Drug Metabolism" (G. G. Gibson, Ed.), pp. 61–98. Taylor and Francis, London.
116. Delaforge, M., Jaouen, M., and Mansuy, D. (1983). Dual effects of macrolide antibiotics on rat liver cytochrome P-450. Induction and formation of metabolite complexes: A structure–activity relationship. *Biochem. Pharmacol.* **32**, 2309–2318.
117. Tinel, M., DeScatoire, V., Larrey, D., Loeper, J., Labbe, G., Letteron, P., and Pessayre, D. (1989). Effect of clarithromycin on cytochrome P-450. Comparison with other macrolides. *J. Pharmacol. Exp. Ther.* **250**, 746–751.
118. Rodin, S. M., and Johnson, B. F. (1988). Pharmacokinetic drug interactions with digoxin. *Clin. Pharmacokinet.* **15**, 227–244.

119. Surjus, A., Tremblay, D., Saint-Salvi, B., Granier, J., and Lefebvre, M. A. (1986). Pharmacokinetic interaction of a new macrolide, roxithromycin (RU 28965) with theophyllin. Presented at the Third World Conference on Clinical Pharmacology and Therapeutics (Stockholm). Abstr. No. 1203.
120. Farrington, E. (1998). Macrolide antibiotics. *Pediatr. Nurs.* **23**, 433–446.
121. Ruff, F., Chu, S. Y., Sonders, R. C., and Sennello, L. T. (1990). Effects of multiple doses of clarithromycin on the pharmacokinetics of theophylline. *In* "Program of the 30th Interscience Conference on Antimicrobial Agents and Chemotherapy," Abstr. No. 761, p. 213. American Society of Microbiology, Washington, DC.
122. Gorski, J. C., Jones, D. R., Haehner-Daniels, B. D., Hamman, M. A., O'Mara, E. M., and Hall, S. D. (1998). Pharmacokinetics and drug disposition. The contribution of intestinal and hepatic CYP3A to the interaction between midazolam and clarithromycin. *Clin. Pharmacol. Ther.* **64**, 133–143.
123. Vance, E., Watson-Bitar, M., Gustavson, L., and Kazanjian, P. (1995). Pharmacokinetics of clarithromycin and zidovudine in patients with AIDS. *Antimicrob. Agents Chemother.* **39**, 1355–1360.
124. Polis, M. A., Piscitelli, S. C., Vogel, S., Witebsky, F. G., Conville, P. S., Petty, B., Kovacs, J. A., Davey, R. T., Walker, R. E., Falloon, J., Metcalf, J. A., Craft, C., Lane, H. C., and Masur, H. (1997). Clarithromycin lowers plasma zidovudine levels in persons with human immunodeficiency virus infection. *Antimicrob. Agents Chemother.* **41**, 1709–1714.
125. Gillum, J. G., Bruzzese, V. L., Israel, D. S., Kaplowitz, L. G., and Polk, R. E. (1996). Effect of clarithromycin on the pharmacokinetics of 2′,3′-dideoxyinosine in patients who are seropositive for human immunodeficiency virus. *Clin. Infect. Dis.* **22**, 716–718.
126. Fraschini, F., Scaglione, F., and Demartini, G. (1993). Clarithromycin clinical pharmacokinetics. *Clin. Pharmacokinet.* **25**, 189–204.
127. Quellet, D., Hsu, A., Granneman, G. R., Carlson, G., Guenther, H., Mukherjee, D., Locke, C., and Leonard, J. M. (1996). Assessment of the pharmacokinetic interaction between ritonavir and clarithromycin. *Clin. Pharmacol. Ther.* **59**, 143.
128. Wallace, R. J., Brown, B. A., Griffith, D. E., Girard, W., and Tanaka, K. (1995). Reduced serum levels of clarithromycin in patients treated with multidrug regimens including rifampin or rifabutin for *Mycobacterium avium–M. intracellulare* infection. *J. Infect. Dis.* **171**, 747–750.
129. Sketris, I. S., Wright, M. R., and West, M. L. (1996). Possible role of the intestinal P-450 enzyme system in a cyclosporine–clarithromycin interaction. *Pharmacotherapy* **16**, 301–305.
130. Honig, P. K., Wortham, D. C., Zamani, K., and Cantilena, L. R. (1994). Comparison of the effect of the macrolide antibiotics erythromycin, clarithromycin and azithromycin on terfenadine steady-state pharmacokinetics and electrocardiographic parameters. *Drug Invest.* **7**, 148–156.
131. Rapeport, W. G., Dewland, P. M., Muirhead, D. C., and Forster, P. L. (1992). Lack of an interaction between azithromycin and carbamazepine. *Br. J. Clin. Pharmacol.* **33**, 551P.
132. Harris, S., Hilligoss, D. M., Colangelo, P. M., Eller, M., and Okerholm, R. (1995). Azithromycin and terfenadine: Lack of drug interaction. *Clin. Pharmacol. Ther.* **58**, 310–315.
133. Chave, J. P., Munafo, A., Chatton, J. Y., Dayer, P., Glauser, M. P., and Biollaz, J. (1992). Once-a-week azithromycin in AIDS patients: Tolerability, kinetics, and effects on zidovudine disposition. *Antimicrob. Agents Chemother.* **36**, 1013–1018.
134. Nahata, M. (1996). Drug interactions with azithromycin and the macrolides: An overview. *J. Antimicrob. Chemother.* **37**, 133–142.
135. Labbe, G., Flor, M., and Lenfant, B. (1999). HMR 3647, a new ketolide antimicrobial, does not inhibit cytochrome P-450 activity *in vitro*. *In* "Proceedings of the 21st International Congress on Chemotherapy," Vol. 44, p. 95. Oxford University Press, Oxford, UK.

Chapter 7

Antimicrobial Macrolides in Clinical Practice*

SALVADOR ALVAREZ-ELCORO
JOSEPH D. C. YAO

*Division of Infectious Diseases, Mayo Clinic
Jacksonville, Florida*

I. Introduction 363
II. Fourteen- and Fifteen-Membered Macrolides 364
   A. Erythromycin 364
   B. Clarithromycin 365
   C. Dirithromycin 369
   D. Roxithromycin 370
   E. Azithromycin 372
III. Sixteen-Membered Macrolides 380
   A. Rokitamycin 380
   B. Spiramycin 380
   C. Midecamycin 381
   D. Miokamycin 381
IV. Concluding Remarks 382
   References 382

I. Introduction

Macrolide antibiotics are a homogeneous group of antimicrobial drugs that have been used to treat clinical infections for several decades. The most clinically useful classification of the macrolides is based on the size of the lactone ring that forms the chemical nucleus of each macrolide molecule [1, 2]. The 14- and 15- membered macrolides include erythromycin, clarithromycin, dirithromycin, roxithromycin, and azithromycin. Erythromycin is the oldest and still the most important of the macrolide antibiotics because it is a useful alternative to penicillin G. It is one of the safest antibiotics available. Clarithromycin and azithromycin have shown some advantages over erythromycin in their antibacterial activity,

*Adapted in part from Alvarez-Elcoro, S., and Enzler, M. J. (1999). The macrolides: Erythromycin, clarithromycin, and azythromycin. *Mayo Clin. Proc.* **74**, 613–634.

pharmacokinetics, and adverse side effects. They also exhibit new indications for possible use in the therapy of certain AIDS-related opportunistic infections. An extensive review of these macrolides has been published in the *Mayo Clinic Proceedings* [3]. Roxithromycin has clinical applications similar to those of erythromycin but its use is largely limited to Europe, South America, and Japan. The 16-member macrolides (leucomycin, rokitamycin, spiramycin, midecamycin, and miokamycin) are still in the investigation stage in the United States and clinical experience with them worldwide is limited. They have *in vitro* antibacterial activities that are similar to those of erythromycin. For the chemical structures of those macrolides, see Chapter 3.

II. Fourteen- and Fifteen-Membered Macrolides

A. Erythromycin

Erythromycin is a natural macrolide derived from *Streptomyces erythreus* that contains a 14-member macrocyclic lactone ring to which is attached two sugar moieties, desosamine and cladinose. As with other macrolides, erythromycin inhibits RNA-dependent protein synthesis by reversibly binding to the 50S ribosomal subunits of susceptible microorganisms [4–6]. They induce dissociation of peptidyl-tRNA from the ribosome during the elongation phase, thereby suppressing RNA-dependent protein synthesis and thus inhibiting bacterial growth [2, 7]. The macrolides are generally bacteriostatic, although bactericidal activity may occur under certain conditions or against particular microorganisms [8, 9]. Erythromycin may interfere with the ribosomal binding of other antibiotics, such as lincomycin and chloramphenicol, suggesting overlapping or common binding sites for these antibiotics.

Erythromycin is the drug of choice for *Mycoplasma pneumoniae* infections, *Legionella* pneumonia, diphtheria, and pertussis, *Chlamydia trachomatis* pneumonia or conjunctivitis, and bacillary angiomatosis. It also has a number of important applications as an alternative drug to penicillin G. Erythromycin is used in the treatment of gastroenteritis caused by *Campylobacter jejuni*, although it does not appear to alter the clinical course of the disease when therapy is started 4 days or more after the onset of the symptoms [10]. Erythromycin is a safer alternative to tetracycline in the treatment of chlamydial pelvic infection during pregnancy [11]. Erythromycin is also effective in eradicating the acute or chronic carrier state of diphtheria.

Prior to 1997, erythromycin was an alternative antibiotic choice to amoxicillin for prophylaxis prior to dental, oral, esophageal, and respiratory procedures to prevent infective endocarditis (IE). Erythromycin is an alternative choice for treating a recurrence of acute rheumatic fever in patients allergic to penicillin [12]. Recently updated recommendations by the American Heart Association for IE prophylaxis have replaced erythromycin with clindamycin, cephalosporins, and the newer macrolides (clarithromycin and azithromycin), due to gastrointestinal intolerance to erythromycin and the unpredictable blood levels that can result from the use of the various erythromycin formulations. The recommended oral dose of erythromycin is 250 mg to 1 g every 6 hr. For intravenous administration, 0.5–1 g every 6 hr is recommended, but use of this route is limited due to the possibility of phlebitis resulting. A summary of some of the clinical indications for erythromycin is given in Table I.

B. Clarithromycin

Clarithromycin is one of the new macrolide antibiotics that is similar in structure to erythromycin [13]. As with other macrolide antibiotics, clarithromycin inhibits the synthesis of proteins by reversible binding of the 50S ribosomal subunits of the 70S ribosome in susceptible bacteria. The resulting blockage effectively halts and inhibits RNA-dependent protein synthesis [4, 5, 14, 15]. Clarithromycin is synthesized by substituting a methoxy group from the C-6 hydroxyl group of erythromycin. Clarithromycin is a 14-member macrolide with a lactone ring attached to two sugar moieties, which confers acid stability and improved antimicrobial and pharmacokinetic properties [2, 9]. A primary metabolite of clarithromycin is the 14R-hydroxy epimer that possesses antimicrobial activity, which is felt to have an additive or synergistic action with the parent compound against a variety of microorganisms.

As with erythromycin and other macrolides, clarithromycin has been used to treat a wide variety of acute upper and lower respiratory tract infections including sinusitis, pharyngitis, acute bronchitis, acute infectious exacerbations of chronic bronchitis (AIECB), and community-acquired pneumonia (CAP) caused by typical and atypical bacterial pathogens [13, 16–22]. Clarithromycin was compared with oral levofloxacin by the Canadian Sinusitis Study Group in a multicenter, double-blind, randomized study in the treatment of acute sinusitis. The study showed that clarithromycin was as effective as levofloxacin in the treatment of acute sinusitis. The group with clarithromycin experienced more adverse events (39.3% vs. 22.5%) when compared with levofloxacin [23]. Clarithromycin was also compared to amoxicillin/clavulanate in the treatment of acute maxillary sinusitis with similar efficacy and fewer side effects [24]. Clarithromycin and other macrolides are frequently used as initial empiric therapy for CAP because of the potential presence of atypical bacterial pathogens. Clarithromycin has been shown to be as effective as erythromycin, penicillin, ampicillin or amoxicillin, and cephalosporins for treatment of a variety of upper and lower respiratory infections [20, 25–30]. Clarithromycin was as effective in the treatment of community-acquired upper and lower respiratory tract infections in hospitals and community settings as beta-lactam agents (with and without beta-lactamase

TABLE I
Comparative Clinical Usage of Selected Macrolides

Organism/Disease	Erythromycin	Clarithromycin	Azithromycin	Dirithromycin	Roxithromycin	Rokitamycin
Mycobacteria						
MAC—prophylaxis	N	Y	Y	N	N	N
MAC—treatment[a]	N	Y	May be alternative	N	N	N
M. abscessus/chelonae	N	Y	N	N	N	N
M. leprae		In vivo		N	N	N
STD						
C. trachomatis[c]	Y	N	Y	N	Y	ND
LGV	Y	N	Promising	N	Y	ND
Chancroid	N	N	Y	N	N	ND
Neisseria gonorrhoeae	N	N	N[b]	N	N	ND
PID	N	N	Y(IV) ± MN	N	N	N
H. pylori	N	Y	Investigational	N	Y	ND
H. influenzae	N	++[c]	++++	++	+++	+++
Trachoma	Topical	N	Promising	N	Topical	ND
IE prophylaxis	N	Y	Y	N	N	N
Campylobacter	Y	Y	Y	Y	Y	Y
Malaria prophylaxis	N	N	Studies in progress	N	N	N
Lyme disease (early)	Y	N	Studies in progress	N	N	N
Bartonellosis	N	N	CR	N	N	N
Rhodococcus equi	N	N	CR	N	N	N
Babesia	N	N	CR	N	N	N
Toxoplasma prophylaxis	N	N	Y	N	Y	N

Source: Adapted in part from Alvarez-Elcoro and Enzler [3] by permission.
Abbreviations: N, no; Y, yes; CR, case reports support activity; IE, infective endocarditis; LGV, lymphgranuloma venereum; PID, pelvic inflammatory disease; MAC, M. avium complex; IV, intravenous; MN, metronidazole; STD, sexually transmitted disease; ND, no published data.
[a]Use in combination with at least one other antibiotic against MAC.
[b]For N. gonorrhoeae, 2 g single dose orally is effective, but excess side effects occur [212].
[c]Based on CDC guidelines [12a].

inhibitor), cephalosporins, and most of the other macrolides. Clarithromycin was similar in efficacy to azithromycin in comparative studies and is as effective as and better tolerated than erythromycin [22, 31, 32]. Clarithromycin appears to be better tolerated than erythromycin, with a lower incidence of gastrointestinal (GI) effects [16]. Clarithromycin has also been compared with the newer fluoroquinolones, sparfloxacin and grepafloxacin, in patients with lower respiratory tract infections with similar clinical and bacteriological responses and excellent tolerance [33, 34].

Clarithromycin has been compared with amoxicillin suspension in the treatment of children with lower respiratory tract infections. No significant differences were seen between the groups with respect to clinical cure rates and incidence and severity of adverse events, which generally were mild [35]. Five days of treatment with clarithromycin suspension was superior to 10 days of penicillin suspension in eradicating *Streptococcus pyogenes* in children with streptococcal pharyngitis [36].

Numerous studies have compared clarithromycin to loracarbef, amoxicillin/clavulanate, amoxicillin suspension, and cefaclor in the treatment of acute otitis media in children with similar clinical outcomes and side effects [37–41]. In mild to moderate skin and skin structure infections, clarithromycin was not significantly better than erythromycin (96–88% eradication rate), with similar GI-related events [42–47].

Clarithromycin was compared to cefadroxil and erythromycin in two large multicenter studies of patients with mild to moderate skin and skin structure infections. Clarithromycin was as effective and safe as cefadroxil and erythromycin, both in adults and children [45, 47].

Mycobacterium avium-intracellulare, also known as *Mycobacterium avium complex* (MAC), is a common infection in advanced HIV-infected patients and is reported to occur in 22% of patients with AIDS [48]. Only *Pneumocystis carinii* pneumonia and Kaposi's sarcoma were more frequent as AIDS defining events. In a randomized, placebo-controlled study Dautzenberg *et al.* [49] demonstrated the ability of clarithromycin alone to decrease MAC bacteremia in patients with AIDS. In another study, the same investigators showed the efficacy of high-dose clarithromycin (1500–2000 mg/day) in the treatment of disseminated MAC infection with bacteriological eradication of 63–98% of the patients in the low- and high-dose groups [50]. Clinical trials in the treatment of disseminated MAC infection in AIDS patients have shown clarithromycin to be effective when combined with other drugs [51]. Clarithromycin may exhibit synergism with ethambutol, rifampin, and an additive effect with clofazimine. Clarithromycin in combination with ethambutol and rifabutin improved functional status, decreased weight loss, and increased survival of AIDS patients with MAC infections, compared with a four-drug non-macrolide-containing regimen [52]. Recent studies have confirmed an increased survival rate with a regimen of clarithromycin, ethambutol, and rifabutin compared with a four-drug non-macrolide-containing regimen [53]. The addition of clofazimine to clarithromycin and ethambutol failed to improve clinical response [54]. Previous studies of a clarithromycin regimen of 1000 mg twice daily (b.i.d.) were associated with a shorter patient survival than a regimen of 500 mg b.i.d. [51, 55]. The 1000-mg b.i.d. regimen had similar efficacy compared with the lower dose of 750 mg b.i.d. [56]. Recent guidelines published by the U.S. Public Health Service for the treatment of disseminated MAC infection in patients with AIDS recommend the inclusion of clarithromycin or azithromycin and at least one more anti-MAC agent with proven activity, such as rifabutin or ethambutol [57–60]. A recent randomized, open-label trial compared azithromycin plus ethambutol versus clarithromycin plus ethambutol for therapy for MAC bacteremia in patients with AIDS. The combination with clarithromycin was significantly more effective in eradicating bacteremia (87.5% vs. 37.5%) [60]. Clarithromycin has been recently compared with azithromycin, each combined with ethambutol in a large double-blind study of 246 patients with disseminated MAC and infected with HIV. The rate of clearance of bacteremia was similar for clarithromycin and azithromycin [61]. Clarithromycin also plays a major role in the treatment of MAC lung infections in HIV-negative patients [62, 63].

Clarithromycin has been shown to be an effective prophylactic agent against MAC infection in patients with advanced HIV infection. In a prospective, double-blind, placebo-controlled trial, clarithromycin prevented 69% of the expected cases of MAC disease [64]. Other studies have demonstrated that clarithromycin alone or in combination with rifabutin prevents MAC infections in AIDS patients and prolongs survival [65, 66]. However, a large prospective study failed to show that the combination of clarithromycin plus rifabutin combination was more effective than clarithromycin alone [67]. Drug-resistant MAC has been reported in 29–58% of patients who developed disseminated infection while taking prophylaxis with clarithromycin [64, 67]. Resistance to clarithromycin and other macrolides is a serious potential problem due to cross-resistance with azithromycin that narrows the therapeutic options available for MAC disease [68].

Clarithromycin has been used successfully in the treatment of infections caused by rapid growing mycobacteria such as *Mycobacterium fortuitum* and *Mycobacterium chelonae* [69–77]. There are several reports of successful treatment of infections caused by these organisms with clarithromycin-containing regimens [78]. Clarithromycin has *in vitro* activity against *Mycobacterium leprae* and was shown to have significant bactericidal effect in the murine footpad model of *M. leprae* infection [79]. In limited human studies, clarithromycin has shown a promising role as a component of a multiple-drug therapy regimen for *M. leprae* infections [80–84].

Helicobacter infection is a common cause of gastritis and its eradication may be associated with cure of peptic ulcers. Clarithromycin has been shown to be an effective therapy for *Helicobacter pylori* infections [85–87]. Clarithromycin monotherapy has shown dose-dependent eradication rates varying from 15% to 54% [88]. Clarithromycin combined with tetracycline and bismuth salts showed eradication of *H. pylori* infection in 72% of patients treated [89]. Clarithromycin has been combined with omeprazole in a recent study resulting in the eradication of the pathogen in 78% of the patients 4 weeks after completion of a 2-week course of therapy [90, 91]. Clarithromycin and metronidazole combined with ranitidine is also an effective regimen for *H. pylori* infection [92]. However, several studies around the world have documented the changing patterns of antibiotic resistance of *H. pylori* with substantial resistance to metronidazole, and emerging resistance to clarithromycin [93–99]. Preliminary data suggest that clarithromycin and other macrolides may have a role in the therapy of other infectious diseases, such as toxoplasmosis, babesiosis, bacillary angiomatosis, and spirochetal infections [100–108].

The dose of clarithromycin is usually 250–500 mg orally twice daily for 7–14 days. There is no need to reduce the dosage for moderate to severe hepatic failure in the presence of normal renal function [109]. Clarithromycin should not be used in severe liver dysfunction because of lack of production of its active metabolite by the liver. As of 2001, no IV formulation of clarithromycin was available in the United States. Clarithromycin is a category C drug in pregnancy and its safety has not been established for children under age 12 [9, 110–114].

C. Dirithromycin

Dirithromycin is a 14-membered macrolide antibiotic approved for clinical use in the United States. Its *in vitro* activities are two- to fourfold less than those of erythromycin against gram-positive bacteria and a variety of clinical isolates, but it is fourfold more active than erythromycin against *Bordetella pertussis* [1, 8, 14]. This antibiotic is equal to erythromycin against *Campylobacter* spp. Its clinical use is limited to treatment of pharyngotonsillitis, mild-to-moderate respiratory tract infections, and skin and soft tissue infections [115]. The major advantage of dirithromycin over erythromycin is the favorable pharmacokinetic properties permitting once-daily oral administration of 500 mg.

Double-blind, randomized, multicenter clinical studies showed dirithromycin to be equally effective as erythromycin base in clinical (92–94% vs. 94–95%) and bacteriological (79–84% vs. 86–88%) responses for 10-day treatment of streptococcal pharyngitis and tonsillitis [116, 117]. Dirithromycin also exhibited comparable efficacies as penicillin V and miokamycin in the treatment of such infections [118, 119]. Very few clinical trials were published on the use of this macrolide in treating acute or chronic sinusitis.

Clinical efficacy and safety of dirithromycin in the treatment of acute bronchitis and acute exacerbations of chronic bronchitis are similar to those of erythromycin

[120–122], miokamycin [123], clarithromycin [124], azithromycin [125], and amoxicillin-clavulanate [126]. Clinical success was found in 84–95% of patients on dirithromycin for 5–7 days, with bacteriological response rates of 75–93%. Despite its relatively poor *in vitro* activity against *H. influenzae*, this drug demonstrated similar eradication rates as other macrolides (70–73% vs. 72–78%) [127]. Comparative clinical trials of this macrolide for CAP showed comparable efficacy to erythromycin, with favorable symptomatic responses of 92–98% and bacteriological eradication rates of 91–93% after 10–14 days of therapy [128–130].

In a double-blind, randomized, multicenter trial [131], dirithromycin (500 mg once daily) showed clinical (96% vs. 99%) and bacteriological (87% vs. 89%) responses that were comparable to those of erythromycin (base 250 mg four times daily) in a 7-day course of therapy for skin and skin structure infections. A larger study comparing 5 days of dirithromycin and 7 days of erythromycin demonstrated equal rates of clinical cure or improvement and bacteriological eradication between the two regimens, but dirithromycin caused significantly fewer GI side effects than erythromycin (4% vs. 8%) [132].

In spite of its *in vitro* activity against *H. pylori*, dirithromycin has not been effective in eradicating this organism *in vivo*, even in combination with metronidazole or proton pump inhibitor [133]. Dirithromycin is well tolerated, with gastrointestinal discomfort as the most frequently reported adverse effect [134]. The incidence and nature of abnormal clinical laboratory evaluation are similar to those of other macrolides, and these infrequent test abnormalities do not result in clinical manifestations, even in elderly patients [135].

D. Roxithromycin

Roxithromycin is a semisynthetic 14-membered ring macrolide in which the erythronolide A lactone ring is modified to prevent inactivation by gastric acid [136]. Therapeutic usage of this antibiotic is found in Europe, South America and Japan, where extensive clinical studies have been done. Its pharmacokinetic profile is characterized by high plasma, tissue, and body fluid concentrations and a long half-life of 15 hr [137]. The drug is well tolerated orally at either a dose of 150 mg twice daily or 300 mg once daily, with peak and trough serum concentrations of 7 or 11 µg and 2.5 or 3 µg, respectively [138, 139]. Roxithromycin has proven clinical efficacy in oral and dental infections, upper and lower respiratory tract infections, skin and soft tissue infections, and urogenital infections.

In a multicenter, double-blind study of acute oral and dental infections in adults, roxithromycin 150 mg twice daily was equally effective as erythromycin ethylsuccinate 1 g twice daily for a mean duration of therapy of 8 days [140]. The success rates of 94% and 91%, respectively, were comparable irrespective of whether surgery was performed, and the trial included patients with dental abscesses.

Given as 10-day therapy, this macrolide antibiotic was as efficacious as amoxicillin-clavulanic acid, azithromycin, and clarithromycin in the treatment of acute pharyngitis, tonsillitis, acute sinusitis, and acute otitis media [141–144]. It is equivalent to doxycycline in achieving successful clinical response (81% vs. 80% response rates) for acute exacerbations of chronic bronchitis [145]. In a large international multicenter, noncomparative, open clinical trial of 40,000 patients, 7- to 14-day therapy with roxithromycin resulted in >95% rates of clinical resolution and improvement in upper and lower respiratory tract infections (LRTI), with side effects reported in only 4% of patients (mainly GI irritability) [146]. Other randomized clinical studies of mild to moderate LRTI, including atypical pneumonia, showed this drug to be comparable to clarithromycin [147, 148] and azithromycin [149–151] in clinical efficacy and tolerability. Compared with erythromycin [152] or amoxicillin-clavulanic acid [153–156], roxithromycin was equally as effective but better tolerated. Although more effective than cefaclor [157, 158] and cefixime [159], this drug was less efficacious than sparfloxacin (79% vs. 94% success rates) [160] in treating patients with acute LRTI. As a macrolide antibiotic, it is useful in the treatment of mycoplasma pneumonia [161].

For skin and soft tissue infections, roxithromycin is an effective and well-tolerated therapy for erysipelas and acne [162, 163]. As with other macrolide antibiotics, its immunomodulatory effects make it useful as an adjunctive therapy of psoriasis vulgaris [164]. Despite its *in vitro* activity against *Borrelia burgdorferi*, monotherapy with this macrolide was not effective for the treatment of Lyme borreliosis [165]. However, a small, nonrandomized, open prospective clinical study of 17 patients with confirmed late Lyme disease (stage II/III) showed a 76% complete recovery rate from a therapeutic combination of roxithromycin 300 mg b.i.d. and co-trimoxazole for 5 weeks [166]. This success rate is similar to that seen with intravenous penicillin and ceftriaxone.

Although having similar clinical cure rates in the treatment of chlamydial conjunctivitis in newborns and adults, roxithromycin achieved a higher microbiological cure rate than erythromycin (92% vs. 57%) after 10 days of oral therapy [167]. However, such therapy may not eradicate chlamydial colonization of the nasopharynx and genital tract [168]. In a single-blind, randomized therapeutic trial of genital chlamydial infection and nonspecific urethritis, roxithromycin was comparable to doxycycline in clinical (81% vs. 85%) and microbiological (92% vs. 100%) cure rates, with fewer drug-related side effects from roxithromycin (19%) than doxycycline (35%) [169]. Similar findings were demonstrated in double-blind, randomized study trials [170, 171].

With increasing evidence for a causative association between *Chlamydia pneumoniae* and coronary heart disease, roxithromycin has been investigated for its anti-chlamydial and anti-inflammatory effects on clinical outcome in patients with unstable angina [172, 173]. Compared with placebo in a double-blind, randomized, prospective, multicenter trial, the drug given at 150 mg b.i.d. for 30 days resulted in a statistically significant reduction in rates of myocardial infarction and ischemic death at day 30 (4% vs. 0%), day 90 (6.3% vs. 0%), and day 180 (7.3% vs. 2.2%) after therapy. While anti-*C. pneumoniae* IgG titers were unchanged, C-reactive protein levels decreased more significantly in the roxithromycin arm than the placebo. The role of roxithromycin and other macrolide antibiotics in the therapy of coronary atherosclerosis remains controversial.

When combined with proton pump inhibitor (PPI) and other antimicrobials for the therapy of duodenal ulcer and chronic active gastritis, roxithromycin showed excellent activity in eradicating *H. pylori*. Eradication rates of 80–100% as determined by urease test, histologic analysis, and microbiologic cultures were achieved in various randomized open clinical trials [174–176]. However, lower rates of cure (50–67%) were found in larger randomized comparative studies [177, 178].

Used as a prophylactic agent for rheumatic fever, 150 mg roxithromycin taken once daily was found to be a suitable alternative to penicillin and erythromycin [179]. In neutropenic patients with acute leukemia and in bone marrow transplant recipients receiving ofloxacin prophylaxis, the addition of roxithromycin resulted in less bacteremia due to viridans group streptococci (0% vs. 9% in ofloxacin alone). However, incidences of bacteremia caused by other organisms, febrile episodes from any cause, infection-associated complications, and antimicrobial usage for therapy were similar between the two groups [180].

When studied as a 4-week therapy at 300 mg b.i.d. for cryptosporidial diarrhea, roxithromycin produced symptomatic improvement and clinical cure rates of 25% and 50%, respectively, in patients with AIDS [181, 182]. The drug also showed efficacy in HIV-infected patients for the prevention of *Pneumocystis carinii* pneumonia and cerebral toxoplasmosis [183]. Given once weekly, 900 mg of roxithromycin was more effective than once monthly aerosolized pentamidine 300 mg therapy in preventing these HIV-related opportunistic infections.

Roxithromycin is usually well tolerated with a lower incidence of side effects than erythromycin. Minor adverse effects include GI upset, abdominal pain, elevated serum transaminase levels, and, rarely, skin rash.

E. Azithromycin

Azithromycin is a newer macrolide that was produced to overcome some of the shortcomings of erythromycin such as GI intolerance, pharmacokinetics, and limited antimicrobial spectrum. Azithromycin has a 15-membered ring, which is derived from the insertion of an amino group into the erythromycin ring. Azithromycin has unique pharmacokinetics which give rise to prolonged tissue levels allowing briefer duration of therapy (3–5 days) for most infections, and a single-dose regimen when treating chlamydial sexually transmitted diseases (STDs). Azithromycin suspension has been safely administered to children in numerous clinical studies. In a recent review of pediatric experience around the world, 2655 children 6 months to 16 years of age received azithromycin. Adverse events were recorded in 8.7% of children treated with azithromycin versus 9.8% who received comparator treatment [183a].

Pediatric studies of pharyngitis and tonsillitis showed that azithromycin was not as effective as a 10-day course of penicillin V. However, it was felt that part of the problem could be attributed to the presence of macrolide resistance (17% in one study) [184–187]. At least two large double-blind, multicenter studies using a higher dose of azithromycin for 5 days in children with streptococcal pharyngitis have shown that azithromycin was clinically superior to a 10-day course of penicillin V [187, 188]. We have no clinical evidence of the efficacy of azithromycin in prevention of rheumatic fever.

The common bacterial pathogens of acute otitis media (OM) in studies of children, based on cultures obtained by diagnostic tympanocentesis, are *S. pneumoniae* (28%), *H. influenzae* (21%), *M. catarrhalis* (20%), group A streptococci (GAS) (4%), *S. aureus* (3%), other bacteria (10%), or sterile (10%) [189]. Numerous studies comparing a short course of azithromycin (3–5 days) with a 10-day course of amoxicillin-clavulanate (amox/clav) in the treatment of children with otitis media demonstrated that azithromycin is equivalent to amox/clav and better tolerated in most studies [190–196]. A recent trial reported 34 cases of OM caused by azithromycin-susceptible *H. influenzae* and treated with azithromycin, a 71% bacteriologic eradication failure rate. Low concentrations of azithromycin in the middle-ear fluid are the possible explanation for a high failure rate [197]. Other trials comparing a short course of azithromycin in children with LRTIs showed no difference when compared with a 10-day course of erythromycin or amox/clav, but significantly fewer side effects were seen in the azithromycin group [198–200].

Several studies have demonstrated no significant difference in the clinical effectiveness of a single dose of azithromycin compared with the standard regimen of 7 days of oral doxycycline in the therapy of *C. trachomatis*-associated STDs both in men and women. Success rates varied from 88% to 100% [201–208]. A single-dose therapy with 1 g of azithromycin is the most cost-effective treatment for *C. trachomatis*, especially in populations where noncompliance is a concern [209–211].

Azithromycin, given in multiple doses over 2–3 weeks, is suggested as an alternate regimen for lymphogranuloma venereum, although clinical studies in this area to support this are limited [212]. Azithromycin was as safe and as effective as erythromycin and better tolerated in two studies treating 61 pregnant women with cervical chlamydial infections [213, 214].

Nongonococcal urethritis (NGU) is a clinical entity caused by either *C. trachomatis* or *Ureaplasma urealyticum* in 35–45% and 15–25% of the cases, respectively, while up to 30% are culture negative. Therapy for NGU has traditionally been 7 days of doxycycline. New studies evaluating the efficacy of azithromycin in men with NGU have demonstrated that a single dose over 3 days was as effective as 7 days of doxycycline, irrespective of the presence or absence of *C. trachomatis* [206, 215–217]. At the present time, doxycycline (7 days) or single-dose azithromycin (1 g) are the recommended regimens for this condition [212].

Multiple pathogens are implicated as causative with pelvic inflamatory disease (PID), although *C. trachomatis* and *N. gonorrhoeae* are the most frequently associated organisms. Intravenous azithromycin (with the addition of metronidazole if anaerobes were suspected) has been shown to be as effective as comparator regimens in the treatment of PID. The recommended regimen is azithromycin 500 mg I.V. for 1 or 2 days followed by 250 mg daily azithromycin orally to complete a 7-day course [218].

Penicillin is no longer effective for the treatment of gonorrhea due to high rates of penicillin-resistant *N. gonorrhoeae*. In the United States, single-dose ceftriaxone 125 mg I.M., or single-dose oral ciprofloxacin (500 mg), ofloxacin (400 gm), or cefixime (400 mg) is recommended for the treatment of *N. gonorrhoeae*, achieving cure rates of 98–100% for uncomplicated genital, rectal, and pharyngeal gonorrhea [212]. Recent clinical trials with 1 g oral azithromycin in the treatment of infections caused by *N. gonorrhoeae* have shown bacteriologic cure rates from 93% to 100% [203, 219, 220]. Most of the strains of *N. gonorrhoeae* isolated from patients with treatment failures were still azithromycin sensitive. The cure rates of 93% were considered unacceptable for this disease. At the present time, a single dose of 1 g azithromycin is not recommended as therapy of *N. gonorrhoeae*. A subsequent study evaluating the clinical efficacy of a larger single dose of oral azithromycin (2 g) in *N. gonorrhoeae* infections revealed cure rates of 98.9% for urethral and cervical infections, 100% for pharyngeal infections, and 96.3% for male and female rectal infections [221]. The larger dose of azithromycin was associated with an unacceptably high rate of side effects (20% gastrointestinal) compared with only 5% reported with the standard dose of 1 g and, therefore, cannot be recommended for treatment of this condition [204]. Azithromycin is active *in vitro* against *Treponema pallidum* and also in the rabbit model for syphilis [222]. However, clinical data are insufficient to support the use of azithromycin for syphilis [223, 224].

Several reports of comparative and noncomparative trials of infections caused by *Haemophilus ducreyi* have shown that a single 1-g dose of azithromycin is 89–100% effective in curing chancroid, and equally effective as 7 days of erythromycin or a single 250-mg dose of I.M. ceftriaxone [225–228].

Tetracyclines, erythromycin, co-trimoxazole, chloramphenicol, and ceftriaxone successfully treat donovanosis if taken daily for several weeks. A recent small pilot study of azithromycin using either 500 mg daily for 7 days (4 cases) or 1 g weekly (7 cases) for 4 weeks resulted in the cure of all cases [229]. An additional 17 cases were cured with azithromycin regimens after completion of the trial.

An open, noncomparative study of azithromycin in sinusitis resulted in an 86% clinical success rate [230]. Comparative trials have shown that azithromycin is as clinically effective as 10 days of phenoxymethylpenicillin, amoxicillin, or erythromycin [231, 232].

Azithromycin compared with 10 days of penicillin V for streptococcal pharyngitis in adults revealed similar clinical success rates (99% for both) and eradication of GAS. No difference was seen in antistreptolysin O (ASO) or anti-DNase B seroconversion rates between the groups.

Numerous studies have compared azithromycin with other antibiotics in the treatment of LRTIs. One study demonstrated that the efficacy of a 5-day azithromycin regimen was comparable to a 7- to 10-day course of erythromycin or amoxicillin [233]. Other studies have shown that a 5-day course of azithromycin was as effective as a 10-day course of cefaclor [234, 235] or co-amoxiclav [236] in the treatment of LRTIs, which included cases of CAP and AIECB. Several studies of AIECB, CAP, and/or acute bronchitis have shown that a short course of azithromycin is equivalent to a 10-day course of co-amoxiclav [237], a 10-day course of clarithromycin [238], or a 10-day course of roxithromycin [151]. A comparative study of weekly azithromycin (500 mg/week) with I.M. benzathine penicillin G (1.2 MU, once) in military recruits revealed that azithromycin was more effective than benzathine penicillin G in the prevention of respiratory infections caused by typical bacterial pathogens [239].

Hammerschlag *et al.* [240] reviewed the effectiveness of azithromycin in CAP caused by *C. pneumoniae* in adult, pediatric, and hospitalized adults in four comparative and three noncomparative CAP trials. Azithromycin was as effective as the comparator antibiotic regimen in all settings with 83–91% experiencing a satisfactory clinical response.

Use of parenteral azithromycin for the treatment of CAP was approved in the United States based on the results of two large clinical trials in hospitalized patients with CAP, including a large comparative trial where azithromycin was shown to be as effective as cefuroxime with or without erythromycin [218]. Parenteral formulation of azithromycin is currently approved in the United States for mild to moderate pneumonia in adults due to *C. pneumoniae, Mycobacterium pneumoniae, Legionella pneumophila, Moraxella catarrhalis, S. aureus,* or *S. pneumoniae* in hospitalized patients who require initial parenteral therapy. The recommended regimen is 500 mg daily for at least 2 days, followed by daily oral azithromycin to complete a 7- to 10-day course. Due to the rising rates of macrolide-resistant *S. pneumoniae* in some areas, physicians should avoid using I.V. azithromycin as monotherapy in patients with severe CAP, particularly in a very ill patient with multiple comorbidities with known risk factors for penicillin-resistant *S. pneumoniae* disease. The safety and effectiveness of I.V. azithromycin in children (<16 years) have not been established.

Numerous comparative studies in adults with skin and soft tissue (SST) infections have shown that azithromycin is as effective as 5–10 days of a first-generation cephalosporin, or 7 days of erythromycin, or dicloxacillin [241–244]. Studies have shown azithromycin to be as effective as dicloxacillin, flucloxacillin, or cefaclor in the treatment of pediatric skin infections [245, 246].

Although susceptibility studies have revealed elevated MIC_{90}s of azithromycin against MAC, experimentally infected mice were treated successfully with azithromycin [247, 248]. A preliminary uncontrolled study treating AIDS patients with daily azithromycin revealed a significant reduction in mycobacteremia in almost 75% of patients treated for at least 20 days with azithromycin [249]. Berry and colleagues [250a] treated a group of AIDS patients with 600 or 1200 mg daily for 6 weeks, and found no significant difference, while the 1200-mg group had significantly more side effects. As noted later in the clarithromycin section, a recent study comparing clarithromycin or azithromycin plus ethambutol therapy for MAC bacteremia revealed that the clarithromycin-containing regimen eradicated bacteremia significantly more quickly and more often than the azithromycin-containing regimen [60]. A major concern in the treatment of MAC disease in AIDS patients is the emergence of resistance. Monotherapy with clarithromycin has led to clinical and microbiological relapse within 12 weeks of the start of therapy [49, 251]. As a consequence, macrolides should be carefully used to treat disseminated MAC disease only in combination with at least one additional agent, such as ethambutol or rifabutin [252, 253]. While rifabutin was the first antibacterial to be shown effective in the prevention of MAC infection in AIDS patients with late-stage disease [250], subsequent studies have shown that clarithromycin was more effective than rifabutin in preventing MAC disease [59]. Azithromycin once weekly, at a dose of 1200 mg, was as effective as rifabutin in preventing MAC disease in late-stage AIDS patients [254]. The combination of azithromycin plus rifabutin was more effective, but this was at the cost of increased side effects. The azithromycin regimen was well tolerated and is currently approved as first-line MAC prophylaxis as is clarithromycin. For some clinicians the lack of significant drug interactions makes azithromycin more appealing than clarithromycin. Emergence of resistance to the macrolides and other drugs is a concern with the MAC prophylaxis regimens in AIDS patients. In recent trials, 9% of patients on azithromycin prophylaxis and 5% of patients on clarithromycin prophylaxis developed breakthrough MAC bacteremia. Eleven percent and 58% of the azithromycin and clarithromycin breakthrough isolates, respectively, were macrolide resistant [254, 255]. Azithromycin has also been used in the treatment of MAC infections in non-AIDS patients. Griffith *et al.* [256] treated 23 patients with MAC lung disease with 600 mg daily azithromycin as monotherapy for 4 months. Cultures of the sputum became negative or were significantly reduced in 38% and 76%, respectively. Unfortunately most of the patients (76%) suffered significant GI side effects and 41% hearing loss. In a subsequent report by the same group, a decreased daily dose of 300 mg azithromycin resulted in resolution of most of the adverse events [257]. Preliminary findings of a 600-mg thrice-weekly azithromycin regimen combined with daily rifabutin, ethambutol, and twice-weekly streptomycin (first weeks) in the treatment of MAC lung disease in non-AIDS patients revealed 13 of 17 converting their sputum by 6 months. Only 4 patients (all less than 50 kg) needed to reduce the azithromycin dose due to GI side effects, and no hearing loss was reported [258].

Butler and colleagues demonstrated that azithromycin was effective in treating 86% of 43 bacteremic adults with typhoid fever and was as effective as chloramphenicol-treated patients (88% of 33 improved). All 83 *Salmonella* strains in this study were susceptible to azithromycin, and all bacteremias were cleared by day 8 of therapy in both treatment groups [259]. On the other hand, Wallace *et al.* reported treating four adults with typhoid fever with azithromycin [3]. All patients were bacteremic with *Salmonella typhi*, and three of four failed therapy. It was postulated that low azithromycin serum levels might be inadequate to treat bacteremic patients with typhoid fever. Additional studies are needed before a recommendation can be made for azithromycin as first-line therapy in typhoid fever.

At the present time, the drugs of choice for enteritis caused by *C. jejuni* are the quinolones. As previously mentioned, erythromycin therapy can shorten the duration of excretion of *C. jejuni* but may fail to provide a significant clinical benefit. With the progressive development of quinolone resistance in some parts of the world and the favorable susceptibility of azithromycin against *C. jejuni*, azithromycin has been suggested as an alternative therapy for the treatment of symptomatic *C. jejuni* enteritis [260].

A recent study of travelers to Thailand comparing azithromycin or ciprofloxacin for the treatment of military personnel who acquired *C. jejuni* enteritis showed that all the azithromycin-treated patients were cured, while there were two clinical failures in the ciprofloxacin group [261]. Almost 50% of the 44 strains were ciprofloxacin resistant. Recently documented increases in quinolone resistance have been associated with rising levels of azithromycin resistance in Thailand: 69% and 31% of 29 *C. jejuni* isolates were found to be resistant to ciprofloxacin and azithromycin, respectively [262]. Rising levels of azithromycin resistance may compromise the effectiveness of azithromycin for *C. jejuni* enteritis.

Azithromycin has been shown to be very active *in vitro* against enteric pathogens such as *Escherichia coli, Shigella*, and *Campylobacter* [263]. One study demonstrated that azithromycin was clinically as effective (82% of 34 cases) as ciprofloxacin for 5 days (89% of 36) in the treatment of moderate to severe shigellosis caused by multidrug-resistant *Shigella* strains [264]. Additionally, azithromycin seems to have a role as prophylaxis in preventing dysentery during an outbreak in a refugee camp in Kenya [265]. Trachoma is a chronic follicular conjunctivitis due to *C. trachomatis* and is the world's leading cause of preventable blindness. The currently recommended treatment of trachoma is topical tetracycline eye ointment for at least 6 weeks, or on 5 consecutive days a month for 6 months. Topical regimens are complicated by the practical and logistical challenges of treating large groups of people in underdeveloped areas where the regimens are often uncomfortable and difficult to apply to the eyes of young children. A single-blind, randomized trial compared a single oral dose of azithromycin with topical tetracycline, with and without erythromycin, for severe cases of trachoma in Gambia [266]. After a 6-month follow-up, trachoma had been resolved in 78% of those treated with azithromycin compared with 72% of the tetracycline-treated subjects, which was not significantly different. Dawson *et al*. [267] showed that treatment of trachoma in children in Egypt with azithromycin, 1–6 doses over 6 months, was similar to that of topical therapy [267].

In vitro studies with susceptibility testing of 97 clinical isolates of *H. pylori* revealed that all were susceptible to erythromycin and azithromycin [268], and an *in vitro* study has also shown that azithromycin is effective against intracellular *H. pylori* [269]. Multiple studies investigating azithromycin in combination with one or two other drugs (amoxicillin, metronidazole, tinidazole, tetracycline, bismuth subsalicylate, and/or a PPI) revealed that the more effective azithromycin-containing regimens involved triple therapy [270–275]. In contrast to clarithromycin, azithromycin is not FDA approved for treatment of *H. pylori* disease. The role of azithromycin in the treatment of *H. pylori* remains to be defined. Several *in vitro* and animal studies have suggested that azithromycin may be clinically active in the treatment of early Lyme disease (LD) [276–278]. Several clinical trials of treatment of early Lyme disease performed in the United States and Europe comparing azithromycin vs. doxycycline, oral penicillin, amoxicillin/probenicid, or oral amoxicillin revealed no significant difference in clinical outcome [279–283]. The precise role and dose of azithromycin in the treatment of early LD remains to be established.

Azithromycin probably has a role as therapy for eradication of *Neisseria meningitidis*. A single dose of azithromycin was as effective as a 2-day regimen of rifampin 600 mg b.i.d. in the eradication of *N. meningitidis* nasopharyngeal carriage in a group of adults (93% vs. 91% eradicated at 2 weeks) [284].

Some *in vitro* studies have demonstrated that azithromycin is active against chloroquine-sensitive and -resistant strains of *Plasmodium falciparum* [285]. A trial using daily azithromycin at a dose of 250 mg as a malaria prophylaxis in volunteers has shown azithromycin to be protective against *P. falciparum* challenge [286, 287]. A randomized, placebo-controlled, double-blind study of malaria prophylaxis compared two regimens of azithromycin versus daily doxycycline for a period of 10 weeks in western Kenya [287]. Both regimens with daily azithromycin and doxycycline provided effective prophylaxis of falciparum malaria in this trial. A potential advantage of azithromycin over doxycycline for malaria prophylaxis is its use in pregnant women and young children, who cannot take tetracyclines but could take azithromycin as an alternative prophylactic agent to doxycycline. Another recent study in Indonesia has shown that daily azithromycin was less effective than daily doxycycline in the prevention of falciparum malaria, although azithromycin was as effective as doxycycline in the prevention of *Plasmodium vivax* malaria [288]. More clinical trials are needed to define the role of azithromycin in malaria prophylaxis.

In vivo studies using a *Babesia microti* hamster model showed that azithromycin alone or with quinine or atovaquone was effective [289, 290]. One of two *B. microti* human cases mentioned in the report by Wittner and colleagues that failed standard therapy was successfully treated with quinine and azithromycin plus atovaquone [290]. Further clinical studies are needed to define the role of azithromycin for this condition.

In vitro studies have shown that azithromycin is active against *Toxoplasma gondii*, and that this activity is additive with pyrimethamine [291]. *In vivo* studies have shown that azithromycin has some prophylactic and therapeutic activity [292–294]. Only a handful of human cases of cerebral toxoplasmosis treated with azithromycin have been described in the literature [295–297]. Early results with escalating doses of oral azithromycin of 900, 1200, or 1500 mg daily plus pyrimethamine in 32 evaluable patients with known or suspected toxoplasma encephalitis have been reported [298]. It was concluded that azithromycin plus pyrimethamine is not as effective as the standard therapy for cerebral toxoplasmosis and should be regarded as second-line or salvage therapy.

Preliminary results of a pilot study of *Cryptosporidium parvum* intestinal infection in children showed that three of four cases were promptly cured within 5 days of azithromycin therapy, and a fourth responded to prolonged azithromycin therapy [299]. On the other hand, another small study in adult AIDS patients with *C. parvum* intestinal infection demonstrated that a 2- to 4-week course of azithromycin 500 mg daily was ineffective in curing this condition [300]. Dupont et al. [301] reported that azithromycin was effective in one of two pulmonary cryptosporidiosis cases.

There is a single case report of a patient with AIDS-related bacillary angiomatosis that responded to azithromycin [302]. Chia and colleagues [303] described the successful treatment of eight of nine cases of cat-scratch disease in normal hosts with azithromycin.

There is a single case report of the successful treatment of *Rhodococcus equi* causing peritoneal dialysis-associated peritonitis with oral azithromycin [304, 305]. Both azithromycin and clarithromycin were as effective as amoxycillin, clindamycin, and erythromycin in a viridans group streptococcal experimental endocarditis prophylaxis model in the prevention of endocarditis in this model [306]. In the 1997 recommendations of the American Heart Association either azithromycin or clarithromycin is recommended as one of the alternatives for IE prophylaxis in people who are penicillin allergic who require antibiotic prophylaxis prior to oral, dental, esophageal, or respiratory tract procedures [12].

III. Sixteen-Membered Macrolides

A. Rokitamycin

Most of the data on rokitamycin have been published in Japanese and Italian journals. As a propionyl ester of leucomycin, rokitamycin has an antimicrobial spectrum similar to that of erythromycin, it is especially potent against *L. pneumophila, M. pneumoniae*, and *Chlamydia*. Like other 16-membered macrolides, it is active against bacteria that are inducibly resistant to erythromycin but inactive against strains that are constitutively resistant to macrolide–lincosamide–streptogramin B antibiotics.

At total daily doses of 600–800 mg, rokitamycin is useful for the treatment of acute odontogenic and oropharyngeal infections, with clinical response rates of 79–85% [307–309]. Clinical efficacy (81–89%) and bacteriological cure (90–94%) rates were achieved in the treatment of acute tonsillitis and other otolaryngological infections [310, 311]. This macrolide is effective in the treatment of CAP, with clinical and bacteriological efficacy of >85% [312].

Various superficial suppurative SST infections are effectively treated with rokitamycin, with clinical improvement or cure in 71–82% of patients, depending on the types of infection [313, 314]. Rokitamycin at 400 mg twice daily for 14 days resulted in excellent clinical and microbiological outcomes (>90% cure rates) in laboratory-proven genital infections due to *Mycoplasma hominis* and *C. trachomatis* [315].

Rokitamycin is a very useful drug for the treatment of *Campylobacter* enteritis, and large comparative studies have shown clinical cure rates of 96–100% and bacteriological eradication rates of 91–95% [316, 317]. As with other macrolides, the main side effects with rokitamycin are related to GI irritability, which occurs in <5% of patients.

B. Spiramycin

Spiramycin is a 16-membered macrolide found in nature and used mainly in Europe. It is prescribed usually at divided doses of 1–2 g/day for the treatment of various odontogenic, upper and lower respiratory tract, and genital infections.

As alternative to penicillin V and tetracycline, spiramycin is as effective as erythromycin in the treatment of acute and chronic periodontitis [318–322]. Clinical trials also indicated spiramycin at 500 mg three times daily for 3 days to be comparable to a 5-day course of penicillin V and erythromycin for the treatment of tonsillo-pharyngitis [323–326].

Spiramycin was equally efficacious as doxycycline in the empirical treatment of acute sinusitis [327]. It is also comparable to erythromycin, clarithromycin, and doxycycline in open, randomized, comparative clinical therapeutic studies of LRTIs [328–330].

This macrolide can be used effectively for the treatment of nongonococcal genital infections [331, 332]. In the therapy of *H. pylori*-associated peptic ulcer disease, combination therapy of spiramycin, metronidazole, and antacid/bismuth resulted in ulcer healing rates of 89–91% and bacteriological eradication rates of 91–95% [333, 334].

The role of spiramycin in the treatment of HIV-associated cryptosporidiosis remains inconclusive and controversial [335, 336]. The drug did hasten clinical recovery and decrease the duration of oocyst excretion in immunocompetent children with cryptosporidial diarrhea [337]. Spiramycin is also used effectively with or without pyrimethamine-sulfadiazine for the treatment of fetal toxoplasmosis [338, 339].

C. Midecamycin

Very little published data exist in the English literature on the clinical use of midecamycin, a naturally occurring macrolide with a spectrum of antimicrobial activities similar to those of other 16-membered macrolides. Administered twice a day, this drug is effective for the treatment of infections in the oral cavity [340], upper and lower respiratory tracts [341, 342], and SST [314]. Clinical use of this macrolide is limited mainly to Japan and Europe.

D. Miokamycin

Much of the information on the clinical efficacy of miokamycin (or miocamycin) has been published in Japanese, Italian, and Spanish medical literature; clinical use of this macrolide is found in these countries and South America. As an acetyl derivative of midecamycin, the drug is administered orally at doses of 600 mg twice daily in adults and 15–25 mg/kg twice daily in children.

Miokamycin is comparable to erythromycin and dirithromycin in clinical and microbiological efficacy for the treatment of streptococcal pharyngitis [119, 343, 344]. However, as possible single-dose therapy in children with culture-proven group A streptococcal pharyngitis, this drug did not eradicate the bacteria as well as benzathine penicillin G (32% vs. 66%) [345]. Favorable clinical experience with miokamycin was found in the treatment of respiratory tract infections, with >95% clinical response and 100% bacteriological eradication in cases in which a bacterial pathogen was cultured [346]. Randomized clinical trials showed this macrolide to be equally efficacious and safe as amoxicillin and dirithromycin in the oral therapy of acute bronchitis, acute exacerbations of chronic bronchitis, and pneumonia in pediatric and adult patients [123, 347]. In a single-blind, randomized clinical study for the treatment of skin and skin structure infections [116, 131], miokamycin was comparable to dirithromycin in achieving clinical (98% vs. 95%) and bacteriological (84% vs. 93%) responses. No studies have been published on the use of miokamycin for the treatment or prophylaxis of other types of infection.

IV. Concluding Remarks

In addition to erythromycin, new macrolides now available for clinical use include azithromycin and clarithromycin as well as other compounds such as roxithromycin. These newer macrolides are more stable and better tolerated than erythromycin, and they have a broader antimicrobial spectrum than erythromycin against *Mycobacterium avium* complex and *Haemophilus influenzae*, nontuberculous mycobacteria, and *Chlamydia trachomatis*. Other macrolides have seen limited use for clinical infections and further evaluation is necessary to determine their antimicrobial role. An area of concern is the increasing number of reports of macrolide resistance in some of the common pathogens, particularly group A streptococci, *Streptococcus pneumoniae*, and *Haemophilus influenzae*. The rapid emergence of macrolide resistance among common respiratory pathogens may limit the usefulness of this group of antimicrobial agents in the future. Judicious use of antibiotics by practicing physicians could help to prevent this growing problem.

References

1. Kirst, H. A., and Sides, G. D. (1989). New directions for macrolide antibiotics: Structural modifications and *in vitro* activity. *Antimicrob. Agents Chemother.* **33**, 1413–1418.
2. Mazzei, T., Mini, E., Novelli, A., and Periti, P. (1993). Chemistry and mode of action of macrolides. *J. Antimicrob. Chemother.* **31**, 1–9.
3. Alvarez-Elcoro, S., and Enzler, M. J. (1999). The macrolides: Erythromycin, clarithromycin, and azithromycin. *Mayo Clin. Proc.* **74**, 613–634.
4. Franklin, T. J., and Snow, G. A. (1981). "Biochemistry of Antimicrobial Action." Chapman and Hall, London.
5. Pestka, S. (1977). Inhibitors of protein synthesis. *In* "Molecular Mechanisms of Protein Biosynthesis" (H. Weissback and S. Pestka, Eds.), p. 467. Academic Press, New York.
6. Oleinick, N. L. (1975). The erythromycins. *In* "Mechanism of Action of Antimicrobial and Antitumor Agents" (J. W. Corcoran and F. E. Hahn, Eds.), p. 396. Springer-Verlag, New York.

7. Kirst, H. A. (1993). Semi-synthetic derivatives of erythromycin. *Prog. Med. Chem.* **30**, 57–88.
8. Fernandes, P. B., and Hardy, D. J. (1988). Comparative *in vitro* potencies of nine new macrolides. *Drugs Exp. Clin. Res.* **14**, 445–451.
9. Piscitelli, S. C., Danziger, L. H., and Rodvold, K. A. (1992). Clarithromycin and azithromycin: New macrolide antibiotics. *Clin. Pharm.* **11**, 137–152.
10. Anders, B. J., Lauer, B. A., Paisley, J. W., and Reller, L. B. (1982). Double-blind placebo controlled trial of erythromycin for treatment of *Campylobacter* enteritis. Lancet **1**, 131–132.
11. Mandell, G. L., Douglas, R. G., and Bennett, J. E. (1992). "Handbook of Antimicrobial Therapy 1992." Churchill Livingstone, New York.
12. Dajani, A. S., Taubert, K. A., Wilson, W., Bolger, A. F., Bayer, A., Ferrieri, P., Gewitz, M. H., Shulman, S. T., Nouri, S., Newburger, J. W., Hutto, C., Pallasch, T. J., Gage, T. W., Levison, M. E., Peter, G., and Zuccaro, Jr., G. (1997). Prevention of bacterial endocarditis. Recommendations by the American Heart Association. *JAMA* **277**, 1794–1801.
12a. Centers for Disease Control and Prevention (1998). 1998 Guidelines for treatment of sexually transmitted disease. *Morb. Mortal. Wkly. Rep.* **47**, 1–111.
13. Bahal, N., and Nahata, M. C. (1992). The new macrolide antibiotics: Azithromycin, clarithromycin, dirithromycin, and roxithromycin. *Ann. Pharmacother.* **26**, 46–55.
14. Kakegawa, T., Ogawa, T., and Hirose, S. (1988). Mode of inhibition of protein synthesis by TE-031 (A-56268). *Chemotherapy* **36**, 123–128.
15. Kakegawa, T., and Hirose, S. (1990). Mode of inhibition of protein synthesis by metabolites of clarithromycin. *Chemotherapy* **38**, 317–323.
16. Anderson, G., Esmonde, T. S., Coles, S., Macklin, J., and Carnegie, C. (1991). A comparative safety and efficacy study of clarithromycin and erythromycin stearate in community-acquired pneumonia. *J. Antimicrob. Chemother.* **27**, 117–124.
17. Hamedani, P., Ali, J., Hafeez, S., Bachand, Jr., R., Dawood, G., Quereshi, S., Raza, R., and Yab, Z. (1991). The safety and efficacy of clarithromycin in patients with *Legionella* pneumonia. *Chest* **100**, 1503–1506.
18. Scaglione, F. (1990). Comparison of the clinical and bacteriological efficacy of clarithromycin and erythromycin in the treatment of streptococcal pharyngitis. *Curr. Med. Res. Opin.* **12**, 25–33.
19. Levenstein, J. H. (1991). Clarithromycin versus penicillin in the treatment of streptococcal pharyngitis. *J. Antimicrob. Chemother.* **27**, 67–74.
20. Bachand, Jr., R. T. (1991). A comparative study of clarithromycin and penicillin VK in the treatment of outpatients with streptococcal pharyngitis. *J. Antimicrob. Chemother.* **27**, 75–82.
21. Marchi, E. (1990). Comparative efficacy and tolerability of clarithromycin and amoxicillin in the treatment of outpatients with acute maxillary sinusitis. *Curr. Med. Res. Opin.* **12**, 19–24.
22. Langtry, H. D., and Brogden, R. N. (1997). Clarithromycin. A review of its efficacy in the treatment of respiratory tract infections in immunocompetent patients. *Drugs* **53**, 973–1004.
23. Lasko, B., Lau, C. Y., Saint-Pierre, C., Reddington, J. L., Martel, A., and Anstey, R. J. (1998). Efficacy and safety of oral levofloxacin compared with clarithromycin in the treatment of acute sinusitis in adults: A multicentre, double-blind, randomized study. The Canadian Sinusitis Study Group. *J. Int. Med. Res.* **26**, 281–291.
24. Dubois, J., Saint-Pierre, C., and Tremblay, C. (1993). Efficacy of clarithromycin vs. amoxicillin/clavulanate in the treatment of acute maxillary sinusitis. *Ear Nose Throat J.* **72**, 804–810.
25. Bachand, Jr., R. T. (1991). Comparative study of clarithromycin and ampicillin in the treatment of patients with acute bacterial exacerbations of chronic bronchitis. *J. Antimicrob. Chemother.* **27**, 91–100.
26. O'Neill, S. J., Millar, E. D., Coles, S. J., and Bachand, R. T. (1991). Safety and efficacy of clarithromycin in the treatment of acute mild to moderate respiratory tract infections. *Ir. Med. J.* **84**, 33–35.

27. Chien, S. M., Pichotta, P., Siepman, N., and Chan, C. K. (1991). Treatment of community acquired pneumonia: A randomized controlled trial comparing clarithromycin (CM) and erythromycin (ER). In "Program of the 31st Interscience Conference on Antimicrobial Agents and Chemotherapy" (Chicago). Abstr. No. 872.
28. Chien, S. M., Pichotta, P., Siepman, N., and Chan, C. K. (1993). Treatment of community-acquired pneumonia. A multicenter, double-blind, randomized study comparing clarithromycin with erythromycin. Canada–Sweden Clarithromycin–Pneumonia Study Group. *Chest* **103**, 697–701.
29. Neu, H. C., and Chick, T. W. (1993). Efficacy and safety of clarithromycin compared to cefixime as outpatient treatment of lower respiratory tract infections. *Chest* **104**, 1393–1399.
30. Adam, D. (1993). Clarithromycin 250 mg b.i.d. for 5 or 10 days in the treatment of adult patients with purulent bronchitis. *Infection* **21**, 265–271.
31. Fong, I. W., Laforge, J., Dubois, J., Small, D., Grossman, R., and Zakhari, R. (1995). Clarithromycin versus cefaclor in lower respiratory tract infections. The Canadian Bronchitis Study Group. *Clin. Invest. Med.* **18**, 131–138.
32. Hatipoglu, O. N., and Tasan, Y. (2000). A comparative efficacy and safety study of clarithromycin, roxithromycin and erythromycin stearate in mild pneumonia. *Yonsei Med. J.* **41**, 340–344.
33. Lipsky, B. A., Unowsky, J., Zhang, H., Townsend, L., and Talbot, G. H. (1999). Treating acute bacterial exacerbations of chronic bronchitis in patients unresponsive to previous therapy: Sparfloxacin versus clarithromycin. *Clin. Ther.* **21**, 954–965.
34. Moola, S., Hagberg, L., Churchyard, G. A., Dylewski, J. S., Sedani, S., and Staley, H. (1999). A multicenter study of grepafloxacin and clarithromycin in the treatment of patients with community-acquired pneumonia. *Chest* **116**, 974–983.
35. Macklin, J. L., James, I., Kearsley, N. J., and Coles, S. J. (1993). A single-blind, randomised, comparative study of clarithromycin and amoxycillin suspensions in the treatment of children with lower respiratory tract infections. *J. Chemother.* **5**, 174–180.
36. McCarty, J., Hedrick, J. A., and Gooch, W. M. (2000). Clarithromycin suspension vs. penicillin V suspension in children with streptococcal pharyngitis. *Adv. Therapy* **17**, 14–26.
37. Gooch III, W. M., Adelglass, J., Kelsey, D. K., Masica, D., Johns, Jr., D., and Weinberg, B. C. (1999). Loracarbef versus clarithromycin in children with acute otitis media with effusion. *Clin. Ther.* **21**, 711–722.
38. McCarty, J. M., Phillips, A., and Wiisanen, R. (1993). Comparative safety and efficacy of clarithromycin and amoxicillin/clavulanate in the treatment of acute otitis media in children. *Pediatr. Infect. Dis. J.* **12**, S122–S127.
39. Coles, S. J., Addlestone, M. B., Kamdar, M. K., and Macklin, J. L. (1993). A comparative study of clarithromycin and amoxycillin suspensions in the treatment of pediatric patients with acute otitis media. *Infection* **21**, 272–278.
40. Pukander, J. S., Jero, J. P., Kaprio, E. A., and Sorri, M. J. (1993). Clarithromycin vs. amoxicillin suspensions in the treatment of pediatric patients with acute otitis media. *Pediatr. Infect. Dis. J.* **12**, S118–S121.
41. Gooch III, W. M., Gan, V. N., Corder, W. T., Khurana, C. M., and Andrews, Jr., W. P. (1993). Clarithromycin and cefaclor suspensions in the treatment of acute otitis media in children. *Pediatr. Infect. Dis. J.* **12**, S128–S133.
42. Hardy, D. J., Guay, D. R., and Jones, R. N. (1992). Clarithromycin, a unique macrolide. A pharmacokinetic, microbiological, and clinical overview. *Diagn. Microbiol. Infect. Dis.* **15**, 39–53.
43. Northcutt, V. J., Craft, J. C., and Pichotta, P. (1990). Safety and efficacy of clarithromycin (C) compared to erythromycin (E) in the treatment (tx) of bacterial skin or skin structure infections (SSTIs). Presented at 30th Interscience Conference on Antimicrobial Agents and Chemotherapy (Atlanta). Abstr. No. 1339.

44. Millikan, L., Coleman, E., Bopp, B., and Northcutt, V. J. (1990). Safety and efficacy of clarithromycin (C) compared with cefadroxil (CEF) in the treatment of mild to moderate bacterial skin or skin structure infections (SSSIs). Presented at 30th Interscience Conference on Antimicrobial Agents and Chemotherapy (Atlanta). Abstr. No. 1340.
45. Parish, L. C. (1993). Clarithromycin in the treatment of skin and skin structure infections: Two multicenter clinical studies. Clarithromycin Study Group. *Int. J. Dermatol.* **32**, 528–532.
46. Gupta, S., and Siepman, H. (1992). Comparative safety and efficacy of clarithromycin vs. standard agents in the treatment of mild to moderate bacterial skin or skin structure infections. Presented at First International Conference on the Macrolides, Azalides and Streptogramins (Santa Fe, NM).
47. Hebert, A. A., Still, J. G., and Reuman, P. D. (1993). Comparative safety and efficacy of clarithromycin and cefadroxil suspensions in the treatment of mild to moderate skin and skin structure infections in children. *Pediatr. Infect. Dis. J.* **12**, S112–S117.
48. Jones, J. L., Hanson, D. L., Chu, S. Y., Fleming, P. L., Hu, D. J., and Ward, J. W. (1994). Surveillance of AIDS-defining conditions in the United States. Adult/Adolescent Spectrum of HIV Disease Project Group. *AIDS* **8**, 1489–1493.
49. Dautzenberg, B., Truffot, C., Legris, S., Meyohas, M. C., Berlie, H. C., Mercat, A., Chevret, S., and Grosset, J. (1991). Activity of clarithromycin against *Mycobacterium avium* infection in patients with the acquired immune deficiency syndrome. A controlled clinical trial. *Am. Rev. Respir. Dis.* **144**, 564–569.
50. Dautzenberg, B., Saint Marc, T., Meyohas, M. C., Eliaszewitch, M., Haniez, F., Rogues, A. M., De Wit, S., Cotte, L., Chauvin, J. P., and Grosset, J. (1993). Clarithromycin and other antimicrobial agents in the treatment of disseminated *Mycobacterium avium* infections in patients with acquired immunodeficiency syndrome. *Arch. Intern. Med.* **153**, 368–372.
51. Chaisson, R. E., Benson, C. A., Dube, M. P., Heifets, L. B., Korvick, J. A., Elkin, S., Smith, T., Craft, J. C., and Sattler, F. R. (1994). Clarithromycin therapy for bacteremic *Mycobacterium avium* complex disease. A randomized, double-blind, dose-ranging study in patients with AIDS. AIDS Clinical Trials Group Protocol 157 Study Team. *Ann. Intern. Med.* **121**, 905–911.
52. Benson, C. A., and Ellner, J. J. (1993). *Mycobacterium avium* complex infection and AIDS: Advances in theory and practice. *Clin. Infect. Dis.* **17**, 7–20.
53. Shafran, S. D., Singer, J., Zarowny, D. P., Phillips, P., Salit, I., Walmsley, S. L., Fong, I. W., Gill, M. J., Rachlis, A. R., Lalonde, R. G., Fanning, M. M., and Tsoukas, C. M. (1996). A comparison of two regimens for the treatment of *Mycobacterium avium* complex bacteremia in AIDS: Rifabutin, ethambutol, and clarithromycin versus rifampin, ethambutol, clofazimine, and ciprofloxacin. Canadian HIV Trials Network Protocol 010 Study Group. *N. Engl. J. Med.* **335**, 377–383.
54. Chaisson, R. E., Keiser, P., Pierce, M., Fessel, W. J., Ruskin, J., Lahart, C., Benson, C. A., Meek, K., Siepman, N., and Craft, J. C. (1997). Clarithromycin and ethambutol with or without clofazimine for the treatment of bacteremic *Mycobacterium avium* complex disease in patients with HIV infection. *AIDS* **11**, 311–317.
55. Cohn, D. L., Fisher, E., Franchino, B., et al. (1996). Comparison of two doses of clarithromycin in a randomized trial of four 3-drug regimens for treatment of disseminated *Mycobacterium avium* complex disease in AIDS: Excess mortality associated with high dose clarithromycin. Presented at 11th International Conference on AIDS (Vancouver, B.C., Canada).
56. Dautzenberg, B., Truffot-Pernot, C., Hazebroucq, J., Legris, S., Guerin, C., Begelman, C., Guermonprez, G., Fievet, M. H., Chastang, C., and Grosset, J. (1997). A randomized comparison of two clarithromycin doses for treatment of *Mycobacterium avium* complex infections. *Infection* **25**, 16–21.
57. Centers for Disease Control and Prevention (1993). Recommendations on prophylaxis and therapy for disseminated *Mycobacterium avium* complex for adults and adolescents infected

with human immunodeficiency virus. U.S. Public Health Service Task Force on Prophylaxis and Therapy for *Mycobacterium avium* Complex. *Morb. Mortal. Wkly. Rep.* **42**, 14–20.
58. Kaplan, J. E., Masur, H., Holmes, K. K., Wilfert, C. M., Sperling, R., Baker, S. A., Trapnell, C. B., Freedberg, K. A., Cotton, D., Powderly, W. G., et al. (1995). USPHS/IDSA guidelines for the prevention of opportunistic infections in persons infected with human immunodeficiency virus: An overview. USPHS/IDSA Prevention of Opportunistic Infections Working Group. *Clin. Infect. Dis.* **21**, S12–S31.
59. Ostroff, S. M., Spiegel, R. A., Feinberg, J., Benson, C. A., and Horsburgh, Jr., C. R. (1995). Preventing disseminated *Mycobacterium avium* complex disease in patients infected with human immunodeficiency virus. *Clin. Infect. Dis.* **21**, S72–S76.
60. Ward, T. T., Rimland, D., Kauffman, C., Hycke, M., Evans, T. G., and Heifets, L. (1998). Randomized, open-label trial of azithromycin plus ethambutol vs. clarithromycin plus ethambutol as therapy for *Mycobacterium avium* complex bacteremia in patients with human immunodeficiency virus infection. *Clin. Infect. Dis.* **27**, 1278–1285.
61. Dunne, M., Fessel, J., Kumar, P., Dickenson, G., Keiser, P., Boulos, M., Mogyros, M., White, Jr., A.C., Cahn, P., O'Connor, M., Lewi, D., Green, S., Tilles, J., Hicks, C., Bissett, J., Schneider, M. M., and Benner, R. (2000). A randomized, double-blind trial comparing azithromycin and clarithromycin in the treatment of disseminated *Mycobacterium avium* infection in patients with human immunodeficiency virus. *Clin. Infect. Dis.* **31**, 1245–1252.
62. Wallace, Jr., R. J., Brown, B. A., Griffith, D. E., Girard, W. M., Murphy, D. T., Onyi, G. O., Steingrube, V. A., and Mazurek, G. H. (1994). Initial clarithromycin monotherapy for *Mycobacterium avium-intracellulare* complex lung disease. *Am. J. Respir. Crit. Care. Med.* **149**, 1335–1341.
63. Wallace, Jr., R. J., Brown, B. A., Griffith, D. E., Girard, W. M., and Murphy, D. T. (1996). Clarithromycin regimens for pulmonary *Mycobacterium avium* complex. The first 50 patients. *Am. J. Respir. Crit. Care. Med.* **153**, 1766–1772.
64. Pierce, M., Crampton, S., Henry, D., Heifets, L., LaMarca, A., Montecalvo, M., Wormser, G. P., Jablonowski, H., Jemsek, J., Cynamon, M., Yangco, B. G., Notario, G., and Craft, J. C. (1996). A randomized trial of clarithromycin as prophylaxis against disseminated *Mycobacterium avium* complex infection in patients with advanced acquired immunodeficiency syndrome. *N. Engl. J. Med.* **335**, 384–391.
65. Datri 001 Study Group (1993). Clarithromycin plus rifabutin for MAC prophylaxis: Evidence for a drug interaction. In "Abstracts of the First National Conference on Human Retrovirus," p. 106. American Society for Microbiology, Washington, DC.
66. Freedberg, K. A., Cohen, C. J., and Barber, T. W. (1997). Prophylaxis for disseminated *Mycobacterium avium* complex (MAC) infection in patients with AIDS: A cost-effectiveness analysis. *J. Acquir. Immune Defic. Syndr. Hum. Retrovirol.* **15**, 275–282.
67. Benson, C. A., Cohn, D. L., and Williams, P. (1996). A phase III prospective, randomized, double-blind study of the safety and efficacy of clarithromycin (CLA) vs. rifabutin (RBT) vs. CLA + RBT for the prevention of *Mycobacterium avium* complex disease in HIV + patients with CD_4 counts <100 cells/μ^3. In "Abstracts of the Third Conference on Retrovirus and Opportunistic Infections," p. 90. American Society of Microbiology, Washington, DC.
68. Heifets, L. B. (1996). Clarithromycin against *Mycobacterium avium* complex infections. *Tuber. Lung Dis.* **77**, 19–26.
69. Bordet, A. L., Machet, L., De Muret, A., Francois-Ramanantsoa, C., Lorette, G., and Vaillant, L. (1997). Mycobacterium chelonae cutaneous infection: Efficacy of prolonged treatment by clarithromycin. *Ann. Dermatol. Venereol.* **124**, 251–253.
70. van Aarem, A., Muytjens, H. L., Smits, M. M., and Cremers, C. W. (1998). Recurrent therapy resistant mastoiditis by Mycobacterium chelonae abscessus, a nontuberculous mycobacterium. *Int. J. Pediatr. Otorhinolaryngol.* **43**, 61–72.

71. Tartaglione, T. (1997). Treatment of nontuberculous mycobacterial infections: Role of clarithromycin and azithromycin. *Clin. Ther.* **19**, 626–638; discussion 603.
72. Saluja, A., Peters, N. T., Lowe, L., and Johnson, T. M. (1997). A surgical wound infection due to Mycobacterium chelonae successfully treated with clarithromycin. *Dermatol. Surg.* **23**, 539–543.
73. Klapper, S. R., Patrinely, J. R., Kaplan, S. L., and Font, R. L. (1995). Atypical mycobacterial infection of the orbit. *Ophthalmology* **102**, 1536–1541.
74. Engelhardt, E., Feldmann, R., Skaria, A., and Salomon, D. (1996). [Postoperative infection with Mycobacterium chelonae]. *Hautarzt* **47**, 863–866.
75. Forslund, T., Rummukainen, M., Kousa, M., Krees, R., Relander, A., and Katila, M. L. (1995). Disseminated cutaneous infection due to Mycobacterium chelonae in a patient with rheumatoid arthritis, amyloidosis, and renal failure. *Nephrol. Dial. Transplant.* **10**, 1234–1236.
76. Mushatt, D. M., and Witzig, R. S. (1995). Successful treatment of Mycobacterium abscessus infections with multidrug regimens containing clarithromycin [letter]. *Clin. Infect. Dis.* **20**, 1441–1442.
77. Butt, A. A. (1998). Cervical adenitis due to *Mycobacterium fortuitum* in patients with acquired immunodeficiency syndrome. *Am. J. Med. Sci.* **315**, 50–55.
78. Murillo, J., Torres, J., Bofill, L., Rios-Fabra, A., Irausquin, E., Isturiz, R., Guzman, M., Castro, J., Rubino, L., and Cordido, M. (2000). Skin and wound infection by rapidly growing mycobacteria: An unexpected complication of liposuction and liposculpture. The Venezuelan Collaborative Infectious and Tropical Diseases Study Group. *Arch. Dermatol.* **136**, 1347–1352.
79. Franzblau, S. G., and Hastings, R. C. (1988). *In vitro* and *in vivo* activities of macrolides against *Mycobacterium leprae*. *Antimicrob. Agents Chemother.* **32**, 1758–1762.
80. Ji, B., Jamet, P., Perani, E. G., Bobin, P., and Grosset, J. H. (1993). Powerful bactericidal activities of clarithromycin and minocycline against *Mycobacterium leprae* in lepromatous leprosy. *J. Infect. Dis.* **168**, 188–190.
81. Chan, G. P., Garcia-Ignacio, B. Y., Chavez, V. E., Livelo, J. B., Jimenez, C. L., Parrilla, M. L., and Franzblau, S. G. (1994). Clinical trial of clarithromycin for lepromatous leprosy. *Antimicrob. Agents Chemother.* **38**, 515–517.
82. Banerjee, D. K., McDermott-Lancaster, R. D., and McKenzie, S. (1997). Experimental evaluation of possible new short-term drug regimens for treatment of multibacillary leprosy. *Antimicrob. Agents Chemother.* **41**, 326–330.
83. Ji, B., Jamet, P., Perani, E. G., Sow, S., Lienhardt, C., Petinon, C., and Grosset, J. H. (1996). Bactericidal activity of single dose of clarithromycin plus minocycline, with or without ofloxacin, against *Mycobacterium leprae* in patients. *Antimicrob. Agents Chemother.* **40**, 2137–2141.
84. Rea, T. H. (2000). Trials of daily, long-term minocycline and rifampin or clarithromycin and rifampin in the treatment of borderline lepromatous and lepromatous leprosy. *Int. J. Lepr. Other Mycobact. Dis.* **68**, 129–135.
85. Graham, D. Y., Opekun, A. R., and Klein, P. D. (1993). Clarithromycin for the eradication of *Helicobacter pylori*. *J. Clin. Gastroenterol.* **16**, 292–294.
86. European *Helicobacter pylori* Study Group (1997). Current European concepts in the management of *Helicobacter pylori* infection. The Maastricht Consensus Report. *Gut* **41**, 8–13.
87. Syngal, S., Wolfe, M. M., Friedman, L. S., and Van Dam, J. (1997). *Helicobacter pylori*: Pathogenic factor in the etiology of gastric diseases. *Infect. Dis. Clin. Pract.* **6**, 17–27.
88. Peterson, W. L., Graham, D. Y., Marshall, B., Blaser, M. J., Genta, R. M., Klein, P. D., Stratton, C. W., Drnec, J., Prokocimer, P., and Siepman, N. (1993). Clarithromycin as monotherapy for eradication of *Helicobacter pylori*: A randomized, double-blind trial. *Am. J. Gastroenterol.* **88**, 1860–1864.

89. Thijs, J. C., Van Zwet, A. A., Moolenaar, W., Oom, J. A., De Korte, H., and Runhaar, E. A. (1994). Short report: Clarithromycin, an alternative to metronidazole in the triple therapy of *Helicobacter pylori* infection. *Aliment. Pharmacol. Ther.* **8**, 131–134.
90. Markham, A., and McTavish, D. (1996). Clarithromycin and omeprazole as *Helicobacter pylori* eradication therapy in patients with *H. pylori*-associated gastric disorders. *Drugs* **51**, 161–178.
91. Logan, R. P., Gummett, P. A., Schaufelberger, H. D., Greaves, R. R., Mendelson, G. M., Walker, M. M., Thomas, P. H., Baron, J. H., and Misiewicz, J. J. (1994). Eradication of *Helicobacter pylori* with clarithromycin and omeprazole. *Gut* **35**, 323–326.
92. Lazzaroni, M., Bargiggia, S., and Porro, G. B. (1997). Triple therapy with ranitidine or lansoprazole in the treatment of *Helicobacter pylori*-associated duodenal ulcer. *Am. J. Gastroenterol.* **92**, 649–652.
93. Mollison, L. C., Stingemore, N., Wake, R. A., Cullen, D. J., and McGechie, D. B. (2000). Antibiotic resistance in *Helicobacter pylori*. *Med. J. Aust.* **173**, 521–523.
94. Katelaris, P. H. (2000). *Helicobacter pylori*: Changing patterns of ulcer disease and antibiotic resistance. *Med. J. Aust.* **173**, 508–509.
95. Lahaie, R. G., and Gaudreau, C. (2000). *Helicobacter pylori* antibiotic resistance: Trends over time. *Can. J. Gastroenterol.* **14**, 895–899.
96. Taylor, D. E. (2000). Pathophysiology of antibiotic resistance: Clarithromycin. *Can. J. Gastroenterol.* **14**, 891–894.
97. Laine, L., Fennerty, M. B., Osato, M., Sugg, J., Suchower, L., Probst, P., and Levine, J. G. (2000). Esomeprazole-based *Helicobacter pylori* eradication therapy and the effect of antibiotic resistance: Results of three U.S. multicenter, double-blind trials. *Am. J. Gastroenterol.* **95**, 3393–3398.
98. Toracchio, S., Cellini, L., Di Campli, E., Cappello, G., Malatesta, M. G., Ferri, A., Ciccaglione, A. F., Grossi, L., and Marzio, L. (2000). Role of antimicrobial susceptibility testing on efficacy of triple therapy in *Helicobacter pylori* eradication. *Aliment. Pharmacol. Ther.* **14**, 1639–1643.
99. Fallone, C. A. (2000). Epidemiology of the antibiotic resistance of *Helicobacter pylori* in Canada. *Can. J. Gastroenterol.* **14**, 879–882.
100. Barza, M., and Eisenberg, E. (1996). Clarithromycin and azithromycin. *Infect. Dis. Clin. Pract.* **5**, 291–302.
101. Lacassin, F., Schaffo, D., Perronne, C., Longuet, P., Leport, C., and Vilde, J. L. (1995). Clarithromycin–minocycline combination as salvage therapy for toxoplasmosis in patients infected with human immunodeficiency virus. *Antimicrob. Agents Chemother.* **39**, 276–277.
102. Fernandez-Martin, J., Leport, C., Morlat, P., Meyohas, M. C., Chauvin, J. P., and Vilde, J. L. (1991). Pyrimethamine–clarithromycin combination for therapy of acute *Toxoplasma* encephalitis in patients with AIDS. *Antimicrob. Agents Chemother.* **35**, 2049–2052.
103. Raffi, F., Struillou, L., Ninin, E., Reliquet, V., Billaud, E., and Milpied, B. (1995). Breakthrough cerebral toxoplasmosis in patients with AIDS who are being treated with clarithromycin [letter]. *Clin. Infect. Dis.* **20**, 1076–1077.
104. Watkins, V. S., Polk, R. E., and Stotka, J. L. (1997). Drug interactions of macrolides: Emphasis on dirithromycin. *Ann. Pharmacother.* **31**, 349–356.
105. Bass, J. W., Vincent, J. M., and Person, D. A. (1997). The expanding spectrum of *Bartonella* infections: II. Cat-scratch disease. *Pediatr. Infect. Dis. J.* **16**, 163–179.
106. Dattwyler, R. J., Grunwaldt, E., and Luft, B. J. (1996). Clarithromycin in treatment of early Lyme disease: A pilot study. *Antimicrob. Agents Chemother.* **40**, 468–469.
107. Zygulska-Mach, H., and Mach, Z. (1975). The effect of gamma rays on glutathion and ascorbic acid content in rabbit lenses [in German]. *Klin. Monatsbl. Augenheilkd.* **166**, 50–55.
108. Braden, B., Helm, B., Fabian, T., and Dietrich, C. F. (2000). Bacillary angiomatosis of the liver, a suspected ultrasound diagnosis? [in German]. *Z. Gastroenterol.* **38**, 785–789.

109. Chu, S. Y., Granneman, G. R., Pichotta, P. J., Decourt, J. P., Girault, J., and Fourtillan, J. B. (1993). Effect of moderate or severe hepatic impairment on clarithromycin pharmacokinetics. *J. Clin. Pharmacol.* **33**, 480–485.
110. Schlossberg, D. (1995). Azithromycin and clarithromycin. *Med. Clin. North. Am.* **79**, 803–815.
111. Hardy, D. J., Swanson, R. N., Rode, R. A., Marsh, K., Shipkowitz, N. L., and Clement, J. J. (1990). Enhancement of the *in vitro* and *in vivo* activities of clarithromycin against *Haemophilus influenzae* by 14-hydroxy-clarithromycin, its major metabolite in humans. *Antimicrob. Agents Chemother.* **34**, 1407–1413.
112. Dabernat, H., Delmas, C., Seguy, M., Fourtillan, J. B., Girault, J., and Lareng, M. B. (1991). The activity of clarithromycin and its 14-hydroxy metabolite against *Haemophilus influenzae*, determined by *in vitro* and serum bactericidal tests. *J. Antimicrob. Chemother.* **27**, 19–30.
113. Garrison, M. W., Malone, C. L., Eiland, J., and Anderson, D. E. (1997). Influence of pH on the antimicrobial activity of clarithromycin and 14-hydroxyclarithromycin against *Haemophilus influenzae* using an *in vitro* pharmacodynamic model. *Diagn. Microbiol. Infect. Dis.* **27**, 139–145.
114. Ives, T. J., Manzewitsch, P., Regnery, R. L., Butts, J. D., and Kebede, M. (1997). *In vitro* susceptibilities of *Bartonella henselae*, *B. quintana*, *B. elizabethae*, *Rickettsia rickettsii*, *R. conorii*, *R. akari*, and *R. prowazekii* to macrolide antibiotics as determined by immunofluorescent-antibody analysis of infected Vero cell monolayers. *Antimicrob. Agents Chemother.* **41**, 578–582.
115. Brogden, R. N., and Peters, D. H. (1994). Dirithromycin. A review of its antimicrobial activity, pharmacokinetic properties and therapeutic efficacy. *Drugs* **48**, 599–616.
116. Derriennic, M., Conforti, P. M., and Sides, G. D. (1993). Dirithromycin in the treatment of streptococcal pharyngitis. *J. Antimicrob. Chemother.* **31**, 89–95.
117. Muller, O., and Wettich, K. (1993). Clinical efficacy of dirithromycin in pharyngitis and tonsillitis. *J. Antimicrob. Chemother.* **31**, 97–102.
118. Watkins, V. S., Smietana, M., Conforti, P. M., Sides, G. D., and Huck, W. (1997). Comparison of dirithromycin and penicillin for treatment of streptococcal pharyngitis. *Antimicrob. Agents Chemother.* **41**, 72–75.
119. Ruggiero, G., Utili, R., Adinolfi, L. E., Attanasio, V., Scarano, M. P., Mazzone, A., Costa, G., Califano, L., and Costa, F. (1993). Clinical efficacy of dirithromycin versus miocamycin in tonsillopharyngitis. *J. Antimicrob. Chemother.* **31**, 103–109.
120. Sides, G. D. (1993). Clinical efficacy of dirithromycin in acute exacerbations of chronic bronchitis. *J. Antimicrob. Chemother.* **31**, 131–138.
121. Gaillat, J. (1993). A multicentre study comparing the safety and efficacy of dirithromycin with erythromycin in the treatment of bronchitis. *J. Antimicrob. Chemother.* **31**, 139–151.
122. Wasilewski, M. M., Johns, D., and Sides, G. D. (1999). Five-day dirithromycin therapy is as effective as seven-day erythromycin therapy for acute exacerbations of chronic bronchitis. *J. Antimicrob. Chemother.* **43**, 541–548.
123. Pozzi, E. (1993). Clinical efficacy of dirithromycin versus miocamycin in the treatment of acute bronchitis or acute exacerbations of chronic bronchitis. *J. Antimicrob. Chemother.* **31**, 153–158.
124. Hosie, J., Quinn, P., Smits, P., and Sides, G. (1995). A comparison of 5 days of dirithromycin and 7 days of clarithromycin in acute bacterial exacerbation of chronic bronchitis. *J. Antimicrob. Chemother.* **36**, 173–183.
125. Cazzola, M., Vinciguerra, A., Di Perna, F., Califano, C., Calderaro, F., Salzillo, A., and Centanni, S. (1999). Comparative study of dirithromycin and azithromycin in the treatment of acute bacterial exacerbations of chronic bronchitis. *J. Chemother.* **11**, 119–125.
126. Van Royen, P., Betz, W., Heyrman, J., Taziaux, P., Van den Haute, M., and Poelman, M. (1997). Dirithromycin versus amoxiclav in the treatment of acute exacerbations of chronic bronchitis. *J. Int. Med. Res.* **25**, 33–40.

127. Cazzola, M., Caputi, M., Santangelo, G., Diu Vinciguerra, A., Perna, F. D., and Polverino, M. (1997). A five-day course of dirithromycin in the treatment of acute exacerbation of severe chronic obstructive pulmonary disease. *J. Chemother.* **9**, 279–284.
128. Jacobson, K. (1993). Clinical efficacy of dirithromycin in pneumonia. *J. Antimicrob. Chemother.* **31**, 121–129.
129. Liippo, K., Tala, E., Puolijoki, H., Bruckner, O. J., Rodrig, J., and Smits, J. P. (1994). A comparative study of dirithromycin and erythromycin in bacterial pneumonia. *J. Infect.* **28**, 131–139.
130. Hernandez, J. M., Sides, G. D., Conforti, P. M., and Smietana, M. G. (1996). Clinical efficacy of dirithromycin in patients with bacteremic pneumonia. *Clin. Ther.* **18**, 1128–1138.
131. Derriennic, M., and Escande, J. P. (1993). Dirithromycin in the treatment of skin and skin structure infections. *J. Antimicrob. Chemother.* **31**, 159–168.
132. Wasilewski, M. M., Wilson, M. G., Sides, G. D., and Stotka, J. L. (2000). Comparative efficacy of 5 days of dirithromycin and 7 days of erythromycin in skin and soft tissue infections. *J. Antimicrob. Chemother.* **46**, 255–262.
133. Laine, L., Stein, C., Garcia, F., Trujillo, M., and Estrada, R. (1996). Prospective evaluation of the macrolide antibiotic dirithromycin for the treatment of *Helicobacter pylori*. *Aliment. Pharmacol. Ther.* **10**, 269–273.
134. Sides, G. D., and Conforti, P. M. (1993). Safety profile of dirithromycin. *J. Antimicrob. Chemother.* **31**, 175–185.
135. Varanese, L. (1993). Pharmacokinetics and safety of dirithromycin in the elderly. *J. Antimicrob. Chemother.* **31**, 169–174.
136. Phillips, I., Pechere, J. C., Davies, A., and Speller, D. (1987). Roxithromycin: A new macrolide. Symposium. Paris, 29–30 May 1987. *J. Antimicrob. Chemother.* **20**, 1–187.
137. Markham, A., and Faulds, D. (1994). Roxithromycin. An update of its antimicrobial activity, pharmacokinetic properties and therapeutic use. *Drugs* **48**, 297–326.
138. Nilsen, O. G., Aamo, T., Zahlsen, K., and Svarva, P. (1992). Macrolide pharmacokinetics and dose scheduling of roxithromycin. *Diagn. Microbiol. Infect. Dis.* **15**, 71S–76S.
139. Pechere, J. C. (1992). Clinical evaluation of roxithromycin 300 mg once daily as an alternative to 150 mg twice daily. *Diagn. Microbiol. Infect. Dis.* **15**, 111S–117S.
140. Deffez, J. P., Scheimberg, A., and Rezvani, Y. (1992). Multicenter double-blind study of the efficacy and tolerance of roxithromycin versus erythromycin ethylsuccinate in acute orodental infection in adults. Odontogenic Infections Study Group. *Diagn. Microbiol. Infect. Dis.* **15**, 133S–137S.
141. Carbon, C., Hotton, J. M., Pepin, L. F., Wohlhuter, C., Souetre, E., Hardens, M., Lozet, H., and Riviera, M. (1996). Economic analysis of antibiotic regimens used in the treatment of pharyngitis: A prospective comparison of azithromycin versus roxithromycin. *J. Antimicrob. Chemother.* **37**, 151–161.
142. Chatzimanolis, E., Marsan, N., Lefatzis, D., and Pavlopoulos, A. (1998). Comparison of roxithromycin with co-amoxiclav in patients with sinusitis. *J. Antimicrob. Chemother.* **41**, 81–84.
143. de Campora, E., Camaioni, A., Leonardi, M., Fardella, P., and Fiaoni, M. (1992). Comparative efficacy and safety of roxithromycin and clarithromycin in upper respiratory tract infections. *Diagn. Microbiol. Infect. Dis.* **15**, 119S–122S.
144. Muller, O. (1996). An open comparative study of azithromycin and roxithromycin in the treatment of acute upper respiratory tract infections. *J. Antimicrob. Chemother.* **37**, 83–92.
145. De Vlieger, A., Druart, M., and Puttemans, M. (1992). Roxithromycin versus doxycycline in the treatment of acute exacerbations of chronic bronchitis. *Diagn. Microbiol. Infect. Dis.* **15**, 123S–127S.
146. Marsac, J. H. (1992). An international clinical trial on the efficacy and safety of roxithromycin in 40,000 patients with acute community-acquired respiratory tract infections. *Diagn. Microbiol. Infect. Dis.* **15**, 81S–84S.

147. Poirier, R. (1991). Comparative study of clarithromycin and roxithromycin in the treatment of community-acquired pneumonia. *J. Antimicrob. Chemother.* **27**, 109–116.
148. Tatsis, G., Tsoukalas, G., Boulbasakos, G., Platsouka, E., Anagnostopoulou, M., Pirounaki, M., Paniara, O., Sioula, E., Raptis, J., and Saroglou, G. (1998). Efficacy and tolerance of roxithromycin versus clarithromycin in the treatment of lower respiratory tract infections. *J. Antimicrob. Chemother.* **41**, 69–73.
149. Morandini, G., Perduca, M., Zannini, G., Foschino, M. P., Miragliotta, G., and Carnimeo, N. S. (1993). Clinical efficacy of azithromycin in lower respiratory tract infections. *J. Chemother.* **5**, 32–36.
150. Schonwald, S., Barsic, B., Klinar, I., and Gunjaca, M. (1994). Three-day azithromycin compared with ten-day roxithromycin treatment of atypical pneumonia. *Scand. J. Infect. Dis.* **26**, 706–710.
151. Laurent, K. (1996). Efficacy, safety and tolerability of azithromycin versus roxithromycin in the treatment of acute lower respiratory tract infections. *J. Antimicrob. Chemother.* **37**, 115–124.
152. Paulsen, O., Christensson, B. A., Hebelka, M., Ljungberg, B., Nilsson-Ehle, I., Nyman, L., Svensson, R., Tull, P., and Varga, Z. (1992). Efficacy and tolerance of roxithromycin in comparison with erythromycin stearate in patients with lower respiratory tract infections. *Scand. J. Infect. Dis.* **24**, 219–225.
153. Dautzenberg, B., Scheimberg, A., Brambilla, C., Camus, P., Godard, P., Guerin, J. C., Lemarie, E., Rezvani, Y., Rosembaum, M., Tuchais, E., et al. (1992). Comparison of two oral antibiotics, roxithromycin and amoxicillin plus clavulanic acid, in lower respiratory tract infections. *Diagn. Microbiol. Infect. Dis.* **15**, 85S–89S.
154. Lousbergh, D., Jochems, G., Everaert, L., and Puttemans, M. (1992). Roxithromycin versus amoxicillin–clavulanic acid in the treatment of respiratory tract infections. *Diagn. Microbiol. Infect. Dis.* **15**, 91S–95S.
155. Karalus, N. C., Garrett, J. E., Lang, S. D., Leng, R. A., Kostalas, G. N., Cursons, R. T., Cooper, B. C., and Ryan, C. J. (1995). Roxithromycin 150 mg b.i.d. versus amoxicillin 500 mg/clavulanic acid 125 mg t.i.d. for the treatment of lower respiratory tract infections in general practice. *Infection* **23**, S15–S20.
156. Scott, W. G., Cooper, B. C., and Scott, H. M. (1995). Pharmacoeconomic evaluation of roxithromycin versus amoxicillin/clavulanic acid in a community-acquired lower respiratory tract infection study. *Infection* **23**, S21–S24.
157. Tilyard, M. W., and Dovey, S. M. (1992). A randomized double-blind controlled trial of roxithromycin and cefaclor in the treatment of acute lower respiratory tract infections in general practice. *Diagn. Microbiol. Infect. Dis.* **15**, 97S–101S.
158. Scott, W. G., Tilyard, M. W., Dovey, S. M., Cooper, B., and Scott, H. M. (1993). Roxithromycin versus cefaclor in lower respiratory tract infection: A general practice pharmacoeconomic study. *Pharmacoeconomics* **4**, 122–130.
159. Salvarezza, C. R., Mingrone, H., Fachinelli, H., and Kijanczuk, S. (1998). Comparison of roxithromycin with cefixime in the treatment of adults with community-acquired pneumonia. *J. Antimicrob. Chemother.* **41**, 75–80.
160. Ortqvist, A., Valtonen, M., Cars, O., Wahl, M., Saikku, P., and Jean, C. (1996). Oral empiric treatment of community-acquired pneumonia. A multicenter, double-blind, randomized study comparing sparfloxacin with roxithromycin. The Scandinavian Sparfloxacin Study Group. *Chest* **110**, 1499–1506.
161. Kaku, M., Kohno, S., Koga, H., Ishida, K., and Hara, K. (1995). Efficacy of roxithromycin in the treatment of mycoplasma pneumonia. *Chemotherapy* **41**, 149–152.
162. Bernard, P., Plantin, P., Roger, H., Sassolas, B., Villaret, E., Legrain, V., Roujeau, J. C., Rezvani, Y., and Scheimberg, A. (1992). Roxithromycin versus penicillin in the treatment of erysipelas in adults: A comparative study. *Br. J. Dermatol.* **127**, 155–159.

163. Akamatsu, H., Nishijima, S., Akamatsu, M., Kurokawa, I., and Asada, Y. (1996). Clinical evaluation of roxithromycin in patients with acne. *J. Int. Med. Res.* **24**, 109–114.
164. Komine, M., and Tamaki, K. (2000). An open trial of oral macrolide treatment for psoriasis vulgaris. *J. Dermatol.* **27**, 508–512.
165. Hansen, K., Hovmark, A., Lebech, A. M., Lebech, K., Olsson, I., Halkier-Sorensen, L., Olsson, E., and Asbrink, E. (1992). Roxithromycin in Lyme borreliosis: Discrepant results of an *in vitro* and *in vivo* animal susceptibility study and a clinical trial in patients with erythema migrans. *Acta Derm. Venereol.* **72**, 297–300.
166. Gasser, R., Wendelin, I., Reisinger, E., Bergloff, J., Feigl, B., Schafhalter, I., Eber, B., Grisold, M., and Klein, W. (1995). Roxithromycin in the treatment of Lyme disease—update and perspectives. *Infection* **23**, S39–S43.
167. Stenberg, K., and Mardh, P. A. (1991). Treatment of chlamydial conjunctivitis in newborns and adults with erythromycin and roxithromycin. *J. Antimicrob. Chemother.* **28**, 301–307.
168. Stenberg, K., and Mardh, P. A. (1993). Treatment of concomitant eye and genital chlamydial infection with erythromycin and roxithromycin. *Acta Ophthalmol. (Copenh.)* **71**, 332–335.
169. Lidbrink, P., Bygdeman, S., Emtestam, L., Gajecki, M., Lapins, J., and Weden, U. (1993). Roxithromycin compared to doxycycline in the treatment of genital chlamydial infection and nonspecific urethritis. *Int. J. STD AIDS*, **4**, 110–113.
170. van Schouwenburg, J., de Bruyn, O., Fourie, E., van Rensburg, J., Rodriques, A., and Pickard, I. (1992). A randomized, comparative study of the efficacy and tolerance of roxithromycin and doxycycline in the treatment of women with positive endocervical cultures for *Chlamydia trachomatis* and *Mycoplasma* spp. in an *in vitro* fertilization program. *Diagn. Microbiol. Infect. Dis.* **15**, 129S–131S.
171. Worm, A. M. (1990). Roxithromycin and erythromycin in chlamydia-negative nongonococcal urethritis. *Acta Derm. Venereol.* **70**, 269–271.
172. Gurfinkel, E., Bozovich, G., Daroca, A., Beck, E., and Mautner, B. (1997). Randomised trial of roxithromycin in non-Q-wave coronary syndromes: ROXIS Pilot Study. ROXIS Study Group. *Lancet* **350**, 404–407.
173. Gurfinkel, E., Bozovich, G., Beck, E., Testa, E., Livellara, B., and Mautner, B. (1999). Treatment with the antibiotic roxithromycin in patients with acute non-Q-wave coronary syndromes. The final report of the ROXIS Study. *Eur. Heart J.* **20**, 121–127.
174. Cellini, L., Marzio, L., Di Girolamo, A., Allocati, N., Grossi, L., and Dainelli, B. (1991). Enhanced clearing of *Helicobacter pylori* after omeprazole plus roxithromycin treatment. *FEMS Microbiol. Lett.* **68**, 255–257.
175. Okada, M., Oki, K., Shirotani, T., Seo, M., Okabe, N., Maeda, K., Nishimura, H., Ohkuma, K., and Oda, K. (1998). A new quadruple therapy for the eradication of *Helicobacter pylori*. Effect of pretreatment with omeprazole on the cure rate. *J. Gastroenterol.* **33**, 640–645.
176. Pohle, T., Stoll, R., Kirchner, T., Heep, M., Lehn, N., Bock, H., and Domschke, W. (1998). Eradication of *Helicobacter pylori* with lansoprazole, roxithromycin and metronidazole—an open pilot study. *Aliment. Pharmacol. Ther.* **12**, 1273–1278.
177. Svoboda, P., Kantorova, I., Ochmann, J., Doubek, J., Kozumplik, L., and Marsova, J. (1997). Pantoprazole-based dual and triple therapy for the eradication of *Helicobacter pylori* infection: A randomized controlled trial. *Hepatogastroenterology* **44**, 886–890.
178. Uygun, A., Ates, Y., Erdil, A., Kadayifci, A., Cetin, C., Gulsen, M., Karaeren, N., and Dagalp, K. (1999). Efficacy of omeprazole plus two antimicrobials for the eradication of *Helicobacter pylori* in a Turkish population. *Clin. Ther.* **21**, 1539–1548.
179. Thamlikitkul, V., Kobwanthanakun, S., Pruksachatvuthi, S., and Lertluknithi, R. (1992). Pharmacokinetics of rheumatic fever prophylaxis regimens. *J. Int. Med. Res.* **20**, 20–26.
180. Kern, W. V., Hay, B., Kern, P., Marre, R., and Arnold, R. (1994). A randomized trial of roxithromycin in patients with acute leukemia and bone marrow transplant recipients receiving fluoroquinolone prophylaxis. *Antimicrob. Agents Chemother.* **38**, 465–472.

181. Sprinz, E., Mallman, R., Barcellos, S., Silbert, S., Schestatsky, G., and Bem David, D. (1998). AIDS-related cryptosporidial diarrhoea: An open study with roxithromycin. *J. Antimicrob. Chemother.* **41**, 85–91.
182. Uip, D. E., Lima, A. L., Amato, V. S., Boulos, M., Neto, V. A., and Bem David, D. (1998). Roxithromycin treatment for diarrhoea caused by *Cryptosporidium* spp. in patients with AIDS. *J. Antimicrob. Chemother.* **41**, 93–97.
183. Durant, J., Hazime, F., Carles, M., Pechere, J. C., and Dellamonica, P. (1995). Prevention of *Pneumocystis carinii* pneumonia and of cerebral toxoplasmosis by roxithromycin in HIV-infected patients. *Infection* **23**, S33–S38.
183a. Treadway, G., and Pontani, D. (1995). Paediatric safety of azithromycin: Worldwide experience. *J. Clin. Pharmacol.* **31**, 164–167.
184. Pacifico, L., Scopetti, F., Ranucci, A., Patarracchia, M., Savignoni, F., and Chiesa, C. (1996). Comparative efficacy and safety of 3-day azithromycin and 10-day penicillin V treatment of group A beta-hemolytic streptococcal pharyngitis in children. *Antimicrob. Agents Chemother.* **40**, 1005–1008.
185. O'Doherty, B. (1996). Azithromycin versus penicillin V in the treatment of paediatric patients with acute streptococcal pharyngitis/tonsillitis. Paediatric Azithromycin Study Group. *Eur. J. Clin. Microbiol. Infect. Dis.* **15**, 718–724.
186. Still, J. G. (1993). Azithromycin suspension vs. penicillin V suspension in treatment of children with streptococcal pharyngitis [abstract PII-101]. *Clin. Pharmacol. Ther.* **53**, 195.
187. Still, J. G. (1995). Management of pediatric patients with group A beta-hemolytic *Streptococcus* pharyngitis: Treatment options. *Pediatr. Infect. Dis. J.* **14**, S57–61.
188. Still, J. G. (1994). Treatment of streptococcal pharyngitis in children with five days of azithromycin suspension. *In* "Program of the 34th Interscience Conference on Antimicrobial Agents and Chemotherapy," Abstr. No. M67. American Society for Microbiology, Washington, DC.
189. Pichichero, M. E. (1992). Therapeutic considerations for management of otitis media, sinusitis, and tonsillopharyngitis. *Pediatr. Asthma Allergy Immunol.* **6**, 167–174.
190. Aronovitz, G. (1996). A multicenter, open label trial of azithromycin vs. amoxicillin/clavulanate for the management of acute otitis media in children. *Pediatr. Infect. Dis. J.* **15**, S15–19.
191. Daniel, R. R. (1993). Comparison of azithromycin and co-amoxiclav in the treatment of otitis media in children. *J. Antimicrob. Chemother.* **31**, 65–71.
192. Khurana, C. M. (1995). Issues concerning antibiotic use in child care settings. *Pediatr. Infect. Dis. J.* **14**, S34–38.
193. Khurana, C. M. (1996). A multicenter, randomized, open label comparison of azithromycin and amoxicillin/clavulanate in acute otitis media among children attending day care or school. *Pediatr. Infect. Dis. J.* **15**, S24–S29.
194. McLinn, S. (1996). A multicenter, double blind comparison of azithromycin and amoxicillin/clavulanate for the treatment of acute otitis media in children. *Pediatr. Infect. Dis. J.* **15**, S20–23.
195. Principi, N. (1995). Multicentre comparative study of the efficacy and safety of azithromycin compared with amoxicillin/clavulanic acid in the treatment of paediatric patients with otitis media. *Eur. J. Clin. Microbiol. Infect. Dis.* **14**, 669–676.
196. Schaad, U. B. (1993). Multicentre evaluation of azithromycin in comparison with co-amoxiclav for the treatment of acute otitis media in children. *J. Antimicrob. Chemother.* **31**, 81–88.
197. Dagan, R., Leibovitz, E., Jacobs, M., Fliss, D., Leiberman, A., and Yagupsky, P. (1997). Bacteriologic response to acute otitis media caused by *Hemophilus influenzae* treated with azithromycin. *In* "Program of the 37th Interscience Conference on Antimicrobial Agents and Chemotherapy," p. 345. American Society for Microbiology, Washington, DC.
198. Roord, J. J., Wolf, B. H., Gossens, M. M., and Kimpen, J. L. (1996). Prospective open randomized study comparing efficacies and safeties of a 3-day course of azithromycin and a 10-day course of erythromycin in children with community-acquired acute lower respiratory tract infections. *Antimicrob. Agents Chemother.* **40**, 2765–2768.

199. Arguedas, A., Rodriguez, A., Loaiza, C., Urruela, R., Rodriguez, F., Bartlett, H., Herrera, M., and Mohs, E. (1997). Azithromycin vs. amox/clav in pediatric community-acquired LRTIs. *Infect. Med.* **14**, 807–809, 813–814.
200. Harris, J., Campbell, M., Kolokathis, A., and The Pediatric Pneumonia Study Group (1996). Safety and efficacy of azithromycin in the treatment of community acquired pneumonia in children. *In* "Program of the 36th Interscience Conference on Antimicrobial Agents and Chemotherapy," p. 286. American Society for Microbiology, Washington, DC.
201. Johnson, R. B. (1991). The role of azalide antibiotics in the treatment of Chlamydia. *Am. J. Obstet. Gynecol.* **164**, 1794–1796.
202. Steingrimsson, O., Olafsson, J. H., Thorarinsson, H., Ryan, R. W., Johnson, R. B., and Tilton, R. C. (1990). Azithromycin in the treatment of sexually transmitted disease. *J. Antimicrob. Chemother.* **25**, 109–114.
203. Lassus, A. (1990). Comparative studies of azithromycin in skin and soft-tissue infections and sexually transmitted infections by *Neisseria* and *Chlamydia* species. *J. Antimicrob. Chemother.* **25**, 115–121.
204. Martin, D. H., Mroczkowski, T. F., Dalu, Z. A., McCarty, J., Jones, R. B., Hopkins, S. J., and Johnson, R. B. (1992). A controlled trial of a single dose of azithromycin for the treatment of chlamydial urethritis and cervicitis. The Azithromycin for Chlamydial Infections Study Group. *N. Engl. J. Med.* **327**, 921–925.
205. Hammerschlag, M. R., Golden, N. H., Oh, M. K., Gelling, M., Sturdevant, M., Brown, P. R., Aras, Z., Neuhoff, S., Dumornay, W., and Roblin, P. M. (1993). Single dose of azithromycin for the treatment of genital chlamydial infections in adolescents. *J. Pediatr.* **122**, 961–965.
206. Whatley, J. D., Thin, R. N., Mumtaz, G., and Ridgway, G. L. (1991). Azithromycin vs. doxycycline in the treatment of nongonococcal urethritis. *Int. J. STD AIDS* **2**, 248–251.
207. Ossewaarde, J. M., Plantema, F. H., Rieffe, M., Nawrocki, R. P., de Vries, A., and van Loon, A. M. (1992). Efficacy of single-dose azithromycin versus doxycycline in the treatment of cervical infections caused by *Chlamydia trachomatis*. *Eur. J. Clin. Microbiol. Infect. Dis.* **11**, 693–697.
208. Nilsen, A., Halsos, A., Johansen, A., Hansen, E., Torud, E., Moseng, D., Anestad, G., and Storvold, G. (1992). A double blind study of single dose azithromycin and doxycycline in the treatment of chlamydial urethritis in males. *Genitourin Med.* **68**, 325-327.
209. Nuovo, J., Melnikow, J., Paliescheskey, M., King, J., and Mowers, R. (1995). Cost-effectiveness analysis of five different antibiotic regimens for the treatment of uncomplicated *Chlamydia trachomatis* cervicitis. *J. Am. Board Fam. Pract.* **8**, 7–16.
210. Magid, D., Douglas, Jr., J. M., and Schwartz, J. S. (1996). Doxycycline compared with azithromycin for treating women with genital *Chlamydia trachomatis* infections: An incremental cost-effectiveness analysis. *Ann. Intern. Med.* **124**, 389–399.
211. Haddix, A. C., Hillis, S. D., and Kassler, W. J. (1995). The cost effectiveness of azithromycin for *Chlamydia trachomatis* infections in women. *Sex. Transm. Dis.* **22**, 274–280.
212. Centers for Disease Control and Prevention (1998). 1998 guidelines for treatment of sexually transmitted diseases. *Morb. Mortal. Wkly. Rep.* **47**, 1–111.
213. Edwards, M., Rainwater, K., Carter, S., Williamson, F., and Newman, R. (1994). Comparison of azithromycin and erythromycin for chlamydia cervicitis in pregnancy [abstract]. *Am. J. Obstet. Gynecol.* **170**, 419.
214. Bush, M. R., and Rosa, C. (1994). Azithromycin and erythromycin in the treatment of cervical chlamydial infection during pregnancy. *Obstet. Gynecol.* **84**, 61–63.
215. Lauharanta, J., Saarinen, K., Mustonen, M. T., and Happonen, H. P. (1993). Single-dose oral azithromycin versus seven-day doxycycline in the treatment of nongonococcal urethritis in males. *J. Antimicrob. Chemother.* **31**, 177–183.
216. Lister, P. J., Balechandran, T., Ridgway, G. L., and Robinson, A. J. (1993). Comparison of azithromycin and doxycycline in the treatment of nongonococcal urethritis in men. *J. Antimicrob. Chemother.* **31**, 185–192.

217. Stamm, W. E., Hicks, C. B., Martin, D. H., Leone, P., Hook III, E. W., Cooper, R. H., Cohen, M. S., Batteiger, B. E., Workowski, K., and McCormack, W. M. (1995). Azithromycin for empirical treatment of the nongonococcal urethritis syndrome in men. A randomized double-blind study. *JAMA* **274**, 545–549.
218. Pfizer (1997). Product monograph: Intravenous azithromycin.
219. Waugh, M. A. (1993). Open study of the safety and efficacy of a single oral dose of azithromycin for the treatment of uncomplicated gonorrhoea in men and women. *J. Antimicrob. Chemother.* **31**, 193–198.
220. Steingrimsson, O., Olafsson, J. H., Thorarinsson, H., Ryan, R. W., Johnson, R. B., and Tilton, R. C. (1994). Single dose azithromycin treatment of gonorrhea and infections caused by *C. trachomatis* and *U. urealyticum* in men. *Sex. Transm. Dis.* **21**, 43–46.
221. Handsfield, H. H., Dalu, Z. A., Martin, D. H., Douglas, Jr., J. M., McCarty, J. M., and Schlossberg, D. (1994). Multicenter trial of single-dose azithromycin vs. ceftriaxone in the treatment of uncomplicated gonorrhea. Azithromycin Gonorrhea Study Group. *Sex. Transm. Dis.* **21**, 107–111.
222. Lukehart, S. A., Fohn, M. J., and Baker-Zander, S. A. (1990). Efficacy of azithromycin for therapy of active syphilis in the rabbit model. *J. Antimicrob. Chemother.* **25**, 91–99.
223. Verdon, M. S., Handsfield, H. H., and Johnson, R. B. (1994). Pilot study of azithromycin for treatment of primary and secondary syphilis. *Clin. Infect. Dis.* **19**, 486–488.
224. Mashkilleyson, A. L., Gromberg, M. A., Kutin, S. A., Kardashenko, B. Y., and Vorobyev, S. V. (1994). Azithromycin in early syphilis [abstract]. *In* "Program Second Int. Conf. Macrolides, Azalides, Streptogramins" (Venice), p. 73.
225. Issoire, C., Casin, I., Perenet, F., Brunat, N., Janier, M., and Perol, Y. (1990). Pilot study of azithromycin in the treament of chancroid caused by *Haemophilus ducreyi*. Presented at First Conf. Infect. Dis. (Montreal).
226. Tyndall, M. W., Agoki, E., Plummer, F. A., Malisa, W., Ndinya-Achola, J. O., and Ronald, A. R. (1994). Single dose azithromycin for the treatment of chancroid: A randomized comparison with erythromycin. *Sex. Transm. Dis.* **21**, 231–234.
227. Martin, D. H., Sargent, S. J., Wendel, G. D., Jr., McCormack, W. M., Spier, N. A., and Johnson, R. B. (1995). Comparison of azithromycin and ceftriaxone for the treatment of chancroid. *Clin. Infect. Dis.* **21**, 409–414.
228. Ballard, R. C., Ye, H., Matta, A., Dangor, Y., and Radebe, F. (1996). Treatment of chancroid with azithromycin. *Int. J. STD AIDS* **7**, 9–12.
229. Bowden, F. J., Mein, J., Plunkett, C., and Bastian, I. (1996). Pilot study of azithromycin in the treatment of genital donovanosis. *Genitourin Med.* **72**, 17–19.
230. Amin, N. M., and Breadon, G. (1995). An open-label, noncomparative study to evaluate the efficacy, safety, and tolerability of azithromycin in the treatment of patients with acute sinusitis. *Clin. Ther.* **17**, 701–707.
231. Haye, R., Lingaas, E., Hoivik, H. O., and Odegard, T. (1996). Efficacy and safety of azithromycin versus phenoxymethylpenicillin in the treatment of acute maxillary sinusitis. *Eur. J. Clin. Microbiol. Infect. Dis.* **15**, 849–853.
232. Felstead, S. J., Daniel, R., and European Azithromycin Study Group (1991). Short-course treatment of sinusitis and other upper respiratory tract infections with azithromycin: A comparison with erythromycin and amoxycillin. *J. Int. Med. Res.* **19**, 363–372.
233. Daniel, R. (1991). Simplified treatment of acute lower respiratory tract infection with azithromycin: A comparison with erythromycin and amoxycillin. European Azithromycin Study Group. *J. Int. Med. Res.* **19**, 373–383.
234. Dark, D. (1991). Multicenter evaluation of azithromycin and cefaclor in acute lower respiratory tract infections. *Am. J. Med.* **91**, 31S–35S.
235. Kinasewitz, G., and Wood, R. G. (1991). Azithromycin versus cefaclor in the treatment of acute bacterial pneumonia. *Eur. J. Clin. Microbiol. Infect. Dis.* **10**, 872–877.

236. Balmes, P., Clerc, G., Dupont, B., Labram, C., Pariente, R., and Poirier, R. (1991). Comparative study of azithromycin and amoxicillin/clavulanic acid in the treatment of lower respiratory tract infections. *Eur. J. Clin. Microbiol. Infect. Dis.* **10**, 437–439.
237. Hoepelman, A. I., Sips, A. P., van Helmond, J. L., van Barneveld, P. W., Neve, A. J., Zwinkels, M., Rozenberg-Arska, M., and Verhoef, J. (1993). A single-blind comparison of three-day azithromycin and ten-day co-amoxiclav treatment of acute lower respiratory tract infections. *J. Antimicrob. Chemother.* **31**, 147–152.
238. Bradbury, F. (1993). Comparison of azithromycin versus clarithromycin in the treatment of patients with lower respiratory tract infection. *J. Antimicrob. Chemother.* **31**, 153–162.
239. Gray, G. C., McPhate, D. C., Leinonen, M., Cassell, G. H., Deperalta, E. P., Putnam, S. D., Karcher, J. A., Sawyer, M. H., Laurila, A., and Connor, J. D. (1998). Weekly oral azithromycin as prophylaxis for agents causing acute respiratory disease. *Clin. Infect. Dis.* **26**, 103–110.
240. Hammerschlag, M. R., Gregory, W. W., Schwartz, D. B., Pistoriius, B. J., Inverso, J. A., and Kolokathis, A. (1997). Azithromycin in the treatment of community-acquired pneumonia due to *Chlamydia pneumoniae*. In "Program of the 37th Interscience Conference on Antimicrobial Agents and Chemotherapy," p. 352. American Society for Microbiology, Washington, DC.
241. Daniel, R. (1991). Azithromycin, erythromycin and cloxacillin in the treatment of infections of skin and associated soft tissues. European Azithromycin Study Group. *J. Int. Med. Res.* **19**, 433–445.
242. Kiani, R. (1991). Double-blind, double-dummy comparison of azithromycin and cephalexin in the treatment of skin and skin structure infections. *Eur. J. Clin. Microbiol. Infect. Dis.* **10**, 880–884.
243. Mallory, S. B. (1991). Azithromycin compared with cephalexin in the treatment of skin and skin structure infections. *Am. J. Med.* **91**, 36S–39S.
244. Amaya-Tapia, G., Aguirre-Avalos, G., Andrade-Villanueva, J., Peredo-Gonzalez, G., Morfin-Otero, R., Esparza-Ahumada, S., and Rodriguez-Noriega, E. (1993). Once-daily azithromycin in the treatment of adult skin and skin-structure infections. *J. Antimicrob. Chemother.* **31**, 129–135.
245. Montero, L. (1996). A comparative study of the efficacy, safety and tolerability of azithromycin and cefaclor in the treatment of children with acute skin and/or soft tissue infections. *J. Antimicrob. Chemother.* **37**, 125–131.
246. Rodriguez-Solares, A., Perez-Gutierrez, F., Prosperi, J., Milgram, E., and Martin, A. (1993). A comparative study of the efficacy, safety and tolerance of azithromycin, dicloxacillin and flucloxacillin in the treatment of children with acute skin and skin-structure infections. *J. Antimicrob. Chemother.* **31**, 103–109.
247. Iseman, M. D. (1989). *Mycobacterium avium* complex and the normal host: The other side of the coin [editorial]. *N. Engl. J. Med.* **321**, 896–898.
248. Inderlied, C. B., Kemper, C. A., and Bermudez, L. E. (1993). The *Mycobacterium avium* complex. *Clin. Microbiol. Rev.* **6**, 266–310.
249. Young, L. S., Wiviott, L., Wu, M., Kolonoski, P., Bolan, R., and Inderlied, C. B. (1991). Azithromycin for treatment of *Mycobacterium avium*–intracellulare complex infection in patients with AIDS. *Lancet* **338**, 1107–1109.
250. Nightingale, S. D., Cameron, D. W., Gordin, F. M., Sullam, P. M., Cohn, D. L., Chaisson, R. E., Eron, L. J., Sparti, P. D., Bihari, B., Kaufman, D. L., et al. (1993). Two controlled trials of rifabutin prophylaxis against *Mycobacterium avium* complex infection in AIDS. *N. Engl. J. Med.* **329**, 828–833.
250a. Berry, A., Koletar, S., and Williams, D. (1993). Azithromycin therapy for disseminited *Mycobacterium avium-intracelluare* in AIDS patients [abstract]. *Nat. Conf. Hum. Retroviruses Relat. Infect.* **1**, 106.
251. Husson, R. N., Ross, L. A., Sandelli, S., Inderlied, C. B., Venzon, D., Lewis, L. L., Woods, L., Conville, P. S., Witebsky, F. G., and Pizzo, P. A. (1994). Orally administered clarithromycin for

the treatment of systemic *Mycobacterium avium* complex infection in children with acquired immunodeficiency syndrome. *J. Pediatr.* **124**, 807–814.
252. Horsburgh, C. R. (1996). Advances in the prevention and treatment of *Mycobacterium avium* disease [editorial]. *N. Engl. J. Med.* **335**, 428–429.
253. Masur, H. (1993). Recommendations on prophylaxis and therapy for disseminated *Mycobacterium avium* complex disease in patients infected with the human immunodeficiency virus. Public Health Service Task Force on Prophylaxis and Therapy for *Mycobacterium avium* Complex. *N. Engl. J. Med.* **329**, 898–904.
254. Havlir, D. V., Dube, M. P., Sattler, F. R., Forthal, D. N., Kemper, C. A., Dunne, M. W., Parenti, D. M., Lavelle, J. P., White, A., Witt, M. D., Bozzette, S. A., and McCutchan, J. A. (1996). Prophylaxis against disseminated *Mycobacterium avium* complex with weekly azithromycin, daily rifabutin, or both. California Collaborative Treatment Group. *N. Engl. J. Med.* **335**, 392–398.
255. Baker, R. L., and Norris, S. A. (1994). Clarithromycin prophylaxis against MAC in AIDS. *In* "Program of the 34th Interscience Conference on Antimicrobial Agents and Chemotherapy," p. 56. American Society for Microbiology, Washington, DC.
256. Griffith, D. E., Brown, B. A., Girard, W. M., Murphy, D. T., and Wallace, Jr., R. J. (1996). Azithromycin activity against *Mycobacterium avium* complex lung disease in patients who were not infected with human immunodeficiency virus. *Clin. Infect. Dis.* **23**, 983–989.
257. Brown, B. A., Griffith, D. E., Girard, W., Levin, J., and Wallace, Jr., R. J. (1997). Relationship of adverse events to serum drug levels in patients receiving high-dose azithromycin for mycobacterial lung disease. *Clin. Infect. Dis.* **24**, 958–964.
258. Wallace, R. J., Griffith, D. E., Brown, B. A., Murphy, D., Girard, W., and Cegielski, P. (1996). Initial results of three times weekly azithromycin in treatment regimens of *Mycobacterium avium*–intracellulare (MAI) lung disease in non-AIDS [abstract]. Presented at the Third International Conference on Macrolides, Azilides, and Streptogramins (Lisbon, Portugal).
259. Butler, T., Sridhar, C. B., Daga, M. K., Jani, K., Pandit, R. B., Khahgria, R., Potkar, C. N., and Johnson, R. B. (1997). Treatment of typhoid fever with azithromycin vs. chloramphenicol in a randomized multicenter trial in India. *In* "Program of the 37th Interscience Conference on Antimicrobial Agents and Chemotherapy," p. 367. American Society for Microbiology, Washington, DC.
260. Rautelin, H., Renkonen, O. V., and Kosunen, T. U. (1993). Azithromycin resistance in *Campylobacter jejuni* and *Campylobacter coli*. *Eur. J. Clin. Microbiol. Infect. Dis.* **12**, 864–865.
261. Kuschner, R. A., Trofa, A. F., Thomas, R. J., Hoge, C. W., Pitarangsi, C., Amato, S., Olafson, R. P., Echeverria, P., Sadoff, J. C., and Taylor, D. N. (1995). Use of azithromycin for the treatment of *Campylobacter* enteritis in travelers to Thailand, an area where ciprofloxacin resistance is prevalent. *Clin. Infect. Dis.* **21**, 536–541.
262. Murphy, Jr., G. S., Echeverria, P., Jackson, L. R., Arness, M. K., LeBron, C., and Pitarangsi, C. (1996). Ciprofloxacin- and azithromycin-resistant *Campylobacter* causing traveler's diarrhea in U.S. troops deployed to Thailand in 1994. *Clin. Infect. Dis.* **22**, 868–869.
263. Gordillo, M. E., Singh, K. V., and Murray, B. E. (1993). In vitro activity of azithromycin against bacterial enteric pathogens. *Antimicrob. Agents Chemother.* **37**, 1203–1205.
264. Khan, W. A., Seas, C., Dhar, U., Salam, M. A., and Bennish, M. L. (1997). Treatment of shigellosis: V. Comparison of azithromycin and ciprofloxacin. A double-blind, randomized, controlled trial. *Ann. Intern. Med.* **126**, 697–703.
265. Shanks, G. D., Ragama, O. B., Aleman, G. M., Andersen, S. L., and Gordon, D. M. (1996). Azithromycin prophylaxis prevents epidemic dysentery. *Trans. R. Soc. Trop. Med. Hyg.* **90**, 316.
266. Bailey, R. L., Arullendran, P., Whittle, H. C., and Mabey, D. C. (1993). Randomised controlled trial of single-dose azithromycin in treatment of trachoma. *Lancet* **342**, 453–456.
267. Dawson, C. R., Schachter, J., Sallam, S., Sheta, A., Rubinstein, R. A., and Washton, H. (1997). A comparison of oral azithromycin with topical oxytetracycline/polymyxin for the treatment of trachoma in children. *Clin. Infect. Dis.* **24**, 363–368.

268. McNulty, C. A., and Dent, J. C. (1988). Susceptibility of clinical isolates of *Campylobacter pylori* to twenty-one antimicrobial agents. *Eur. J. Clin. Microbiol. Infect. Dis.* **7**, 566–569.
269. Hulten, K., Cars, O., Hjelm, E., and Engstrand, L. (1996). In vitro activity of azithromycin against intracellular *Helicobacter pylori*. *J. Antimicrob. Chemother.* **37**, 483–489.
270. Bertoni, G., Sassatelli, R., Nigrisoli, E., Tansini, P., Bianchi, G., Della Casa, G., Bagni, A., and Bedogni, G. (1996). Triple therapy with azithromycin, omeprazole, and amoxicillin is highly effective in the eradication of *Helicobacter pylori*: A controlled trial versus omeprazole plus amoxicillin. *Am. J. Gastroenterol.* **91**, 258–263.
271. al-Assi, M. T., Genta, R.M., Karttunen, T. J., Cole, R. A., and Graham, D. Y. (1995). Azithromycin triple therapy for *Helicobacter pylori* infection: Azithromycin, tetracycline, and bismuth. *Am. J. Gastroenterol.* **90**, 403–405.
272. Tursi, A., Cammarota, G., Montalto, M., Papa, A., Fedeli, G., and Gasbarrini, G. (1996). The use of azithromycin in short-term low-dose triple therapies for *Helicobacter pylori* infection [letter]. *Am. J. Gastroenterol.* **91**, 817–818.
273. Pilotto, A., Di Mario, F., Franceschi, M., Leandro, G., Soffiati, G., Scagnelli, M., Bozzola, L., and Valerio, G. (1996). Cure of *Helicobacter pylori* infection in the elderly: Effects of eradication on gastritis and serological markers. *Aliment. Pharmacol. Ther.* **10**, 1021–1027.
274. Caselli, M., Ruina, M., Fabbri, P., Balhous, W., and Alvisi, V. (1996). A comparative trial of short term therapy with omeprazole plus either amoxicillin or azithromycin for *Helicobacter pylori* eradication [letter]. *J. Antimicrob. Chemother.* **37**, 849–850.
275. Caselli, M., Trevisani, L., Tursi, A., Sartori, S., Ruina, M., Luzzi, I., Gaudenzi, P., Alvisi, V., and Gasbarrini, G. (1997). Short-term low-dose triple therapy with azithromycin, metronidazole and lansoprazole appears highly effective for the eradication of *Helicobacter pylori*. *Eur. J. Gastroenterol. Hepatol.* **9**, 45–48.
276. Johnson, R. C., Kodner, C., Russell, M., and Girard, D. (1990). In vitro and in vivo susceptibility of *Borrelia burgdorferi* to azithromycin. *J. Antimicrob. Chemother.* **25**, 33–38.
277. Preac-Mursic, V., Wilske, B., Schierz, G., Suss, E., and Gross, B. (1989). Comparative antimicrobial activity of the new macrolides against *Borrelia burgdorferi*. *Eur. J. Clin. Microbiol. Infect. Dis.* **8**, 651–653.
278. Dever, L. L., Jorgensen, J. H., and Barbour, A. G. (1993). Comparative in vitro activities of clarithromycin, azithromycin, and erythromycin against *Borrelia burgdorferi*. *Antimicrob. Agents Chemother.* **37**, 1704–1706.
279. Massarotti, E. M., Luger, S. W., Rahn, D. W., Messner, R. P., Wong, J. B., Johnson, R. C., and Steere, A. C. (1992). Treatment of early Lyme disease. *Am. J. Med.* **92**, 396–403.
280. Strle, F., Ruzic, E., and Cimperman, J. (1992). Erythema migrans: Comparison of treatment with azithromycin, doxycycline and phenoxymethylpenicillin. *J. Antimicrob. Chemother.* **30**, 543–550.
281. Weber, K., Wilske, B., Preac-Mursic, V., and Thurmayr, R. (1993). Azithromycin versus penicillin V for the treatment of early Lyme borreliosis. *Infection* **21**, 367–372.
282. Luft, B. J., Dattwyler, R. J., Johnson, R. C., Luger, S. W., Bosler, E. M., Rahn, D. W., Masters, E. J., Grunwaldt, E., and Gadgil, S. D. (1996). Azithromycin compared with amoxicillin in the treatment of erythema migrans. A double-blind, randomized, controlled trial. *Ann. Intern. Med.* **124**, 785–791.
283. Strle, F., Maraspin, V., Lotric-Furlan, S., Ruzic-Sabljic, E., and Cimperman, J. (1996). Azithromycin and doxycycline for treatment of *Borrelia* culture-positive erythema migrans. *Infection* **24**, 64–68.
284. Girgis, N. I., Frenck, R. W., and Sultan, Y. (1997). Single dose azithromycin compared to a four dose rifampin regimen for the eradication of meningococcal nasopharyngeal carriage. *In* "Program of the 37th Interscience Conference on Antimicrobial Agents and Chemotherapy," p. 364. American Society for Microbiology, Washington, DC.

285. Gingras, B. A., and Jensen, J. B. (1992). Activity of azithromycin (CP-62,993) and erythromycin against chloroquine-sensitive and chloroquine-resistant strains of *Plasmodium falciparum* in vitro. *Am. J. Trop. Med. Hyg.* **47**, 378–382.
286. Kuschner, R. A., Heppner, D. G., Andersen, S. L., Wellde, B. T., Hall, T., Schneider, I., Ballou, W. R., Foulds, G., Sadoff, J. C., Schuster, B., et al. (1994). Azithromycin prophylaxis against a chloroquine-resistant strain of *Plasmodium falciparum*. *Lancet* **343**, 1396–1397.
287. Andersen, S. L., Ager, A., McGreevy, P., Schuster, B. G., Wesche, D., Kuschner, R., Ohrt, C., Ellis, W., Rossan, R., and Berman, J. (1995). Activity of azithromycin as a blood schizonticide against rodent and human plasmodia *in vivo*. *Am. J. Trop. Med. Hyg.* **52**, 159–161.
288. Taylor, W. R., Richie, T. L. F., Fryauff, D. F., Picarima, H., Ohrt, C., Tang, D., Braitman, D., Murphy, G. S., Widjaja, H., Tjitra, E., Ganjar, A., Jones, T. R., Basri, H., and Berman, J. (1999). Malaria prophylaxis using azithromycin: A double-blind, placebo-controlled trial in Iraian, Jaya, Indonesia. *Clin. Infect. Dis.* **28**, 74–81.
289. Weiss, L. M., Wittner, M., Wasserman, S., Oz, H. S., Retsema, J., and Tanowitz, H. B. (1993). Efficacy of azithromycin for treating *Babesia microti* infection in the hamster model. *J. Infect. Dis.* **168**, 1289–1292.
290. Wittner, M., Lederman, J., Tanowitz, H. B., Rosenbaum, G. S., and Weiss, L. M. (1996). Atovaquone in the treatment of *Babesia microti* infections in hamsters. *Am. J. Trop. Med. Hyg.* **55**, 219–222.
291. Cantin, L., and Chamberland, S. (1993). *In vitro* evaluation of the activities of azithromycin alone and combined with pyrimethamine against *Toxoplasma gondii*. *Antimicrob. Agents Chemother.* **37**, 1993–1996.
292. Araujo, F. G., Guptill, D. R., and Remington, J. S. (1988). Azithromycin, a macrolide antibiotic with potent activity against *Toxoplasma gondii*. *Antimicrob. Agents Chemother.* **32**, 755–757.
293. Dumas, J. L., Chang, R., Mermillod, B., Piguet, P. F., Comte, R., and Pechere, J. C. (1994). Evaluation of the efficacy of prolonged administration of azithromycin in a murine model of chronic toxoplasmosis. *J. Antimicrob. Chemother.* **34**, 111–118.
294. Derouin, F., Almadany, R., Chau, F., Rouveix, B., and Pocidalo, J. J. (1992). Synergistic activity of azithromycin and pyrimethamine or sulfadiazine in acute experimental toxoplasmosis. *Antimicrob. Agents Chemother.* **36**, 997–1001.
295. Saba, J., Morlat, P., Raffi, F., Hazebroucq, V., Joly, V., Leport, C., and Vilde, J. L. (1993). Pyrimethamine plus azithromycin for treatment of acute toxoplasmic encephalitis in patients with AIDS. *Eur. J. Clin. Microbiol. Infect. Dis.* **12**, 853–856.
296. Farthing, C., Rendel, M., Currie, B., and Seidlin, M. (1992). Azithromycin for cerebral toxoplasmosis [letter]. *Lancet* **339**, 437–438.
297. Wiselka, M. J., Read, R., and Finch, R. G. (1996). Response to oral and intravenous azithromycin in a patient with toxoplasma encephalitis and AIDS. *J. Infect.* **33**, 227–229.
298. Hardy, W. D., Bozzette, S., Safrin, S., Black, J., Farthing, C., and Saag, M. (1994). Results from recent therapeutic trials for opportunistic infections from the United States. *AIDS* **8**, S15.
299. Hicks, P., Zwiener, R. J., Squires, J., and Savell, V. (1996). Azithromycin therapy for *Cryptosporidium parvum* infection in four children infected with human immunodeficiency virus. *J. Pediatr.* **129**, 297–300.
300. Blanshard, C., Shanson, D. C., and Gazzard, B. G. (1997). Pilot studies of azithromycin, letrazuril and paromomycin in the treatment of cryptosporidiosis. *Int. J. STD AIDS* **8**, 124–129.
301. Dupont, C., Bougnoux, M. E., Turner, L., Rouveix, E., and Dorra, M. (1996). Microbiological findings about pulmonary cryptosporidiosis in two AIDS patients. *J. Clin. Microbiol.* **34**, 227–229.

302. Guerra, L. G., Neira, C. J., Boman, D., Ho, H., Casner, P. R., Zuckerman, M., and Verghese, A. (1993). Rapid response of AIDS-related bacillary angiomatosis to azithromycin. *Clin. Infect. Dis.* **17**, 264–266.
303. Chia, J. K., Nakata, M. M., Lami, J. L., Park, S. S., and Ding, J. C. (1998). Azithromycin for the treatment of cat-scratch disease. *Clin. Infect. Dis.* **26**, 193–194.
304. Hoque, S., Weir, A., Fluck, R., and Cunningham, J. (1996). *Rhodococcus equi* in CAPD-associated peritonitis treated with azithromycin. *Nephrol. Dial. Transplant.* **11**, 2340–2341.
305. Mascellino, M. T., Iona, E., Ponzo, R., Mastroianni, C. M., and Delia, S. (1994). Infections due to *Rhodococcus equi* in three HIV-infected patients: Microbiological findings and antibiotic susceptibility. *Int. J. Clin. Pharmacol. Res.* **14**, 157–163.
306. Rouse, M. S., Steckelberg, J. M., Brandt, C. M., Patel, R., Miro, J. M., and Wilson, W. R. (1997). Efficacy of azithromycin or clarithromycin for prophylaxis of viridans group streptococcus experimental endocarditis. *Antimicrob. Agents Chemother.* **41**, 1673–1676.
307. Sasaki, J., Yamada, Y., Takai, H., Ohmura, H., Abe, H., Shiki, K., Narita, Y., Michi, K., Ohno, K., Kawanishi, I., et al. (1985). Clinical studies on TMS-19-Q.O tablets, the preparation of a new macrolide antibiotic, in the field of oral surgery [in Japanese]. *Jpn. J. Antibiot.* **38**, 615–633.
308. Sasaki, J., Yamada, Y., Morihana, K., Kaneko, A., Takai, H., Ohmura, H., Abe, H., Ikeshima, K., Sesimo, Y., Mishina, M., et al. (1985). Clinical evaluation of the TMS-19-Q.GC tablet in odontogenic infections. A comparative double-blind study with josamycin [in Japanese]. *Jpn. J. Antibiot.* **38**, 1389–1419.
309. Fornaseri, C., Carbone, V., Giangrandi, D., Giordano, M., and Mortellaro, C. (1997). Rokitamycin in odontostomatology. Controlled study of doses [in Italian]. *Minerva Stomatol.* **46**, 693–699.
310. Baba, S., Wada, K., Hatano, T., Murai, K., Kinoshita, H., Kawamura, S., Sugita, R., Fujimaki, Y., Sanbe, B., Ueda, R., et al. (1985). Clinical evaluation of TMS-19-Q, a new macrolide antibiotic, in otorhinolaryngological infections [in Japanese]. *Jpn. J. Antibiot.* **38**, 595–614.
311. Baba, S., Kinoshita, H., Mori, Y., Sanbe, B., Ueda, R., Kawamura, S., Sugita, R., Fujimaki, Y., Nomura, Y., Kawabata, I., et al. (1985). Clinical evaluation of the TMS-19-Q.GC tablet in acute tonsillitis. A comparative double blind study with josamycin [in Japanese]. *Jpn. J. Antibiot.* **38**, 1368–1388.
312. Hara, K., Izumikawa, K., Suzuyama, Y., Shigeno, Y., Komori, M., Tomita, H., Sai, M., Ikebe, A., Iwasaki, H., Saito, A., et al. (1985). Clinical studies on TMS-19-Q.O tablet in respiratory tract infection [in Japanese]. *Jpn. J. Antibiot.* **38**, 553–574.
313. Watanabe, S., Takizawa, K., Shimada, S., Yamada, K., Nakagawa, H., Kukita, A., Miura, Y., Tsukinaga, I., Tagami, H., Tanita, Y., et al. (1985). Clinical and bacteriological evaluation of TMS-19-Q in superficial suppurative skin and soft tissue infection [in Japanese]. *Jpn. J. Antibiot.* **38**, 575–594.
314. Watanabe, S., Kukita, A., Miura, Y., Tsukinaga, I., Tagami, H., Tanita, Y., Nonami, E., Shishiba, T., Fujita, K., Shigeno, Y., et al. (1985). Clinical evaluation of the TMS-19-Q.GC tablet on superficial suppurative disease. A comparative double blind study with midecamycin [in Japanese]. *Jpn. J. Antibiot.* **38**, 1331–1354.
315. Colombo, U., Pifarotti, G., Amidani, M., Viezzoli, T., and Pifarotti, P. (1998). Rokitamycin in the treatment of female genital *Chlamydia* and *Mycoplasma* infections. Comparative study vs. josamycin [in Italian]. *Minerva Ginecol.* **50**, 491–497.
316. Motohiro, T., Aramaki, M., Oda, K., Kawakami, A., Tanaka, K., Koga, T., Sakata, Y., Yamashita, F., Suzuki, K., Ishii, M., et al. (1990). Evaluation of effectiveness of rokitamycin dry syrup in acute enteritis in pediatrics. A comparative study on rokitamycin and fosfomycin dry syrups [in Japanese]. *Jpn. J. Antibiot.* **43**, 257–284.
317. Obana, M., Tomizawa, I., Takizawa, Y., Nitta, Y., Sagara, H., Seo, T., Sato, J., Tsunoda, T., Ota, S., Machii, A., et al. (1991). Comparison of clinical efficacy of rokitamycin (RKM) and

ofloxacin (OFLX) for the treatment of *Campylobacter* enteritis by a double-blind method. The Research Committee for the Effect of Rokitamycin, Research Group for Infectious Enteritis [in Japanese]. *Kansenshogaku Zasshi* **65**, 1165–1182.
318. Mills, W. H., Thompson, G. W., and Beagrie, G. S. (1979). Clinical evaluation of spiramycin and erythromycin in control of periodontal disease. *J. Clin. Periodontol.* **6**, 308–316.
319. Chin Quee, T., Al-Joburi, W., Lautar-Lemay, C., Chan, E. C., Iugovaz, I., Bourgouin, J., and Delorme, F. (1988). Comparison of spiramycin and tetracycline used adjunctively in the treatment of advanced chronic periodontitis. *J. Antimicrob. Chemother.* **22**, 171–177.
320. Al-Joburi, W., Quee, T. C., Lautar, C., Iugovaz, I., Bourgouin, J., Delorme, F., and Chan, E. C. (1989). Effects of adjunctive treatment of periodontitis with tetracycline and spiramycin. *J. Periodontol.* **60**, 533–539.
321. Lo Bue, A. M., Sammartino, R., Chisari, G., Gismondo, M. R., and Nicoletti, G. (1993). Efficacy of azithromycin compared with spiramycin in the treatment of odontogenic infections. *J. Antimicrob. Chemother.* **31**, 119–127.
322. Bain, C. A., Beagrie, G. S., Bourgouin, J., Delorme, F., Holthuis, A., Landry, R. G., Roy, S., Schuller, P., Singer, D., and Turnbull, R. (1994). The effects of spiramycin and/or scaling on advanced periodontitis in humans. *J. Can. Dent. Assoc.* **60**, 209, 212–207.
323. Suprihati, Noersinggih, and Hoedijono, R. (1984). Treatment of acute tonsillo-pharyngitis: A comparative study of spiramycin and erythromycin. *Curr. Med. Res. Opin.* **9**, 192–196.
324. Zulkifli, Fachri, H., Efiaty, S., Yunisaf, M. H., and Iskandar, N. (1984). A comparative study of spiramycin and erythromycin in acute tonsillo-pharyngitis. *Curr. Med. Res. Opin.* **8**, 708–713.
325. Soekrawinata, T., Ibrahim, T., and Driyatno, E. (1984). Spiramycin and erythromycin in the treatment of acute tonsillo-pharyngitis: A comparative study. *Curr. Med. Res. Opin.* **9**, 296–300.
326. Manolopoulos, L., Adamopoulos, C., Tzagaroulakis, A., Maragoudakis, P., and Kaffes, T. (1989). Spiramycin versus penicillin V in the empiric treatment of bacterial tonsillitis. *Br. J. Clin. Pract.* **43**, 94–96.
327. Boezeman, A. J., Kayser, A. M., and Siemelink, R. J. (1988). Comparison of spiramycin and doxycycline in the empirical treatment of acute sinusitis: Preliminary results. *J. Antimicrob. Chemother.* **22**, 165–170.
328. Biermann, C., Loken, A., and Riise, R. (1988). Comparison of spiramycin and doxycycline in the treatment of lower respiratory infections in general practice. *J. Antimicrob. Chemother.* **22**, 155–158.
329. De Cock, L., and Poels, R. (1988). Comparison of spiramycin with erythromycin for lower respiratory tract infections. *J. Antimicrob. Chemother.* **22**, 159–163.
330. Rocha, R. T., Awad, C. E., Ali, A., Matyas, R., Vital, A. C., Silva, C. O., Dainesi, S. M., Salazar, M. S., and Nakatani, J. (1999). Comparison of spiramycin and clarithromycin for community-acquired lower respiratory tract infections. *Int. J. Clin. Pract.* **53**, 433–436.
331. Segev, S., Samra, Z., Eliav, E., Rosen, N., and Rubinstein, E. (1988). The efficacy and safety of spiramycin in the treatment of nongonococcal urethritis in men. *J. Antimicrob. Chemother.* **22**, 183–187.
332. Dylewski, J., Clecner, B., Dubois, J., St-Pierre, C., Murray, G., Bouchard, C., and Phillips, R. (1993). Comparison of spiramycin and doxycycline for treatment of *Chlamydia trachomatis* genital infections. *Antimicrob. Agents Chemother.* **37**, 1373–1374.
333. Berstad, A., Berstad, K., Wilhelmsen, I., Hatlebakk, J. G., Nesje, L. B., and Hausken, T. (1995). Spiramycin in triple therapy of *Helicobacter pylori*-associated peptic ulcer disease. An open pilot study with 12-month follow-up. *Aliment. Pharmacol. Ther.* **9**, 197–200.
334. Olafsson, S., Berstad, A., Bang, C. J., Nysaeter, G., Coll, P., Tefera, S., Hatlebakk, J. G., Hausken, T., and Olafsson, T. (1999). Spiramycin is comparable to oxytetracycline in eradicating

H. pylori when given with ranitidine bismuth citrate and metronidazole. *Aliment. Pharmacol. Ther.* **13**, 651–659.
335. Woolf, G. M., Townsend, M., and Guyatt, G. (1987). Treatment of cryptosporidiosis with spiramycin in AIDS. An "*N* of 1" trial. *J. Clin. Gastroenterol.* **9**, 632–634.
336. Connolly, G. M., Dryden, M. S., Shanson, D. C., and Gazzard, B. G. (1988). Cryptosporidial diarrhoea in AIDS and its treatment. *Gut* **29**, 593–597.
337. Saez-Llorens, X., Odio, C. M., Umana, M. A., and Morales, M. V. (1989). Spiramycin vs. placebo for treatment of acute diarrhea caused by *Cryptosporidium*. *Pediatr. Infect. Dis. J.* **8**, 136–140.
338. Couvreur, J., Thulliez, P., Daffos, F., Aufrant, C., Bompard, Y., Gesquiere, A., and Desmonts, G. (1993). *In utero* treatment of toxoplasmic fetopathy with the combination pyrimethamine-sulfadiazine. *Fetal Diagn. Ther.* **8**, 45–50.
339. Couvreur, J., Thulliez, P., Daffos, F., Aufrant, C., Bompard, Y., Gesquiere, A., and Desmonts, G. (1991). Fetal toxoplasmosis. *In utero* treatment with pyrimethamine sulfamides [in French]. *Arch. Fr. Pediatr.* **48**, 397–403.
340. Pappalardo, G., Caltabiano, M., and Mattina, R. (1979). Controlled clinical trial of a new antibiotic "CM 9164" (midecacin) in dental and stomatological practice [in Italian]. *Minerva Stomatol.* **28**, 167–186.
341. Soejima, R., Niki, Y., Hino, J., Kishimoto, T., Nakagawa, Y., Sumi, M., Saito, A., Nakayama, I., Tomizawa, M., Hiraga, Y., et al. (1989). Double-blind comparative study of roxithromycin (RU 28965) and midecamycin acetate (MOM) in the treatment of pneumonia [in Japanese]. *Kansenshogaku Zasshi* **63**, 501–529.
342. Sacristan, J. A., Elviro, J., Garcia de Lomas, J., Palomino-Nicas, J., Sobradillo, V., Sanchez Gascon, F., Valencia, A., Calbo, F., Viejo, J. L., Zamarron, C., Coca, E., Casado, M. A., Garcia Perez, L. E., and Hernandez, J. M. (1997). Double-blind clinical trial comparing 5 days of dirithromycin versus 7 days of diacetylmidecamycin in acute bronchitis and acute exacerbations of chronic bronchitis [in Spanish]. *Enferm. Infecc. Microbiol. Clin.* **15**, 357–360.
343. Borzani, M., Varotto, F., Garlaschi, L., Conio, F., Dell'Olio, M., and Careddu, P. (1989). Clinical and microbiological evaluation of miocamycin activity against group A beta-hemolytic streptococci in pediatric patients. Three years' incidence of erythromycin-resistant group A streptococci. *J. Chemother.* **1**, 35–38.
344. Principi, N., Marchisio, P., Calanchi, A., Onorato, J., Plebani, A., Reali, E., Rancilio, L., Grasso, E., Magni, L., Caramia, G., et al. (1990). Streptococcal pharyngitis in Italian children: Epidemiology and treatment with miocamycin. *Drugs Exp. Clin. Res.* **16**, 639–647.
345. Lagos, R., Topelberg, S., Herrera, P., Dattas, J. P., Vallejos, M., and Aguilera, A. L. (1993). Benzathine penicillin G and miocamycin in the treatment of children with streptococcal pharyngitis: A controlled therapeutic trial [in Spanish]. *Rev. Med. Chil.* **121**, 1274–1279.
346. Rimoldi, R., Fioretti, M., and Bandera, M. (1985). Clinical experience with miocamycin in the treatment of respiratory tract infections. *Drugs Exp. Clin. Res.* **11**, 263–268.
347. Agostoni, C., Giovannini, M., Fraschini, F., Scaglione, F., Galluzzo, C., Riva, E., and Ferrara, F. (1988). Comparison of miocamycin versus amoxycillin in lower respiratory tract infections in children. Clinical response and effect on natural killer activity. *J. Int. Med. Res.* **16**, 305–311.

Chapter 8

Ivermectin in Clinical Practice

OSAMU ZAHA
TETSUO HIRATA
FUKUNORI KINJO
ATSUSHI SAITO

*First Department of Internal Medicine, Faculty of Medicine
University of the Ryukyus, Okinawa, Japan*

I. Introduction 403
II. Novel Activity of Ivermectin in Clinical Practice 405
   A. Onchocerciasis 405
   B. Strongyloidiasis 407
   C. Bancroftian Filariasis 413
   D. Scabies 414
III. Concluding Remarks 414
    References 416

I. Introduction

In 1979, as a result of a joint research between the Kitasato Institute, Tokyo, Japan, headed by S. Ōmura, and Merck, USA, headed by E. O. Stapley, avermectins, which have a unique mode of action and potent anthelmintic activity, were discovered in culture mycelia of *Streptomyces avermitilis*. Ivermectin is a dihydro derivative of avermectin B_1, ensuring its efficacy and safety. It has an extremely broad-spectrum nematocidal activity.

The mode of action of ivermectin is described in detail in Chapter 5. Specific avermectin binding sites have been identified and characterized in the nematodes including *Caenorhabditis elegans* and *Haemonchus contortus* and anthropods such as *Drosophila melanogaster* and *Schistocerca americana* [1–3]. Electrophysiological studies have demonstrated that the avermectin binding site is associated with a glutamate-gated chloride channel in nematodes and arthropods [4, 5].

Ivermectin has been used to an extraordinary extent in veterinary medicine against a wide range of parasites, including insects and nematodes, in cattle, sheep, and horses. Ivermectin exhibits potent microfilaricidal activity against the other major filarial parasites of humans, *Wuchereria bancrofti, Brugia malayi, Loa loa,* and *Mansonella ozzardi* but not against *M. perstans*. Ivermectin also has excellent efficacy against human cutaneous larva migrans, for which good alternative treatments have not been available. Activity of ivermectin has also been demonstrated against the intestinal nematodes *Ascaris lumbricoides, Trichuris trichiura,* and *Enterobius vermicularis* in humans.

Ivermectin, however, has not been approved for the treatment of any human diseases other than onchocerciasis and strongyloidiasis. The treatment of onchocerciasis has been radically improved by the introduction of ivermectin. Large field trials in Africa in the 1980s established ivermectin as the treatment of choice for onchocerciasis. Strongyloidiasis, which is widely distributed in the tropics and subtropics, is an intestinal parasitic disease. Thiabendazole, which has been commonly used, is most effective for strongyloidiasis. However, because of its adverse effects, thiabendazole is not considered to be suitable as a therapeutic drug for strongyloidiasis. On the other hand, the anthelmintic effect of ivermectin for strongyloidiasis is as excellent as that of thiabendazole, and the incidences of adverse effects with the use of ivermectin are low and their severity is very mild. Ivermectin is currently the most useful drug for strongyloidiasis in mass treatment.

A combination of ivermectin and albendazole exhibits more activity against *W. bancrofti* than ivermectin alone. Because albendazole has broad potency against intestinal helminths, combination therapy with ivermectin and albendazole would carry the added advantage of reducing the prevalence and intensity of filarial parasites as well as of intestinal nematodes.

Preliminary studies indicated that ivermectin has the potential to be the drug of choice for ectoparasitic infestations (mites, lice) of humans as well. Recently, there is growing evidence that scabies can also be treated by oral administration of ivermectin. Treatment of scabies is very difficult, and relapse is frequent after topical scabicidal therapy. Ivermectin has been proven to be a safe and effective alternative for the therapy of severe *Sarcoptes scabiei* infestation unresponsive to conventional treatment.

The efficacy of ivermectin for human onchocerciasis was established in the 1980s and is summarized in this chapter briefly. Clinical study of ivermectin in human strongyloidiasis has not yet been completed. Here we describe the efficacy of ivermectin for human strongyloidiasis as observed in our institution during the past decade. Furthermore, we would like to describe the combination therapy with ivermectin and albendazole for bancroftian filariasis and also the efficacy of ivermectin in the treatment of human scabies.

II. Novel Activity of Ivermectin in Clinical Practice

A. Onchocerciasis

Onchocerciasis is prevalent in tropical Africa, parts of Latin America, and the Arabian peninsula. Of 37 countries where onchocerciasis is endemic, 30 are in sub–Sahara Africa, 6 are in Latin America, and the last is Yemen. According to World Health Organization (WHO) information in 2000, 120 million people are at the risk worldwide of contracting this disease. A total of 18 million people are infected with this disease, 99% of whom are in Africa. Of those infected, more than 6.5 million suffer from impaired visual acuity and 270,000 are blind. This disease is a public health and economic development problem in Africa, where it constitutes a serious obstacle to socioeconomic progress.

Onchocerciasis is transmitted by blackflies belonging to the genus *Simulium*, usually *S. damnosum*. When the blackfly bites human skin, the larvae of *Onchocerca volvulus* escape through the membranous labrum into the wound and penetrate the tissues to develop into adult filaria, which are found clinically in nodules scattered around the body. Development of the adult worm takes about 1 year. The females can live for about 9–14 years. During this period they produce many millions of living embryos known as microfilariae, which migrate to the skin and eyes of their human host.

Microfilariae are the main cause of the clinical manifestations of the disease. These include dermatitis, resulting in very severe itching, depigmentation and atrophy of the skin, and lymphadenitis, which may lead to hanging groin and elephantiasis of the genitals. The microfilariae enter the eye, where they can often be seen in the cornea and anterior chamber with a slit lamp; eventually they induce blindness. Impaired visual acuity is the most serious complication of this disease and usually affects those persons with more than moderate infections.

1. Older Drugs Used against the Parasite O. volvulus

Suramin is effective for treatment of the hemolymphatic stage of african trypanosomiasis. It has been shown to have macrofilaricidal effects against *O. volvulus*. Difficulties associated with its mode of administration and its toxicity have limited its usefulness. Suramin inhibits a number of enzymes that play roles in the metabolism of the filariae and also of the host. Because of its extreme toxicity, including a fatal progressive wasting syndrome, exfoliative dermatitis, progression of chorioretinitis, and development of optic atrophy, suramin is not recommended for treatment of onchocerciasis by WHO [6].

Diethylcarbamazine was introduced in the late 1940s and was the drug of choice for human onchocerciasis. Diethylcarbamazine has microfilaricidal effects and rapidly kills microfilariae of *O. volvulus* in the skin and eye, but also results in cellular cytotoxicity with severe damage to the human host [7]. Ocular complications may result in impairment of visual capacity, including visual field constriction, anterior uveitis, and punctate keratitis [8]. During treatment with diethylcarbamazine, systemic reactions include pruritus, edema of the skin, fever, hypotension, increased eosinophilia, lymphadenopathy, splenomegaly, and proteinuria. These drawbacks have prevented its use in the routine treatment of onchocerciasis. Thus, diethylcarbamazine is no longer recommended for the treatment of human onchocerciasis by WHO [6].

2. Effect of Ivermectin against the Parasite O. volvulus

The introduction of ivermectin has for the first time provided a feasible method of chemotherapy for large-scale treatment of onchocerciasis. At present, ivermectin is being used by more than 33 million patients per year. The first reports of the use of ivermectin for human *O. volvulus* infection were prepared by Aziz and coworkers in 1982 [9, 10]. Since then, extensive large community trials in hyperendemic communities have convincingly established its effectiveness and safety. In phase II studies, detailed ocular examinations were performed serially over a 12-month period. The local inflammatory responses in ocular tissues are less severe with ivermectin therapy than with diethylcarbamazine [11, 12]. This is particularly important in the eye, where microfilariae disappear slowly after ivermectin therapy. In 1987, ivermectin was approved for use in humans by the French government authorities, and Merck & Co., the manufacturer of ivermectin, has supplied it free of charge through WHO for as long as it is needed for treatment of onchocerciasis.

A single dose of ivermectin not only reduces microfilariae of *O. volvulus* in the skin and eye, but also induces long-lasting suppression of microfilaremia. It appears to have only limited effects on adult worms. The reduction of skin microfilaria counts it induced persisted for at least 12 months [13–16]. A few months after treatment, when the adult female worm recommences its reproduction, skin microfilarial loads will begin to increase again. However, the buildup of microfilarial loads in each individual is successively less after each treatment. According to various assessments of safety and efficacy for treatment of onchocerciasis, the optimal dose of ivermectin is 150 µg/kg, and a single oral dose every 6–12 months is sufficient to prevent the development of ocular lesions and blindness [17–20]. Ivermectin apparently does not have macrofilaricidal effects [21], and treatment with it is therefore suppressive rather than curative, and must be continued for at least the life span of adult worms (up to 15 years) [16]. Repeated, more frequent administration may have longer term sterilizing effects on adult females, which may lead to a shorter adult life span [11].

Rarely, patients with onchocerciasis who are also heavily infected with *L. loa* may develop a serious or even fatal encephalopathy either spontaneously or following treatment with an effective microfilaricide. Ivermectin is not approved for use in children under 5 years of age or 15-kg body weight, pregnant women, breast-feeding mothers within 1 week of delivery, and individuals with impaired hepatic function or impaired renal function.

B. Strongyloidiasis

Strongyloidiasis is an intestinal parasitic disease caused by *Strongyloides stercoralis*. Strongyloidiasis is widely distributed in the tropics and subtropics. It is estimated that more than 10 million people are infected with this disease. *S. stercoralis* is distinguished by a capacity that permits an ongoing cycle of autoinfection. This peculiar autoinfection is the main cause of long-lasting intractable infection without further exposure to exogenous infective larvae. In immunocompromised hosts, large numbers of invasive *S. stercoralis* larvae can disseminate widely in the body. Moreover, large amounts of intestinal bacteria are dispersed through the bloodstream by autoinfecting larvae, causing sepsis, pneumonia, and purulent meningitis as fatal complications.

Once strongyloidiasis has become severe, however, complete eradication is difficult. Even in the asymptomatic state, strongyloidiasis must be treated because of the potential for fatal hyperinfection. In particular, for patients requiring the use of immunosuppressants, and for those scheduled to undergo surgery of the digestive tract or other major surgery, early diagnosis and treatment are imperative.

The peculiar life cycle of *S. stercoralis* should be taken into consideration in treatment. A two-course treatment with an interval of 2 weeks is most effective. We assume that ivermectin does not act on filariform larvae when mobile within the human body, but instead acts only after the larvae have settled into the small intestine. Because the cycle of autoinfection with *S. stercoralis* is 3 to 4 weeks long, it is important to provide at least two courses of treatment at intervals of 2 weeks in order to completely eradicate the parasite. Studies from other countries, however, have recommended the use of ivermectin in a single dose of 200 µg/kg.

For disseminated strongyloidiasis, an antibacterial drug should be given concomitantly with an anthelmintic. Efficacy, as measured by eradication rate, is defined as the absence of larvae in follow-up post-treatment stool examinations. Because larvae of *S. stercoralis* are usually excreted in the stools in far less abundance than are ova of other parasites, it is necessary to repeat the stool examination after a sufficient period of time. In patients infected with human T-lymphotropic virus type I (HTLV-I), the capability of eradication is decreased. Although concentration techniques such as use of the Baermann apparatus are generally recommended, the ordinary agar plate culture method developed in our department is one in which larvae are detected by tracing crawling marks on agar medium. This method is much simpler to perform and much more sensitive than conventional methods such as the direct smear method, formalin ether concentration method, and filter paper culture method [22–27].

TABLE I
Incidence of Adverse Reactions in the Patients Treated with Thiabendazole[a]

Adverse reactions	After one course $n = 151$ (%)	After two courses $n = 112$ (%)	After three courses $n = 96$ (%)	After four courses $n = 33$ (%)
Body as whole				
Asthenia/fatigue	24.5 (37)[b]	24.1 (27)	27.1 (26)	24.2 (8)
Gastrointestinal				
Nausea and vomiting	32.5 (49)	32.1 (36)	18.8 (18)	15.2 (5)
Anorexia	21.2 (32)	22.3 (25)	20.8 (20)	18.2 (6)
Heartburn	11.9 (18)	13.4 (15)	10.4 (10)	9.1 (3)
Abdominal pain	6.0 (9)	9.8 (11)	5.2 (5)	0.0 (0)
Constipation	2.6 (4)	3.6 (4)	2.1 (2)	0.0 (0)
Diarrhea	0.7 (1)	3.6 (4)	0.0 (0)	6.1 (2)
Nervous system/psychiatric				
Dizziness or vertigo	23.8 (36)	19.6 (22)	12.5 (12)	9.1 (3)
Headache	15.9 (24)	14.3 (16)	9.4 (9)	3.0 (1)
Numbness of extremities	3.3 (5)	5.4 (6)	1.0 (1)	0.0 (0)
Skin				
Itching	2.0 (3)	4.5 (5)	0.0 (0)	0.0 (0)
Exanthema	1.3 (2)	6.3 (7)	0.0 (0)	0.0 (0)
TOTAL	67.5 (102)	58.0 (65)	59.4 (57)	45.5 (15)

[a]Thiabendazole: 1500 mg t.i.d. for 5 days, repeated 3–4 courses every 2 weeks.
[b]Parentheses indicate number of patients.

1. Effect of Benzimidazole Derivatives against the Parasite S. stercoralis

Thiabendazole, which has strong activities against *S. stercoralis*, was the first benzimidazole to be introduced in the 1960s. However, the incidence of severe side effects with it is known to be extremely high. The precise mode of action of thiabendazole against parasites is unknown, but it may inhibit the helminth-specific enzyme fumarate reductase. In our study of 162 patients, thiabendazole (via a 500-mg tablet) was given 3 times a day for 5 days. Three or four courses were given at intervals of 2 weeks. After one course of treatment, some adverse effects, such as general fatigue, dizziness, nausea, anorexia, and headache, were observed in 67.5% of the patients (Table I). Consequently, 45.1% of the patients dropped out during the course of treatment, and dose decrease was required for 32.1% of patients. Only 22.8% (37 patients) could be given a full course of treatment. Liver dysfunction was observed in 33.8% of patients; the incidence of this increased with dose dependence, and its severity also intensified (Table II). In our patients treated with more than three courses, the rates of eradication after 2 weeks, 6 months, and 1 year were each 100% (Table III). These results suggest that treatment should be repeated at least twice for complete eradication of *S. stercoralis*. Although thiabendazole has strong activity, it is not a practical drug suitable for mass treatment of strongyloidiasis [28]. Adverse reactions to thiabendazole have also been reported

TABLE II
Relationship between Dosage of Thiabendazole and Severity of Liver Dysfunction[a]

ALT (IU/liter)	<36–50	<100	<200	<500	<1000	>1001	Total
Thiabendazole							
After 1 course (n = 148)	7 (4.7)[b]	5 (3.4)					12 (8.1)
After 2 courses (n = 59)	5 (8.5)	3 (5.1)	8 (13.6)	5 (8.5)	1 (1.7)	1 (1.7)	23 (39.0)
After 3 courses (n = 33)	2 (6.1)	7 (21.2)	4 (12.1)	2 (6.1)			15 (45.4)

[a] The criteria for abnormal changes were set as an increase exceeding 120% of the upper boundary of the normal range in case the change was from normal to abnormal, and as an increase exceeding 200% of the previous value in case the change was from abnormal.
[b] Parentheses indicate percentage of patients with liver dysfunction.

TABLE III
Eradication Rate of *S. stercoralis* in the Patients Treated with Thiabendazole

Regimen	1 week (%)	6 months (%)
One course only (n = 29)	91.7 (22/24)[a]	89.5 (17/19)
Two courses (n = 26)	95.8 (23/24)	100.0 (14/14)
Three courses (n = 64)	100.0 (59/59)	100.0 (48/48)
Four courses (n = 33)	100.0 (27/27)	100.0 (29/29)

[a] Parentheses indicate number of eradicated cases/number of patients examined.

by Franz [29], Grove [30], and Pelletier [31]. In their studies, from 20% to 89% of patients treated with thiabendazole sustained adverse reactions. In the other prospective trials using thiabendazole, eradication rates ranged from 50% to 100% [32, 33]. Variability in eradication rate may be due to differences in dosing schedules, methods of stool examination, and follow-up period.

Mebendazole is known to be active against *S. stercoralis*. In our study of 96 patients with *S. stercoralis*, the efficacy of and tolerance to mebendazole were assessed for various dose schedules (Table IV). The rates of eradication after 3–7 months and after 8–15 months of treatment with 200 mg/day for 28 days were 77.3% (17/22) and 85.0% (17/20), respectively. There have been reports that even when mebendazole is used for the treatment of hydatid disease, adverse effects and liver dysfunction rarely occur. However, liver dysfunction occurred more often than anticipated. Because the incidence of liver dysfunction appeared to be dose

TABLE IV
Administration Methods, Anthelmintic Effects, and Adverse Effects of Mebendazole and Albendazole

Regimen	Duration of follow-up	Eradication rate	Adverse effects		Liver dysfunction
Mebendazole (33 cases)	200 mg b.i.d. for 28 days (powder)	2–24 days 3–7 months 8–15 months 24 months	81.8% (18/22)[a] 77.3% (17/22) 85.0% (17/20) 93.8% (15/16)	Headache 11.1% Constipation 11.1% General fatigue 7.4% Abdominal pain 7.4% Overall 40.7%	71.4%
Mebendazole (47 cases)	200 mg b.i.d. for 5 days, repeated 1, 3, and 4 weeks later (powder)	2–24 days 3–7 months 8–15 months	100.0% (46/46) 80.0% (24/30) 87.1% (27/31)	Diarrhea 10.6% Abdominal pain 6.4% Dizziness 6.4% Appetite loss 4.3% Overall 34.0%	51.1%
Mebendazole (16 cases)	200 mg b.i.d. for 4 days, repeated 1 week later (tablet)	2–24 days 8–15 months	93.8% (15/16) 69.2% (9/13)	Diarrhea 6.3% Headache 6.3% Overall 12.5%	25.0%
Albendazole (27 cases)	400 mg b.i.d. for 3 days, repeated 2 weeks later (tablet)	14 days 28 days	70.4% (19/27) 66.7% (16/24)	Abdominal pain 14.8% Headache 7.4% Nausea 3.7% Overall 18.5%	33.3%

[a] Parentheses indicate number of eradicated cases/number of patients examined.

dependent, we administered 200 mg/day for 5 days in another trial, and repeated this regimen 1, 3, and 4 weeks later. The rates of eradication after 3–7 months and after 8–15 months were 80.0% (24/30) and 87.1% (27/31), respectively. However, a high incidence of liver dysfunction was observed (51.1%). Finally, two courses of 4-day administration were given at an interval of 1 week. Although the incidence of liver dysfunction was decreased (25.0%), the rate of eradication after 8–15 months was also decreased (69.2%) [34–38].

Albendazole is also being tried for strongyloidiasis. In our study, with two courses of 3-day administration at an interval of 2 weeks, the eradication rate at 2 weeks after initial treatment was 70.4% (19 of 27 patients) and was 66.7% (16 of 24 patients) 2 weeks after second treatment. Although the degree of liver dysfunction was low, the rate of eradication was not satisfactory [39] (Table IV). Some trials using albendazole 400 mg/day for 3 days have been reported by Pene *et al.* [40] and Pungpak *et al.* [41]. In their studies, eradication rates ranged from 36% to 73%.

2. Effect of Ivermectin against the Parasite S. stercoralis

The efficacy of ivermectin for human *S. stercoralis* infection was reported by Naquira *et al.* [42] and Freedman *et al.* [43] in 1989. In the former study, which included 110 patients, the eradication rates at 30 days varied from 67% (10/15) for a single 50 μg/kg dosage to 94% (16/17) for a single 200 μg/kg dosage. In heavily infected patients, larval output after ivermectin therapy was remarkably reduced and the clinical condition was also improved. In the latter study, 10 patients with light *S. stercoralis* infection were given a single 200 μg/kg dose of ivermectin. All stool examinations were negative 1 month later.

As already mentioned, the adverse effects of thiabendazole are severe. On the other hand, mebendazole also features a high incidence of liver dysfunction, and albendazole has an unsatisfactory eradication rate. Although in the literature a single 150–200 μg/kg dose of ivermectin is given, in our department a two-course administration with an interval of 2 weeks is used [44–46]. This regimen was established considering the time from autoinfection until appearance of parasites in the intestines, and because ivermectin would be ineffective for larvae and eggs in body tissues. Table V shows the eradication rate of ivermectin with administration of two tablets (6 mg) at an interval of 2 weeks regardless of body

TABLE V
Eradication Rate of *S. stercoralis* in the Patients Treated with Ivermectin[a]

Treatment	13 days after second (%)	10–12 months (%)	16–18 months (%)	22–24 months (%)
Two courses	91.6 (109/119)[b]	98.2 (54/55)	100.0 (48/48)	97.3 (36/37)

[a]Dose of ivermectin: Approximately 110 μg/kg. Eradication rate at 13 days after first dose: 84.4% (95/112).

[b]Parentheses indicate number of eradicated cases/number of patients examined.

weight (about 100 μg/kg). Although the literature suggests this regimen may be inadequate in patients weighing more than 40 kg, satisfactory results were obtained in 109 of the 119 patients. The rate of anti-HTLV-I antibody positivity in the resistant group was significantly higher (80.8%) than that in the group with eradication (29.2%). It is known that cellular immunity, in which eosinophils and IgE play an important role against parasitic infection, is reduced in patients infected with HTLV-I. HTLV-I infection provides suitable conditions for *S. stercoralis* infection (or autoinfection).

Adverse reactions accompanying the first administration of ivermectin occurred in nine patients (7.2%), and included dizziness (2.4%), nausea (1.6%), and diarrhea (1.6%). Each of these adverse reactions was mild and transient and none required particular treatment. After the second administration, adverse reactions occurred in five patients (4.0%), but this frequency was lower and adverse reactions were mild and transient. Mazzotti-type reactions associated with the treatment of onchocerciasis did not occur in these patients with strongyloidiasis treated with ivermectin (Table VI).

The values of AST (aspartate aminotransferase) and ALT (alanin aminotransferase) changed from normal to abnormal in 13 patients (10.4%) after first administration. Twelve of the 13 patients had mild and transient increases in AST and ALT prior to treatment, while in one patient with fatty liver ALT increased

TABLE VI
Incidence of Adverse Reactions in the Patients Treated with Ivermectin

Adverse reactions	After first administration $n = 125$ (%)	After second administration $n = 125$ (%)
Body as whole		
Asthenia/fatigue	1.6 (2)[a]	0.8 (1)
Myalgia	0.8 (1)	—
Gastrointestinal		
Nausea	1.6 (2)	—
Anorexia	0.8 (1)	—
Borborygmus	0.8 (1)	0.8 (1)
Diarrhea	1.6 (2)	—
Nervous system/psychiatric		
Dizziness or vertigo	2.4 (3)	—
Blurred vision	0.8 (1)	—
Skin		
Exanthema	—	0.8 (1)
Itching	—	1.6 (2)
Others	3.2 (4)	—
TOTAL	7.2 (9)	4.0 (5)

[a]Parentheses indicate number of patients.

from 49 to 198 IU/liter. After second administration, elevation of AST and ALT values occurred in 4 patients (3.2%). Three of the 4 patients also had mild and transient increases in AST and ALT prior to treatment, while in 1 patient with liver cirrhosis AST increased from 66 to 170 IU/liter. During the treatment period, liver dysfunction occurred in 17 patients (13.6%). In these 17 patients, symptoms such as fatiguability were absent and no significant clinical problems were noted. Our study indicated that ivermectin is currently the most useful agent against *S. stercoralis* [47].

In two controlled clinical studies using albendazole and thiabendazole as comparative agents, the efficacy of ivermectin (a single dose of 170–200 µg/kg) was significantly greater than that of albendazole (200 mg twice daily for 3 days) [48, 49]. Ivermectin administered as a single dose of 200 µg/kg for 1 day was as efficacious as thiabendazole administered at 50 mg/kg twice daily for 3 days [50].

C. Bancroftian Filariasis

Wuchereria bancrofti, a mosquito-transmitted parasite, is the major cause of human lymphatic filariasis in tropic areas. Current estimates suggest that about 120 million people are infected. Infection often leads to microfilaremia without clinical manifestations. The most common clinical presentation of this disease are asymptomatic microfilaremia, filarial fever, and lymphatic obstruction. Ivermectin rapidly reduces microfilaremia concentration in peripheral blood and may inhibit larval development in mosquitoes. Because the drug does not kill adult worms, microfilaremia concentration begins to increase gradually 3–12 months after treatment.

Efficacy of combined ivermectin and albendazole therapy for bancroftian filariasis was reported by Addiss *et al.* [51] in 1997 and Ismail *et al.* [52] in 1998. One hundred and thirteen schoolchildren infected with *W. bancrofti* were randomly assigned treatment with placebo, a single 200–400 µg/kg dose of ivermectin (mean, 273 µg/kg), 400 mg albendazole, or a combination of 200–400 µg/kg ivermectin and 400 mg albendazole. Four months after treatment, the proportion of children who remained positive for microfilariae was significantly lower in the ivermectin plus albendazole group, but there were no significant changes in the other three groups. Systemic adverse reactions did not differ significantly between children who received ivermectin alone and those who were treated with ivermectin and albendazole [51]. In a trial on 50 asymptomatic patients with *W. bancrofti*, the efficacy of a single dose of albendazole 600 mg alone or in combination with ivermectin 400 µg/kg or diethylcarbamazine 6 mg/kg was compared with a single dose of the combination diethylcarbamazine 6 mg/kg and ivermectin 400 µg/kg over a period of 15 months after treatment. Albendazole plus ivermectin was the most effective regimen for clearing microfilariae. Nine of 13 subjects (69%) were amicrofilaremic 15 months after treatment compared to 1 of 12 (8%), 3 of 11 (27%), and 3 of 10 (30%) in the groups treated with albendazole, albendazole plus diethylcarbamazine, and diethylcarbamazine plus ivermectin, respectively. All four regimens were well tolerated and clinically safe, although mild, self-limited systemic reactions were observed in all treatment groups [52].

D. Scabies

Scabies is a skin disease caused by infection with the ectoparasitic mite *Sarcoptes scabiei*. Its most common symptom is widespread itchy lesions associated with allergic reaction. Scabies occurs worldwide, affecting all races and all ages. Scabies often occurs within families, hospitals, and nursing homes for the aged. *Scabies* mites are transmitted by close human-to-human contact with infected individuals. The female mites burrow just under the surface of the skin, laying eggs in burrows. Treatment with topical medications (scabicide) is usually effective, and the medication should be applied thoroughly to all skin from the neck down for at least 12 hr. Such treatment requires much time and labor, and it is often difficult to eradicate massively occurring scabies.

The efficacy of ivermectin for human scabies infection was reported by Glaziou *et al.* [53] and Macotela-Ruiz and Pena-Gonzalez [54] in 1993. Forty-four patients with the diagnosis of scabies were randomly assigned to receive a single 100 µg/kg oral dose of ivermectin or benzyl benzoate 10% application in French Polynesia. At day 30 after treatment, the cumulative recovery rates were 70% (16/23) in the group receiving ivermectin and 48% (10/21) in the group given benzyl benzoate [53]. Moreover, there have been many reports of the efficacy of ivermectin for human scabies since 1995. Oral administration of ivermectin once or twice at 100–200 µg/kg yields favorable effects, without safety problems or serious side effects [55–62].

III. Concluding Remarks

The principal methods of treatment for onchocerciasis have been vector control and early diagnosis and treatment. Repeated annual treatments may interrupt transmission. Combining ivermectin treatment with vector control maximized the effect of vector control, leading to more rapid decline in the transmission of infection.

The treatment of onchocerciasis has changed radically since the introduction of ivermectin. Ivermectin is better tolerated and more effective than previously approved treatments (including diethylcarbamazine and suramin) for onchocerciasis. Although ivermectin is still far from being the final solution for onchocerciasis, it has greatly simplified and improved the quality of patient treatment and decreased the burden of this disease.

The Onchocerciasis Control Programme has made considerable progress through both vector control and ivermectin distribution in West Africa during the last 25 years. This success led to launching of the African Programme for Onchocerciasis Control in 1995. In Latin America, the Onchocerciasis Elimination Programme in the Americas is successfully using ivermectin distribution. Furthermore, as of 2001, 30 million people had been protected from infection by the disease, 100,000 have been prevented from going blind, and 1.25 million have been freed of onchocercal infection.

Because manifestations of strongyloidiasis are milder than those of other tropical diseases, strongyloidiasis may be overlooked in regions where parasitic diseases such as those caused by hookworms and roundworms are endemic. With further improvement in hygiene and the resulting decrease in parasitic diseases, increasing attention will be focused on strongyloidiasis.

Among chemotherapeutic agents, thiabendazole, a representative benzimidazole compound, is most effective for strongyloidiasis. However, it has problems with safety, because its use can cause adverse effects and liver dysfunction at a high incidence that can be severe. High incidences of liver dysfunction were observed with mebendazole, and the rate of eradication was not sufficient with albendazole [63, 64].

In our clinical use of ivermectin, more than 400 patients were studied. Administration was discontinued in only 2 patients due to liver dysfunction. Few subjective symptoms that were clinically problematic were observed. Thus, the rate of eradication obtained with ivermectin was excellent, and its incidences of adverse effects and liver dysfunction were low and their severity was mild.

Ivermectin is indicated for intestinal strongyloidiasis in France (1987) and the USA (1996). With a view to obtaining approval for treatment with ivermectin, we are proceeding with a clinical trial at 200 µg/kg. With administration by this method, a high rate of eradication and safety are being confirmed. Ivermectin is as effective as, and better tolerated than, thiabendazole for treating strongyloidiasis.

Single-dose ivermectin greatly reduces microfilarial concentration. An estimated 300 million Africans are at risk of lymphatic filariasis. Merck plans to donate ivermectin (Mectizan) for the disease through the same system successfully established for the treatment of onchocerciasis.

Combined treatment with ivermectin and albendazole for bancroftian filariasis was more effective than treatment with ivermectin alone. In tropical countries where bancroftian filariasis is endemic, infection with intestinal helminths is also an important public health concern. Additional benefits of this combination are its potent, broad-spectrum activity against intestinal helminths and potential relative safety.

Under the present method of treatment with external preparations only, oral administration of ivermectin once or twice may reduce time and labor and be effective for prophylaxis of outbreaks of scabies. Ivermectin is effective for the treatment of human scabies. It is simpler to use and, therefore, may improve compliance and control of scabies infestations. As of 2001, ivermectin had not been approved for treatment of human scabies. We suggest that the effects of ivermectin on human scabies be studied in detail in more patients as soon as possible.

References

1. Schaeffer, J. M., and Haines, H. W. (1989). Avermectin binding in *Caenorhabditis elegans:* A two-state mode for the avermectin binding site. *Biochem. Pharmacol.* **38**, 2329–2338.
2. Rohrer, S. P., Birzin, E. T., Eary, C. H., Schaeffer, J. M., and Shoop, W. L. (1994). Ivermectin binding sites in sensitive and resistant *Haemonchus contortus. J. Parasitol.* **80**, 493–497.
3. Rohrer, S. P., Bizin, E. T., Costa, S. D., Arena, J. P., Hayes, E. C., and Schaeffer, J. M. (1995). Identification of neuron-specific ivermectin binding sites in *Drosophila melanogaster* and *Schistocerca americana. Insect Biochem. Mol. Biol.* **25**, 11–17.
4. Arena, J. P., Liu, K. K., Paress, P. S., Schaeffer, J. M., and Cully, D. F. (1992). Expression of a glutamate-activated chloride current in *Xenopus* oocytes injected with *Caenorhabditis elegans* RNA: Evidence for modulation by avermectin. *Mol. Brain Res.* **15**, 339–348.
5. Cully, D. F., Paress, P. S., Liu, K. K., Schaeffer, J. M., and Arena, J. P. (1996). Identification of a *Drosophila melanogaster* glutamate-gated chloride channel sensitive to the antiparasitic agent avermectin. *J. Biol. Chem.* **271**, 20,187–20,191.
6. World Health Organization (1995). Report of a WHO expert committee on onchocerciasis control, Technical Report 852. World Health Organization, Geneva.
7. Taylor, H. R., and Greene, B. M. (1981). Ocular changes with oral and transepidermal diethylcarbamazine therapy of onchocerciasis. *Br. J. Ophthamol.* **65**, 494–502.
8. Francis, H., Awadzi, K., and Ottesen, E. A. (1985). The Mazzotti reaction following treatment of onchocerciasis with diethylcarbamazine: Clinical severity as a function of infection intensity. *Am. J. Trop. Med. Hyg.* **34**, 529–536.
9. Aziz, M. A., Diallo, S., Diop, I. M., Lariviere, M., Porta, M., and Gaxotte, P. (1982). Efficacy and tolerance of ivermectin in human onchocerciasis. *Lancet* **2**, 171–173.
10. Aziz, M. A., Diallo, S., Diop, I. M., Lariviere, M., Porta, M., and Gaxotte, P. (1982). Ivermectin in onchocerciasis. *Lancet* **2**, 1456–1457.
11. Taylor, H. R., Murphy, R. P., Newland, H. S., White, A. T., D'Anna, S. A., Keyvan-Larijani, E., Aziz, M. A., Cupp, E. W., and Greene, B. M. (1986). Comparison of the treatment of ocular onchocerciasis with ivermectin and diethylcarbamazine. *Arch. Ophthalmol.* **104**, 863–870.
12. Dadzie, K. Y., Bird, A. C., Awadzi, K., Schulz-Key, H., Gilles, H. M., and Aziz, M. A. (1987). Ocular findings in a double-blind study of ivermectin versus diethylcarbamazine versus placebo in the treatment of onchocerciasis. *Br. J. Ophthalmol.* **71**, 78–85.
13. Greene, B. M., Taylor, H. R., Cupp, E. W., Murphy, R. P., White, A. T., Aziz, M. A., Schulz-Key, H., D'Anna, S. A., Newland, H. S., Goldschmidt, L. P., Auer, C., Hanson, A. P., Freeman, S. V., Reber, E. W., and Williams, P. N. (1985). Comparison of ivermectin and dietylcarbamazine in the treatment of onchocerciasis. *N. Engl. J. Med.* **313**, 133–138.
14. Lariviere, M., Vingtain, P., Aziz, M., Beauvais, B., Weimann, D., Derouin, F., Ginoux. J., Schulz-Key, H., Gaxotte, P., Basset, D., and Sarfati, C. (1985). Double-blind study of ivermectin and diethylcarbamazine in African onchocerciasis patients with ocular involvement. *Lancet* **2**, 174–177.

15. Awadzi, K., Dadzie, K. Y., Schulz-Key, H., Haddock, D. R., Gilles, H. M., and Aziz, M. A. (1985). The chemotherapy of onchocerciasis X. An assessment of four single dose treatment regimes of MK-933 (ivermectin) in human onchocerciasis. *Ann. Trop. Med. Parasitol.* **79**, 63–78.
16. Diallo, S., Aziz, M. A., Lariviere, M., Diallo, J. S., Diop-Mar, I., N'Dir, O., Badiane, S., Py, D., Schulz-Key, H., Gaxotte, P., and Victorius, A. (1986). A double-blind comparison of the efficacy and safety of ivermectin and diethylcarbamazine in a placebo controlled study of Senegalese patients with onchocerciasis. *Trans. R. Soc. Trop. Med. Hyg.* **80**, 927–934.
17. White, A. T., Newland, H. S., Taylor, H. R., Erttmann, K. D., Keyvan-Larijani, E., Nara, A., Aziz, M. A., D'Anna, S. A., Williams, P. N., and Greene, B. M. (1987). Controlled trial and dose-finding study of ivermectin for treatment of onchocerciasis. *J. Infect. Dis. Sep.* **156**, 463–470.
18. Greene, B. M., White, A. T., Newland, H. S., Keyvan-Larijani, E., Dukuly, Z. D., Gallin, M. Y., Aziz, M. A., Williams, P. N., and Taylor, H. R. (1987). Single dose therapy with ivermectin for onchocerciasis. *Trans. Assoc. Am. Physicians* **100**, 131–138.
19. Newland, H. S., White, A. T., Greene, B. M., D'Anna, S. A., Keyvan-Larijani, E., Aziz, M. A., Williams, P. N., and Taylor, H. R. (1988). Effect of single-dose ivermectin therapy on human *Onchocerca volvulus* infection with onchocercal ocular involvement. *Br. J. Ophthalmol.* **72**, 561–569.
20. Greene, B. M., Dukuly, Z. D., Munoz, B., White, A. T., Pacque, M., and Taylor, H. R. (1991). A comparison of 6-, 12-, and 24-monthly dosing with ivermectin for treatment of onchocerciasis. *J. Infect. Dis.*, **163**, 376–380.
21. Awadzi, K., Dadzie, K. Y., Schulz-Key, H., Gilles, H. M., Fulford, A. J., and Aziz, M. A. (1986). The chemotherapy of onchocerciasis XI. A double-blind comparative study of ivermectin, diethylcarbamazine and placebo in human onchocerciasis in northern Ghana. *Ann. Trop. Med. Parasitol.* **80**, 433–442.
22. Arakaki, T., Hasegawa, H., Asato, R., Ikeshiro, T., Kinjo, F., and Saito, A. (1988). A new method to detect *Strongyloides stercoralis* from human stool. *Jpn. J. Trop. Med. Hyg.* **16**, 11–17.
23. Arakaki, T., Iwanaga, M., Kinjo, F., Saito, A., Asato, R., and Ikeshiro, T. (1990). Efficacy of agar-plate culture in detection of *Strongyloides stercoralis* infection. *J. Parasitol.* **76**, 425–428.
24. Koga, K., Kasuya, S., and Ohtomo, H. (1992). How effective is the agar plate method for *Strongyloides stercoralis*? *J. Parasitol.* **78**, 155–156.
25. Sato, Y., Kobayashi, J., Toma, H., and Shiroma, Y. (1995). Efficacy of stool examination for detection of *Strongyloides* infection. *Am. J. Trop. Med. Hyg.* **53**, 248–250.
26. Salazar, S. A., Gutierrez, C., and Berk, S. L. (1995). Value of the agar plate method for the diagnosis of intestinal strongyloidiasis. *Diagn. Microbiol. Infect. Dis.*, **23**, 141–145.
27. Kobayashi, J., Hasegawa, H., Forli, A. A., Nishimura, N. F., Yamanaka, A., Shimabukuro, T., and Sato, Y. (1995). Prevalence of intestinal parasitic infection in five farms in Holambra, Sao Paulo, Brazil. *Rev. Inst. Med. Trop. Sao Paulo* **37**, 13–18.
28. Oyakawa, T., Kuniyoshi, T., Arakaki, T., Higashionna, A., Shikiya, K., Sakugawa, H., Kadena, K., Kitsukawa, K., Kinjo, F., Saito, A., Oyadomari, Y., Ishihara, M., Miyagi, M., Morishima, A., Kinjo, Y., and Asato, R. (1991). New trial with thiabendazole for treatment of human strongyloidiasis. *Kansenshogaku Zasshi.* **65**, 304–310.
29. Franz, K. H. (1963). Clinical trials with thiabendazole against human strongyloidiasis. *Am. J. Trop. Med. Hyg.* **12**, 211–214.
30. Grove, D. I. (1982). Treatment of strongyloidiasis with thiabendazole: An analysis of toxicity and effectiveness. *Trans. R. Soc. Trop. Med. Hyg.* **76**, 114–118.
31. Pelletier, L. L. (1984). Chronic strongyloidiasis in World War II Far East ex-prisoners of war. *Am. J. Trop. Med. Hyg.* **33**, 55–61.
32. Berk, S. L., Verghese, A., Alvarez, S., Hall, K., and Smith, B. (1987). Clinical and epidemiologic features of strongyloidiasis. A prospective study in rural Tennessee. *Arch. Intern. Med.* **147**, 1257–1261.

33. Beus, A. (1989). Comparative study of thiabendazole and mebendazole in strongyloidiasis. *Lijec. Vjesn.* **111**, 98–101.
34. Shikiya, K., Kuniyoshi, T., Higashionna, A., Arakaki, T., Oyakawa, T., Kadena, K., Kinjo, F., Saito, A., and Asato, R. (1990). Treatment of strongyloidiasis with mebendazole and its combination with thiabendazole. *Kansenshogaku Zasshi.* **64**, 1408–1415.
35. Shikiya, K., Kuniyoshi, T., Uechi, H., Oyakawa, T., Kinjo, F., Saito, A., Ikema, M., Nakamura, H., Yamashiro, M., and Asato, R. (1991). Treatment of strongyloidiasis with mebendazole—long-term eradication and new trials. *Kansenshogaku Zasshi.* **65**, 433–441.
36. Shikiya, K., Kinjo, N., Ikema, M., Yamashiro, A., Uechi, H., Oyakawa, T., Kinjo, F., Saito, A., Nakamura, H., Ohwan, T., Yamashiro, M., and Asato, R. (1991). Comparison of efficacy of powder and tablet of mebendazole in the treatment of strongyloidiasis. *Kansenshogaku Zasshi.* **65**, 681–686.
37. Shikiya, K., Zaha, O., Niimura, S., Ikema, M., Nakamura, H., Nakayoshi, T., Uechi, H., Kinjo, F., Saito, A., Ohwan, T., Yamashiro, M., and Asato, R. (1992). Long term eradication rate of mebendazole therapy for strongyloidiasis. *Kansenshogaku Zasshi.* **66**, 354–359.
38. Higa, F., Kitsukawa, K., Gaja, M., Tateyama, M., Shikiya, K., Shigeno, Y., Kinjo, F., and Saito, A. (1992). Cytotoxicity of mebendazole against established cell lines from the human, rat, and mouse liver. *Arch. Toxicol.* **66**, 224–227.
39. Niimura, S., Hirata, T., Zaha, O., Nakamura, H., Kouchi, A., Uehara, T., Uechi, H., Ohshiro, J., Shikiya, K., Kinjo, F., and Saito, A. (1992). Clinical study of albendazole therapy for strongyloidiasis. *Kansenshogaku Zasshi.* **66**, 1231–1235.
40. Pene, P., Mojon, M., Garin, J. P., Coulaud, J. P., and Rossignol, J. F. (1982). Albendazole: A new broad spectrum anthelmintic. Double-blind multicenter clinical trial. *Am. J. Trop. Med. Hyg.* **31**, 263–266.
41. Pungpak, S., Bunnag, D., Chindanond, D., and Radmoyos, B. (1987). Albendazole in the treatment of strongyloidiasis. *Southeast Asian J. Trop. Med. Public Health* **18**, 207–210.
42. Naquira, C., Jimenez, G., Guerra, J. G., Bernal, R., Nalin, D. R., Neu, D., and Aziz, M. (1989). Ivermectin for human strongyloidiasis and other intestinal helminths. *Am. J. Trop. Med. Hyg.* **40**, 304–309.
43. Freedman, D. O., Zierdt, W. S., Lujan, A., and Nutman, T. B. (1989). The efficacy of ivermectin in the chemotherapy of gastrointestinal helminthiasis in humans. *J. Infect. Dis.* **159**, 1151–1153.
44. Shikiya, K., Kinjo, N., Uehara, T., Uechi, H., Ohshiro, J., Arakaki, T., Kinjo, F., Saito, A., Iju, M., and Kobari, K. (1992). Efficacy of ivermectin against *Strongyloides stercoralis* in humans. *Intern. Med.* **31**, 310–312.
45. Shikiya, K., Uehara, T., Uechi, H., Ohshiro, J., Arakaki, T., Oyakawa, T., Sakugawa, H., Kinjo, F., Saito, A., and Asato, R. (1991). Clinical study on ivermectin against *Strongyloides stercoralis*. *Kansenshogaku Zasshi.* **65**, 1085–1090.
46. Shikiya, K., Zaha, O., Niimura, S., Nakamura, H., Nakayoshi, T., Kochi, A., Uehara, T., Uechi, H., Ohshiro, J., Kinjo, F., and Saito, A. (1992). Clinical study of eradicated and resistant patients to treatment with ivermectin for strongyloidiasis. *Kansenshogaku Zasshi.* **66**, 935–943.
47. Shikiya, K., Zaha, O., Niimura, S., Uehara, T., Ohshiro, J., Kinjo, F., Saito, A., and Asato, R. (1994). Clinical study on ivermectin against 125 strongyloidiasis patients. *Kansenshogaku Zasshi.* **68**, 13–20.
48. Datry, A., Hilmarsdottir, I., Mayorga-Sagastume, R., Lyagoubi, M., Gaxotte, P., Biligui, S., Chodakewitz, J., Neu, D., Danis, M., and Gentilini, M. (1994). Treatment of *Strongyloides stercoralis* infection with ivermectin compared with albendazole: Results of an open study of 60 cases. *Trans. R. Soc. Trop. Med. Hyg.* **88**, 344–345.
49. Marti, H., Haji, H. J., Savioli, L., Chwaya, H. M., Mgeni, A. F., Ameir, J. S., and Hatz, C. (1996). A comparative trial of a single-dose ivermectin versus three days of albendazole for treatment of *Strongyloides stercoralis* and other soil-transmitted helminth infections in children. *Am. J. Trop. Med. Hyg.* **55**, 477–481.

50. Gann, P. H., Neva, F. A., and Gam, A. A. (1994). A randomized trial of single- and two-dose ivermectin versus thiabendazole for treatment of strongyloidiasis. *J. Infect. Dis.* **169**, 1076–1079.
51. Addiss, D. G., Beach, M. J., Streit, T. G., Lutwick, S., LeConte, F. H., Lafontant, J. G., Hightower, A. W., and Lammie, P. J. (1997). Randomised placebo-controlled comparison of ivermectin and albendazole alone and in combination for *Wuchereria bancrofti* microfilaraemia in Haitian children. *Lancet* **350**, 480–484.
52. Ismail, M. M., Jayakody, R. L., Weil, G. J., Nirmalan, N., Jayasinghe, K. S., Abeyewickrema, W., Rezvi Sheriff, M. H., Rajaratnam, H. N., Amarasekera, N., de Silva, D. C. L., Michalski, M. L., and Dissanaike, A. S. (1998). Efficacy of single dose combinations of albendazole, ivermectin and diethylcarbamazine for the treatment of bancroftian filariasis. *Trans. R. Soc. Trop. Med. Hyg.* **92**, 94–97.
53. Glaziou, P., Cartel, J. L., Alzieu, P., Briot, C., Moulia-Pelat, J. P., and Martin, P. M. (1993). Comparison of ivermectin and benzyl benzoate for treatment of scabies. *Trop. Med. Parasitol.* **44**, 331–332.
54. Macotela-Ruiz, E., and Pena-Gonzalez, G. (1993). The treatment of scabies with oral ivermectin. *Gac. Med. Mex.* **129**, 201–205.
55. Kar, S. K., Mania, J., and Patnaik, S. (1994). The use of ivermectin for scabies. *Natl. Med. J. India* **7**, 15–16.
56. Currie, B. J., Connors, C. M., and Krause, V. L. (1994). Scabies programs in aboriginal communities. *Med. J. Aust.* **161**, 636–637.
57. Lawrence, G. W., Sheridan, J. W., and Speare, R. (1994). We can get rid of scabies: New treatment available soon. *Med. J. Aust.* **161**, 232.
58. Marty, P., Gari-Toussaint, M., Le Fichoux, Y., and Gaxotte, P. (1994). Efficacy of ivermectin in the treatment of an epidemic of sarcoptic scabies. *Ann. Trop. Med. Parasitol.* **88**, 453.
59. Meinking, T. L., Taplin, D., Hermida, J. L., Pardo, R., and Kerdel, F. A. (1995). The treatment of scabies with ivermectin. *N. Engl. J. Med.* **333**, 26–30.
60. Youssef, M. Y., Sadaka, H. A., Eissa, M. M., and el-Ariny, A. F. (1995). Topical application of ivermectin for human ectoparasites. *Am. J. Trop. Med. Hyg.* **53**, 652–653.
61. Currie, B. J., Maguire, G. P., and Woodm, Y. K. (1995). Ivermectin and crusted (Norwegian) scabies. *Med. J. Aust.* **163**, 559–560.
62. Aubin, F., and Humbert, P. (1995). Ivermectin for crusted (Norwegian) scabies. *N. Engl. J. Med.* **332**, 612.
63. Saito, A. (1995). Strongyloidiasis: Epidemiology, clinical manifestation and new methods for diagnosis and treatment. *J. Infect. Chemother.* **1**, 98–106.
64. Zaha, O., Hirata, T., Kinjo, F., and Saito, A. (2000). Strongyloidiasis: Progress in diagnosis and treatment. *Intern. Med.* **39**, 695–700.

Chapter 9

Tacrolimus and Other Immunosuppressive Macrolides in Clinical Practice

TADAHIRO AMAYA

Development Division
Fujisawa Pharmaceutical Co.
Osaka, Japan

JUN HIROI

Medical Science Research
Fujisawa Pharmaceutical Co.
Osaka, Japan

IRA D. LAWRENCE

Research and Development
Fujisawa Healthcare, Inc.
Chicago, Illinois 60015-2548

I. Introduction 421
II. Tacrolimus, a Brief Developmental History 424
III. Novel Activity of Tacrolimus and Other Immunosuppressive
    Macrolides in Clinical Practice 425
    A. Organ Transplantation 426
    B. Autoimmune Diseases 431
IV. Concluding Remarks 442
    References 444

I. Introduction

The introduction of tacrolimus into the armamentarium of immunosuppressant agents in 1993 ushered in a new era in the prevention of rejection following organ transplantation. Tacrolimus (FK506) was the first in a new class of immunosuppressant drugs that were macrolide derivatives. It was initially discovered in 1984, by Fujisawa Pharmaceutical Company, utilizing a mixed lymphocyte reaction as a screening system for compounds isolated from the broth of *Streptomyces tsukubaensis*, a bacterium found in the soil at the base of Mount Tsukuba [1, 2]. Initially given the company research compound designation of FK506, the generic name of tacrolimus is an acronym based on the following rationale: *t* for Mount Tsukuba, *acrol* for macrolide, and *imus* for immunosuppressant.

Tacrolimus was originally found to inhibit T-cell activation at an early stage of the cell cycle (G_0 to G_1) through a calcium-mediated signal [3, 4]. Tacrolimus is similar in its mechanism to cyclosporin, in that it exerts its activity via an intracellular binding protein, known collectively as the immunophilins. Although similar in their actions, the drugs bind to unique immunophilins, tacrolimus to FKBP-12 (for FK binding protein with a 12-kDa molecular weight) while cyclosporin binds to cyclophilin [5, 6]. It was initially believed that the enzymatic activity of FKBP as a peptidylprolyl isomerase (PPIase, rotamase) was the mechanism by which tacrolimus exerted its immunosuppressive effects, based on the observation that the enzymatic activity of each immunophilin is inhibited by tacrolimus or cyclosporin [5, 7]. This initial theory was disproved, however, when a derivative of tacrolimus, which possessed the enzymatic activity but not the immunosuppressant activity, was discovered [8]. It is now known that both immunophilins interact with a specific intracellular protein, calcineurin, a calmodulin-dependent serine/threonine phosphatase [9, 10]. It is the interaction of this tacrolimus–FKBP complex and calcineurin that exerts the immunosuppressive effects of the drug. This tacrolimus–FKBP complex inhibits the dephosphorylation of NFAT by calcineurin, which permits the translocation of this transcription factor into the nucleus of T lymphocytes [11–14] and results in the inhibition of the production of a large number of cytokines such as IL-2, -3, -4, -5, -8, TNF-α, GM-CSF, and INF-γ [4, 15]. The inhibitory effect of tacrolimus on other transcription factors such as AP-1 or NFκB is also known [16] but felt to be less significant [17]. Using X-ray crystallography of the tacrolimus–FKBP complex and two fragments of calcineurin, it has been determined that the drug's activity is dependent on the three-dimensional structure of this binding event [18]. Of particular importance is that the immunosuppressive mechanisms of tacrolimus are different from that of glucocorticoids (Fig. 1), this may explain why tacrolimus is clinically effective in instances where steroids have failed or demonstrated suboptimal efficacy.

Rapamycin (sirolimus) is another macrolide antibiotic that possesses potent immunosuppressant activity. Rapamycin has a chemical structure partially similar to that of tacrolimus (Fig. 2). It was first isolated from *Streptomyces hygroscopicus* strains found in soil obtained on Rapa Nui (Easter Island), hence the name *rapamycin* [19, 20]. This compound was initially investigated as an antifungal agent and later found to have immunosuppressive activity [21]. Rapamycin also binds to FKBP, but its immunosuppressive mechanisms are distinct from those of tacrolimus and cyclosporin in that it does not act via the calcineurin pathway [22, 23]. The immunosuppressive effects of rapamycin result from its inhibition of T-cell [23, 24] and B-cell [25] proliferation. The key effect on those cells results from the blocking of the signals of several cytokines (IL-2 and IL-4), leading to interruption of the cell cycle from the G_1 to the S phase. Unlike tacrolimus, the complex of rapamycin and FKBP-12 does not inhibit the dephosphorylase

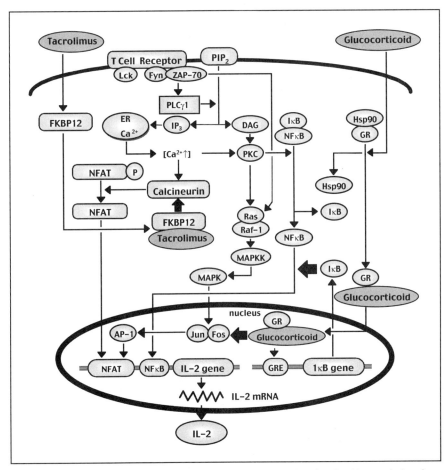

Fig. 1. The immunosuppressive mechanism of tacrolimus. (Reprinted with permission from Amaya, T., Hiroi, J., and Lawrence, I., *Folia Pharmacologica Japonica* **117**[5], Fig. 2, p. 353.)

activity of calcineurin [9] but rather binds to and modulates the function of a target protein, mTOR (mammalian target of rapamycin). The exact mode of action of the further molecular interaction has not been well elucidated but RAPA-FKBP is known to regulate p70 S6 kinase and inhibit protein synthesis including elongation factors [26–28]. Rapamycin has also been reported to inhibit cyclin-dependent kinase activity [29, 30] and the initiation factor of translational activity [31, 32]. This evidence suggests that rapamycin regulates the cell cycle progression mainly by inhibiting the translational phase of protein synthesis.

Based on the different mechanism of action from cyclosporin and tacrolimus, rapamycin has been investigated as an adjunctive immunosuppressant agent for prevention of rejection after organ transplantation in combination with these agents.

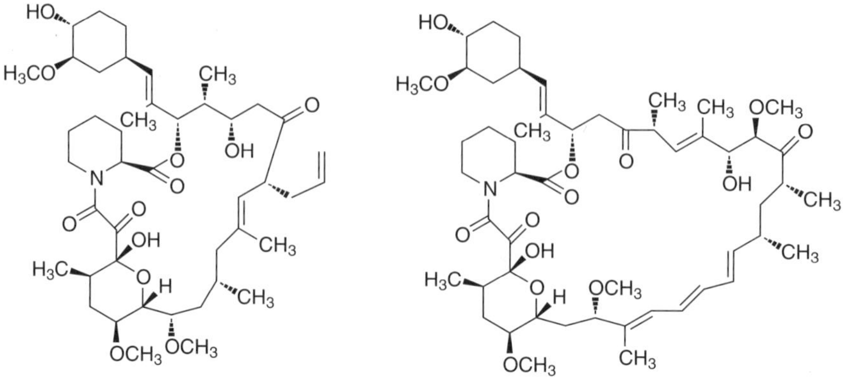

Fig. 2. Chemical structures of tacrolimus and rapamycin.

II. Tacrolimus, a Brief Developmental History

Given its potent immunosuppressant actvity and its mechanistic similarity to cyclosporin, tacrolimus was initially targeted for development as an agent to prevent the rejection of transplanted organs. Based on a paper by Dr. Takenori Ochiai of Chiba University who initiated collaboration on tacrolimus development with Fujisawa's Tsukuba Institute, Dr. Thomas Starzl, then chairman of transplantation surgery at the University of Pittsburgh, became interested in tacrolimus as the first real alternative to cyclosporin. Unfortunately, initial studies in animals revealed significant dose-related toxicities that raised serious concerns over the ability of tacrolimus to be used in a clinical setting. These data—with their very negative impact—discouraged the company from proceeding with further development of tacrolimus. However, Dr. Starzl felt strongly that given the potency of tacrolimus (30–100 times more potent than cyclosporin on a milligram-to-milligram basis), an appropriate dose reduction could alleviate many of the toxicity concerns without compromising the desired immunosuppressant activity. The first clinical experience with tacrolimus occurred in a liver transplant patient with severe refractory rejection who had several times failed conventional, cyclosporin-based antirejection treatment. The patient responded to treatment with tacrolimus, saving the transplanted organ and the patient, without significant adverse events. Subsequent similar positive experiences with tacrolimus in patients who were not adequately controlled with conventional cyclosporin-based regimens led Dr. Starzl and the group at the University of Pittsburgh to strongly encourage Fujisawa to develop tacrolimus as a clinical agent, initially for the prevention of organ rejection following transplantation. Thus, he rescued tacrolimus and tacrolimus has since rescued many patients. We discuss the results of these programs later in this chapter.

As clinical experience grew with the use of tacrolimus, establishing both its pharmacological actions and safety profile, the possibility of developing the drug for uses other than transplantation became more feasible. Of particular importance was the finding that, unlike cyclosporin, tacrolimus was active when used topically on the skin to treat immune-mediated dermatologic diseases. Topical application would seem to provide local benefit while obviating many of the concerns associated with the systemic toxicity of this class of agents [33]. We discuss the results of clinical studies using tacrolimus topically in greater detail later.

III. Novel Activity of Tacrolimus and Other Immunosuppressive Macrolides in Clinical Practice

Since its introduction in 1993 in Japan and 1994 in the United States, tacrolimus has become a primary immunosuppressant for the prevention of rejection after organ transplantation in many types of solid organ transplants [34–38] as well as for the prevention of graft-versus-host disease (GVHD) following bone marrow transplantation [39, 40]. Tacrolimus is now well established for both primary and rescue immunosuppression in solid organ transplantation such as liver and kidney. Additionally, tacrolimus has been successfully used for other types of solid organ transplantation including heart [35], pancreas [36], lung [37], and intestine [38], frequently providing effective rejection prophylaxis when cyclosporin-based regimens have been ineffective or suboptimal.

The role of both T and B lymphocytes in a variety of disease states beyond transplantation has become increasingly important in the past decade. This is especially true of those diseases frequently referred to as "autoimmune" in their etiology, such as rheumatoid arthritis, nephrotic syndrome, systemic lupus erythematosus, inflammatory bowel disease, and so on. In addition, several other major diseases are also known to have a component of T- or B-cell-mediated pathogenesis, for example, atopic dermatitis, psoriasis, and asthma. Until very recently, the mainstay of therapy for these diseases was the corticosteroids, which were often less than satisfactory in efficacy and often associated with undesirable side effects, especially in growing children and the elderly. Thus, the search for new agents with different mechanisms of action and which did not have the same adverse event profile as conventional corticosteroids led to the subsequent evaluation of drugs such as tacrolimus and sirolimus to treat several of these diseases.

Very recently topical tacrolimus (ointment) has been shown to be a very effective therapy for the treatment of moderate to severe atopic dermatitis, making it the first true alternative to steroids for treating this widespread disease. It seems logical that if topical tacrolimus is effective in atopic dermatitis, it is likely that it might show similar benefit in other inflammatory skin diseases such as psoriasis and alopecia areata, as well as other diseases such as ocular allergies, and vernal conjunctivitis and diseases of the respiratory tract. All of these disease states are particularly amenable to treatment with topical agents.

Recent investigations into the pathophysiology of ischemic brain injury and neural regeneration following central nervous system (CNS) and peripheral nerve injury have demonstrated that tacrolimus may provide benefit both in protecting against irreversible CNS injury following ischemic injury and in promoting nerve regeneration following injury [41–43].

We now turn to a discussion of the role of these agents in the treatment of several of these specific disease states in greater detail.

A. Organ Transplantation

1. Tacrolimus

The efficacy of tacrolimus as a primary immunosuppressant for the prophylaxis of rejection and for rescue therapy following failure of conventional cyclosporin-based rejection prophylaxis has been demonstrated in numerous clinical studies in adults and pediatrics using various types of combination therapy since 1989. Tacrolimus is now well established not only as a primary immunosuppressant in organ transplantation but also an excellent rescue agent for patients experiencing post-transplant rejection while on cyclosporin-based regimens [44].

a. Liver Transplantation. Many randomized comparative studies have clearly demonstrated that tacrolimus-based immunosuppression is effective in the prevention of rejection following liver transplantation based on patient survival, graft survival, and/or prevention of acute rejection. In many instances, tacrolimus was shown to be numerically and/or statistically significantly superior to cyclosporin-based immunosuppression in these parameters of success. Large-scale randomized multicenter studies, conducted in the United States and Europe in adult liver transplantation, showed favorable patient and graft survival rates for tacrolimus and statistically significantly lower rates for acute rejection in tacrolimus-based immunosuppression [45–48]. Meta analysis for a sample of 1000 pooled patients who were enrolled in two multicenter clinical studies revealed a statistically significant advantage for tacrolimus over cyclosporin in terms of patient and graft survival at 3 years post-transplantation [49]. The European study also demonstrated that patients receiving tacrolimus-based immunosuppression showed a statistically significantly lower incidence of chronic rejection than did those receiving cyclosporin-based immunosuppression at 1 year (1.5 vs. 5.3%, $P < 0.05$) and 3 years (2.0 vs. 6.9%, $P < 0.05$).

In rescue therapy, tacrolimus also demonstrated excellent results. Patients initially receiving tacrolimus who were subsequently converted to cyclosporin due to rejection were fewer than those who required conversion from cyclosporin to tacrolimus in a U.S. study. It would appear that tacrolimus might have a unique capacity to reverse acute refractory rejection [50]. These multicenter studies also showed significant steroid sparing or withdrawal benefits with tacrolimus when compared to conventional cyclosporin-based immunosuppressive regimens.

The clinical benefit of tacrolimus in liver transplantation seen in the large-scale comparative studies in the United States and Europe have been corroborated by many subsequent clinical studies. Plosker and Foster [44] have provided an excellent review of several additional randomized comparative studies comparing tacrolimus- and cyclosporin (microemulsion)-based regimens, demonstrating very similar results in patient and graft survival rates, and in most cases favoring the tacrolimus-based regimen in terms of the percentage of patients with acute rejection. The results of an interim analysis of a study presented at the Transplant 2000 meetings suggested that tacrolimus was superior to cyclosporin (microemulsion) in preventing acute rejection following liver transplantation [51].

Results of a randomized comparative study in pediatric liver transplantation between tacrolimus and cyclosporin showed equal patient survival (80% vs. 81%), graft survival (70% vs. 71%), and retransplantation rate (17% vs. 19%) at 1 year after transplantation. Of note, however, was a trend favoring tacrolimus in the prevention of acute rejection (52% vs. 79%) [52]. An additional benefit of tacrolimus in children is its steroid-sparing effect allowing for them to achieve more normal growth. In addition, the lack of gingival hyperplasia and hirsutism has made tacrolimus particularly valuable in the treatment of children undergoing liver transplantation.

The favorable therapeutic effect of tacrolimus on rejection after liver transplantation may be due, in part, to its unique action both as an hepatotrophic agent [53] and an inhibitor of iNOS induced by IL-1 during NFκB activation in hepatocytes [54]. These beneficial effects have not been observed with cyclosporin.

Positive clinical results in preventing rejection following living related donor liver transplantation for children and adults have also been reported by a Kyoto University group in Japan [55]. Living related liver transplantation is a new technique that may provide a partial solution to the shortage of organ donors and is gaining increasing acceptance by many transplant surgeons.

b. Kidney Transplantation. Several comparative studies have reported on tacrolimus versus conventional cyclosporin-based regimens in renal transplantation. In two large, multicenter randomized studies conducted in the United States and Europe [56, 57], the tacrolimus-based immunosuppressive regimen was shown to be equal to the cyclosporin-based immunosuppressive regimens in terms of patient and graft survival at 1-year post-transplantation. These studies also demonstrated that tacrolimus had statistically significantly lower rates of acute rejection (26% vs. 46% presumed; 31% vs. 46%, biopsy confirmed). Follow-up data at 3 or 4 years in the United States [58] and Europe [59] showed similar patient and graft survival rates, but the tacrolimus regimen in the U.S. data showed significantly fewer graft failures than did the cyclosporin-based regimen. Conversion of one drug to the alternative therapy occurred in 8.3% of tacrolimus-treated patients and 24.2% of cyclosporin-treated patients during the 3-year follow-up period.

A multicenter randomized comparative trial of tacrolimus in combination with azathioprine or mycofenolate mofetil (MMF) versus cyclosporin (microemulsion) with MMF after cadaveric kidney transplantation demonstrated that all regimens yielded similar acute rejection and graft survival rates at 1 year. The tacrolimus-MMF regimen was associated with the lowest rate of steroid-resistant rejection requiring antilymphocyte therapy. In addition, the tacrolimus-treated patients had lower incidence of hyperlipidemia, a side effect of particular concern in these patients [60].

In an abstract presented at Transplant 2000 (AST/ASTS), a multicenter randomized comparative study of a tacrolimus-based immunosuppressive regimen versus a cyclosporin (microemulsion)-based immunosuppressive regimen was reported. Both regimens showed similar patient and graft survival rates at 6 months post-transplantation, but the tacrolimus regimen showed a statistically significantly lower rate of biopsy-confirmed acute rejection and clinically suspected acute rejection. Additionally, tacrolimus showed a significantly better safety profile with regard to the incidence of hypertension, elevated cholesterol, and hypertrichosis, all of which have significantly negative impact on the patients' quality of life. Only tremor was observed to be significantly higher in the tacrolimus-treated group [61]. At the same conference, data from a 5-year follow-up study of a multicenter, randomized, comparative study between tacrolimus- and cyclosporin-based immunosuppressive therapies showed that tacrolimus-treated patients experienced significantly better graft survival at 5 years (when crossover due to refractory rejection was taken into account), and were also more likely to remain on their initially randomized therapy when compared to cyclosporin-treated patients. Additionally, treatment failure due to discontinuation for rejection, toxicity or graft loss, use of antihypertensive medication, and use of lipid reducing agents were all significantly lower in tacrolimus-treated patients when compared to the cyclosporin-treated patients [62].

The benefit of tacrolimus in rescue therapy following kidney transplantation has also been demonstrated in many clinical trials. Note that earlier conversion (within 6 months of transplant) to tacrolimus from cyclosporin is generally more successful than is later conversion (after 6 months from transplant) [63].

In pediatric renal transplantation, the tacrolimus regimen showed very favorable results in terms of patient and graft survival at 1 and 4 years, as well as in the acute rejection rate and the rate of steroid reduction or withdrawal [64].

   c. Other Organ Transplantation. Although fewer clinical studies on the use of tacrolimus in other types of organ transplantation exist, there is a consistent trend in patient survival rates at 1 or 2 years post-transplantation, a trend that is very similar to or better than a cyclosporin-based regimen. As has been seen in both liver and kidney transplant studies, more patients receiving tacrolimus were free of acute rejection than were those receiving a cyclosporin-based regimen. Tacrolimus has been particularly successful when used in transplantation of solid organs such as pancreas, kidney/pancreas, heart, lung, and intestine where intensive immunosuppression might be necessary. This might be due in part to the greater immunosuppressive potency of tacrolimus [44].

A recent report has suggested that cyclosporin, but not tacrolimus, induces the synthesis of transforming growth factor β(TGF-β) that has fibrogenic properties. It is believed that this effect may contribute to the development of the fibrotic changes frequently seen in transplanted organs with chronic rejection [65] and may also be associated with bronchiolitis obliterans, which is frequently seen following lung transplantation [66].

In bone marrow transplantation, it has been demonstrated that grade II–IV acute GVHD is observed significantly less often with a tacrolimus and methotrexate (MTX) regimen than with a cyclosporin and MTX regimen. Both regimens have similar survival rates in nonadvanced disease, in matched sibling bone marrow transplantation [39]. In advanced cases, further investigation is necessary to better define the benefit of tacrolimus to prevent GVHD and to improve overall survival rates. In bone marrow transplantation from matched, unrelated donors, the combination of tacrolimus and MTX significantly decreased the risk for acute GVHD when compared to the combination of cyclosporin and MTX, with no significant increase in toxicity, infections, or leukemia relapse rate [40].

In general, the tendency for tacrolimus use in organ transplantation in comparison to conventional immunosuppressants can be summarized as follows:

- Tacrolimus is now an established first-line agent for the prevention of post-transplant rejection in liver and kidney transplantation.
- Tacrolimus-based immunosuppression provides a better therapeutic regimen in terms of patient and/or graft survival rates.
- Tacrolimus-based immunosuppression demonstrates lower rates of acute rejection in transplantation of almost all other organs evaluated to date.
- Tacrolimus has a significant steroid-sparing effect, resulting in the reduction of adverse effects due to long-term steroid treatment.
- Tacrolimus shows excellent efficacy as rescue therapy for refractory acute rejection and treatment failure due to the serious adverse effects associated with cyclosporin-based treatment.
- Tacrolimus can reverse ongoing refractory acute rejection in any type of transplantation, which does not appear to be the case with cyclosporin.
- Both tacrolimus and cyclopsorin are associated with nephrotoxicity.

- Tacrolimus appears to have a higher incidence of diabetes mellitus, and some forms of neurotoxicity like tremor, although these tend to be dose related and can often be reversed by lowering the dose of tacrolimus or discontinuing the drug. Cyclosporin is associated with a greater incidence of hypertension, hypercholesterolemia, hirsutism, and gingival hyperplasia.
- Mycofenolate mofetil is replacing azathioprine as the adjunctive agent of choice in combination with both of these primary immunosuppressive regimens.

2. Rapamycin

Rapamycin (sirolimus) is a potent immunosuppressant with a different mechanism of action from tacrolimus and cyclosporin. Rapamycin forms a complex with FKBP-12, the receptor for tacrolimus in the cytosol, but the rapamycin–FKBP-12 complex does not interact with calcineurin, a key enzyme for activation of NFAT. Although tacrolimus and rapamycin share a common intracellular receptor, their mechanisms of action are quite distinct. Tacrolimus and cyclosporin inhibit the early stage in the signal transduction pathway for immunological responses, whereas rapamycin blocks the late stage of immunological activation. This different mode of pharmacological actions between tacrolimus and rapamycin allows for their different usage in clinical practices for the prophylaxis of organ rejection. In fact, given their different mechanisms of action, rapamycin and tacrolimus (or cyclosporin) are often used in combination.

Rapamycin is currently indicated for the prophylaxis of organ rejection for patients undergoing renal transplantation; it is recommended that rapamycin be used in a combination regimen with cyclosporin and steroids. Studies are currently ongoing on the concurrent use of tacrolimus and rapamycin. Several randomized comparative studies have been performed to investigate the clinical role of rapamycin in renal transplantation [67, 68], but there are not enough data to draw a conclusion for its positioning as a single agent vis-à-vis conventional immunosuppressants. Rapamycin does appear to have some advantages over calcineurin inhibitors. Those differences are most likely due to the different mechanisms of action of the agents.

The most remarkable advantage for rapamycin is its lack of significant toxicities in the kidney, pancreas, and CNS frequently seen with cyclosporin and tacrolimus and which are believed to occur, in part, due to effects on calcineurin inhibition via the tacrolimus–FKBP or cyclosporin–cyclophilin complex. On the other hand, because rapamycin exerts its effect not only on T cells but also on other cells that are stimulated by the growth factors it inhibits, its effects are not T-cell specific. This means that rapamycin affects several unique biochemical events, giving rise to its own unique pattern of adverse events [69]. Compared to tacrolimus or cyclosporin, rapamycin is associated with a significantly higher incidence of adverse effects such as hypertriglyceridemia, hypercholesterolemia, thrombocytopenia, and leukopenia. Acute rejection was seen with equal frequency to cyclosporin after renal transplantation [67, 68]. Currently, rapamycin is not used as a base immunosuppressant after organ transplantation, but rather should be used in conjunction with cyclosporin or tacrolimus.

B. Autoimmune Diseases

1. Tacrolimus (Systemic)

It is quite reasonable to postulate that tacrolimus, which has specific effects on the activation of T cells, would have a potential role in the treatment of autoimmune diseases, which are frequently mediated by an abnormal immune response of T cells. The benefit of tacrolimus in a variety of nontransplant indications has been most extensively studied by the University of Pittsburgh, which has evaluated its activity in such diseases as pyoderma gangrenosum, psoriasis, Behçet's syndrome, nephrotic syndrome, primary sclerosing cholangitis, polymyositis, and other immunologically mediated diseases [70, 71]. In most of these studies, doses of about 0.15 mg/kg twice a day were used, based on the initial recommended doses of tacrolimus for organ transplantation. As the dose of tacrolimus in transplantation has decreased over time with increasing experience, these initially evaluated doses would seem too high. New studies evaluating the benefit of lower doses on the treatment of these diseases are currently available for some diseases, as we now discuss.

In rheumatoid arthritis (RA), T cells play a critical role in both the cause and long-term maintenance of the disease. Tacrolimus, as a T-cell-specific immunosuppressant, should be effective against RA based on the pharmacological evidences that tacrolimus both inhibits the activation of T cells and blocks cross-talk between T cells and antigen-presenting cells [72].

A double-blind, comparative clinical study conducted in the United States to determine the optimal daily dosage of tacrolimus in methotrexate failure patients with advanced RA demonstrated that tacrolimus was significantly more effective at daily doses of 3 mg/patient using the criteria of improvement in the ACR 20 when compared to an inactive placebo [73]. Similar results were obtained in a Japanese double-blind comparative study in RA patients failing one disease modifying antirheumatic drug (DMARD). A daily dosage of 3 mg/patient of tacrolimus was effective in improving the ACR criteria of pain and swelling in the joints and the erythrocyte sedimentation rate with no difference in the safety profile between tacrolimus groups and the placebo group [74]. The daily dosage of 3 mg/patient was more effective than 1.5 mg/patient.

A preliminary, open-label study also demonstrated promising antirheumatic efficacy of tacrolimus at a daily dose of 2–6 mg/patient in 12 patients with severe, recalcitrant RA who had failed, on average, 5.3 DMARD and all who had failed methotrexate treatment due either to toxicities ($N = 5$) or lack of efficacy ($N = 7$) [75].

The positive clinical responses in these nonresponders to the most commonly used oral DMARD, methotrexate, are particularly encouraging. It is also noteworthy that 4 of the 5 patients who had an ACR 50 response, and both completers who had ACR 20 responses after 6 months, had previously failed cyclosporin. Note also that the required prednisone dose showed a significant reduction, and analgesic use decreased in 7 patients who completed 6 months of treatment. Dyspepsia, tremor, and headache were seen but all were reported as mild and transient. Laboratory parameters were stable throughout the 6 months of the treatment period [75]. No cases of renal toxicity were reported in any of these patients.

In the U.S. double-blind study in which NSAID use was unrestricted, a dose-dependent increase in serum creatinine from baseline was observed. In contrast, although slight increases were seen in the Japanese study, most patients' creatinine levels remained within the normal range. Whether the concomitant use of NSAIDs was a confounding factor in the development of increased creatinine is not clear. It is important to note that the antirheumatic effect of tacrolimus becomes clear within the first 4 weeks after treatment initiation and, therefore, the potential exists to reduce the dose of NSAIDs at that time. Further clinical studies are currently under way to confirm these results in a larger cohort of patients who have failed at least one DMARD in the United States and Japan, using a double-blind, placebo-controlled paradigm. This study should also better define the safety profile of tacrolimus in this patient population.

We want to stress that the effective daily dosage of 3 mg/patient in RA is less than one-half of the daily dosage currently used in organ transplantation. If shown to be effective and safe, tacrolimus, with its unique mode of action, would be a significant addition to the treatment choices available to patients with rheumatoid arthritis [72].

Clinical experiences in the treatment of Crohn's disease and ulcerative colitis with tacrolimus have been reported by both European and American investigators. They have shown that tacrolimus was effective in both ulcerative colitis and Crohn's disease, especially severe cases that were resistant to conventional therapies [76–78]. Clinical studies in several other autoimmune diseases including systemic lupus erythematosus, multiple sclerosis, and myasthenia gravis are currently under way in the United States, Japan, and Europe.

Psoriasis is a skin disease that involves the hyperproliferation of epidermal cells but its exact pathogenesis is not completely understood. Recently new insights into the pathogenesis of psoriasis suggest that T cells, rather than the keratinocyte, may play a central role in this disease as the trigger of primary importance and that several pro-inflammatory cytokines like IL-6, IL-8, and TNF-α are overproduced in psoriasis. T-cell-triggered hyperproliferation of keratinocytes in psoriasis also suggests some defect in the feedback mechanism for down-regulation in the cell cycle [79–82]. Tacrolimus inhibits T-cell activation and biosynthesis of pro-inflammatory cytokines, which are important in the pathogenesis of psoriasis.

The effectiveness of systemic tacrolimus on psoriasis has been demonstrated in a double-blind, placebo-controlled study [83].

The most exciting pharmacological actions of tacrolimus to be identified in recent years are its neuroprotective and neurotrophic actions. These actions may be elicited by inhibition of calcineurin or FKBP, large quantities of which colocalize in the brain [84]. Although the roles of FKBP and calcineurin in the CNS are not well understood, tacrolimus has been shown to have inhibitory effects on these factors in focal cerebral ischemic injury [85]. The mechanisms of the neuroprotective action of tacrolimus in an experimental stroke model are under study, and several possibilities have been suggested. Tacrolimus is known to increase the phosphorylation of nitric oxide synthase (NOS) and protect against neurotoxicity induced by glutamate [86]. These findings suggest that tacrolimus reduces the NOS enzymatic activity by inhibiting calcineurin [87]. Moreover it has also been shown that tacrolimus might protect against apoptosis in neuronal cells [88]. It has also been reported that brain damage resulting from ischemia may be induced by the infiltration of inflammatory cells including T cells [89]. Tacrolimus may protect neuronal cells by inhibition of cytokine production by these inflammatory cells. Studies to further evaluate this agent in ischemic stroke are currently being planned.

Another interesting effect of tacrolimus is its ability to induce neurite outgrowth *in vitro* [42] and nerve regeneration *in vivo* [90]. Because this effect is shown by rapamycin as well, which does not inhibit calcineurin, participation of FKBP, but not calcineurin, seems to be important in the neurotrophic action of tacrolimus. A beneficial role for tacrolimus in a variety of neurological diseases in addition to stroke such as diabetic neuropathy, spinal cord injury, amyotrophic lateral sclerosis, Alzheimer's disease, and Parkinson's disease have been suggested [43].

2. Tacrolimus (Topical)

The use of cyclosporin systemically to treat recalcitrant inflammatory skin diseases was a significant advancement in treatment in the early 1990s. The clinical benefit of oral cyclosporin for severe, recalcitrant atopic dermatitis and psoriasis was confirmed in several well-designed clinical studies. Unfortunately, due to concerns over systemic toxicity, its usage was largely limited to the most severe cases. All efforts to develop an alternative formulation, especially a topical formulation, failed, in large part due to the physicochemical properties of cyclosporin, especially its heavy molecular weight, which led to very poor cutaneous penetration. Tacrolimus appeared to possess several advantages for topical use, including a smaller molecular weight of 822 Da (two-thirds of the molecular weight of cyclosporin), and a clinical and *in vitro* potency 30–100 times greater than that of cyclosporin.

Based on these facts, in 1989, Fujisawa began to develop a topical formulation of tacrolimus to allow for the local treatment of immune-mediated dermatologic diseases while avoiding potential systemic side effects. The development of a topical immunomodulator is of great clinical relevance since, to date, the only topically active immunomodulating drugs were topical corticosteroids, which are associated with a number of undesirable side effects. These side effects have resulted in strict limitations on the length and locale of treatment, especially with regard to the more potent steroid classes. This is especially true in children, where the side effects of steroids have the greatest potential for long-term harm. The resultant product, an ointment formulation of tacrolimus, known by the trade name Protopic®, has become available in two concentrations (0.03% and 0.1%) in the United States and 0.1% in Japan for the treatment of moderate to severe atopic dermatitis. It is currently under review for marketing authorization in Europe. It has been hailed "as the most important advance for the treatment of atopic dermatitis since the introduction of topical corticosteroids over 40 years ago" [91].

   a. Atopic Dermatitis. Atopic dermatitis (AD) is a chronic and recalcitrant inflammatory disease of the skin, characterized by chronic eczematous skin lesions and severe pruritus. The pathogenesis of AD involves a complex immunologic process, superimposed on a variety of environmental and psychological factors. It is commonly believed that AD is induced by the activation of inflammatory cells such as T cells, mast cells, and eosinophils [92–95]. Recent studies have revealed that the initiation of AD is driven by activation of Th2-type cells in response to an antigen, followed by a Th1-type reaction in the chronic inflammatory phase of AD. Th1- and Th2-type cytokines contribute to different sequential stages of the development of AD skin lesions [96].

Topical tacrolimus suppresses cytokine and costimulatory molecule expression in epidermal and local draining lymph node cells during the initial skin immune response [97]. The inhibitory effect of tacrolimus on the production of cytokines in T cells has been demonstrated in both Th1 and Th2 cells [98]. This is coincident with reports that the transcription factor NFAT, a target for the calcium-regulated phosphatase calcineurin, mediates transcription of both Th1- and Th2-derived cytokines [99]. The effects of tacrolimus on other inflammatory cells such as skin mast cells, basophils, eosinophils, and Langerhans' cells have also been studied extensively. Tacrolimus has been shown to inhibit histamine release and cytokine production from human skin, lung, and cord blood-derived cultured mast cells [100–102]. Tacrolimus has also been reported to have a direct inhibitory activity on eosinophil activation [103, 104].

The role of epidermal dendritic cells, that is, Langerhans' cells, in pathogenesis of AD has been well elucidated; recently, interactions between tacrolimus and the antigen-presenting capacity of Langerhans' cells, specifically the down-regulation of the expression of high-affinity FCεRI, has been clarified [105–108]. From these *in vitro* studies, the potential efficacy of tacrolimus in AD, due to its effects on a large number of inflammatory cells whose role in the pathogenesis of atopic dermatitis has been established, has been postulated. This potential benefit was first confirmed in an NC mouse model of atopic dermatitis [109]. An additional benefit of tacrolimus is its antifungal activity against *Malassezia furfur* (*pityrosporum ovale*). This fungus has been implicated in the pathogenesis of AD as well as seborrheic dermatitis and may provide an additional rationale for the clinical benefit of tacrolimus in these cutaneous diseases. The growth of all strains of this fungus clinically isolated from healthy adults and patients with atopic dermatitis was inhibited at 0.5–32 mg/liter, concentrations that would appear to be easily obtained locally by topical application with 0.1% tacrolimus ointment [110].

Cyclosporin has been shown to have the same effects as tacrolimus on T cells [111] and mast cells [112, 113], in addition to a clinical effect on AD when given orally [114, 115]. Unfortunately, none of these effects is seen when cyclosporin is applied topically, in marked contrast to tacrolimus. It is this unique property of tacrolimus to exert its effects following topical application that has raised so much interest in its potential use as a topical agent in the clinical setting.

The traditional mainstays of topical treatment for AD have been the topical corticosteroids. It is interesting to note, however, that steroids are not believed to have any effects on histamine release from human mast cells [116, 117]. A direct comparative study of the mechanism of steroids versus tacrolimus has not been performed. However, in preliminary work done by Fujisawa Laboratories in the United States utilizing sequential punch biopsies from patients with atopic dermatitis treated with either steroids or tacrolimus topically, some differences were noted in the patterns of cytokine expression between the two treatments (M. Kobayashi, personal communication, 2000). Additional studies will need to be performed before clear conclusions can be drawn.

In aggregate, these results provide new insights into the novel mechanisms of action of topical tacrolimus treatment. This would suggest that topically applied tacrolimus represents a significant advance in the treatment of atopic dermatitis and potentially other immune-mediated diseases of the skin [91]. We now review the available clinical data to confirm this postulate.

Topical tacrolimus has been successfully developed as an ointment. An open trial in Japan as the first human experience highlighted the therapeutic potential of topical tacrolimus [33]. Three different concentrations of tacrolimus ointment (0.03%, 0.1%, and 0.3%) were evaluated. All concentrations provided excellent clinical improvements with minimal adverse effects, primarily local burning and pruritus. The benefits were most profound on the face and neck, areas traditionally most difficult to control due to restrictions on the use of topical steroids due to side effects such as skin atrophy and telangiectasia formation. Following these initially promising results, a large number of clinical trials have been conducted to further establish the clinical efficacy and safety of topically applied tacrolimus. These trials have been conducted in Japan [118–120], North America [121–125], and Europe [126–128]. To date, more than 8000 patients in 28 clinical studies have been treated with tacrolimus ointment (data on file, Fujisawa Pharmaceutical Company, Ltd).

In a phase 2, double-blind study in Japan either 0.03% and 0.1% tacrolimus ointment or its vehicle was applied twice a day for 3 weeks to adult patients with moderate to severe atopic dermatitis. Using the endpoint of the physician's global assessment of patient improvement, the number of patients achieving "moderate and marked improvement" was significantly higher in the tacrolimus ointment groups when compared to vehicle (0.03%, 71.6%; 0.1%, 91.9%; and vehicle, 49.2%). The only adverse event reported was local skin irritation at the site of application in 40% of patients in the tacrolimus groups. From this dose finding study, it was determined that 0.1% appeared to be the optimal concentration for tacrolimus ointment in Japan. Subsequently, two randomized, double-blind studies were conducted in Japan to compare tacrolimus ointment 0.1% with betamethasone valerate, 0.12% in patients with AD involving the trunk and extremities or alclometasone dipropionate, 0.1% in patients with acute AD in face and neck. Patients received treatment for 3 weeks in the first study and 1 week in the second study. In the assessment of the Physician's Global Improvement rating at the end of the first (trunk and extremities) study, tacrolimus ointment 0.1% was shown to be equal to or better than betamethasone valerate 0.12% (Table I), with a greater number of patients achieving "cure" (cleared) at the end of 3 weeks of treatment.

In the second study, involving the face and neck, tacrolimus ointment also demonstrated statistically significantly better efficacy when compared to alclometasone dipropionate 0.1% (Table II), again with more patients achieving a rating of "cured" (cleared). Moreover statistically significantly greater improvements of several secondary parameters were also seen in the tacrolimus group such as reduction of erythema and swelling; reduction in the number of papules,

TABLE I
Final Global Improvement Rating in Trunk/Extremities Study

Drugs	Cured	Markedly improved	Moderately improved	Slightly improved	No change	TOTAL	U test
FK[a]	13 (16.7)[c]	41 (69.2)	19 (93.6)	3	2	78	N.S.
BV[b]	9 (10.7)	43 (61.9)	24 (90.5)	7	1	84	

[a]FK: 0.1% tacrolimus ointment.

[b]BV: 0.12% betamethasone valerate.

[c]Parentheses indicate cumulative percent. 90% confidence intervals for difference in improvement rate: −3.9% ~ 10.1%.

TABLE II
Final Global Improvement Rating in Face and Neck Study

Drugs	Cured	Markedly improved	Moderately improved	Slightly improved	No change	Total	U test
FK[a]	10 (13.7)[c]	53 (86.3)	8	2		73	P < 0.001
AD[b]	3 (4.3)	22 (35.7)	24	18	3	70	

[a]FK: 0.1% tacrolimus ointment.
[b]AD: 0.1% alclometasone dipropionate ointment.
[c]Parentheses indicate cumulative percent.

serous papules, phlyctenule; and improvements in prurigo nodularis, lichenification, desquamation, and itching, when compared to the alclometasone treated group. In both studies, more patients in the tacrolimus-treated group experienced burning sensation at the site of application than in the vehicle group. These sensations subsided in a few days as the atopic dermatitis improved, as might be anticipated, given the highly sensitive nature of the skin of patients during an acute flare of atopic dermatitis. There were no other differences in overall safety between the tacrolimus and steroid groups.

A long-term safety study in adults was performed in patients with atopic dermatitis, treated regardless of the location of the atopic dermatitis, for periods of up to 2 years, in Japan. A total of 568 patients were enrolled; more than 400 completed 1 year of treatment and more than 300 completed 2 years. The study demonstrated that tacrolimus ointment was effective and well tolerated for the long-term treatment of AD in adults. Rapid clinical response was achieved within the first few months of treatment and was maintained throughout the 2-year study period [120]. Adverse events were primarily restricted to application site reactions and the incidence decreased rapidly over time as the skin healed. Similar results for safety and efficacy in adult patients with atopic dermatitis were obtained from a 1-year study of 0.1% tacrolimus ointment twice a day monotherapy [128]. Local irritations such as a burning sensation and pruritus were common but tended to occur only when initiating treatment. Systemic absorption was minimal with the maximum blood concentration of tacrolimus being less than 1 ng/ml. Marked and excellent or cleared improvement rates in investigators' global assessment were 54%, 81%, and 86% at week 1, month 6, and month 12, respectively. These results show tendencies similar to those of the short-term trials reviewed here.

Large-scale, multicenter, double-blind comparative studies have also been conducted in the United States and Europe. Hanifin et al. [121] and the Tacrolimus Ointment Study Group reported the results focused on the efficacy of tacrolimus

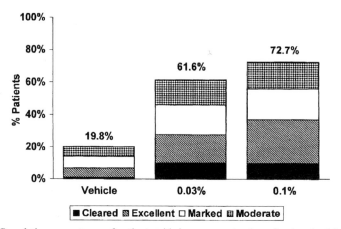

Fig. 3. Cumulative percentages of patients with improvement ratings for the physician's global evaluation of clinical response ($N = 632$). Cleared (100% improvement, *solid bars*), excellent (90–99% improvement, *hatched bars*), marked (75–89% improvement, *open bars*), or moderate (50–74% improvement, *speckled bars*). (Reprinted with permission from Hanifin *et al.* [121], Fig. 1, p. S32.)

ointment in a pooled group of 632 adult patients with moderate to severe atopic dermatitis. Patients were treated with tacrolimus ointment (0.03% or 0.1%) or vehicle twice daily for 12 weeks [121] and could treat any portion of their body involved with atopic dermatitis, up to 100% body surface area (BSA). The primary endpoint for this study was the physician's global assessment of improvement at the end of treatment. The study demonstrated that both tacrolimus ointment concentrations were significantly more effective than vehicle in the treatment of moderate to severe atopic dermatitis in adult patients, and that the 0.1% ointment was significantly more effective than the 0.03% concentration, as shown in Fig. 3. The additional benefit of 0.1% tacrolimus was most apparent in those adults with the most severe disease, in women, and in African-Americans. In the evaluation of the individual signs and symptoms of atopic dermatitis, which were collected as secondary endpoints, such as edema, erythema, and excoriation, a statistically greater improvement was seen in both tacrolimus treatment groups compared with vehicle. In the case of edema, excoriation, and oozing, the 0.1% concentration exerted a significantly greater improvement when compared to the 0.03% concentration (Fig. 4).

With respect to safety in a total of 631 adult patients who applied tacrolimus ointment (0.03% or 0.1%) or vehicle twice daily for up to 12 weeks, a sensation of skin burning was observed more often in the tacrolimus ointment groups than in the vehicle group (25% vs. 45% and 57%, vehicle vs. 0.03% and 0.1%, respectively). These application site adverse events were generally of short duration and their prevalence decreased after a few days of treatment. No clinically meaningful

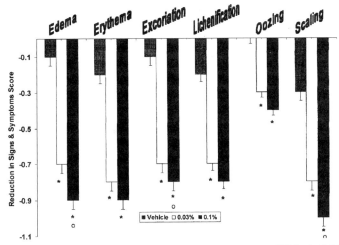

Fig. 4. Change from baseline to end of treatment (least squares mean ± SE) for individual signs of atopic dermatitis. Vehicle (*shaded bars*), 0.03% tacrolimus ointment (*open bars*), and 0.1% tacrolimus ointment (*solid bars*). *, significantly greater improvement compared with vehicle group ($P < 0.001$); O, significantly greater improvement for the 0.1% concentration compared with the 0.03% concentration of tacrolimus ointment ($P = 0.05$). (Reprinted with permission from Hanifin et al. [121], Fig. 3, p. S33.)

changes in any laboratory parameters were noted. There was minimal absorption of tacrolimus through affected skin; mean and median tacrolimus whole-blood concentrations were below the limit of quantification throughout the study in the majority of patients evaluated [123].

Paller et al. [125] reported the safety and efficacy of tacrolimus ointment for the treatment of 351 children ages 2–15 years with moderate to severe atopic dermatitis in a randomized, double-blind, vehicle controlled study. Patients were allowed to treat up to 100% BSA twice daily for 12 weeks with either 0.03%, 0.1% tacrolimus ointment, or vehicle. Although there were no differences in efficacy between the two concentration groups, significantly more patients achieved clinical improvement of 90% or better with 0.03% or 0.1% tacrolimus ointment compared with vehicle. Adverse events seen with significantly greater frequency in the 0.03% tacrolimus ointment-treated patients compared with vehicle were transient local adverse events, especially skin burning sensation and pruritus. No statistically significant differences were seen between either tacrolimus groups and the vehicle group with respect to the overall incidence of nonapplication adverse events, and no clinically meaningful changes in mean or median values for any laboratory parameters were noted.

Because atopic dermatitis is a chronic and recalcitrant disease, which usually requires prolonged treatment, the long-term safety of a new therapy especially for

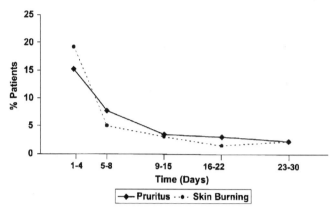

Fig. 5. Prevalence of application site adverse events of pruritus (*solid line*) and skin burning (*dashed line*) during first month of the study. (Reprinted with permission from Kang *et al.* [124], Fig. 1, p. S61.)

pediatrics, such as tacrolimus ointment, must be established. A long-term study in 255 pediatric patients (ages 2–15 years) was conducted by Kang *et al.* [124] using 0.1% tacrolimus ointment applied twice daily for up to 1 year. The results of that study, with regard to efficacy, were consistent with the results of the 12-week double-blind comparative studies described above. From the perspective of safety, no significant long-term events were seen, and the pattern of adverse events was consistent with, and similar to, that seen in the 12-week studies.

As shown in Fig. 5, the prevalence of skin burning decreased after the first few days of treatment, ostensibly as the skin healed and the therapeutic barrier function was restored. Of particular note, no increase was observed in the rate of infections during the 1-year study. Improvement in the patient's atopic dermatitis was seen early in the study and was sustained throughout the remainder of the study [124].

We emphasize that, in terms of systemic safety for children, tacrolimus whole-blood concentrations were consistent with minimal absorption of tacrolimus through the affected skin; in more than 90% of blood samples collected, the level of tacrolimus was below the limit of detection of the assay used.

Drake *et al.* [122], using a validated quality-of-life (QOL) measurement instrument, assessed changes in QOL for all three U.S. phase 3 studies mentioned above. In that assessment, tacrolimus ointment therapy with either 0.1% or 0.03% was associated with a statistically significantly greater QOL improvement than placebo treatment in adults, children, and toddlers. As illustrated in Fig. 6, the patients' preference for continuing on the study therapy supports the QOL benefits of tacrolimus ointment for atopic dermatitis from the patients' perspectives.

One of the most feared complications of conventional treatment with topical steroids to treat atopic dermatitis has been the development of cutaneous atrophy.

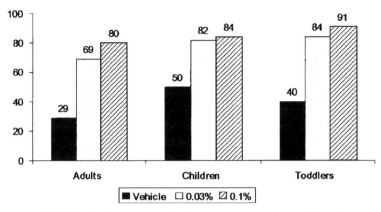

p ≤ 0.001 for differences among treatment groups for each age category.

Fig. 6. Preference for continuing with study therapy (adults, children, and toddlers).

This complication of topical steroid use is particularly serious on the face and the intertriginous areas of the body, where the skin is thinnest. The resultant atrophy can lead to scarring and disfigurement, especially when used chronically. This side effect of steroids is presumed to be due, in large part, to steroidal effects on collagen synthesis. Tacrolimus ointment has been shown to differ significantly from steroids, in that it does not reduce collagen synthesis in one published report [127]. This provides a significant advantage for tacrolimus, especially for use on the face and other areas of the skin where steroids are usually contraindicated. Another concern of chronic steroid use has been its effect on growth and maturation in children. Thus far, there is no evidence that tacrolimus affects growth or maturation in children, although longer studies will be required to confirm this initial observation.

Tacrolimus ointment is the first topical, nonsteroidal, effective and safe immunomodulatory agent to offer the potency of a corticosteroid without its attendant adverse effects. It truly represents the first therapeutic advance in the treatment of atopic dermatitis since the introduction of topical steroids more than 40 years ago [105, 129].

b. Other Skin Diseases The clinical use of topical tacrolimus has been extended to other inflammatory skin diseases such as pyoderm gangrenosum [130], rheumatic skin ulcers [131], and oral/perineal Crohn's disease [132]. Additional diseases where topical tacrolimus may be of clinical benefit include the cutaneous manifestations of GVHD after bone marrow transplantation, psoriasis, contact dermatitis, leichen planus, hand/foot eczema, alopecia areata, and seborrheic dermatitis. Different topical formulations may optimize the benefits of topically applied tacrolimus in these diseases. The role of topical tacrolimus as a therapeutic option to conventional topical corticosteroids in the treatment of a variety of

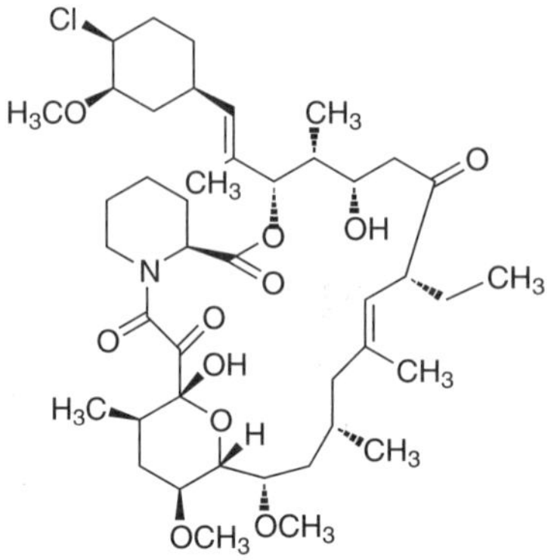

Pimecrolimus(ASM-981)

Fig. 7. Chemical structure of pimecrolimus.

clinical diseases of the skin remains to be fully realized. As data continue to be collected, now that this agent is available commercially in the United States and Japan, its role in the treatment armamentarium of physicians should continue to increase.

3. Rapamycin

Because rapamycin exerts its immunosuppressive effects by inhibiting T- and B-cell functions that are, however, different from those of tacrolimus or cyclosporin, it might also have a therapeutic potential in rheumatoid arthritis and the other autoimmune diseases. Unfortunately, few reports exist on the clinical benefit of this agent in such diseases, with the exception of psoriasis [133, 134], where its efficacy was clearly shown, when used systemically.

4. Pimecrolimus

Recently, Novartis has begun development of a new macrolide immunosuppressant, an ascomycin derivative (pimecrolimus; ASM 981; Fig. 7), in a topical formulation of a 1% cream. Preliminary data presented at several international conferences suggest that it may have benefit in mild atopic dermatitis with a good safety profile, but definitive studies have not yet been published.

IV. Concluding Remarks

Both tacrolimus and rapamycin (and more recently pimecrolimus or ascomycin) are classified in immunosuppressive macrolides. In the case of tacrolimus and rapamycin, the mechanisms by which they exert immunosuppression are quite different, although both are ligands of the same immunophilin, FKBP-12. Tacrolimus impedes the early phases of the immune signal transduction pathway in the cell by inactivating the dephosphatase activity of calcineurin. Rapamycin blocks the later stages of the immune pathway by inhibiting cell cycle progression. In addition, tacrolimus inhibits calcineurin activity, whereas rapamycin does not. Preliminary data would suggest that ascomycin acts in a way more similar to tacrolimus, although, again, minimal data are available for review.

As we have outlined briefly in the discussion above, tacrolimus has become a significant agent in the treatment of a large number of immunologically mediated diseases. The role of tacrolimus-based primary immunosuppression in the prevention of rejection following solid organ transplants, especially liver and kidney transplants, is now established throughout the world. In addition to the essential role in the primary immunosuppression, its unique properties when compared to cyclosporin have led to its ability to rescue patients who have experienced rejection while on conventional cyclosporin-based immunosuppressive regimens. Moreover, several key safety advantages, including a lack of gingival hyperplasia, no hypertrichosis, and a lower incidence of hypertension and hyperlipidemia have made it a reasonable alternative for many patients experiencing these unwanted side effects of cyclosporin-based regimens, especially children and women. Rapamycin, in its own right, offers a significant alternative as an adjunctive agent in the prevention of rejection in transplantation. Its role as a primary prophylaxis agent remains to be established, however.

The potential benefit of this class of agents in other clinical disease states, such as rheumatoid arthritis and other autoimmune diseases, has begun to be established and promises to provide physicians with reasonable new choices when treating patients with these challenging and often difficult to manage diseases. More recently, the potential of tacrolimus to provide benefit to patients with neurologic injury has raised the possibility of a new avenue of clinical benefit for this class of drugs. The studies to evaluate this therapeutic potential will be of great interest and importance to all patients and physicians.

The unique ability of tacrolimus to be effective when used topically has ushered in a new era in the treatment of immune-mediated skin diseases. As experience is gained both with longer term use of this agent in atopic dermatitis as well as in treating other dermatologic conditions, the macrolides, especially tacrolimus, will no doubt become a major therapeutic class for this important and common group of conditions.

The development of the macrolide immunosuppressants was the result of careful basic and clinical research conducted by the pharmaceutical industry in close collaboration with the medical and scientific communities from around the world. The resultant drugs that have been developed to date and those still on the horizon promise to continue to benefit patients and provide physicians with valuable and novel therapeutic choices.

Finally, we should emphasize again that tacrolimus unlocked the black box of the intracellular signal transduction pathway to immunity by explanation of interactions among tacrolimus, immunophilin, calcineurin, and the nuclear factor of activated T cells. Thus tacrolimus has been playing the immensely important roles of both providing a novel therapeutic option for recalcitrant immune diseases and the evolution of cell biology by elucidation of an immune signal transduction pathway in cells as well as the new understandings in nerve physiology by elucidating the unique function of FKBP and calcineurin in the brain. We now know that many possibilities exist for controlling immunosuppression by analyzing the mode of actions of tacrolimus and rapamycin. Hereafter, intensive investigation of the pharmacological action of tacrolimus in the neuronal system should provide a distinct explanation of the role of FKBP in CNS function, and the chance to develop a novel therapy for nerve degenerative diseases such as Alzheimer's disease, ischemic stroke, and Parkinson's disease probably after catching the natural ligand(s) of FKBPs and other immunophilins in the brain (so specially called neurophilins).

References

1. Kino, T., Hatanaka, H., Hashimoto, M., Nishiyama, M., Goto, T., Okuhara, M., Kohsaka, M., Aoki, H., and Imanaka, H. (1987). FK506, a novel immunosuppressant isolated from a *Streptomyces*. I. Fermentation, isolation, and physico-chemical and biological characteristics. *J. Antibiot.* **40**, 1249–1255.
2. Kino, T., Hatanaka, H., Miyata, S., Inamura, N., Nishiyama, M., Yajima, T., Goto, T., Okuhara, M., Kohsaka, M., Aoki, H., and Ochiai, T. (1987). FK506, a novel immunosuppressant isolated from a *Streptomyces*. II. Immunosuppressive effect of FK506 *in vitro*. *J. Antibiot.* **40**, 1256–1265.
3. Sawada, S., Suzuki, G., Kawase, Y., and Takaku, F. (1987). Novel immunosuppressive agent, FK506. *In vitro* effects on the cloned cell activation. *J. Immunol.* **139**, 1797–1803.
4. Tocci, M. J., Matkovich, D. A., Collier, K. A., Kwok, P., Dumont, F., Lin S., Degudicibus, S., Siekierka, J. J., Chin, J., and Hutchinson, N. I. (1989). The immunosuppressant FK506 selectively inhibits expression of early T cell activation genes. *J. Immunol.* **143**, 718–726.
5. Siekierka, J. J., Hung, S. H. Y., Poe, M., Lin, C. S., and Sigal, N. H. (1989). A cytosolic binding protein for the immunosuppressant FK506 has peptidyl-prolyl isomerase activity but is distinct from cyclophilin. *Nature* **341**, 755–757.
6. Siekierka, J. J., Staruch, M. J., Hung, S. H. Y., and Sigal, N. H. (1989). FK506, a potent novel immunosuppressive agent, binds to a cytosolic protein which is distinct from the cyclosporin A binding protein, cyclophilin. *J. Immunol.* **143**, 1580–1583.
7. Harding, M. W., Galat, A., Uehling, D. E., and Schreiber, S. L. (1989). A receptor for the immunosuppressant FK506 is a cis-trans peptidyl-prolyl isomerase. *Nature* **341**, 758–760.
8. Schreiber, S. L., and Crabtree, G. R. (1992). The mechanism of action of cyclosporin A and FK506. *Immunol. Today* **13**, 136–142.
9. Liu, J., Farmer, Jr., J. D., Lane, W. S., Friedman, J., Weissman, I., and Schreiber, S. L. (1991). Calcineurin is a common target of cyclophilin-cyclosporin A and FKBP-FK506 complexes. *Cell* **66**, 807–815.
10. Lane, B. C., Miller, L. N., Kawai, M., Or, Y.-S., Wiedeman, P., Holzman, T. F., and Luly, J. R. (1993). Evaluation of calcineurin's role in the immunosuppressive activity of FK506, related macrolactams, and cyclosporine. *Transpl. Proc.* **25**, 644–646.

11. McCaffrey, P. G., Perrino, B. A., Soderling, T. R., and Rao, A. (1993). NF-ATp, a T lymphocyte DNA-binding protein that is a target for calcineurin and immunosuppressive drugs. *J. Biol. Chem.* **268**, 3747–3752.
12. Fruman, D. A., Klee, C. B., Bierer, B. E., and Burakoff, S. J. (1992). Calcineurin phosphatase activity in T lymphocytes is inhibited by FK506 and cyclosporin A. *Proc. Natl. Acad. Sci. USA* **89**, 3686–3690.
13. Ruff, V. A., and Leach, K. L. (1995). Direct demonstration of NFATp dephosphorylation and nuclear localization in activated HT-2 cells using a specific NF-ATp polyclonal antibody. *J. Biol. Chem.* **270**, 22,602–22,607.
14. O'Keefe, S. J., Tamura, J., Kincaid, R. L., Tocci, M. J., and O'Neill, E. A. (1992). FK-506- and CsA-sensitive activation of the interleukin-2 promoter by calcineurin. *Nature* **357**, 692–694.
15. Spencer, C. M., Goa, K. L., and Gillis, J. C. (1997). Tacrolimus: An update of its pharmacology and clinical efficacy in the management of organ transplantation. *Drugs* **54**, 925–975.
16. Baksh, S., and Burakoff, S. J. (2000). The role of calcineurin in lymphocyte activation. *Semin. Immunol.* **12**, 405–415.
17. Mattila, P. S., Ullman, K. S., Fiering, S., Emmel, E. A., McCutcheon, M., Crabtree, G. R., and Herzenberg, L. A. (1990). The action of cyclosporin A and FK506 suggest a novel step in the activation in T lymphocytes. *EMBO J.* **9**, 4425–4433.
18. Griffith, J. P., Kim, J. L., Kim, E. E., Sintchak, M. D., Thomson, J. A., Fitzgibbon, M. J., Fleming, M. A., Caron, P. R., Hsiao, K., and Navia, M. A. (1995). X-ray structure of calcineurin inhibited by the immunophilin-immunosuppressant FKBP12–FK506 complex. *Cell* **82**, 507–522.
19. Vezina, C., Kudelski, A., and Sehgal, S. N. (1975). Rapamycin (AY22,989), a new antifungal antibiotic. I. Taxonomy of the producing streptomycete and isolation of the active principle. *J. Antibiot.* **28**, 721–726.
20. Sehgal, S. N., Baker, H., and Vezina, C. (1975). Rapamycin (AY22,989), a new antifungal antibiotic. II. Fermentation, isolation and characterization. *J. Antibiot.* **28**, 727–732.
21. Martel, R. R., Klicius, J., and Galet, S. (1977). Inhibition of the immune response by rapamycin, a new antifungal antibiotic. *Can. J. Physiol. Pharmacol.* **55**, 48–51.
22. Sehgal, S. N., and Banasbach, C. C. (1993). Rapamycin: *In vitro* profile of a new immunosuppressive macrolide. *Annal NY Acad. Sci.* **685**, 58–67.
23. Dumont, F. J., Strauch, M. J., Koprak, S. L., Melino, M. R., and Sigal, N. H. (1990). Distinct mechanisms of suppression of murine T cell activation by the related macrolides FK506 and rapamycin. *J. Immunol.* **144**, 251–258.
24. Terada, N., Lucas, J. J., Szepesi, A., Franklin, R. A., Domenico, J., and Gelfand, E. W. (1993). Rapamycin blocks cell cycle progression of activated T cells prior to events characteristic of the middle to late G_1 phase of the cycle. *J. Cell Physiol.* **154**, 7–15.
25. Aaguaard-Tillery, K. M., and Jelinek, D. F. (1994). Inhibition of human B lymphocyte cell cycle progression and differentiation by rapamycin. *Cell Immunol.* **156**, 493–507.
26. Price, D. J., Grove, J. R., Calvo, V., Avruch, J., and Bierer, B. E. (1992). Rapamycin-induced inhibition of the 70-kilodalton S6 protein kinase. *Science* **257**, 973–977.
27. Chung, J., Kuo, C. J., Crabtree, G. R., and Blenis, J. (1992). Rapamycin-FKBP specifically blocks growth-dependent activation of and signalling by the 70 kd S6 protein kinases. *Cell* **69**, 1227–1236.
28. Terada, N., Patel, H. R., Takase, K., Kohno, K., Nairn, A. C., and Gelfand, E. W. (1994). Rapamycin selectively inhibits translation of mRNAs encoding elongation factors and ribosomal proteins. *Proc. Natl. Acad. Sci. USA* **91**, 11477–11481.
29. Albers, M. W., Williams, R. T., Brown, E. J., Tanaka, A., Hall, F. L., and Schreiber, S. L. (1993). FKBP-Rapamycin inhibits a cyclin-dependent kinase activity and a cyclin D1-cdk association in early G_1 of an osteosarcoma cell line. *J. Biol. Chem.* **268**, 22825–22829.

30. Nourse, J., Firpo, E., Flanagan, W. M., Coats, S., Polyak, K., Lee, M.-H., Massague, J., Crabtree, G. R., and Roberts, J. M. (1994). Interleukin-2-mediated elimination of the p27^{Kip1} cyclin-dependent kinase inhibitor prevented by rapamycin. *Nature* **372**, 570–573.
31. Beretta, L., Gingras, A.-C., Svitkin, Y. V., Hall, M. N., and Sonenberg, N. (1996). Rapamycin blocks the phosphorylation of 4E-BP1 and inhibits cap-dependent initiation of translation. *EMBO J.* **15**, 658–664.
32. Brunn, G. J., Hudson, C. C., Sekulic, A., Williams, J. M., Hosoi, H., Houghton, P. J., Lawrence, Jr., J. C., and Abraham, R. T. (1997). Phosphorylation of the translational repressor PHAS-1 by the mammalian target of rapamycin. *Science* **277**, 99–101.
33. Nakagawa, H., Etoh, T., Ishibashi, Y., Higaki, Y., Kawashima, M., Torii, H., and Harada, S. (1994). Tacrolimus ointment for atopic dermatitis. *Lancet* **344**, 883.
34. Starzl, T., Todo, S., Fung, J., Demetris, A. J., Venkataramanan, R., and Jain, A. (1989). FK506 for liver, kidney, and pancreas transplantation. *Lancet* **II**(8670), 1000–1004.
35. Meiser, B. M., Pfeiffer, M., Schmidt, D., Ueberfuhr, P., Reichenspurner, H., Paulus, D., Scheidt, W. V., Kreuzer, E., Seidel, D., and Reichart, B. (1999). The efficacy of the combination of tacrolimus and mycophenolate mofetil for prevention of acute myocardial rejection is dependent on routine monitoring of mycophenolic acid trough acid levels. *Transpl. Proc.* **31**, 84–87.
36. Gruessner, R. W. G., Bartlett, S. T., Burke, G. W., and Stock, P. G. (1998). Suggested guidelines for the use of tacrolimus in pancreas/kidney transplantation. *Clin. Transpl.* **12**, 260–262.
37. Reichenspurner, H., Kur, F., Treede, H., Meiser, B. M., Welz, A., Vogelmeier, C., Schwaiblmeier, M., Müller, C., Fürst, H., Briegel, J., and Reichat, B. (1999). Tacrolimus-based immunosuppressive protocols in lung transplantation. *Transpl. Proc.* **31**, 171–172.
38. Goulet, O., Jan, D., Lacaille, F., Colomb, V., Michel, J. L., Damotte, D., Jouvet, P., Brousse, N., Faure, C., Cézard, J. P., Sarnacki, S., Peuchmaur, M., Hubert, P., Ricour, C., and Révillon, Y. (1999). Intestinal transplantation in children: Preliminary experience in Paris. *J. Parenter. Enter. Nutr.* **23**, S121–S125.
39. Ratanatharathorn, V., Nash, R. A., Przepiorka, D., Devine, S. M., Klein, J. L., Weisdorf, D., Fay, J. W., Nademanee, A., Antin, J. H., Christiansen, N. P., Jagt, R. V. D., Herzig, R. H., Litzow, M. R., Wolff, S. N., Longo, W. L., Petersen, F. B., Karanes, C., Avalos, B., Storb, R., Buell, D. N., Maher, R. M., Fitzsimmons, W. E., and Wingard, J. R. (1998). Phase III study comparing methotrexate and tacrolimus (prograf, FK506) with methotrexate and cyclosporine for graft-versus-host disease prophylaxsis after HLA: Identical sibling bone marrow transplantation. *Blood* **92**(7), 2303–2314.
40. Nash, R. A., Antin, J. H., Karanes, C., Fay, J. W., Avalos, B. R., Yeager, A. M., Przepiorka, D., Davies, S., Petersen, F. B., Bartels, P., Buell, D., Fitzsimmons, W., Anasetti, C., Storb, R., and Ratanatharathorn, V. (2000). Phase III study comparing methotrexate and tacrolimus with methotrexate and cyclosporine for prophylaxis of acute graft-versus-host disease after marrow transplantation from unrelated donors. *Blood* **96**(6), 2062–2068.
41. Sharkey, J., Jones, P. A., McCarter, J. F., and Kelly, J. S. (2000). Calcineurin inhibitors as neuroprotectants. *CNS Drugs* **13**, 1–13.
42. Lyons, W. E., George, E. B., Dawson, T. M., Steiner, J. P., and Snyder, S. H. (1994). Immunosuppressant FK 506 promotes neurite outgrowth in cultures of PC 12 cells and sensory ganglia. *Proc. Natl. Acad. Sci. USA* **91**, 3191–3195.
43. Snyder, S. H., Lai, M. M., and Burnett, P. E. (1998). Immunophilins in the nervous system. *Neuron* **21**, 283–294.
44. Plosker, G. L., and Foster, R. H. (2000). Tacrolimus. *Drugs* **59**, 323–389.
45. U.S. Multicenter FK506 Liver Study Group (1994). A comparison of tacrolimus (FK506) and cyclosporine for immunosuppression in liver transplantation. *N. Engl. J. Med.* **331**, 1110–1115.
46. European FK506 Multicenter Liver Study Group (1994). Randomised trial comparing tacrolimus (FK506) and cyclosporin in prevention of liver allograft rejection. *Lancet* **344**, 423–428.

47. Wiesner, R. H. (1998). A long-term comparison of tacrolimus (FK506) versus cyclosporine in liver transplantation: A report of the United States FK506 Study Group. *Transplantation* **66**, 493–499.
48. Pichlmayr, R., Winkler, M., Neuhaus, P., McMaster, P., Calne, R., Otto, G., Williams, R., Groth, C. G., and Bismuth, H. (1997). Three-year follow-up of the European Multicenter Tacrolimus (FK506) Liver Study. *Transpl. Proc.* **29**, 2499–2502.
49. Busuttil, R. W., and Holt, C. D. (1997). Tacrolimus (FK506) is superior to cyclosporine in liver transplantation. *Transpl. Proc.* **29**, 534–538.
50. Denton, M. D., Magee, C. C., and Sayegh, M. H. (1999). Immunosuppressive strategies in transplantation. *Lancet* **353**, 1083–1091.
51. Ogrady, J. G. (2000). Tacrolimus vs microemulsified cyclosporine in liver transplantation: Preliminary results of the TMC trial. Presented at Transplant (May 13–17, Chicago). Abstr. No. 198.
52. McDiarmid, S. V., Busuttil, R. W., Ascher, N. L., Burdick, J., D'Alessandro, A. M., Esquivel, C., Kalayoglu, M., Klein, A. S., Marsh, J. W., Miller, C. M., Schwartz, M. E., Shaw, B. W., and So, S. K. (1995). FK506 (tacrolimus) compared with cyclosporine for primary immunosuppression after pediatric liver transplantation: Results from the U.S. Multicenter Trial. *Transplantation* **59**, 530–536.
53. Peters, D. H., Fitton, A., Plosker, G. L., and Faulds, D. (1993). Tacrolimus, a review of its pharmacology, and therapeutic potential in hepatic and renal transplantation. *Drugs* **46**, 746–794.
54. Kaibori, M., Sakitani, K., Oda, M., Kamiyama, Y., Masu, Y., Nishizawa, M., Ito, S., and Okumura, T. (1999). Immunosuppressant FK506 inhibits inducible nitric oxide synthase gene expression at a step of NF-κ-B activation in rat hepatocytes. *J. Hepatol.* **30**, 1138–1145.
55. Minamiguchi, S., Sakurai, T., Fujita, S., Okuno, T., Haga, H., Mino, M., Kanehira, K., Matsushiro, H., Nakashima, Y., Inomata, Y., Tanaka, K., and Yamabe, H. (1999). Living related liver transplantation: Histopathologic analysis of graft dysfunction in 304 patients. *Human Pathol.* **30**, 1479–1487.
56. Prisch, J. D., Miller, J., Deierhoi, M. H., Vincenti, F., and Filo, R. S. (1997). A comparison of tacrolimus (FK506) and cyclosporine for immunosuppression after cadaveric renal transplantation. FK506 Kidney Transplant Study Group. *Transplantation* **63**(7), 977–983.
57. Mayer, A. D., Dmitrewski, J., Squifflet, J. P., Besse, T., Grabensee, B., Klein, B., Eigler, F. W., Heemann, U., Pichlmayr, R., Behrend, M., Vanrenterghem, Y., Donck, J., Hooff, J. V., Christiaans, M., Morales, J. M., Andres, A., Johnson, R. W. G., Short, C., Buchholz, B., Rehmert, N., Land, W., Schleibner, S., Forsythe, J. L. R., Talbot, D., Neumayer, H. H., Hauser, I., Ericzon, B. G., Brattström, C., Claesson, K., Mühlbacher, F., and Pohanka, E. (1997). Multicenter randomized trial comparing tacrolimus (FK506) and cyclosporine in the prevention of renal allograft rejection: A report of the European Tacrolimus Multicenter Renal Study Group. *Transplantation* **64**, 436–443.
58. Jensik, S. C., and the FK506 Kidney Transplant Study Group (1998). Tacrolimus (FK506) in kidney transplantation: Three-year survival results of the U.S. multicenter, randomized, comparative trial. *Transpl. Proc.* **30**, 1216–1218.
59. Mayer, A. D., for the European Tacrolimus Multicenter Renal Study Group (1999). Four-year follow-up of the European Tacrolimus Multicentre Renal Study. *Transpl. Proc.* **31** (Suppl. 7A), 27S–28S.
60. Johnson, C., Ahsan, N., Gonwa, T., Halloran, P., Stegall, M., Hardy, M., Matzger, R., Shield, C., Rocher, L., Scandling, J., Sorensen, J., Mulloy, L., Light, J., Corwin, C., Danovitch, G., Wachs, M., VanVeldhuisen, P., Salm, K., Tolzman, D., and Fitzsimmons, W. F. (2000). Randomized trial of tacrolimus (prograf) in combination with azathioprine or mychophenolate mofetil versus cyclosporine (neoral) with mychophenolate mofetil after cadaveric kidney transplantation. *Transplantation* **69**, 834–841.

61. Margareiter, R., and the European Tacrolimus vs Cyclosporin-Microemulsion Renal Transplantation Study Group (2000). A prospective, randomized multicentre study to compare the efficacy and safety of tacrolimus and cyclosporin-microemulsion in renal transplantation. Presented at Transplant (May 13–17, Chicago). Abstr. No. 8.
62. Pirsch, J. D., and the FK506 Kidney Transplant Study Group (2000). Tacrolimus versus cyclosporin in kidney transplantation: Five-year survival results of the U.S. multicenter, randomized, comparative study. Presented at Transplant (May 13–17, Chicago). Abstr. No. 10.
63. Jordan, M. L., Naraghi, R., Shapiro, R., Smith, D., Vivas, C. A., Scantlebury, V. P., Gritsch, H. A., McCauley, J., Randhawa, P., Demetris, A. J., McMichael, J., Fung, J. J., and Starzl, T. E. (1998). Tacrolimus for rescue of refractory renal allograft rejection. *Transpl. Proc.* **30**, 1257–1260.
64. Shapiro, R., Scantlebury, V. P., Jordan, M. L., Vivas, C., Ellis, D., Lombardozzi-Lane, S., Gilboa, N., Albin, H., Irish, W., McCauley, J., Fung, J. J., Hakala, T. R., Simmons, R. L., and Starzl, T. E. (1999). Pediatric renal transplantation under tacrolimus-based immunosuppression. *Transplantation* **67**, 299–303.
65. Hutchinson, I. V. (1999). The role of transforming factor-β in transplant rejection. *Transpl. Proc.* **31** (Suppl. 7A), 9S–13S.
66. Zhang, J.-G., Walmsley, M. W., Moy, J. V., Cunningham, A. C., Talbot, D., Dark, J. H., and Kirby, J. A. (1998). Differential effects of cyclosporin A and tacrolimus on the production of TGF-β: Implications for the development of obliterative bronchiolitis after lung transplantation. *Transpl. Int.* **11** (Suppl. 1), S325–S327.
67. Kahan, B. D., Podbielsky, J., Napoli, K. L., Katz, S. M., Meier-Kriesch, H.-U., and Von Buren, C. T. (1998). Immunosuppressive effects and safety of a sirolimus/cyclosporine combination regimen for renal transplantation. *Transplantation* **66**, 1041–1046.
68. Groth, C. G., Bäckman, L., Morales, J.-M., Calne, R., Kreis, H., Lang, P., Touraine, J.-L., Claesson, K., Campistol, J. M., Duran, D., Warmner, L., Brattström, C., and Charpentier, B. (1999). Sirolimus (rapamycin)-based therapy in human renal transplantation. *Transplantation* **67**, 1036–1042.
69. Sehgal, S. N. (1998). Rapamune® (RAPA, rapamycin, sirolimus): Mechanism of action immunosuppressive effect results from blockade of signal transduction and inhibition of cell cycle progression. *Clin. Biochem.* **31**, 335–340.
70. Lekto, E., Bhol, K., Pinar, V., Foster, V. P., and Ahmed, A. R. (1999). Tacrolimus (FK506). *Ann. Allergy Asthma Immunol.* **83**, 179–189.
71. Oddis, C. V., Sciurba, F. C., Elmagd, K. A., and Starzl, T. E. (1999). Tacrolimus in refractory polymyositis with interstitial lung disease. *Lancet* **353**, 1762–1763.
72. Sakuma, S., Kato Y., Nishigaki, F., Sasakawa, T., Magari, K., Miyata, S., Ohkubo, Y., and Goto, T. (2000). FK506 potently inhibits T cell activation induced TNF-α and IL-1β production *in vitro* by human peripheral blood mononuclear cells. *Br. J. Pharmacol.* **130**, 1655–1663.
73. Furst, D. E., Sherrer, Y., Fleischmann, R. M., Block, J., Rutstein, J., and Stamler, D. (1999). Efficacy of FK506 in rheumatoid arthritis (RA). A 6 month dose-ranging study in RA patients failing methotrexate (MTX). Presented at 1999 annual scientific meeting, American College of Rheumatology, Boston. Internal medicine (in submission).
74. Kondo, H., Irimaziri, S., Sugawara, S., Hashimoto, H., Uchida, S., Hara, M., and Abe, T., for the Japanese FK506 RA Study Group (2000). Efficacy and safety of FK506 in rheumatoid arthritis (RA): A 16 week, double blind, randomized study. Presented at 64th ACR (Oct. 29–Nov. 2, Philadelphia).
75. Gremillion, R. B., Poserver, J. O., Manek, N., West, J. P., and Van Vollenhoven, R. F. (1999). Tacrolimus (FK506) in the treatment of severe, refractory rheumatoid arthritis: Initial experience in 12 patients. *J. Rheumatol.* **26**, 2332–2336.
76. Fellermann, K., Ludwing, D., Stahl, M., David-Walek, T., and Stange, E. F. (1998). Steroid-unresponsive acute attacks of inflammatory bowel disease: Immunomodulation by tacrolimus (FK506). *Am. J. Gastroenterol.* **93**, 1860–1866.

77. Lowry, P. W., Weaver, A. L., Tremaine, W. J., and Sandborn, W. J. (1999). Combination therapy with oral tacrolimus (FK506) and azathioprine or 6-mercaptopurine for treatment-refractory Crohn's disease perianal fistulae. *Inflam. Bowel Dis.* **5**, 239–245.
78. Bousvoros, A., Kirschner, B., Werlin, S., Balint, J., Kendall, R., Griffiths, A., and Leichtner, A. M. (1997). Oral tacrolimus treatment of severe colitis in children. *Gastroenterology* **112** (Suppl. A), 491.
79. Wrone-Smith, T., and Nickoloff, B. J. (1996). Dermal injection of immunocytes induces psoriasis. *J. Clin. Invest.* **98**, 1878–1887.
80. Schulz, B. S., Michel, G., Wagner, S., Süss, R., Beets, A., Peter, R. U., Kemény, L., and Ruzicka, T. (1993). Increased expression of epidermal IL-8 receptor in psoriasis. *J. Immunol.* **151**, 4399–4406.
81. Lemster, B. H., Carroll, P. B., Rilo, H. R., Johnson, N., Nikaein, A., and Thomson, A. W. (1995). IL-8/IL-8 receptor expression in psoriasis and the response to systemic tacrolimus (FK506) therapy. *Clin. Exp. Immunol.* **99**, 148–154.
82. Michel, G., Auer, H., Kemény, L., Böcking, A., and Ruzicka, T. (1996). Antioncogene P53 and mitogenic cytokine interleukin-8 aberrantly expressed in psoriatic skin are inversely regulated by the antipsoriatic drug tacrolimus (FK506). *Biochem. Pharmacol.* **51**, 1315–1320.
83. European FK506 Multicentre Psoriasis Study Group (1996). Systemic tacrolimus (FK506) is effective for the treatment of psoriasis in a double-blind placebo controlled study. *Arch. Dermatol.* **132**, 419–423.
84. Steiner, J. P., Dawson, T. M., Fotuhi, M., Glatt, C. E., Snowman, A. M., Cohen, N., and Snyder, S. H. (1992). High brain densities of the immunophilin FKBP colocalized with calcineurin. *Nature* **358**, 584–587.
85. Sharkey, J., and Butcher, S. P. (1994). Immunophilins mediate the neuroprotective effects of FK506 in focal cerebral ischemia. *Nature* **371**, 336–339.
86. Dawson, T. M., Steiner, J. P., Dawson, V. L., Dinerman, J. L., Uhl, G. R., and Snyder, S. H. (1993). Immunosuppressant FK506 enhances phosphorylation of nitric oxide synthase and protects against glutamate neurotoxicity. *Proc. Natl. Acad. Sci. USA* **90**, 9808–9812.
87. Sabatini, D. M., Lai, M. M., and Snyder, S. H. (1997). Neural roles of immunophilins and their ligands. *Mol. Neurobiol.* **15**, 223–239.
88. Wang, H.-G., Pathan, N., Ethell, I. M., Krajewsky, S., Yamaguchi, Y., Shibasaki, F., McKeon, F., Bobo, T., Franke, T. F., and Reed, J. C. (1999). Ca^{2+}-induced apoptosis through calcineurin dephosphorylation of BAD. *Science* **284**, 339–343.
89. Stoll, G., Jander, S., and Schrocter, M. (1998). Inflammation and glial responses in ischemic brain lesions. *Progr. Neurobiol.* **56**, 149–171.
90. Wang, M.-S., Zeleny-Pooley, M., and Gold, B. G. (1997). Comparative dose-dependence study of FK506 and cyclosporin A on the rate of axonal regeneration in rat sciatic nerve. *J. Pharmacol. Exp. Ther.* **282**, 1084–1093.
91. Bieber, T. (1998). Topical tacrolimus (FK506): A new milestone in the management of atopic dermatitis. *J. Allergy Clin. Immunol.* **102**, 555–557.
92. Juhlin, L. (1967). Localization and content of histamine in normal and diseased skin. *Acta Derm.Venereol.* **47**, 383–391.
93. Mihm, Jr., M. C., Soter, N. A., Dvorak, H. F., and Austen, K. F. (1976). The structure of normal skin and the morphology of atopic eczema. *J. Invest. Dermatol.* **67**, 305–312.
94. Zachary, C. B., Allen, M. H., and Macdonald, D. M. (1985). *In situ* quantification of T-lymphocyte subsets and Langerhans cells in the inflammatory infiltrate of atopic eczema. *Br. J. Dermatol.* **112**, 149.
95. Leiferman, K. M., Ackerman, S. J., Sampson, H. A., Haugen, H. S., Venecie, P. Y., and Gleich, G. J. (1985). Dermal deposition of eosinophil-glanular major basic protein in atopic dermatitis. *N. Engl. J. Med.* **313**, 282–285.

96. Grewe, M., Bruijnzeel-Koomen, C. A. F. M., Schöpf, E., Thepen, T., Langeveld-Wildschut, A. G., Ruzicka, T., and Krutmann, J. (1998). A role for Th1 and Th2 cells in the immunopathogenesis of atopic dermatitis. *Immunol. Today* **19**, 359–361.
97. Homey, B., Assmann, T., Vohr, H.-W., Ulrich, P., Lauerma, A. I., Ruzicka, T., Lehmann, P., and Schuppe, H.-C. (1998). Topical FK506 suppresses cytokine and costimulatory molecule expression in epidermal and local draining lymph node cells during primary immune response. *J. Immunol.* **160**(11), 5331–5340.
98. Sakuma, S., Higashi, Y., Sato N., Sasakawa, T., Sengoku, T., Ohkubo, Y., Amaya, T., and Goto, T. (2001). Tacrolimus suppressed the production of cytokines involved in atopic dermatitis by direct stimulation of human PBMC system. *Int. Immunopharmacol.* **1**, 1219–1226.
99. Rao, A. (1994). NF-ATp: A transcription factor required for the co-ordinate induction of several cytokine genes. *Immunol. Today* **15**, 274–281.
100. De Paulis, A., Cirillo, R., Ciccarelli, A., De Crescenzo, G., Oriente, A., and Marone, G. (1991). Characterization of the anti-imflammatory effect of FK506 on human mast cells. *J. Immunol.* **147**, 4278–4285.
101. Tachimoto, H., Ebisawa, M., Kimata, M., Mori, K., Iikura, Y., and Saito, H. (1997). Effect of immunosuppressive drugs on cytokine production from cultured human mast cells and basophils. *J. Allergy Clin. Immunol.* **99** (Part 2), S123.
102. De Paulis, A., Stellato, C., Cirillo, R., Ciccarelli, A., Oriente, A., and Marone, G. (1992). Anti-inflammatory effect of FK506 on human skin mast cells. *J. Invest. Dermatol.* **99**, 723–728.
103. Hom, J. T., and Estridge, T. (1993). FK506 and rapamycin modulate the functional activities of human peripheral blood eosinophils. *Clin. Immunol. Immunopathol.* **68**, 293–300.
104. Yamashita, N., Akimoto, Y., Minoguchi, K., Sekine, K., Nakajima, M., Okano, Y., Ohta, K., and Sakane, T. (1999). Inhibitory effect of pemirolast potassium and FK506 on degranulation and IL-8 production of eosinophils. *Allergol. Int.* **48**, 37–42.
105. Bieber, T. (1997). Fc εRI-expressing antigen-presenting cells: New players in the atopic game. *Immunol. Today* **18**(7), 311–313.
106. Wollenberg, A., Sharma, S., Regele, D., Geiger, E., Haberstok, J., and Bieber, T. (2001). Topical tacrolimus (FK506) leads to profound phenotypic and functional alterations of epidermal antigen presenting dendritic cells in atopic dermatitis. *J. Allergy Clin. Immunol.* **107**, 519–525.
107. Panhans-Gross, A., Novak, N., Kraft, S., and Bieber, T. (2001). Human epidermal Langerhans cells are targets for the immunosuppressive macrolide tacrolimus (FK506). *J. Allergy Clin. Immunol.* **107**, 345–352.
108. Panhans-Gross, A., and Bieber, T. (1998). Differential regulation of the phenotypic and functional maturation of human epidermal Langerhans cells by FK506 (tacrolimus) and bethamethasone valerate *in vitro*. *J. Invest. Dermatol.* **110**, 630.
109. Hiroi, J., Sengoku, T., Morita, K., Kishi, S., Sato, S., Ogawa, T., Tsuduki, M., Matsuda, H., Wada, A., and Esaki, K. (1998). Effect of tacrolimus hydrate (FK506) ointment on spontaneous dermatitis in NC/Nga mice. *Jpn. J. Pharmacol.* **76**, 175–183.
110. Nakagawa, H., Etoh, T., Yokota, Y., Ikeda, F., Hatano, K., Teratani, N., Shimomura, K., Mine, Y., and Amaya, T. (1966). Tacrolimus has antifungal activities against *Malassezia furfur* isolated from healthy adults and patients with atopic dermatitis. *Clin. Drug. Invest.* **12**, 244–250.
111. Liu, J. (1993). FK506 and cyclosporin, molecular probes for studying intracellular signal transduction. *Immunol. Today* **14**, 290–295.
112. Stellato, C., de Paulis, A., Ciccarelli, A., Cirillo, R., Patella, V., Casolaro, V., and Marone, G. (1992). Anti-inflammatory effect of cyclosporin A on human skin mast cells. *J. Invest. Dermatol.* **98**, 800–804.
113. Hultsch, T., Albers, M. W., Schreiber, S. L., and Hohman, R. J. (1991). Immunophiline ligands demonstrate common features of signal transduction leading to exocytosis or transcription. *Proc. Natl. Acad. Sci. USA* **88**, 6229–6233.

114. Camp, R. D., Reitamo, S., Friedmann, P. S., and Heule, F. (1993). Cyclosporin A in severe, therapy-resistant atopic dermatitis: Report of an international workshop. *Br. J. Dermatol.* **129**, 217–220.
115. Salek, M. A., Finlay, A. Y., Luscombe, D. K., Allen, B. R., Berth-Jones, J., Kahn, G. K., Marks, R., Motley, R. J., Ross, J. S., and Sowden, J. M. (1993). Cyclosporin greatly improves the quality of life of adults with severe atopic dermatitis: A randomized, double-blind placebo controlled trial. *Br. J. Dermatol.* **129**, 422–430.
116. Cohan, V. L., Undem, B. J., Fox, C. C., Adkinson, Jr., N. F., Lichtenstein, L. M., and Schleimer, R. P. (1989). Dexamethasone does not inhibit the release of mediators from human mast cells residing in airway, intestine or skin. *Am. Rev. Respir. Dis.* **140**, 951–954.
117. Ebisawa, M., Tachimoto, H., Iikura, Y., and Saito, H. (1997). Corticosteroids inhibit cytokine production but not histamine release from cultured human mast cells. *J. Allergy Clin. Immunol.* **99** (Part 2), S123.
118. Nakagawa, H., and the Japanese FK506 Ointment Study Group (1998). Comparative study of 0.1% FK506 (tacrolimus) ointment vs 0.12% betamethasone valerate ointment in atopic dermatitis (trunk and extremities lesions). Presented at Clinical Dermatology 2000 (June 18–20, Singapore). Abstr. No. 271.
119. Nakagawa, H., and the Japanese FK506 Ointment Study Group (1998). Comparative study of FK506 (tacrolimus) ointment vs alclometasone dipropionate ointment in atopic dermatitis (face and neck lesions). *J. Invest. Dermatol.* **110**(4), 683.
120. Kawashima, M., and the Japanese FK506 Ointment Study Group (2000). Long-term treatment with FK506 (tacrolimus) ointment in patients with atopic dermatitis—analysis at the time of 2-year observation. Presented at Clinical Dermatology 2000 (May 17–20, Vienna). Abstr. No. 235.
121. Hanifin, J. M., Ling, M. R., Langley, R., Breneman, D., and Rafal, E., and the Tacrolimus Ointment Study Group (2001). Tacrolimus ointment for the treatment of atopic dermatitis in adult patients: Part 1, Efficacy. *J. Am. Acad. Dermatol.* **44**, S28–S38.
122. Drake, L., Prendergast, M., Maher, R., Breneman, D., Korman, N., Satoi, Y., Beusterien, K. M., and Lawrence, I. (2001). The impact of tacrolimus ointment on health related quality of life of adult and pediatric patients with atopic dermatitis. *J. Am. Acad. Dermatol.* **44**, S65–S72.
123. Soter, N. A., Fleischer, A. B., Webster, G. F., Monroe, E., and Lawrence, I., and the Tacrolimus Ointment Study Group (2001). Tacrolimus ointment for the treatment of atopic dermatitis in adult patients: Part II, Safety. *J. Am. Acad. Dermatol.* **44**, S39–S46.
124. Kang, S., Lucky, A. W., Pariser, D., Lawrence, I., and Hanifin, J. M., and the Tacrolimus Ointment Study Group (2001). Long-term safety and efficacy of tacrolimus for the treatment of atopic dermatitis in children. *J. Am. Acad. Dermatol.* **44**, S58–S64.
125. Paller, A., Eichefield, L. F., Leung, D. Y. M., Stewart, D., and Appell, M., and the Tacrolimus Ointment Study Group (2001). A 12-week study of tacrolimus ointment for the treatment of atopic dermatitis in pediatric patients. *J. Am. Acad. Dermatol.* **44**, S47–S57.
126. Ruzicka, T., Bieber, T., Schöpf, E., Rubins, A., Dobozy, A., Bos, J. D., Jablonska, S., Ahmed, I., Thestrup-Pedersen, K., Daniel, F., Finzi, A., and Reitamo, S. (1997). A short-term trial of tacrolimus ointment for atopic dermatitis. *N. Engl. J. Med.* **337**, 816–821.
127. Reitamo, S., Rissanen, J., Remitz, A., Granlund, H., Erkko, P., Elg, P., Autio, P., and Lauerma, A. I. (1998). Tacrolimus ointment does not affect collagen synthesis: Results of a single-center randomized trial. *J. Invest. Dermatol.* **111**, 396–398.
128. Reitamo, S., Wollenberg, A., Schöpf, E., Perrot, J. L., Marks, R., Ruzicka, T., Christophers, E., Kapp, A., Lahfa, M., Rubins, A., Jablonska, S., and Rustin, M., for the European Tacrolimus Ointment Study Group (2000). Safety and efficacy of 1 year of tacrolimus ointment monotherapy in adults with atopic dermatitis. *Arch. Dermatol.* **136**, 999–1006.
129. Ruzicka, T., Assmann, T., and Homey, B. (1999). Tacrolimus, the drug for the turn of the millenium? *Arch. Dermatol.* **135**, 574–580.

130. Richter-Hintz, D., Schuppe, H. C., Homey, B., and Ruzicka, T. (2000). Topical tacrolimus (FK506) is effective in the treatment of pyoderma gangrenosum. *J. Am. Acad. Dermatol.* **42**, 304–305.
131. Schuppe, H.-C., Richter-Hintz, D., Stierle, H. E., Homey, B., Ruzicka, T., and Lehman, P. (2000). Topical tacrolimus for recalcitrant leg ulcer in rheumatoid arthritis. *Rheumatology* **39**, 105–106.
132. Casson, D. H., Eltumi, M., Tomlin, S., Walker, J. A., and Murch, S. H. (2000). Topical tacrolimus may be effective in the treatment of oral and perineal Crohn's disease. *Gut* **47**, 436–440.
133. Ormerod, A. D., Copeland, P., Shah, S. A. A., and Winfield, A. (1999). Penetration, safety and efficacy of the topical immunosuppressive sirolimus in psoriasis. *Br. J. Dermatol.* **141**, 975.
134. Reitamo, S., Spuls, P., Sassolas, B., Lahfa, M., Claudy, A., and Griffith, C. (1999). A double-blind study in patients with severe psoriasis to assess the clinical activity and safety of rapamycin (sirolimus) alone or in association with a reduced dose of cyclosporin. *Br. J. Dermatol.* **141**, 978–979.

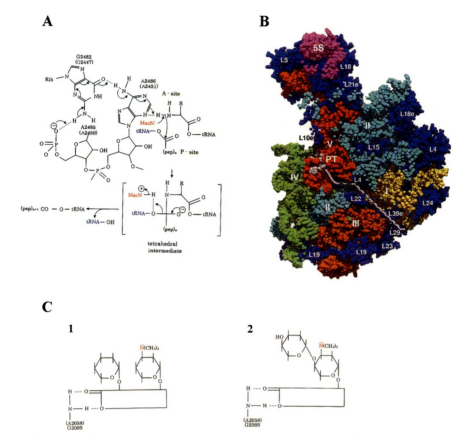

Chapter 10, Fig. 3. Hypothetical mode of action of a macrolide antibiotic that is present near peptidyltransferase (PT). (A) On the basis of the peptidyltransferase reaction mechanism proposed by Nissen *et al.* [27] and Muth *et al.* [28], a putative mode for a 14- or 16-membered macrolide representative of the antibiotics (red) is shown. The N linked to the C3′ position of macrolide antibiotics (red) could accept a proton more easily than the N3 of A2486 (A2451 for *E. coli*) base could. Consequently, the drugs appear to inhibit translocation or peptide bond formation. See the text for details. (B) The 50S subunit from *H. marismortui* has been cut in half, roughly bisecting its central protuberance and its peptide tunnel along the entire length. Two halves have been opened like the pages of a book. The tunnel surface is shown with backbone atoms of the RNA color coded by domain: domains I, yellow; II, light blue; III, orange; IV, green; V, light red; and 5S, pink. Proteins (indicated by L) are blue. [From Nissen, P., Hansen, J., Ban, N., Moore, P., and Steitz, T. A. (2000). *Science* **289**, 927; reprinted by permission of the American Association for the Advancement of Science.] (C) Schematic structures of representative 14- and 16-membered macrolide antibiotics. (C1) erythromycin and (C2) leucomycin. The N (red) of dimethyl amino sugar, bearing a lone pair of electrons (indicated by a colon, :), corresponds to that in the MacN of macrolide antibiotics previously described in part A. Rectangular forms represent lactone moieties. Numbers and those in parentheses indicate the base positions in 23S rRNA from *H. marismortui* and *E. coli*, respectively (see the text) [26, 27].

Chapter 10

Mode of Action and Resistance Mechanisms of Antimicrobial Macrolides

YOSHINORI NAKAJIMA

Division of Microbiology
Hokkaido College of Pharmacy
Hokkaido, Japan

I. Introduction 453
II. Mode of Action of Macrolide Antibiotics... 454
   A. Inhibition of *in Vitro* Protein Synthesis 454
   B. Binding of Macrolide Antibiotics to Ribosomes 455
   C. Effect of Macrolides on Polypeptide Synthesis 459
   D. Interaction of Macrolides with Peptidyltransferase 468
III. Mechanisms of Resistance to Antimicrobial Macrolides 472
   A. General Concept 472
   B. Resistance in Clinical Isolates... 473
   C. Alteration of the MLS Target Site 476
   D. Decreased Macrolide Accumulation.. 479
   E. Translational and Transcriptional Attenuation 483
   F. Enzymatic Inactivation of Macrolide Antibiotics 484
IV. Important Developments in Macrolide Antibiotics 485
V. Concluding Remarks... 486
VI. Addendum.. 487
   References 488

I. Introduction

Macrolide antibiotics, including analogues synthesized recently, each consist of a large lactone ring (aglicone of 12- to 16-membered rings) to which one or more sugars (which can be amino sugars, non-nitrogenous sugars or both) are linked [1–3]. The first macrolide antibiotic discovered was picromycin, a 14-membered-lactone-ring macrolide (isolated in 1950). Successively, it was later found that erythromycin (1952) and oleandomycin (1954), which are 14-membered macrolides, as well as carbomycin (1952), leucomycin (1953), and spiramycin (1954), which are 16-membered macrolides, have antibiotic activity.

Macrolides, including semisynthetic derivatives such as roxithromycin, clarithromycin, azithromycin, and telithromycin (formerly HMR3647), usually exhibit potent activity against most gram-positive bacteria such as *Staphylococcus* spp. and *Streptococcus* spp. but exhibit various degrees of activity against gram-negative bacteria such as *Haemophilus, Neisseria, Legionella*, and *Mycoplasma* spp. [3].

In a broad sense, the term *macrolides* is used in the same way as the term *MLS antibiotics*, which include macrolide, lincosamide, and type B streptogramin antibiotics. A single group of macrolide antibiotics is hereafter referred to as *macrolide antibiotics*.

It is widely accepted that MLS antibiotics inhibit protein synthesis by binding to closely related sites on the 50S subunit of the 70S ribosome of bacteria [4], despite being structurally different from each other (see Figs. 1 and 2 in a later section). That is the reason why, when inducible resistant *Staphylococcus aureus* cells are exposed to a low concentration of the drug (0.05 μg erythromycin/ml ~ 6.8×10^{-8} M), they show resistance against not only erythromycin but also other macrolide antibiotics as well as lincosamide and type B streptogramin antibiotics. Erythromycin has been widely used and has been the object of extensive molecular and biological studies.

MLS antibiotics have a narrow spectrum of activity against bacteria that include gram-positive cocci (including staphylococci, streptococci, and enterococci) and bacilli and against gram-negative cocci. After the use of β-lactam antibiotics, macrolide antibiotics are often used as a safe remedy against infection by one of these bacteria, because they do not have severe adverse effects. Gram-negative bacilli such as *Pseudomonas aeruginosa* and *Escherichia coli*, which occasionally cause opportunistic infection, are usually intrinsically resistant to MLS antibiotics. However, certain gram-negative bacteria, including *Haemophilus, Bordetella, Legionella, Campylobacter, Chlamydia*, and *Treponema* spp., are susceptible to MLS antibiotics. Antibiotics such as β-lactams and quinolones are impotent against *Mycoplasma* spp., whereas macrolide antibiotics such as erythromycin are effective against the bacteria.

In this chapter, the molecular–biological mode of action of macrolide antibiotics and the biochemical and genetic mechanisms of resistance to MLS antibiotics are reviewed. Based on a recent X-ray crystallographic study on a 50S ribosomal subunit from *Haloarcula marismortui* and the finding of intracellular macrolide accumulation, the mode of action from the viewpoint of a new hypothetical concept, deposition binding, and mechanisms of drug resistance in clinically isolated bacteria are discussed. In addition, recent major developments in macrolide antibiotics are briefly described.

II. Mode of Action of Macrolide Antibiotics

A. Inhibition of *in Vitro* Protein Synthesis

At a high concentration (1 mg/ml, equivalent to 1.4×10^{-3} M), erythromycin inhibits the growth of *E. coli* strain K cells and hardly interferes with the synthesis of nucleic acids but greatly interferes with the production of proteins in cells [5].

It has been found that macrolide-susceptible bacteria such as *Bacillus subtilis* and *S. aureus* accumulate about 10 times more erythromycin than do resistant bacteria, namely, clinically isolated *S. aureus* [6], spontaneously mutated *B. subtilis* [7], and intrinsically resistant *E. coli* [8].

Observation of the inhibition of protein synthesis [5] and increased macrolide accumulation [6, 9] due to the binding of the antibiotics to ribosome prompted researchers to investigate more detailed mechanisms of the actions of drugs. Many well-organized reviews on the action mechanisms of the drugs have been published from the1960s until the 1990s [2, 3, 10–19].

The decreased accumulation of macrolide antibiotics in resistant gram-positive bacteria is discussed in detail in Section III.

B. Binding of Macrolide Antibiotics to Ribosomes

Despite the fact that macrolide antibiotics have displayed great ability as growth inhibitors against gram-positive bacteria, cell-free extracts from a gram-negative bacterium *E. coli* strain have been used in almost all *in vitro* studies on the mode of actions of the drugs. In the 1960s, extensive studies on cell-free protein synthesis were carried out to determine which process of the synthesis is inhibited by macrolide antibiotics. Cell-free systems for protein synthesis contain ribosome, natural or synthetic m-RNA such as polyuridylic and polyadenylic acids [poly (U) and poly (A), respectively], tRNAs, enzymes, amino acids, an ATP-generating system, GTP, buffer, certain inorganic salts, and some organic compounds. Moreover, fruitful discoveries in bacterial genetics have inspired investigators to engage in numerous studies in the molecular biological field [20, 21].

On the other hand, studies on the structure and function of ribosomes have extensively been performed [22–25]. Quite recently, Ban *et al.* [26] and Nissen *et al.* [27] have published reports on the first observation that peptidyltransferase activity in 50S ribosomal subunit is carried out by a 23S ribosomal RNA, the so-called ribozyme, present in the same subunit. The 50S ribosomal subunit was obtained from a gram-negative halophilic Archaean bacterium, *H. marismortui* or from *E. coli* [28]. Moreover, Wimberly *et al.* [29] and Carter *et al.* [30] reported that the 30S ribosomal subunit originates from a gram-negative heat-loving bacterium, *Thermus thermophilus*.

Erythromycin selectively binds to a single site on the 50S ribosomal subunit of 70S bacterial ribosome in the presence of Mg^{2+} and K^+ ions but does not bind to mammalian 80S ribosome [4, 31].

Such selective binding occurs with 1:1 stoichiometry [32]. Erythromycin-binding affinities to *E. coli* and *S. aureus* ribosomes give apparent dissociation constant K_d values of 1×10^{-7} and 1×10^{-8} M, respectively [33, 34].

Despite having functional differences, all macrolide antibiotics appear to compete mutually for binding to ribosome, suggesting the presence of almost the same site or an overlapping site of interaction, since they are mutually exclusive in their binding to ribosome [35–38].

It is well known that nonprotonated molecules of a macrolide antibiotic are the active species of an inhibitor of protein synthesis [39]. In our experimental condition (pH 7.6), nonprotonated molecules of rokitamycin can be calculated to be about 10 times as numerous as those of erythromycin from pK_a values of 7.6 and 8.7, respectively (Table I) [34]. Despite having the advantage in the number of nonprotonated molecules, the ratio of rokitamycin bound per ribosome (41%) is only about half as much as that of bound erythromycin (76%), when the molecules of the antibiotics each are incubated with the same amount of ribosome molecules. This suggests that rokitamycin's mode of binding to ribosomes may be noticeably different from that of erythromycin.

For example, it is evident that antibiotic molecules are distributed randomly among ribosomes if the ribosomes are not clumped. Such an event would be expected to follow the Poisson distribution: $Pn\,(m) = m^n e^{-m}/n!$, in which P is the probability that any antibiotic molecule (n) at a binding site of the ribosome will be detected, and m is the mean number of drug molecules bound per ribosome. When the ratio of erythromycin molecules to ribosomal particles is one to one, it follows from the Poisson series that the proportion of ribosomes:

binding to 0 molecular species of the antibiotic = e^{-1} = 0.368
binding to 1 molecular species of the antibiotic = e^{-1} = 0.368
binding to 2 molecular species of the antibiotic = $e^{-1}/2!$ = 0.184
binding to 3 molecular species of the antibiotic = $e^{-1}/3!$ = 0.061
...

binding to n (more than four) molecular species of the antibiotic = $e^{-1}/n! \sim 0$.

As shown in this example, even if the ratio between drug and ribosomemolecules stands at one to one (344 pmol/ml), only about 63% of erythromycin molecules can be expected to bind to any ribosome particles present in a reaction mixture. The percentage of the molecules of erythromycin capable of being bound to ribosome particles agrees approximately with that of our practical result (76%) [34]. The difference between the value obtained from the Poisson distribution (63%) and that from our experimental result (76%) might arise from contamination in that the molecules of ribosome probably congregate with each other to a small extent, forming dimers, trimers, and so on.

Assuming that the volume of a single 50S subunit, observed by means of low-angle and X-ray scattering, is about 3×10^{-18} ml and that the subunit gives a hydration value of 58% in g water/g dry ribosomes [50], the internal-water volume of the subunit is $(3 \times 10^{-18} \times 0.58)$ 1.7×10^{-18} ml per ribosomal subunit.

In the above example, 262 and 158 pmol in 344 pmol (ribosomal subunit/ml) can be bound to erythromycin and rokitamycin, respectively. On the other hand,

TABLE I
Solubility and pK_a Values Representative of Some Macrolide, Lincosamide,
and Streptogramin Type B Antibiotics[a]

Antibiotic	Solubility (mg/ml)	pK_a
14-membered macrolide		
Erythromycin	2	8.6–8.8
Oleandomycin	—[b]	8.5
Megalomicin A	2–5[c]	8.6–8.8[c]
Clarithromycin	0.1–0.5	8.5[d]
Dirithromycin	—	9.0 (in aq. 66% DMF[e])
Telithromycin	0.6[f]	3.0, 5.1, 8.7[f]
ABT-773	0.04	4.2, 8.6[g]
15-membered macrolide		
Azithromycin	0.1	8.1, 8.8[g]
16-membered macrolide		
Mycinamicin I	—	8.6[h]
Spiramycin I	≦1	7.7[h] (in methanol)
Tylosin	5	7.0[h]
Carbomycin A	0.3	6.8
Niddamycin	—	7.2 (in aq. 75% DMF)
Josamycin (leucomycin A$_3$)	≦0.1	6.8 (in aq. 50% ethanol)
Rokitamycin[i]	≦0.1	7.6
Midecamycin A$_1$	—	6.9
YM133[j]	0.0005	6.8 (in aq. 50% ethanol)
Lincosamide		
Lincomycin	≦0.1	7.6
Streptogramin type B		
Mikamycin B	≦0.1[k]	9.0[l] (in aq. 40% methanol)

[a]Budavari, S., O'Neil, M. J., Smith, A., Heckelman, P. E., and Kinneary, J. F. (1996). "The Merck Index: An Encyclopedia of Chemicals, Drugs, and Biologicals," 12th ed. Merck & Co., Whitehouse Station, NJ, was consulted for citing the values indicated in this table unless otherwise stated.

[b]Not known, although it is known that the solubility of mycinamicin I in water is fairly soluble (about 1 mg/ml).

[c]Megalomicin A shows approximately the same water solubility as erythromycin but not megalomicin C$_2$, which has the most minimal water solubility and the greatest inhibitory activity against gram-positive bacteria among the four complexes (A, B, C$_1$, and C$_2$). Complex A would have a pK_a value of at least 8.7, like erythromycin, because the antibiotic contains desosamine in addition to megosamine [40, 41].

[d]The pK_a was obtained from an abstract for a new macrolide TE-031; presented at Japanese Society of Chemotherapy 35th Japanese General Meeting on Chemotherapy. Morioka, Iwate, pp. 5 (in Japanese) [42].

[e]DMF, dimethylformamide.

[f]This value was quoted from "HMR3647: Chemistry and Pharmacy." (2000). *Japan Investigator's Brochure HMR3647*, 7th ed., p. 7. Aventis Pharma Ltd., Tokyo. Telithromycin (formerly HMR3647) has three pK_a values: pK_1 for the pyridinium ring confers 3.0; pK_2 for the imidazolium ring, 5.1; and pK_3 for the D-desosamine, 8.7 [43].

[g]The pK_a values, 4.2 and 8.6, in ABT-773 depend on the N in quinoline and C3'-dimethylamino group, respectively [44]. Those of azithromycin, 8.1 and 8.8, given by the 9a-imido in the lactone ring and C3'-dimethylamino group, respectively, were obtained from "Zithromac: Chemistry and Pharmacy." (2000). *In* "Interviewform," 2nd ed., pp. 11–12. Pfizer Pharmaceuticals, Tokyo.

[h–l]For superscripts h, i, j, k, and l, see Refs. 45, 46, 47, 48, and 49, respectively.

the total water volume present in 344 pmol of 50S ribosome is 3.6×10^{-4} ml. The concentration of erythromycin and rokitamycin in the internal water of the 50S ribosome goes up to 9.6×10^{-4} M (0.7 mg/ml) and 4.4×10^{-4} M (0.36 mg/ml), respectively (Table I). Consequently, erythromycin binds to macrolide-sensitive ribosome in a water-soluble state but rokitamycin in a precipitated state. The macrolide antibiotic, like rokitamycin, is hereafter referred to as an agent capable of causing "deposition binding."

This putative new concept can account for the reason why rokitamycin acts as a bactericidal agent against susceptible cells despite the fact that ^{14}C-labeled rokitamycin ($K_d = 3.4 \times 10^{-8}$ M), a 3″-O-propionyl derivative of leucomycin A_5, has about half as much affinity for the ribosome as erythromycin ($K_d = 1.6 \times 10^{-8}$ M) does. In fact, ^{14}C-labeled rokitamycin is hardly replaced by nonlabeled rokitamycin, when the labeled drug had previously been bound to ribosomes from susceptible S. aureus [34, 51].

However, the question still arises as to why rokitamycin can bind only about half as much as can erythromycin. A possible answer is that, assuming that a single rokitamycin molecule can play two functional roles (namely, as if one molecule of the drug corresponds to two molecules of erythromycin), it may be enough to explain an experimental result [34]. That is why the former drug, a 16-membered-ring drug (except for bisglycosides such as mycinamicin), has a disaccharide-monoglycoside residue [52].

This suggests that one macrolide molecule is able to arrive at a binding site of ribosome and that, after the drug is bound to the site of the ribosome, we can determine, in terms of the water solubility of macrolide antibiotics, whether reversible or cohesive (probably irreversible) binding occurs. In other words, the solubility of an antibiotic depends primarily on the difference between intracellular pH and pK_a values and could determine the extent of the antibiotic's ability to deposit at a 50S ribosomal binding site.

The solubilities of erythromycin, spiramycin, carbomycin A, and rokitamycin are ~2, ~1.5, 0.3, and ~0.1 mg/ml at $20 \pm 5°C$, respectively (Table I) [53]. In addition, the higher the temperature in an aqueous solution of the antibiotic is, the more difficult it is for the drug to be dissolved [42].

Chloramphenicol, a bacteriostatic agent, inhibits growth of gram-positive and gram-negative bacteria. This drug has a water solubility of about 1 mg/ml at 28°C [53] and a probable pK_a value of about 8.7.

In contrast, in S. aureus carrying the erm gene responsible for MLS resistance—the so-called N^6,N^6-dimethylation of a specific adenine residue in 23rRNA, drug accumulation is easily decreased to about one-tenth of that in susceptible cells when they are washed five times with chilled salt solution. That is why a macrolide antibiotic present in their cells can hardly bind to their 50S ribosomes even at the highest permissible concentration of the drug [6, 54].

Studies based on the binding of erythromycin and its derivatives to 50S ribosome from *S. aureus* have revealed that seven hydrogen bonds associate with six adjacent nitrogenous bases (including two pairs of hydrogen bonds, 2'-hydroxyl and 3'-dimethylamino groups, with one of the bases) of the nucleotides (probably in 23S rRNA). The residues capable of forming hydrogen bonds, the 2'-hydroxyl, 3'-dimethylamino, 11-,12-hydroxyl, 9-carbonyl, 3"-methoxy, and 6-hydroxyl groups, are all required to form an erythromycin–ribosome complex [37].

In addition to the seven hydrogen bonds concerned with the interaction between erythromycin and ribosome, another ketone residue (−O−CO−) present in lactone of the antibiotic may be required to bind to ribosome. That is why dimethylation or a mutation at A2058 in 23S rRNA gives rise to the inability of ribosome to bind to the drug, rendering the ribosome resistant to not only macrolide antibiotics but also to lincosamide and type B streptogramin antibiotics (Table II). The inability is likely to be brought about by hydrogen-bond destruction caused by a conformational change due to the mono- or dimethylation at the N6 position of A2058 in 23S rRNA.

This suggests that a ketone residue must be common to all MLS antibiotic structures. This residue might be available for forming a complex between any one of the antibiotic residues and the nitrogenous base of ribosomal RNA; that is, residues: −CO−O− for macrolides (Figs. 1 and 2); −CO−NH− for lincosamides; and −CO−O− or an unspecified ketone residue, for which −CO−NH− or −CO−NR− is required, for type B streptogramins (Fig. 2).

C. Effect of Macrolides on Polypeptide Synthesis

The synthesis of most peptides in living cells depends on the presence of ribosomes, although many biologically active peptides, such as gramicidin S, tyrocidines, and polymixins, all of which are produced as antibiotics, are not synthesized by means of ribosomes.

It has been shown that the first step in ribosome-dependent peptide synthesis is activation of amino acids to form amino acid adenylates. The amino acids are then transferred to RNA present in the soluble extract of the cell, the so-called transfer RNA (tRNA) to which the amino acids become fixed by an ester linkage. These two steps are usually referred to as the formation of aminoacyl-tRNA. The next step, the translation step of codons in messenger RNA (mRNA), which is associated with ribosomes, to provide a polypeptide includes three stages: (1) chain initiation by mutual coordination with initiation factors, (2) chain elongation in aid of elongation factors, and (3) chain termination in support of release factors.

Most macrolide antibiotics are not able to inhibit the formation of aminoacyl-tRNA. Jordan [78], however, has shown that incorporation of ^{14}C-glycine into

TABLE II
MLS Resistance Due to Target Modification, Mediation by *erm* Genes, and Mutation at Several Sites in the Peptidyl Transferase Circle of 23S rRNA Domain V from Major Clinical Isolates

Resistance mechanism (resistant phenotype) Host	Gene[a]	References
I. Target modification (MLS[b])		
Staphylococcus aureus	ermA [erm(A)]	55
	ermB [erm(B)]	56
	ermC [erm(C)]	57, 58
	ermGM [erm(Y)]	59
Streptococcus pyogenes	ermTR [erm(A)]	60
Staphylococcus epidermidis	ermM [erm(C)]	61
Streptococcus sanguis	ermAM [erm(B)]	62
Enterococcus faecalis	ermAMR [erm(B)]	63
Corynebacterium diphtheriae	ermCD [erm(X)]	64, 65
Clostridium perfringens	ermP [erm(B)]	66
	ermQ [erm(Q)]	67
Clostridium difficile	ermZ [erm(B)]	68
Escherichia coli	ermBC [erm(B)]	69
Bacteroides fragilis	ermF [erm(F)]	70
	ermFS [erm(F)]	71
	ermFU [erm(F)]	72
II. Target mutation (MLS or ML)[b]		
Helicobacter pylori	A2058G[c]	73
	A2059G	74
Propionibacteria spp	A2058G	75
	A2059G	
Mycoplasma pneumoniae	A2059G	76

[a] Genetic names in brackets are given according to a new paradigm by assigning two genes of 80% amino acid identity to the same class and same letter designation [77].

[b] MLS, macrolide, lincosamide, and streptogramin type B antibiotics; ML, macrolide and lincosamide antibiotics.

[c] For example, A2058G corresponds to a transition (A→G) mutation at a position related to *E. coli* in 23 rRNA position 2058. Position 2058 corresponds to *H. pylori* position 2142 [52].

soluble RNA, so-called tRNA, from *S. aureus* cells was inhibited by chalcomycin more than protein synthesis. Because the effect of chalcomycin has not been tested in other steps or stages using a cell-free protein synthesizing system, it remains to be seen whether this represents the primary target of chalcomycin.

Macrolide antibiotics neither restrain the binding of aminoacyl-tRNA to ribosomes, nor do they inhibit ribosome-dependent hydrolysis of GTP in the presence

Symbolic macrolide ring system	Typical macrolide antibiotic	Activity (MIC; μg/ml)	
		S. aureus	E. coli
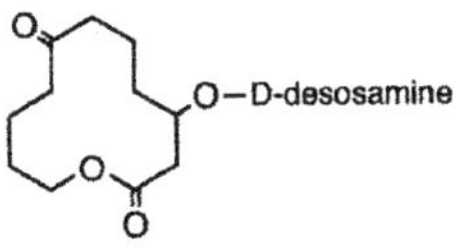 (12-membered ring with O-D-desosamine)	methymycin neomethymycin	4-7	1,875
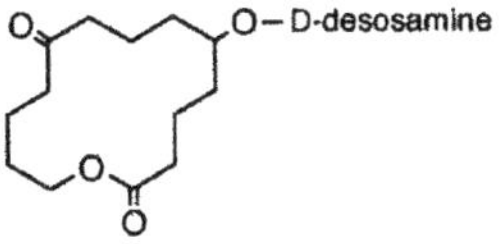 (14-membered ring with O-D-desosamine)	narbomycin picromycin	0.3-10	—[a]
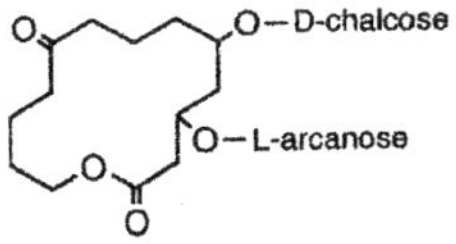 (14-membered ring with O-D-chalcose, O-L-arcanose)	kujimycin A lankamycin	3-6.3	>100
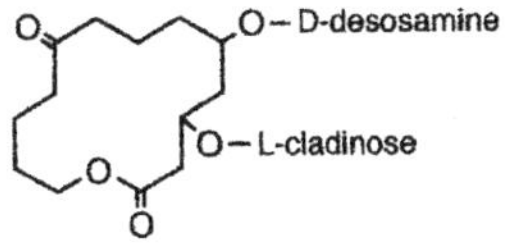 (14-membered ring with O-D-desosamine, O-L-cladinose)	erythromycin A oleandomycin [b] clarithromycin [c] dirithromycin roxithromycin flurithromycin	0.03-3.8	100

Fig. 1. Schematic structures of macrolide antibiotics and their antibiotic activities against bacteria representative of gram-positive *S. aureus* or gram-negative *E. coli*.

[a] No available drugs against indicated bacteria.

[b] Oleandomycin, rosamicin, mycinamicin I, aldgamycin E, and angolamycin contain oleandrose, desosamine, desosamine, aldgarose, and angolosamine in place of cladinose, mycaminose, mycaminose, chalcose, and mycaminose, respectively.

[c] Semisynthetic antibiotics are underlined.

[d] Telithromycin incorporates in its structure as residues: R_1 (CH_3-) and R_2 (pyridinyl-imidazolyl-butyryl-); ABT-773: R_1 (quinolylallyl-) and R_2 (H-).

(Continued)

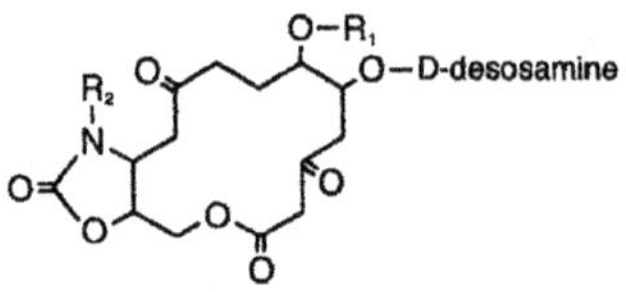 (structure with O-R₁, O-D-desosamine, R₂, N, O)	telithromycin [d] ABT-773	0.05-0.4	6.3-25
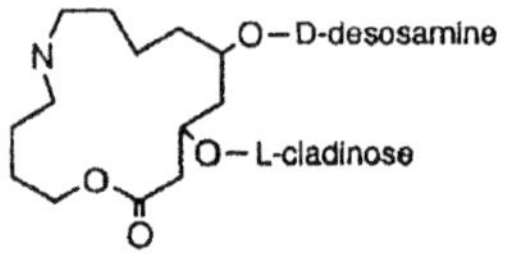 (structure with O-D-desosamine, O-L-cladinose, N)	azithromycin	0.1-0.8	0.5-2
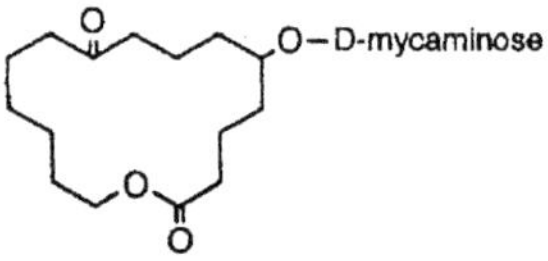 (structure with O-D-mycaminose)	rosamicin [b] cirramycin A₁	0.03-1.6	3-12
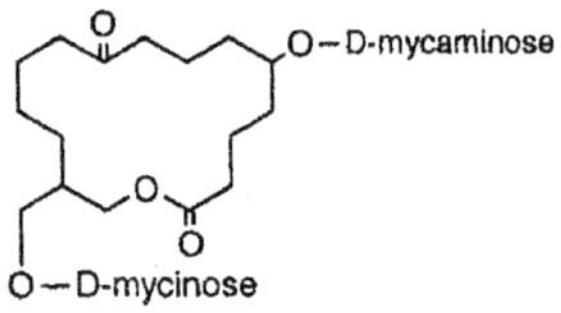 (structure with O-D-mycaminose, O-D-mycinose)	mycinamicin I [b] lactenocin tylosin B	0.05-0.8	16-100
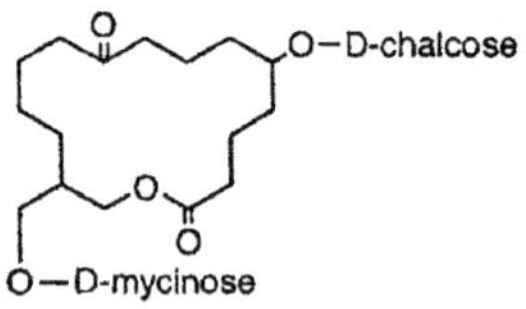 (structure with O-D-chalcose, O-D-mycinose)	neutramycin chalcomycin aldgamycin E [b]	0.2-31	12-50

Fig. 1. (*Continued*)

Structure	Antibiotic		
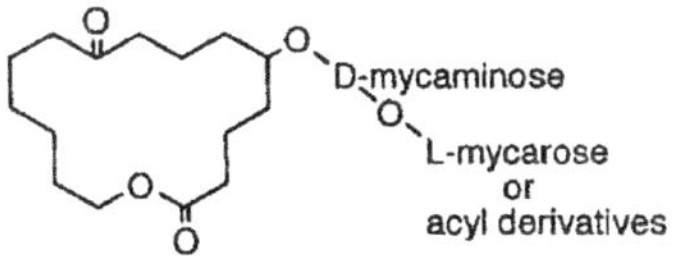 D-mycaminose, L-mycarose or acyl derivatives	carbomycin A niddamycin leucomycin A₁ josamycin maridomycin I deltamycin A₁ midecamycin A₁ miokamycin rokitamycin	0.04-1.6	25-100
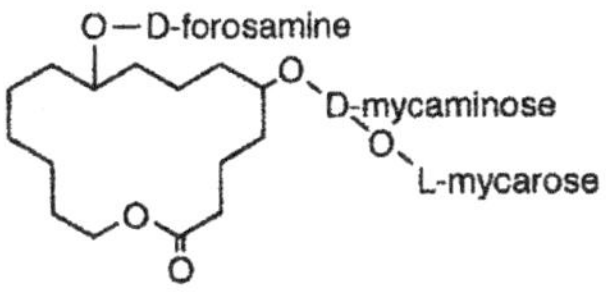 O—D-forosamine, D-mycaminose, L-mycarose	spiramycin I	0.4-1.6	33
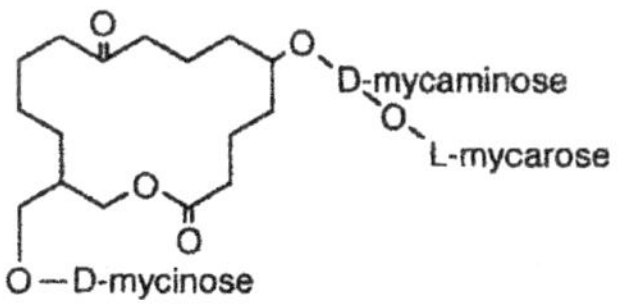 D-mycaminose, L-mycarose, O—D-mycinose	tylosin angolamycin [b] relomycin	0.4-12.5	>100
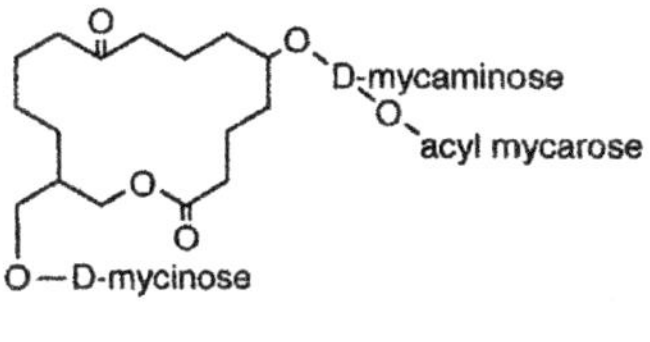 D-mycaminose, acyl mycarose, O—D-mycinose	YM133	0.4-1.6	—[a]

Fig. 1. (*Continued*)

of elongation factor G (EF-G) or with elongation factor (EF-Tu) and aminoacyl-tRNA [79–81].

More detailed information on the mechanism of protein synthesis in eukaryotes as well as on the structures and functions of their ribosomes can be found in several reviews [18, 82–85].

As shown in Fig. 1, macrolide antibiotics have such similar chemical characteristics and physiological properties that the drugs are able to cause them to accumulate into susceptible *S. aureus* cells [6, 8, 9]. They show, however, various

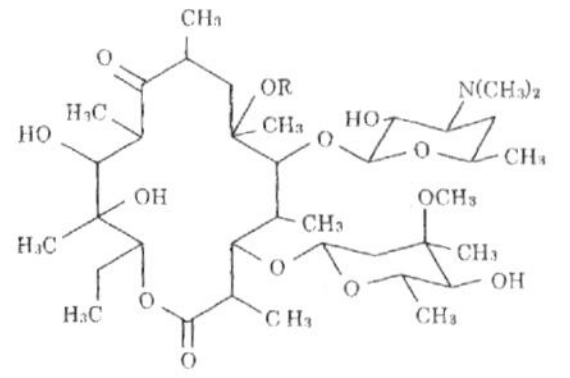

	R
erythromycin A (EM)	H
clarithromycin (CAM)	CH_3

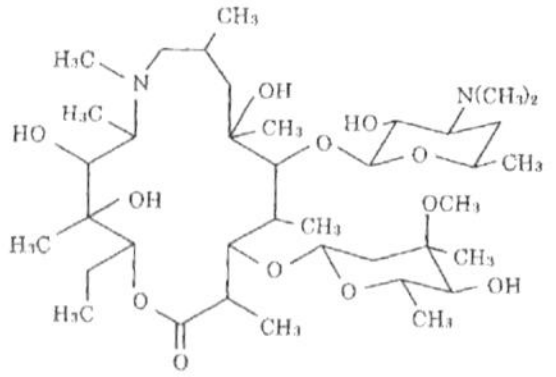

azithromycin (AZM)

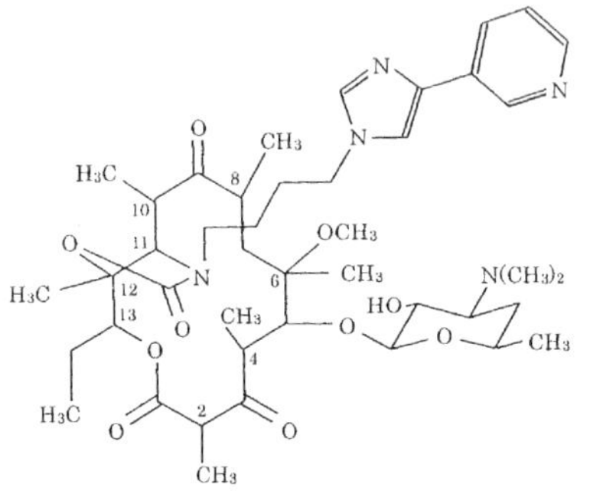

telithromycin (HMR 3647)

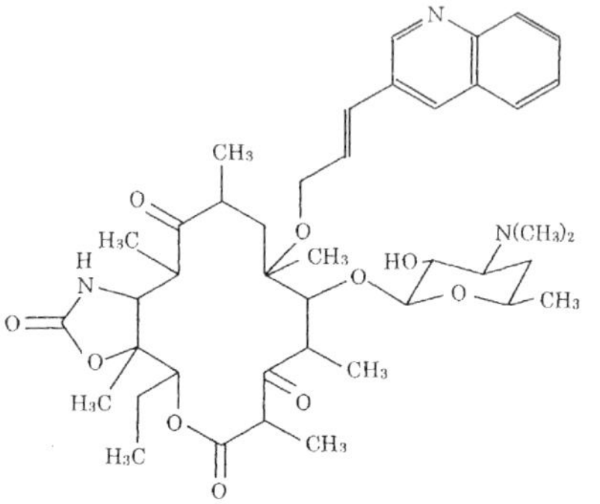

ABT-773

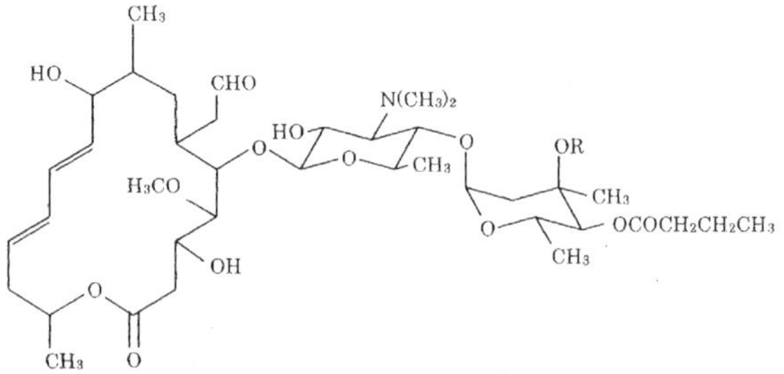

	R
leucomycin A_5 (LM A_5)	H
rokitamycin (RKM)	$COCH_2CH_3$

Fig. 2. Chemical structures of macrolide, lincosamide, and streptogramin type B antibiotics. *Macrolide antibiotics (M):* EM, CAM, AZM, HMR 3647, ABT-773, LM A_5, RKM, TL, and YM133. *Lincosamide antibiotics (L)*: LCM and CLDM. *Streptogramin type B antibiotic (S):* mikamycin B.

(*Continued*)

inhibitory features in susceptibility tests using bacterial cells and in sensitivity tests using cell-free protein-synthesizing systems [3, 15, 16, 86, 87].

In addition to showing various inhibitory features, the 50S ribosomal subunit assembly in bacterial cells (*E. coli*, *S. aureus*, and *B. subtilis*) has recently been found to be inhibited by some macrolides (chiefly 14-membered macrolides).

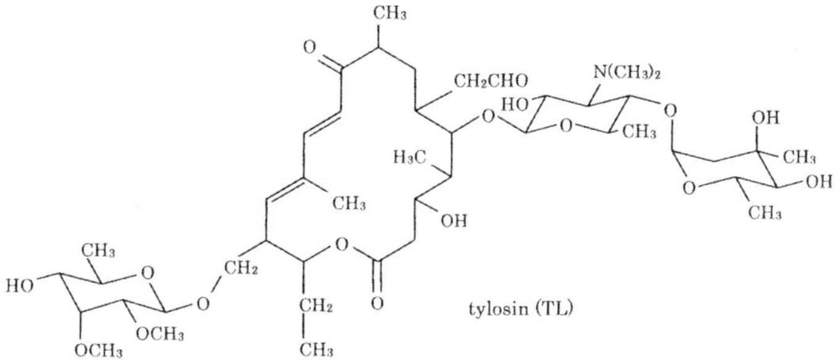

tylosin (TL)

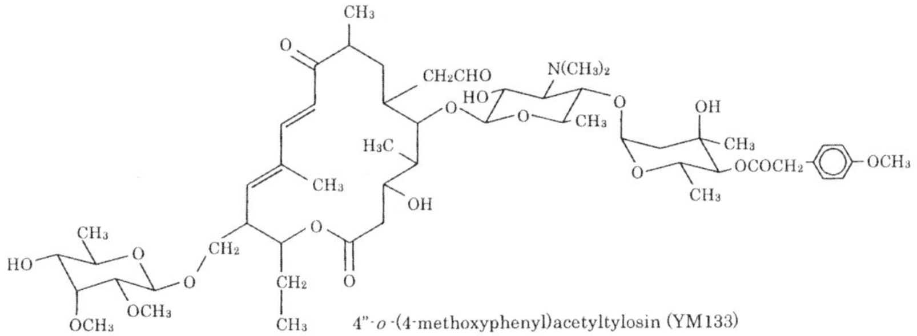

4"-*o*-(4-methoxyphenyl)acetyltylosin (YM133)

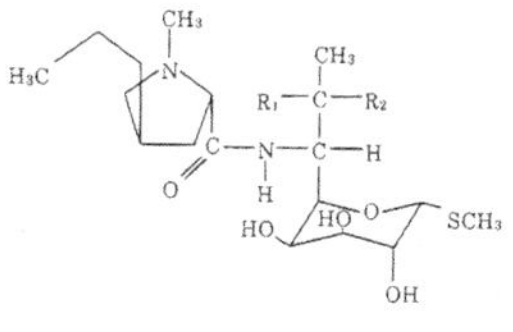

	R_1	R_2
lincomycin (LCM)	OH	H
clindamycin (CLDM)	H	Cl

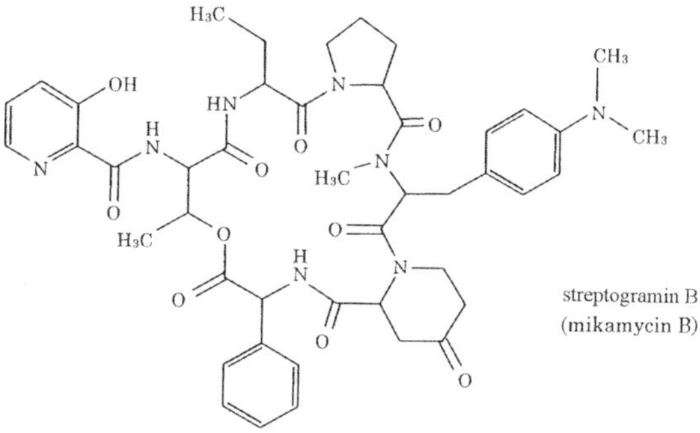

streptogramin B (mikamycin B)

Fig. 2. (*Continued*)

Telithromycin, a derivative of erythromycin, the so-called ketolide antibiotics, blocks the formation of not only 50S but also 30S ribosomal subunits [87]. Although there is no logical explanation for their complex inhibitory action, such inhibition is required for long exposures (from 4.25 to 6 hr) to the ketolide.

Taubman et al. [7] and Mao [8] have observed that the concentration of intracellular erythromycin in drug-susceptible bacteria (*B. subtilis* and *S. aureus*) is 44- to 90-fold greater than that in an extracellular medium (about 1 μg erythromycin/ml ~1.35×10^{-6} M). It seems that the amount of the intracellularly accumulated drug can be determined by the amount of ribosome present in the bacterial cells, because erythromycin binds to ribosome at a ratio of one to one (in the case of *S. aureus*, corresponding to about 2.6×10^{4} ribosomes/cell).

Until about 20 years ago, macrolide antibiotics were usually divided into three practical groups in terms of the extent of their inhibitory action in a peptidyltransferase reaction. Macrolide antibiotics in the first group, such as carbomycin and niddamycin, strongly inhibit the peptidyltransferase reaction in systems employing *N*-acylated amino acids, dipeptides, or oligopeptides as donated substrate [79, 88, 89]. Therefore, these macrolides, so to speak, "inhibit" the peptidyltransferase. Macrolide antibiotics in the second group, such as spiramycin and tylosin, usually inhibit the same reactions as those inhibited by macrolides in the first group, but are not able to effectively inhibit peptidyltransferase activity in the reaction of puromycin with fMet-tRNA^{fMet} [79, 90]. Macrolide antibiotics in the final group, such as erythromycin and oleandomycin, have either an inhibitory or stimulatory effect on the reaction depending on the nature of the transferred substrate (for example, oligolysylpeptide or *N*-acylaminoacyl residues [89, 91] or [88, 89, 92], respectively).

In the late 1980s, macrolide antibiotics were newly divided structurally and functionally into two groups. One group consists of 14-membered-ring macrolides such as erythromycin and including methymycin, a 12-membered-ring macrolide, and azithromycin, a 15-membered-ring macrolide. The other group consists of 16-membered-ring macrolides such as leucomycin, spiramycin, and tylosin.

It is usually accepted that the 16-membered-ring macrolides inhibit peptidyltransferase activity [79, 88, 89], because they inhibit puromycin reaction although poorly. The reaction is used as an assay system for peptidyltransferase activity, because puromycin characteristically interrupts peptide bond formation by virtue of its structural similarity to the 3′ end of aminoacyl-tRNA. Puromycin enters the A site (the so-called aminoacyl site) on the ribosome and is incorporated into either a nascent polypeptide or into *N*-acylaminoacylate, consequently causing premature release of puromycinyl polypeptide or *N*-acylaminoacyl-puromycin from the ribosome.

Evidence also indicates that 16-membered-ring macrolides, as peptidyltransferase inhibitors, hinder the polyuridylic acid-dependent polymerization of phenylalanine, despite the fact that 14-membered-ring macrolides are not able to inhibit polyphenylalanine synthesis. In particular, 16-membered-ring macrolides containing at least one disaccharide-monoglycoside in their structures, such as leucomycin, spiramycin, carbomycin, and tylosin, may cause degradation of polyribosome [93, 94].

On the other hand, the inhibitory effect of erythromycin, a 14-membered-ring macrolide, on such a peptidyltransferase reaction is markedly diminished in terms of the character of a substrate. Erythromycin inhibits poly(A)-dependent polymerization of a transferred substrate such as lysine residue linked to tRNA but not other oligonucleotide-dependent polymerization of an amino acid linked either to tRNA or to oligonucleotides such as CACCA and UACCA. It has been shown that the transfer of *N*-acylaminoacyl residues to puromycin (puromycin reaction) is usually stimulated by erythromycin [88, 89, 95]. Igarashi *et al.* [96] have also confirmed these findings. That is to say, they found that erythromycin inhibits the release of a deacylated tRNA from the P site of ribosome. The release of such a deacylated tRNA from the P site and the translocation of peptidyl-tRNA from the A site to the P site of ribosome occurs concomitantly when EF-G catalyzes the GTP-dependent movement of the ribosome and the codon-anticodon-linked mRNA–peptidyl-tRNA complex.

In this connection, erythromycin, spiramycin, and rokitamycin, at low concentrations of these antibiotics, stimulated polyadenylate-directed polylysine synthesis in systems employing cell-free extract containing ribosome from *S. aureus* and S100 from *E. coli* Q13 [97]. The amounts of macrolides that cause stimulation of polylysine synthesis seem to be about the same or a little less than those of bacterial ribosomes (unpublished data).

Noller and coworkers [98] showed that protein-depleted 23S rRNA had peptidyltransferase activity but it was difficult to eliminate the last traces of protein without losing the transferase activity. Recently, in terms of omission and addition tests using six domains of 23S rRNA synthesized individually by T7 RNA polymerase, Nitta *et al.* [99, 100] demonstrated conclusively that domain V is inevitably required for peptide bond formation.

Menninger and Otto [101] proposed a major inhibitory mechanism common to probably all macrolide antibiotics. In *E. coli* mutants with temperature-sensitive peptidyl-tRNA hydrolase (aminoacyl-tRNA hydrolase; EC 3.1.1.29), they observed that peptidyl-tRNA accumulates at a nonpermissive temperature (40°C) and that the cells die. The accumulation at a high temperature was enhanced when the cells were pretreated with erythromycin, carbomycin, or spiramycin at doses sufficient to inhibit protein synthesis in wild-type cells but not sufficient to kill either mutant or wild-type cells at the permissive temperature (30°C). Based on their observations, they suggested that stimulated dissociation of peptidyl-tRNA from ribosomes is the major mechanism of action of macrolide antibiotics. Their observations agree with recent results showing that a macrolide antibiotic binds to peptidyltransferase in ribosome.

More detailed study of the stimulated dissociation of peptidyl-tRNA from ribosomes and of the bactericidal effect of the semisynthetic antibiotics on susceptible bacteria is needed to determine whether there is a relationship between

D. Interaction of Macrolides with Peptidyltransferase

1. Is Peptidyltransferase Composed of Protein?

As stated in Section II.B, macrolide antibiotics interact with peptidyltransferase by forming hydrogen bonds usually but occasionally hydrophobic bonds present in a 50S ribosomal subunit. That is why noncovalent interactions appear to be the key to the antibiotic flexibility and specificity observed between different macrolides and their derivatives and also probably to the solubility of the drugs in water.

In the 1970s, much attention was focused on the proteins from an *E. coli* 50S ribosomal subunit to try to determine the binding site of macrolide antibiotics. First, Teraoka and Nierhaus [102] found that core particles in which the L15 and L16 proteins were absent from 50S ribosomal subunits incubated with 1.3 *M* LiCl were unable to bind erythromycin or to exhibit peptidyltransferase activity. When the particles were mixed with L15 and L16 proteins, they recovered not only their drug-binding activity but also their enzyme activity. Protein L15 alone retained a weak binding capability ($K_d = 2 \times 10^{-5}$ *M*). Based on their observations, they suggested that L15 and L16 proteins are required for both antibiotic binding and peptidyltransferase activity. Second, macrolide-resistant bacteria were obtained by exposure to either various mutagens such as nitrosoguanidine and ethyl methanesulfonate or a high concentration of antibiotics.

Strains of *E. coli* resistant to erythromycin and other macrolides have been found to undergo mutations concerned with either 50S ribosomal protein L4 encoded by the *rplD* (formerly *eryA*) gene or protein L22 encoded by the *rplV* (formerly *eryB*) gene [103–105]. Ribosomes including L4 proteins in which one of the amino acids of a wild-type L4 protein had been replaced with any other amino acid do not bind erythromycin, and the formation of *N*-acetylphenylalanyl puromycin is markedly inhibited [105].

Although it is not clear that alteration of protein L22 brings about resistance to erythromycin [105], protein L22 from wild-type *H. marismortui*, a species of the genus *Haloarcula* belonging to Archaea, is considered to play an important role in the surface of the nascent polypeptide exit tunnel (Fig. 3B; see also color plate). In addition, the nonglobular region of L22, the long β-hairpin loop, lies between 23S rRNA segments of domains I through VI [26, 27]. Hence, it seems natural that the protein can undergo a mutation that makes it more resistant to erythromycin than do other 50S ribosomal proteins.

Ribosomal proteins are abundant everywhere on the surface of 23S rRNA except in the active site where peptide bond formation occurs and where it

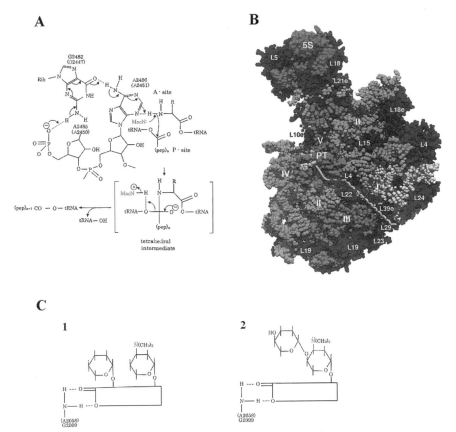

Fig. 3. Hypothetical mode of action of a macrolide antibiotic that is present near peptidyltransferase (PT). (A) On the basis of the peptidyltransferase reaction mechanism proposed by Nissen et al. [27] and Muth et al. [28], a putative mode for a 14- or 16-membered macrolide representative of the antibiotics (red) is shown. The N linked to the C3′ position of macrolide antibiotics (red) could accept a proton more easily than the N3 of A2486 (A2451 for *E. coli*) base could. Consequently, the drugs appear to inhibit translocation or peptide bond formation. See the text for details. (B) The 50S subunit from *H. marismortui* has been cut in half, roughly bisecting its central protuberance and its peptide tunnel along the entire length. Two halves have been opened like the pages of a book. The tunnel surface is shown with backbone atoms of the RNA color coded by domain: domains I, yellow; II, light blue; III, orange; IV, green; V, light red; and 5S, pink. Proteins (indicated by L) are blue. [From Nissen, P., Hansen, J., Ban, N., Moore, P., and Steitz, T. A. (2000). *Science* **289**, 927; reprinted by permission of the American Association for the Advancement of Science.] (C) Schematic structures of representative 14- and 16-membered macrolide antibiotics. (C1) erythromycin and (C2) leucomycin. The N (red) of dimethyl amino sugar, bearing a lone pair of electrons (indicated by a colon, :), corresponds to that in the MacN of macrolide antibiotics previously described in part A. Rectangular forms represent lactone moieties. Numbers and those in parentheses indicate the base positions in 23S rRNA from *H. marismortui* and *E. coli*, respectively (see the text) [26, 27].

makes contact with the 30S ribosomal subunit. Most of the proteins stabilize the structure of 50S ribosome by interacting with several of the six RNA domains, often using uniquely folded extensions that intrude into the 50S subunit's inside [26].

According to Nissen et al. [27], proteins L4, L22, and L39e (the letter e represents a protein in a bacterial 50S ribosomal subunit, a protein that belongs to one of the homologs in eukaryotic 60S ribosomal subunit) have been shown to be present in the polypeptide exit tunnel [average diameter, about 15 Å; length of the tunnel, 100 Å (Fig. 3B)] present in the 50S ribosomal subunit from H. marismortui. Six other proteins (L19, L22, L23, L24, L29, and L31e) are known to be located in the exit area of the polypeptide tunnel [26, 27]. If these proteins are involved in the inhibitory actions of macrolide antibiotics, mutant bacteria resistant to macrolides will develop in the future.

2. Is Peptidyltransferase Composed of RNA?

Ban et al. [26] used X-ray crystallography in a recent study to determine the crystal structure (i.e., to obtain a 2.4-Å atomic resolution map) of the 50S ribosomal subunit from H. marismortui. They found that the subunit includes almost all of the entire chain of 23S rRNA (2711 out of 2923 nucleotides) and 5S rRNA (all 122 nucleotides) as well as 27 of its 31 proteins.

Nissen et al. [27] revealed the atomic structure of the 50S ribosomal subunit from the bacteria and its complexes with two substrate analogs: Yarus analog, CCdA-p-puromycin, or a mini-helix analog of an aa-tRNA, 5′-CCGGCGGGCU-GGUUCAAACCGGCCCGCCGGACC-3′-puromycin. The former analog was used as an A-site substrate analog, but its CCdA moiety is expected to bind at the P site, and the latter analog was used as the imitated acceptor stem of a $tRNA^{tyr}$ so as to bind to the A site. Both substrate analogs bound to the site of conserved rRNA residues in domain V are composed entirely of 23S rRNA. The site is the so-called "peptidyltransferase loop."

Based on their observations, they concluded that the ribosome is ribozyme and directs the catalytic properties of its all-RNA active site. The secondary structures of both 5S and 23S rRNA from H. marismortui are remarkably close to those deduced for them by phylogenetic comparison.

The mechanism of peptide bond synthesis is thought to resemble the reverse of the acylation step in the serine protease, with the base of A2486 (A2451 in E. coli) playing the same general base role as histidine-57 in chymotrypsin. This A is universally conserved within the central loop of domain V [28].

Nissen et al. [27] have suggested that the N3 of A2486 (A2451 in E. coli) residue is most likely to be involved in catalysis, presumably as a general base in the first step. There are two reasons for this supposition. One reason is that base N3 is about 3 Å away from the phosphoramide oxygen of the Yarus analog, which is similar to the carbonyl oxygen of a nascent peptide bond, and base N3 is about

4 Å away from the amide, which corresponds to the amide nitrogen of the nascent peptide bond. The other reason is that there is no other titratable RNA functional group closer than 5 Å to the nascent peptide bond and no other group available to function as a general base.

Similarly, a single conserved A (number 2451) within the loop of domain V in *E. coli* 23S rRNA has about a neutral pK_a (where K_a is the acid dissociation constant) of 7.6 ± 0.2, which is about the same as that reported for the peptidyltransferase reaction [106]. The neutral pK_a at position N3 of adenosine was determined by the pH dependence of dimethylsulfate (DMS) modification.

Moreover, the most sensitive nucleoside to the same DMS modification as that just stated was identified as A2451 within domain V of *E. coli* 23S rRNA by means of an autoradiogram (of reverse transcription products) developed in 6% denaturing polyacrylamide gel electrophoresis [28]. This result also agrees with our observation that poly(A)-directed polylysine synthesis by cell-free extracts carried out in medium SA (pH 7.6) is promoted the most [97].

No conclusive evidence concerning the functional mode of all macrolide antibiotics has yet been obtained. However, based on the X-ray crystallographic structure for the 50S ribosomal subunit from *H. marismortui*, a new catalytic reaction was established as a ribozyme. The catalytic reaction is called the charge relay system and consists of A2485 (*E. coli* number A2450) phosphate, G2482 (G2447) including G2102 (G2061) (but a base G2102 is not shown in Fig. 3A), and A2486 (A2451) linked by hydrogen bonds [27, 28] (Fig. 3A).

Wilhelm *et al.* [36] have presumed that macrolide antibiotics and lincomycin may mimic peptidyl-tRNA by binding to some ribosomal region through not only the sugar moieties but also the natural ester region of the lactone ring or the peptidyl linkage. In lincomycin, the linkage is similar to the juncture between tRNA chains and polypeptides rather than to the ester region present in the macrolides (Fig. 2). The model proposed by Wilhelm *et al.*, in that the macrolides were considered to be analogues to acyl-tRNA, agrees with our interpretation of the drug function.

However, our erythromycin and leucomycin A_5 model is different from that of Wilhelm *et al.* [36]. Our speculation that aminoacylated ribose at the 3' end of tRNA may resemble a 3'-dimethylaminoglycoside of macrolides (or, probably, the 3'-methoxyglucoside of lankamycin) disagrees with Wilhelm and coworkers' concept that a substituted N9-ribosyl glycoside (R_2 = N-sugar), at position N9 of the purine ring present in the 3'-end ribonucleotide, may mimic the amino sugar [$(CH_3)_2$ = N-sugar] of the usual macrolides [52, 107]. How do macrolide antibiotics act as inhibitory agents to protein synthesis?

Macrolide antibiotics may be permitted to arrive near the active center of peptidyltransferase [the N3 atom position of A2486 (A2451)] by virtue of the structural similarity of the drug to an acyl-tRNA (Fig. 3A) [52]. The antibiotics usually have a dimethylamino residue in the C3' position in each drug structure

(Fig. 2). Macrolide antibiotic probably takes up its position adjacent to the amino group of the amino acyl tRNA, compared with the position of the N3 atom of A2486 (A2451).

Dimethylamino residue in the C3′ position of a macrolide antibiotic has a pK_a value ranging from 6.9 to 8.8 (Table I). The dimethylamino residue of the drug, like nucleophilic atom N3 in the position of A2486 (A2451) ($pK_a = 7.6 \pm 0.2$), also may act as a general base catalysis (Fig. 3A). A macrolide with such a pK_a, like the N3 atom of A2486 (A2451), can easily accept a proton from the nucleophilic amino group of the A-site-bound aminoacyl-tRNA during formation of the tetrahedral intermediate. The protonated drug, except for deposited drug, could no longer stay in a space adjacent to the nucleophilic amino group of the A-site-bound aminoacyl-tRNA because only a nonprotonated molecule of macrolide antibiotic is able to bind to 50S ribosome. However, immediately, the protonated-drug molecule could in turn transfer its proton (general acid catalysis) to the 3′-oxyanion-leaving group of the P-site tRNA without its being released from near the position of A2486 in domain V (Fig. 3A). In addition, such immediate proton transfer might account for stimulation of peptide bond formation in the presence of a low concentration of macrolide antibiotics. That is likely to be the reason why the protonated dimethylamino residue of the drug exists quite near the 3′-oxyanion-leaving group of the P-site tRNA. Cooperatively, the tetrahedral intermediate is resolved into an amide linkage (Fig. 3A). This suggests that a neutral pK_a and a low degree of solubility in water of the drug are closely related to the inhibition of catalytic peptidyltransferase and to its conformation change due to deposition binding, respectively.

Deposition binding might cause a great conformational change in the overall three-dimensional structure essential for peptidyltransferase activity and the introduction of nascent peptide into the exit tunnel. Such modification also probably renders correct alignment of the peptidyl and aminoacyl substrates at the ribosome catalytic site, hindering peptide formation. In this connection, additional investigation is needed to obtain clear direct evidence.

Further research should eventually shed light on the molecular basis of the detailed total translation step and on the function of the 30S ribosomal subunit [29, 30].

III. Mechanisms of Resistance to Antimicrobial Macrolides

A. General Concept

Antibiotics have been used for the treatment of bacterial infections for more than 50 years. However, since that time, we have learned that antibiotics did not kill certain bacteria. It is generally accepted that there are two major classes of antibiotic resistance: intrinsic resistance and acquired resistance.

Intrinsic resistance in gram-negative bacteria such as *Pseudomonas aeruginosa* and *E. coli*, which occasionally cause opportunistic infection, is usually expressed by a chromosomal gene. Intrinsic resistance means that inherent features of the cell are responsible for preventing antibiotic action. For example, the cell envelope present in such organisms gives rise to low permeability of the cells to many agents, which are effective against other bacteria.

In contrast, acquired resistance may be the result of mutations in chromosomal genes or of the acquisition of plasmids and transposons. An acquired resistance arises from the selective pressure exerted on bacteria in terms of the application, including incorrect use or wrong prescriptions of antibiotics for chemotherapy.

Genes for resistance are not new creations. There seems to be no exception to the rule in genes responsible for resistance to macrolide antibiotics. In fact, many kinds of clinical isolates that carry resistant determinant(s) to macrolide antibiotics rarely develop the same mechanism as drug-resistant mutants, which arise *in vitro* from treatment with mutagens.

In the case of inducible MLS-resistant bacteria, their exposure to uninducible macrolides such as 16-membered-ring macrolides or certain 14-membered-ring macrolides (oleandomycin for a strain bearing the *ermA* gene, for example) gives rise to constitutive MLS-resistant mutants.

This section reviews mechanisms resistant to macrolides in major clinical isolates, especially acquired resistance bacteria.

In the late 1950s, antagonistic action between erythromycin and spiramycin was found in clinically isolated staphylococci [108, 109]. This observation promoted investigations of the biochemical mechanism and led to our current understanding of inducible or constitutive MLS resistance that is mediated by *erm* genes containing, respectively, a functional regulation mechanism or constitutively mutated regulatory region.

B. Resistance in Clinical Isolates

Erythromycin was first obtained from *Saccharopolyspora erythreus* (formerly *Streptomyces erythreus*) [110] in 1952. Shortly after unsuccessful erythromycin treatment (for only 7–10 days) in two patients with acute bacterial endocarditis [111], resistance to erythromycin emerged in two strains of bacteria, particularly in *S. aureus*.

In the two bacteria with acquired erythromycin resistance, obtained in 1952 clinical isolates in the United States, it is too late to determine retrospectively whether the kind of resistant genotype (*erm* or *msr*) in the two *S. aureus* strains could be specified. The *erm* gene encodes a methyltransferase (Erm protein) that catalyzes dimethylation of a specific adenine residue in 23S rRNA and requires *S*-adenosylmethionine as a cofactor [112]. The *msr* gene encodes a cell-membrane protein, which acts as an enhanced erythromycin-efflux pump.

Although the specific genotype in erythromycin-resistant *S. aureus* could not be determined, it has recently been reported that the *erm* gene was present in 98% of the erythromycin-resistant strains isolated from blood between 1959 and 1988 in Denmark [113]. Accordingly, the resistant strains isolated from the blood of the two endocardial patients must therefore have borne the *erm* gene. These epidemiological observations suggest that the *erm* gene responsible for macrolide resistance resides in most clinically isolated bacteria. If this is the case, where did the *erm* gene come from?

It is generally recognized that such a determinant responsible for resistance to macrolide-lincosamide-streptogramin type B antibiotics was probably acquired by pathogenic bacteria from a pool of resistance genes in other microbial genera, including the antibiotic-producing organisms Actinomycetes. The sequences of the resistance genes would then subsequently have been integrated by site-specific recombination into several classes of naturally occurring gene-expression cassettes such as integrons or similar structures. The cassettes or other multidrug-resistant structures formed by nonhomologous recombination would further have been incorporated into respective replicon genes that control the initiation and replication of DNA. Such acquisition of antibiotic resistance would have resulted in the development of R plasmids and conjugative transposons. Consequently, the resistance genes are disseminated within the bacterial population through a variety of gene transfer processes [114, 115].

In addition, note that antibiotic preparations themselves are often contaminated with DNA encoding antibiotic-resistant genes. Fluorescence spectroscopy and PCR amplification have shown that a number of antibiotic preparations are contaminated with significant amounts of DNA from the antibiotic-producing bacterium (for example, the gene *ermSF* of *Streptomyces fradiae*, which is responsible for resistance to MLS antibiotics) [115].

Three mechanisms of resistance to macrolides in bacteria are known: (1) modification of the target of the antibiotics mediated by the so-called *erm* gene (Table II) [116]; (2) enhanced efflux mediated by *msr*, *mef*, and *mre* genes (Table III) [61, 117, 118]; and (3) inactivation of macrolide antibiotics by the erythromycin esterase encoded by the *ere* gene [119, 120], by streptogramin B hydrolase encoded by the *vgb* gene [121], by macrolide phosphotransferase encoded by the *mph* gene [122, 123], and by lincosamide nucleotidyltransferase encoded by the *lin* gene [124] (Table IV).

MLS resistance due to modification of the drug target, a specific adenine residue of 23S rRNA, is widespread at present and has been found in *Staphylococcus* spp. [108, 109, 140–142], *Streptococcus* spp. [56, 62], *Corynebacterium diphtheriae* [143], *Clostridium* spp. [66–68], *Bacillus* spp. [144–148], *Lactobacillus* spp. [149], *Propionibacterium* spp. [150], *Bacteroides fragilis* [70–72], *E. coli* [69], and *Klebsiella* spp. [151].

TABLE III
Macrolide Resistance Due to Decreased Accumulation (Enhanced Efflux) in Staphylococci and Streptococci

Resistant phenotype[a]	Host	Genotype	References
MS	Staphylococcus epidermidis	msrA [msr (A)]	125, 126
PMS	Staphylococcus aureus	msrSA [msr (A)]	127
PMS	Staphylococcus aureus	msrSA' [msr (A)]	122, 128
PMS	Staphylococcus xylosus	msrB	129
M	Enterococcus faecium	msrC	130
M	Streptococcus pyogenes	mefA [mef (A)]	117, 131
M	Streptococcus pneumoniae	mefE [mef (A)]	132
M	Streptococcus agalactiae	mreA	118

[a]PMS, partial macrolide and streptogramin B antibiotics; MS, macrolide and streptogramin B antibiotics; M, macrolide antibiotics.

TABLE IV
Macrolide, Lincosamide, and Type B Streptogramin Resistance Due to Inactivation, Hydrolytic Degradation, or Modification by a Certain Transferase of the Drugs in Clinically Isolated Bacteria

Resistant phenotype[a]	Enzyme	Gene	Host	References
Degradation				
M_d	Erythromycin esterase type I	ereA [ere(A)]	E. coli	119
M_d	Erythromycin esterase type II	ereB [ere(B)]	E. coli	133
S_d	Streptogramin B hydrolase	vgb [vgb(A)]	S. aureus	121
		vgbB [vgb(B)]	S. aureus	134
Modification				
M_m	Macrolide 2′-phosphotransferase (14-membered-ring macrolides only)	mphA [mph(A)]	E. coli	123
M_m	Macrolide 2′-phosphotransferase (14-membered, rather than 16-membered-ring macrolides)	mphK[b] [mph(A)]	E. coli	135
M_m	Macrolide 2′-phosphotransferase (14- and 16-membered-ring macrolides)	mphB [mph(B)]	E. coli	136
M_m	Phosphotransferase[c]	mphBM [mph(C)]	S. aureus	122
L_m	3-Lincomycin, 4-clindamycin-O-nucleotidyltransferase	linA [lnu(A)]	S. haemolyticus	137, 138
		linA' [lnu(A)]	S. aureus	137
		linB [lnu(B)]	E. faecium	139

[a]Subscript letters d and m represent degradation and modification, respectively.
[b]Gene mphK differs from mphA at only five amino acid positions.
[c]Phosphotransferase-like enzyme.

Macrolide resistance is mediated by the *erm* gene in other antibiotic-producing bacteria such as *Streptomyces* spp. [152–159], *Micromonospora* sp. [160], *Saccharopolyspora* sp. [161], and *Arthrobacter* sp. [162], all of which have, together with cooperative determinants, coding for a transport ATPase. The resistance in all of these bacteria has been described in detail [116].

Ribosomal RNA, a ribozyme that is able to act as a peptidyltransferase, is one of the important components of living cells [26, 27]. Most mutational changes of the conserved base sequence in 23S rRNA would be unfavorable for the essential function of ribosome and, in the course of evolutionary events, would lead to cessation of cell growth and probably before long to cell death. Thus, the modification of the target site (in 23 rRNA) of the drugs by methyltransferase emerges in clinically isolated bacteria resistant to macrolides rather than other target site mutations. For the latter example, mutations such as deletion, inversion, and translocation as well as point mutation affecting ribosomal protein L4 or L12 (which also gives rise to macrolide resistance) are included. Such a modification appears to be related to an evolutionary event from the prebiotic RNA world [163–165].

C. Alteration of the MLS Target Site

1. Methylation of Domain V

In terms of the basic principle of the secondary structure, *E. coli* 23S rRNA has six domains [166]. Macrolide antibiotics interact with two regions (in domains II and V) of 23S rRNA [166–168], and domain V plays a particularly important role on the occasion of translation, that is, during the peptidyltransferase reaction [26, 27, 98–100, 169].

In clinically isolated strains of *S. aureus* and some other bacteria, modification involves methylation of A2058, which corresponds to an adenine residue at position 2058 based on the *E. coli* numbering system (*E. coli* number) [170, 171]. Alterations in domain V have been accomplished by adenine-N^6-methyltransferase specified by an *erm* gene.

Resistance to erythromycin can also result from an adenine-to-guanine transition at position 2058 of 23S rRNA from *E. coli* or mitochondria in yeast [172, 173]. In addition, *H. marismortui* has guanine (G) at a homologous site 2099 to *E. coli* number A2058 [26]. This microorganism should show resistance to erythromycin.

The adenine residue at position 2058 underwent dimethylation by methyltransferase, which was produced transcriptionally (in the case of *ermK* [*erm*(D), where the gene in brackets hereafter indicates the corresponding gene in the new naming group) [77, 148] or posttranscriptionally (in the case of *ermC* [*erm*(C)]) [57, 174] or both [146, 175], or mutationally (Table II) [169, 176].

Regardless of whether regulation of *erm* gene expression, in terms of MLS resistance, is sustained inducibly or constitutively, the gene codes for an enzyme, Erm (erythromycin resistance methylase). Methylation prevents MLS antibiotics from binding to the internal loop in domain V in the 23S rRNA. The dimethylated alteration of A2058 probably gives rise to a conformational (ill-defined) change in the RNA, leading to high resistance to MLS antibiotics, because binding sites of these drugs overlap [116, 177–179]. This kind of resistant mechanism also appears to create a phenotypically decreased accumulation of macrolide antibiotics in resistant cells [6, 9].

As well as clinical isolates, *erm* genes have been isolated from soil bacteria including antibiotic-producing microbes such as *Bacillus licheniformis* (*ermD* [*erm*(D)]), *B. sphaericus* (*ermG* [*erm*(G)]), *B. subtilis* (*ermIM* [*erm*(C)]), *B. anthracis* (*ermJ* [*erm*(D)]), and *B. licheniformis* (*ermK* [*erm*(D)]) as well as *Saccharopolyspora erythreus* (*ermE*; erythromycin producer [*erm*(E)]), *Arthrobacter luteus* (*ermR* or *ermA'*, AR; erythromycin producer [*erm*(R)]), and *Streptomyces fradiae* (*ermSF* [*erm*(S)]).

Extensive studies on *erm* alleles and their regulation of macrolide resistance have been reviewed [107, 116, 180–183]. Recently, Roberts *et al.* [77] suggested a new nomenclature for naming MLS genes and proposed its use in conjunction with the rules developed for identifying and naming new tetracycline-resistant genes [184, 185]. According to their proposal, if a new gene sequence shows ≧80% amino acid homology to any member of a gene class and confers a similar phenotype to the host, the new gene (designated with a single letter) should be placed in the existing group and not given a new letter or number designation.

The new guidelines seem to have the advantage of organizing the *N*-methyltransferases well, regardless of whether the gene was originally identified in pathogenic, opportunistic, or normal floral bacteria, in nosocomial infectious microorganisms, or in an antibiotic-producing species.

It is noteworthy that in the eye lens of the duck, gene duplication and separation of function may have occurred with one gene product acting primarily as a δ-crystallin and one acting primarily as an argininosuccinate lyase, although the two proteins specified by the respective cDNA clones are 94% identical in predicted amino acid sequence [186].

The focus here is on mechanisms in some clinically isolated bacteria that are representative of those that have acquired resistance to MSL antibiotics but we do not discuss to what extent a relationship exists between the similarity of amino acid sequences and that of their function.

2. Mutation of Domain V

In addition to the base methylation, transitional mutations (A2142G and A2143G) near the peptidyltransferase region in domain V of the RNA from clarithromycin-resistant *Helicobacter pylori* have clinically been found to give rise to a high level of MLS resistance. The residues at positions 2142 and 2143 are homologous to adenine residues at positions 2058 and 2059 of *E. coli* 23S rRNA, respectively [73, 169]. Mutants of the transitional type have been obtained during use of clarithromycin for therapy. Despite the same therapy as the drug, neither transversional mutants [such as A2142C (A2058C) or A2142T (A2058T)], in which gene (23S rDNA) coding for 23s rRNA has been altered, nor other acquired (or insertional) mutants such as *erm*-like or *msr*-like gene, have yet been found in clinical isolates of *H. pylori*. Every mutant gives rise to resistance to macrolides. It is curious that, in *H. pylori*, both transitional (A2142G and A2143G) and transversional (A2142C, A2142T, and A2143C) mutations are generated by the use of cDNA technology to produce a point mutation at a predetermined position in the 23S rDNA, the so-called site-directed mutagenesis [73, 187].

A class of mutation in 23S rRNA genes such as A2142G and A2142C was linked with high-level resistance to MLS antibiotics, and A2143G mutation was linked with high-level resistance to macrolide and clindamycin but not to streptogramin B antibiotic. Another class of mutation in the RNA genes such as A2142T and A2143C caused an intermediate level of resistance to clarithromycin, clindamycin, and streptogramin B antibiotic (quinupristin), except that the latter mutant showed susceptibility to quinupristin [73]. This class of resistance has been reported in 23S rRNAs obtained from clinical isolates of *Mycobacterium intracellular* and *Propionibacterium* spp. [75, 176].

3. Mutation of Domain II

The contribution of domain II to erythromycin resistance (dependent on the amount of E-peptide encoded by nucleotides at positions 1248 to 1265 in 23S rRNA) has been confirmed in the domain of *E. coli* 23S rRNA using deletion and point mutants [188].

The deletion and point mutations in the 23S rRNA domain II were responsible for erythromycin resistance in *E. coli*: A clone that mediates erythromycin resistance has been obtained from a certain plasmid containing the *rrnB* operon of the bacteria, the plasmid that was exposed to a hydroxylamine mutagen [167, 188, 189].

A 12-nucleotide deletion (positions 1219–1230) observed within an rRNA hairpin structure between nucleotides 1198 and 1247 in domain II of the *E. coli* 23S rRNA gene conferred erythromycin resistance. This 12-nucleotide sequence is located upstream of an open reading frame (ORF), which encodes a peptide MVLFV, E-peptide [190]. The conserved amino acids required for forming specific contacts with rRNA or ribosomal proteins may be N-terminal formyl methionine, the third Leu or Ile, and C-terminal hydrophobic amino acid commonly represented by Val [190]. The expression of the pentapeptide *in vivo* renders *E. coli* cells resistant to erythromycin [167]. Curiously, such a deletion and other engineered deletions did not affect binding of erythromycin to the mutant ribosomes as assayed by footprinting *in vivo* [191]. In contrast to this, point mutations at the central loop in domain V gave rise to a marked decrease in the ribosome–drug interaction [178, 191]. Because synthetic E-peptide did not affect the sensitivity of the cell-free translation system to erythromycin, E-peptide enters the site of its action cotranslationally and acts in a *cis* arrangement [190].

These findings are interpreted to indicate that erythromycin resistance mutation in domain II caused an increase in the peptide and disrupted an indirectly functional interaction between domains II and V, because such a mutation could affect alteration of the stability of a secondary rRNA structure (hairpin sequence structure) in domain II. In addition, the Shine-Dalgarno (SD) sequence of the rRNA-encoded E-peptide ORF is sequestered in the hairpin structure. Thereby, SD and E-peptide codon are not accessible to ribosomes of wild-type *E. coli*. The conformational change of the hairpin structure by erythromycin resistance mutation can be recognized by ribosomes for the initiation of translation of E-peptide. Thus, the increase of the peptide is expected to show resistance to macrolide antibiotics such as erythromycin, oleandomycin, and spiramycin but not clindamycin and chloramphenicol without preventing their binding to the target.

Although a mini-gene encoding a five-amino-acid peptide, MVLFV, has permanently been introduced into *B. subtilis* [192], an erythromycin-resistant mechanism due to production of E-peptide has not yet been found in clinical bacteria.

D. Decreased Macrolide Accumulation

In recent years, new resistance phenotypes (macrolide and streptogramin B antibiotics [MS] or partial macrolide and streptogramin B antibiotics [PMS], and macrolide antibiotics [M]) have been observed in clinical isolates of staphylococci and streptococci. Ross *et al*. [61, 125, 193] and Goldman and Capobianco [126] reported that MS-resistant strains of *Staphylococcus epidermidis* were resistant to 14- or 15-membered-ring macrolides and streptogramin B, but were susceptible to 16-membered-ring macrolide and lincosamide antibiotics.

Two types of resistance, M type and MS or PMS type, due to decreased macrolide accumulation, have been recognized in clinical isolates of staphylococci and streptococci (Table III).

Transporters, which mediate multidrug efflux, have usually been divided into four classes depending on (1) whether they are conducted by *proton motive force* (PMF) or by *ATP*, and (2) whether they consist of *a single protein* that has any one of the 4-, 12-, or 14-transmembrane-spanning domains [194, 195] or whether they are a *more complex multicomponent transporter* [183]. The more complex transporters, in addition to a multidrug efflux protein (MexB, for example), contain a membrane fusion protein such as MexA and an outer membrane protein such as OprM [194].

The four classes of transporters (PMF-dependent single or complex and ATP-dependent single or complex transporters) do not correspond to the two types of macrolide efflux transporters observed in clinical isolates, that is, M type and MS or PMS type. These groupings are based on diverseness in the profile of susceptibility to macrolide antibiotics. First, the M phenotype is usually characterized as having resistance to 14- and 15-membered macrolides (erythromycin and azithromycin, respectively) and occasionally, to a lesser extent, to 16-membered macrolides (spiramycin and josamycin) but being susceptible to lincosamide and streptogramin type B antibiotics [196]. The respective genes, *mefA* from *Streptococcus pyogenes* [117], *mefE* from *S. pneumoniae* [132], and *mreA* from *S. agalactiae* [118], were cloned, functionally expressed, and sequenced.

A comparison of the deduced amino acid sequences between *mefA* and *mefE* genes revealed that the two genes were 90% identical. Further analysis of their amino acid sequences disclosed the presence of 12-transmembrane domains. However, the amino acid sequence deduced from the *mreA* base sequence was significantly different from both of them. Although MreA had short recurrent hydrophobic regions of about 10 amino acids, it might be associated transiently with the cell membrane or perhaps with specific membrane proteins [118].

Proteins MefA, MefE, and MreA are thought to be driven by PMF, because decreased macrolide accumulation by their mediating was increased to the same accumulation level as that in susceptible streptococci in the presence of some uncouplers, such as carbonylcyanide-*m*-chlorophenylhydrazone (CCCP), 2,4-dinitrophenol (DNP), and arsenate.

Another resistance group, MS phenotype *Staphylococcus epidermidis*, has been studied by Ross *et al.* [125]. This MS resistance group showed inducible resistance to 14-membered-ring macrolides and to streptogramin type B antibiotics but susceptibility to 16-membered-ring macrolides and lincosamides. A subcloned 1.9-kb DNA sequence from one strain of *S. epidermidis*, which was located on a 31.5-kb plasmid, contained the *msrA* gene, which conferred MS resistance. The sequence revealed an ORF that encoded a 488-amino-acid protein (MsrA) whose regulation is mediated by translational attenuation [61], a mechanism of regulation similar to that which regulates *ermC* [57, 174].

The production of MsrA protein is regulated by the process that occurs after completion of transcription from the template strand of the *msrA* gene. That is to say, it is governed by about a 300-nt leader sequence that encodes, in sequence, an upstream ribosome-binding sequence (RBS or SD1), GGAGG, a putative eight-amino-acid leader peptide, MTASMRLK, and a noncoding region that contains four inverted complementary repeat sequences. Finally, the MsrA-ORF is preceded by its own RBS, AGGAG, which could be sequestered by the secondary structure of the leader region. If a stalled erythromycin–ribosome complex occupies the upstream leader peptide sequence, the ORF codon is translated.

On the other hand, although PMS-resistant *S. aureus* was first described as an inducible MS-like resistant strain [127], it was later found that the strain conferred inducible resistance to mycinamicin, a 16-membered-ring macrolide, in addition to 14-membered-ring macrolides and streptogramin type B antibiotics. Matsuoka *et al.* disclosed that the gene product of *S. auerus msrSA'* including *msrSA*, MsrSA' protein, is 98% similar to the *S. epidermidis* MsrA protein [45, 59, 122, 128, 197].

A cloned 5.04-kb DNA sequence contains the *msrSA* gene, which is responsible for PMS resistance in a constitutive mutant of *S. aureus* (pMC38), revealing the same ORF that encodes a 488-amino-acid protein, MsrSA. Moreover, 44 nucleotides including the leader region that is located upstream of the ORF have been deleted in the mutant. Thus, the regulation of PMS resistance seems to be mediated by translational attenuation of ORF in the *msrSA* sequence [197]. Consequently, it is highly likely that the MS phenotype *S. epidermidis* strain that confers *msrA* also gives resistance to mycinamicin, when MS resistance is induced by an optimum concentration of erythromycin (for example, 1.35 μg/ml in the case of *S. aureus*) [45].

Thus, the name of PMS (partial macrolide and streptogramin B) resistance may be more suitable than that of MS resistance, because of including resistance to 16-membered macrolides, mycinamicin I and II. An inducible PMS-resistant strain of *S. aureus* PM2104 carries a plasmid pEP2104, the gene (*msrSA*) of which is responsible for the resistance to erythromycin, oleandomycin, mycinamicin, and mikamycin B antibiotic. The PMS resistance conferred a similar phenotype to other host strains of *S. aureus* NCTC8325, which was transformed with the same pEP2104 [197]. In contrast, an inducible PM-resistant strain of *S. epidermidis* carries a plasmid pUL5050, the genes (*msrA*) of which are responsible for the resistance to erythromycin, oleandomycin, and streptogramin type B antibiotic only. The MS resistance conferred a different phenotype from other host strains of *S. aureus* RN4220, which was transformed with a high-copy-number plasmid pUL5054 [61]. The host strain of *S. aureus* RN4220 (pUL5054) showed constitutive resistance to erythromycin but resistance to streptogramin type B antibiotic remained inducible. If a high copy of the *msrA* operon causes constitutive MS resistance, it is unclear why streptogramin B resistance remains inducible.

A decreased accumulation in terms of enhanced efflux of the antibiotics accounts for the resistance mechanism to erythromycin in PMS- or MS-resistant staphylococci, but the amount of erythromycin (EM) accumulated in induced or constitutive EM-resistant cells, nevertheless, was about 8 pmol/0.76 μl (assuming 76% intracellular fluid volume [7]) per bacterium $\sim 11 \times 10^{-6}$ M. At an external erythromycin concentration of 1 μg/ml (1.36×10^{-6} M), even the resistant bacteria take up 11×10^{-6} M of EM. The accumulation of the drug in resistant *S. aureus* cells is 8 times greater than the extracellular level. The accumulation of the drug in

EM-susceptible or resistance-uninduced cells of *S. aureus* strains, on the other hand, is 12 to 15 times greater than the same extracellular level. The addition of CCCP to the culture of these strains increases erythromycin accumulation from 34×10^{-6} to 37×10^{-6} M. The increase in the bacterial cells in the presence of CCCP is estimated to be 25 to 27 times higher than the extracellular concentration. The amounts that accumulate in the presence of an uncoupler (0.1 m*M*) are only about twice the level of that in the absence of an inhibitor [197, 198]. In addition, ribosome has a high affinity for erythromycin (K_d: 1×10^{-8} to 3×10^{-8} *M*), regardless of whether it is sensitive or resistant *S. aureus* ribosome [199].

Assuming that the major accumulation of erythromycin in *S. aureus* cells is determined by uptake dependent on the amounts of ribosome present in the cells, these observations of drug accumulation can be easily accounted for. As described in a previous section, erythromycin molecules can bind to ribosome at a certain equilibrium state expressed as a dissociation constant (K_d) [39]. For example, on the supposition that an intracellular concentration of erythromycin increases by 30 times the level of that in an extracellular medium (1 μg/ml), the nonprotonated molecules of erythromycin (pK_a = 8.8) are able to occupy no more than 1% of a drug concentration present in *S. aureus* cell. That is why it is expected that intracellular pH would be 6.8 or less [200] and that erythromycin (even as a free base) with pK_a = 8.8 is one of the most water-soluble (i.e., deposition-resisting) drugs in the macrolide antibiotics (Table I).

Therefore, most molecules of erythromycin that exist as protonated erythromycin in intracellular fluid could easily be discharged into the extracellular medium through the help of an efflux protein such as MsrSA. That is why positive charged molecules of the drug cannot pass across the lipid layer present in the cell membrane despite a drug accumulation that results from an intracellular concentration gradient that is dependent on an affinity of ribosome to erythromycin. The efflux of protonated erythromycin molecules will continue until the difference in the concentration between both the outside and inside of the cell membrane disappears.

In contrast, most 16-membered-ring macrolides including 14-membered ketolides such as telithromycin and probably ABT-773, which can cause deposition at their binding sites of 50S ribosomes, can hardly be discharged across a cell membrane into the extracellular medium. That is why the drugs, owing to their slight solubility in water, can deposit on the binding site of the ribosomes at 0.6 or less mg/ml and why only the macrolides dissolved in intracellular cell fluid must be discharged into extracellular medium (Table I). These macrolide antibiotics exhibit outstanding inhibitory activity against PMS- or MS- and M-phenotype-resistant bacteria.

In this connection, a 16-memberd macrolide, YM 133, has a pK_a value of around seven and shows the most powerful inhibitory activity against even constitutively MLS-resistant *S. aureus*. The lowest concentrations (MICs) of YM 133 and tylosin that inhibit the constitutive-resistant strain of *S. aureus* carrying the *erm* gene were 12.5 and >800 μg/ml, respectively (unpublished data). That is why YM 133 probably has a pK_a value similar to that of intracellular pH and a strong deposition affinity for ribosome from the cell that bears constitutive MLS resistance specified by the *erm* gene (Fig. 2 and Table I) [47].

Further work is needed to determine the real role of Msr, Mef, and Mre proteins in enhanced drug efflux.

E. Translational and Transcriptional Attenuation

MLS or PMS resistance, usually induced by erythromycin or occasionally by oleandomycin, follows a negative *cis* regulation by attenuators. In the absence of erythromycin, for example, the mRNA required for ErmC, methyltransferase, is previously synthesized on a template of the *ermC* operon in an inactive conformation, which permits the initiation of efficient translation [57, 58].

In the case of the *ermK* operon, the synthesis of mRNA required for ErmK is not previously achieved because of termination independent of the ρ factor, transcriptional attenuation, in the absence of the drug [175].

With regards to translational attenuation, the induction of MLS and MS or PMS resistance by the presence of erythromycin is generally thought to result from stalling of an erythromycin–ribosome complex on RNA synthesized without previous preparation on the template DNA sequence of the leader (control) peptide.

Two tests to determine whether the ternary complex, protein-synthesizing ribosome, mRNA, and peptidyl-tRNA, is destabilized or stabilized were conducted in the presence of erythromycin. The stabilization of the ternary complex was confirmed by the profile of the footprint of the ribosome occupying the *ermC* message covering the leader peptide coding region [183].

Lovett and Rogers [201], in a recent review, discussed cotranslational regulation of protein synthesis. That is to say, a translational process conducted by the ribosome of the leader peptides (of the *catA86* specifying chloramphenicol acetyltransferase and *cmlA* encoding a membrane protein that probably alters chloramphenicol transport operons) appears to take place concomitantly with the synthesis of the nascent peptide that is capable of binding to rRNA and inhibiting the peptidyltransferase. In addition, a nascent 5- or 8-residue leader peptide is found to inhibit ribosomal peptidyltransferase.

On the other hand, in the case of the *ermC* operon responsible for inducible MLS resistance, the *ermC* leader peptide is insoluble in the buffer systems commonly used to study peptide interaction with ribosomes or rRNA [201]. Mayford and Weisblum [202], however, have shown that the first nine amino acids of the *ermC* leader peptide, MGIFSIFVI, were sufficient to specify induction and that, out of these, IFVI were the most significant, because mutations at these amino acids completely abolished induction.

Peptidyltransferase as a ribozyme may be inhibited by nascent-leader peptides capable of binding (probably in a deposition-like manner) to the center of the enzyme, because the nascent peptides repress the enzyme activity *in vitro* [201].

F. Enzymatic Inactivation of Macrolide Antibiotics

Enzymatic mechanisms that inactivate macrolides appear to be rare among clinical isolates compared with mechanisms of phenotypically decreased macrolide accumulation due to either a target site alteration (i.e., ribosome modification) or enhanced macrolide efflux.

As shown in Table IV the enzymes that are capable of inactivation of macrolide, lincosamide, or streptogramin type B antibiotics can specifically distinguish only one or a class of the drugs as a corresponding substrate in acquired resistance bacteria.

According to the systematic arrangement and naming of enzymes by the Enzyme Commission (EC), enzymes concerned with inactivation of the antibiotics all belong to either of two divisions out of six major divisions. The first type of enzymatic inactivation is due to degradation of macrolides and belongs to one of the EC 3 division. This includes any enzyme that catalyzes the hydrolysis of various bonds, for example, C–O, C–N, C–C, and phosphoric anhydride bonds. Another type of enzymatic inactivation due to modification of the antibiotics belongs to one of the EC 2 division and includes any enzyme that catalyzes transfer of a phosphate or a nucleotidyl group from a donor such as ATP to an antibiotic.

Recently, a plasmid, pMS97, residing in a strain of *S. aureus* clinically isolated in Japan in 1971 and investigated by Matsuoka *et al.* [59, 122, 128], was found to carry not only *msrSA′* and *ermGM* genes but also, unexpectedly, the *mphBM* gene (GenBank Q9ZNK8), probably encoding macrolide phosphotransferase. Practically, it may be difficult to isolate *S. aureus* strains that are able to inactivate a macrolide antibiotic by the enzyme, because they produce phosphatase that may render the drug active. If the enzymatic activity of MphBM protein differs in substrate specificity from that of *S. aureus* phosphatase, a phosphorylated macrolide antibiotic is expected to be obtained.

The emergence of such a multiresistant isolate may be a result of exposure of the originally susceptible *S. aureus* strain to various macrolide antibiotics, because several varieties of macrolide antibiotics had been used for chemotherapy in Japan during the late 1960s to 1970s.

Our interest has focused on individual origin, assemblage of these three resistant genes, and the regulation mechanism of macrolide resistance in the plasmid pMS97: Where has each of them come from? Have they ever constituted a cluster themselves? How do they regulate the expression of drug resistance, independently or cooperatively?

IV. Important Developments in Macrolide Antibiotics

Based on improvements such as increased acid stability, improved pharmacokinetics, a broader spectrum of action, and increased effectiveness against erythromycin-resistant strains, four generations of macrolides are distinguishable [183], as discussed next.

First-generation macrolide antibiotics include 14-membered-ring macrolides such as erythromycin, oleandomycin, and megalomicin A [40, 41]. They generally have high water solubility (1–5 mg/ml) and a high pK_a value of 8 to 9 (Table I). Erythromycin and oleandomycin have usually been considered to show bacteriostatic action due to translocation inhibition and probably due to their causing the stabilization of polyribosomes [93, 203]. They differ in the extent of inducer ability, but they all are able to act as an inducer of MLS resistance in most inducible-macrolide-resistant isolates of *S. aureus*. Quantitative values for induction of resistance (100% for erythromycin, 28% for oleandomycin, and 77% for megalomicin, for example) refer to the capability of resistant cells to grow in the presence of a high concentration of a 16-membered-ring macrolide (rokitamycin) (unpublished data). In addition, these macrolides can act as inducers toward efflux-based resistant strains bearing *msr* genes.

Second-generation drugs, semisynthetic and natural 16-membered-ring macrolides (carbomycin, leucomycin, spiramycin, and rokitamycin), show potent activities against inducible resistant *S. aureus* strains. The strains show susceptibility to MLS antibiotics, unless the bacteria are exposed to any of the 14-membered-ring macrolides, such as erythromycin, oleandomycin, and megalomicin. Second-generation drugs, however, give rise to a mutation into constitutive MLS or PMS resistance in *S. aureus* strains that show inducible resistance to MLS or PMS antibiotics. They generally have low water solubility (0.1–1 mg/ml) and a pK_a value of about 7 except for mycinamicin and rokitamycin (Table I).

Rokitamycin, a semisynthetic derivative of leucomycin A_5 (Fig. 2), has a unique property that enables the drug to bind irreversibly to ribosome, probably owing to deposition at a binding site near the active center of the peptidyltransferase. The solubility in water and pK_a values of the drug are less than 0.1 mg/ml and 7.6, respectively. In addition, this drug has a greater bactericidal effect on susceptible *S. aureus* strains despite its lower affinity to ribosome than that of erythromycin [34].

In terms of the bactericidal effect, note that a 16-membered-ring macrolide, which bears at least one disaccharide-monoglycoside residue related to a stable binding (probably deposition binding) to ribosome, causes polyribosome degradation [93, 94].

Third-generation macrolides, which include a 15-membered-ring macrolide, azithromycin, increase the acid stability of 14-membered-ring macrolides. Their pharmacokinetics is thus improved. These macrolides have high pK_a value but low water solubility (Table I), so they may be bactericidal against susceptible bacteria.

The antibacterial spectrum of these macrolides is broader. For example, the clinical modification at C6 [e.g., O-methylation (clarithromycin)] or at C9 [e.g., some 9-ether oxime derivatives (roxithromycin and dirithromycin)] stabilizes 14-membered lactone rings in acidic media, even when the modified drugs have been orally administered. Drugs with an expanded erythromycin A-lactone ring (e.g., azithromycin) are also more stable in acidic media and display better anti-gram-negative activity than does erythromycin A.

In terms of *fourth-generation macrolides*, the substitution of L-cladinose at C3 in erythromycin A with a keto produces ketolides, such as telithromycin (formerly HMR3647) and ABT-773, and also causes an increase in acid stability and a new characteristic different from erythromycin. For example, the semisynthetic drugs do not induce MLS resistance, unlike 14-membered-ring macrolides produced naturally by antibiotic producers [204], but they do not inhibit staphylococcal strains carrying erythromycin-induced or constitutive MLS resistance specified by the *erm* gene.

Despite having the same spectrum of action as that of other natural macrolides, ketolides have better *in vitro* activity against streptococci and other gram-positive microorganisms, including oxacillin-resistant *Staphylococcus* spp. and vancomycin-resistant enterococci [204–206], because a remarkable decrease takes place in the water solubility of those ketolides (Table I).

These studies on modifications to erythromycin A have provided a great deal of important and interesting pharmacokinetic improvements, broadening an antibacterial spectrum and developing potency. For details, see Weisblum [183] and Bryskier *et al.* [207, 208]. Further work on structure–activity relationships in macrolide antibiotics with due consideration for their water solubility will provide a sufficient opportunity to improve antibacterial macrolides.

V. Concluding Remarks

The molecular biological mode of action of macrolide antibiotics and the biochemical and genetic mechanism of resistance to macrolide, lincosamide, and type B streptogramin antibiotics were reviewed in this chapter.

X-ray crystallographic study of the 50S ribosomal subunit of *H. marismortui* has shown that macrolide antibiotics can act chiefly as inhibitors of peptidyltransferase, that is, a ribozyme bearing a charge-relay system. It is likely that the poor water solubility of the macrolide antibiotic could give rise to drug deposition at a binding site near the peptidyltransferase, and that the extent of the solubility of the drug could determine bacteriostatic or bactericidal action.

The resistance mechanisms are divided into three major groups: (1) target modification, (2) decreased macrolide accumulation due to enhanced drug efflux, and (3) inactivation of the antibiotics.

The phenotypically decreased accumulation of macrolide antibiotics is usually observed in not only MLS-resistant but also PMS-resistant *S. aureus*, and probably in other clinical isolates.

In connection to mechanism 3, inactivation of macrolide antibiotics by pathogenic *Nocardia* spp. was not described in this review. For example, we did not consider the inactivation due to phosphorylation (class EC 3), glycosylation (EC 3), reduction (EC 1), deacylation (EC 3), or a combination of phosphorylation and reduction [209]. The organisms are usually found in soil and water, and most cases are opportunistic, occurring in immunosuppressed patients.

Because it is impossible at present to predict when and where a gene responsible for antibiotic resistance will develop and be disseminated, we do not know what kind of acquired resistant mechanism has worked previously.

Two major strategies are used to prevent the emergence of macrolide-resistant bacteria. First, efforts should be made to preserve the effectiveness of the antibiotics by determining which drugs are still available for causal bacteria in infectious diseases with a rapid accurate diagnosis. A global surveillance of resistance is needed as soon as possible to determine what is happening in the world. Second, overreliance on the antimicrobial agents must be removed from the minds of the people who use or purchase a drug and the medical doctors who advise the use of an antibiotic.

Important developments in terms of new derivatives of erythromycin and ketolides are proceeding steadily.

VI. Addendum

Recently, the structural basis for the interaction of antibiotics with the peptidyl transferase center in eubacteria was reported. Schlünzen *et al.* [210] have determined the high-resolution X-ray structure of the 50S ribosomal subunit of the eubacterium *Deinococcus radiodurans*, complexed with the antibiotic clindamycin and three macrolides (erythromycin, clarithromycin, and roxithromycin), including another peptidyl transferase inhibitor such as chloramphenicol. They found that the antibiotic binding sites are composed exclusively of segments of 23S rRNA at the peptidyl transferase cavity—clindamycin target A and P sites, and the macrolides bind at the entrance to the tunnel where the drugs sterically block the progression of nascent peptide. The authors' paper should give a mass of valuable information on interaction of drugs with target sites. Whether their work on ribosome–antibiotic interactions in soaked crystals can explain the overall biochemical and physiological effects of the drugs in an intracellular aquatic state, however, remains to be seen.

References

1. Woodward, R. B. (1957). Structure and biogenesis of the macrolides. *Angew. Chem.* **69**, 50–58.
2. Greenwood, D. (1995). Inhibitors of bacterial protein synthesis. *In* "Antimicrobial Chemotherapy," 3rd ed. (D. Greenwood, Ed.), pp. 32–48. Oxford University Press, Oxford.
3. Bryskier, A., and Butzler, J. P. (1997). Macrolides. *In* "Antibiotic and Chemotherapy," 7th ed. (F. O'Grady, H. P. Lambert, R. G. Finch, and D. Greenwood, Eds.), pp. 377–393. Churchill Livingstone, New York.
4. Taubman, S. B., Jones, N. R, Young, F. E., and Corcoran, J. W. (1966). Sensitivity and resistance to erythromycin in *Bacillus subtilis* 168: The ribosomal binding of erythromycin and chloramphenicol. *Biochem. Biophys. Acta* **123**, 438–440.
5. Brock, T. D., and Brock, L. M. (1959). Similarity in mode of action of chloramphenicol and erythromycin. *Biochem. Biophys. Acta* **33**, 274–275.
6. Nakajima, Y., Inoue, M., Oka, Y., and Yamagishi, S. (1968). A mode of resistance to macrolide antibiotics in *Staphylococcus aureus*. *Jpn. J. Microbiol.* **12**, 248–250.
7. Taubman, S. B., Young, F. E., and Corcoran, J. W. (1963). Antibiotic glycosides, IV. Studies on the mechanism of erythromycin resistance in *Bacillus subtilis*. *Proc. Natl. Acad. Sci USA* **50**, 955–962.
8. Mao, J. C.-H., and Putterman, M. (1968). Accumulation in gram-positive and gram-negative bacteria as a mechanism of resistance to erythromycin. *J. Bacteriol.* **95**, 1111–1117.
9. Yamagishi, S., Nakajima, Y., Inoue, M., and Oka, Y. (1971). Decrease in accumulation of macrolide antibiotics as a mechanism of resistance in *Staphylococcus aureus*. *Jpn. J. Microbiol.* **15**, 39–52.
10. Vazquez, D. (1967). Macrolide antibiotics—Spiramycin, carbomycin, angolamycin, methymycin, and lankamycin. *In* "Antibiotics—Mechanism of Action" (D. Gottlieb and P. D. Shaw, Eds.), Vol. I, pp. 366–377. Springer-Verlag, Berlin.
11. Hahn, F. E. (1967). Erythromycin and oleandomycin. *In* "Antibiotics—Mechanism of Action" (D. Gottlieb and P. D. Shaw, Eds.), Vol. I, pp. 378–386. Springer-Verlag, Berlin.
12. Oleinick, N. L. (1975). The erythromycins. *In* "Antibiotics—Mechanism of Action of Antimicrobial and Antitumor Agents" (J. W. Corcoran and F. E. Hahn, Eds.), Vol. III, pp. 396–419. Springer-Verlag, Berlin.
13. Vazquez, D. (1975). The macrolide antibiotics. *In* " Antibiotics—Mechanism of Action of Antimicrobial and Antitumor Agents" (J. W. Corcoran and F. E. Hahn, Eds.), Vol. III, pp. 459–479. Springer-Verlag, Berlin.
14. Vazquez, D. (1979). Macrolide antibiotics of the spiramycin and carbomycin groups. *In* "Molecular Biology, Biochemistry and Biophysics—Inhibitors of Protein Biosynthesis" (A. Kleinzller, G. F. Springer, and H. G. Wittmann, Eds.), Vol. 30, pp. 120–126. Springer-Verlag, Berlin.
15. Vazquez, D. (1979). Macrolide antibiotics of the spiramycin and carbomycin groups. *In* "Molecular Biology, Biochemistry and Biophysics—Inhibitors of Protein Biosynthesis" (A. Kleinzller, G. F. Springer, and H. G. Wittmann, Eds.), Vol. 30, pp. 169–175. Springer-Verlag, Berlin.
16. Gale, E. F., Cundliffe, E., Reynolds, P. E., Richmond, M. H., and Waring, M. J. (1981). Antibiotic inhibitors of ribosome function. *In* "Molecular Basis of Antibiotic Action," 2nd ed., pp. 402–547. John Wiley & Sons, London.
17. Pestka, S. (1977). Inhibitor of protein synthesis. *In* "Molecular Mechanism of Protein Biosynthesis" (H. Weisbach and S. Pestka, Eds.), pp. 468–533. Academic Press, New York.
18. Corcoran, J. W. (1984). Mode of action and resistance mechanisms of macrolides. *In* "Macrolide Antibiotics: Chemistry, Biology, and Practice (S. Ōmura Ed.), pp. 231–259. Academic Press, Orlando, FL.

19. Russell, A. D., and Chopra, I. (1996). Mode of action of antibiotics and their uptake into bacteria. *In* "Understanding Antibacterial Action and Resistance," 2nd ed., pp. 22–95. Ellis Horwood, London.
20. Nirenberg, M. W., and Matthaei, J. H. (1961). The dependence of cell-free protein synthesis in *E. coli* upon naturally occurring or synthetic polynucleotides. *Proc. Natl. Acad. Sci. USA* **47**, 1588–1602.
21. Mao, J. C.-H. (1967). Protein synthesis in a cell-free extract from *Staphylococcus aureus*. *J. Bacteriol.* **94**, 80–86.
22. Lengyel, P., and Söll, D. (1969). Mechanism of protein biosynthesis. *Bacteriol. Rev.* **33**, 246–301.
23. Nomura, M., Mizushima, S., Ozaki, M., Traub, P., and Lowry, C. V. (1969). Structure and function of ribosomes and their molecular components. *In* "The Mechanism of Protein Synthesis," The Cold Spring Harbor Symposia on Quantitative Biology, Vol. XXXIV, pp. 49–61. Cold Spring Harbor Laboratory, Cold Spring Harbor, NY.
24. Nomura, M., and Held, W. A. (1974). Reconstitution of ribosomes: Studies of ribosome structure, function and assembly. *In* "Ribosomes" (M. Nomura, A. Tissières, and P. Lengyel, Eds.), pp. 193–223. Cold Spring Harbor Laboratory, Cold Spring Harbor, NY.
25. Noller, H. F., Moazed, D., Stern, S., Powers, T., Allen, P. N., Robertson, J. M., Weiser, B., and Triman, K. (1990). Structure of rRNA and its functional interactions in translocation. *In* "The Ribosome—Structure, Function, and Evolution" (W. E. Hill, A. Dahlberg, R. A. Garrett, P. B. Moore, D. Schlessinger, and J. R. Warner, Eds.), pp. 73–92. American Society for Microbiology, Washington, DC.
26. Ban, N., Nissen, P., Hansen, J., Moor, P. B., and Steitz, T. A. (2000). The complete atomic structure of the large ribosomal subunit at 2.4Å resolution. *Science* **289**, 905–920.
27. Nissen, P., Hansen, J., Ban, N., Moor, P. B., and Steitz, T. A. (2000). The structural basis of ribosome activity in peptide bond synthesis. *Science* **289**, 920–930.
28. Muth, G. W., Ortoleva-Donnelly, L., and Strobel, S. A. (2000). A single adenosine with a natural pK_a in the ribosomal peptidyl transferase center. *Science* **289**, 947–950.
29. Wimberly, B. T., Brodersen, D. E., Clemons Jr., W. M., Morgan-Warren, R. J., Carter, A. P., Vonrhein, C., Hartsch, T., and Ramakrishnan, V. (2000). Structure of the 30S ribosomal subunit. *Nature* **407**, 327–339.
30. Carter, A. P., Clemons, W. M., Broderson, D. E., Morgan-Warren, R. J., Wimberly, B. T., and Ramakrishnan, V. (2000). Functional insights from the structure of the 30S ribosomal subunit and its interactions with antibiotics. *Nature* **407**, 340–348.
31. Mao, J. C.-H., Putterman, M., and Wiegand, R. G. (1970). Biochemical basis for the selective toxicity of erythromycin. *Biochem. Pharmacol.* **19**, 391–399.
32. Mao, J. C.-H. (1967). The stoichiometry of erythromycin binding to ribosomal particles of *Staphylococcus aureus*. *Biochem. Pharmacol.* **16**, 2441–2443.
33. Fernandez-Muñoz, R., and Vázquez, D. (1973). Quantitative binding of ^{14}C-erythromycin to *E. coli* ribosomes. *J. Antibiot. (Tokyo)* **26**, 107–108.
34. Endou, K., Matsuoka, M., and Nakajima, Y. (1990). Adhesive binding of rokitamycin by *Staphylococcus aureus* ribosomes. *FEMS Microbiol. Lett.* **72**, 93–96.
35. Vazquez, D. (1967). Inhibitors of protein synthesis at the same ribosome level: Studies on their site of action. *Life Sci.* **6**, 381–386.
36. Wilhelm, J. M., Oleinick, N. L., and Corcoran, J. W. (1968). Interaction of antibiotics with ribosomes: Structure–function relationships and a possible common mechanism for the antibacterial action of the macrolides and lincomycin. *Antimicrob. Agents Chemother.* **1967**, 236–249.
37. Mao, J. C.-H., and Putterman, M. (1969). The intermolecular complex of erythromycin and ribosome. *J. Mol. Biol.* **44**, 347–361.

38. Fernandez-Muñoz, R., Monro, R. E., Torres-Pinedo, R., and Vázquez, D. (1971). Substrate- and antibiotic-binding sites at the peptidyltransferase centre of *E. coli* ribosomes: Studies on the chloramphenicol, lincomycin and erythromycin sites. *Eur. J. Biochem.* **23**, 185–193.
39. Mao, J. C-H., and Wiegand, R. G. (1968). Mode of action of macrolides. *Biochim. Biophys. Acta* **157**, 404–413.
40. Weinstein, M. J., Wagman, G. H., Marquez, J. A., Testa, R.T., Oden, E., and Waitz, J. A. (1969). Megalomicin, a new macrolide antibiotic complex produced by *Micromonospora*. *J. Antibiot.* **22**, 253–258.
41. Marqez, J., Murawski, A., Wagman, G. H., Jaret, R. S., and Reimann, H. (1969). Isolation, purification and preliminary characterization of megalomicin. *J. Antibiot.* **22**, 259–264.
42. Nakagawa, Y., Itai, S., Yoshida, T., and Nagai, T. (1992). Physicochemical properties and stability in the acidic solution of a new macrolide antibiotic, clarithromycin, in comparison with erythromycin. *Chem. Pharm. Bull.* **40**, 725–728.
43. Vazifeh, B., Preira, A., Bryskier, A., and Labro, M. T. (1998). Interactions between HNR3647, a new ketolide, and human polymorphonuclear neutrophils. *Antimicrob. Agents Chemother.* **42**, 1944–1951.
44. Hernandez, L., Sadrzadeh, N., Krill, S., Ma, Z., and Marsh, K. (1999). Preclinical pharmacokinetic profile of ABT-773 in mouse, rat, monkey, and dog. Presented at American Society for Microbiology 39th Interscience Conf. on Antimicrobial Agents and Chemotherapy (San Francisco). Poster No. 2148.
45. Matsuoka, M., Endou, K., Saitoh, S., Katoh, M., and Nakajima, Y. (1995). A mechanism of resistance to partial macrolide and streptogramin B antibiotics in *Staphylococcus aureus* clinically isolated in Hungary. *Biol. Pharm Bull.* **18**, 1482–1486.
46. Morohoshi, T. Personal communication.
47. Yoshioka, T. Personal communication.
48. Waksman, S. A. (1962). Description of the various antibiotics produced by Actinomycetes: Mikamycin B. *In* "The Actinomycetes," Vol. III, Part B, pp. 302–303. William and Wilkins Co., Baltimore.
49. Korzybski, T., Kowszyk-Gindifer, Z., and Kurylowicz, W. (1978) Mikamycin. *In* "Antibiotics: Origin, Nature, and Properties" (T. Korzybski, Z. Kowszyk-Gindifer, and W. Kurylowicz, Eds.), Vol. I, pp. 367–371. American Society for Microbiology, Washington, DC.
50. Van Holde, K. E., and Hill, W. E. (1974). General physical properties of ribosomes. *In* "Ribosomes" (M. Nomura, A. Tissières, and P. Lengyel, Eds.), pp. 53–91. Cold Spring Harbor Laboratory, Cold Spring Harbor, NY.
51. Endou, K., Matsuoka, M., Taniguchi, H., and Nakajima, Y. (1993). Implication of cohesive binding of a macrolide antibiotic, rokitamycin, to ribosomes from *Staphylococcus aureus*. *J. Antibiot.* **46**, 478–485.
52. Nakajima, Y. (1999). Mechanisms of bacterial resistance to macrolide antibiotics. *J. Infect. Chemother.* **5**, 61–74.
53. Weiss, P. J., Andrew, M. L., and Wright, W. W. (1957). Solubility of antibiotics in 24 solvents: Use in analysis. *Antibiot. Chemother.* **7**, 374–377.
54. Saito, T., Hashimoto, H., and Mitsuhashi, S. (1968). Drug resistance of staphylococci, formation of erythromycin–ribosome complex, decrease in the formation of erythromycin–ribosome complex in erythromycin-resistant strains of *Staphylococcus aureus*. *Jpn. J. Microbiol.* **13**, 119–121.
55. Murphy, E. (1985). Nucleotide sequence of *ermA*, a macrolide-lincosamide-streptogramin B determinant in *Staphylococcus aureus*. *J. Bacteriol.* **162**, 633–640.
56. Shaw, J. H., and Clewell, D. B. (1985). Complete nucleotide sequence of macrolide-lincosamide-streptogramin B-resistance transposon Tn917 in *Streptococcus faecalis*. *J. Bacteriol.* **164**, 782–796.

57. Horinouchi, S., and Weisblum, B. (1980). Posttranscriptional modification of mRNA conformation: Mechanism that regulates erythromycin-induced resistance. *Proc. Natl. Acad. Sci. USA* **77**, 7079–7083.
58. Gryczan, T., Grandi, G., Hahn, J., Grandi, R., and Dubnau, D. (1980). Conformational alteration of mRNA structure and the posttranscriptional regulation of erythromycin-induced drug resistance. *Nucleic Acids Res.* **8**, 6081–6097.
59. Matsuoka, M., Inoue, M., and Nakajima, Y. (1998). A new class of *erm* genes mediating MLS-coresistance in *Staphylococcus aureus*: It resides on plasmid pMS97 together with *msrSA'* gene coding for an active efflux pump. Presented at American Society for Microbiology 38th Interscience Conf. on Antimicrobial Agents and Chemotherapy (San Diego, CA). Abstr. No. 14352.
60. Kataja, J., Huovinen, P., and Seppala, H. (2000). Erythromycin resistance genes in group A streptococci of different geographical origins. The macrolide resistance study group. *J. Antimicrob. Chemother.* **46**, 789–792.
61. Ross, J. I., Eady, E. A., Cove, J. H., Cunliffe, W. J., Baumberg, S., and Wootton, J. C. (1990). Inducible erythromycin resistance in staphylococci is encoded by a member of the ATP-binding transport super-gene family. *Mol. Microbiol.* **4**, 1207–1214.
62. Horinouchi, S., Byeon, W. H., and Weisblum, B. (1983). A complex attenuator regulates inducible resistance to macrolides, lincosamides, and streptogramin type B antibiotics in *Streptococcus sanguis*. *J. Bacteriol.* **154**, 1252–1262.
63. Oh, T-G., Kwon, A-R., and Choi, E-C. (1998). Induction of *ermAMR* from a clinical strain of *Enterococcus faecalis* by 16-membered-ring macrolide antibiotics. *J. Bacteriol.* **180**, 5788–5791.
64. Serwold-Davis, T. M., and Groman, N. B. (1986). Mapping and cloning of *Corynebacterium diphtheriae* plasmid pNG2 and characterization of its relatedness to plasmid from skin coryne forms. *Antimicrob. Agents Chemother.* **30**, 69–72.
65. Hodgson, A. L. M., Krywult, J., and Radford, A. J. (1990). Nucleotide sequence of the erythromycin resistance gene *Corynebacterium* plasmid pNG2. *Nucleic Acids Res.* **18**, 1891.
66. Berryman, D. I., and Rood, J. I. (1989). Cloning and hybridization analysis of *ermP*, a macrolide-lincosamide-streptogramin B resistance determinant from *Clostridium perfringens*. *Antimicrob. Agents Chemother.* **33**, 1346–1353.
67. Berryman, D. I., Lyristis, M., and Rood, J. I. (1994). Cloning and sequence analysis of *ermQ*, the predominant macrolide-lincosamide-streptogramin B resistance gene in *Clostridium perfringens*. *Antimicrob. Agents Chemother.* **38**, 1041–1046.
68. Hächler, H., Berger-Bächi, B., and Kayser, F. H. (1987). Genetic characterization of a *Clostridium difficile* erythromycin-clindamycin resistance determinant that is transferable to *Staphylococcus aureus*. *Antimicrob. Agents Chemother.* **31**, 1039–1045.
69. Brisson-Noël, A., Arthur, M., and Courvalin, P. (1988). Evidence for natural gene transfer from gram-positive cocci to *Escherichia coli*. *J. Bacteriol.* **170**, 1739–1745.
70. Rasmussen, J. L., Odelson, D. A., and Macrina, F. L. (1986). Complete nucleotide sequence and transcription of *ermF*, a macrolide-lincosamide-streptogramin B resistance determinant from *Bacteroides fragilis*. *J. Bacteriol.* **168**, 523–533.
71. Smith, C. J. (1987). Nucleotide sequence analysis of Tn*4551*: Use of *ermFS* operon fusions to detect promoter activity in *Bacteroides fragilis*. *J. Bacteriol.* **169**, 4589–4596.
72. Halula, M. C., Manning, S., and Macrina, F. L. (1991). Nucleotide sequence of *ermFU* a macrolide-lincosamide-streptogramin (MLS) resistance gene encoding an RNA methylase from the conjugal element of *Bacteroides fragilis* V503. *Nucleic Acids Res.* **19**, 3453.
73. Wang, G., and Taylor, D. E. (1998). Site-specific mutations in the 23S rRNA gene of *Helicobacter pylori* confer two types of resistance to macrolide-lincosamide-streptogramin B antibiotics. *Antimicrob. Agents Chemother.* **42**, 1952–1958.

74. Occhialini, A., Urdaci, M., Doucet-Populaire, F., Bébéar, C. M., Lamouliatte, H., and Mégraud, F. (1997). Macrolide resistance in *Helicobacter pylori*: Rapid detection of point mutations and assays of macrolide binding to ribosomes. *Antimicrob. Agents Chemother.* **41**, 2724–2728.
75. Ross, J. I., Eady, E. A., Cove, J. E., Jones, C. E., Ratyal, A. H., Miller, Y. W., Vyakrnam, S., and Cunliffe, W. J. (1997). Clinical resistance to erythromycin and clindamycin in cutaneous *Propionibacteria* isolated from acne patients is associated with mutations in 23S rRNA. *Antimicrob. Agents Chemother.* **41**, 1162–1165.
76. Lucier, T. S., Heitzman, K., Liu, S. K., and Hu, P. C. (1995). Transition mutation in the 23S rRNA of erythromycin-resistant isolates of *Mycoplasma pneumoniae*. *Antimicrob. Agents Chemother.* **39**, 2770–2773.
77. Roberts, M. C., Sutcliffe, J., Courvalin, P., Jensen, L. B., Rood, J., and Seppala, H. (1999). Nomenclature for macrolide and macrolide-lincosamide-streptogramin B resistance determinants. *Antimicrob. Agents Chemother.* **43**, 2823–2830.
78. Jordan, D. C. (1963). Effect of chalcomycin on protein synthesis by *Staphylococcus aureus*. *Can. J. Microbiol.* **9**, 129–132.
79. Mao, J. C-H., and Robishaw, E. E. (1971). Effects of macrolides on peptide-bond formation translocation. *Biochemistry* **10**, 2054–2061.
80. Hill, R. N. (1969). The effect of antibiotics on the interaction of T-factor, aminoacyl-tRNA and ribosomes. *J. Gen. Microbiol.* **58**, viii.
81. Cerná, J., Rychlík, I., and Pulkrábek, P. (1969). The effect of antibiotics on corded binding of peptidyl-tRNA to the ribosome and on the transfer of the peptidyl residue to puromycin. *Eur. J. Biochem.* **9**, 27–35.
82. Adamson, S. D., Howard, G. A., and Herbert, E. (1969). The ribosome cycle in a reconstituted cell-free system from reticulocytes. *In* "Symposia on Quantitative Biology," Vol. XXXIV, pp. 547–554. Cold Spring Harbor Laboratory, Cold Spring Harbor, NY.
83. Moldave, K. (1985). Eukaryotic protein synthesis. *Ann. Rev. Biochem.* **54**, 1109–1149.
84. Merrik, W. C. (1990). Eukaryotic mRNAs: Strange solutions require unusual problems. *In* "The Ribosome—Structure, Function, and Evolution" (W. E. Hill, A. Dahlberg, R. A. Garrett, P. B. Moore D. Schlessinger, and J. R. Warner, Eds.), pp. 203–214. American Society for Microbiology, Washington, DC.
85. Wool, I. G., Endo, Y., Chan, Y-L., and Glück, A. (1990). Structure, function, and evolution of mammalian ribosome. *In* "The Ribosome—Structure, Function, and Evolution" (W. E. Hill, A. Dahlberg, R. A. Garrett, P. B. Moore D. Schlessinger, and J. R. Warner, Eds.), pp. 203–214. American Society for Microbiology, Washington, DC.
86. Champney, W. S., and Burdine, R. (1995). Macrolide antibiotics inhibit 50S ribosomal subunit assembly in *Bacillus subtilis* and *Staphylococcus aureus*. *Antimicrob. Agents Chemother.* **39**, 2141–2144.
87. Champney, W. S. (1998). Inhibition of translation and 50S ribosomal subunit formation in *S. aureus* all by 11 different ketolide antibiotics. *Curr. Microbiol.* **37**, 418–425.
88. Monro, R. E., and Vázquez, D. (1967). Ribosome-catalyzed peptidyl transfer: Effect of some inhibitors of protein synthesis. *J. Mol. Biol.* **28**, 161–165.
89. Cerná, J., Jonák, J., and Rychlík, I. (1971). Effect of macrolide antibiotics on the ribosomal peptidyl transferase in cell-free systems derived from *Escherichia coli* B and erythromycin-resistant mutant of *Escherichia coli* B. *Biochim. Biophys. Acta* **240**, 109–121.
90. Kubota, K., Okuyama, A., and Tanaka, N. (1972). Differential effects of antibiotics on peptidyl transferase reactions. *Biochem. Biophys. Res. Commun.* **47**, 1196–1202.
91. Jayaraman, J., and Goldberg, I. H. (1968). Localization of sparsomycin action to the peptide-bond-forming step. *Biochemistry* **7**, 418–421.
92. Mao, J. C.-H., and Robishaw, E. E. (1972). Erythromycin, a peptidyltransferase effector. *Biochemistry* **11**, 4864–4872.

93. Ennis, H. L. (1972). Polysome metabolism in *Escherichia coli*. Effect of antibiotics on polysome stability. *Antimicrob. Agents Chemother.* **1**, 197–203.
94. Cundliffe, E. (1969). Antibiotics and polyribosomes. II. Some effects of lincomycin, spiramycin, and streptogramin A *in vivo*. *Biochemistry* **8**, 2063–2066.
95. Vogel, Z., Vogel, T., and Elson, D. (1971). The effect of erythromycin on peptide bond formation and termination reaction. *FEBS Lett.* **15**, 249–253.
96. Igarashi, K., Ishitsuka, H., and Kaji, A. (1969). Comparative studies on the mechanism of action of lincomycin, streptomycin, and erythromycin. *Biochem. Biophys. Res. Commun.* **37**, 499–504.
97. Nakajima, Y., Takeda, R., Tani, K., Endou, K., Matsuoka, M., and Yamagishi, S. (1990). Greatly improved activity of staphylococcal ribosomes in polyadenylate directed polylysine synthesis: As an assay system for investigating their sensitivity to macrolide antibiotics. *J. Pharmacobio. Dyn.* **13**, 378–383.
98. Noller, H. F., Hoffarth, V., and Zimniak, L. (1992). Unusual resistance of peptidyl transferase to protein extraction procedures. *Science* **256**, 1416–1419.
99. Nitta, I., Ueda, T., and Watanabe K. (1998). Possible involvement of *Escherichia coli* 23S ribosomal RNA in peptide bond formation. *RNA* **4**, 257–267.
100. Nitta, I., Kamada, Y., Noda, H., Ueda, T., and Watanabe, K. (1998). Reconstitution of peptide bond formation with *Escherichia coli* 23S rRNA ribosomal RNA domains. *Science* **281**, 666–669.
101. Menninger, J. R., and Otto, D. P. (1982). Erythromycin, carbomycin, and spiramycin inhibit protein synthesis by stimulating the dissociation of peptidyl-tRNA from ribosomes. *Antimicrob. Agents Chemother.* **21**, 811–818.
102. Teraoka, H., and Nierhaus, K. H. (1978). Protein from *Escherichia coli* ribosomes involved in the binding of erythromycin. *J. Mol. Biol.*, **126**, 185–193.
103. Otaka, E., Teraoka, H., Tamaki, M., Tanaka, K., and Osawa, S. (1970). Ribosomes from erythromycin-resistant mutants of *Escherichia coli* Q 13. *J. Mol. Biol.* **48**, 499–510.
104. Dekio, S., Takata, R., Osawa, S., Tanaka, K., and Tamaki, M. (1970). Genetic studies of the ribosomal proteins in *Escherichia coli*. IV. Pattern of the alteration of ribosomal protein components in mutants resistant to spectinomycin or erythromycin in different strains of *Escherichia coli*. *Mol. Gen. Genet.* **107**, 39–49.
105. Wittmann, H., Stöffler, G., Apirion, D. Rosen, L., Tanaka, K., Tamaki, M., Takata, R., Dekio, S., Otaka, E., and Osawa, S. (1973). Biochemical and genetic studies on two different types of erythromycin-resistant mutants of *Escherichia coli* with altered ribosomal proteins. *Mol. Gen. Genet.* **127**, 175–189.
106. Pestka, S. (1972). Transfer ribonucleic acid–ribosome complexes. XVIII. Peptidyl-puromycin synthesis on polyribosomes from *Escherichia coli*. *Proc. Natl. Acad. Sci. USA* **69**, 624–628.
107. Nakajima, Y. (1995). Macrolides, the attractive antibiotics—from a structural and functional aspect. *Jpn. J. Bacteriol.* **50**, 717–736 (in Japanese).
108. Chabbert, Y., and Hervé, J. (1956). *In vitro* antagonism between erythromycin and spiramycin. *Ann. Inst. Pasteur* **90**, 787–790.
109. Garrod, L. P. (1957). The erythromycin group of antibiotics. *Br. Med. J.* **2**, 57–63.
110. McGuire, J. M., Bunch, R. L., Anderson, R. C., Boaz, H. E., Flynn, E. H., Powell, H. M., and Smith, J. W. (1952). "Ilotycin," a new antibiotic. *Antibiot. Chemother.* **2**, 281–283.
111. Haight, T. H., and Finland, M. (1952). Resistance of bacteria to erythromycin. *Proc. Soc. Exp. Biol. Med.* **81**, 183–188.
112. Skinner, R., Cundliffe, E., and Schmidt, F. J. (1983). Site of action of a ribosomal RNA methylase responsive to erythromycin and other antibiotics. *J. Biol. Chem.* **258**, 12702–17206.
113. Westh, H., Hougaard, D. M., Vuust, J., and Rosdahl, V. T. (1995). Prevalence of *erm* gene classes in erythromycin-resistant *Staphylococcus aureus* strains isolated between 1959 and 1988. *Antimicrob. Agents Chemother.* **39**, 369–373.

114. Davis, J. (1994). Inactivation of antibiotics and the dissemination of resistance genes. *Science* **264**, 375–382.
115. Webb, V., and Davis, J. (1993). Antibiotic preparations contain DNA: A source of drug resistance genes? *Antimicrob. Agents Chemother.* **37**, 2379–2384.
116. Weisblum, B. (1995). Erythromycin resistance by ribosome modification. *Antimicrob. Agents Chemother* **39**, 577–585.
117. Clancy, J., Petitpas, J., Dib-Hajj, F., Yuan, W., Cronan, M., Kamath, A. V., Bergeron, J., and Retsema, J. A. (1996). Molecular cloning and functional analysis of a novel macrolide-resistance determinant, *mefA*, from *Streptococcus pyogenes*. *Mol. Microbiol.* **22**, 867–879.
118. Clancy, J., Dib-Hajj, F., Petitpas, J., and Yuan, W. (1997). Cloning and characterization of a novel macrolide efflux gene, *mreA*, from *Streptococcus agalactiae*. *Antimicrob. Agents Chemother.* **41**, 2719–2723.
119. Ounissi, H., and Courvalin, P. (1985). Nucleotide sequence of the gene *ereA* encoding the erythromycin esterase in *Escherichia coli*. *Gene* **35**, 271–278.
120. Wondrack, L., Massa, M., Yang, B. V., and Sutcliffe, J. (1996). Clinical strain of *Staphylococcus aureus* inactivates and causes efflux of macrolides. *Antimicrob. Agents Chemother.* **40**, 992–998.
121. Allignet, J., Loncle, V., Mazodier, P., and El Solh, N. (1988). Nucleotide sequence of a staphylococcal plasmid gene, *vgb*, encoding a hydrolase inactivating the B components of virginiamycin-like antibiotics. *Plasmid* **20**, 271–275.
122. Matsuoka, M., Endou, K., Kobayashi, H., Inoue, M., and Nakajima, Y. (1998). A plasmid that encodes three genes for resistance to macrolide antibiotics in *Staphylococcus aureus*. *FEMS Microbiol. Lett.* **167**, 221–227.
123. O'Hara, K., Kanda, T., Ohmiya, K., Ebisu, T., and Kono, M. (1989). Purification and characterization of macrolide 2′-phosphotransferase from a strain of *E. coli* that is highly resistant to erythromycin. *Antimicrob. Agents Chemother.* **33**, 1354–1357.
124. Brisson-Noël, A., and Courvalin, P. (1986). Nucleotide sequence of gene *linA* encoding resistance to lincosamide in *Streptococcus haemolyticus*. *Gene* **43**, 247–253.
125. Ross, J. I., Farrell, A. M., Eady, E. A., Cove, J. H., and Cunliffe, W. J. (1989). Characterization and molecular cloning of the novel macrolide-streptogramin B resistance determinant from *Staphylococcus epidermidis*. *J. Antimicrob. Chemother.* **24**, 851–862.
126. Goldman, R. C., and Capobianco, J. O. (1990). Role of an energy-dependent efflux pump in plasmid pNE24-mediated resistance to 14- and 15-membered macrolides in *Staphylococcus epidermidis*. *Antimicrob. Agents Chemother.* **34**, 1973–1980.
127. Jánosi, L., Nakajima, Y., and Hashimoto, H. (1990). Characterization of plasmids that confer inducible resistance to 14-membered macrolide and streptogramin type B antibiotics in *Staphylococcus aureus*. *Microbiol. Immunol.* **34**, 723–735.
128. Matsuoka, M., Endou, K., Kobayasi, H., Inoue, M., and Nakajima, Y. (1997). A dyadic plasmid that shows MLS and PMS resistance in *Staphylococcus aureus*. *FEMS Microbiol. Lett.* **148**, 91–96.
129. Milton, I. D., Hewitt, C. L., and Harwood, C. R. (1992). Cloning and sequencing of a plasmid-mediated erythromycin resistance determinant from *S. xylosus*. *FEMS Microbiol. Lett.* **97**, 141–148.
130. Portillo, A., Ruiz-Larrea, F., Zarazaga, M., Alonso, A., Martinez, J. L., and Torres, C. (2000). Macrolide resistance genes in *Enterococcus* spp. *Antimicrob. Agents Chemother.* **44**, 967–971.
131. Sutcliffe, J., Tait-Kamradt, A., and Wondrack, L. (1996). *Streptococcus pneumoniae* and *Streptococcus pyogenes* resistant to macrolides but sensitive to clindamycin: A common resistance pattern mediated by an efflux system. *Antimicrob. Agents Chemother.* **40**, 1817–1824.
132. Tait-Kamradt, A., Clancy, J., Cronan, M., Dib-Hajj, F., Wondrack, L., Yuan, W., and Sutcliffe, J. (1997). *mefE* is necessary for erythromycin-resistant M phenotype in *Streptococcus pneumoniae*. *Antimicrob. Agents Chemother.* **41**, 2251–2255.

133. Authur, M., Autissier, D., and Courvalin, P. (1986). Analysis of the nucleotide sequence of the *ereB* gene encoding the erythromycin esterase type II. *Nucleic Acids Res.* **14**, 4987–4999.
134. Allignet, J., Liassine, N., and El Solh, N. (1998). Characterization of a staphylococcal plasmid related to pUB110 and carrying two novel genes, *vatC* and *vgbB*, encoding resistance to streptogramins A and B and similar antibiotics. *Antimicrob. Agents Chemother.* **42**, 1794–1798.
135. Kim, S. K., Baek, M. C., Choi, S. S., Kim, B. K, and Choi, E. C. (1996). Nucleotide sequence expression and transcriptional analysis of the *Escherichia coli mphK* gene encoding macrolide phosphotransferase. *Molecules Cells* **6**, 153–160.
136. Noguchi, N., Katayama, J., and O'hara, K. (1996). Cloning and nucleotide sequence of the *mphB* gene for macrolide 2′-phosphotransferase II in *Escherichia coli*. *FEMS Microbiol. Lett.* **144**, 197–202.
137. Brisson-Noël, A., Delrieu, P., Samain, D., and Courvalin, P. (1988). Inactivation of lincosamide antibiotics in *Staphylococcus*. Identification of lincosamide O-nucleotidyltransferases and comparison of the corresponding resistance genes. *J. Biol. Chem.* **263**, 15880–15887.
138. Brisson-Noël, A., and Courvalin, P. (1986). Nucleotide sequence of gene *linA* encoding resistance to lincosamide in *Streptococcus haemolyticus*. *Gene* **43**, 247–253.
139. Bozdogan, B., Berrezouga, L., Kuo, M.-S., Yurek, D. A., Farley, K. A., Stockman, B. J., and LeClercq, R. (1999). A new resistance gene, *linB*, conferring resistance to lincosamide by nucleotidylation in *Enterococcus faecium* HM1025. *Antimicrob. Agents Chemother.* **43**, 925–929.
140. Jones, W. F., Nichols, R. L., and Finland, M. (1966). Development of resistance and cross-resistance *in vitro* to erythromycin, carbomycin, oleandomycin and streptogramin. *Proc. Soc. Exp. Biol. Med.* **93**, 388–393.
141. Lampson, B. C., and Parisi, J. T. (1986). Naturally occurring *Staphylococcus epidermidis* plasmid expressing constitutive macrolide-lincosamide-streptogramin B resistance contains a deleted attenuator. *J. Bacteriol.* **166**, 479–483.
142. Thakker-Varia, S., Jenssen, W. D., Moon-McDermott, L., Weinstein, M. P., and Dubin, D. T. (1987). Molecular epidemiology of macrolide-lincosamide-streptogramin B resistance in *Staphylococcus aureus* and coagulase-negative staphylococci. *Antimicrob. Agents Chemother.* **31**, 735–743.
143. Serwold-Davis, T. M., and Groman, N. B. (1988). Identification of a methylase gene for erythromycin resistance within the sequence of a spontaneously deleting fragment of *Corynebacterium diphtheriae* plasmid pNG2. *FEMS Microbiol. Lett.* **46**, 7–14.
144. Monod, M., Denoya, C., and Dubnau, D. (1986). Sequence and properties of pIM13, a macrolide-lincosamide-streptogramin B resistance plasmid from *Bacillus subtilis*. *J. Bacteriol.* **167**, 138–147.
145. Monod, M., Mohan, S., and Dubnau, D. (1987). Cloning and analysis of *ermG*, a new macrolide-lincosamide-streptogramin B resistance element from *Bacillus sphaericus*. *J. Bacteriol.* **169**, 340–350.
146. Gryczen, T., Israelic-Reches, M., Del Bue, M., and Dubnau, D. (1984). DNA sequence and regulation of *ermD*, a macrolide-lincosamide-streptogramin B resistance element from *Bacillus licheniformis*. *Mol. Gen. Genet.* **194**, 349–356.
147. Kim, H-S., Choi, E-C., and Kim, B-K. (1993). A macrolide-lincosamide-streptogramin B resistance determinant from *Bacillus anthracis* 590: Cloning and expression of *ermJ*. *J. Gen. Microbiol.* **139**, 601–607.
148. Kwak, J-H., Choi, E-C., and Weisblum, B. (1991). Transcriptional attenuation control of *ermK*, a macrolide-lincosamide-streptogramin B resistance determinant from *Bacillus licheniformis*. *J. Bacteriol.* **173**, 4725–4735.
149. Klaenhammer, T. R. (1994). Molecular characterization of a plasmid-born (pGT633) erythromycin resistance determinant (*ermGT*) from *Lactobacillus reuteri* 100–63. *Plasmid* **31**, 60–71.

150. Eady, E. A., Ross, J. I., Cove, J. H., Holland, K. T., and Cundliffe, W. J. (1989). Macrolide-lincosamide-streptogramin B (MLS) resistance in cutaneous propionibacteria: Definition of phenotype. *J. Antimicrob. Chemother.* **23**, 493–502.
151. Arthur, M., Andremont, A., and Courvalin, P. (1987). Distribution of erythromycin esterase and RNA methylase gene in members of the family Enterobacteriaceae highly resistant to erythromycin. *Antimicrob. Agents Chemother.* **31**, 404–409.
152. Kamimiya, S., and Weisblum, B. (1988). Translational attenuation control of *ermSF*, an inducible resistance determinant encoding rRNA N-methyltransferase from *Streptomyces fradiae*. *J. Bacteriol.* **170**, 1800–1811.
153. Zalacain, M., and Cundliffe, E. (1989). Methylation of 23S rRNA by *tlrA* (*ermSF*), a tylosin resistance determinant from *Streptomyces fradiae*. *J. Bacteriol.* **171**, 4254–4260.
154. Zalacain, M., and Cundliffe, E. (1991). Cloning of *tlrD*, a fourth resistance gene, from the tylosin producer, *Streptomyces fradiae*. *Gene* **97**, 137–142.
155. Epp, J. K., Burgett, S. G., and Schoner, B. E. (1987). Cloning and nucleotide sequence of carbomycin-resistance gene from *Streptomyces thermotolerans*. *Gene* **53**, 73–83.
156. Calcutt, M. J., and Cundliffe, E. (1990). Cloning of a lincosamide resistance determinant from *Streptomyces caelestis*, the producer of celesticetin and characterization of the resistance mechanism. *J. Bacteriol.* **172**, 4710–4714.
157. Zhang, H.-Z., Schmidt, H., and Piepersberg, W. (1992). Molecular cloning and characterization of two lincomycin-resistance genes, *lmrA* and *lmrB*, from *Streptomyces lincolnensis* 78-11. *Mol. Microbiol.* **6**, 2147–2157.
158. Jenkins, G., and Cundliffe, E. (1991). Cloning and characterization of two genes from *Streptomyces lividans* that confer inducible resistance to lincomycin and macrolide antibiotics. *Gene* **108**, 55–62.
159. Hara, O., and Hutchinson, C. R. (1990). Cloning of midecamycin (MLS)-resistance genes from *Streptomyces mycarofaciens*, *Streptomyces lividans* and *Streptomyces coelicolor* A3(2). *J. Antibiot.* **43**, 977–991.
160. Inouye, M., Morohoshi, T., Horinouchi, S., and Beppu, T. (1994). Cloning and sequences of two macrolide-resistance-encoding genes from mycinamicin-producing *Micromonospora griseorubida*. *Gene* **141**, 39–46.
161. Uchiyama, H., and Weisblum, B. (1985). *N*-Methyl transferase of *Streptomyces erythreus* that confers resistance to the macrolide-lincosamide-streptogramin B antibiotics: Amino acid sequence and its homology to cognate R-factor enzymes from pathogenic bacilli and cocci. *Gene* **38**, 103–110.
162. Roberts, A. N., Hudson, G. S., and Brenner, S. (1985). An erythromycin-resistance gene from an erythromycin-producing strain of *Arthrobacter* sp. *Gene* **35**, 259–270.
163. Davis, J. (1990). What are antibiotics? Archaic functions for modern activities. *Mol. Microbiol.* **4**, 1227–1232.
164. Crick, F. H. C. (1968). The origin of the genetic code. *J. Mol. Biol.* **38**, 367–379.
165. Wittman, H. G., Stöfler, G., Apirion, D., Rosen, L., Tanaka, K., Tamaki, L., Tanaka, R., Dekio, S., Otaka, E., and Osawa, S. (1973). Biochemical and genetic studies on two different types of erythromycin resistant mutants of *Escherichia coli* with altered ribosomal proteins. *Mol. Gen. Genet.* **127**, 175–189.
166. Noller, H. F., Kop, J., Wheaton, V., Brosius, J., Gutell, R. R., Kopylov, A. M., Dohme, F., Herr, W., Stahl, D. A., Gopta, R., and Wpese, C. R. (1981). Secondary structure model for 23S ribosomal RNA. *Nucleic Acids Res.* **9**, 6167–6189.
167. Douthwaite, S., Prince, J. B., and Noller, H. F. (1985). Evidence for functional interaction between domains II and V of 23S ribosomal RNA from an erythromycin mutant. *Proc. Natl. Acad. Sci. USA* **82**, 8330–8334.
168. Douthwaite, S. (1992). Functional interactions within 23S rRNA involving the peptidyltransferase center. *J. Bacteriol.* **174**, 1333–1338.

169. Versalovic, J., Shortridge, D., Kibler, K., Griffy, M. V., Beyer, J., Flamm, R. K., Tanaka, S. K., Graham, D. Y., and Go, M. F. (1996). Mutations in 23S rRNA are associated with clarithromycin resistance in *Helicobacter pylori*. *Antimicrob. Agents Chemother.* **40**, 477–480.
170. Lai, C. J., and Weisblum, B. (1971). Altered methylation of ribosomal RNA in an erythromycin-resistant strain of *Staphylococcus aureus*. *Proc. Natl. Acad. Sci. USA* **68**, 856–860.
171. Egebjerg, J., Larsen, N., and Garrett, R. A. (1990). Structural map of 23S rRNA. *In* "The Ribosome—Structure, Function, and Evolution" (W. E. Hill, A. Dahlberg, R. A. Garrett, P. B. Moore, D. Schlessinger, and J. R. Warner, Eds.), pp. 168–179. American Society for Microbiology, Washington, DC.
172. Vester, B., and Garrett, R. A. (1987). A plasmid-coded and site-directed mutation in *Escherichia coli* 23 S RNA that confers resistance to erythromycin: Implications for the mechanism of action of erythromycin. *Biochimie* **69**, 891–900.
173. Sor, F., and Fukuhara, H. (1982). Identification of two erythromycin resistance mutations in the mitochondrial gene coding for the large ribosomal RNA in yeast. *Nucleic Acids Res.* **10**, 6571–6577.
174. Shivakumar, A. G., Hahn, J., Grandi, G., Kozlov, Y., and Dubnau, D. (1980). Posttranscriptional regulation of an erythromycin resistance protein specified by plasmid pE194. *Proc. Natl. Acad. Sci. USA* **7**, 3903–3907.
175. Choi, S-S., Kim, S-K., Oh, T-G., and Choi, E-C. (1997). Role of mRNA termination in regulation of *ermK*. *J. Bacteriol.* **179**, 2065–2067.
176. Meier, A., Kirschner, P., Springer, V. A., Steingrub, V. A., Brown, B. A., Wallace, R. J., and Bötger, E. C. (1994). Identification of mutations in the 23S ribosomal RNA gene of clarithromycin resistant *Mycobacterium intracellulare*. *Antimicrob. Agents Chemother.* **38**, 381–384.
177. Moazed, D., and Noller, H. F. (1987). Chloramphenicol, erythromycin, and vernamycin B protect overlapping sites in the peptidyl transferase region of 23S ribosomal RNA. *Biochemie* **69**, 879–884.
178. Douthwaite, S., and Aagaard, C. (1993). Erythromycin binding is reduced in ribosomes with conformational alterations in the 23S rRNA peptidyl transferase loop. *J. Mol. Biol.* **232**, 725–731.
179. Fernandez-Muñoz, R., Monro, R. E., Torres-Pinedo, R., and Vázquez, D. (1971). Substrate- and antibiotic-binding sites at the peptidyltransferase center of *Escherichia coli* ribosomes. Studies on the chloramphenicol, lincomycin and erythromycin sites. *Eur. J. Biochem.* **23**, 185–193.
180. Weisblum, B. (1985). Inducible resistance to macrolides, lincosamides and streptogramin type B antibiotics: The resistance phenotype, its biological diversity, and structural elements that regulate expression—a review. *J. Antimicrob. Chemother.* **16** (Suppl. A), 63–90.
181. Weisblum, B. (1995). Insights into erythromycin action from studies of its activity as inducer of resistance. *Antimicrob. Agents Chemother.* **39**, 797–805.
182. Leclercq, R., and Courvalin, P. (1991). Bacterial resistance to macrolide, lincosamide, and streptogramin antibiotics by target modification. *Antimicrob. Agents Chemother.* **35**, 1267–1272.
183. Weisblum, B. (1998). Macrolide resistance. *Drug Resistance Updates* **1**, 29–41.
184. Levy, S. B., McMurry, L. M., Burdett, V., Courvalin, P., Hillen, W., Roberts, M. C., and Taylor, D. E. (1989). Nomenclature for tetracycline resistance determinants. *Antimicrob. Agents Chemother.* **33**, 1373–1374.
185. Levy, S. B., McMurry, L. M., Barbosa, T. M., Burdett, V., Courvalin, P., Hillen, W., Roberts, M. C., Rood, J. I., and Taylor, D. E. (1999). Nomenclature for new tetracycline resistance determinants. *Antimicrob. Agents Chemother.* **43**, 1523–1524.
186. Wistow, G. J., and Piatigorsky, J. (1990). Gene conversion and splice-site slippage in the argininosucccinate lyases/δ-crystallins of the duck lens: Members of an enzyme superfamily. *Gene* **96**, 263–270.

187. Sigmund, C. D., and Morgan, E. A. (1982). Erythromycin resistance due to a mutation in a ribosomal RNA operon of *Escherichia coli*. *Proc. Natl. Acad. Sci. USA* **79**, 5602–5606.
188. Dam, M., Douthwaite, S., Tenson, T., and Mankin, A. S. (1996). Mutation in domain II of 23S rRNA facilitates translation of a 23S rRNA-encoded pentapeptide conferring erythromycin resistance. *J. Mol. Biol.* **259**, 1–6.
189. Tenson, T., DelBlasio, A., and Mankin, A. (1997). A functional peptide encoded in the *Escherichia coli* 23S rRNA. *Proc. Natl. Acad. Sci. USA* **93**, 5641–5646.
190. Tenson, T., Xiong, L., Kloss, O., and Mankin, A. S. (1997). Erythromycin resistance peptides selected from random peptide libraries. *J. Biol. Chem.* **272**, 17425–17430.
191. Douthwaite, S., Powers, T., Lee, J. Y., and Noller, H. F. (1989). Defining the structural requirements for a helix in 23S ribosomal RNA that confers erythromycin resistance. *J. Mol. Biol.* **209**, 655–665.
192. Novikova, S. I., Bushueva, A. M., Trachuk, L. A., Konstantinova, G. E., Serkina, A. V., Hoischen, C., Gumpert, J., Chestukhina, G. G., Mankin, A., and Shevelev, A. B. (2000). Introduction of a mini-gene encoding a five-amino acid peptide confers erythromycin resistance on *Bacillus subtilis* and provides temporary erythromycin protection in *Proteus mirabilis*. *FEMS Microbiol. Lett.* **182**, 213–218.
193. Ross, J. I., Eady, E. A., Cove, J. H., and Baumberg, S. (1996). Minimal functional system required for expression of erythromycin resistance by msrA in *Staphylococcus aureus* RN4220. *Gene* **183**, 143–148.
194. Paulsen, I. T., Brown, M. H., and Skurray, R. A. (1996). Proton-dependent multidrug efflux systems. *Microbiol. Rev.* **60**, 575–608.
195. Lewis, K. (1994). Multidrug resistance pumps in bacteria: Variations on a theme. *Trends Biochem. Sci.* **19**, 119–123.
196. Endou, K., Matsuoka, M., and Nakajima, Y. (1987). A mode of resistance to macrolide antibiotics: Regarding *Staphylococcus aureus* that does not show MLS-coresistance. *J. Pharmacobio. Dyn.* **10**, S–32.
197. Matsuoka, M., Jánosi, L., Endou, K., and Nakajima, Y. (1999). Cloning and sequences of inducible and constitutive macrolide resistance genes in *Staphylococcus aureus* that correspond to an ABC transporter. *FEMS Microbiol. Lett.* **181**, 91–100.
198. Matsuoka, M., and Nakajima, Y. (1996). A distinctive effect of CCCP on the transfer of erythromycin to 1-octanol: As a possible model for playing a role in promoting the intracellular antibiotic-accumulation through lipid in a staphylococcal cytoplasmic membrane. *Res. Commun. Mol. Pathol. Pharmacol.* **92**, 85–93.
199. Matsuoka, M. (2000). Study of macrolide, lincosamide, and streptogramin B antibiotics resistance in *Staphylococcus aureus*. *Yakugaku Zasshi* **120**, 374–386 (in Japanese).
200. Gutstein, M. (1932). Determination of hydrogen-ion concentration in living yeast and bacterial cells. *Protoplasma* **17**, 454–470 (in German).
201. Lovett, P. S., and Rogers, E. J. (1996). Ribosome regulation by the nascent peptide. *Microbiol. Rev.* **60**, 366–385.
202. Mayford, M., and Weisblum, B. (1989). ermC leader peptide, amino acid sequence critical for induction by translational attenuation. *J. Mol. Biol.* **206**, 69–79.
203. Cundliffe, E., and McQuillen, K. (1967). Bacterial protein synthesis: The effect of antibiotics. *J. Mol. Biol.* **30**, 137–146.
204. Agouridas, C., Bonnefoy, A., and Chantot, J-F. (1998). HMR3647: Antibacterial activity and resistance. *In* "Abstracts of the 4th International Conference on the Macrolides, Azalides, Streptogramins & Ketolides," p. 25 (Abstr. 1.24). ICMAS, Barcelona, Spain.
205. Jones, R. N., and Biedenbach, D. J. (1997). Antimicrobial activity of RU66647, a new ketolide. *Diagn. Microbiol. Infect. Dis.* **27**, 7–12.

206. Ma, Z., Clark, R. F., Wang, S., Nilius, A. M., Flamm, R. K., and Or, Y. S. (1999). Design, synthesis and characterization of ABT-773: A novel ketolide highly active against multidrug resistant pathogens. Presented at American Society for Microbiology 39th Interscience Conf. on Antimicrobial Agents and Chemotherapy (San Francisco). Poster No. 2133.
207. Bryskier, A., Agouridas, C., and Chantot, J. F. (1995). New insights into the structure–activity relationship of macrolides and azalides. In "New Macrolides, Azalides, and Streptogramins in Clinical Practice." (H. C. Neu, L. S. Young, S. H. Zinner, and J. F. Acar, Eds.), pp. 3–30. Marcel Dekker, NY.
208. Bryskier, A., Agouridas, C., and Chantot, J. F. (2000). Ketolide: Novel antibacterial agent designed to overcome erythromycin A resistance. In "New Considerations for Macrolides, Azalides, Streptogramins, and Ketolides" (S. H. Zinner, L. S. Young, J. F. Acar, and C. Ortiz-Neu, Eds.), pp. 79–102. Marcel Dekker, NY.
209. Yazawa, K., Mikami, Y., Sakamoto, T., Ueno, Y., Morisaki, N., Iwasaki, S., and Furihata, K. (1994). Inactivation of the macrolide antibiotics erythromycin, midecamycin, and rokitamycin by pathogenic *Nocardia* species. *Antimicrob. Agents Chemother.* **38**, 2197–2199.
210. Schlünzen, F., Zarivach, R., Harms, J., Bashan, A., Tocilj, A., Albrecht, R., Yonath, A., and Franceschi, F. (2001). Structural basis for the interaction of antibiotics with the peptidyl transferase centre in eubacteria. *Nature* **413**, 814–821.

Chapter 11

Mode of Action of Macrolides with Motilin Agonistic Activity—Motilides

NOBUHIRO INATOMI
Discovery Research Laboratory III
Takeda Chemical Industries, Ltd.
Osaka, Japan

FUMIHIKO SATO
Pharmacology Laboratory II
Takeda Chemical Industries, Ltd.
Osaka, Japan

ZEN ITOH
5-10, Chiyodamachi 1-chome
Maebashi, Gunma, Japan

SATOSHI ŌMURA
Kitasato Institute for Life Sciences,
Kitasato University
and The Kitasato Institute
Tokyo, Japan

I. Introduction 501
II. Mode of Action of Motilin 504
   A. *In Vivo* Study 504
   B. *In Vitro* Study 506
III. Invention of Motilides 507
   A. Discovery of GMS Activity of Erythromycin A 507
   B. Chemical Modification of Erythromycin A 508
IV. Biological Activity of Motilides 510
   A. Gastrointestinal Motor-Stimulating Action *in Vivo* .. 510
   B. *In Vitro* Study on Rabbit Gastrointestinal Tract 521
V. Clinical Trials of Motilides 526
   A. Effects of Motilin and Erythromycin A on Gastric Emptying in Humans 526
   B. Other Clinical Trials of Motilides 527
VI. Concluding Remarks 527
   References 528

I. Introduction

The pattern of gastrointestinal motility in dogs and humans under normal conditions is well known to be quite different between the digestive (fed) and interdigestive (fasted) states. A typical example of the two contractile patterns in a conscious dog is shown in Fig. 1. In the digestive state, relatively weak repetitive contractions without quiescence are observed. By contrast, during the fasted state,

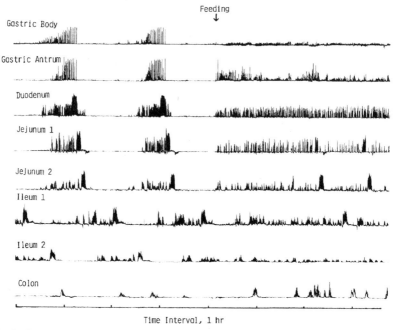

Fig. 1. Gastrointestinal contractile pattern before and after feeding in a dog. (From Inatomi et al. [21] with permission of the editor.)

bursts of strong contractions occur periodically and migrate to the colon. In the stomach, a group of intense phasic contractions occurs at about 100-min intervals and lasts for about 25 min before abrupt and spontaneous termination. These strong contractions migrate along the small intestine in the caudad direction at a constant velocity. A similar contractile pattern has also been shown in the human gastrointestinal tract [1]. The burst of strong contractions during the fasted state is called the interdigestive migrating complex (IMC). The interdigestive contractile activity is divided into four phases, I to IV. Phase I corresponds to quiescence without action potential in electrical activity. Phase II means irregular contractions of weaker contractile force forming segmentations preceding phase III. Phase III is a group of strong contractions with maximum contractile force. Phase IV indicates contractions seen in the short period of transition from phase III to I and is often missing, especially in the stomach.

Motilin is a 22-amino-acid polypeptide isolated by Brown et al. [2] in the search for duodenal factors stimulating gastric motility in dogs by intraduodenal alkalization. Although differences in the amino acid sequence among animal species are seen, the 6-amino-acid sequence of the N terminal is common to the species. Pearse et al. [3] reported that, using an indirect immunofluorescence technique, motilin is produced by a type of endocrine cell that is abundant in the duodenum and jejunum,

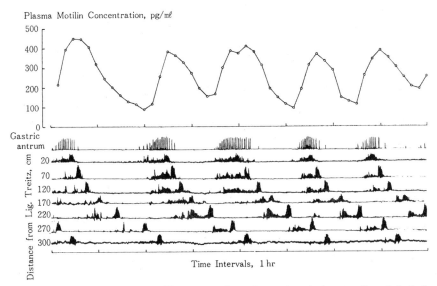

Fig. 2. Changes in the plasma motilin concentration and gastrointestinal contractile activity in the interdigestive state in a dog. (From Itoh and Sekiguchi [4] by permission of Taylor & Francis AS.)

present in the ileum, but absent from the stomach and large intestine. Itoh and Sekiguchi [4] found that a cyclic change in the plasma immunoreactive motilin concentration occurs in fasted dogs. Peaks of plasma motilin level coincide well with the occurrence of phase III in the stomach (Fig. 2). Similar changes in the plasma motilin concentration during fasting were observed in humans [5]. The periodic changes of plasma motilin concentration were not observed in the digestive state. The intravenous administration of porcine motilin has been shown to induce strong migrating contractions mimicking the natural IMC in dogs [6]. These findings suggest that motilin is involved in the occurrence of IMC in humans and dogs.

Erythromycin A (EM) has been used clinically as an antibiotic, and one of its side effects is observed as symptoms related to the gastrointestinal tract, such as nausea, vomiting, and diarrhea, especially when injected intravenously [7]. We found that EM induces strong gastrointestinal contractions in dogs [8]. Intravenous infusion of EM lactobionate at 50–100 µg/kg hr^{-1} induced a group of strong contractions in the stomach and the duodenum, and the contractions migrated along the intestinal tract (Fig. 3). The EM-induced contractions were quite similar to the naturally occurring IMC and the motilin-induced contractions. It was of interest that, although EM and motilin differ in chemical structure, they showed quite similar gastrointestinal motor-stimulating (GMS) activity. We conducted chemical modification of EM and found many derivatives that are more potent than EM without antimicrobial activities. We then designated this family of macrolide compounds *motilides* for the macrolides that have motilin-like activity.

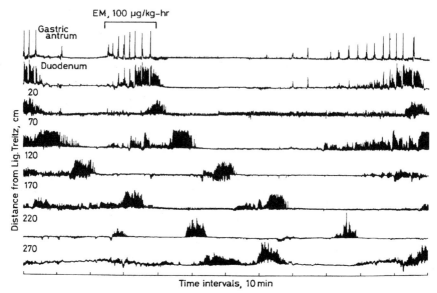

Fig. 3. Effect of intravenous infusion of erythromycin A (EM) on gastrointestinal motor activity in a dog. (From Itoh et al. [17] with permission of the editor.)

II. Mode of Action of Motilin

A. *In Vivo* Study

1. Effect in the Fasted State

Motilin does not stimulate gastrointestinal contractile activity in all species of experimental animals, even when it is administered in the fasted state. The rat, guinea pig, and rabbit do not respond to motilin even in the conscious state. Among experimental animals, the dog is the most suited for the study of motilin.

We examined the effect of exogenous motilin on the gastrointestinal motility of conscious dogs in both the digestive and interdigestive states. We used dogs weighing 10–15 kg, and implanted strain gauge force transducers permanently in the stomach and small intestine under pentobarbital anesthesia [9]. After a 2-week recovery period, the experiments were started. For intravenous administration of motilin and withdrawal of blood samples, a silicone tube was permanently implanted in the superior vena cava through the right external jugular vein. This tube was filled with heparin-containing saline, tipped with a plastic stopper. The tube and the lead wires of the force transducers were passed out of the abdominal cavity and sutured onto the back of the dog. After the operation, a jacket protector was placed on each dog to protect the lead wires and the silicone tubing from the

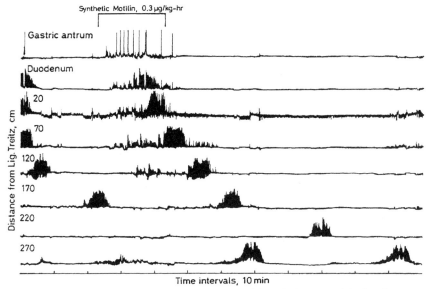

Fig. 4. Effect of intravenous infusion of motilin on gastrointestinal motor activity in a dog.

dog's scratching or biting. Gastrointestinal contractile activity was recorded on multichannel recorders through amplifiers by connecting the lead wires of the force transducers to the cable leads from the amplifiers under the jacket protector.

Figure 4 shows the effect of intravenous infusion of motilin on gastrointestinal contractile activity in fasted dogs. Motilin at 0.1 and 0.3 µg/kg hr^{-1} induced a series of strong contractions in the stomach and duodenum, which migrated through the small intestine toward the terminal ileum. When the dose of motilin was increased to 3 µg/kg hr^{-1}, strong contractions were evoked simultaneously in the stomach and upper small intestine. If a larger dose was infused or 3 µg/kg of motilin was given in a single bolus injection, intense contractions were induced followed by a couple of retrograde migrating contractions originating in the mid–small intestine, and vomiting was usually observed [10]. Exogenous motilin also induces contractile activity in the lower esophageal sphincter in association with motilin-induced strong contractions in the stomach. Infusion of synthetic motilin starting 10 min after the termination of natural IMC induced gallbladder contractions very similar to those observed during IMC.

Both naturally occurring IMC and motilin-induced contractions are inhibited by feeding. Feeding promptly changed the contractile patterns in the lower esophageal sphincter and stomach. The immediate change from the fasted to the digestive pattern suggests the involvement of neuronal regulation. Gastrin and cholecystokinin (CCK), which are gastrointestinal hormones released after feeding, show inhibitory effects on motilin-induced stomach contractions [11, 12]. Secretin, vasoactive intestinal polypeptide (VIP), glucagon and gastrin inhibitory polypeptide (GIP), which are members of the secretin family, do not show inhibition of motilin-induced IMC-like contractions. Instillation of acid solution into the stomach inhibited both IMC and motilin-induced gastric contractions [13].

Motilin-induced contractions in the gastrointestinal tract as well as the naturally occurring IMC are completely inhibited by atropine (a muscarinic receptor antagonist) and hexamethonium (a ganglion blocker). These findings indicate that motilin-induced contractions are brought about through the cholinergic pathway. Dopamine, a neurotransmitter in the brain, showed inhibition of motilin-induced contractions probably due to inhibition of acetylcholine release from the cholinergic nerve terminals. Ondansetron and granisetron, serotonin-3 (5-HT_3) receptor antagonists, inhibited spontaneous IMC and motilin-induced contractions in the stomach and altered the contractile pattern in the duodenum and small intestine. However, the motilin-induced contractions in the denervated fundic pouch were not affected at all by 5-HT_3 receptor antagonists [14]. As described in a later section, ondansetron completely inhibited the gastric contractions induced by a low dose (1 µg/kg hr^{-1}) of motilin, but did not completely inhibit the contractions induced by a high dose (3 µg/kg hr^{-1} or more) of motilin [15]. The results suggest that the 5-HT_3 receptor antagonist blocks the transmission of signals in the vagal cholinergic neurons.

2. Effect in the Fed State

Intravenous motilin at 0.5 µg/kg hr^{-1}, which induces IMC-like contractions in the fasted state, does not influence contractile activity in the gastrointestinal tract in the fed state. When given at a supr60physiological dose (3 µg/kg hr^{-1}) in a single bolus injection (0.6 µg/kg), motilin stimulated contractions in the stomach, duodenum, and small intestine. The migrating contractions as observed in the fasted state were not observed in the fed state.

The increased contractile activity induced by exogenous motilin in the fed state was strongly inhibited by atropine and hexamethonium, but was little affected by ondansetron. These results suggest that cholinergic neurons are involved in the stimulatory effect of motilin at high doses in the fed state as well as at low doses in the fasted state.

B. *In Vitro* Study

Motilin induces contractions in the isolated small intestine of the rabbit and human, but not the isolated intestine of the rat, guinea pig, and dog. We investigated the effect of motilin on rabbit intestinal segments by means of the Magnus method. The duodenum and small intestine were isolated, and the specimens were suspended longitudinally in Kreb's solution at 37°C and continuously gassed with 95% O_2–5% CO_2. Isotonic contractions were recorded on a polygraph with an isotonic transducer and amplifier. Motilin was added cumulatively to the organ bath, and the mean of the increase was expressed relative to that induced by 10^{-4} M acetylcholine.

The rabbit intestinal segments showed rhythmic contractions superimposed on the basal tonus. Porcine motilin caused an increase in the basal tonus in a concentration-dependent manner without affecting the frequency of phasic contractions. The sensitivity of the response to motilin decreased aborally; the minimum effective concentrations for the duodenum, jejunum, and ileum were 1×10^{-10}, 3×10^{-9}, and 1×10^{-8} M, respectively. The maximal contractile responses of the duodenum, jejunum, and ileum were 89%, 50%, and 11%, respectively.

Similar contractile activity of motilin was observed in the isolated muscle strip of human gastric antrum and isolated myocytes [16]. Motilin (10^{-8}–10^{-6} M) caused concentration-dependent contractions of human antral muscle strips with an EC_{50} value of 1.4×10^{-7} M. Motilin caused concentration-dependent contractions of human antral myocytes with maximal contraction of 25.2% and an EC_{50} value of 4.9×10^{-12} M, indicating a direct contractile effect of motilin on smooth-muscle cells.

III. Invention of Motilides

A. Discovery of GMS Activity of Erythromycin A

Z. Itoh, one of the authors, found incidentally in 1984 that EM stimulates canine gastrointestinal motility when injected into dogs recovering from a surgical operation to implant the force transducers in the gastrointestinal tract to prevent infection. At first, the dose of injected EM was so high that gastrointestinal motility was intensely stimulated and the dog vomited. As the dosage was decreased to 100 µg/kg hr^{-1}, a group of strong contractions was observed in the fundus, body, and antrum of the stomach, and the contractions migrated along the intestine as shown in Fig. 3. The pattern of these contractions was very similar to that of motilin [17]. Many substances were known to stimulate gastrointestinal motility, but no substance except motilin was known to simultaneously stimulate contractile activity in the proximal and distal regions of the stomach and cause migrating contractions along the gastrointestinal tract. Tomomasa et al. [18] investigated the effect of EM on the gastrointestinal motility of healthy human subjects and found that infusion of 1 and 3 mg/kg hr^{-1} of EM induces strong migrating contractions that mimic naturally occurring IMC.

The effects of other macrolide compounds on gastrointestinal motility were also investigated in conscious dogs. Oleandomycin, which has a 14-membered ring like EM, showed about one-tenth of the activity of EM in stimulating gastrointestinal motility. But 16-membered ring macrolides such as leukomycin, acetylspiromycin, and tyrocine showed no GMS activity [7]. These results suggested that the GMS activity of macrolides is due to the specific feature contained in the chemical structure of EM.

These findings prompted us to chemically modify EM in an attempt to find a new prokinetic drug (a drug that stimulates gastrointestinal motility and enhances propulsion of the luminal contents) to treat gastrointestinal disorders. Motilin has a unique GMS activity, but cannot be used as an orally prescribed prokinetic drug, because it is not absorbed from the gastrointestinal tract, and furthermore motilin degrades rapidly in the blood. Because EM is a low molecular weight compound and orally active, EM or its derivatives were expected to be developed as a novel prokinetic agent. The problem was that, because EM has potent antibiotic activities, it might affect the intestinal bacterial flora and induce EM-resistant bacteria after repeated administrations. Therefore, we planned systematic chemical modification of EM to obtain EM derivatives having stronger GMS activity without antibacterial activities.

B. Chemical Modification of Erythromycin A

EM's chemical structure consists of a 14-membered macrocyclic lactone ring (erythronolide) connected to a deoxyamino sugar (desosamine) and a deoxy sugar (cladinose) as shown in Fig. 5. We synthesized about 250 EM derivatives and examined their GMS and antibacterial activities [19, 20]. GMS activity was tested by intravenous injection of the test compounds to fasted conscious dogs with permanently implanted force transducers in the stomach, and antibacterial activity was estimated as minimum inhibitory concentration (MIC) by agar dilution method. The EM derivatives shown in Fig. 5 exhibited higher GMS activities with less antibacterial activities compared with those of EM (Table I).

EM201 obtained by mild acid treatment of EM showed GMS activity 10 times as strong as that of EM with much weaker antibacterial activities. Based on this result, a variety of *N*-methylalkyl analogues of EM201 were prepared from EM by sequential reactions with AcOH, I_2/NaOAc, and alkylhalide/diisopropylethylamine. Among them, EM523 (*N*-methylethyl) and EM574 (*N*-methylisopropyl) showed their respective GMS activities to be 18 and 248 times as strong as that of EM without antibacterial activity.

Quaternary *N*-substituted derivatives, EM485, EM491, EM511, and EM536, were prepared by *N*-alkylation of EM201 with corresponding alkyl halides. Similarly, quaternary *N*-substituted dihydroderivatives, EM502, EM506, and EM507, were obtained by *N*-alkylation of 9,9-dihydroerythromycin A 6,9-epoxide (EM501). EM485 and EM536 proved to have 21 and 2890 times more potent GMS activities than EM, respectively, and no or little antibacterial activities.

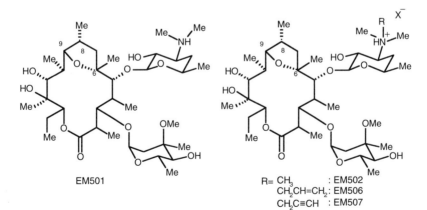

Fig. 5. Chemical structures of erythromycin A derivatives. (Reprinted with permission from Ōmura et al. [19]. Copyright 1987 American Chemical Society.)

TABLE I
Antibacterial Activities and Gastric Motor Stimulant Activities
of Erythromycin A and Its Derivatives

Compound	Antibacterial activity (MIC; μg/ml)[a]					Motor stimulant activity[b] (EM = 1)
	SA	BS	BC	EC	KP	
EM	0.2	0.1	0.1	12.5	6.25	1
EM201	50	25	25	>100	>100	10
EM523	>100	>100	>100	>100	>100	18
EM574	>100	>100	>100	>100	>100	248
EM485	>100	>100	>100	>100	>100	21
EM491	>100	>100	>100	>100	>100	111
EM511	100	>100	>100	>100	>100	256
EM536	100	100	100	>100	>100	2890
EM502	>100	>100	>100	>100	>100	65
EM506	100	100	100	>100	>100	115
EM507	>100	>100	>100	>100	>100	202

Source: Reprinted with permission from Ōmura *et al.* [19]. Copyright 1987 American Chemical Society.

[a]SA, *Staphylococcus aureus* ATCC6358P; BS, *Bacillus subtilis* ATCC6633; BC, *Bacillus cereus* IFO3001; EC, *Escherichia coli* NIHJ; KP, *Klebsiella pneumoniae* ATCC10031.

[b]Motor stimulant activity: Gastric antral motor index was measured after intravenous injection of each compound in conscious dogs.

Although EM536 showed the most potent GMS activity among the EM derivatives tested, the compound was poorly absorbed from intestine when it was given orally or intraduodenally. The reason is presumably due to its ionic nature. Thus, EM523 and EM574 were selected to be candidates for clinical trials. EM523 was initially chosen to be developed, but later it was replaced by EM574, which has more potent GMS activity than the former.

IV. Biological Activity of Motilides

A. Gastrointestinal Motor-Stimulating Action *in Vivo*

1. Effect in the Fasted State

The GMS activity of EM574 was studied in fasted conscious dogs using permanently implanted force transducers. To measure the motility quantitatively, the antral signals were input into a signal processor every 100 msec, and the motor index of drug-induced contractions was calculated [21]. Intravenous EM574 (0.3 μg/kg or more) induced phase III–like contractions starting from the stomach (Fig. 6), and the motor index increased in a dose-dependent manner (Fig. 7). Intraduodenal EM574 (1 μg/kg or more) induced IMC-like contractions within

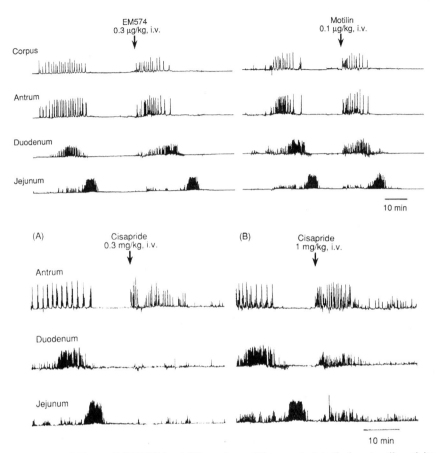

Fig. 6. *Top:* Effects of (A) EM574 and (B) porcine motilin on gastrointestinal contractile activity in the interdigestive state in the conscious dog. Both agents induced IMC-like migrating contractions. *Bottom:* Effect of cisapride on gastrointestinal contractile activity in the fasting dog [(A), 0.3 mg/kg; (B), 1 mg/kg]. The contractile pattern of cisapride-evoked contractions was different from those induced by EM574 and motilin. (From Inatomi *et al.* [15] with permission of the editor.)

5 min of being administered. EM574 was 322 and 186 times more potent than EM at inducing gastric contractions after intravenous and intraduodenal administration, respectively. Intravenous motilin (0.01 μg/kg or more) induced phase III–like contractions (Figs. 6 and 7). The maximum response was obtained with 0.3 μg/kg, and the gastric motility did not increase further at higher doses. Intraduodenal motilin (10 μg/kg) did not induce gastrointestinal contractions. A 5-HT$_4$ receptor agonist, cisapride given intravenously (100 μg/kg or more), induced contractions of the stomach, but the contractile pattern differed from those induced by motilin and EM574 (Fig. 6).

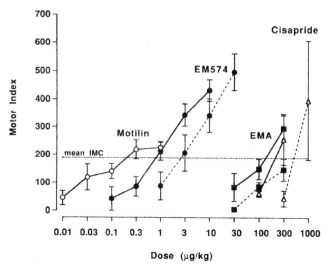

Fig. 7. Dose-related effects of motilin, EM574, erythromycin, and cisapride on gastric motility in the conscious dog. –○– motilin, i.v.; –●– EM574, i.v.; –●– EM574, i.d.; –■– erythromycin A, i.v.; –■– erythromycin A, i.d.; –△– cisapride, i.v.; –△– cisapride, i.d. The mean IMC indicates the mean gastric motor index during IMC. Data are expressed as means ±SE of four to six dogs. (From Inatomi et al. [15] with permission of the editor.)

Because the action of EM574 on gastrointestinal motility resembled that of motilin, the effect of EM574 on endogenous motilin release was investigated. Intraduodenal EM574 (30 μg/kg) increased antral motility markedly; the motor index exceeded the value during IMC 10 min after administration and then gradually decreased (Fig. 8). The mean plasma motilin concentration increased significantly in response to EM574, but the peak value was less than that during phase III. The results suggest that EM574 causes contractions mainly by acting on motilin receptors and that endogenous motilin released in association with EM574-induced contractions is involved, in part, in the motor-stimulating action.

The effects of pretreatment with various receptor antagonists on intraduodenal EM574 (10 μg/kg) and 30-min-infused motilin (1 μg/kg hr^{-1})-induced gastrointestinal motility were examined. Each antagonist was administered intravenously. Pretreatment with atropine (0.1 mg/kg) inhibited EM574- and motilin-induced gastric motility significantly by 95% and 94%, respectively (Table II). Hexamethonium (10 mg/kg) significantly inhibited the responses evoked by EM574 and motilin by 76% and 70%, respectively. Phentolamine and propranolol did not affect EM574- and motilin-induced gastric motility significantly. Mepyramine, famotidine, methysergide, and ketanserine inhibited neither EM574- nor motilin-induced gastric motility. The 5HT$_3$-receptor antagonist ondansetron (0.1 mg/kg) inhibited EM574- and motilin-induced gastric motility by 91% and 94%, respectively. Naloxone,

TABLE II
Effects of Receptor Antagonists on Gastric Motility Induced by EM574 and Motilin in Conscious Dogs[a]

Antagonist	Receptor	Dose (mg/kg, i.v.)	Gastric motor index EM574[b]		Motilin[c]	
Saline	—		266 ± 36		250 ± 22	
Atropine	Muscarinic	0.1	12 ± 5[d]	(95)[e]	15 ± 10[d]	(94)
Hexamethonium	Nicotinic	10	65 ± 25[d]	(76)	75 ± 17[d]	(70)
Phentolamine	α	2	221 ± 51	(17)	182 ± 41	(27)
Propranolol	β	1	247 ± 29	(7)	198 ± 28	(21)
Mepyramine	H_1	3	332 ± 62	(−25)	367 ± 41	(−47)
Famotidine	H_2	1	322 ± 56	(−21)	246 ± 32	(2)
Methysergide	5–HT_1, 5–HT_2	2	262 ± 34	(2)	253 ± 46	(−1)
Ketanserin	5–HT_2	1	358 ± 61	(−35)	300 ± 56	(−20)
Ondansetron	5–HT_3	0.1	23 ± 6[d]	(91)	15 ± 4[d]	(94)
Naloxone	Opioid	1	240 ± 23	(10)	213 ± 26	(15)
Saline	—		313 ± 41		361 ± 22	
CP-99994	NK_1	0.3	278 ± 78	(11)	320 ± 49	(11)
SR48968	NK_2	0.3	292 ± 25	(7)	391 ± 42	(−8)

Source: From Inatomi *et al.* [15] with permission of the editor.
[a]Data are expressed as means ± SE of three or five dogs.
[b]10 μg/kg, i.d.
[c]1 μg/kg hr^{-1}, i.v.
[d]$P < 0.01$ compared to saline treatment.
[e]Numbers in parentheses represent a percentage inhibition relative to saline treatment.

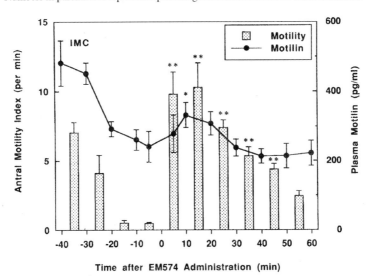

Fig. 8. Effect of intraduodenal EM574 (30 μg/kg) on gastric motility (□) and plasma motilin concentration (–●–). Data are expressed as means ± SE of five dogs. $^*P < 0.05$, $^{**}P < 0.01$ compared to the value just before EM574 administration. (From Inatomi *et al.* [15] with permission of the editor.)

CP-99994, and SR48968 had no significant effect on EM574- or motilin-induced gastric motility. The gastric motility stimulated by intravenous cisapride (1 mg/kg) was inhibited strongly by atropine and hexamethonium by 97% and 98%, respectively, but no inhibition was observed after treatment with ondansetron. These findings indicate that EM574 and motilin exert their actions by stimulating cholinergic neurons and that 5-HT_3 receptors are involved in their actions.

Intravenous infusion of EM574 (3 µg/kg hr^{-1}) and motilin (1 µg/kg hr^{-1}) induced phase III-like contractions, and the effect was sustained during the infusion for at least 1 hr. Ondansetron (0.1 mg/kg) abolished the gastric motility induced by EM574 at 3 µg/kg hr^{-1} or less for more than 60 min. However, gastric contractions were induced when the dose of EM574 was increased to 10 µg/kg hr^{-1} and higher under ondansetron (0.1 mg/kg) treatment, and the contractions were not inhibited even by 3 mg/kg of ondansetron, but were abolished by atropine (Fig. 9). The same responses to ondansetron were also observed with motilin-induced gastric contractions; the gastric motility induced by intravenous motilin

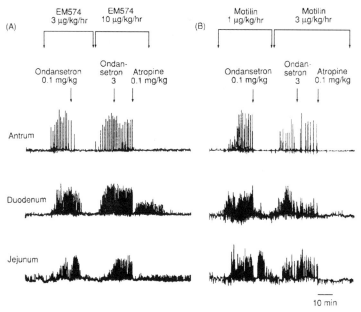

Fig. 9. Effects of ondansetron and atropine on gastrointestinal contractions induced by (A) EM574 and (B) motilin. Ondansetron (0.1 mg/kg) inhibited gastric contractions induced by EM574 at 3 µg/kg hr^{-1} and motilin at 1 µg/kg hr^{-1}. These inhibitory effects lasted for more than 1 hr (data not shown). Under the ondansetron treatment, increasing the doses of EM574 and motilin to 10 and 3 µg/kg hr^{-1}, respectively, resulted in prominent gastric contractions, which were not inhibited by ondansetron at 3 mg/kg, but were abolished by atropine. (From Inatomi et al. [15] with permission of the editor.)

(1 µg/kg hr^{-1}) was abolished by ondansetron at 0.1 mg/kg, but not the gastric contractions induced by intravenous infusion of motilin at 3 µg/kg hr^{-1}. The contractions were not inhibited by a higher dose of ondansetron, but were abolished by atropine (Fig. 9). The results suggest that 5-HT$_3$ receptors are present on the neuronal pathway stimulated by low doses, but not on that stimulated by high doses of EM574 and motilin.

The role of the vagus nerve in EM574- and motilin-induced gastrointestinal contractions was investigated using a cooling technique for temporary blockade of the vagus nerve. The experiment was performed on dogs whose vagosympathetic nerves had been previously isolated in skin loops on each side of the neck. Transient nerve blockade was accomplished by circulating 50% v/v alcohol at −20°C through copper cooling jackets placed around the skin loops [15]. Intravenous EM574 (3 µg/kg hr^{-1}) induced phase III-like contractions of the stomach and small intestine. Unilateral vagal blockade did not inhibit EM574-induced contractions. But bilateral vagal blockade almost completely inhibited the contractions in each region (Fig. 10). Under bilateral vagal blockade, increasing the dose of EM574 to 10 µg/kg hr^{-1} and higher induced strong contractions. The gastrointestinal contractions induced by intravenous motilin at 1 µg/kg hr^{-1} were not affected by unilateral vagal blockade, but they were abolished by bilateral blockade, under

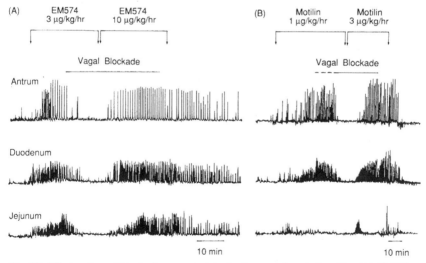

Fig. 10. Effects of vagal blockade on gastrointestinal contractions induced by (A) EM574 and (B) motilin. The period of bilateral vagal blockade is indicated by an unbroken bar and that of unilateral vagal blockade by a dashed bar. Bilateral blockade inhibited gastrointestinal contractions induced by EM574 (3 µg/kg hr^{-1}) and motilin (1 µg/kg hr^{-1}). After increasing the doses of EM574 and motilin, strong contractions were induced. Unilateral vagal blockade did not affect motilin-induced contractions. (From Inatomi *et al.* [15] with permission of the editor.)

which conditions motilin, at 3 µg/kg hr^{-1} or more, induced contractions in each region (Fig. 10). The results indicate that EM574 and motilin induce gastrointestinal contractions by stimulating vagal cholinergic nerves; and at higher doses, they evoke contractions via stimulation of a nonvagal pathway.

Because EM574 affects gastric motility by stimulating vagal cholinergic neurons, the possibility that EM574 stimulates gastric secretion was considered. The effect of EM574 on gastric acid secretion was investigated in dogs with gastric fistulae and permanently implanted force transducers. In the phase I state, intraduodenal EM574 (10 µg/kg) induced a marked increase in gastric motility; a peak was observed 15 min after administration, and then the motility gradually decreased. The gastric acid secretion was not affected by EM574, whereas subcutaneous tetragastrin (5 µg/kg) increased the acid secretion markedly. The results indicate that EM574 and motilin have little or no effect on basal gastric secretion, and that gastric motility and acid secretion are stimulated via different vagal cholinergic neurons. EM574 seems to stimulate the vagal neurons associated with gastrointestinal motility selectively.

Figure 11 is a diagram showing the sites of action of EM574 and motilin. At relatively low doses, EM574 and motilin stimulate the vagal cholinergic neurons;

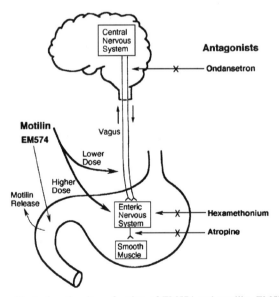

Fig. 11. Diagram illustrating the sites of action of EM574 and motilin. EM574 and motilin at lower doses seem to stimulate motilin receptors on vagal afferent nerves, and the signals are transmitted via vagal efferent cholinergic neurons. 5-HT$_3$-receptor antagonists probably block the stimulation of this pathway. At higher doses, EM574 and motilin also stimulate cholinergic neurons in the enteric nervous system. Hexamethonium and atropine inhibited the actions of EM574 and motilin by blocking the stimulation of both pathways. (From Inatomi et al. [15] with permission of the editor.)

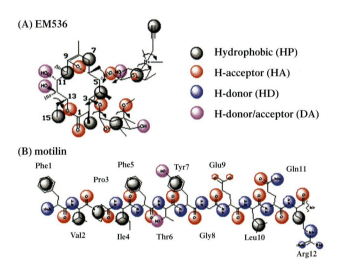

Chapter 11, Fig. 18. Chemical structures and functional atoms of (A) EM536 and (B) motilin [44].

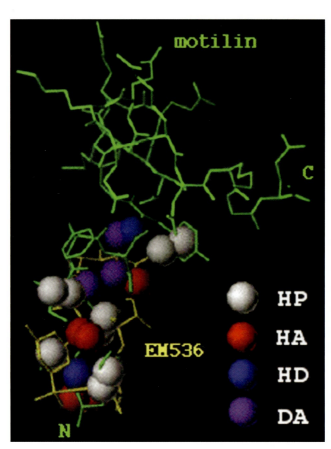

Chapter 11, Fig. 19. Stereopairs of the resulting alignment between the NMR structure in SDS micelles of motilin (green) and the selected conformation of EM536 (yellow) [44].

and at higher doses, they also stimulate the cholinergic neurons of the enteric nervous system. 5-HT$_3$ receptors are likely to be associated with the transmission of signals in the vagal pathway.

2. Effect in the Fed State

The effect of EM574 and motilin on gastric emptying was investigated in conscious dogs with a permanently implanted force transducer in the gastric antrum [22]. After an 18-hr fast, dogs were given 2.4 kJ/10 ml kg^{-1} caloric semisolid meals thoroughly mixed with 30 mg/kg acetaminophen suspended in 0.2 ml/kg 0.5% methyl cellulose. Venous blood samples were withdrawn every 15 min, and plasma acetaminophen concentrations were determined by reversed-phase HPLC according to the method of Adriaenssens and Prescott [23]. This method of measuring gastric emptying using acetaminophen is based on the finding of Heading et al. [24] that the peak plasma acetaminophen concentration correlates significantly with the half-time of gastric emptying as measured with radioisotopes in humans. Harasawa et al. [25] reported the usefulness and validity of the combination of caloric semisolid meals and acetaminophen as a test meal for gastric emptying.

Feeding of the test meal caused slight contractions in the gastric antrum (Fig. 12). Intraduodenal EM574 at 30 µg/kg induced marked antral contractions. Gastric emptying, expressed as the elevation in the plasma acetaminophen concentration, was also accelerated by EM574 at 30 µg/kg. The gastric emptying as well as the antral motor index was enhanced by EM574 in a dose-dependent manner; a significant effect was observed at 30 µg/kg (Fig. 13). Intravenous infusion of porcine motilin at 3 µg/kg hr^{-1} during the 30-min postprandial period stimulated antral contractile activity (Fig. 12) and accelerated gastric emptying. The effect of motilin was dose dependent; the dose of 3 µg/kg hr^{-1} enhanced both the motor index and emptying process significantly (Fig. 13). Cisapride (1 mg/kg) given intraduodenally evoked marked antral contractions and slightly enhanced gastric emptying (Figs. 12 and 13). At 3 µg/kg, however, gastric emptying was not enhanced despite a further increase in the motor index (Fig. 13). These results indicated that the dose of EM574 and motilin needed to stimulate antral motility in the digestive state was about 10 times higher than the dose needed in the interdigestive state. Cisapride did not enhance normal gastric emptying despite the increase in motility as reported previously [26, 27]. This might have been due to the transient increase in antropyloroduodenal resistance evoked by cisapride [28]. The prominent enhancing effect of EM574 on gastric emptying could suggest that it induces more coordinated gastroduodenal contractions than does cisapride.

We then examined the effect of EM574 on gastric emptying in two gastroparesis models, induced by intraduodenal instillation of oleic acid and intravenous infusion of dopamine. Lipid or free fatty acid in the lumen of the intestine is a potent inhibitor of gastric emptying [29], and this inhibition is thought to be

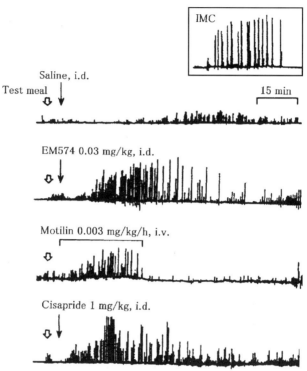

Fig. 12. Effects of EM574, porcine motilin, and cisapride on gastric antral motility after a semi-solid meal. The meal was ingested within 5 min. EM574 and cisapride were administered intraduodenally immediately after feeding. Motilin was infused intravenously during 30-min postprandial period. A representative pattern of interdigestive migrating complex (IMC) is shown in the inset to indicate the magnitude of contractions. (From Sato et al. [22] with permission from Elsevier Science.)

mediated mainly by CCK [30], which relaxes the proximal stomach and elicits contractions of the pylorus. Dopamine acts as an inhibitory modulator of gastrointestinal motility by reducing both tonic and phasic contractile activity and blocking antroduodenal coordination [31]. Dopamine suppresses the release of acetylcholine from the guinea pig stomach [32]. The instillation of 10 ml of 48% oleic acid emulsion into the duodenum decreased antral motor index and delayed gastric emptying (Table III). Intraduodenal EM574 at 30 µg/kg 15 min after feeding significantly increased the antral motor index and reversed delayed gastric emptying almost completely. Cisapride at 1 mg/kg induced significant antral contractions and tended to ameliorate the delayed emptying (Table III). Postprandial infusion of dopamine (0.6 mg/kg hr^{-1}) tended to suppress antral motility and impaired gastric emptying significantly (Table III). EM574 (30 µg/kg) given intraduodenally 1 min after the meal tended to stimulate antral motility and

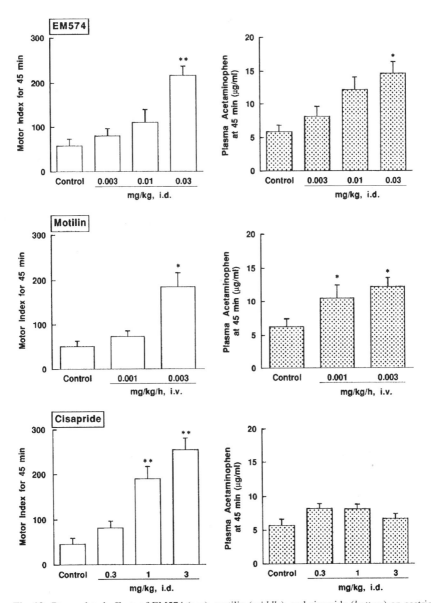

Fig. 13. Dose-related effects of EM574 (*top*), motilin (*middle*), and cisapride (*bottom*) on gastric antral motor index (*left*) and gastric emptying indicated as plasma acetaminophen concentration (*right*). EM574 and cisapride were administered intraduodenally immediately after the meal. Motilin was infused intravenously during 30-min postprandial period. Data are expressed as means ± SE of five dogs. *$P < 0.05$, **$P < 0.01$ compared to each control by paired t test with Bonferroni's correction. (From Sato *et al.* [22] with permission from Elsevier Science.)

TABLE III
Effects of EM574 and Cisapride on Experimentally Induced Gastroparesis[a]

	Antral motor index	Plasma acetaminophen (μg/ml)
Oleic acid-induced gastroparesis	(15–60 min)	(60 min)
Normal	53.6 ± 9.1[f]	12.2 ± 1.3[f]
Oleic acid[b]	39.3 ± 10.2	3.1 ± 0.7
+EM574[c] 0.01 mg/kg	70.0 ± 12.9	4.0 ± 1.2
+EM574[c] 0.03 mg/kg	188.1 ± 36.5[g]	11.4 ± 0.7[f]
+Cisapride[c] 1 mg/kg	109.5 ± 30.7[g]	4.5 ± 0.8
Dopamine-induced gastroparesis	(0–45 min)	(45 min)
Normal	82.9 ± 13.2	7.3 ± 0.1[f]
Dopamine[d]	51.0 ± 8.4	3.3 ± 0.8
+EM574[e] 0.01 mg/kg	84.4 ± 9.2	6.4 ± 1.5
+EM574[e] 0.03 mg/kg	241.0 ± 50.8	12.3 ± 1.1[f]
+Cisapride[e] 1 mg/kg	98.7 ± 26.1	5.3 ± 1.6

Source: From Sato et al. [22] with permission from Elsevier Science.
[a]Data are expressed as means ± SE of five or six dogs.
[b]Ten milliliters of 48% oleic acid was instilled into the duodenum after feeding.
[c]Drugs were administered intraduodenally 15 min after the meal.
[d]Dopamine hydrochloride at 0.6 mg/kg hr^{-1} was infused intravenouly during the postprandial period.
[e]Drugs were given intraduodenally 1 min after the meal.
[f]$P < 0.05$ compared to each gastroparesis control by paired t test with or without Bonferri's correction.
[g]$P < 0.01$ compared to each gastroparesis control by paired t test with or without Bonferri's correction.

reversed the delayed emptying significantly. EM574 (10 μg/kg) as well as cisapride (1 mg/kg) tended to induce moderate antral contractions and partially ameliorated the delayed emptying.

Atropine pretreatment (0.1 mg/kg), which suppressed test-meal-elicited antral contractions almost completely for 60 min, inhibited the increase in the antral motor index by EM574 only during the first 30 min. The accelerating effect of EM574 on gastric emptying was abolished by atropine. The plasma acetaminophen concentrations in atropinized dogs treated with EM574 were as low as in saline-treated atropinized dogs. These results suggest that EM574-induced antral contractions occurring via a cholinergic neural pathway play a very important role in enhancement of gastric emptying. Shiba et al. [33] reported that atropine-resistant contractions in the postprandial stomach are evoked by EM523, an effect that is eliminated by co-administration of an NK_1-receptor antagonist and a $5-HT_3$-receptor antagonist in conscious dogs. The atropine-resistant component alone might be insufficient for antroduodenal coordination.

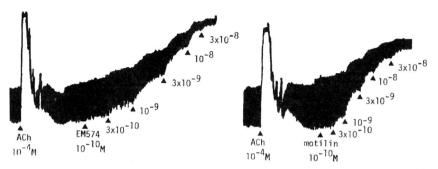

Fig. 14. Responses of isolated rabbit duodenum to acetylcholine, EM574, or porcine motilin. After supramaximal dosing of acetylcholine followed by washing out, EM574 or motilin was applied in cumulative concentrations. EM574 and motilin produced an increase in the basal tonus of the segment in a dose-dependent manner. (From Sato et al. [34] with permission from Elsevier Science.)

B. *In Vitro* Study on Rabbit Gastrointestinal Tract

1. Contractile Activity

The effect of EM574 on isolated rabbit gastrointestinal tract was investigated to clarify whether EM574 acts as a motilin receptor agonist [34]. EM574 did not induce contractions in the guinea pig or rat intestinal segments even at 10^{-5} M. But it caused an increase in the basal tonus of the rabbit isolated duodenum in a concentration-dependent manner; the contractile pattern was very similar to that induced by motilin (Fig. 14). The sensitivity to EM574 differed according to the site in the intestine; the minimum effective concentrations for the duodenum, jejunum, and ileum were 3×10^{-10}, 1×10^{-8}, and 1×10^{-8} M, respectively. The maximum contractile responses of the duodenum, jejunum, and ileum were 90%, 47%, and 10%, respectively, of the response induced by 10^{-4} M acetylcholine. EM, EM502, EM523, and EM536 also induced contractions in the isolated rabbit duodenal segments. Their dose–response curves were almost parallel, with pEC_{50} values of 6.52, 7.21, 7.95, and 7.84, respectively. The pEC_{50} value for EM574 was 8.26.

Porcine motilin did not induce contractions in the guinea pig or rat intestinal segments. Motilin induced concentration-dependent contractions in the rabbit intestinal segments (Fig. 14). As with the results obtained for EM574, the sensitivity of the response to motilin decreased aborally; the minimum effective concentrations for the duodenum, jejunum, and ileum were 1×10^{-10}, 3×10^{-9}, and 1×10^{-8} M, respectively. The maximum contractile responses of the duodenum, jejunum, and ileum were 89%, 50%, and 11%, respectively. The pEC_{50} value in the duodenum for porcine motilin was 8.69; the pEC_{50} value for [Tyr^{23}]canine motilin was 8.58. The contractile responses of duodenal segments to EM574 and porcine motilin were not sensitive to atropine, tetrodotoxin, mepyramine, or CP-99994. Verapamil (10^{-5} M) suppressed the response to EM574 as well as motilin. Removal of Ca^{2+} from the media also suppressed the response to EM574 and motilin markedly. These results indicate that EM574, like motilin, exerts a direct effect on intestinal smooth muscle.

2. Receptor Binding

Because EM574 and motilin showed very similar contractile responses in the isolated rabbit intestinal segments, receptor binding studies were performed to clarify whether EM574 acts on the motilin receptors. ^{125}I-Motilin binding to the smooth-muscle tissue of the rabbit gastric antrum was examined according to the method of Bormans *et al.* [35] with minor modifications. ^{125}I-labeled [Tyr^{23}]canine motilin was used as a radioligand.

Nonspecific binding to smooth-muscle tissue homogenate of rabbit gastric antrum accounted for less than 20% of total binding as determined from the radioligand bound in the presence of excess (10^{-6} M) unlabeled motilin. Porcine and [Tyr^{23}]canine motilin displaced labeled motilin bound to the antral homogenate with pIC$_{50}$ values of 9.20 and 8.97, respectively. Scatchard analysis of the self-displacement curves revealed linear relationships consistent with single binding sites with a pK_d value of 8.92 and a maximal binding capacity of 24 ± 4 fmol/mg protein. Pentagastrin and substance P (each 10^{-6} M) did not reduce labeled motilin binding. EM574 displaced labeled motilin bound to the antral homogenates with a pIC$_{50}$ of 8.21; EM, EM502, EM523, and EM536 also displaced labeled motilin with pIC$_{50}$ values of 7.37, 7.77, 8.25, and 8.17, respectively (Fig. 15). The displacement curves were almost parallel. A plot of binding affinity (pIC$_{50}$ value) versus contractile activity (pEC$_{50}$ value) demonstrated that changes in receptor binding were reflected by changes in contractile activity (Fig. 16). The pIC$_{50}$ and pEC$_{50}$ values for each motilin and erythromycin derivative were strongly correlated ($r = 0.94$, $P < 0.01$). The results suggest that EM574 stimulates contractile activity via motilin receptors on the surface of smooth muscle. These findings,

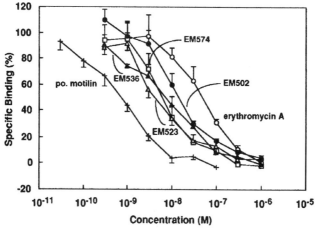

Fig. 15. Displacement studies with porcine (po.) motilin (+) and erythromycin A (○) and its derivatives EM502 (●), EM523 (△), EM536 (▲), and EM574 (□). The binding displacement curves are almost parallel. Data are expressed as means ± SE of three to five experiments. (From Sato *et al.* [34] with permission from Elsevier Science.)

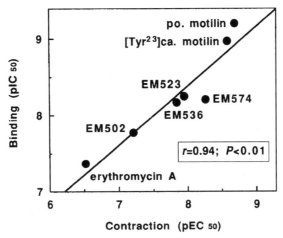

Fig. 16. Correlation ($r = 0.94$; $P < 0.01$) between contractile activity (pEC_{50}) and binding affinity (pIC_{50}) for porcine (po.) motilin, [Tyr^{23}]canine (ca.) motilin, and erythromycin derivatives. (From Sato et al. [34] with permission from Elsevier Science.)

obtained in rabbit antral tissue, are consistent with the report that EM574 is a motilin receptor agonist in the human gastric antrum [16].

3. Autoradiographic Analysis

The distribution of motilin receptors in the rabbit gastric antrum was investigated by the method of Sheikh et al. [36] with minor modifications. Frozen sections (7 μm) of rabbit antral tissue were mounted on gelatin-coated slides and incubated for 2 hr at 25°C with 2×10^{-10} M ^{125}I-motilin in the presence or absence of 10^{-6} M unlabeled [Tyr^{23}]canine motilin or 3×10^{-6} M EM574 in 0.25 ml of Tris-HCl buffer (pH 7.4). After incubation, the mounted sections were washed repeatedly, then placed for 1 hr in fixative containing 0.5% glutaraldehyde and 4% paraformaldehyde in 100 mM phosphate buffer (pH 7.4) at 4°C. They were then rinsed in chilled distilled water, dried, coated with NR-M2 photographic emulsion, and exposed for 20 days at 4°C. The slides were developed and counterstained with Kernechtrot. The distribution of autoradiographic grains at 10 sites each (1000 μm^2) in the mucosal, submucosal, longitudinal muscle, and circular muscle layer and in the myenteric plexus sections was counted using an image analyzer.

High densities of autoradiographic grains were observed in the circular muscle layer and myenteric plexus, and these were significantly reduced in the presence of excess unlabeled motilin or EM574 (Table IV). The densities of grains in the mucosal, submucosal, and longitudinal layers were less than those in the circular muscle layers and myenteric plexus, and were not affected by the addition of excess unlabeled motilin or EM574 (Table IV).

The predominant distribution of motilin binding sites in the circular muscle layer has already been reported by Sakai et al. [37], but the present findings are

TABLE IV
Distribution of Autoradiographic Grains in the Rabbit Antrum after Incubation with ^{125}I-Motilin[a]

Competitor added to ^{125}I-motilin	Grains/100 μm^{2} [b]				
	Mucosal layer	Submucosal layer	Circular muscle layer	Myenteric plexus	Longitudinal muscle layer
Control	5.6 ± 0.7	4.5 ± 1.0	22.2 ± 5.2	19.4 ± 6.0	4.4 ± 0.4
Unlabeled motilin	5.3 ± 1.1	3.8 ± 1.0	6.2 ± 0.8[c]	7.3 ± 1.5[c]	3.9 ± 0.7
EM574	5.1 ± 1.0	3.8 ± 1.2	5.9 ± 1.2[c]	8.8 ± 2.4[c]	3.9 ± 0.7

Source: From Sato *et al.* [22] with permission from Elsevier Science.

[a] Values were obtained by incubating tissue sections with 2×10^{-10} M [^{125}I-Tyr^{23}]canine motilin in the presence or absence of 1×10^{-6} M unlabeled [Tyr^{23}]canine motilin or 3×10^{-6} M EM574 for 2 hr. Data are expressed as means ± SD of 10 sites.

[b] Grain densities were calculated from light microscope autoradiographs using an image analyzer.

[c] $P < 0.01$ compared to control (Dunnett's test).

the first direct morphological evidence of specific myenteric plexus binding sites. Scatchard plot analysis detected only a single class of binding sites in the smooth-muscle tissue contained in the myenteric plexus. The motilin binding sites are probably those located in the circular muscle layer, because the total number of binding sites in the circular muscle layer is apparently much greater than in the myenteric plexus due to the difference in area. Thus, motilin receptors in the circular muscle layer, but not in the myenteric plexus, seem to be responsible for the pIC$_{50}$ values obtained in the binding studies, and this is in line with the significant correlation between the binding affinity and contractile activity, which is insensitive to neural blocking agents.

4. Three-Dimensional Structure of Motilin and the Motilide

From the perspective of structure–activity relationships, it is interesting that both motilin and motilides act as agonists for the motilin receptor, even though they have very different chemical structures (Fig. 17). Studies on the structure–activity relationship of motilin revealed that four hydrophobic residues, Phe 1, Val 2, Ile 4, and Tyr 7, of the motilin N-terminal region are most essential for its activity [38–41].

The three-dimensional (3D) structure of motilin in sodium dodecyl sulfate (SDS) micelles was determined using high-resolution NMR spectroscopy [42].

Recently, Hirono *et al.* applied a rational computational procedure consisting of conformational analysis and a novel superposing method, SUPERPOSE [43], to investigate the 3D structure–activity relationship between motilin and a motilide [44]. The conformational analysis was performed to obtain a number of energetically stable conformations of the motilide EM536, based on the structure of EM523 on which X-ray analysis has been performed. Next, an alignment between the NMR structure of motilin and a total of 714 distinct conformations for EM536

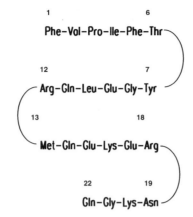

Fig. 17. Structure of porcine (human) motilin.

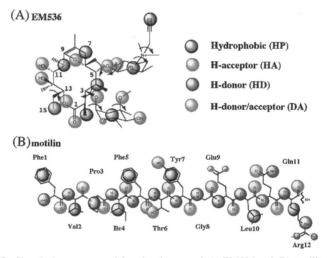

Fig. 18. Chemical structures and functional atoms of (A) EM536 and (B) motilin [44].

was carried out using SUPERPOSE and assuming that the NMR structure in SDS micelles of motilin is its active conformation. In the SUPERPOSE program the properties of the functional atoms in the motilide and motilin were divided into four types, comprising hydrophobic atoms (HP), hydrogen-bonding donors (HD), hydrogen-bonding acceptors (HA), and hydrogen-bonding donors/acceptors (DA). The functional atoms for EM536 and motilin were defined as shown in Figs. 18A and B (see also color plate), respectively.

The best alignment suggested for the superposing of motilin and EM536 is shown in Fig. 19 (see also color plate). In this alignment the four HP atoms located on the

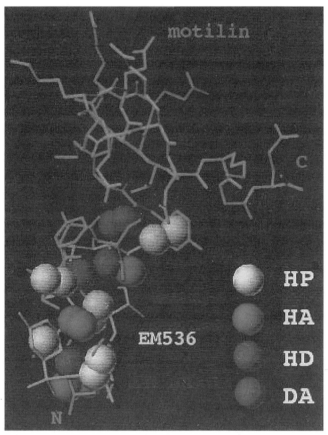

Fig. 19. Stereopairs of the resulting alignment between the NMR structure in SDS micelles of motilin (green) and the selected conformation of EM536 (yellow) [44].

propargyl group, the amino sugar ring, C7, and C15 in EM536 are superposed on those corresponding to Phe 1, Val 2, Ile 4, and Tyr 7 of motilin, respectively.

V. Clinical Trials of Motilides

A. Effects of Motilin and Erythromycin A on Gastric Emptying in Humans

After the discovery of motilin, its unique ability to stimulate gastric motility led to clinical studies on the effect of motilin on gastric emptying. Although earlier data on the effects of motilin had been inconclusive [45, 46], Peeters *et al.* [47] clearly showed that motilin accelerated delayed gastric emptying in patients with diabetic gastroparesis.

Because EM induces motilin-like contractions and is active by oral administration, it would be more suitable as a prokinetic agent for clinical use than motilin. In 1990, Janssens *et al.* [48] demonstrated that EM improved delayed gastric emptying in diabetic gastroparesis. Thereafter, several clinical studies on the enhancing effects of EM on gastric emptying were reported in postvagotomy gastroparesis [49, 50], idiopathic gastroparesis [51], primary anorexia nervosa [52], and gastroparesis after cancer chemotherapy [53].

B. Other Clinical Trials of Motilides

Intravenous EM523 at 2 mg significantly accelerated delayed gastric emptying in diabetic gastroparesis [54]. However, orally administered EM523 was less effective than expected; therefore, EM523 was replaced by EM574, which has about 10 times more potency. In healthy subjects, oral EM574 significantly shortened gastric half-emptying time in a dose-related manner [55]. Mean gastric half-emptying times were 173, 158, 147, and 149 min at doses of 5, 10, 20, and 30 mg, respectively (placebo, mean 189 min). Several motilides developed by other companies have also been entered into clinical trials. In healthy subjects, orally given ABT-229, an erythromycin B derivative synthesized by Abbott Laboratories, at 4 and 16 mg increased the gastric emptying rate to a similar extent [56]. Terazepide (KC-11458) is a motilide developed by Solvay Pharmaceuticals. It has been reported that oral terazepide at 4–64 mg accelerated gastric emptying in a dose-related manner in healthy subjects [57]. GM-611 is a relatively acid-stable motilide developed by Chugai Pharmaceutical Co. [58], the clinical trial of which is under way.

From the studies described above, it is evident that motilides not only stimulate gastric motility but also accelerate delayed gastric emptying in certain clinical conditions. As the next step, the clinical benefit to patients with gastrointestinal disorders (e.g., symptom relief) needs to be proved in double-blind and placebo-controlled studies. Motilides seem to offer an additional therapeutic possibility for the treatment of patients with impaired gastric motor function.

VI. Concluding Remarks

Recently, various seed compounds have been found using *in vitro* experimental systems, such as a high-throughput screening system using target molecules of receptors, enzymes, and so on. And it is very rare that seed compounds are found *in vivo*. The discovery of motilides is a good example of serendipity. The authors incidentally found that the effect of EM on canine gastrointestinal motility was very similar to that of motilin. Based on this finding, a number of EM derivatives were synthesized and their biological activities were evaluated, resulting in the discovery of very unique and potent peptide receptor agonists, called motilides. But this kind of serendipity is brought about not only by chance but also by having enough knowledge, experience, and deep insight in the areas of physiology, pharmacology, and pharmaceutics.

It is very interesting that motilin and the motilides, despite the fact that the chemical structures of the two agents are apparently different, share the same binding site on the receptor. Recently, the receptor for motilin was identified in the human gastrointestinal system [59]. The human motilin receptor is a family of G-protein-coupled receptors, and shows 52% overall amino acid sequence identity to the receptor for growth hormone secretagogues. The identification of the human motilin receptor will promote the analysis of the interaction between the receptor and its ligands. It is also expected that the motilin receptors of experimental animals such as dogs and rabbits will be identified and the localization will be studied. These studies will bring us a clearer understanding of the physiological role of motilin in humans and animals. The identification of human motilin receptors may also facilitate the development of selective agonists and antagonists with chemical structures different from that of EM.

Recently, many nonpeptide antagonists for peptide receptors have been reported. For example, low molecular weight antagonists for angiotensin II receptors, CCK receptors, endothelin receptors, neurokinin receptors, vasopressin receptors, and luteinizing hormone-releasing hormone (LH-RH) receptors have been developed or are under development as therapeutic agents. Although nonpeptide agonists for peptide receptors have also been found, the number of nonpeptide agonists is less than that of antagonists. The first example of the famous low molecular weight peptide receptor agonist is morphine. The motilides should become the second example.

References

1. Vantrappen, G., Janssens, J., Hellemans, J., and Ghoos, Y. (1977). The interdigestive motor complex of normal subjects and patients with bacterial overgrowth of the small intestine. *J. Clin. Invest.* **59**, 1158–1166.
2. Brown, J. C., Cook, M. A., and Dryburgh, J. R. (1973). Motilin, a gastric motor activity stimulating polypeptide: The complete amino acid sequence. *Can. J. Biochem.* **51**, 533–537.
3. Pearse, A. G. E., Polak, J. M., Bloom, S. R., Adams, C., Dryburgh, J. R., and Brown, J. C. (1974). Enterochromaffin cells of the mammalian small intestine as the source of motilin. *Virchows Arch. B Cell Pathol.* **16**, 111–120.
4. Itoh, Z., and Sekiguchi, T. (1983). Interdigestive motor activity in health and disease. *Scand. J. Gastroenterol.* **18**(Suppl. 82), 121–134.
5. Peeters, T. L., Vantrappen, G., and Janssens, J. (1980). Fasting plasma motilin levels are related to the interdigestive motility complex. *Gastroenterology* **79**, 716–719.
6. Itoh, Z., Honda, R., Hiwatashi, K., Takeuchi, S., Aizawa, I., and Takayanagi, R. (1976). Motilin-induced mechanical activity in the canine alimentary tract. *Scand. J. Gastroenterol.* **11**(Suppl. 39), 93–110.
7. Itoh, Z., Suzuki, T., Nakaya, M., Inoue, M., and Mitsuhashi, S. (1984). Gastrointestinal motor-stimulating activity of macrolide antibiotics and analysis of their side effects on the canine gut. *Antimicrob. Agents Chemother.* **26**, 863–869.

8. Itoh, Z., Suzuki, T., Nakaya, M., Inoue, M., Arai, H., and Wakabayashi, K. (1985). Structure–activity relation among macrolide antibiotics in initiation of interdigestive migrating contractions in the canine gastrointestinal tract. *Am. J. Physiol.* **248**, G320–G325.
9. Itoh, Z., Honda, R., Takeuchi, S., Aizawa, I., and Takayanagi, R. (1977). An extraluminal force transducer for recording contractile activity of the gastrointestinal smooth muscle in conscious dogs: Its construction and implantation. *Gastroenterol. Jpn.* **12**, 275–283.
10. Aizawa, I., Negishi, K., Suzuki, T., and Itoh, Z. (1984). Gastrointestinal contractile activity associated with vomiting in the dog. In "Gastrointestinal Motility" (C. Roman, ed.), pp. 159–165. MTP Press Ltd., Lancaster, UK.
11. Itoh, Z., Takeuchi, S., Aizawa, I., and Couch, E. F. (1977). Inhibitory effect of pentagastrin and feeding on natural and motilin induced interdigestive contractions in the stomach of conscious dogs. *Gastroenterol. Jpn.* **12**, 284–288.
12. Lee, K. Y., Kim, M. S., and Chey, W. Y. (1980). Effect of a meal and gut hormones on plasma motilin and duodenal motility in dog. *Am. J. Physiol.* **238**, G280–283.
13. Yamamoto, O., Matsunaga, Y., Haga, N., and Itoh, Z. (1994). Vagovagal inhibition of motilin-induced phase III contractions by antral acidification in dog stomach. *Am. J. Physiol.* **267**, G129–134.
14. Itoh, Z., Mizumoto, A., Iwanaga, Y., Yoshida, N., Torii, K., and Wakabayashi, K. (1991). Involvement of 5-hydroxytryptamine 3 receptors in regulation of interdigestive gastric contractions by motilin in the dog. *Gastroenterology* **100**, 901–908.
15. Inatomi, N., Sato, F., Marui, S., Itoh, Z., and Ōmura, S. (1996). Vagus-dependent and vagus-independent mechanism of action of the erythromycin derivative EM574 and motilin in dogs. *Jpn. J. Pharmacol.* **71**, 29–38.
16. Satoh, M., Sakai, T., Sano, I., Fujikura, K., Kayama, H., Oshima, K., Itoh, Z., and Ōmura, S. (1994). EM-574, an erythromycin derivative, is a potent motilin receptor agonist in human gastric antrum. *J. Pharmacol. Exp. Ther.* **271**, 574–579.
17. Itoh, Z., Nakaya, M., Suzuki, T., Arai, H., and Wakabayashi, K. (1984). Erythromycin mimics exogenous motilin in gastrointestinal contractile activity in the dog. *Am. J. Physiol.* **247**, G688–G694.
18. Tomomasa, T., Kuroume, T., Arai, H., Wakabayashi, K., and Itoh, Z. (1986). Erythromycin induces migrating motor complex in human gastrointestinal tract. *Dig. Dis. Sci.* **31**, 157–161.
19. Ōmura, S., Tsuzuki, K., Sunazuka, T., Marui, S., Toyoda, H., Inatomi, N., and Itoh, Z. (1987). Macrolides with gastrointestinal motor stimulating activity. *J. Med. Chem.* **30**, 1941–1943.
20. Tsuzuki, K., Sunazuka, T., Marui, S., Toyoda, H., Ōmura, S., Inatomi, N., and Itoh, Z. (1989). Motilides, macrolides with gastrointestinal motor stimulating activity. I. *O*-Substituted and tertiary *N*-substituted derivatives of 8,9-anhydroerythromycin A 6,9-hemiacetal. *Chem. Pharm. Bull.* **37**, 2687–2700.
21. Inatomi, N., Satoh, H., Maki, Y., Hashimoto, N., Itoh, Z., and Ōmura, S. (1989). An erythromycin derivative, EM-523, induced motilin-like gastrointestinal motility in dogs. *J. Pharmacol. Exp. Ther.* **251**, 707–712.
22. Sato, F., Marui, S., Inatomi, N., Itoh, Z., and Ōmura, S. (2000). EM574, an erythromycin derivative, improves delayed gastric emptying of semi-solid meal in conscious dogs. *Eur. J. Pharmacol.* **395**, 165–172.
23. Adriaenssens, P. I., and Prescott, L. F. (1978). High performance liquid chromatographic estimation of paracetamol metabolites in plasma. *Br. J. Clin. Pharmacol.* **6**, 87–88.
24. Heading, R. C., Nimmo, J., Prescott, L. F., and Tothill, P. (1973). The dependence of paracetamol absorption on the rate of gastric emptying. *Br. J. Pharmacol.* **47**, 415–421.
25. Harasawa, S. Tani, N., Suzuki, S., Miwa, M., Sakita, R., Nomiyama, T., and Miwa, T. (1979). Gastric emptying in normal subjects and patients with peptic ulcer: A study using the acetaminophen method. *Gastroenterol. Jpn.* **14**, 1–10.

26. Edwards, C. A., Holden, S., Brown, C., and Read, N. W. (1987). Effect of cisapride on the gastrointestinal transit of a solid meal in normal human subjects. *Gut* **28**, 13–16.
27. Fraser, R., Horowitz, M., Maddox, A., and Dent, J. (1993). Dual effects of cisapride on gastric emptying and antropyloroduodenal motility. *Am. J. Physiol.* **264**, G195–G201.
28. Malbert, C. H., Serthelon, J. P., and Dent, J. (1992). Changes in antroduodenal resistance induced by cisapride in conscious dogs. *Am. J. Physiol.* **263**, G202–G208.
29. Hunt, J. N., and Knox, M. T. (1968). A relation between the chain length of fatty acids and the slowing of gastric emptying. *J. Physiol. (Lond.)* **194**, 327–336.
30. Kleibeuker, J. H., Beekhuis, H., Jansen, J. B. M., Pipers, D. A., and Lambers, C. B. H. W. (1988). Cholecystokinin is a physiological hormonal mediator of fat-induced inhibition of gastric emptying in man. *Eur. J. Clin. Invest.* **18**, 173–177.
31. Valezuela, J. E. (1976). Dopamine as a possible neurotransmitter in gastric relaxation. *Gastroenterology* **71**, 1019–1022.
32. Kusunoki, K., Taniyama, K., and Tanaka, C. (1985). Dopamine regulation of [^{3}H]acetylcholine release from guinea pig stomach. *J. Pharmacol. Exp. Ther.* **234**, 713–719.
33. Shiba, Y., Mizumoto, A., Inatomi, N., Haga, N., Yamamoto, O., and Itoh, Z. (1995). Stimulatory mechanism of EM523-induced contractions in postprandial stomach of conscious dogs. *Gastroenterology* **109**, 1513–1521.
34. Sato, F., Sekiguchi, M., Marui, S., Inatomi, N., Shino, A., Itoh, Z., and Ōmura, S. (1997). EM574, an erythromycin derivative, is a motilin receptor agonist in the rabbit. *Eur. J. Pharmacol.* **322**, 63–71.
35. Bormans, V., Peeters, T. L., and Vantrappen, G. (1986). Motilin receptors in rabbit stomach and small intestine. *Regul. Pept.* **15**, 143–153.
36. Sheikh, S. P., Roach, E., Fuhlendorff, J., and Williams, J. A. (1991). Localization of Y$_1$ receptors for NPY and PYY on vascular smooth muscle cells in rat pancreas. *Am. J. Physiol.* **260**, G250–G257.
37. Sakai, T., Satoh, M., Sonobe, K., Nakajima, M., Shiba, Y., and Itoh, Z. (1994). Autoradiographic study of motilin binding sites in the rabbit gastrointestinal tract. *Regul. Pept.* **53**, 249–257.
38. Poitras, P., Gagnon D., and St-Pierre, S. (1992). Autophosphorylation of β-connectin (TITIN 2) in vitro. *Biochem. Biophys. Res. Commun.* **183**, 34–40.
39. Peeters, T. L., Macielag, M. J., Depoortere, I., Konteatis, Z. D., Florance, J. R., Losser, R. A., and Galdes, A. (1992). D-Amino acid and alanine scans of the bioactive portion of porcine motilin. *Peptides* **13**, 1103–1107.
40. Miller, P., Gagnon, D., Dickner, M., Aubin, P., St-Pierre, S., and Politres, P. (1995). Structure–function studies of motilin analogs. *Peptides* **16**, 11–18.
41. Haramura, M., Tsusuki, K., Okamachi, A., Yogo, K., Ikuta, M., Kozono, T., Takanashi, H., and Murayama, E. (1999). Structure–activity study of intact porcine motilin. *Chem. Pharm. Bull.* **47**, 1555–1559.
42. Javet, J., Zdunek, J., Damberg, P., and Gaslund, A. (1997). Three-dimensional structure and position of porcine motilin in sodium dodecyl sulfate micelles determined by ^{1}H-NMR. *Biochemistry* **36**, 8153–8163.
43. Iwase, K., and Hirono, S. (1999). Estimation of active conformations of drugs by a new molecular superposing procedure. *J. Comput.-Aided Mol. Des.* **13**, 499–512.
44. Gouda, H., Sunazuka, T., Ōmura, S., and Hirono, S. (2000). Three-dimensional structure–activity relationship analysis between motilin and motilide using conformational analysis and a novel molecular superposing method. *Chem. Pharm. Bull.* **48**, 1835–1837.
45. Ruppin, H., Domschke, S., Domschke, W., Wunsch, E., Jaeger, F., and Demling, L. (1975). Effects of 13-nle-motilin in man—inhibition of gastric evacuation and stimulation of pepsin secretion. *Scand. J. Gastroenterol.* **10**, 199–202.
46. Christofides, N. D., Modlin, I. M., Fitzpatrick, M. L., and Bloom, S. R. (1979). Effect of motilin on the gastric emptying and gut hormone release during breakfast. *Gastroenterology* **76**, 903–907.

47. Peeters, T. L., Muls, E., Janssens, J., Urbain, J. L., Bex, M., Van Cutsem, E., Depoortere, I., De Roo, M., Vantrappen, G., and Bouillon, R. (1992). Effect of motilin on gastric emptying in patients with diabetic gastroparesis. *Gastroenterology* **102**, 97–101.
48. Janssens, J., Peeters, T. L., Vantrappen, G., Tack, J., Urbain, J. L., De Roo, M., Muls, E., and Bouillon, R. (1990). Improvement of gastric emptying in diabetic gastroparesis by erythromycin. *N. Engl. J. Med.* **322**, 1028–1031.
49. Mozwecz, H., Pavel, D., Pitrak, D., Orellana, P., Schlesinger, P. K., and Layden, T. J. (1990). Erythromycin stearate as prokinetic agent in postvagotomy gastroparesis. *Dig. Dis. Sci.* **35**, 902–905.
50. Xynos, E., Mantides, A., Papageorgiou, A., Fountos, A., Pechlivanides, G., and Vassilakis, J. S. (1992). Erythromycin accelerates delayed gastric emptying of solids in patients after truncal vagotomy and pyloroplasty. *Eur. J. Surg.* **158**, 407–411.
51. Richards, R. D., Davenport, K. G., Hurm, K. D., and McCallum, R. W. (1993). The treatment of idiopathic and diabetic gastroparesis with acute intravenous and chronic oral erythromycin. *Am. J. Gastroenterol.* **88**, 203–207.
52. Stacher, G., Peeters, T. L., Bergmann, H., Wiesnagrotzki, S., Schneider, C., Granser-Vacariu, G. V., Gaupmann, G., and Kugi, A. (1993). Erythromycin effects on gastric emptying, antral motility and plasma motilin and pancreatic polypeptide concentrations in anorexia nervosa. *Gut* **34**, 166–172.
53. Maliakkal, B. J., Polodori, G., Gordon, C., Davis, L., and Desai, T. K. (1991). Severe gastroparesis following cancer chemotherapy and prokinetic response to erythromycin. *Gastroenterology* **100**, A466.
54. Nakamura, T., Ishii, M., Arai, Y., Tandoh, Y., Terada, A., and Takebe, K. (1994). Effect of intravenous administration of EM523L on gastric emptying and plasma glucose levels after a meal in patients with diabetic gastroparesis: A pilot study. *Clin. Ther.* **16**, 989–999.
55. Choi, M.-G., Camilleri, M., Burton, D. D., Johnson, S., and Edmonds, A. (1998). Dose-related effects of N-demethyl-N-isopropyl-8,9-anhydroerythromycin A 6,9-hemiacetal on gastric emptying of solids in healthy human volunteers. *J. Pharmacol. Exp. Ther.* **285**, 37–40.
56. Verhagen, M. A., Samsom, M., Maes, B., Geypens, B. J., Ghoos, Y. F., and Smout, A. J. (1997). Effects of a new motilide, ABT-229, on gastric emptying and postprandial antroduodenal motility in healthy volunteers. *Aliment. Pharmacol. Ther.* **11**, 1077–1086.
57. Steinbrede, H., Aygen, S., and Steinborn, C. (1997). KC 11458, a new motilin agonist, is effective in the acceleration of gastric emptying in healthy male volunteers. *Gut* **41**(Suppl. 3), A155.
58. Yamada, K., Chen, S., Abdullah, N. A., Tanaka, M., Ito, Y., and Inoue, R. (1996). Electrophysiological characterization of a motilin agonist, GM611, on rabbit duodenal smooth muscle. *Am. J. Physiol.* **271**, G1003–1016.
59. Feighner, S. D., Tan, C. P., McKee, K. K., Palyha, O. C., Hreniuk, D. L., Pong, S. S., Austin, C. P., Figueroa, D., MacNeil, D., Cascieri, M. A., Nargund, R., Bakshi, R., Abramovitz, M., Stocco, R., Kargman, S., O'Neill, G., Van Der Ploeg, L. H., Evans, J., Patchett, A. A., Smith, R. G., and Howard, A. D. (1999). Receptor for motilin identified in the human gastrointestinal system. *Science* **284**, 2184–2188.

Chapter 12

Novel Activity of Erythromycin and Its Derivatives

SHOJI KUDOH
ARATA AZUMA

Fourth Department of Internal Medicine
Nippon Medical School
Tokyo, Japan

JYUN TAMAOKI

Tokyo Women's Medical University
Tokyo, Japan

HAJIME TAKIZAWA

Graduate School of Medicine
University of Tokyo
Tokyo, Japan

KOH NAKATA

Research Institute
International Medical Center of Japan
Tokyo, Japan

HAJIME GOTO

School of Medicine
Kyorin University
Tokyo, Japan

I. Erythromycin Treatment in Diffuse Panbronchiolitis 534
   A. Discovery of Erythromycin Treatment 534
   B. Clinical Efficacy and Improvement of Prognosis 535
   C. Mechanism of Action 537
   D. Conclusion 540
II. Inhibition of Chloride Channel 541
   A. Effect of Macrolide Antibiotics on Airway Secretions 541
   B. Effect on Water Secretions by the Chlorine Channel Block 542
   C. Clinial Studies on Sputum Production 545
   D. Conclusion 546
III. Effects of Macrolides on Cytokine/Chemokine Expression 546
   A. Clinical Evidence of Macrolide Antibiotics Effect
      on Cytokine Production 547
   B. Inhibitory Action of Macrolides on Cytokine Production:
      In Vitro Findings 548
   C. Molecular Mechanisms of the Anti-Inflammatory Actions
      of Macrolides 550
   D. Future Directions 552
IV. Modulation of Bacterial Function 553
   A. Enhancement of Susceptibility of *P. aeruginosa*
      to Serum by Macrolides 554
   B. Inhibitory Action of Macrolides on Biofilm Formation
      of *P. aeruginosa* 554

    C. Inhibitory Action of Macrolides on the Expression
       of Virulence Factors of *P. aeruginosa* 555
    D. Suppression of Aminoglycoside-Inactivating Enzyme
       by Macrolides 555
    E. Suppressive Action of Macrolides on the Production
       of Verotoxin by *E. coli* O157 556
    F. Effect of Erythromycin on *Shigella* Invasion into Host Cells 556
    G. Conclusion 557
 V. New Challenge for Novel Action 557
    A. Mechanism of Structure–Activity Relationship 557
    B. Challenge for New Field of Inflammatory Diseases 558
    C. Conclusion 563
       References 564

I. Erythromycin Treatment in Diffuse Panbronchiolitis

Diffuse panbronchiolitis is a chronic airway disease predominantly affecting East Asians. It is pathologically characterized by chronic inflammation diffusely located in the region of respiratory bronchioles and is clinically counted as a sinobronchial syndrome with severe lower airway infection [1]. An unidentified gene in the human leukocyte antigen class I region is assumed to predispose Asians to the disease [2]. Prognosis of diffuse panbronchiolitis has recently been improved dramatically over the past 20 years with the use of low-dose long-term erythromycin (EM) treatment. The beneficial effect of EM and other 14-membered ring macrolides in this condition is considered to be due to an anti-inflammatory rather than anti-infective mechanism. Recent investigations have revealed many new actions to the epithelial cells and inflammatory cells, e.g., neutrophils, lymphocytes, and macrophages. Furthermore, erythromycin treatment of diffuse panbronchiolitis provides a new understanding of the pathophysiology and the treatment of chronic infectious airway disease.

A. Discovery of Erythromycin Treatment

During the past 20 years, diffuse panbronchiolitis (DPB) has changed from a fatal disease to a curable disease. Before EM treatment, the prognosis of patients with DPB was very poor. According to a study by a research group of the Ministry of Health and Welfare of Japan in 1981 [3], the 5-year survival rate was 42% from the first medical examination. Major bacterial species infecting the airway often changed from *Haemophilus influenzae* to *Pseudomonas aeruginosa* as the disease advanced. After the disease changed to a *Pseudomonas* infection, the prognosis became extremely poor, i.e., the 5-year survival rate was only 8%. In 1982, EM was first used to treat DPB by us. We first considered the use of EM following an encounter with one patient with DPB who had markedly improved after the treatment. At that time it was not possible to cure patients with DPB; thus we suspected that the medication he was taking (600 mg daily EM for >2 years) was responsible for his cure. Therefore, a clinical trial using low-dose EM was started, and in 1984 the clinical efficacy of EM therapy in patients with DPB was first reported [4]. In 1987, a paper on 4 years of experience with EM therapy was published [5].

B. Clinical Efficacy and Improvement of Prognosis

From the early Japanese reports from individual institutes, at least three important points are accepted today. First, the clinical effects of the therapy were excellent. Second, no significant changes in bacterial species are seen before or after therapy. Third, even in cases involving *Pseudomonas* infection, a certain clinical effect could be shown. According to a retrospective study of 52 patients treated with low-dose erythromycin, 37 patients had received various conventional treatments using β-lactams, aminoglycosides, new quinolones, or steroids for an average of 43 months. However, the result was, of course, poor. Conversely, clinical improvement was excellent after erythromycin therapy for an average of 20 months [6]. After this study, the Ministry of Health and Welfare research group did another retrospective study comparing EM therapy with long-term administration of new quinolones. In this study also, EM therapy revealed far better effects than treatment using new quinolones [7].

The Ministry of Health and Welfare research group then did a prospective double-blind, placebo-controlled study involving the use of 600 mg daily of EM for 3 months. In this study, clinical efficacy was evaluated based on a scoring system using these items: dyspnea on exertion, amount of sputum, chest radiology findings, arterial oxygen tension, forced expiratory volume in 1 sec, and serum C-reactive protein level. As shown in Fig. 1, a "moderately improved" rating was noted for 57% in the EM group and for 15% in the placebo group. Conversely, the rates of aggravation were 6% and 38% in the treatment and placebo groups, respectively [8]. This double-blind study established the clinical efficacy of EM treatment for DPB in Japan.

Figure 2 shows survival curves of DPB patients based on initial diagnosis [9]. In the 1970s, the 5-year survival rate was 63%. Between 1980 and 1984, the survival rate was 72%. However, after 1985 (after EM treatment was described), the survival curve significantly improved, i.e., the 5-year survival rate was 92%. From the survival rates based on patient age, it was found that EM therapy is more beneficial for older rather than younger patients. Furthermore, additional analysis of whether EM therapy was useful revealed that there was no significant difference between patients treated in the 1970s before EM therapy was established and patients treated after 1984 who had not received EM. Therefore, EM therapy certainly contributed to the recent improvement in the prognosis for DPB.

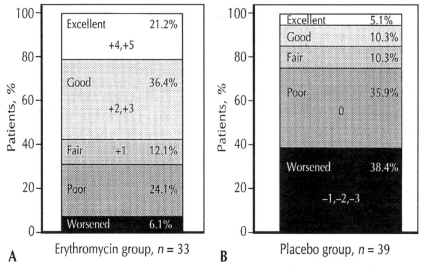

Fig. 1. Therapeutic effects of erythromycin (EM) treatment on diffuse panbronchiolitis: a double-blind study of EM, 600 mg/d (A) and placebo (B) for 3 months. Number values indicate total scores on six variables (degree of dyspnea on exertion, chest radiology findings, PaO_2, forced expiratory volume in 1 sec, C-reactive protein and sputum volume). 1 point, improved; 0 point, unchanged; −1 point, worsened.

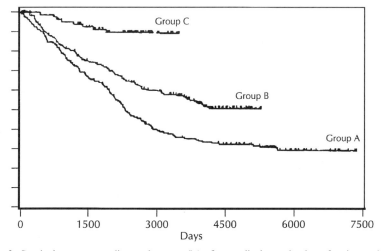

Fig. 2. Survival curves according to the year of the first medical examination of patients with diffuse panbronchiolitis (A, 1970–1979; B, 1980–1984; C, 1985–1990). (Adapted from Kudoh et al. [9] with permission. From AJRCCM, official journal of the American Thoracic Society. © American Lung Association.)

C. Mechanism of Action

Macrolides including EM are, of course, antibiotics with a bactericidal effect. In EM treatment for DPB, however, the mechanism of action cannot, from the clinical viewpoint, be attributed to the bactericidal effect. First, the disease can improve without eliminating bacteria. Second, improvement can be found even in patients with *P. aeruginosa* infection. Third, the maximal concentration of EM in serum or sputum, which is approximately 1 µg/ml, is lower than the minimum inhibitory concentration for major species of bacteria [10]. Thus, many recent Japanese investigations have focused on the action to cells relating to the airway inflammation of DPB, i.e., airway epithelial cells, neutrophils, lymphocytes, or macrophages.

1. Inhibition of Hypersecretion

A large amount of sputum is a characteristic manifestation of DPB. Sputum volume markedly decreases after EM therapy. The reduction of sputum volume is the most sensitive parameter as shown by the double-blind study mentioned here. Goswami *et al.* [11] first reported that EM dose-dependently inhibited mucous secretion from airway mucosa with a marker of glycoconjugate *in vitro*. Tamaoki *et al.* [12] first observed *in vitro* ion transport on epithelial cells and reported that EM dose-dependently inhibited ion transport when erythromycin attached to the serosa. Furthermore, they found this inhibition was due to blocking of the chloride channel. The inhibition of the chloride channel also means that erythromycin inhibits the secretion of water that moves with the chloride ion. Inhibition of mucus and water secretion from the epithelial cells can be interpreted as an important mechanism of action in improving hypersecretion in patients with DPB.

2. Inhibition of Neutrophil Accumulation

The most important advance of the recent study concerns the effect on neutrophil accumulation in the inflammatory region of the airway. Numerous neutrophils are found in the broncoalveolar lavage fluid (BALF), frequently reaching 70 to 80%. After EM treatment, neutrophil elastase decreases in sputum [13] and also in BALF [14]. Kadota *et al.* [15] reported that the percentage of neutrophils in BALF markedly decreased after erythromycin therapy with a decrease in neutrophil chemotactic activity. Oishi *et al.* [16] reported that interleukin-8 in BALF markedly decreases with the number of neutrophils and with the concentration of neutrophil elastase. Takizawa *et al.* [17, 18] showed that EM dose-dependently inhibited interleukin-8, interleukin-6, and granulocyte-macrophage colony-stimulating factor secretion from the epithelial cells *in vitro* using a cell line of human airway epithelial cells. A similar inhibition was also found on epithelial cells stimulated by *H. influenzae* endotoxin [19]. Recently, Desaki *et al.* [20] reported that EM suppressed nuclear factor-κ B and activator protein-1 activation in human bronchial epithelial cells in relation to mRNA expression of interleukin-8 on epithelial cells.

3. The Action on Lymphocytes and Macrophages

Pathological characteristics of DPB include chronic inflammation with lymphocytes, plasma cells, and foamy macrophages in the region of respiratory bronchioles. These foci disappeared after EM treatment. According to a study on BALF of a patient with DPB, memory T cells and activation of CD8+ cells, mainly cytotoxic T cells, significantly increase in DPB and decrease after the erythromycin treatment [21]. Keicho et al. [22] from our group reported that EM dose-dependently inhibited proliferation of lymphocytes but did not inhibit the expression of interleukin-2 and CD25. This is a different mechanism from that of tacrolimus. They concluded that the inhibitory effect of EM on T cells existed in the later activation process based on the inhibition of the T-cell response to interleukin-2. Furthermore, Keicho et al. [23] and other Japanese investigators have concurred that erythromycin accelerates both the differentiation and proliferation of the monocyte–macrophage system. EM inhibits lipopolysaccharide production of human monocyte-stimulated tumor necrosis factor-α in vitro [24]. However, we have not yet clarified the role of these actions in lymphocytes and macrophages in the improvement of the region of the respiratory bronchiole in DPB.

4. Action Points on the Airway Inflammation

Figure 3 shows a summary of the action points of the airway inflammation according to recent papers. First, EM inhibits hypersecretion due to inhibition of mucus and water secretion from the epithelial cells. Second, EM inhibits neutrophil

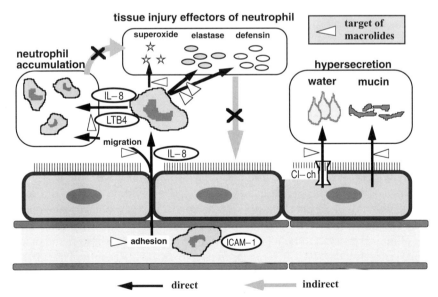

Fig. 3. Airway inflammation and action points of erythromycin. ICAM-1, intercellular adhesion molecule-1; IL-8, interleukin-8; LTB-4, leukotriene B4.

accumulation on the inflammatory region due to inhibition of neutrophil attachment to the capillary vessels, interleukin-8 secretion from the epithelial cells, and secretion of interleukin-8 and leukotriene B4 from the neutrophil itself [25]. These actions result in a reduction of the substances injuring mucosa, such as elastase or superoxide anion [26]. These actions obviously play important roles in the improvement of airway inflammation, although there are controversies concerning the action of EM in the neutrophil activity itself [27–30] and its actions in lymphocytes and macrophages.

Today, EM treatment has been established as a basic treatment for DPB. Furthermore, erythromycin has been widely applied in treating chronic airway inflammatory disorders, not only in lower airway diseases (DPB, chronic bronchitis, bronchiectasis, or bronchial asthma [31–33]) but also upper airway diseases (chronic sinusitis or exudative otitis media). Other forms of 14-membered ring macrolides, i.e., clarithromycin (CAM) [34–36] and roxithromycin (RXM) [37] have also been used for a similar purpose.

5. A Role for the Anti-Inflammatory Actions of Erythromycin in Chronic Airway Infection

Figure 4 shows the change of bacterial species in the sputum of patients with DPB before and after treatment. With the conventional treatment *H. influenzae* decreased and *P. aeruginaosa* increased after therapy. EM therapy reduced both *Haemophilus* and *Pseudomonas* organisms and increased normal flora.

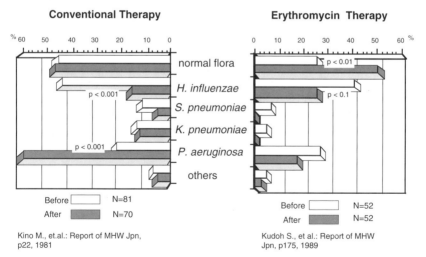

Fig. 4. Bacterial flora in sputum before and after treatment: conventional therapy versus erythromycin treatment. (Left panel adapted from Kino [3] with permission. Right panel adapted from Kudoh *et al.* [5] with permission.)

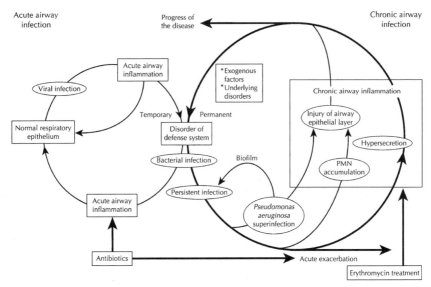

Fig. 5. Infection and inflammation of respiratory tract. EM may block the "vicious circle" of chronic airway infection. PMN, polymolphonucear cells.

We believe that treatment with antibiotics for chronic airway infection is naturally limited, because chronic airway infection is different from acute airway infection in terms of the disease process. First, chronic airway infection exists for certain reasons, i.e., a defect of the defense mechanism of the airway. Second, chronic airway infection is accompanied by an inflammatory process, which results in the "vicious circle" proposed by Cole [38]. Presently, we think that EM blocks the "vicious circle" in chronic airway infection as it inhibits the inflammatory process (Fig. 5).

Furthermore, it has become clear recently that even the subminimum inhibitory concentration of 14-membered ring macrolides has the inhibitory effects of biofilm formation and expression of virulence factors (piocyanin, elastase, proteases) of *P. aeruginosa* (see following section). Thus, an anti-inflammatory agent such as EM should be used as basic treatment in the therapy of chronic airway infection.

D. Conclusion

In Japan, a movement to clarify the mechanism of action of EM treatment for DPB was formed. A study group on the novel action of macrolides was established in 1994 and many clinical and experimental investigations have been reported on, e.g., the inhibitory effect on experimental bleomycin lung injury, experimental reperfusion lung injury, experimental pulmonary metastasis of cancer cells, and so on. Fourteen-membered ring macrolides are known to have a motilin-like action that is a gastrointestinal hormone stimulating peristalsis of the gastrointestine [39, 40]. The anti-inflammatory properties mentioned here should be defined as the third action of 14-membered ring macrolides. It is expected that the anti-inflammatory action of erythromycin will be widely applied for many diseases in the future.

II. Inhibition of Chloride Channel

Airway hypersecretion is one of the characteristic features of chronic airway diseases including chronic bronchitis, asthma, bronchiectasis, and DPB, and a large amount of secretions stagnated in the respiratory lumen may cause airflow limitation, impairment of mucociliary transport, and recurrent respiratory infection [41]. Airway secretions consist of the mucus synthesized and released by submucosal glands and goblet cells, and the water transported across airway mucosa [42].

It is known that secretion of water from the submucosa toward the lumen and absorption of water in the opposite direction are generally correlated with secretion of Cl and absorption of Na, respectively, by airway epithelial cells [43]. In fact, at least four types of Cl channels are located on the apical membrane of airway epithelium, i.e., cystic fibrosis transmembrane conductance regulator (CFTR), outwardly rectifying Cl channel (ORCC), Ca^{2+}-activated Cl channel, and volume-sensitive Cl channel [44]; certainly the epithelium can actively secrete Cl through these channels, thereby making the lumen electrically negative. This transepithelial potential difference will drive net diffusion of Na toward the lumen across the tight junctions and other leak pathways within the epithelium. The resulting transfer of salt creates an osmotic pressure difference that moves fluid into the lumen. Similar reasoning explains the stimulation of water absorption with increases in active Na absorption. Thus, the direction and amount of water moved across the epithelium will depend, in part, on the balance between these two opposing transport processes [45], and it is possible that airway epithelial Cl secretion is upregulated in patients with airway hypersecretion and, hence, the drugs capable of inhibiting the Cl channel function could be of value in the treatment of such patients.

A. Effect of Macrolide Antibiotics on Airway Secretions

There is increasing evidence that long-term administration of macrolide antibiotics provides a marked reduction in the volume of airway secretions in some patients with asthma [31], bronchorrhea [46], chronic bronchitis, and DPB [15, 35]. One explanation for the mechanism of efficacy would be macrolides' direct effects on airway secretory cells. Indeed, Goswami et al. [11] first studied nasal mucous glycoconjugate secretion from healthy nonsmoking adults before and after treatment with erythromycin base, penicillin, ampicillin, tetracycline, or cephalosporins. They found that erythromycin at a concentration of 10 μM reduced nasal secretion by 35% in both the resting state and when the nose was stimulated with methacholine or histamine, but other antibiotics had no effect on glycoconjugate secretion. There is evidence that erythromycin has anti-inflammatory properties with direct effects on neutrophil migration and its activation and secondary effects on mucous secretion. It has been shown that inhalation of lipopolysaccharide from *Eschericia coli* in guinea pig secretions, as assessed histologically by mucous scores, which are inversely related to the magnitude of mucous discharge [47], is dose-dependently inhibited by orally administered erythromycin and clarithromycin, but not by amoxicillin or cefaclor [48]. Similarly, erythromycin and roxithromycin inhibit interleukin-8-induced recruitment of neutrophils and the associated goblet secretion in the guinea pig trachea [49]. However, whether macrolides exert their antisecretory effect by solely inhibiting neutrophil function or by acting on goblet cells and neutrophils independently remains uncertain.

B. Effect on Water Secretions by the Chlorine Channel Block

Regarding the effects of macrolides on water transport, the experiment with electrical properties of cultured canine tracheal epithelium [12] showed that erythromycin added to the submucosal side at concentrations of $10 M$ and higher dose-dependently decreased short-circuit current, an electrical parameter that reflects net value of actively transported ions across airway epithelium (Table I). Moreover, this effect was not altered by the Na channel blocker amiloride but was abolished by the Cl channel blocker diphenylamine-2-carboxylate or substitution

TABLE I
Effect of Erythromycin on Bioelectric Properties of Canine Tracheal Epithelium in Culture

	Isc ($\mu A/cm^2$)	PD (mV)	G (mS/cm^2)
Baseline	8.2 ± 0.8	2.6 ± 0.4	3.2 ± 0.5
Erythromycin (M)	8.5 ± 1.2	2.4 ± 0.4	3.5 ± 0.5
Erythromycin (S)	2.5 ± 0.4[a]	1.2 ± 0.3[b]	2.1 ± 0.3[c]

Source: Reprinted from Tamaoki et al. [12] with permission from European Respiratory Society Journals, Ltd.

Note: Erythromycin (100 μM) was added to the mucosal (M) or submucosal solution (S) in an Ussing chamber. Values are means ± SE; $n = 12$. Isc, short-circuit current; PD, potential difference; G, conductance.

[a]$P < 0.001$, significantly different from baseline values.
[b]$P < 0.01$, significantly different from baseline values.
[c]$P < 0.05$, significantly different from baseline values.

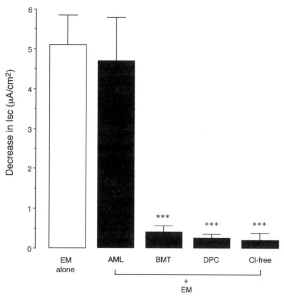

Fig. 6. Effects of amiloride (AML, 100 μM), bumetanide (BMT, 100 μM), diphenylamine-2-carboxylate (DPC, 100 μM), and substitution of Cl in the medium with gluconate (Cl-free) on the decrease in short-circuit current (Isc) induced by erythromycin (EM, 100 μM) in canine cultured tracheal epithelium. Responses are expressed as means ± SE; $n = 9$ for each group. ***$P < 0.001$, significantly different from the response to EM alone. (Reprinted from Tamaoki *et al.* [12] with permission from European Respiratory Society Journals, Ltd.)

of Cl in the bathing medium with gluconate, an anion that cannot be transported by airway epithelium (Fig. 6). These *in vitro* findings suggest that erythromycin may reduce water secretion through a selective inhibition of airway epithelial Cl channel. A discrepancy seems to exist in the concentrations of erythromycin required to produce its *in vitro* and *in vivo* effects. The mean serum concentration following the ingestion of 500 mg erythromycin by the adult volunteers has been reported to be 1.6 μM [50], whereas *in vitro* experiments showed that at least 10 μM erythromycin was required to decrease mucous production and Cl transport. However, because of the species difference, the findings may not necessarily negate its clinical significance. In addition, the serum concentration of erythromycin does not accurately reflect the local concentration since this drug has been shown to concentrate intracellularly more than 10-fold [51].

Ikeda and associates [52] have subsequently shown the effects of antibiotics on Cl secretion by acinar cells isolated from guinea pig nasal glands using a microfluorimetric imaging method and a patch-clamp whole-cell recording. In this experiment Cl current evoked by acetylcholine was inhibited by roxithromycin and erythromycin but not by josamycin (Figs. 7 and 8), indicating that among

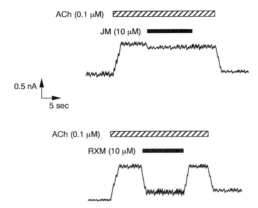

Fig. 7. Effects of josamycin (JM) and roxithromycin (RXM) at concentrations of 10 μM on isolated Cl currents in acinar cells of guinea pig nasal gland. After eliciting a control response to acetylcholine (ACh, 0.1 μM), the cells were exposed to each macrolide. The pipette solution was Na-gluconate and the external solution was NaCl without K. A distinct Cl current was isolated when the membrane potential was clamped at 0 mV. (Reprinted from Ikeda et al. [52] with permission from AJRCCM, official journal of the American Thoracic Society. © American Lung Association.)

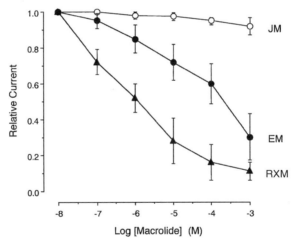

Fig. 8. Concentration-dependent effects of josamycin (JM), erythromycin (EM), and roxithromycin (RXM) on acetylcholine (0.1 μM) induced inward currents at −90 mV in acinar cells of guinea pig nasal gland. All currents were normalized to the control response induced by acetylcholine alone. (Reprinted from Ikeda et al. [52] with permission from AJRCCM, official journal of the American Thoracic Society. © American Lung Association.)

macrolides the drugs having 14-membered lactone ring are capable of directly inhibiting Cl channel functions.

Because these experiments have been conducted on excised tissues and cultured cells, they may not accurately reflect *in vivo* ion transport due to the lack of innervation and blood supply. Therefore, the *in vivo* effects of macrolides on the Cl channel were investigated by measuring Cl diffusion potential difference across rabbit tracheal mucosa using a high-impedance voltmeter under open-circuit conditions [53]. Consequently, intravenous administration of clarithromycin reduced the Cl diffusion potential difference in a dose-dependent manner, whereas aminobenzyl penicillin, cefazolin and amikacin had no effect, thus confirming the specific inhibition of Cl secretion by a 14-membered macrolide.

C. Clinial Studies on Sputum Production

In clinical studies, Tamaoki and colleagues [35] examined the effect of clarithromycin on sputum production and its rheological properties in patients with chronic lower respiratory tract infections. Clarithromycin was given at 100 mg twice daily for 8 weeks and compared with placebo. They showed that clarithromycin almost halved sputum volume, and that the percent solids of the sputum increased, with no effect of placebo. Elastic modulus (G′) significantly increased (at 10 Hz), whereas dynamic viscosity (h′) remained unchanged (Fig. 9). The reduction of sputum production and the corresponding increase in

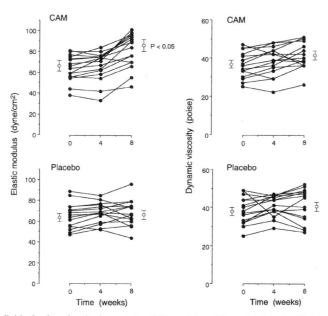

Fig. 9. Individual values for elastic modulus (left panels) and dynamic viscosity (right panels) of the sputum obtained from patients with chronic airway hypersecretion before (0 week), during (4 weeks), and after (8 weeks) the treatment with clarithromycin (CAM, upper panels) or placebo (lower panels). Open circles indicate means ± SE values at 0 and 8 weeks of the treatment. (Reprinted from Tamaoki *et al.* [35] with permission from the American Society for Microbiology.)

solid composition of the secretions may be associated with the inhibition of airway epithelial Cl secretion. More recently, Rubin *et al.* [36] also showed that in patients with purulent rhinitis clarithromycin (500 mg twice daily) for 2 weeks did not significantly alter sputum viscoelasticity but substantially decreased secretion volume and increased mucociliary transportability. The increased transportability of mucus may be attributed to diminished secretion adhesiveness and/or direct stimulation of airway ciliary motility [13].

D. Conclusion

Fourteen-membered ring macrolides can inhibit Cl secretion across the airway epithelial Cl channel. Although the subcellular mechanism of this action warrants further studies, the inhibition of Cl secretion may lead to the reduction of liquid secretion across the airway mucosa toward the lumen. It is thus likely that the favorable effects of 14-membered macrolides on chronic airway hypersecretion might be related, at least in part, to the action on airway epithelial Cl channel.

III. Effects of Macrolides on Cytokine/Chemokine Expression

Erythromycin is a macrolide antibiotic widely used for the treatment of upper and lower respiratory tract infections. Recent reports further showed that EM and its analogues are effective for the treatment of chronic airway diseases such as DPB, bronchial asthma, and chronic sinusitis [5, 15, 32]. This effectiveness is considered to be apart from their antimicrobial actions, because they are effective at half of the recommended dosage and even in cases without concomitant infection. Its precise mechanisms, however, remain unclear. Several cytokines including IL-1, TNF-α, and IL-8 have been reported to be elevated in BALF from patients with such airway inflammatory diseases (Table II), and to be decreased

TABLE II
Cytokines and Chemokines Important in the Pathogenesis of Chronic Airway Inflammatory Diseases

Diffuse Panbronchiolitis and Chronic Bronchitis/Emphysema:	
IL-1 β	IL-8
TNF-α	RANTES
IL-6	
Bronchial Asthma:	
IL-1 β	IL-8
TNF-α	IL-13
GM-CSF	IL-16
IL-3	RANTES
IL-4	Eotaxin
IL-5	TARC
IL-6	MDC

after appropriate therapy, suggesting roles in airway inflammatory processes [15, 54]. There is increasing evidence that macrolide antibiotics show modulating effects on cytokine expression in clinical and experimental settings. *In vitro* studies further indicated that these drugs have inhibitory actions on cytokine production and/or expression in various cells. This section focuses on the effects of the macrolides on cytokine/chemokine production.

A. Clinical Evidence of Macrolide Antibiotics Effect on Cytokine Production

Kadota and his associates [15] demonstrated an increase of neutrophil chemotactic activity (NCA) in BALF, which showed a clear correlation with neutrophil numbers. They further showed that inflammatory cytokines such as IL-8, IL-1β, and TNF-α were also increased in BALF from patients with chronic airway inflammatory diseases such as DPB and bronchiectasis [25]. They showed that treatment with 14-membered ring macrolide antibiotics such as EM caused a decline in both neutrophil number and these inflammatory cytokines and chemokines. These cytokines are potent activators of neutrophils, among which IL-8 is one of the most potent chemotactic factors in the airways [55, 56]. Therefore, it is probable that EM attenuates airway inflammatory responses by decreasing the local cytokine/chemokine levels and thus decreasing the recruitment of inflammatory cells such as neutrophils.

Kadota and associates [57] found that IL-8 mRNA expression was detected in airway epithelium, endothelium, and alveolar macrophages by an *in situ* hybridization technique in the lungs with DPB. Among these resident cells, airway epithelial cells are one of the potent sources of cytokines and chemokines [58, 59], and their anatomic location suggests their proviral role in the regulation of cell recruitment into the airways. Therefore, we studied whether or not macrolides had any effect on cytokine expression and production by human bronchial epithelial cells. First, we evaluated the changes in IL-8 mRNA levels and IL-8 protein release by airway epithelial cells before and after macrolide therapy. In five patients with chronic airway diseases (DPB, chronic bronchitis, and diffuse bronchiectasis), we obtained airway epithelial cells before and after macrolide therapy [60]. All of the patients received oral EM or clarithromycin (CAM) therapy for more than 3 months with no side effects. Clinical signs and symptoms such as dyspnea on exertion, daily amount of sputa, chest radiographic findings, and arterial blood gas analysis were improved in four patients. In accordance with these clinical changes, IL-8 mRNA levels corrected by β-actin transcripts were decreased in the four patients (patients 1 through 4) when assessed by reverse transcription–polymerase chain reaction (RT-PCR) (Fig. 10). Spontaneous IL-8 release from epithelial cells was also decreased by macrolide therapy in four patients (before therapy: 442 ± 34.5 pg/24 hr/10^5 cells; after therapy: 209 ± 18.0 pg/24 hr/10^5 cells, $n = 4$, $p < 0.05$, Student's *t* test), but not in patient 5.

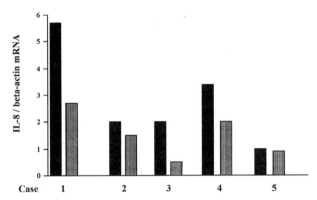

Fig. 10. Evaluation of IL-8 mRNA levels in human bronchial epithelial cells before (left bar) and after (right bar) as assessed by semiquantitative RT-PCR analysis. The vertical bar indicates the magnitude of IL-8 mRNA corrected by β-actin transcripts. Patients 1 to 4 showed a decline in IL-8 mRNA levels, whereas patient 5 did not show any change.

B. Inhibitory Action of Macrolides on Cytokine Production: *In Vitro* Findings

1. Airway Epithelial Cells

It is well documented that normal human bronchial epithelial cells release a variety of cytokines and chemokines, and proinflammatory cytokines such as IL-1α, IL-1β, and TNF-α stimulate their production *in vitro* (Fig. 11) [61]. We cultured human normal and transformed bronchial epithelial cells, and studied the effect of EM, CAM, and RXM on IL-6 [17], IL-8 [60, 62] and GM-CSF [63] production. Among the antimicrobes tested, only 14-membered-ring macrolides EM, CAM, and RXM showed an inhibitory action on cytokine release by unstimulated and cytokine-stimulated human bronchial epithelial cells (Fig. 12). In contrast, a 16-membered ring macrolide josamycin (JM) failed to show such effects. LDH release assay, a trypan blue dye exclusion test, and a colorimetric MTT assay showed that this effect was not due to cytotoxicity. Percent inhibition of IL-8 protein release in human primary bronchial epithelial cells was $25.0 \pm 5.67\%$ at $10^{-6}\,M$, which is comparable with clinically observed serum concentration.

To assess the effect of macrolide antibiotics on IL-8 production by inflamed airway epithelium, bronchial epithelial cells were obtained from 10 patients (3 with DPB, 5 with sinobronchial syndrome, 1 with nonatopic asthma associated with chronic sinusitis, and 1 with diffuse bronchiectasis, mean age of 54.8, all were non- or ex-smokers) under fiber optic bronchoscopy as previously reported [64, 65]. Spontaneous IL-8 release by airway epithelial cells from inflamed airways was significantly suppressed with the addition of EM and CAM, but not with ABPC *in vitro* [60]. Khair *et al.* [19] reported that EM inhibited release of IL-8 as well as of IL-6 from *H. influenzae* endotoxin-stimulated normal bronchial epithelial cells.

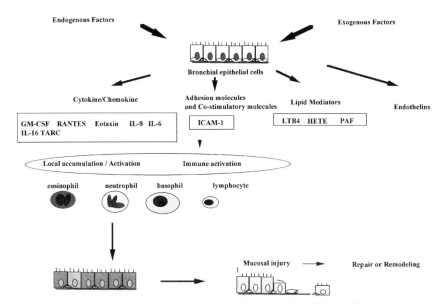

Fig. 11. Airway epithelial cells as sources of cytokines and chemokines in the airways. Airway epithelial cells express and release a variety of cytokines/chemokines, adhesion molecules, and lipid mediators, and thereby participate in the regulation of inflammatory responses in the airways.

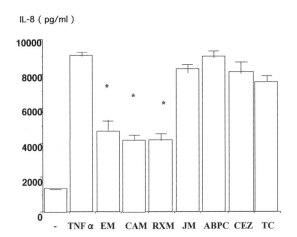

ABPC: aminobenzyl penicillin, CEZ: cefazolin, TC: tetracyclin

Fig. 12. Effect of macrolide antibiotics on IL-8 production by normal human bronchial epithelial cells in culture. The primary human bronchial epithelial cells were cultured until confluence, and the culture supernatants were harvested 24 hr after the stimulation of TNF-α with and without EM, CAM, or RXM. The three macrolide antibiotics showed a significant inhibitory effect on IL-8 release by bronchial epithelial cells. *$p < 0.01$ as compared to the cells with TNF alone (ANOVA).

2. Alveolar Macrophages and Neutrophils

Alveolar macrophages are another important source of cytokines in the lung. Iino and coworkers [24] reported that EM suppressed IL-1β and TNF-α production by human peripheral blood monocytes. Fujii and coworkers [66] demonstrated that 14-membered ring macrolides uniquely inhibited IL-8 production by a vitamin D-differentiated macrophage cell line THP-1 cells. Sugiyama and associates showed that chronic administration of EM in rats induced an inhibitory change in cytokine production such as GRO/CINC-1 and CINC-2α, homologues of human IL-8 and MIP-2, respectively [67]. Oishi and his coworkers [16] studied the production of IL-8 by neutrophils, which stimulated inactivated *P. aeruginosa* bacilli, and they found that IL-8 release from the neutrophils was inhibited by EM treatment.

3. Lymphocytes and Eosinophils

Treatment with EM has long been described in the literature as an alternative therapy for intractable bronchial asthma [68]. It was reported that EM has a corticosteroid-sparing effect. Simultaneous administration of EM with theophylline causes increased levels of the latter drug [69], and thus augments its bronchodilator effect. However, it has been demonstrated that EM alone decreased airway hyperresponsiveness and asthma severity [32]. Konno *et al.* [33] found that RXM inhibited production of IL-2 and IL-4 by peripheral lymphocytes. Nakahara and associates studied the effects of macrolides on cytokine production by peripheral lymphocytes from patients with asthma [70]. Production of Th$_2$-derived cytokines IL-4 and IL-5 was significantly suppressed by EM, whereas that of Th$_1$-derived cytokines such as IFN-γ rather increased. Therefore, it is probable that macrolides exert inhibitory effects on Th$_2$ cytokines in asthma patients. Kohyama *et al.* [71] showed that EM significantly suppressed IL-8 release from human peripheral blood eosinophils from atopic donors. It appeared that this process was at posttranscriptional stages; however, further mechanisms of such immunomodulatory effects of macrolides have not been investigated thus far.

C. Molecular Mechanisms of the Anti-Inflammatory Actions of Macrolides

1. Fourteen-Membered-Ring Macrolide Antibiotics Suppress Gene Expression of Inflammatory Cytokines in Human Bronchial Epithelial Cells

We evaluated the effects of macrolides on steady-state levels of IL-6 and IL-8 mRNA by Northern blot analysis [17, 60]. As shown in Fig. 13, bronchial epithelial cells expressed constitutive IL-8 mRNA, which was significantly upregulated by the cytokines such as IL-1α, IL-1β, and TNF-α. We reported that EM, CAM, and RXM inhibited steady-state levels of IL-6 and IL-8 expression in normal and immortalized bronchial epithelial cells [60, 62, 63]. This action appeared to be unique, because other antibiotics including a 16-membered ring macrolide JM

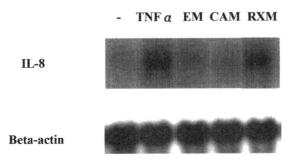

Fig. 13. Northern blot analysis showing the inhibitory effect of macrolide antibiotics on IL-8 mRNA expression in normal human bronchial epithelial cells in culture. Total cellular RNAs were isolated, and equivalent doses of RNA (10 µg/lane) were electrophoresed on formaldehyde-denatured agarose gels. Northern blot analysis showed that IL-8 mRNA levels in cells stimulated by TNF-α were suppressed by pretreatment with EM, CAM, or RXM.

did not show any effect. Therefore, it is probably one mechanism of the clinical beneficial effect of these macrolide antibiotics.

Second, we evaluated the transcriptional rates of IL-8 in the presence or absence of EM by using a run-on transcription assay. EM clearly decreased the transcriptional rates of IL-8 mRNA. We evaluated the time course of IL-8 mRNA levels after treatment with actinomycin D, an RNA synthesis inhibitor; there was no difference in the presence or absence of EM, showing that EM does not affect the decay speed of IL-8 mRNA. These findings indicated that EM suppresses IL-8 mRNA at the transcriptional levels.

2. *Fourteen-Membered Ring Macrolide Antibiotics Suppress Activation of Transcription Factors of Inflammatory Cytokines in Human Bronchial Epithelial Cells*

Many inflammatory cytokines including IL-8 are regulated at transcriptional levels, and a variety of transcription factors such as nuclear factor-κ B (NFκB) and activator protein-1 (AP-1) play important roles in such processes. Abe and coworkers [71] demonstrated that CAM repressed TNF-α-induced AP-1 activation in human bronchial epithelial cells. We studied the effect of EM and CAM on the phorbol myristate acetate (PMA)-induced activation of NFκB and AP-1. Pretreatment of EM and CAM before the PMA treatment showed an inhibitory effect on both of the transcription factors as assessed by electrophoretic mobility shift assay (EMSA) (Fig. 14) [20]. In contrast, the macrolides showed no effect on the activation of cyclic AMP-responsive element binding protein (CREB), suggesting that the suppressive effect on some transcription factors is specific. We further evaluated the effect of EM on the phosphorylation of inhibitor of NFκB (IκB), which is a crucial step for transactivation of NFκB. EM did not influence the phosphorylation processes *in vitro* (Okazaki *et al.*, unpublished data, January 2001). These data suggest that EM acts at the process of nuclear translocation of

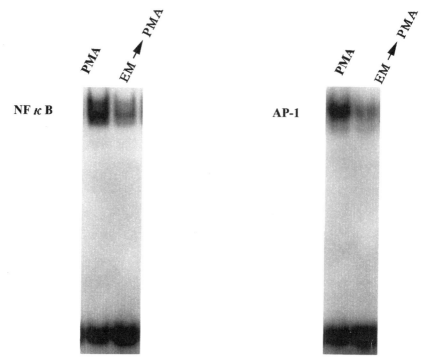

Fig. 14. Electrophoretic mobility shift assay for the detection of NFκB and AP-1 in human bronchial epithelial cell line Bet1A. The human bronchial epithelial cells were preincubated with macrolide antibiotics (10^{-6} M) 24 hr before stimulation by phorbol myristate acetate (PMA, 10^{-7} M). The nuclear extracts were isolated and EMSAs were performed for the detection of (left) NFκB and (right) AP-1. EM and CAM suppressed the activation of both of the transcription factors crucial in the regulation of the IL-8 gene.

NFκB, or at the stages of DNA binding within the nucleus (Fig. 15). Further studies are necessary to elucidate the molecular events important for their potentials.

D. Future Directions

EM also has a motilin-like stimulating activity on gastrointestinal smooth muscles [39]. Therefore, the inhibitory effect on cytokine expression in human cells summarized here may be a third bioactivity of the macrolide antibiotic. We recently reported that some of these derivatives have inhibitory effect on IL-8 production by human airway epithelial cells [72]. These analogues also showed inhibitory action on the activation of NFκB and AP-1 assessed by EMSA (M. Desaki *et al.*, unpublished observations, January 2001). Characterization of the chemical structure responsible for its potential would be important to pursue and further investigation for the molecular mechanism would be necessary for a possible new type of anti-inflammatory agent.

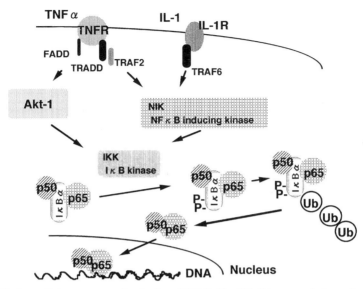

Fig. 15. Intracellular processes of activation of NFκB. The NFκB is present in latent form in the cytoplasm as a result of binding to the inhibitor protein IκB. Cytokine-induced signal transduction results in selective IκB phosphorylation, which is in turn ubiquitinated and degraded by the proteasome pathway. Free NFκB migrates to the mucleus by several localization signals. Binding of NFκB to its specific site of genes induces transcription of several NFκB-dependent genes. NFκB is then inactivated by newly synthesized IκB both in the cytoplasm and nucleus.

IV. Modulation of Bacterial Function

Macrolides are lipophilic molecules composed of a characteristic 12- to 16-membered lactone ring as the aglycon and several amino and/or neutral sugars. Erythromycin, clarithromycin, and roxithromycin with 14-membered rings; azithromycin with a 15-membered ring; and rokitamycin, josamycin, and midecamycin (MDM) with 16-membered rings have been so far used in clinical practice.

These macrolides bind to the 23S ribosomal RNA in the 50S subunit of the 70S ribosome of bacteria with a dissociation constant (K_d) of 10^{-8} M in the presence of Mg^{2+} and K^+, and block the translocation step in protein synthesis, thereby preventing the release of tRNA after peptide bond formation. When both the 50S and 30S subunits are formed as the receptors, the drug recognizes only the large ribosomal subunit, with 1:1 stoichiometry and a K_d of 10^{-7} M, indicating that it has less affinity to the subunit than for the ribosome itself. The 30S subunits may modify the conformation of the 50S subunit [73].

Although macrolides can bind to 23S ribosomal RNAs of both gram-positive and gram-negative bacteria, incorporation and accumulation of sufficient amounts of macrolide into the cells are necessary to exhibit bactericidal or bacteriostatic effect.

Gram-positive bacteria have a cell wall of relatively simple structure consisting of peptideglycan as a major component, whereas gram-negative bacteria have a surrounding outer membrane including lipopolysaccharide through which many drugs of up to 400 Da can diffuse relatively freely through a variety of channel-forming proteins known as porins. Macrolides are large compounds with a molecular weight of approximately 800; therefore, gram-negative bacteria are generally resistant to macrolides. There is increasing evidence, however, that subinhibitory levels of macrolides can affect bacteria in ways distinct from their inhibitory effect of protein synthesis described above. These subinhibitory effects have been studied mainly on *P. aeruginosa* infection. Administration of 14- and 15-membered ring macrolides for periods of several months to years have been reported to improve the clinical symptoms and prognosis of patients with chronic *P. aeruginosa* infections [9]. This may be considered paradoxical, since laboratory susceptibility tests of these antibiotics have demonstrated them to be neither bactericidal nor bacteriostatic against *P. aeruginosa* at their usual therapeutic doses. It has been speculated, therefore, that the clinical efficacies of these macrolides may be due to an indirect result of the effects on the organism's virulence factors, host defense systems, or both.

A. Enhancement of Susceptibility of *P. aeruginosa* to Serum by Macrolides

The mechanism of the bactericidal activity of serum against gram-negative bacteria is complex and involves the participation of complement, antibodies, and lysozyme. In general, *P. aeruginosa* sp. is resistant to serum [74]. However, subinhibitory levels of EM, CAM, and AZM enhance the susceptibility of *P. aeruginosa* to serum bactericidal activity by altering cell membrane structure [75]. The susceptibility is dependent on the time of exposure to the macrolides. Serum susceptibility is significantly enhanced by 36 hr of exposure, but not 12 or 24 hr of exposures. It is plausible that the accumulation of macrolides in *P. aeruginosa* is time dependent and, thus, they affect the rate of protein synthesis in the microorganism.

B. Inhibitory Action of Macrolides on Biofilm Formation of *P. aeruginosa*

P. aeruginosa forms a bacterial biofilm by producing alginate, when it adheres to mucosa or various medical devices. Ichimiya *et al.* reported the effect of certain macrolides on biofilm formation by *P. aeruginosa* [76, 77]. Using scanning electron microscopy, it was observed that biofilms on a Teflon sheet with CAM decreased markedly in a dose-dependent manner compared with a control Teflon sheet without CAM. Interestingly, this effect was observed despite that CAM has no direct bactericidal activity against *P. aeruginosa*. It seemed likely that CAM inhibited the biofilm synthesis via inhibition of polysaccharide synthesis. CAM inhibited with 1/256-1/64 MIC production of alginate, a major component of biofilm, whereas RKM, piperacillin, ceftazidime, and ofloxacin did not inhibit alginate production.

C. Inhibitory Action of Macrolides on the Expression of Virulence Factors of *P. aeruginosa*

Inhibitory effects of macrolides against the expression of virulence factors in *P. aeruginosa* have been studied at subinhibitory drug concentrations.

P. aeruginosa infection is preceded by selective adhesion of bacteria to the host target cells via adhesins, including lectins. The production of both lectins and many of the virulence factors is positively controlled by transcription activators including signaling autoinducers (*N*-acyl-L-homoserine lactones). Sofer *et al.* reported that EM at sub-MIC concentrations suppressed the production of *P. aeruginosa* hemagglutinins (including lectins) [78]. On the other hand, AZM at sub-MIC concentrations strongly suppressed the synthesis of elastase, proteases, lecithinase, and DNase [79]. CAM and EM were far less effective. Gelatinase was reduced to almost the same level by these three macrolides, while hemolysins and lipase were only marginally affected. In these virulence factors, pyocianin, a pigment, is known to suppress superoxide anion production from neutrophils, differentiation and proliferation of lymphocytes, cilliary beating of bronchial epithelial cells, and nitrogen intermediate and cytokine production from alveolar macrophages. EM suppresses the production of pyocianin dose-dependently *in vitro* [80]. *In vivo*, the concentration of pyocianin in sputum from patients with chronic lower respiratory tract infection is reduced after administration of EM.

EM also modulates the effect of pyocianin indirectly. Denning *et al.* reported that EM suppressed IL-8 production from airway epithelial cells induced by pyocianin stimulation [81]. Thus, it is likely that EM prevents the lesion infected with *P. aeruginosa* from tissue injury caused by pyocianin in both direct and indirect manners.

D. Suppression of Aminoglycoside-Inactivating Enzyme by Macrolides

Although aminoglycosides are an effective antibiotic groups against *P. aeruginosa*, the resistant clinical isolates producing aminoglycoside-inactivating enzyme (aminoglycoside 3′-phosphorylating enzyme: APH(3′)-Ib) are increasing. Nakamura *et al.* reported that the activity of APH(3′)-Ib was remarkably suppressed in *P. aeruginosa* after incubation with macrolides [82]. They also demonstrated that this effect was not observed when *P. aeruginosa* was treated with 11 other antibiotics including protein synthesis inhibitors, clindamycin and minocycline, suggesting that the effect was specific for macrolides. Moreover, this effect was conserved even in macrolide-resistant *P. aeruginosa* strain with macrolide 2′ phosphotransferase gene, but abrogated after incubation with macrolides with hydrolyzed lactone ring, indicating that lactone ring was crucial for the effect.

E. Suppressive Action of Macrolides on the Production of Verotoxin by *E. coli* O157

Enterohemorrhagic *Escherichia coli* (EHEC) O157:H7 causes serious hemorrhagic colitis, thrombotic thrombocytopenic purpura, and hemolytic uremic syndrome in humans. Verotoxins produced by EHEC are the pathogens of these diseases. Two different verotoxins have been reported, VT1 and VT2, that destroy mucosa and renal epithelium. The effect of macrolides at sub-MIC concentration was reported on the production of verotoxin (VT) by *E. coli* O157. Nakata *et al.* observed that the production of VT1 was suppressed up to 10 hr when bacteria were incubated with 1/100 of MIC of CAM [83]. The production of VT1 reached the control level after 22 hr even with 1/10 of MIC of CAM. On the other hand, production of VT2 was completely suppressed by 22 hr with 1/10 of MIC. This effect was observed when *E. coli* O157 was incubated with EM. In contrast to the macrolides, ampicillin at sub-MIC levels did not inhibit the production of both VT1 and VT2. To clarify the mechanism of suppression of VT production by macrolides, a macrolide-resistant gene of a *S. pneumonia* strain, the *ermAM* gene, encoding an enzyme of methylation of 23S RNA was transformed into an *E. coli* O157 strain. The suppressive effect of both VT1 and VT2 production from the transformed strain by CAM was remarkably decreased compared with the wild-type strain, indicating that inhibition of protein synthesis by macrolides is responsible for the suppression of VT production [84].

F. Effect of Erythromycin on *Shigella* Invasion into Host Cells

Shigella invades epithelial cells by inducing cytoskeletal reorganization localized at the site of bacteria–host cell interaction. During entry, the *Shigella* type III secretion apparatus allows the insertion of a pore that contains the IpaB and IpaC proteins into cell membranes. Honma *et al.* reported the effect of EM on *Shigella flexineri* invasion of Caco-2 cells even at concentrations less than the MIC [85]. When a *S. flexineri* strain was treated with sub-MIC levels of EM, the invasion efficiency decreased. The carrier rate of the invasion plasmid containing virulence genes was reduced by EM treatment. The presence of the invasion plasmid was found to increase the susceptibility of the organisms to EM. The growth of virulent organisms carrying the invasion plasmid was inhibited at 25 µg/ml of

EM, whereas the growth of organisms without the plasmid was inhibited at 100 μg/ml of EM. It is likely that EM passed through the type III apparatus, encoded by the invasion plasmid, and suppressed the growth of invasive organisms selectively.

G. Conclusion

Challenges against bacterial infections are continuously evolving. Macrolides present excellent antibacterial effects that allow for a simple dosing regimen with minimal side effects. Current macrolide usage includes a variety of community-acquired respiratory tract, skin, and soft tissue, and sexually transmitted disease infections. Macrolides have also substantial activity against atypical organisms such as *Mycobacterium avium* complex (MAC) and *Chlamydia trachomatis*. Due to never-ending need for new antibiotic therapies, several other potential indications have been studied for macrolides. This section presented various current research associated with the potential use of macrolides for modulation of bacterial functions at sub-MIC concentrations. This research indicated that macrolides might be clinically useful for the treatment of patients with some gram-negative bacteria infections. As bacterial resistance patterns fluctuate globally, macrolides may be an alternative therapy for the previously mentioned indications, which will also enhance patient compliance and therefore effectively eradicate infection worldwide.

V. New Challenge for Novel Action

Fourteen-membered ring macrolides such as erythromycin and oleandomycin exhibit antimicrobial activity and are characterized by the following: (1) Instability of EM to acid is characterized by aglycon structure with the carbonyl group at the C-9 position, and the hydroxyl group at C-6 and C-12 positions to form an inactive enolether derivative. (2) A dimethyl amino group at the C-3' position in desosamine is required for antimicrobial activity. (3) Antimicrobial activity is a result of cladinose as a neutral sugar [86]. Based on the means of disintegration by acid, clarithromycin was designed to be stable to acid by the methylation of the hydroxyl group at the C-6 position. Clarithromycin demonstrated 16-fold higher antimicrobial activity in comparison with EM, which was characterized by extreme stability in acid (95% of CAM remained stable in an acidic medium in which erythromycin was completely disintegrated). The serum concentration of CAM was retained 4- to 5-fold higher and longer than that of erythromycin.

A. Mechanism of Structure–Activity Relationship

Anti-inflammatory activities of these 14-membered ring macrolides have not yet been clarified in relation to their structure. To study the structure–activity relationships of macrolides, we examined six macrolides including three new

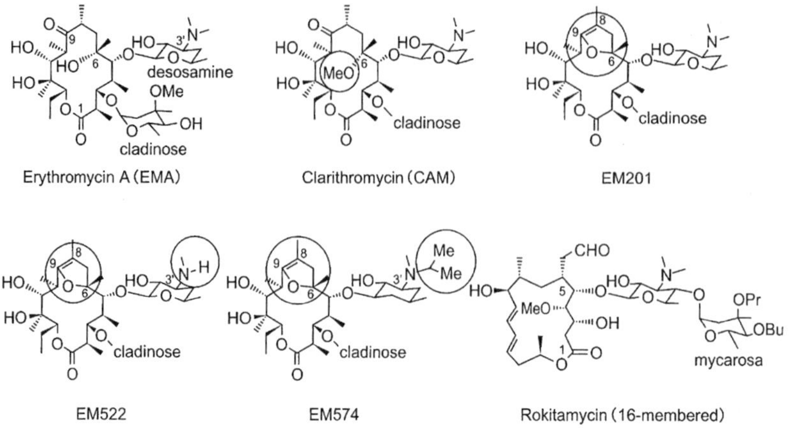

Fig. 16. Comparison of the structures of erythromycin derivatives.

compounds, EM201, EM522, and EM574. They were screened as the acceleration of gastrointestinal motor stimulating activity (GMSA), but not with bactericidal effect (Fig. 16). EM201, EM522, and EM574 strongly inhibited the proliferation of T lymphocytes and differentiation of monocytic cells, but did not commonly inhibit water and mucous secretion of epithelial cells. These results suggest that the antimicrobial effect, GMSA, the inhibition of some cytokines releasing from lymphocyte and/or macrophage, and the inhibition of water and mucous secretions from epithelial cells might be a result of the different structural factors of macrolides [87] (Table III). Structural factors associated with anti-inflammatory activities remain unclear. Further investigation will be needed.

B. Challenge for New Field of Inflammatory Diseases

1. New Field of Chronic Airway Diseases

Today in Japan, 14-membered ring macrolides (EM, CAM, and RXM) are widely applied for treatment of chronic airway diseases including DPB, chronic bronchitis, bronchiectasis, and chronic sinusitis, as mentioned in previous sections. In 1998, Jaffe reported on a pilot study of patients with cystic fibrosis treated with azythromycin. AZM caused improvement of pulmonary functions. The patients were in an advanced stage of the disease and had been awaiting lung transplantation in the United Kingdom [88]. We have sometimes experienced the improvement of mimic airway inflammation, i.e., folicular bronchiolitis, in patients with rheumatoid arthritis and ulceritive colitis by treatment with EM. We actually experienced a patient with bare lymphocyte syndrome (BLS), an extremely rare disease showing progressive bronchiectasis, whose pulmonary functions improved by treatment with long-term and low-dose erythromycin [89] (Table IV).

TABLE III
Biological Activities of Erythromycin A Derivatives and Rokitamycin

Derivatives	Antimicrobial	Motilide	Hematopoietic cells			Epithelioid cells		
			Inhibition of T-cell proliferation	Differentiation of monocytes	Inhibition of IL-8	Inhibition of Cl^- channels		Inhibition of mucous secretion
Erythromycin A	++	+	+	+	+	+	+	+-
Clarithromycin	++	-	++	++	++	+-	++	+-
EM201	-	++	+++	+++	+-	+	+-	-
EM522	-	++	+++	+++	+	++	-	+-
EM574	-	+++	+++	+++	++	+	n.d.	+-
Rokitamycin	++	-	++	++	-	+	n.d.	-
Concentration	$-7 \sim -5$	$-8 \sim -5$	$-6 \sim -4$	$-6 \sim -4$	-6	$-8 \sim -6$	$-5 \sim -4$	-5
Contributor	Sunazuka	Sunazuka	Akagawa	Akagawa	Takizawa	Tamaoki	Miwa	Shimizu Irokawa

Note: Concentration: 10^n *M*; n.d., not determined.

TABLE IV
Lung Function Tests of Patient with Bare Lymphocyte Syndrome[a]

Date (y.m)	VC (L)	%VC	FEV1 (L)	FEV1%	TLC (L)	RV (L)	RV/TLC(%)	V50/V25	DLco (ml/min/mmHg)	PaO$_2$ (mmHg)	PaCO$_2$ (mmHg)
81.4	2.25	75	1.51	67	4.51	2.65	58.7		23.4		
82.4										68	42
Start of EM therapy at 82.8											
82.10	1.79	60	1.12	71	4.45	2.66	59.8	2.11			
83.2	2.19	73	1.33	61	5.11	2.69	52.6	2.72	25.9	72	38
83.10	2.60	81	1.47	64	5.08	2.48	48.8	3.00		71	37
84.2	2.60	87	1.70	65	5.23	2.63	50.3			74	38
84.3	2.73	92	1.71	62	5.23	2.50	47.8	3.33			
85.3	2.82	95	1.79	62	4.62	1.60	39.0	3.28	24.2	66	37
85.12	3.12	106	1.81	60	5.02	1.90	37.6	3.79		74	37
87.7	2.34	80	1.48	69	4.36	1.66	38.1	3.12		75	37
89.7	3.07	107	1.87	61	4.55	1.41	31.0	3.58	18.7	79	39
90.2										86	44
91.3										75	39
91.12											
92.4											
94.9										87	37
95.9										86	35
96.10										86	37
99.1	2.78	98	1.83	66	4.89	1.67	41.3	3.39	24.0		

Source: Reprinted from Azuma et al. [89] with permission from the *Sarcoidosis Vasculitis and Diffuse Lung Disease*.
[a]PaO$_2$ levels increased to over 85 mmHg, as compared to 68 mmHg before EM therapy. RV and RV/TLC decreased, while FEV1 and VC increased. %VC recovered and V50/V25 changed from 2.11 to 3.58.

These results suggest that the process of bacterial infection and neutrophil inflammation in chronic airway inflammation will be commonly improved by treatment with 14- and 15-membered ring macrolides. Modes of action include antineutrophil activities, inhibition of IL-8, decreasing oxygen radicals, inhibitions of LTB4 and defensin, and further improvement of mucociliary clearance [90].

2. New Challenge for Anti-Inflammation Therapy

Since erythromycin and its derivatives exhibit several activities including motilin-like activity and anti-inflammatory actions, we expected them to be useful for treatment of systemic inflammatory diseases including chronic lower and upper airway diseases, chronic inflammatory bowel diseases, skin diseases of unknown origin, and autoimmune diseases. These inflammatory diseases are known to be associated with inflammatory cell activity.

Neutrophils and their injurious substances have been reported to play important roles in the progression of lung fibrosis induced by bleomycin (BLM) [91] (Fig. 17). We found that E-selectin played an essential role in neutrophil adhesion to endothelial cells and subsequent migration into lung tissue and progression of lung fibrosis [92]. A new structural macrolide, macrosphelide-A, which is an inhibitor for sialyl-Le^x-dependent adhesion of HL60 cells to the vascular endothelial cells, successfully inhibited BLM-induced pulmonary fibrosis. The 14-membered ring macrolides have been reported to improve survival time of patients with DPB as a result of several anti-inflammatory mechanisms [9]. We investigated the mechanisms by which 14-membered ring macrolides prevented BLM-induced lung fibrosis in mice, and found that these macrolides inhibited VCAM-1 mRNA induction by BLM in a mice model [93]. This finding indicates that 14-membered ring macrolides decrease the ability of inflammatory cells, especially neutrophils, to migrate into injurious tissue, and thus inhibit tissue

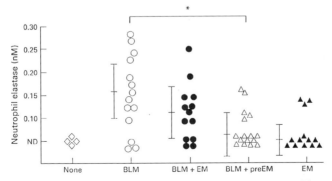

Fig. 17. Comparison of neutrophil elastase activity in BALF between groups. The activity of the BLM + pre-EM group was significanitly inhibited compared with that of the BLM alone group (*$p < 0.10$). ND = not detectable (<0.10 nM). (Reprinted from Azuma *et al.* [91] with permission from the BMJ Publishing Group.)

damage by injurious substances released by inflammatory cells. The inhibitory effect of 14-membered ring macrolides on inflammatory cell infiltration is prominent for neutrophils but less so for lymphocytes, whereas corticosteroid inhibits inflammation rather than neutrophil inflammation. The inhibition by macrolides on cell-to-cell interaction is an important mechanism of their anti-inflammatory action. Similar findings were obtained in the field of dermatology. RXM inhibited the expression of adhesion molecule ICAM-1 in a dose-dependent manner, resulting in attenuation of the activity of inflammatory dermal disorders [94].

In the production of injurious substances, i.e., oxidants, EM was proved to inhibit HADPH/NADH oxidase induction, resulting in the reduction of superoxide anion synthesis [26]. Sato *et al.* reported that EM exhibited an antioxidant effect in the influenza virus-infected mice model. Although mice infected with influenza virus usually die, EM dramatically improved the survival ratio of infected mice. This improvement is due to the antioxidant effect, which showed inhibition of IFN-γ causing attenuation of inducible nitric oxide synthase (iNOS) transcription by erythromycin [95] (Fig. 18).

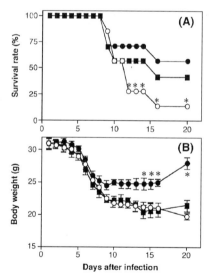

Fig. 18. Effect of EM on survival rate (A) and body weight (B) of mice infected with influenza virus. At each time point after influenza virus infection [1.5 × LD50 of influenza virus (A/Kumamoto/Y5/H2N2)], the survival rate of infected mice was evaluated and the body weight of mice was measured. The mice were given EM intraperitoneally (solid squares: 1.0 mg/kg/d; solid circles: 3.3 mg/kg/d in saline/0.5% DMSO) every 24 h from Day 1 to Day 6 after virus infection. The control group (open circles) was injected intraperitoneally with 0.5 ml saline/0.5% DMSO. Fourteen mice were used in each experimental group: $p < 0.05$, control versus EM-treated groups. (Reprinted from Sato *et al.* [95] with permission from AJRCCM, official journal of the American Thoracic Society. © American Lung Association.)

Control

CAM pre-treated

Fig. 19. Marked reduction of experimental metastatic lung cancer [96]. Marked metastasis in control groups (upper panel). However, marked attenuation of metastasis in pre-treated with clarithromycin (lower panel).

3. Modulation of Cancer Biology

In the field of cancer metastasis, investigators are interested in the antimetastatic effects of macrolides in various strains of cancer cells *in vivo*. We found that metastasis of B16 mouse melanoma cells to the lung was inhibited by CAM with inhibition of integrin function *in vivo* [96] (Fig. 19). Some investigators reported the marked attenuation of cancer metastasis to the lung by inhibition of angiogenesis treated with RXM [97]. These findings suggest the clinical usefulness of 14-membered ring macrolides for reduction of progressive metastasis of cancer cells in cancer-bearing patients. Well-designed and controlled clinical trials are needed to prove these preclinical effects.

4. New Strategy for Dermatitis Associated with Neutrophil Inflammation

In the field of dermatology, new potentials for application of macrolide were recently reported. Preliminary results show 14-membered ring macrolides improved (1) pustulosis palmaris et plantaris, which is usually involved in hand and/or foot with aseptic pustuls; (2) acne vulgaris; and (3) psoriasis vulgaris [98]. A common feature of the skin diseases is neutrophil-associated inflammation. A large-scale clinical study to clarify the effect of macrolides on inflammatory skin diseases is scheduled for the near future. Furthermore, the mechanisms of action of macrolides on subcutaneous inflammatory cells, including neutrophils, lymphocytes, dendritic cells, macrophages, and fibroblasts, are now under investigation.

C. Conclusion

A study group to examine novel actions of macrolides was established in 1994 in Japan. Many clinicians and basic investigators have performed studies in many biological fields related to chronic inflammation. Some of these studies have been considered in this paper. Macrolides are likely to be useful for not only DPB but also other chronic neutrophil-induced inflammation, such as cystic fibrosis [88]. These effects were associated with both modulation of host defense mechanisms and modulation of bacterial activity. New derivatives of erythromycin that showed stronger motilide activity than erythromycin but no bactericidal effects have been developed. Development of new derivatives exhibiting anti-inflammatory actions without bactericidal activity will be needed, because low-dose and long-term erythromycin therapy can induce the formation of macrolide-resistant bacterial strains. We propose new clinical trials with new derivatives for the evaluation of anti-inflammatory effects. The anti-inflammatory effects noted here can be considered to be the third action of macrolides. We expect the anti-inflammatory activity of macrolides to be widely applied in the treatment of chronic inflammatory diseases in the future.

References

1. Homma, H., Yamanaka, A., Tanimoto, S., Tamura, M., Chijimatsu, Y., Kira, S., and Izumi, T. (1983). Diffuse panbronchiolitis. A disease of the transitional zone of the lung. *Chest* **83**, 63–69.
2. Keicho, N., Ohashi, J., Tamiya, G., Nakata, K., Taguchi, Y., Azuma, A., Ohishi, N., Emi, M., Park, M. H., Inoko, H., Tokunaga, K., and Kudoh, S. (2000). Fine localization of a major disease-susceptibility locus for diffuse panbronchiolitis. *Am. J. Hum. Gen.* **66**, 501–507.
3. Kino, M. (1981). A study on patients with diffuse panbronchiolitis diagnosed by pathological examination. *In* "Annual Report of the Research Committee on Interstitial Lung Disease." Ministry of Health and Welfare, Japan.
4. Kudoh, S., and Kimura, H. (1984). Clinical effect of low-dose long-term administration of macrolides on diffuse panbronchiolitis. *Jpn. J. Thorac. Dis.* **22** (Suppl.), 254.
5. Kudoh, S., Uetake, T., Hagiwara, K., Hirayama, M., Hus, L. H., Kimura, H., and Sugiyama, Y. (1987). Clinical effects of low-dose long-term erythromycin chemotherapy on diffuse panbronchiolitis. *Jpn. J. Thorac. Dis.* **25**, 632–642.
6. Kudoh, S., Yamaguchi, T., and Kurashima, A. (1988). Long-term therapeutic effects of erythromycin on diffuse panbronchiolitis—A retrospective study. *In* "Annual Report of the Research Committee on Interstitial Lung Disease," pp. 25–29. Ministry of Health and Welfare, Japan.
7. Yamamoto, M., Kondo, A., Tamura, M., Izumi, T., Ina, Y., and Noda, M. (1990). Long-term therapeutic effects of erythromycin and new quinolone antibacterial agents on diffuse panbronchiolitis. *Jpn. J. Thorac. Dis.* **28**, 1305–1313.
8. Yamamoto, M., Kudoh, S., Ina, Y., and Tamura, A. (1990). Clinical efficacy of erythromycin for patients with diffuse panbronchiolitis—A double blind study. *Saishin Igaku* **45**, 103–108.
9. Kudoh, S., Azuma, A., Yamamoto, M., Izumi, T., and Ando, M. (1998). Improvement of survival in patients with diffuse panbronchiolitis treated with low-dose erythromycin. *Am. J. Respir. Crit. Care Med.* **157**, 1829–1832.
10. Nagai, H., Shishido, H., Yoneda, R., Yamaguchi, E., Tamura, A., and Kurashima, A. (1991). Long-term low-dose administration of erythromycin to patients with diffuse panbronchiolitis. *Respiration* **58**, 145–149.
11. Goswami, S. K., Kivity, S., and Marom, Z. (1990). Erythromycin inhibits respiratory glycoconjugate secretion from human airway *in vitro*. *Am. Rev. Respir. Dis.* **141**, 72–78.
12. Tamaoki, J., Isono, K., Sakai, N., Kanemura, T., and Konno, K. (1992). Erythromycin inhibits Cl secretion across canine tracheal epithelial cells. *Eur. Respir. J.* **5**, 234–238.
13. Mikami, M. (1992). Clinical and pathophysiological significance of neutrophil elastase in sputum and the effect of erythromycin in chronic respiratory diseases. *Jpn. J. Thorac. Dis.* **29**, 72–83.

14. Ichikawa, Y., Ninomiya, H., Koga, Tanaka, M., Kinoshita, M., Tokunaga, N., Yano, T., and Oizumi, K. (1992). Erythromycin reduces neutrophils and neutrophil-derived elastolytic-like activity in the lower respiratory tract of bronchiolitis patients. *Am. Rev. Respir. Dis.* **146**, 196–203.
15. Kadota, J., Sakito, O., Kohno, S., Sawa, H., Mukae, H., Oda, H., Kawakami, K., Fukushima, K., Hiratani, K., and Hara, K. (1993). A mechanism of erythromycin treatment in patients with diffuse panbronchiolitis. *Am. Rev. Respir. Dis.* **147**, 153–159.
16. Oishi, K., Sonoda, F., Kobayashi, S., and Matsumoto, K. (1994). Role of interleukin-8 (IL-8) and an inhibitory effect of erythromycin on IL-8 release in the airways of patients with chronic airway diseases. *Infect. Immun.* **62**, 4145–4152.
17. Takizawa, H., Desaki, M., Ohtoshi, T., Kawasaki, S., Kohyama, T., Sato, M., Tanaka, M., Kasama, T., Kobayashi, K., Nakajima, J., and Ito, K. (1995). Erythromycin suppresses interleukin 6 expression by human bronchial epithelial cells: A potential mechanism of its anti-inflammatory action. *Biochem. Biophys. Res. Commun.* **210**, 781–786.
18. Takizawa, H., Desaki, M., Ohtoshi, T., Kawasaki, S., Kohyama, T., Sato, M., Tanaka, M., Kasama, T., Kobayashi, K., Nakajima, J., and Ito, K. (1997). Erythromycin modulates IL-8 expression in normal and inflamed human bronchial epithelial cells. *Am. J. Respir. Crit. Care Med.* **156**, 266–271.
19. Khair, O. A., Devalia, J. L., Abdelaziz, M. M., Sapsford, R. J., and Davis, R. J. (1995). Effect of erythromycin on *Hemophilus influenzae* endotoxin-induced release of IL-6, IL-8 and sICAM-1 by cultured human bronchial epithelial cells. *Eur. Respir. J.* **8**, 1451–1457.
20. Desaki, M., Takizawa, H., Ohtoshi, T., Kasama, T., Kobayashi, K., Sunazuka, T., Omura, S., Yamamoto, K., and Ito, K. (2000). Erythromycin suppresses nuclear factor-kappa B and activator protein-1 activation in human bronchial epithelial cells. *Biochem. Biophys. Res. Commun.* **267**, 124–128.
21. Kawakami, K., Kadota, J., Iida, K., Fujii, T., Shirai, R., Matsubara, Y., and Kohno, S. (1997). Phenotypic characterization of T cells in bronchoalveolar lavage fluid (BALF) and peripheral blood of patients with diffuse panbronchiolitis: The importance of cytotoxic T cells. *Clin. Exp. Immun.* **107**, 410–416.
22. Keicho, N., Kudoh, S., Yotsumoto, H., and Akagawa, K. S. (1993). Antilymphocytic activity of erythromycin distinct from that of FK506 or cyclosporin A. *J. Antibiot.* **46**, 1406–1413.
23. Keicho, N., Kudoh, S., Yotsumoto, H., and Akagawa, K. S. (1994). Erythromycin promotes monocyte to macrophage differentiation. *J. Antibiot.* **47**, 80–89.
24. Iino, Y., Toriyama, M., Kudo, K., Natori, Y., and Yuo, A. (1992). Erythromycin inhibition of lipopolysaccharide-stimulated tumour necrosis factor-alpha production by human monocytes *in vitro*. *Ann. Oto. Rhino. Laryngol.* **101**, 16–20.
25. Sakito, O., Kadota, J., Kohno, S., Abe, K., Shirai, R., and Hara, K. (1996). Interleukin 1 beta, tumor necrosis factor alpha, and interleukin 8 in bronchoalveolar lavage fluid of patients with diffuse panbronchiolitis: A potential mechanism of macrolide therapy. *Respiration* **63**, 42–48.
26. Umeki, S. (1993). Anti-inflammatory action of erythromycin. Its inhibitory effect on neutrophil NADPH oxidase activity. *Chest* **104**, 1191–1193.
27. Nelson, S., Summer, W. R., Terry, P. B., Warr, G. A., and Jakab, G. J. (1987). Erythromycin induced suppression of pulmonary antibacterial defenses. *Am. Rev. Respir. Dis.* **136**, 1207–1212.
28. Hirata, T., Matsunobe, S., Matsui, Y., Kado, M., Mikiya, K., and Oshima, S. (1990). Effect of erythromycin on the generation of neutrophil chemiluminescence *in vitro*. *Jpn. J. Thorac. Dis.* **28**, 1066–1071.
29. Oda, H., Kadota, J. I., Kohno, S., and Hara, K. (1994). Erythromycin inhibits neutrophil chemotaxis in bronchoalveoli of diffuse panbronchiolitis. *Chest* **106**, 1116–1123.
30. Hojo, M., Fujita, I., Hamasaki, Y., Miyazaki, M., and Miyazaki, S. (1994). Erythromycin does not directly affect neutrophil functions. *Chest* **105**, 520–523.

31. Suez, D., and Szefler, S. J. (1986). Excessive accumulation of mucus in children with asthma: A potential role for erythromycin? A case discussion. *J. Allergy Clin. Immunol.* **77**, 330–334.
32. Miyatake, H., Taki, F., Taniguchi, H., Suzuki, R., Takagi, K., and Satake, T. (1991). Erythromycin reduces the severity of bronchial hyperresponsiveness in asthma. *Chest* **99**, 670–673.
33. Konno, S., Asano, K., Kurokawa, M., Ikeda, K., Okamoto, K., and Adachi, K. (1994). Antiasthmatic activity of a macrolide antibiotic, roxithromycin: Analysis of possible mechanisms *in vitro* and *in vivo*. *Int. Arch. Allergy Immunol.* **105**, 308–316.
34. Yanagihara, K., Tomono, K., Sawai, T., Hirakata, Y., Kadota, J., Koga, H., Tashiro, T., and Kohno, S. (1997). Effect of clarithromycin on lymphocytes in chronic respiratory *Pseudomonas aeruginosa* infection. *Am. J. Respir. Crit. Care Med.* **156**, 337–342.
35. Tamaoki, J., Takeyama, K., Tagaya, E., and Konno, K. (1995). Effect of clarithromycin on sputum properties and its rheological properties in chronic respiratory tract infections. *Antimicrob. Agents Chemother.* **39**, 1688–1690.
36. Rubin, B. K., Druce, H., Ramirez, O. E., and Palmer, R. (1997). Effect of clarithromycin on nasal mucus properties in healthy subjects and in patients with purulent rhinitis. *Am. J. Respir. Crit. Care Med.* **155**, 2018–2023.
37. Suzuki, H., Shimomura, A., Ikeda, K., Oshima, T., and Takasaka, T. (1997). Effects of long-term low-dose macrolide administration on neutrophil recruitment and IL-8 in the nasal discharge of patients. *Tohoku J. Exp. Med.* **182**, 115–124.
38. Cole, P. C. A. (1986). "Vicious circle" hypothesis of the pathogenesis of bronchiectasis. *Eur. J. Respir. Dis.* **69** (Suppl.), 6–15.
39. Kondo, Y., Torii, K., Omura, S., and Itoh, Z. (1988). Erythromycin and its derivatives with motilin-like biological activities inhibit the specific binding of 125I-motilin to duodenal muscle. *Biochem. Biophys. Res. Commun.* **150**, 877–882.
40. Peeters, T. L. (1993). Erythromycin and other macrolides as prokinetic agents. *Gastroenterology* **105**, 1886–1899.
41. Wanner, A. (1977). State of the art: Clinical aspects of mucociliary transport. *Am. Rev. Respir. Dis.* **116**, 73–125.
42. Nadel, J. A., Widdicombe, J. H., and Peatfield, A. C. (1985). Regulation of airway secretions, ion transport, and water movement. In "The Respiratory System" (A. P. Fishman, Ed.), pp. 419–445. American Physiological Society, Bethesda, MD.
43. Welsh, M. J. (1987). Electrolyte transport by airway epithelia. *Physiol. Rev.* **67**, 1143–1184.
44. Boucher, R. C. (1994). State of the art: Human airway ion transport. *Am. J. Respir. Crit. Care Med.* **150**, 581–593.
45. Widdicombe, J. H., Kondo, M., and Mochizuki, H. (1991). Regulation of airway mucosal ion transport. *Int. Arch. Allergy Appl. Immunol.* **94**, 56–61.
46. Marom, Z. M., and Goswami, S. K. (1991). Respiratory mucus hypersecretion (bronchorrhea): A case discussion. Possible mechanism(s) and treatment. *J. Allergy Clin. Immunol.* **87**, 1050–1055.
47. Tokuyama, K., Kuo, H. P., Rohde, J. A. L., Barnes, P. J., and Rogers, D. F. (1990). Neural control of goblet cell secretion in guinea pig airways. *Am. J. Physiol.* **259**, L108–L115.
48. Tamaoki, J., Takeyama, K., Yamawaki, I., Kondo, M., and Konno, K. (1997). Lipopolysaccharide-induced goblet cell hypersecretion in the guinea-pig trachea: Inhibition by macrolides. *Am. J. Physiol.* **272**, L15–L19.
49. Tamaoki, J., Nakata, J., Tagaya, E., and Konno, K. (1996). Effects of roxithromycin and erythromycin on interleukin 8-induced neutrophil recruitment and goblet cell secretion in guinea pig tracheas. *Antimicob. Agents Chemother.* **40**, 1726–1728.
50. Anderson, R., Fernandes, A. C., and Eftychis, H. E. (1984). Studies on the effects of ingestion of a single 500 mg oral dose of erythromycin stearate on leucocyte motility and transformation and on release *in vitro* of prostaglandin E_2 by stimulated leucocytes. *J. Antimicrob. Chemother.* **14**, 41–50.

51. Johnson, J. D., Hand, W. L., Francis, J. B., King-Thompson, N., and Corwin, R. W. (1980). Antibiotic uptake by alveolar macrophages. *J. Lab. Clin. Med.* **95**, 429–439.
52. Ikeda, K., Wu, D., and Takasaka, T. (1995). Inhibition of acetylcholine-evoked Cl-currents by 14-membered macrolide antibiotics in isolated acinar cells of the guinea pig nasal gland. *Am. J. Respir. Cell Mol. Biol.* **13**, 449–454.
53. Tamaoki, J., Takemura, H., Tagaya, E., and Konno, K. (1995). Effect of clarithromycin on transepithelial potential difference in rabbit tracheal mucosa. *J. Infect. Chemother.* **1**, 112–115.
54. Mattoli, S., Mattoso, V. L., Soloperto, M., Allegra, L., and Fasoli, A. (1991). Cellular and biochemical characteristics of bronchoalveolar lavage fluid in symptomatic nonallergic asthma. *J. Allergy Clin. Immunol.* **87**, 794–802.
55. Yoshimura, T., Matsushima, K., Oppenheim, J. J., and Leonard, E. J. (1987). Neutrophil chemotactic factor produced by lipopolysaccharide (LPS)-stimulated human blood mononuclear leukocytes: Partial characterization and separation from interleukin-1 (IL 1). *J. Immunol.* **139**, 3474–3483.
56. Carre, P. C., Mortenson, R. L., King, T. E., Noble, P. W., Sable, C. L., and Riches, D. W. H. (1991). Increased expression of the interleukin-8 gene by alveolar macrophages in idiopathic pulmonary fibrosis. *J. Clin. Invest.* **88**, 1802–1810.
57. Kadota, J., Ikeda, K., Kawakami, K., Matsubara, Y., Shirai, R., Abe, K., Taniguchi, H., Fujii, T., Kawamoto, S., Kaseda, M., and Kohno, S. (1997). Analysis of inflammatory cell infiltration and its related factors in the lung of patients with diffuse panbronchiolitis (in Japanese). *Jpn. J. Inflammation* **17**, 261–267.
58. Nakamura, H., Yoshimura, K., Jaffe, H. A., and Crystal, R. G. (1991). Interleukin-8 gene expression in human bronchial epithelial cells. *J. Biol. Chem.* **266**, 19,611–19,617.
59. Takizawa, H., Ohtoshi, T., Kikutani, T., Okazaki, H., Akiyama, N., Sato, M., Shoji, S., and Ito, K. (1995). Histamine activates bronchial epithelial cells to release inflammatory cytokines *in vitro*. *Int. Arch. Allergy Immunol.* **108**, 260–267.
60. Takizawa H., Ohtoshi, T., Kawasaki, S., Kohyama, T., Sato, M., Tanaka, M., Kasama, T., Kobayashi, K., Nakajima, J., and Ito, K. (1997). Erythromycin modulates IL-8 expression in human bronchial epithelial cells: Studies with normal and inflamed airway epithelium. *Am. J. Respir. Crit. Care Med.* **156**, 266–271.
61. Takizawa, H. (1998). Airway epithelial cells as regulators of airway inflammation. *Int. J. Mol. Med.* **1**, 367–378.
62. Kawasaki S., Takizawa, H., Ohtoshi, T., Takeuchi, N., Kohyama, T., Nakamura, H., Kasama, T., Kobayashi, K., Nakahara, K., Morita, Y., and Yamamoto, K. (1998). Roxithromycin inhibits cytokine production and neutrophil attachment with human bronchial epithelial cells *in vitro*. *Antimicrob. Agents Chemother.* **42**, 1499–1502.
63. Takizawa, H., Ohtoshi, T., Takeuchi, N., and Ito, K. (1996). Effect of macrolide antibiotics on the expression and release of inflammatory cytokines in human bronchial epithelial cells [in Japanese]. *J. Jpn. Bronchoesophagol. Soc.* **47**, 185–188.
64. Kelsen, S. G., Mardini, I. A., Zhou, S., Benovic, J. L., and Higgins, N. C. (1992). A technique to harvest viable tracheobronchial epithelial cells from living human donors. *Am. J. Respir. Cell Mol. Biol.* **7**, 66–72.
65. Tanaka, M., Takizawa, H., Satoh, M., Okada, Y., Yamasawa, F., and Umeda, A. (1994). Assessment of an ultrathin bronchoscope which allows cytodiagnosis of small airways. *Chest* **106**, 1443–1447.
66. Fujii, T., Kadota, J., Morikawa, T., *et al.* (1996). Inhibitory effect of erythromycin on interleukin 8 production by 1 alpha,25-dihydroxyvitamin D3-stimulated THP-1 cells. *Antimicrob. Agents Chemother.* **40**, 1548–1551.
67. Sugiyama, Y., Yanagisawa, K., Tominaga, S.-I., and Kitamura, S. (1999). Effects of long-term administration of erythromycin on cytokine production in rat alveolar macrophages. *Eur. Respir. J.* **14**, 1113–1116.

68. Kamada, A. K., Hill, M. R., Ikle, D. N., Brenner, A. M., and Szefler, S. J. (1993). Efficacy and safety of low-concentration troleandomycin therapy in children with severe, steroid-requiring asthma. *J. Allergy Clin. Immunol.* **91**, 873–882.
69. Nakahara, H., Higashida, A., Nogami, J., Iwanaga, K., Ueshima, H., Sawaguchi, H., Haraguchi, R., Muraki, M., Kubo, Y., and Nakajima, S. (1997). Effect of roxithromycin on cytokine production by peripheral monocytes derived from patients with bronchial asthma [in Japanese]. *Jpn. J. Antibiot.* **50** (Suppl.), 113–115.
70. Kohyama, T., Takizawa, H., Kawasaki, S., Akiyama, N., Sato, M., and Ito, K. (1999). Fourteen-membered ring macrolides inhibit IL-8 release by human eosinophils from atopic donors. *Antimicrob. Agents Chemother.* **43**, 907–911.
71. Abe, S., Nakamura, H., Inoue, S., Takeda, H., Saito, H., Kato, S., Mukaida, N., Matsushima, K., and Tomoike, H. (2000). Interleukin-8 gene repression by clarithromycin is mediated by the activator protein-1 binding site in human bronchial epithelial cells. *Am. J. Respir. Cell Mol. Biol.* **22**, 51–60.
72. Sunazuka, T., Takizawa, H., Desaki, M., Suzuki, K., Obata, R., Otoguro, K., and Omura, S. (1999). Effects of erythromycin and its derivatives on interleukin-8 release by human bronchial epithelial cell line BEAS-2B cells. *J. Antibiot.* **52**, 71–74.
73. Aumercier, M., and Goffic, F. (1993). Mechanism of action of the macrolide and streptogramin antibiotics. *In* "Macrolides: Chemistry, Pharmacology and Clinical Uses" (A. J. Bryskier, Ed.), pp. 115–123. Arnette Blackwell, Paris.
74. Wolska, K., Bukowski, K., Anusz, Z., and Jakubczak, A. (1999). Bactericidal activity of human, swine and cattle serum against *Pseudomonas aeruginosa* strains. *Med. Dosw. Mikrobiol.* **51**, 339–345.
75. Tateda, K., Hirakata, Y., Furuya, N., Ohno, A., and Yamaguchi, K. (1993). Effects of sub-MICs of erythromycin and other macrolide antibiotics on serum sensitivity of *Pseudomonas aeruginosa*. *Antimicrob. Agents. Chemother.* **37**, 675–680.
76. Kobayashi, H. (1995). Airway biofilm disease: Clinical manifestations and therapeutic possibilities using macrolides. *J. Infect. Chemother.* **1**, 1–15.
77. Ichimiya, T., Yamasaki, T., and Nasu, T. (1994). *In vitro* effects of antimicrobial agents on *Pseudomonas aeruginosa* biofilm formation. *J. Antimicrob. Chemother.* **34**, 331–341.
78. Sofer, D., Gilboa-Garber, N., Belz, A., and Garber, N. C. (1999). Subinhibitory erythromycin represses production of *Pseudomonas aeruginosa* lectins, autoinducer and virulence factors. *Chemotherapy* **45**, 335–341.
79. Molinari, G., Guzman, C. A., Pesce, A., and Schito, G. C. (1993). Inhibition of *Pseudomonas aeruginosa* virulence factors by subinhibitory concentrations of azithromycin and other macrolide antibiotics. *J. Antimicrob. Chemother.* **31**, 681–688.
80. Sato, K., Suga, M., Nishimura, J., Kushima, Y., Muranaka, H., and Ando, M. (1997). Pyocyanine synthesis by *Pseudomonas aeruginosa* in chronic airway infection and the effect of erythromycin on its biological activity. *Jpn. J. Antibiot.* **50** (Suppl.), 89–91.
81. Denning, G. M., Wollenweber, L. A., Railsback, M. A., Cox, C. D., Stoll, L. L., and Britigan, B. E. (1998). *Pseudomonas* pyocyanine increases interleukin-8 expression by human airway epithelial cells. *Infect. Immun.* **66**, 5777–5784.
82. Nakamura, A., Naito, Y., Nakazawa, K., Ohara, K., and Sawai, T. (2000). Selective suppression of aminoglycoside 3′-phosphorylating enzyme activities by macrolides—A study by using macrolide-resistant bacteria. *Jpn. J. Antibiot.* **53** (Suppl. A), 28–31.
83. Nakata, K., Tozu, T., Hoshikawa, Y., Sakai, A., Tanaka, N., Akashi, T., and Kanegasaki, S. (1997). Suppressive effect of clarithromycin on the production of verotoxin by *E. coli* O157. *Kansenshogaku Zasshi.* **71**, 437–442.
84. Nakata, K., Totsu, T., Tanaka, N., Akashi, T., and Kanegasaki, S. (1998). Inhibitory effects of macrolide antibiotics at low concentrations on vero toxin production by pathogenic *E. coli* O157. *Jpn. J. Antibiot.* **51** (Suppl.), 146–150.

85. Honma, Y., Sasakawa, C., Tsuji, T., and Iwanaga, M. (2000). Effect of erythromycin on *Shigella* infection of Caco-2 cells. *FEMS Immunol. Med. Microbiol.* **27**, 139–145.
86. Nakagawa, A., and Omura, S. (1984). Structure and stereochemistry of macrolides. *In* "Macrolide Antibiotics," Chap. 2, pp. 37–84. Academic Press, Inc., San Diego.
87. Azuma, A. (2001). Novel activity of erythromycin derivatives on inflammatory lung diseases. *Recent Res. Devel. Respir. & Crit. Care Med.* (in press).
88. Jaffe, A., Francis, J., Rosenthal, M., and Bush, A. (1998). Long-term azithromycin may improve lung function in children with cystic fibrosis. *Lancet* **351**, 420.
89. Azuma, A., Keicho, N., Furukawa, H., Yabe, T., and Kudoh, S. (2001). Prolonged survival of bare lymphocyte syndrome Type I patient with diffuse panbronchiolitis treated with erythromycin. *Sarcoidosis Vasculitis and Diffuse Lung Disease* **18**, 312–313.
90. Tanaka, E., Kanthakumar, K., Cundell, D. R., Tsang, K. W., Taylor, G. W., Kuze, F., Cole, P. J., and Wilson, R. (1994). The effect of erythromycin on *Pseudomonas aeruginosa* and neutrophil mediated epithelial damage. *J. Antimicrob. Chemother.* **33**(4), 765–775.
91. Azuma, A., Furuta, T., Enomoto, T., Hashimoto, Y., Uematsu, K., Nukariya, N., Murata, A., and Kudoh, S. (1998). Preventive effect of erythromycin on experimental bleomycin-induced acute lung injury in rats. *Thorax* **53**, 186–189.
92. Azuma, A., Takahashi, S., Nose, M., Araki, K., Araki, M., Takahashi, T., Hirose, M., Kawashima, H., Miyasaka, M., and Kudoh, S. (2000). Role of E-selectin in bleomycin induced lung fibrosis in mice. *Thorax* **55**, 147–152.
93. Azuma, A., Li, Y. J., Usuki, J., Aoyama, A., Enomoto, T., and Kudoh, S. (2001). Fourteen membered-ring macrolides inhibit the VCAM-1 mRNA induction preventing neutrophil induced lung injury and fibrosis in bleomycin challenged mice. *Chest* **120**S, 20–22.
94. Yamawaki, M., Akamatsu, H., and Horio, T. (1998). Influence of roxithromycin to the expression of adhesion molecule on dermal endothelial cells. *Jpn. J. Antibiot.* **52**, 128–130.
95. Sato, K., Suga, M., Akaike, T., Fujii, S., Muranaka, H., Doi, T., Maeda, H., and Ando, M. (1998). Therapeutic effect of erythromycin on influenza virus-induced lung injury in mice. *Am. J. Respir. Crit. Care Med.* **157**, 853–857.
96. Shibuya, M., Kokubo, Y., Takechi, T., Hibino, T., Ono, Y., Shinoda, K., Suh, D., Matsuda, K., Azuma, A., and Kudoh, S. (1998). Inhibitory effect of clarithromycin on experimental lung metastasis of melanoma cells by inhibition of cell adhesion and migration into the tissue. *Jpn. J. Antibiot.* **52**, 121–127.
97. Yatsunami, J., Tsuruta, N., Fukuno, Y., Kawashima, M., Taniguchi, S., and Hayashi, S. (1999). Inhibitory effects of roxithromycin on tumor angiogenesis, growth and metastasis of mouse B16 melanoma cells. *Clin. Exp. Metastasis* **17**, 119–124.
98. Komiyane, M., Tokura, S., Matsunaga, Y., Akamatsu, H., and Tamaoki, K. (2000). Symposium 2: Novel activities of macrolides in dermatology. *Jpn. J. Antibiot.* **54** (Suppl. A), 100–112.

Chapter 13

Mode of Action of Avermectin

SATOSHI ŌMURA

Kitasato Institute for Life Sciences
Kitasato University
and The Kitasato Institute
Tokyo, Japan

I. Introduction 571
II. Target of Avermectin Action 571
III. Cloning and Structure of Avermectin Binding Protein 572
IV. Concluding Remarks 575
    References 575

I. Introduction

Since the discovery of avermectin (AVM), its outstanding antiparasitic activity has attracted great attention from many people, and investigations into the mechanism of its specific biological activity have been actively conducted. From the beginning, AVM was presumed to be a blocker of neurotransmitters, such as ibotenic acid, because of its immediate antiparasitic and anthelmintic effects.

Studies of the mode of action of this antibiotic have progressed in expectation of providing useful information for developing other superior antiparasitic agents. Because an excellent review [1] of the studies up to 1989 exists, only mode of action achievements are reviewed in this chapter.

II. Target of Avermectin Action

The study of the mechanism of AVM's action began around the time when this drug was introduced for practical use. Fritz *et al.* [2] and Mellin *et al.* [3] reported that AVM is a blocker for neurotransmission, and the target causing the inhibition was considered to be the γ-aminobutyric acid (GABA)-gated chloride ion channel [4]. However, Duce and Scott [5] later indicated the existence of multiple sites of action for AVM—one of which is independent of the GABA receptor–chloride ion channel complex—by means of an experiment using both the GABA-sensitive and -insensitive muscle fibers in *Schistcerca greguria*.

In addition, it turns out that AVM specifically inhibits the binding of GABA and benzodiazepines to rat brain membrane and that specific binding sites exist in canine and rat brains. This was shown by an experiment that used [^{3}H]-ivermectin (22, 23-dihydroavermectin B1$_a$, IVM) [6, 7]. Schaeffer and Haines [8] reported a parasitic worm sensitive to AVM. They demonstrated that the binding affinity of [^{3}H]-IVM to membranes isolated from *Caenorhabditis elegans* is approximately 100-fold higher than that observed in rat brain and elucidated the mechanism of high selective toxicity of this anthelmintic. Arena *et al.* [9, 10] found that AVM remarkably increased inward membrane current in *Xenopus laevis* oocytes injected with poly(A)$^+$RNA from *C. elegans* and the AVM-induced current was blocked by picrotoxin. This made way for the cloning of AVM-binding proteins, a chloride ion channel. They further demonstrated that the AVM-sensitive current expressed in *Xenopus* oocytes is modulated by glutamate. These results are consistent with the findings previously reported by Duce and Scott [5]. Thus, the target of AVM was defined at last.

Based on a series of studies of the mechanism of action of AVM, Rohrer *et al.* [11] identified specific AVM-binding proteins by means of a photoaffinity labeling technique. Azido-AVM(I) was confirmed to retain anthelmintic activity at the same level as AVM and to bind competitively with [^{3}H]-IVM to membranes from *C. elegans*. ^{125}I-azido-AVM(II) was incubated with *C. elegans* membranes and became covalently linked to membrane proteins by UV exposure. Three *C. elegans* polypeptides of 53, 47, and 8 kDa bound specifically to [^{3}H]-IVM were detected in a sodium dodecyl sulfate/5–20% polyacrylamide gradient gel electrophoresis. On the other hand, a single major polypeptide of ~47 kDa was labeled in a membrane fraction of an insect, *Drosophila melanogaster*. Azido-AVM did not bind to rat brain membrane and therefore the AVM-binding protein was identified [11].

III. Cloning and Structure of Avermectin Binding Protein

Cully *et al.* [12] cloned cDNAs encoding an AVM-sensitive glutamate-gated chloride channel from *C. elegans* to elucidate its structure and properties. Screening of the genes was carried out with a cDNA library prepared by using a poly(A)$^+$RNA fraction, 1.8 kilobasepairs (kbp) in length, which was sensitive to IVM PO$_4^{2+}$ and to glutamate in *Xenopus* oocytes. Two kinds of cDNAs, GluCl-α and GluCl-β, were consequently isolated, and the pharmacologic properties of the GluCl-α and -β channels expressed in the oocytes were investigated (Table I). These data were entirely consistent with the characteristics that had been partially reported.

TABLE I
Pharmacology

Compound	Agonists Concentration (μM)		Percent of glutamate (10 mM)
IVMPO$_4$	1		117 ± 13
Ibotenate	500		18 ± 2
D-Glutamate | 10 | | 1 ± 2 |

Compound	Antagonists Concentration (μM)	Ligand (mM)	Inhibition (%)
Picrotoxin	100	Glutamate 1	68 ± 4
Picrotoxin	100	IVMPO$_4$ 1	61 ± 2
Flufenamic acid	200	IVMPO$_4$ 1	60 ± 4
Strychnine	100	Glutamate 1	0
Bicuculine	100	Glutamate 1	0
CNQX	10	Glutamate 1	0

Compound	Avermectins Potentiation of glutamate (%)	IVMPO$_4$ activation (%)
Ivermectin	582 ± 113	49 ± 10
Milbemycin D	499 ± 155	60 ± 4
L648548	298 ± 38	10 ± 3
Octahydro-AVM	70 ± 15	Inactive

The nucleotide sequences of cDNA clones, GluCl-α and GluCl-β, revealed single large open reading frames of 1383 and 1302 base pairs, respectively. From predicted animo acid sequences, the molecular weights of GluCl-α protein and GluCl-β protein were estimated to be 52,550 and 49,900, respectively.

An N-terminal extracellular domain and four hydrophobic transmembrane domains M1 to M4 were found in the GluCl-α and GluCl-β sequences. Both proteins were also observed to contain the conserved cysteine residues found in the extracellular domain of all ligand-gated chloride channels, positioned at amino acids 252 and 263 in GluCl-α and amino acids 233 and 234 in GluCl-β. The GluCl-α protein contained a strong consensus sequence for a protein kinase C phosphorylation site located between putative membrane-spanning domains M3 and M4. In comparison with the structures of other chloride ion channels, both proteins were shown to contain putative N-linked glycosylation sites in the proposed extracellular domain as found in the GABA receptor sequence. Glutamate-gated chloride channels have only been reported in invertebrates and

Dros GluClα	MGSGHYFWAI	LYFASLCSAS	LANNAKVNFR	E---------	---KEKKVLD	38
C. elegans GluClα	--MATWIVGK	LIIASLILGI	QAQQARTKSQ	DIFEDDNDNG	TSPIHIPIEQ PQTSDSKILA	68
C. elegans GluClβ1	----------	----------	----MTTP	SSFSILLLLL	LMPVVTNGEY SMQSEQEILN	34
Dros GluClα	QILGAG----	---KYDARLR	PSGIN-GTDG	PAIVRINLFV	MEYSVQLTFR EQWTDERLKF	100
C. elegans GluClα	HLFTSG----	---YDFRVR	PPTDN-G--G	PVVVSVNMLL	RTISKIDKVN ESWIDKRLSY	127
C. elegans GluClβ	ALLKN-----	---YDMRVR	PPANSSTEG	AVNVRVNIMI	RMISKIDVUN EQWIDPRLAY	95
Dros GluClα	D--DIQGRLK	YLTLTEANR-	VWMPDLFFSN	EKEGHFHNII	MPNVYIRIFP NGSVKYSURI SLTLACPMNL	167
C. elegans GluClα	G-VKGDGQPD	FVILTVGHQ-	UWMPDTFFPN	EKQAYKHTID	KPNVLLIRHN DGTVLYSVRI SIVLSCPMYL	95
C. elegans GluClβ	ENLGFYNPPA	FLTVPHVKKS	LMIPDTFFPT	EKAAHRHLID	MENMFLRIYP DGKILYSSRI SLTSSCPMNL	165
Dros GluClα	KLYPLDROUK	SLRMASYGWT	TNDLVFLWKR	G-DPVQVVKN	L--HLPRFTL EKFLTDY-CN SKTNTGEYSC	233
C. elegans GluClα	QYYPMDVQQC	SIDLASYAYT	TKDIEYLMKE	H-SPLQLKVG	LSSSLPSFQL TNTSTTY-CT INSQLPPVSY	263
C. elegans GluClβ	QLYPLDYQSC	NFDLVSYAHT	MNDIMYEWDP	S-TPVQLKPG	VGSDLPNFIL KNYTNADCT SHTNTGSYGC	234
Dros GluClα	LKVDLLERRE	FSYYLIQIYI	PCCMLVIVSW	VSFWLDQGAV	PARVSLGVTT LLTMATQTSG INASLPPVSY	303
C. elegans GluClα	LRTTIQLKRE	FSEYYLIQLYI	PSCMLVIVSW	VSFWFDRTAI	EPARVTLGVTT LLTMATAQSAG INSQLPPVSY	333
C. elegans GluClβ	LRMQLLEKRQ	ESYYLVQLYA	PTTMIVIVSW	VSFWIDLHST	AGRVALGVTT LLTMTTMQSA INAKLPPVSY	304
			M3		M1 M2	
Dros GluClα	TKAIDVWTGV	CLTFVFGALL	EFALVNEASR	SGSNKANMHK	ENMKFKRRDL EQA--SLDA ASDLLDTDSN	370
C. elegans GluClα	IKAIDVWIGA	CMTFIFGALL	EFALVNHI--	-ANKQGVE-	--RKARTER EKA--EIPL LQNLHNDVPT	392
C. elegans GluClβ	VKVVDVWLGA	CQTFVFGALL	EYAFVSEQDS	VRQNDRSREK	AA--RKAQRRR EKL--EMVD AEVYQPPCTC	370
Dros GluClα	ATFAMKPLVR	HPGDPLALEK	R-L---QCEV	HMQAPKRPNC	CKTWLSKFPT RQCS---RSK RLDVISRITF	433
C. elegans GluClα	KVFNQEEKVR	D--TVPLNR	RQM---NSFL	NLLETK---	EWND---ISK RVDLISRALF	439
C. elegans GluClβ	HTEARETFR	D--------	--TVPLNR	-----	---KVRRYFT KPDY---LPA KLDFYARFVV	405
Dros GluClα	PLVFALFNLV	YWSTY----	LFREEEDE--	--		456
C. elegans GluClα	PVLFFVFNIL	YWSRFGQONV	LF-------	--		461
C. elegans GluClβ	ELAELAFNVI	YWVSCLIMSA	NASTPESLV-	--		434
	M4					

Fig. 1. Alignment of the deduced amino acid sequence of DrosGluCl-α with *C. elegans* GluCl [13]. Boxes indicate the residues that are identical. Closed circles indicate the four conserved cysteine residues in GluCl. Open circles indicate the putative N-linked glycosylation sites in DrosGluCl-α. The four transmembrane regions (M1–M4) are underlined. Gaps have been introduced to give optimal alignments.

are found on neuronal somata, crustacean muscle, and are expressed in oocytes injected with the insect muscle poly(A)$^+$RNA.

In 1996, a *Drosophila melanogaster* glutamate-gated chloride channel, DrosGluCl-α, was cloned by Cully *et al.* and was compared with those of *C. elegans* [13]. The alignment of the deduced amino acid sequences reported in both papers [12, 13] is shown in Fig. 1. Several properties of the DrosGluCl-α channel differ from those of the nematode *C. elegans* GluCl-α and -β channels; for instance, whereas the latter was sensitive to AVM potentiation of the glutamate response, the former was not.

IV. Concluding Remarks

As mentioned, the screening of potent antiparasitic agents using the expression system of the glutamate-gated chloride channel became possible only after the target of AVM was elucidated to be the chloride channels peculiar to nematodes, insects, ticks, etc. [1].

Because it possesses such specific activity and because there is not much serious drug resistance to be considered, AVM will continue to be widely used as a superior antiparasitic agent for human and animals. Moreover, the biological activity of the antibiotic is thought to be exceedingly useful for detailed studies on the mechanism of neurotransmission such as ibotenic acid and picrotoxin.

References

1. Turner, M., and Schaeffer, J. M. (1989). Mode of action of ivermectin. *In* "Ivermectin and Avermectin" (W. C. Campbell, Ed.), pp. 73–88. Springer-Verlag, New York.
2. Fritz, L. C., Wang, C. C., and Gorio, A. (1979). Avermectin B1a irreversibly blocks postsynaptic potentials at the lobster neuromuscular junction by reducing muscle membrane resistance. *Proc. Natl. Acad. Sci. USA.* **76**, 2062–2066.
3. Mellin, T. N., Busch, R. D., and Wang, C. C. (1983). Postsynaptic inhibition of invertebrate neuromuscular transmission by avermectin B1a. *Neuropharmacology* **22**, 89–96.
4. Kass, I. S., Wang, C. C., Walrond, J. P., and Stretton, A. O. W. (1983). Avermectin B1a, a paralyzing anthelmintic that affects interneurons and inhibitory motoneurons in *Ascaris*. *Proc. Natl. Acad. Sci. USA.* **77**, 6211–6215.
5. Duce, I. R., and Scott, R. H. (1985). Actions of dihydroavermectin B1a on insect muscle. *Br. J. Pharmacol.* **85**, 395–401.
6. Pong, S. S., and Wang, C. C. (1980). The specificity of high affinity binding of avermectin B1a to mammalian brain. *Neuropharmacology* **19**, 311–317.
7. Drexler, G., and Sieghart, W. (1984). Properties of a high affinity binding site for [^{3}H]avermectin B1a. *Eur. J. Pharmacol.* **99**, 269–277.
8. Schaeffer, J. M., and Haines, H. W. (1989). Avermectin binding in *Caenorhabditis elegans*. A two-state mode for the avermectin binding site. *Biochem. Pharmacol.* **38**, 2329–2338.

9. Arena, J. P., Liu, K. K., Paress, P. S., and Cully, D. F. (1991). Avermectin-sensitive chloride currents induced by *Caenorhabditis elegans* RNA in *Xenopus* oocytes. *Mol. Pharmacol.* **40**, 368–374.
10. Arena, J. P., Liu, K. K., Paress, P. S., Schaeffer, J. M., and Cully, D. F. (1992). Expression of a glutamate-activated chloride current in *Xenopus* oocytes injected with *Caenorhabditis elegans* RNA: Evidence for modulation by avermectin. *Mol. Brain Res.* **15**, 339–348.
11. Rohrer, S. P., Meinke, P. T., Hayes, E. C., Mrozik, H., and Schaeffer, J. M. (1992). Photoaffinity labeling of avermectin binding sites from *Caenorhabditis elegans* and *Drosophila melanogaster*. *Proc. Natl. Acad. Sci. USA.* **89**, 4168–4172.
12. Cully, D. F., Vassilatis, D. K., Liu, K. K., Paress, P. S., Van der Ploeg, L. H. V., Schaeffer, J. M., and Arena, J. P. (1994). Cloning of an avermectin-sensitive glutamate-gated chloride channel from *Caenorhabditis elegans*. *Nature* **371**, 707–711.
13. Cully, D. F., Paress, P. S., Liu, K. K., Schaeffer, J. M., and Arena, J. P. (1996). Identification of a *Drosophila melanogaster* glutamate-gated chloride channel sensitive to the antiparasitic agent avermectin. *J. Biol. Chem.* **271**, 20,187–20,191.

Chapter 14

Mode of Action of FK506 and Rapamycin

NOBUHIRO TAKAHASHI
Tokyo University of Agriculture and Technology
Tokyo, Japan

I.	Introduction	577
II.	Initial Cellular Target for FK506 and Rapamycin; Peptidyl Prolyl *cis-trans* Isomerases (Rotamases, Immunophilins)	586
	A. Role of PPIases as Foldase in Protein Folding	587
	B. Specific Roles of PPIases in Highly Differentiated Cells	587
	C. Presence of a Third Class of PPIase Insensitive to Immunosuppressants	588
	D. Role of PPIases in Stabilization of Protein Complexes and Regulation of Their Functions	590
	E. Isoforms of FKBP12	599
III.	Target of FK506–FKBP12 Complex: Calcineurin	599
IV.	Target of Rapamycin–FKBP12 Complex: mTOR/FRAP/RFAT	604
V.	Intervention of Intracellular Signaling Pathways by FK506 and Rapamycin	607
	A. Calcineurin Signaling Pathway	607
	B. mTOR Signaling Pathway	610
	References	611

I. Introduction

FK506 (tacrolimus), a natural product of *Streptomyces tukubaensis*, is a 23-membered macrolide antibiotic that contains a unique hemiketal-masked, α,β-diketo amide moiety (Fig. 1) [1]. It has a mode of action and toxicity profile similar to those of cyclosporin A (CsA), an immunosuppressive cyclic undecapeptide produced by *Tolypocladium inflatum* (Fig. 1), which has been used clinically for two decades to prevent rejection on transplantation of organs, such as kidney, heart, liver, and bone marrow. The use of CsA has led to a remarkable increase of survival of all organ allografts and has fueled a dramatic increase in the number of kidney, liver, cardiac, and bone marrow transplants [2]. Thus, CsA has revolutionized transplantation surgery, both in terms of efficiency and quality of life of the patient. FK506 is said to be a more potent immunosuppressant than CsA and was approved for the treatment and prevention of kidney transplant rejection.

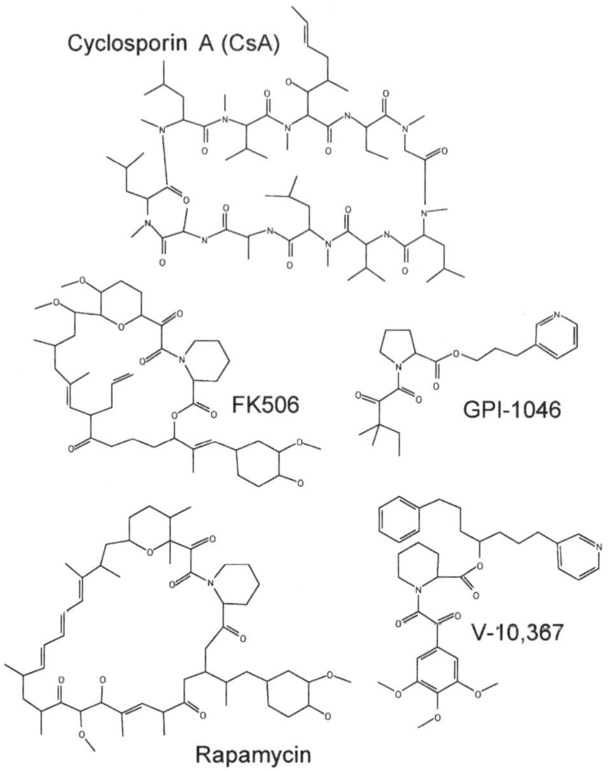

Fig. 1. The structure of cyclosporin A (CsA), FK506, rapamycin, and FK506's derivatives. Note the common structural features of FK506, rapamycin, GPI-1046 and V-10,367, while CsA has an entirely different chemical structure. Despite their structural differences, CsA and FK506, as immunosuppressants, exhibit a nearly identical spectrum of action on T lymphocytes. In contrast, regardless of their structural similarity, FK506 and rapamycin exhibit quite different spectra of action on T lymphocytes, though both FK506 and rapamycin are powerful immunosuppressants. GPI-1046 and V-10,367 do not have any immunosuppressive activity; however, they have neurotrophic activities as do FK506 and rapamycin.

Although the chemical structures of CsA and FK506 are quite different, it was not surprising that they showed a very similar mode of action at cellular levels, because FK506 was screened as an immunosuppressant by mimicking CsA's inhibitory effects on cytokine production by and the proliferation of T cells [1]. Thus, both FK506 and CsA block events occurring in T-cell activation, such as those apparently leading to activation of the transcription of early genes, including interleukins-2, -3, -4, -5 (IL-2, etc.), tumor necrosis factor α (TNF-α), granulocyte macrophage colony-stimulating factor (GM-CSF), interferon-γ, etc., which function in coordinating the various cells involved in the immune response [3]. Because of their powerful immunomodulating activities on T lymphocytes, FK506

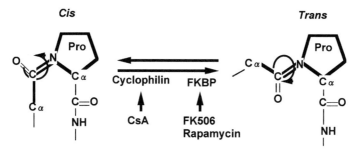

Fig. 2. *Cis-trans* isomerization of the prolyl peptide bond (Modified from Takahashi [3] with permission.) Peptide bonds exist mainly in the *trans* configuration in proteins and peptides. However, depending on amino acid composition and sequence, the prolyl residues of naturally occurring proteins may be stabilized in the *cis* configuration. Prolyl isomerization is known as slow step in protein folding. Both cyclophilin and FK506-binding protein are able to catalyze the *cis-trans* isomerization of the imide bond in proteins and, thus, to accelerate the folding of some proteins *in vitro* and *in vivo*. CsA inhibits the enzymatic activity of cyclophilin, whereas FK506 and rapamycin inhibit that of FK506-binding protein. However, CsA and FK506 (rapamycin) do not cross-inhibit each other's PPIase activity.

and CsA are also used in treating many autoimmune diseases and allergies, including psoriasis, Behcet's uveitis, rheumatoid arthritis, asthma and atopic dermatitis. However, they are not devoid of side effects, such as hypertension, neurotoxicity, and especially nephrotoxicity, which is one of the most worrisome toxicities; therefore, the full therapeutic potential of FK506 and CsA has been limited [3].

Despite the similarity of their action mode and toxicity profile at clinical and cellular levels, what was surprising is the fact that even at the molecular level CsA and FK506 bind to and inhibit different members—but of the same family—of proteins, cyclophilin (CyP) and FK506-binding protein (FKBP), respectively [4, 5]. Both catalyze the *cis-trans* interconversion around peptidyl prolyl amide bonds in peptides and accelerate the folding of proline-containing proteins (Fig. 2). However, the peptidyl prolyl *cis-trans* isomerase (PPIase) activity is not involved directly in T-cell activation. Instead, when bound to their cognate PPIases (FKBP12 and CyPA, respectively), FK506 and CsA form a ternary complex with calcineurin, a calcium calmodulin-dependent serine/threonine protein phosphatase [6], and inhibit its ability to dephosphorylate the cytoplasmic subunit of nuclear factor of activated T cells (NF-ATC). Dephosphorylation of NF-ATC is required for its translocation to the nucleus and consequently for association with a newly synthesized nuclear subunit of NF-AT (NF-ATN), which is essential to transcribe the IL-2 and other cytokine genes in T lymphocytes [7, 8] (Fig. 3). Thus, the inhibition of the translocation of NF-ATC is the main event caused by CsA and FK506 at the molecular level for inducing their immunosuppressive effects.

Besides the immunosuppressive activities, FK506 and CsA have been reported to have neurotrophic activities [9–11]. Although CsA is demonstrated to

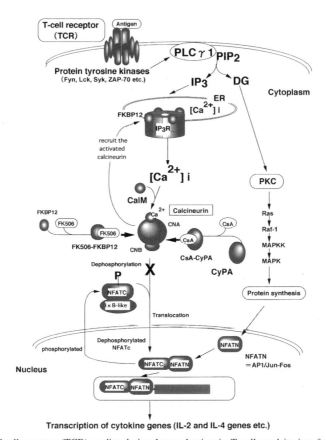

Fig. 3. T-cell receptor (TCR)-mediated signal transduction in T cells and its interference by the immunosuppressants (Modified from Takahashi [3] with permission.) Binding of antigen to TCR induces the activation of various protein tyrosine kinases, such as Lyn, Fyn, Syk, and ZAP-70, which lead to the production of inositol-3-phosphate (IP3) and diacylglycerol (DG) by phospholipase Cγ1 (LPCγ1). IP3 induces the mobility of Ca^{2+} from endoplasmic reticulum (ER) to cytoplasm through its binding to the Ca^{2+}-releasing channel comprising four IP3 receptors (IP3R) on the membrane of ER. The Ca^{2+}-releasing channel is modulated by FKBP12, which binds to each of the four IP3R subunits. The increase of $[Ca^{2+}]i$ in the cytoplasm activates calcineurin [CN: alpha-subunit (CNA) and beta-subunit (CNB)] in a manner dependent on calmodulin (CalM), and induces dephosphorylation of the cytoplasmic subunit of nuclear factor of T cells (NF-ATC) by CN. This causes the translocation of NF-ATC from cytoplasm to nucleus. On the other hand, DG induces the activation of protein kinase C (PKC), which activates the MAP kinase cascade (including Ras, Raf-1, MAPKK, etc.), resulting in *de novo* synthesis of the nuclear subunit of NF-AT (NF-ATN). Consequently, NF-ATC and NF-ATN associate in the nucleus to transcribe the cytokine genes, such as those of IL-2 and IL-4. When the sustained $[Ca^{2+}]i$ influx is turned off, the translocated NF-ATC is phosphorylated by nuclear kinases and the transcription of the cytokine genes is terminated. CsA and FK506 interrupt the cytokine productions when complexed with their respective receptors, CyPA and FKBP12, through inhibiting the activity of calcineurin. This prevents NF-ATC from translocation to the nucleus. The binding to IP3R enables FKBP12 to interact with the activated CN, thus FKBP12 anchors the phosphatase to the Ca^{2+}-releasing channel and modulates the receptor's phosphorylation state. In addition, FKBP12 also modulates by itself the open and closed states of the Ca^{2+} channel. Therefore, FK506 interrupts the function of the Ca^{2+} channel as well.

have only neuroprotective activity, FK506 and its derivatives exhibit neuroregenerative as well as neuroprotective activities [12]. They seem to hold great promise for the treatment of nerve injuries and neurological diseases because these compounds readily cross the blood–brain barrier, being orally effective in a variety of animal models of ischemia, traumatic nerve injury, and human neurodegenerative disorders [13]. This is in contrast to neurotrophins, protein factors such as nerve growth factors that cannot cross the blood–brain barrier, thus presenting many difficulties in their administration to the patient with neuronal injuries. Because calcineurin is highly localized in the central nervous system, especially in those neurons vulnerable to ischemic and traumatic insults, it is considered to play important roles in neuron-specific functions. In fact, calcineurin has been reported to be involved in many neuronal functions such as neurotransmitter release, neurite outgrowth, and neuronal cell death [14, 15]. Thus, some of the neuroprotective activities of FK506 and CsA may be attributed to their ability to inhibit calcineurin. However, the effect of CsA is different from that of FK506 on nerve regeneration, thus, some neurotrophic activities seem to arise via different mechanisms by which calcineurin is not involved. In addition, FK506-related compounds, such as GPI-1046 and V-10,367 (Fig. 1) that bind to FKBP12 but do not inhibit calcineurin are also able to increase nerve regeneration [16, 17]. Thus, FK506's or its derivatives' abilities to increase nerve regeneration arise via a calcineurin-independent mechanism. This character is important because most of the unwanted side effects produced by these drugs arise via calcineurin inhibition, opening up the development of this class of compounds—devoid of calcineurin-inhibiting activity—for the treatment of human neurological disorders. Currently, two possible calcineurin-independent mechanisms might explain how FK506 and its derivatives induce neuroregenerative effects. One mechanism is that mediated by the Ca^{2+}-releasing channel because FKBP12, not CyP, regulates the function of Ca^{2+}-releasing channels, such as ryanodine receptors (RyR) and inositol 1,4,5-triphosphate receptors (IP3R), which may play crucial roles in Ca^{2+} signaling in the nerve system [12]. Another mechanism is that mediated by isotypes of FKBP12, which are not involved in calcineurin inhibition even when complexed with FK506 [18]. One such candidate of FKBP isotypes is FKBP52, which is a component of steroid receptor complexes. Ca^{2+}-releasing channels and FKBP52 may represent novel targets for new drugs for the treatment of a variety of neurological disorders. Current clinical interest in FK506 and its derivatives revolve around its neuroprotective and neuroregenerative properties.

One of the analogues of FK506, rapamycin (sirolimus) produced by *Streptomyces hygroscopicus*, is a 31-membered macrocyclic lactone that contains a masked α- and β-diketo amide L-pipecolic acid residue similar to that found in FK506, and a unique triene segment including a distinctive shikimic acid-derived trisubstituted cyclohexane moiety (Fig. 1) [19]. It was originally isolated as an antifungal agent with potent anticandida activity, and was later shown to have antitumor activities. Rapamycin is also known as an inhibitor of the growth factor and/or nutrient-dependent proliferation of most of the cells originated from yeast to mammalians [20, 21]. In addition, as expected because of its structural similarity with FK506, rapamycin also blocks early events in the activation of T lymphocytes, thus exhibiting immunosuppresive activity [22]. In fact, recent clinical trials have demonstrated a decrease in acute rejection episodes in renal transplant patients receiving rapamycin compared with controls. Rapamycin is now in phase III clinical trials in renal transplantation as a novel immunosuppressant [19]. However, some preliminary results show that rapamycin is also not devoid of side effects. The major toxicities associated with rapamycin treatment include thrombocytopenia and hyperlipidemia [23]. Nonetheless, it is of great interest that its action on lymphocytes is quite different from those of FK506 and CsA. FK506 and CsA mediate their inhibition of cytokine transcription early in the G_1 phase during T-cell activation, whereas rapamycin exhibits its inhibitory effects on the proliferation of T cells at the G_1/S transition of the cell cycle induced by cytokines. Thus, the events that are blocked by FK506/CsA and rapamycin are independent, but sequentially related in the signaling pathway that follows T-cell activation (Fig. 4) [24].

While no antagonism between CsA and rapamycin was detected, FK506 and rapamycin in molar excess are selective reciprocal antagonists, indicating that a common intracellular receptor is involved in mediating the immunosuppressive activity of these drugs. Rapamycin, like FK506, binds to FKBP12 and inhibits FKBP's PPIase activity. However, when complexed with FKBP12, it does not bind to calcineurin, but binds to mTOR/FRAP (mammalian target of rapamycin/FKBP and rapamycin-associated protein), which is one of the homologues of phosphatidylinositol-3-phosphate (PI3P) kinase [25] (Fig. 4). The mTOR protein kinases have been shown to regulate cellular proliferation in response to nutrients (especially branched amino acids, such as leucine) and growth factors [including IL-2, insulin, epidermal growth factor (EGF), and other growth factors] [20, 26–28] (Fig. 5). The rapamycin–FKBP12 complex blocks cellular proliferation by inhibiting the function of mTOR, and consequently by inhibiting downstream targets essential for protein synthesis and cell cycle progression. One such downstream target is p70S6k, a serine/threonine protein kinase that has been linked to translational control by virtue of its ability to phosphorylate the ribosomal protein S6. mTOR can phosphorylate p70S6k directly. Another well-known target is PHAS-I/4E-BP1 (protein, heat, and acid stable-I/eukaryotic initiation factor 4E binding protein 1). It inhibits assembly of the eukaryotic translation initiation factor (eIF4F) complex with the mRNA 5′ cap structure by binding directly to eIF4E, one of the components of eIF4F that recruits 40S ribosomal subunits to the 5′ end of mRNA [24]. Phosphorylation of PHAS-I/4E-BP1 by mTOR regulates its binding to eIF4E: Hypophosphorylated PHAS-I/4E-BP1 interacts strongly with eIF4E, whereas hyperphosphorylated isoforms do not. PHAS-I/4E-BP1 is

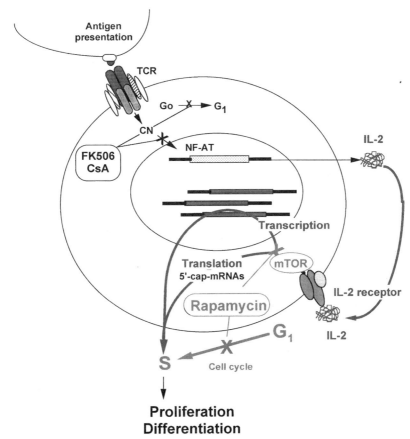

Fig. 4. The sites of inhibitory action by FK506/CsA and rapamycin in the T-cell activation cascade. T cells are activated by the complex series of events that take place over several days after antigen presentation to T-cell receptors (TCR). FK506/CsA and rapamycin act within the first hours of the process. FK506 and CsA are potent inhibitors of the Ser/Thr protein phosphatase, calcineurin (CN), which comprises the main stream of the TCR-mediated signal transduction pathway leading to the translocation of NF-ATC and the transcription of early T-cell activation gene, i.e., the IL-2 gene (G_0/G_1 transition of the cell cycle). Rapamycin does not have any effect on this process, though it reverses the inhibitory effect of FK506 but not that of CsA. The subsequent T-cell activation involves expression of the IL-2 receptor (IL-2R) on the surface of the cell. By binding IL-2 to IL-2R, the IL-2R signal transduction pathway is activated and induces translation of many different proteins and transcription of a different set of genes, leading to T-cell proliferation and differentiation (G_1/S transition of the cell cycle). Rapamycin, but not FK506, inhibits the action of mTOR, thus interrupting the response of the T cells to IL-2. (Modified from Takahashi [24] with permission.)

hypophosphorylated in quiescent cells, but is hyperphosphorylated on multiple sites following exposure to a variety of extracellular stimuli, such as those from growth factors or nutrients. Thus, inhibition of mTOR by rapamycin blocks the initial steps of translation and leads to growth arrest of the cell [29, 30]. In addition,

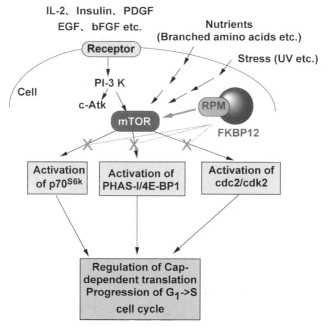

Fig. 5. Activation of mTOR signal transduction pathways in various cells and the downstream factors affected by mTOR. The mTOR protein kinase is activated through its upstream kinases (such as PI3K and c-Atk) by a number of different stimuli, such as growth factors (IL-2, insulin, EGF, PDGF, bFGF, etc.), nutrients (amino acids; Leu, Asn, etc.), and stress (UV, etc.) in various cells. On stimulation, mTOR activates its downstream pathways by phosphorylating directly translational regulators, such as p70S6k and PHAS-I/4E-BP1. mTOR is also able to activate the signaling pathway that involves cell cycle regulators, such as cdc2 and cdk2. Thus, the activation of the mTOR signal pathway leads to cap-dependent translation and G_1/S cell progression, and eventually allows cells to proliferate. Rapamycin complexed with FKBP12 binds to mTOR and interrupts its action to the downstream factors. (Modified from Takahashi [24] with permission.)

mTOR (TOR) has also been reported to be involved in other core biological processes such as DNA damage checkpoint/repair, cell cycle checkpoint, and autophagy [31–33]. Rapamycin certainly affects those processes. Furthermore, rapamycin is reported to increase neurite outgrowth and to inhibit growth of certain cancer cells [34–36].

Thus, because of the additional clinical value of rapamycin as well as FK506, many medicinal chemistry efforts have been made to generate their analogues (and also those of CsA) with the hope of identifying compounds with an improved or new therapeutic index that could have broader therapeutic utility than the parent drugs. These efforts yielded several compounds with unique biochemical attributes, some showing evidence for dissociation between clinically beneficial property

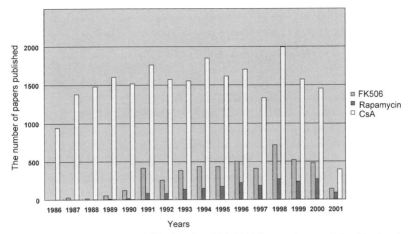

Fig. 6. The number of papers published during 1986–2001 for certain macrolides. (Retrieved with PubMed via the Internet.)

and toxic property, which may pave the way toward designing safer FK506- and rapamycin- (and CsA-) related drugs.

Besides being therapeutic drugs, FK506 and rapamycin (and CsA) have been used extensively to solve a number of problems as experimental probes in basic research and offered many answers to the mechanisms of intracellular signal transduction and of other cellular processes. Because of their clinical value and usefulness in basic research, the pace of publications on these macrolides has increased currently to approximately 1000 papers (2500 for CsA) per year, and during the last two decades their total reached more than 8000 (50,000 for CsA) [3] (Fig. 6). Those investigations have uncovered many new aspects of their modes of actions. It has become clear that, as a result of calcineurin or mTOR inhibition, FK506 or rapamycin alters multiple biochemical processes in a variety of cells. In addition, their actions are also reflected from the broad functions of the downstream factors (NF-AT, or p70S6k and PHAS-I/4E-BP1, etc.) of calcineurin or mTOR, and of their direct target protein (FKBP12) in many biological systems. Moreover, a number of additional members (such as FKBP52 in steroid hormone receptor complexes) of the FKBP as well as the other target and downstream protein families have been identified, providing further insights into the complexity of FK506's and rapamycin's biological effects. These may account for the adverse useful effects as well as unwanted side effects of the drugs [3]. Thus, their unique mechanisms of action have created great interest in their use as biochemical probes of signal transduction pathways and of the other cellular processes. This chapter reviews the current state of our understanding of the mechanism of action of FK506 and rapamycin.

II. Initial Cellular Target for FK506 and Rapamycin; Peptidyl Prolyl cis-trans Isomerases (Rotamases, Immunophilins)

The initial target proteins for CsA and FK506/rapamycin are shown to be CyPA (MW = 17,735; pI = 9.6) and FKBP12 (MW = 11,819, pI = 8.9), respectively. Both are cytosolic peptidyl prolyl *cis-trans* isomerases (PPIases) catalyzing the interconversion of peptidyl prolyl amide bonds in peptide and protein substrates [4, 5, 37] (Fig. 2). Interestingly, both FK506 and rapamycin inhibit the activity of FKBP12, but CsA does not; conversely, CsA inhibits that of CyPA, but FK506 or rapamycin does not. The inhibition constants (K_i) of FK506 and rapamycin are 0.2 and 0.4 nM, respectively, which are 15- to 30-fold smaller than that observed in inhibition of the CyPA's PPIase activity by CsA (K_i = 6 nM) [38]. While FK506 and CsA do not cross-bind to each other's target protein, rapamycin competes with FK506 in binding to FKBP12, and has a slightly higher affinity for FKBP12 than does FK506 (rapamycin K_d = 0.2 nM, FK506 K_d = 0.4 nM) [39]. Although both FKBP12 and CyPA can catalyze the same chemical reaction as PPIase, neither has structural similarity by means of amino acid sequence [40] and three-dimensional structure [41, 42]. Despite these differences, the binding of the drugs to PPIase occurs at the active site of their respective PPIases. Thus, all of the drugs seem to be mimicking a twisted Xaa-Pro bond in the peptide substrate of PPIases [43, 44]. Based on an isomer-specific proteolytic assay with a variety of synthetic chromogenic Succinyl-Ala-Xaa-Pro-Phe-p-nitroanilide, CyPA is shown to accept any amino acid at the Xaa position in the synthetic peptide as substrate, whereas FKBP12 prefers peptides in which Xaa is a hydrophobic amino acid, such as phenylalanine, tyrosine, and leucine [45] (Table I). Differences in the substrate specificity between FKBP12 and CyPA observed in the synthetic peptides *in vitro* probably reflect on their functions *in vivo*.

TABLE I
Substrate Specificity of PPIases[a]

Substrate	CyPA	FKBP12	Pin1
AA<u>S</u>PF-pNA	5400	29	9
AAp<u>S</u>PF-pNA	50	1	3760
AA<u>T</u>PF-pNA	5200	27	4
AAp<u>T</u>PF-pNA	9	2	1370
Ac-A<u>Y</u>PY-pNA	2420	880	5
Ac-Ap<u>Y</u>PY-pNA	3270	64	3

Source: Modified with permission from Yaffe *et al.* [63]. Copyright © 1997 American Association for the Advancement of Science.
Key: A = Ala, S = Ser, T = Thr, P = Pro, F = Phe, Y = Tyr, pS = phosphorylated Ser, pT = phosphorylated Thr, pY = phosphorylated Tyr, Ac = Acetyl, pNA = p-nitroanilide.
[a]The value of the rate constant is given as that of k_{cat}/K_m mM^{-1}s^{-1}.

A. Role of PPIases as Foldase in Protein Folding

Paralleled with their ability to isomerize peptide substrates, both FKBP12 and CyPA are able to act on protein substrates as well and thus are able to accelerate the slow, rate-limiting steps in the refolding of many proteins *in vitro*. This foldase activity of PPIases is also shown to be necessary for accelerating refolding of a fusion protein that was synthesized in a reticulocyte lysate and imported into the matrix of isolated yeast mitochondria [46] and for protein folding and subunit assembly of heterodimeric luciferase from *Vibrio harveyi* after *de novo* synthesis of subunits in rabbit reticulocyte lysate [47]. These foldase activities of PPIases are certainly sensitive to the immunosuppressants. These observations, coupled with their ubiquitous presence in almost all organisms and in every cellular compartment, suggest that PPIases have some general cellular functions as protein foldase in the cell [48, 49]. However, all mutants with the defect of single and multiple PPIase (CyPs and FKBPs) genes were viable with no remarkable phenotype in yeast. Thus, despite the accumulating evidence for a general role of PPIases as protein foldase, the nature of this role in cells has been difficult to elucidate *in vivo* and is still controversial.

B. Specific Roles of PPIases in Highly Differentiated Cells

While its general role as a protein foldase remains elusive, a number of physiological roles for PPIases have been reported in many specialized phenomena that are observed in highly differentiated cells of higher organisms. For example, Nina A, an isotype of CyPA, is involved in the trafficking and the maturation of specific isoforms of rhodopsins in *Drosophila* photoreceptor cells, for which the PPIase activity was shown to be absolutely required [50]. A similar function is reported for RanBP2, the other CyP isotype, which is found in cone photoreceptors of the vertebrate retina. It probably acts as a catalyst for proper folding of red/green opsin [51]. The necessity of PPIase activity is further reported in modulating human HIV-1 infectivity [52], in which CyPA regulates cleavage of Gag polypeptide by HIV-1 protease through its ability to isomerize the proline–peptide bond adjacent to the cleavage site of Gag polypeptide [53]. CsA treatment inhibits Gag processing in extracellular virus-like particles and causes accumulation of particles, which display mostly immature virion morphology and lack condensed capsids [54]. The other examples showing that PPIase activity is required for exhibiting FKBP's or CyP's function include those for (1) the dephosphorylation of connexin43 by protein-phosphatase 2B (PP2B), which is involved in modulation of gap junctional intercellular communication in hamster fibroblasts [55]; (2) the suppression of epidermal growth factor (EGF) receptor autophosphorylation, which regulates negative EGF receptor signal transduction in intact A431 cells [56]; (3) the inhibition of interferon regulatory factor-4 (IRF-4)-PU.1 binding to the immunoglobulin light chain enhancer E (lambda2-4) as well as IRF-4-PU.1

transactivation [57]; and (4) the modulation of the activity of the Sin3-Rpd3 histone deacetylase, which regulates transcriptional repression and gene silencing in yeast and mammalian cells [58]. In those examples, the PPIase activity was shown to be definitively required for achieving PPIase's physiological functions, which are certainly affected by the immunosuppressants.

C. Presence of a Third Class of PPIase Insensitive to Immunosuppressants

Along with the CyPs and FKBPs, a third class of PPIase, the parvulin, has been found [59]. In the case of CyPs and FKBPs they interact with immunosuppressants and are therefore termed immunophilins. However, parvulin neither interacts with nor is inhibited by FK506 and CsA. Thus, parvulin is not an immunophilin. An inhibitor for parvulin, juglone (5-hydroxy-1, 4-naphthoquinone), which has no immunosuppressive activity, is reported, however, its specificity to parvulin is questionable because juglone is known to modify the sulfhydryl groups of some other proteins than parvulin [60]. Parvulins, like immunophilins, are conserved from bacteria to human. Therefore, as the third PPIase that complements the enzymatic activities of immunophilins, it was expected to have some essential function for viability of the cell. In fact, in yeast cells, the one-parvulin homologue, Ess1, was found to be essential, while all other CyPs and FKBPs are dispensable for growth [61]. In addition, Pin1, a human parvulin homologue, was also shown to be essential for mitosis, and more interestingly it was reported to have a unique substrate specificity, i.e., it isomerizes only phosphorylated Ser/Thr-Pro bonds, which restrain the already slow *cis-trans* isomerization [62, 63] (Table I).

A number of Ser/Thr protein kinases are involved in cell cycle regulation and intracellular signal transduction (such as cdc kinases and MAP kinases). Many of those are so-called proline-directed protein kinases, which phosphorylate Ser or Thr preferentially adjacent to a proline in a protein substrate. Therefore, the substrate specificity of Pin1 is extremely important for us to understand how the function of phosphorylated proteins is regulated, because it predicts the presence of a two-step process for controlling the function of phosphorylated proteins (Fig. 7). Namely, adding phosphates to proteins involved in, for example, cell cycle or signal transduction creates binding sites for Pin1, which can then latch onto them and twist the peptide bond next to the prolines it contacts. That might, in turn, change the shape of the whole protein, altering its ability to interact with still other proteins, its location in the cell, its life span, etc. [64]. A few, but significant protein substrates of Pin1 have been identified, including cdc25 phosphatase, RNA polymerase II, and Alzheimer-associated tau protein [65-67]. The former two substrates are well-known key factors that regulate cell cycle progression and proliferation. Furthermore, since phosphorylation renders phosphorylated Ser/Thr-Pro bonds resistant to the catalytic action of FKBPs and CyPs

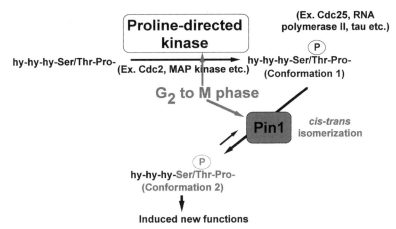

Fig. 7. Possible two-step regulation of protein function by proline-directed protein kinases and Pin1 (a hypothesis) [63–65]. Phosphorylation (P) of some proteins on Ser/Thr-Pro sites by proline-directed kinases, such as cdc2 and MAP kinases, creates binding sites for Pin1. Pin1 prefers to recognize the phosphorylated Ser/Thr-Pro with a hydrophobic amino acid (hy) stretch at the amino terminus, and it isomerizes the peptide bond adjacent to proline in the phosphorylated Ser/Thr-Pro sequence. This action of Pin1 changes the shape of the whole protein from conformation 1 to conformation 2, which might have the ability to interact with other proteins, to translocate to other cellular compartments, or to change its life span. This regulation may be critical for the coordinated G_1/M progression in mitosis.

(Table I), other mechanisms may be involved by the prolyl-isomerization in regulating protein function; briefly, immunophilins and parvulins may act in coordinated fashion on the same substrates *in vivo*. In fact, both CyPA and Ess1 have been shown to interact with Sin3-Rpd3 histone deacetylase complex and regulate transcriptional silencing by their interaction during chromatin remodeling in yeast cells [58]. In the same yeast system, Ess1 interacts with the carboxy-terminal domain (CTD) of RNA polymerase II and isomerizes the CTD, thus altering its interaction with proteins, such as Sin3-Rpd3 histone deacetylase, that are required for transcription of essential cell cycle genes [66]. These observations suggest that PPIases have some general roles in the regulation of the transcriptional machinery. The unusually high content of Ser/Thr-Pro motif among transcription factors may be indicative of their significance in such transcriptional processes [68a,b]. The precise mechanism of how PPIases are involved in controlling the function of those transcription factors remains unsolved. Thus, the compounds that bind to and/or inhibit specifically parvulin-type PPIases may be very useful for elucidating PPIase's roles in transcription, mitosis, and the other various physiological processes in the cell, and for developing new drugs to treat cancer, Alzheimer disease, etc.

D. Role of PPIases in Stabilization of Protein Complexes and Regulation of Their Functions

Other than the ability to act as an enzyme that interacts with and releases its substrate after completion of its reaction, immunophilins have the ability to recognize specific peptide conformations and to form stable multiprotein complexes, such as those of Ca^{2+} channels, steroid receptors, growth factor receptors, etc. In those protein complexes, it is proposed that the binding of protein ligands to immunophilins stabilizes specific conformations in surface loops of the proteins that play a critical role in regulating the function of immunophilin–protein complexes. Such roles of PPIases in protein complexes may be comparable to that of immunophilin in an immunosuppressant–immunophilin complex, in which immunophilin triggers to render a stable conformation for immunosuppressant by its binding, creating a new surface on the complex to interact with another target, i.e., calcineurin or mTOR. This may also be similar to the role predicted for Pin1-type parvulin in regulating its phosphorylated-protein substrate.

1. Ca^{2+}-Releasing Channel Complexes

The Ca^{2+}-releasing channels play crucial roles in Ca^{2+}-mediated signaling that triggers excitation–contraction coupling, T-lymphocyte activation, fertilization, and many other cellular functions. There are two types of intracellular Ca^{2+}-releasing channels, one on the sarcoplasmic reticulum of striated muscle [ryanodine receptors (RyRs)] and another on the endoplasmic reticulum of almost all types of cells [inositol 1,4,5-trisphosphate receptors (IP3Rs)]. The Ca^{2+}-releasing channel is a tetramer that is composed of four receptor subunits, each of which binds one FKBP molecule. Among the isoforms reported for these receptors (for RyRs: RyR1, expressed predominantly in skeletal muscle; RyR2, expressed predominantly in cardiac muscle; and RyR3, expressed in specialized muscles and nonmuscle tissues including the brain; for IP3Rs: types 1, 2, and 3, which have been characterized by cDNA cloning; most cells have at least one form of IP3R, and some other cells express all three types), RyR1 and IP3R1 are found to bind to FKBP12, and RyR2 binds to FKBP12.6, one of the isotypes of FKBP12 [69]. In all cases, the binding of FKBP molecule has been reported to be required for coordinated gating of the channel comprising four-receptor subunits.

In the case of an RyR1 channel, binding of FKBP12 to RyR1 modulates channel gating by increasing channels with full conductance levels, decreasing open probability after caffeine activation, and increasing mean open time [70]. In addition, the binding of FKBP12 is also known to cause so-called "coupled gating," an event in which a number of individual RyR1 channels exhibit simultaneous openings and closings in planar lipid bilayers [71]. In the absence of FKBP12, RyR1 channels exhibit increased gating frequency and cannot induce coupled gating, suggesting that FKBP12 stabilizes and synchronizes the channels in the open and closed states. FK506 or rapamycin interrupts the FKBP12–RyR1 interaction, thus repealing the stabilizing effects, which are again recovered on rebinding of FKBP12 [70]. This indicates that the FKBP12–RyR1 complex is probably in equilibrium with the cytosolic pool of FKBP12 in maintaining its Ca^{2+}-releasing activity. The action of FK506 or rapamycin is to dissociate FKBP12 from RyR1 and thus to disturb the equilibrium. This action of the drugs is linked to their reported effects that increase the sensitivity to agonists such as caffeine and is suspected to be the cause of cardiac dysfunction associated with high-dose immunosuppressant therapy by promoting leakage of Ca^{2+} from the sarcoplasmic reticulum [72]. The role of FKBP12 in regulating RyR-type Ca^{2+} channels is also confirmed by constructing FKBP12-deficient mice that have normal skeletal muscle but have severe dilated cardiomyopathy and ventricular septal defects that mimic a human congenital heart disorder, noncompaction of left ventricular myocardium. In this animal model, FKBP12 is shown to be indispensable for modulating the Ca^{2+} release activity of both skeletal and cardiac RyR-type Ca^{2+} channels [73].

A similar function of FKBP12 has also been reported for the IP3R1-Ca^{2+} channel whose function is regulated by at least two major cellular signaling pathways: the second messenger IP3 activates the channel, and phosphorylation by nonreceptor protein tyrosine kinases (e.g., Fyn) increases its open probability (Fig. 8). Namely, disrupting the IP3R1-FKBP12 interaction by FK506 or rapamycin increases Ca^{2+} flux through IP3R, an effect that is reversed by added FKBP12 [74]. The binding of FKBP12 to IP3R1 occurs at residues 1400–1401 in IP3R, a leucyl-prolyl dipeptide epitope that structurally resembles FK506 [75]. Interestingly, binding to IP3R1 at this site enables FKBP12 to interact with calcineurin, presumably to anchor the phosphatase to IP3R and modulate the receptor's phosphorylation status [76]. Since FKBP12 does not interact with calcineurin in the absence of IP3R, these findings indicate that the binding of FKBP12 to IP3R induces its ability to interact with calcineurin (Fig. 8). This, in turn, suggests that FK506 promotes an FKBP12–calcineurin interaction by mimicking a structurally similar dipeptide epitope present within IP3R that uses FKBP12 to anchor calcineurin to IP3R. Thus, the binding of FKBP12 seems to cause not only conformational change of its ligand but also that of FKBP12 itself, and stabilizes the conformation of the complex as a whole, creating a new surface on the complex to interact with another target, calcineurin.

Although FKBP12.6 binds selectively to RyR2, FKBP12.6 differs from FKBP12 by only 18 of 108 amino acids in their amino acid sequences. In addition, only FKBP12.6 is able to exchange with bound FKBP12.6 of RyR2, whereas both FKBP isoforms bind to RyR1 and exchange with bound FKBP12 of RyR1. This binding specificity of human FKBP12.6 is determined by only three amino acid residues (Gln31, Asn32, and Phe59) and accounts for the selective binding to cardiac RyR2. Mutations of FKBP12 to the three critical amino acids of FKBP12.6 conferred selective binding to RyR2, while those mutations

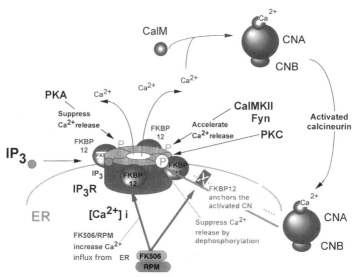

Fig. 8. Regulation of the function of a Ca^{2+}-releasing channel (IP3R) by FKBP12, calcineurin (CN), and protein kinases. The Ca^{2+}-releasing channel is composed of four IP3R subunits, each of which binds one FKBP12 molecule. On the binding of IP3, the IP3R-Ca^{2+} channel releases Ca^{2+} from endoplasmic reticulum (ER). FKBP12 stabilizes and synchronizes the channels in the open and closed states. The gating of the Ca^{2+} channel is also regulated by several protein kinases, such as protein kinase A (PKA) and C (PKC), Ca^{2+}-calmodulin-dependent protein kinase (CalMK), and Fyn. The binding of FKBP12 to IP3R enables FKBP12 to interact with the activated CN (Ca^{2+}–CalM–CAN–CNB complex). This facilitates the CN to dephosphorylate the phosphorylated Ca^{2+} channel. FK506 and rapamycin (RPM) destabilize the channel by dissociating FKBP12 from IP3R subunit, resulting in the increase of the gating frequency and thus the loss of $[Ca^{2+}]i$ influx from ER.

still retained binding to the skeletal muscle RyR1 [77]. This specificity of FKBP12.6 provides it with a unique role in pancreatic beta cells. In the case of islet microsomes from pancreatic cells, cyclic ADP-ribose (cADPR) induces Ca^{2+} release from the endoplasmic reticulum, leading to insulin secretion from pancreatic beta cells. By its unique binding specificity, FKBP12.6 is rendered the ability to bind cADPR. Therefore, cADPR is able to dissociate FKBP 12.6 from RyR2 and to induce Ca^{2+} release by RyR2 Ca^{2+}-releasing channel. Mimicking the effect of cADPR, the binding of FK506 to FKBP12.6 also frees the RyR2 from FKBP12.6, causing it to release Ca^{2+}. Thus, this example shows a mechanism by which the gating of Ca^{2+}-releasing channel is regulated. The binding of FKBP12.6 to RyR2 probably keeps closing the Ca^{2+} channel by stabilizing its conformation. In contrast, the dissociation of FKBP12.6 from RyR2 by either cADPR or FK506 causes relaxation of the conformation of Ca^{2+} channel and opens up the gate to release Ca^{2+}. In fact when islet microsomes were treated with cADPR, FKBP12.6 dissociated from the microsomes and moved to the supernatant, releasing Ca^{2+} from the intracellular stores. The microsomes that were then devoid of FKBP12.6 did not show Ca^{2+} release by cADPR [78].

In addition to the regulation by cADPR, the gating of RyR2 is also regulated by protein kinase A (PKA), which phosphorylates RyR2 and leads to dissociation of FKBP12.6. This regulation is proposed to link to human heart dysfunction because RyR2 is hyperphosphorylated by PKA in failing heart [79], which is shown to result in defective channel function due to increased sensitivity to Ca^{2+}-induced activation. PKA seems to be one of the essential factors for regulating the RyR2 Ca^{2+}-releasing channel. In fact, PKA as well as FKBP12.6 is reported to be the main component of RyR2 channel complex. The other factors present in the RyR2 channel complex include the protein phosphatases PP1 and PP2A, and an anchoring protein, mAKAP, which may also be involved in the regulation of RyR2 Ca^{2+} channel function.

Regarding involvement of PPIase activity in regulating Ca^{2+}-releasing channel, two models are proposed. One of the models is that the opening and closing of Ca^{2+}-releasing channels are achieved by *cis-trans* isomerization of prolyl peptidyl bonds located around the gate of Ca^{2+}-releasing channel. FKBP forces the conformational transition of Ca^{2+}-releasing channel by its PPIase activity. Another model is against this and proposes that modulation of the Ca^{2+}-releasing channel be achieved by FKBP's ability to bind to each receptor subunit. The binding of FKBP just stabilizes the transitional conformation of Ca^{2+}-releasing channel, which is induced by events other than the *cis-trans* isomerization, such as the binding of IP3 or ryanodine to the channel. The latter model is proposed based on the result that substitution of wild-type FKBP on RyR complex with FKBP mutants that are devoid of any PPIase activity measurable by the protease-coupled assay with a peptide substrate did not alter Ca^{2+} flux by Ca^{2+}-releasing channel [80]. However, it has been shown that the PPIase mutants devoid of any PPIase activity measured by the protease-coupled assay could still have the PPIase activity when measured by an assay based on the catalysis of a proline-limited protein folding reaction [81]. Thus, the involvement of PPIase activity in the mechanism by which FKBP modulates gating of Ca^{2+} channels remains uncertain. Anyway, regardless of how PPIase activity is involved in regulating the function of Ca^{2+}-releasing channels, it would be certain to say that the binding of FKBP to the receptors would stabilize the conformation of Ca^{2+} channel complex either by fixing its structure on or by inducing a stable configuration.

2. *Steroid-Hormone Receptor Complexes*

The immunosuppressants had been well known to exhibit steroid-sparing effects in clinical therapy, which further enhanced the effect and reduced the amount of the steroid used with the immunosuppressant during treatment for preventing rejection after organ transplantation. Regarding the mechanism of action of steroid hormones, there have been the frequent controversies and equivocations of earlier studies due to the fact that the native, hormone-free state of their receptors is a large multiprotein complex that resisted description for many years because of its unstable and dynamic nature. Thus, how the immunosuppressants achieve the steroid-sparing effect had remained a mystery for a long period until the steroid-hormone receptor complexes were found to contain immunophilins and their nature came to be understood [82]. It is now believed that the steroid-sparing effect of FK506 and CsA are due to their ability to potentiate the steroid-induced transcriptional activity of the steroid-hormone receptor [83–86]. However, the mechanism by which the immunosuppressants potentiate the transcriptional activity of the steroid-hormone receptor is very complicated.

In various cells, the steroid-hormone receptors, such as those for glucocorticoid (GC) and progesterone (PG), undergo rapid steroid-mediated translocation from the cytoplasm to the nucleus and eventually perform transcription of the target gene. In nonstimulated cells, the inactivated steroid receptors are chaperoned into a conformation that is optimal for binding hormone by chaperone molecules. Upon steroid stimulation, a sequential and dynamic series of protein–protein interactions is required for forming a stable steroid–steroid receptor heterocomplex in their inactive, ligand-friendly state that are assembled by a multiprotein chaperone system comprising hsp90, hsp70, Hip (p48), hsp40, and p23 [87]. In addition to chaperones, steroid-receptor heterocomplexes contain Hop (p60) or a high molecular weight immunophilin, i.e., FKBP51, FKBP52 (FKBP59/hsp56), or CyP40. These proteins, which have tetratricopeptide repeat (TPR) motifs, bind to hsp90 via a region containing TPR motifs. Hsp90 is present as dimer in and as a basic constituent of the steroid receptor complexes. An hsp90 dimer provides one binding site located in its carboxy-terminal region containing a MEEVD motif for one of the TRP-containing proteins; thus, this accounts for the common heterotetrameric structure of native receptor heterocomplexes being one molecule of receptor, two molecules of hsp90, and one molecule of a TPR-containing protein [88–90] (Fig. 9). Thus, any one of the TPR-containing proteins competes with each other for binding to hsp90, indicating that every one of the steroid-hormone receptor heterocomplexes contains—if any—only one of the TRP motif-containing proteins. For example, in human breast cancer cells treated with PG and rapamycin, the nuclear PG–PG receptor complex contains hsp90, hsp70, and FKBP52 but not FKBP51, although the latter was part of an unliganded PR heterocomplex associated with hsp90. In the same cells, CyP40 was associated only with hsp90 in the cytosol [91].

This nature of competitive binding for hsp90 determines the sensitivity of the steroid-hormone receptor complexes to steroid. This is exemplified for the GC-receptor complex from squirrel monkeys, which are known to have a low-binding affinity to GC. This character is associated with incorporation of FKBP51 into GC-receptor heterocomplexes. The exchange of FKBP52 by FKBP51 renders the high-affinity GC-receptor complex from the other species GC resistant [92].

14. Mode of Action of FK506 and Rapamycin 595

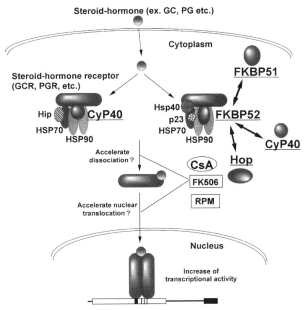

Fig. 9. Regulation of the function of steroid-hormone receptor complexes by TRP-containing proteins and molecular chaperones (hypothesis). Abbreviations are given in the text.

Thus, the control of the relative amounts of TRP-containing proteins in the cell is one of the mechanisms by which the cell controls its sensitivity to steroid hormones. Actually, there have been a number of reports that the amounts of TRP-containing proteins are controlled by changing their cellular localization, transcriptional state, or posttranslational modification, situations which are induced by various stimulus, such as growth factors, steroids, and stress [93–95].

Because of these regulatory natures for steroid-hormone complexes, any disturbance of protein–protein interactions present in the steroid-hormone complexes affects the action of steroid hormones. Immunosuppressants, FK506, rapamycin, and CsA are certainly able to interrupt such interactions. This is why the immunosuppressants have effects on the action of steroid hormones in various cells. At least three members, CyP40, FKBP51, and FKBP52, of immunophilin families play important roles in the maturation processes of the steroid-hormone receptor complexes (Fig. 9). This, in turn, predicts some differences in the mode of action among the immunosuppressants because the three immunophilins act in a coordinated way with hsp90 but modulate receptor activity differently. For example, in the case of rabbit progesterone receptor (PGR) in Lc13 cells, CsA as well as the hormone is able to translocate PGR, which is cytoplasmic in the absence of progesterone, into the nucleus, but FK506 or rapamycin cannot do so. On the other hand, growth factors influence the localization of

FKBP52 (FKBP59) but not that of PGR or Cyp40, thus this phenomenon is affected by FK506 or rapamycin but not by CsA [96]. In the case of human rheumatoid synovial fibroblasts, FK506 potentiates downregulation of synovial COX-2 mRNA expression induced by GC. This FK506-induced potentiation of GR-mediated repression is the result of increased translocation of GR to the nucleus and subsequent repression of NFκB transactivation [97]. CsA does not cause this potentiation. Thus, FK506 and CsA differ in their effects on translocation of steroid-hormone receptor in different cells.

In addition to the binding activity of hsp90 to TRP-containing proteins, it mediates ATP-independent chaperone activity via the same region as that binding the TRP-containing proteins. The region also overlaps the hsp90 dimerization domain, and includes structural elements important for steroid receptor interaction [98]. Therefore, the competitive binding of specific immunophilins and the other TPR-containing proteins provides an additional regulatory mechanism for hsp90 chaperone activity and thus is also a critical step for receptor maturation, determining the protein composition and the three-dimensional conformation of steroid receptors.

Other than hsp90 and TRP-containing proteins, some additional players influence the process of receptor maturation. These include p23, which binds hsp60 and stabilizes a high-affinity ligand-binding state of steroid-hormone receptor [99, 100]; hsp70, which is bridged by Hop (p60) in the binding to hsp90; and Hip (p48), which interacts with hsp70. The interaction of these proteins with hsp90 is regulated by the presence or absence of other proteins or substances. For example, the presence of ATP modulates formation of hsp90-Hop (p60) complex, and stabilizes it at low protein concentrations [101]. However, binding of p23 to hsp90 is independent from the affinity of the hsp90-Hop (p60) complex and the stabilizing effect of ATP. The binding of those proteins including TRP-containing proteins to hsp90 leads to dissociation of some proteins from the receptor complex while others are recruited. In principle, for several steroid receptors, binding to hsp90 is required for the receptor to be in a native hormone-binding state, and for all of the receptors, hormone binding promotes dissociation of the receptor from hsp90 and conversion of the receptor to the DNA-binding, transcriptionally active state (Fig. 9).

Finally, note that in addition to the control of protein–protein interaction based on a multichaperone-complex system, the receptor maturation process is also regulated by phosphorylation–dephosphorylation of the proteins involved. For example, phosphorylation of FKBP52 at Thr143 in the hinge I region by casein kinase II (CK2) interrupts its binding to hsp90 that is also able to associate with CK2 [102]. Although this phosphorylation does not affect the ability of FKBP52 to bind FK506, CK2 can regulate the protein composition of chaperone-containing steroid-receptor complex. Other than CK2 Ca^{2+}/calmodulin kinase is also reported to be involved in PG- and GC-receptor-mediated transcription in human breast cancer cells [91].

Thus, the maturation and translational activity of the steroid-hormone receptor complexes are regulated by many factors including immunophilins. FK506/rapamycin and CsA certainly affect the functions of steroid-hormone receptor by interrupting the roles of immunophilins in the maturation and transcriptional activity of steroid-hormone receptor complexes.

3. Growth-Hormone Receptor Complexes

Growth arrest in many cell types is triggered by transforming growth factor beta (TGF-β), which signals through two TGF-β receptors (TGF-β RI, and TGF-β RII) that consist of heteromeric serine-threonine kinase in their cytosolic carboxyl-terminal domains. In a yeast genetic screen, FKBP12 was found to interact with TGF-β RI [103]. In mammalian cells FKBP12 also binds to ligand-free TGF-β RI, from which it is released on phosphorylation of TGF-β RI mediated by TGF-β RII that is activated by binding of TGF-β to either of TGF-β RI or RII (Fig. 10).

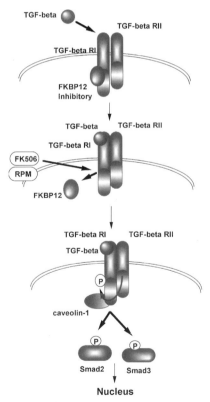

Fig. 10. Role of FKBP12 in TGF-β receptor signal transduction. See text for explanation of the figure.

The phosphorylation occurs within a glycine/serine (GS)-rich box that is located upstream of the serine/threonine kinase domain in the cytoplasmic portion of TGF-β RI. Interrupting the FKBP12–TGF-β RI interaction with FK506 enhances the TGF-β activity; thus, the FKBP12 binding is inhibitory to the signaling pathway of TGF-β [104]. The binding of FKBP12 to TGF-β RI occurs at a Leu-Pro sequence that is located next to the activating phosphorylation sites in the GS-rich box of TGF-β RI. According to the results of X-ray analysis of the FKBP12–TGF-β RI fragment complex, TGF-β RI adopts an inactive conformation that is maintained by the unphosphorylated GS-rich region, to which FKBP12 binds and is capping the TGF-β RII phosphorylation sites and further stabilizing the inactive conformation of TGF-β RI [105]. The binding of FKBP12, therefore, prevents TGF-β RI from being approached by TGF-β RII for phosphorylation. However, FKBP12 does not inhibit TGF-β RI association with TGF-β RII, which occurs at a region apart from the GC-rich box of TGF-β RI. As expected, rapamycin as well as FK506 reverses the inhibitory effect of FKBP12 on TGF-β RI phosphorylation, but does not affect the interaction between TGF-β RI and RII [106].

On dissociation of FKBP12 from TGF-β RI by binding of TGF-β, TGF-β RI is phosphorylated by TGF-β RII and is activated to phosphorylate two downstream effectors, Smad2 and Smad3, leading to their translocation into the nucleus (Fig. 10). The phosphorylation of Smad2 is suppressed by caveolin-1, which interacts with the TGF-β RI [107]. FKBP12 is no longer involved in these downstream events. However, FK506 and rapamycin cannot directly affect those events, but indirectly eventually cause inhibition of growth arrest of the cell, though as an exception they have no such effect on embryonic fibroblasts and thymocytes [108].

A similar role for FKBP12 is also reported in the signal transduction system of epidermal growth factor (EGF) receptor, which also interacts with FKBP12. FK506 and rapamycin induce a significant stimulation of EGF receptor autophosphorylation, thus, FKBP12 again acts negatively on EGF receptor autophosphorylation [56]. In addition, in the case of the EGF receptor, the inhibitory effect on autophosphorylation is shown to be dependent on only FKBP's PPIase activity, but not CyPA's PPIase activity. Thus, both PPIase activity and substrate specificity of FKBP12 seem to be indispensable for exhibiting an inhibitory effect on the EGF receptor function and probably on the TGF-β RI function as well.

In addition to the role of FKBP12 in modulating growth factor receptors, FKBP12 is reported to have a role in endocytosis. Recently, using a chimeric TGF-β receptor system in which TGF-β RI and FKBP12 are covalently fused, a specific enhancement of internalization is observed when FKBP12 binding to TGF-β RI is prevented with rapamycin. Thus, FKBP12 seems to act also as a negative regulator on the endocytosis of TGF-β receptor [109]. However, its detailed mechanism of action is not well understood.

E. Isoforms of FKBP12

Because FK506 (and rapamycin) and CsA bind with high affinities to their respective PPIases, the binding criterion allow us to classify the PPIases for two superfamilies: the CyP and the FKBP families. In addition to these families, we now know that there is one additional PPIase family, the parvulin family. All three families of PPIases are found in virtually every living organism from human to bacteria. Many organisms express multiple members of each of these families in different cellular compartments, from the cytosol to the nucleus, from the endoplasmic reticulum to the mitochondria, from the cell surface to the extracellular space. Although there have been a number of reports regarding the identification of new members and their functions for each of the three PPIase families, only those belonging to the FKBP family are described in this section because this review focuses on the action of FK506 and rapamycin.

So far, at least 14 unique FKBP open reading frames (ORFs) are recognized in the genome sequences of humans [110]. Their corresponding proteins are not necessarily identified as actual translation products. However, the number represents that of a probable unique protein species belonging to the FKBP family that humans are able to express in their entire body. Although many additional pseudogenes for FKBP12 are recognized in the genomes, they are not included in the number described. The expected members of the FKBP family range in size from 12 to 135 kDa; some have domains other than the FKBP domain as part of a different protein; and some contain a variable number of FKBP domains ranging from two to four. Table II summarizes the FKBP isotypes that have been identified so far as the expressed transcripts and/or proteins in mammalians. The table includes 10 FKBP isotypes isolated from mammalians and one from chicken: FKBP12, FKBP12.6, FKBP13, FKBP23, FKBP25, FKBP51, FKBP52/FKBP59, FKBP60, FKBP65, IPBP12, and cFKBP/SMAP (chicken). We have not yet found all of the isotypes corresponding to those predicted to be present in the human genome; i.e., isotypes that contain two FKBP domains and that are higher than 65 kDa in size are not found. They must be waiting to be found. Although the roles of FKBP12 have been well described in various biological events, most of the other proteins have not been well characterized yet. Solving the cellular functions of those PPIases will be critical for understanding the mode of action of FK506 and rapamycin in detail.

III. Target of FK506–FKBP12 Complex: Calcineurin

The inhibition of PPIase activity was shown to be an insufficient requirement for mediating immunosuppressive effects of FK506 and CsA. However, FKBP12 and CyPA were proposed to induce a "gain-of-function" transformation of their ligand by binding to their active sites as PPIase, by which FK506 and CsA create

TABLE II
Summary of Isotypes of FKBP-Type PPIases Isolated as Transcript and/or Protein[a]

Isotype	MW (kDa)	Tissue distribution	Intracellular localization	Expected function	Others	Ref. No.
FKBP12	12	Nonspecific	Cytosol, nucleus, ER membrane	Modulator of RyRs and IP3Rs; suppressor of TGF-beta R signaling; probable modulator of calcineurin (in yeast)	Receptor for FK506 and rapamycin; complex with FK506 binds to calcineurin; complex with rapamycin binds to mTOR	3, 5, 6, 40, 56, 70, 74, 138, 142, 143
FKBP12.6	12.6	Cardiac muscle, pancreas	ER membrane	Modulator of RyR2	Binding to cADPR causes dissociation from RyR2	69, 78
FKBP13	13	Nonspecific	ER membrane, erythrocyte membrane	Protein folding and trafficking; function as a component of membrane cytoskeletal scaffolds	Contains ER retention signal; interacts with a homologue of erythrocyte membrane cytoskeletal protein 4.1	111, 112
FKBP23	23	Heart, lung, testis, most expressed early in embryonic development	ER	?	Contains ER retention signal, and two EF-hand motifs; binds to Ca^{2+}	113
FKBP25	25	Thymus, brain, and spleen	Nucleus	Regulation of transcriptional activity of YY1; associates with histone deacetylases HDAC1 and HDAC2 and with YY1	Higher sensitivity to rapamycin than FK506 (rapamycin $K_i = 0.9$ nM and FK506 $K_i = 160$ nM); contains nuclear localization signal, and binds to DNA, casein kinase II, and nucleolin; downregulated following p53 induction	114–118
FKB51	51	Nonspecific (T-cell specific in mouse)	Cytosol, nucleus	Negative modulator of steroid-hormone receptor; interacts with hsp90	Complex with FK506 binds to calcineurin	119–121

FKBP52/59	52	Nonspecific	Cytosol, nucleus	Modulator of steroid-hormone receptor; binds with hsp90, GC receptor and dynein at different sites; phosphorylated by casein kinase II.	Contains phosphorylation-site motifs, a potential calmodulin-binding site, three-unit TPR and three FKBP12 domains; interacts with interferon regulatory factor-4 (IRF-4) and FKBP-associated protein 48	57, 88, 102, 122–126
FKBP60	60	Heart, skeletal muscle, lung, liver, and kidney	ER	?	Has homology to FKBP65 and SMAP; contains ER retention signal, EF hand motif, and 4 FKBP12 domains; binds to Ca^{2+} and phosphorylated	127
FKBP65	65	Lung, spleen, heart, brain, and testis	ER	Protein folding and trafficking	Contains four FKBP12 domains glycosylated and phosphorylated; associated with c-Raf-1, tropoelastin	128–132
IPBP12	12	Erythrocyte	Membranes	?	Inhibited not only by FK506 but also by inositol 1,4,5-triphosphate, inositol 1,3,4,5-tetrabisphosphate, and phosphatidylinositol 4- and 4,5-phosphates	134
cFKBP/ SMAP	65	Smooth muscle (chicken)		Smooth-muscle differentiation	Has homology to FKBP65 and FHBP60	133

[a]See text for definitions of abbreviations.

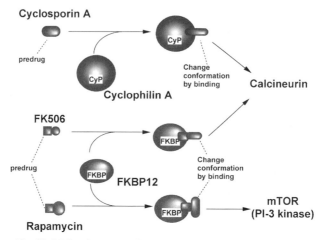

Fig. 11. Molecular targets of PPIase-immunosuppressant complexes.

a new surface on the FK506–FKBP12 and CsA–CyPA complexes to interact with calcineurin [6] (Fig. 11). Calcineurin is an eukaryotic Ca^{2+}- and calmodulin-dependent serine/threonine protein phosphatase that is conserved from yeast to human. It is a heterodimeric protein consisting of a catalytic subunit, calcineurin A, which contains an active site binuclear metal (Ca^{2+}) center, and a tightly associated, myristoylated, Ca^{2+}-binding subunit, calcineurin B. The binding of the complexes to and inhibition of calcineurin depends on the presence of calcineurin B and Ca^{2+}-calmodulin [135]; that is, the FK506–FKBP12 and CsA–CyPA complexes interact tightly with calcineurin only when the enzymatic activity of a heteromeric dimer is amplified by the binding of calmodulin in the increasing concentration of Ca^{2+} that is released from endoplasmic reticulum through activation of IP3R (Figs. 3 and 8). On the other hand, any one of the drugs and PPIases alone does not inhibit the phosphatase activity even in the presence of Ca^{2+} and calmodulin. The structural basis of gain-of-function in transformation of the immunosuppressants in their immunophilin complexes has been given by their NMR and X-ray analyses, by which the structures of FK506 and CsA in solution are shown to be quite different from those in the FKBP12–FK506 and CsA–CyPA complexes, respectively. Thus, the gain-of-function concept became widely accepted, though there are some alterations from the original concept. Namely, the gain-of-function is achieved by the specific binding of PPIases to the twisted *trans* isomer of the drugs and by subsequent induction of additional drugs' conformational changes in equilibrium [136]. It is not induced in the active conformational change of the drugs forced by the isomerase activity of CyPA or FKBP12 as might have been expected [137].

Comparison of the three-dimensional structures of the CsA–CyP and FK506–FKBP complexes showed no apparent similarity between them. These findings raise the question as to how, given their structural differences, these two complexes can both bind and inhibit calcineurin. While a crystal structure of the FK506–FKBP12–calcineurin complex has been reported [138], no structure for a CsA–CyP–calcineurin complex has been determined. However, based on the results from various molecular modeling studies [139], as well as kinetics studies for inhibition of calcineurin [140], both classes of drug–PPIase complexes are thought to interact with a common locus on the calcineurin molecule. Namely, both complexes bind in a hydrophobic groove between the calcineurin A catalytic and the regulatory B subunit. Thus, the two drug–PPIase complexes prevent substrate from binding in the calcineurin's active site. Interestingly, significant protein–protein interactions between FKBP12 and a heterodimeric calcineurin are present in the crystal structure of the FK506–FKBP12–calcineurin complex, suggesting that some interactions occur between the two proteins in the absence of the drugs *in vivo*. Supporting this suggestion, the presence of direct interactions between FKBP12 and calcineurin A subunit has been observed in the yeast two-hybrid system in the absence of any exogenous drug, though this interaction is not as robust as the FK506-dependent interaction [141]. Unlike the FK506–FKBP12–calcineurin complex for which calcineurin B is required, calcineurin B subunit is not required for formation of the FKBP12–calucineurin A complex. Therefore, direct interaction of FKBP12 with calcineurin A is not affected by FK506 or rapamycin in the absence of calcineurin B. In addition, neither PPIase activity nor FKBP12 active site is required to form the ligand-independent FKBP12–calcineurin A complex. These results suggest that calcineurin B and FKBP12 may compete for binding to calcineurin A. In contrast, in the presence of calcineurin B and calmodulin, the FKBP12–calcineurin A interaction is dramatically potentiated by FK506. These findings indicate that surface residues of FKBP12 that are required for FKBP12–FK506 to bind calcineurin are completely dispensable when FKBP12 alone binds to calcineurin [80]. In an analysis of the CyPA–calcineurin interaction with a yeast two-hybrid system, CyPA is also found to bind calcineurin A in the absence of CsA. In this case, the addition of CsA potentiates binding of CyPA to both calucineurin subunits. Compared with the FKBP12–calcineurin A complex, the ligand-independent CyPA–calcineurin A complex is more robust and is potentiated to a lesser degree by CsA binding. Although differences are slight between the nature of FKBP12–calcineurin and CyPA–calcineurin interactions, PPIases and calcineurin, either alone or in complex with endogenous ligands, functionally interact with each other. In this yeast model, CsA and FK506 no longer serve to bring two completely unrelated proteins together, but rather may capitalize on the inherent ability of PPIases to interact with calcineurin [141]. Thus, it is probable that PPIases normally bind to and regulate calcineurin. Additional yeast genetic studies have suggested that FKBP12 functions normally to inhibit calcineurin, whereas CyPA normally activates calcineurin and, if so, then the two PPIases would reciprocally modulate calcineurin [141]. In mammalian cells, the direct interaction between PPIase and calcineurin has not yet been reported, however, the presence of such interactions *in vivo* would explain why the action of the two structurally unrelated immunosuppressants converged on the same target, calcineurin (Fig. 11).

IV. Target of Rapamycin–FKBP12 Complex: mTOR/FRAP/RFAT

When bound to the same binding protein (FKBP12) as that for FK506, rapamycin has an ability to interact with mTOR, a homologue of PI3P kinase, which is related to the ataxia-telangiectasia gene product [25] (Fig. 11). Initially, the target of rapamycin was identified by yeast genetic analyses in that binding of rapamycin to FKBP12 was shown to generate a toxic complex that inhibits cell growth in the G_1 phase. These analyses implicated two related proteins, target of rapamycin (TOR) 1 and TOR2, as the targets of the FKBP12–rapamycin complex in yeast [142]. Thus, mTOR is a mammalian counterpart of yeast TORs. The mTOR protein from humans has a molecular weight of 289 kDa and is composed of 2549 amino acid residues. It shows 42% and 45% identity in their amino acid sequences with TOR1 and TOR2, respectively [142, 143]. Like yeast TORs, the mTOR molecule consists of a carboxyl-terminal repressor (CTR) domain (2430–2459), a PI3K domain (residues 2255–2450), an FKBP12–rapamycin binding (FRB) domain (residues 2015–2114), and a large amino-terminal (AT) domain (residues 1–2014) (Fig. 12). As expected from the origin of its name, the rapamycin–FKBP12

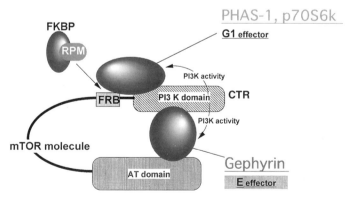

Fig. 12. Molecular model for mTOR function and action of rapamycin (RPM)–FKBP12 complex (hypothesis). See text for explanation. (Modified from Zheng *et al.* [146] with permission from Elsevier Science.)

complex binds to mTOR via the FRB domain. According to the result of X-ray analysis, FKBP12 and the FRB domain bind together as a result of the ability of rapamycin to occupy two different hydrophobic binding pockets simultaneously. The structure shows extensive interactions between rapamycin and both proteins, but fewer interactions between the proteins [43, 44]. This is in contrast to the presence of significant protein–protein interactions observed in the FK506–FKBP12–calcineurin complex [138]. The absence of the protein–protein interaction between FKBP12 and the FRB domain allows us to construct two chimeric human-derived proteins that are reconstituted by rapamycin into a transcription factor complex, which can transcribe a therapeutic protein gene, such as erythropoietin, under pharmacological control [144].

The FRB domain, with its upstream activator sequence of the mTOR molecule, is required for activation of the PI3K domain. It, as a fragment that does not have the other domains of mTOR, induces a drastic blockage of the G_1 to S cell cycle progression in human cells; thus, the FRB domain alone is toxic to the cell [145]. However, the binding of the rapamycin–FKBP12 complex to the FRB domain of mTOR does not inhibit PI3K activity directly. Instead, it probably interferes selectively with mTOR binding to or with phosphorylation of effectors that are required for functioning mTOR [146] (Fig. 12). Candidates for such effectors can be substrates for mTOR kinase or possibly the FRB domain's upstream activator sequence of mTOR, or the other proteins. As the direct substrates of mTOR kinase, p70S6k1, p70S6k2, PHAS-I/4E-BP1, and 4E-BP2 have been identified [30, 147–152]. For example, the mTOR kinase phosphorylates p70S6k1 on Thr-389, a residue whose phosphorylation is rapamycin sensitive *in vivo* and necessary for S6 kinase activity. In the case of PHAS-I/4E-BP1, its phosphorylation occurs at multiple sites and is controlled by several kinases. mTOR phosphorylates it on Thr36, Thr45, Ser64, Thr69, and Ser82 *in vitro*; all of these sites are adjacent to Pro residue, thus, mTOR seems to be a kind of proline-directed protein kinase, such as cdc and MAP kinases [153]. Of these Ser/Thr sites, phosphorylation on Thr36 and Thr45 is essential for induction of the subsequent phosphorylations of PHAS-I/4E-BP1 *in vivo*, which lead to its dissociation from the cap-binding protein, eIF-4E [154] (Fig. 13). The phosphorylation on these sites is affected by rapamycin *in vivo*. The subsequent phosphorylation on the other sites on PHAS-I/4E-BP1 is probably mediated by PKC delta that interacts directly with mTOR. This causes dissociation of 4E-BP1 from eIF4E, and then cap-dependent translation is stimulated [148]. Thus, phosphorylation of PHAS-I/4E-BP1 on five Ser/Thr-Pro sites is controlled by multiple mechanisms by which translational repression is governed [155, 156] (Fig. 13).

The kinase activity of mTOR is regulated intramolecularly by its internal domains; i.e., the FRB domain and amino-terminal domain are required for the acceleration of the activity and the carboxy-terminal repressor domain is

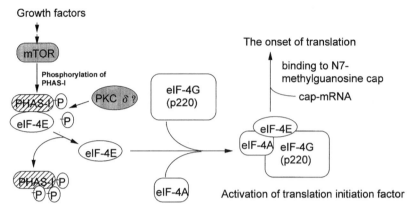

Fig. 13. Regulation of cap-dependent translation by mTOR. See text for explanation. (Modified from Takahashi [24] with permission.)

required for the suppression of the activity (Fig. 12). In addition to the intramolecular modulation, mTOR activity is also regulated intermolecularly by other proteins. For example, mTOR interacts with gephyrin via its large amino-terminal domain. Gephyrin is a widely expressed protein necessary for the clustering of glycine receptors at the cell membrane of neurons. This interaction is necessary for mTOR to signal to downstream molecules, including the p70S6k and the PHAS-I/4E-BP1 [157]. However, the binding of mTOR to gephyrin is not affected by rapamycin. Identification of gephyrin as a regulator of mTOR activity seems to support a model proposed for the dual functions of TOR2 based on the results of yeast genetic analyses [146]. In this yeast model, the TOR2 molecule has at least two functions that are required for regulation of the intact kinase domain. One function is that for the G_1 phase progression in the yeast cell cycle that is sensitive to the action of rapamycin. Another function is that required for cell viability that is resistant to the action of rapamycin. Accordingly, the presence of the two effectors is assumed. One effector essential for the G_1 progression binds to the carboxyl-terminal portion of TOR2 corresponding to that containing the FRB and PI3 kinase domains of mTOR (Fig. 12, G_1 effector). On the other hand, another effector essential for cell viability binds to the amino-terminal portion of TOR2 corresponding to the region excluding the FRB and PI3 kinase domains of mTOR (Fig. 12, E effector). In mammalian cells, the former (G_1 effector) fits the substrate of mTOR kinase, such as p70S6k or PHAS-I. The latter (E effector) fits the regulatory factor of mTOR, such as gephyrin. Thus, the mode of action of mTOR on p70S6k and PHAS-I/4E-BP1 seems to correspond with that of yeast TOR2. Recently, an additional mTOR homologue has been found. This protein, termed human SMG-1 (hSMG-1), has a carboxyl-terminal domain unique to the phosphatidylinositol kinase-related kinases and a FRB-like domain similar to that found in mTOR [158]. This suggests that mammals may also have a second signaling system regulated by TOR1-like mTOR, in which gephyrin-like factor is not involved.

The mTOR kinase activity is also modulated by its upstream protein kinases other than gephyrin in the cell. Protein kinase B (PKB)/Akt is one such kinase. PKB directly phosphorylates mTOR at Ser2448, which is induced by insulin, but attenuated by amino acid starvation. The phosphorylation of Ser2448 is now believed to be a point of convergence for the counteracting regulatory effects of growth factors and amino acid levels [159]. The other candidates for the upstream regulator of mTOR include c-Abl protein-tyrosine kinase that is activated by ionizing radiation and certain other DNA-damaging agents [160], and PI3 kinases that are sensitive to wortmannin or LY294002 [161, 162a]. In addition, on upstream stimulation, such as that inhibited by wortmannin or that increased by calyculin A treatment in the case of T cells, mTOR is known to autophosphorylate Ser2481 in the region located in a hydrophobic region near the conserved carboxyl-terminal mTOR tail. This carboxyl-terminal tail is required for mTOR kinase activity and for signaling to the translational regulator p70S6k [162b, 163]. However, those upstream events are not affected by rapamycin. Thus, rapamycin is very useful to draw a line between the events occurring upstream and downstream of the mTOR function.

V. Intervention of Intracellular Signaling Pathways by FK506 and Rapamycin

A. Calcineurin Signaling Pathway

Although calcineurin is found to be a target of the immunosuppressants, it is more abundant in the brain than the immune system. In addition, calcineurin is also present in almost all other tissues throughout the body. Thus, FK506 and CsA have the potential to affect all of those tissues and cells.

In the case of the activated T cells, both FK506 and CsA specifically block T-cell receptor (TCR)-mediated transcription of many cytokine genes including the IL-2 gene (Fig. 3). When looking at the transcription of the IL-2 gene in activated T cells, of the many enhancer-binding elements present in the IL-2 promoter, both immunosuppressants affect specifically the binding site for the NF-AT associated protein. The binding activity of NF-AT correlates with the level of IL-2 transcription in T cells. Thus, NF-AT is a crucial transcription factor in T-cell activation induced by antigen [164, 165]. NF-AT is a complex factor made of at least two subunits, NF-ATC and NF-ATN. The former is T-cell specific and preexists in cytoplasm, while the latter is ubiquitous, rapidly induced by the activation of a PKC-dependent pathway and predominantly localized in the nucleus. Thus, the functional NF-AT requires the association of the two subunits, NF-ATC and NF-ATN (Fig. 3). Because an increase in intracellular Ca^{2+} concentration $[Ca^{2+}]i$ leads to the nuclear translocation of NF-ATC and the assembly of a functional transcription factor NF-AT [166], the existence of a subunit of NF-AT in the cytoplasm offers a connection between calcineurin and IL-2 transcription on the two sides of the nuclear membrane. Many lines of evidence offer a signaling pathway leading to IL-2 transcription upon the binding of antigen to TCR as follows [3] (Fig. 3):

Step 1: Binding of antigen to TCR triggers the activation of a number of signaling factors including protein tyrosine kinases, such as Fyn and ZAP70, and induces the activation of phospholipase C.

Step 2: Phospholipase C hydrolyzes phosphatidylinositol diphosphate and releases diacyl glycerol (DG) and inositol-3-phosphate (IP3).

Step 3: The released IP3 increases $[Ca^{2+}]i$ in the cytosol by opening an IP3R-Ca^{2+}-releasing channel on ER.

Step 4: By the elevated $[Ca^{2+}]i$, calcineurin is activated in a calmodulin-dependent manner.

Step 5: Calcineurin dephosphorylates NF-ATC directly and induces translocation of NF-ATC to the nucleus.

Step 6: The translocated NF-ATC associates with NF-ATN in the nucleus, which is translated via the PKC signaling pathway that is activated by DG, and transcribes IL-2-gene.

Thus, inhibition of calcineurin by the FK506–FKBP12 and CsA–CyPA complexes causes blocking of the translocation of NF-ATC from the cytoplasm to the nucleus, resulting in a failure to activate the IL-2 gene.

Besides the IL-2 gene, ones regulated by transcription factor NF-AT include those of IL-3, IL-4, GM-CSF, TNF-α, CD-40 ligand, Fas(CD95) ligand, and those involved in differentiation of $CD4^+CD8^+$ T cells and in the B-cell receptor response of self-tolerant B cells etc [8, 167]. In the signal transduction leading to the transcriptional activation of these genes, NF-AT is probably activated by calcineurin because both FK506 and CsA have been shown to inhibit most of the transcription of these genes. However, the regulatory mechanism by which NF-AT is activated seems to be much more complicated than expected. There are at least two reasons for this. First, calcineurin has at least three isoforms for the A subunit (A-1, A-2, A-3) and two for the B subunit (B-1 and B-2) [168]. In mouse, one of the A forms of calcineurin was knocked out, but no defect was observed in the TCR-mediated IL-2 production, indicating that the other isoforms mediate the TCR signal transduction pathway [169]. There are also some other reports in which different isoforms of calcineurin have been described to exhibit distinct expression patterns in different cells, indicating that each of the isoforms has a distinct cellular function. Second, so far, four genes encode proteins belonging to the cytoplasmic subunit of NF-AT (NF-ATC): NF-AT1, NF-AT2 (NF-ATp), NF-AT3 (NF-AT4/NF-ATx), and NF-AT4 (NF-AT3). In addition, there are many theoretical molecular species for the nuclear subunit of NF-AT (NF-ATN, AP-1). Since NF-ATN is made of two protein constituents, Jun and Fos proteins, each of which has multiple isoforms, their combination makes at least 12 different forms of NF-ATN [170]. Furthermore, four more candidates (NIP45, c-Maf, NF45, and NF90) of NF-ATN have been reported [171–173].

Regarding the specific role of a particular combination of the NF-AT's isoform, an interesting result is reported for an experiment in transgenic mice bearing a disruption in the NF-AT1 gene [174, 175]. Despite the expected role of NF-AT1 in the immune system, unexpectedly, the pulmonary and aortic valves in heart completely and specifically failed to form as a result of the disruption of the gene. These valves control the flow of blood from ventricles of the heart into the arteries leading to the lungs and the main circulation, respectively. An interesting point of this experiment is that although NF-AT2 (NF-ATp) is biochemically and genetically similar in many ways to NF-AT1, genetic disruption of the NF-AT2 gene did not lead to an obvious defect in heart development. However, it does disrupt the immune system, thus the expected role of NF-AT in the immune system is attributed to NF-AT2 [176]. In addition FK506 or CsA can inhibit the translocation of NF-AT1 from the cytoplasm to the nucleus in cells from the pulmonary and aortic valves of normally developing heart [174, 175]. A comparable activation system of NF-AT1 with the immune system may be used in the developing fetal heart; namely, endocardial cells activate NF-AT1 through a calcineurin-specific pathway [177]. Therefore, FK506 and CsA probably affect the formation of heart structure. In contrast to the effect on heart development, there has been a report that administration of FK506 and CsA prevented disease in mice that were genetically predisposed to develop hypertrophic cardiomyopathy [178]. Despite the critical role of the calcineurin pathway in heart formation and hypertrophy, genetic disruption of the gene for a calcineurin A subunit did not show any disturbance in mice [179]. These results coupled with the presence of isoforms of calcineurin indicate that there may be a heart-specific form of calcineurin in heart. Those observations point out the importance of answering the question of which one or combination of NF-ATN and NF-ATC (or of isoforms of calcineurin A and B) is playing a role in the transcription of a particular gene (or in leading to the particular gene transcription) in a specific subset of immune or other type of cells.

There are a number of reports that FK506 and CsA have affected various phenomena mediated by T cells in the immune system as well as by the other cells from many other tissues, indicating that calcineurin-signaling pathways are involved in various phenomena throughout the body.

B. mTOR Signaling Pathway

Rapamycin inhibits the mTOR kinases, which regulate cell proliferation and mRNA translation induced by various external signals, including those by growth factors, nutrients, and stresses, such as UV (Fig. 5). mTOR kinases and their sensitivity to rapamycin are well conserved from yeast to man. Thus, the target of rapamycin is emerging as a general effector for signals, which indicate to the cell whether the external environment is conducive for growth. Use of rapamycin has been instructive in identifying potential signaling components downstream of mTOR, leading to the observation that both protein synthesis and turnover are under mTOR control. The central issues concerning mTOR are the identification of the proliferative and antiproliferative signals that mediate its function and the mechanisms by which these signals are transduced to downstream molecules. However, emerging evidence indicates that mTOR signaling is highly complex and is involved in a variety of cellular processes. Therefore, it is not an easy task to identify the molecules working downstream of mTOR as well as the signaling that controls the function of mTOR.

For overcoming the difficulties, a genome-wide genetic linkage map analysis for TOR was tried by using systematically generated yeast deletion strains for rapamycin sensitivity [20]. In yeast cells, TOR signaling is also highly complex and is involved in starvation responses, including G_1 cell cycle arrest, glycogen accumulation, autophagocytosis, reduced protein synthesis, and sporulation. In addition, TOR is shown to regulate transcription of genes involved in ribosome biogenesis and nutrient responses. Thus, in many senses the function of yeast TOR is comparable to that of mTOR. According to the result of the genomic-wide analysis, the genes related to the TOR function are clustered into eight functional groups: protein synthesis, carbon and nitrogen catabolic repression, metabolic biosyntheses, mitochondria biogenesis and functioning, general transcription, vacuolar biogenesis and functioning, ubiquitin-dependent proteolysis, and spindle stability and functioning [20]. Although the majority of the genes are already known to function with TOR, some other genes have never been connected to TOR function. Thus, these results may provide important links to future mechanistic studies of the roles of mTOR.

Finally, so far, the use of FK506 and rapamycin as well as CsA as experimental probes has offered substantial answers to the mechanisms of intracellular signal transduction and of other important cellular processes in basic biology. On the other hand, it also led to more questions than answers. One of the most notable questions is "What are the physiological roles of FKBPs (and CyPs)?" They are involved not only in common physiological processes occurring in cells from human to bacteria, but also in specified processes occurring in highly differentiated cells of higher organisms. FK506 and rapamycin affect many of those functions because these drugs are able to bind with high affinity to most FKBPs. Solving the cellular functions of FKBPs will be critical for understanding in detail the mode of action of these drugs. Nonetheless, we cannot emphasize enough the usefulness of these macrolide compounds for clinical application and basic research. The availability of new macrolides will accelerate the pace of basic research as well as development of more specific and less toxic drugs.

References

1. Kino, T., and Goto, T. (1993). Discovery of FK506 and update. In "Immunomodulating Drugs" (V. Georgiev and H. Yamaguchi, Eds.), pp. 13–21. New York Academy of Science, New York.
2. Borel, J. F. (1989). Pharmacology of cyclosporine (sandimmune) IV. Pharmacological properties in vivo. Pharmacol. Rev. **41**, 259–371.
3. Takahashi, N. (1999). Pharmacodynamics of cyclosporin A. Clin. Exp. Nephrol. **3**, S16–S26.
4. Takahashi, N., Hayano, T., and Suzuki, M. (1989). Peptidyl-prolyl cis-trans isomerase is the cyclosporin A-binding protein cyclophilin. Nature **337**, 473–475.
5. Siekierka, J. J., Hung, S. H. Y., Poe, M., Lin, C. S., and Sigal, N. H. (1989). A cytosolic binding protein for the immunosuppressant FK506 has peptidyl-prolyl isomerase activity but is distinct from cyclophilin. Nature **341**, 755–757.
6. Liu, J., Farmer, Jr., J. D., Jane, W. S., Friedman, J., Weissman, I., and Schreiber, S. L. (1991). Calcineurin is a common target of cyclophilin–cyclosporin A and FKBP–FK506 complexes. Cell **66**, 807–815.
7. Timmerman, L. A., Clipstone, N. A., Ho, S. N., Northrop, J. P., and Crabtree, G. R. (1996). Rapid shuttling of NF-AT in discrimination of Ca^{2+} signals and immunosuppression. Nature **383**, 837–840.
8. Rao, A. (1994). NF-ATp: A transcription factor required for the co-ordinate induction of several cytokine genes. Immunol. Today **15**, 274–281.
9. Sharkey, J., and Butcher, S. P. (1994). Immunophilins mediate the neuroprotective effects of FK506 in focal cerebral ischaemia. Nature **371**, 336–339.
10. Steiner, J. P., Connolly, M. A., Valentine, H. L., Hamilton, G. S., Dawson, T. M., Hester, L., and Snyder, S. H. (1997). Neurotrophic actions of nonimmunosuppressive analogues of immunosuppressive drugs FK506, rapamycin and cyclosporin A. Nat. Med. **3**, 421–428.
11. Gold, B. G. (2000). Neuroimmunophilin ligands: Evaluation of their therapeutic potential for the treatment of neurological disorders. Exp. Opin. Invest. Drugs **9**, 2331–2342.
12. Snyder, S. H., Sabatini, D. M., Lai, M. M., Steiner, J. P., Hamilton, G. S., and Suzdak, P. D. (1998). Neural actions of immunophilin ligands. Trends Pharmacol. Sci. **19**, 21–26.
13. Gold, B. G., Zeleny-Pooley, M., Chaturvedi, P., and Wang, M. S. (1998). Oral administration of a nonimmunosuppressant FKBP-12 ligand speeds nerve regeneration. Neuroreport **9**, 553–558.
14. Gold, B. G. (1997). FK506 and the role of immunophilins in nerve regeneration. Mol. Neurobiol. **15**, 285–306.
15. Morioka, M., Hamada, J., Ushio, Y., and Miyamoto, E. (1999). Potential role of calcineurin for brain ischemia and traumatic injury. Progr. Neurobiol. **58**, 1–30.
16. Steiner, J. P., Hamilton, G. S., Ross, D. T., Valentine, H. L., Guo, H., Connolly, M. A., Liang, S., Ramsey, C., Li, J. H., Huang, W., Howorth, P., Soni, R., Fuller, M., Sauer, H., Nowotnik, A. C., and Suzdak, P. D. (1997). Neurotrophic immunophilin ligands stimulate structural and functional recovery in neurodegenerative animal models. Proc. Natl. Acad. Sci. USA **94**, 2019–2024.
17. Costantini, L. C., and Isacson, O. (2000). Immunophilin ligands and GDNF enhance neurite branching or elongation from developing dopamine neurons in culture. Exp. Neurol. **164**, 60–70.

18. Gold, B. G., Densmore, V., Shou, W., Matzuk, M. M., and Gordon, H. S. (1999). Immunophilin FK506-binding protein 52 (not FK506-binding protein 12) mediates the neurotrophic action of FK506. *J. Pharmacol. Exp. Ther.* **289**, 1202–1210.
19. Sehgal, S. N. (1998). Rapamune (RAPA, rapamycin, sirolimus): Mechanism of action immunosuppressive effect results from blockade of signal transduction and inhibition of cell cycle progression. *Clin. Biochem.* **31**, 335–340.
20. Chan, T. F., Carvalho, J., Riles, L., and Zheng, X. F. (2000). A chemical genomics approach toward understanding the global functions of the target of rapamycin protein (TOR). *Proc. Natl. Acad. Sci. USA* **97**, 13,227–13,232.
21. Cardenas, M. E., Cutler, N. S., Lorenz, M. C., Di Como, C. J., and Heitman, J. (1999). The TOR signaling cascade regulates gene expression in response to nutrients. *Genes Dev.* **13**, 3271–3279.
22. Sehgal, S. N., and Bansbach, C. C. (1993). Rapamycin: *In vitro* profile of a new immunosuppressive macrolide. *In* "Immunomodulating Drugs" (V. Georgiev and H. Yamaguchi, Eds.), pp. 58–67. New York Academy of Science, New York.
23. Meier-Kriesche, H. U., and Kaplan, B. (2000). Toxicity and efficacy of sirolimus: Relationship to whole-blood concentrations. *Clin. Ther.* Suppl. B, B93–B100.
24. Takahashi, N. (1998). The mechanism of action of rapamycin in the mTOR mediated signal transduction [in Japanese]. *Clin. Immunol.* **30**, 487–496.
25. Sabatini, D. M., Erdjument-Bromage, H., Lui, M., Tempst, P., and Snyder, S. H. (1994). RAFT1: A mammalian protein that binds to FKBP12 in a rapamycin-dependent fashion and is homologous to yeast TORs. *Cell* **78**, 35–43.
26. Beck, T., and Hall, M. N. (1999). The TOR signalling pathway controls nuclear localization of nutrient-regulated transcription factors. *Nature* **402**, 689–692.
27. Xu, G., Kwon, G., Marshall, C. A., Lin, T. A., Lawrence, J. C., and McDaniel, M. L. (1998). Branched-chain amino acids are essential in the regulation of PHAS-I and p70 S6 kinase by pancreatic beta-cells. A possible role in protein translation and mitogenic signaling. *J. Biol. Chem.* **273**, 28,178–28,184.
28. Schmelzle, T., and Hall, M. N. (2000). TOR, a central controller of cell growth. *Cell* **103**, 253–262.
29. Gingras, A. C., Gygi, S. P., Raught, B., Polakiewicz, R. D., Abraham, R. T., Hoekstra, M. F., Aebersold, R., and Sonenberg, N. (1999). Regulation of 4E-BP1 phosphorylation: A novel two-step mechanism. *Genes Dev.* **13**, 1422–1437.
30. Brunn, G. J., Hudson, C. C., Sekulic, A., Williams, J. M., Hosoi, H., Houghton, P. J., Lawrence, Jr., J. C., and Abraham, R. T. (1997). Phosphorylation of the translational repressor PHAS-I by the mammalian target of rapamycin. *Science* **277**, 99–101.
31. Tee, A. R., and Proud, C. G. (2000). DNA-damaging agents cause inactivation of translational regulators linked to mTOR signalling. *Oncogene* **19**, 3021–3031.
32. Hashemolhosseini, S., Nagamine, Y., Morley, S. J., Desrivieres, S., Mercep, L., and Ferrari, S. (1998). Rapamycin inhibition of the G1 to S transition is mediated by effects on cyclin D1 mRNA and protein stability. *J. Biol. Chem.* **273**, 14,424–14,429.
33. Kamada, Y., Funakoshi, T., Shintani, T., Nagano, K., Ohsumi, M., and Ohsumi, Y. (2000). TOR-mediated induction of autophagy via an Apg1 protein kinase complex. *J. Cell Biol.* **150**, 1507–1513.
34. Hosoi, H., Dilling, M. B., Liu, L. N., Danks, M. K., Shikata, T., Sekulic, A., Abraham, R. T., Lawrence, J. C., and Houghton, P. J. (1998). Studies on the mechanism of resistance to rapamycin in human cancer cells. *Mol. Pharmacol.* **54**, 815–824.
35. Hosoi, H., Dilling, M. B., Shikata, T., Liu, L. N., Shu, L., Ashmun, R. A., Germain, G. S., Abraham, R. T., and Houghton, P. J. (1999). Rapamycin causes poorly reversible inhibition of mTOR and induces p53-independent apoptosis in human rhabdomyosarcoma cells. *Cancer Res.* **59**, 886–894.

36. Parker, E. M., Monopoli, A., Ongini, E., Lozza, G., and Babij, C. M. (2000). Rapamycin, but not FK506 and GPI-1046, increases neurite outgrowth in PC12 cells by inhibiting cell cycle progression. *Neuropharmacology* **39**, 1913–1919.
37. Freedman, R. B. (1989). A protein with many functions? *Nature* **337**, 407–408.
38. Fischer, G., Wittmann-Liebold, B., Lang, K., Kiefhaber, T., and Schmid, F. X. (1989). Cyclophilin and peptidyl-prolyl *cis-trans* isomerase are probably identical proteins. *Nature* **337**, 476–478.
39. Parson, W. H., Sigal, N. H., and Wyvratt, M. J. (1993). FK506—A novel immunosuppressant. *In* "Immunomodulating Drugs" (V. Georgiev and H. Yamaguchi, Eds.), pp. 22–36. New York Academy of Science, New York.
40. Maki, N., Sekiguchi, F., Nishimaki, J., Miwa, K., Hayano, T., Takahashi, N., and Suzuki, M. (1990). Complementary DNA encoding the human T-cell FK506-binding protein, a peptidyl-prolyl *cis-trans* isomerase distinct from cyclophilin. *Proc. Natl. Acad. Sci. USA* **87**, 5440–5443.
41. Fesik, S. W., Gampe, R. T., Holzman, T. F., Egan, D. A., Edalji, R., Luly, J. R., Simmer, R., Helfrich, R., Kioshore, V., and Rich, D. H. (1990). Isotope-edited NMR of cyclosporin A bound to cyclophilin: Evidence for a *trans* 9,10 amide bond. *Science* **250**, 1406–1409.
42. Michnick, S. W., Rosen, M. K., Wandless, T. J., Karplus, M., and Schreiber, S. L. (1991). Solution structure of FKBP, a rotamase enzyme and receptor for FK506 and rapamycin. *Science* **252**, 836–839.
43. Rosen, M. K., Standaert, R. F., Galat, A., Nakatsuka, M., and Schreiber, S. L. (1990). Inhibition of FKBP rotamase activity by immunosuppressant FK506: Twisted amide surrogate. *Science* **248**, 863–866.
44. Choi, J., Chen, J., Schreiber, S. L., and Clardy, J. (1996). Structure of the FKBP12-rapamycin complex interacting with the binding domain of human FRAP. *Science* **273**, 239–242.
45. Harrison, R. K., and Stein, R. L. (1990). Substrate specificities of the peptidyl prolyl *cis-trans* isomerase activities of cyclophilin and FK506-binding protein: Evidence for the existence of a family of distinct enzymes. *Biochemistry* **29**, 3813–3816.
46. Matouschek, A., Rospert, S., Schmid, K., Glick, B. S., and Schatz, G. (1995). Cyclophilin catalyzes protein folding in yeast mitochondria. *Proc. Natl. Acad. Sci. USA* **92**, 6319–6323.
47. Tyedmers, J., Kruse, M., Lerner, M., Demand, J., Hohfeld, J., Solsbacher, J., Volkmer, J., and Zimmermann, R. (2000). Assembly of heterodimeric luciferase after *de novo* synthesis of subunits in rabbit reticulocyte lysate involves hsc70 and hsp40 at a post-translational stage. *Eur. J. Biochem.* **267**, 3575–3582.
48. Lang, K., Schmid, F. X., and Fischer, G. (1987). Catalysis of protein folding by prolyl isomerase. *Nature* **329**, 268–270.
49. Pennisi, E. (1996). Expanding the eukaryote's cast of chaperones. *Science* **274**, 1613–1614.
50. Shieh, B.-H., Stamnes, M. A., Seavello, S., Harris, G. L., and Zuker, C. S. (1989). The nineA gene required for visual transduction in *Drosophila* encodes a homologue of cyclosporin A-binding protein. *Nature* **338**, 67–70.
51. Ferreira, P. A., Nakayama, T. A., Pak, W. L., and Travis, G. H. (1996). Cyclophilin-related protein RanBP2 acts as chaperone for red/green opsin. *Nature* **383**, 637–640.
52. Thali, M., Bukovsky, A., Kondo, E., Rosenwirth, B., Walsh, C. T., Sodroski, J., and Gottlinger, H. G. (1994). Functional association of cyclophilin A with HIV-1 virions. *Nature* **372**, 363–365.
53. McCornack, M. A., Kakalis, L. T., Caserta, C., Handschumacher, R. E., and Armitage, I. M. (1997). HIV protease substrate conformation: Modulation by cyclophilin A. *FEBS Lett.* **414**, 84–88.
54. Streblow, D. N., Kitabwalla, M., Malkovsky, M., and Pauza, C. D. (1998). Cyclophilin A modulates processing of human immunodeficiency virus type 1 p55Gag: Mechanism for antiviral effects of cyclosporin A. *Virology* **245**, 197–202.

55. Cruciani, V., Kaalhus, O., and Mikalsen, S. O. (1999). Phosphatases involved in modulation of gap junctional intercellular communication and dephosphorylation of connexin43 in hamster fibroblasts: 2B or not 2B? *Exp. Cell Res.* **252**, 449–463.
56. Lopez-Ilasaca, M., Schiene, C., Kullertz, G., Tradler, T., Fischer, G., and Wetzker, R. (1998). Effects of FK506-binding protein 12 and FK506 on autophosphorylation of epidermal growth factor receptor. *J. Biol. Chem.* **273**, 9430–9434.
57. Mamane, Y., Sharma, S., Petropoulos, L., Lin, R., and Hiscott, J. (2000). Posttranslational regulation of IRF-4 activity by the immunophilin FKBP52. *Immunity* **12**, 129–140.
58. Arevalo-Rodriguez, M., Cardenas, M. E., Wu, X., Hanes, S. D., and Heitman, J. (2000). Cyclophilin A and Ess1 interact with and regulate silencing by the Sin3-Rpd3 histon deacetylase. *EMBO J.* **19**, 3739–3749.
59. Rahfeld, J. U., Rucknagel, K. P., Schelbert, B., Ludwig, B., Hacker, J., Mann, K., and Fischer, G. (1994). Confirmation of the existence of a third family among peptidyl prolyl *cis-trans* isomerases. Amino acid sequence and recombinant production of parvulin. *FEBS Lett.* **352**, 180–184.
60. Hennig, L., Christner, C., Kipping, M., Schelbert, B., Rucknagel, K. P., Grabley, S., Kullertz, G., and Fischer, G. (1998). Selective inactivation of parvulin-like peptidyl-prolyl *cis/trans* isomerase by juglon. *Biochemistry* **37**, 5953–5960.
61. Dolinski, K., Muri, S., Cardenas, M., and Heitman, J. (1997). Cyclophilins and FKBPs are dispensible for viability in *Saccharomyces cerevisiae*. *Proc. Natl. Acad. Sci. USA* **94**, 13,093–13,098.
62. Lu, K. P., Hanes, S. D., and Hunter, T. (1996). A human peptidyl-prolyl isomerase essential for regulation of mitosis. *Nature* **380**, 544–547.
63. Yaffe, M. B., Schutkowski, M., Shen, M., Zhou, X. Z., Stukenberg, P. T., Rahfeld, J.-U., Xu, J., Kuang, J., Kischner, M. W., Fischer, G., Cantley, L. C., and Lu, K. P. (1997). Sequence-specific and phosphorylation-dependent proline isomerization: A potential mitotic regulatory mechanism. *Science* **278**, 1957–1960.
64. Vogel, G. (1997). Pinning down cell division. *Science* **278**, 1883–1884.
65. Zhou, X. Z., Kops, O., Werner, A., Lu, P.-J., Shen, M., Stoller, G., Kullertz, G., Stark, M., Fischer, G., and Lu, K. P. (2000). Pin1-dependent prolyl isomerization regulates dephosphorylation of cdc25C and tau protein. *Mol. Cell* **6**, 873–883.
66. Wu, X., Wilcox, C. B., Devasahayam, G., Hackett, R. L., Arevalo-Rodriguez, M., Cardenas, M. E., Heitman, J., and Hanes, S. D. (2000). The Ess1 prolyl isomerase is linked to chromatin remodeling complexes and the general transcription machinery. *EMBO J.* **19**, 3727–3738.
67. Lu, P.-J., Wulf, G., Zhou, X. Z., Davies, P., and Lu, K. P. (1999). The prolyl isomerase Pin1 restores the function of Alzheimer-associated phosphorylated tau protein. *Nature* **399**, 784–788.
68. (a) Takahashi, N. (1990). Mechanism of action of immunosuppressant receptors [in Japanese]. *Med. Immunol.* **20**, 59-67. (b) Suzuki, M. (1989). SPXX, a frequent sequence motif in gene regulatory proteins. *J. Mol. Biol.* **207**, 61–84.
69. Lam, E., Martin, M. M., Timerman, A. P., Sabers, C., Fleischer, S., Lukas, T., Abraham, R. T., O'Keefe, S. J., O'Neill, E. A., and Wiederrecht, G. J. (1995). A novel FK506 binding protein can mediate the immunosuppressive effects of FK506 and is associated with the cardiac ryanodine receptor. *J. Biol. Chem.* **270**, 26,511–26,522.
70. Brillantes, A. B., Ondrias, K., Scott, A., Kobrinsky, E., Ondriasova, E., Moschella, M. C., Jayaraman, T., Landers, M., Ehrlich, B. E., and Marks, A. R. (1994). Stabilization of calcium release channel (ryanodine receptor) function by FK506-binding protein. *Cell* **77**, 513–523.
71. Marx, S. O., Ondrias, K., and Marks, A. R. (1998). Coupled gating between individual skeletal muscle Ca^{2+} release channels (ryanodine receptors). *Science* **281**, 818–821.
72. Marks, A. R. (1997). Intracellular calcium-release channels: Regulators of cell life and death. *Am. J. Physiol.* **272**, H597–H605.

73. Shou, W., Aghdasi, B., Armstrong, D. L., Guo, Q., Bao, S., Charng, M. J., Mathews, L. M., Schneider, M. D., Hamilton, S. L., and Matzuk, M. M. (1998). Cardiac defects and altered ryanodine receptor function in mice lacking FKBP12. *Nature* **391**, 489–492.
74. Cameron, A. M., Steiner, J. P., Sabatini, D. M., Kaplin, A. I., Walensky, L. D., and Snyder, S. H. (1995). Immunophilin FK506 binding protein associated with inositol 1,4,5-trisphosphate receptor modulates calcium flux. *Proc. Natl. Acad. Sci. USA* **92**, 1784–1788.
75. Cameron, A. M., Nucifora, Jr., F. C, Fung, E. T., Livingston, D. J., Aldape, R. A., Ross, C. A., and Snyder, S. H. (1997). FKBP12 binds the inositol 1,4,5-trisphosphate receptor at leucine-proline (1400–1401) and anchors calcineurin to this FK506-like domain. *J. Biol. Chem.* **272**, 27,582–27,588.
76. Cameron, A. M., Steiner, J. P., Roskams, A. J., Ali, S. M., Ronnett, G. V., and Snyder, S. H. (1995). Calcineurin associated with the inositol 1,4,5-trisphosphate receptor–FKBP12 complex modulates Ca^{2+} flux. *Cell* **83**, 463–472.
77. Xin, H. B., Rogers, K., Qi, Y., Kanematsu, T., and Fleischer, S. (1999). Three amino acid residues determine selective binding of FK506-binding protein 12.6 to the cardiac ryanodine receptor. *J. Biol. Chem.* **274**, 15,315–15,319.
78. Noguchi, N., Takasawa, S., Nata, K., Tohgo, A., Kato, I., Ikehata, F., Yonekura, H., and Okamoto, H. (1997). Cyclic ADP-ribose binds to FK506-binding protein 12.6 to release Ca^{2+} from islet microsomes. *J. Biol. Chem.* **272**, 3133–3136.
79. Marx, S. O., Reiken, S., Hisamatsu, Y., Jayaraman, T., Burkhoff, D., Rosemblit, N., and Marks, A. R. (2000). PKA phosphorylation dissociates FKBP12.6 from the calcium release channel (ryanodine receptor): Defective regulation in failing hearts. *Cell* **101**, 365–376.
80. Timerman, A. P., Wiederrecht, G., Marcy, A., and Fleischer, S. (1995). Characterization of an exchange reaction between soluble FKBP-12 and the FKBP–ryanodine receptor complex. Modulation by FKBP mutants deficient in peptidyl-prolyl isomerase activity. *J. Biol. Chem.* **270**, 2451–2459.
81. Scholz, C., Schindler, T., Dolinski, K., Heitman, J., and Schmid, F. X. (1997). Cyclophilin active site mutants have native prolyl isomerase activity with a protein substrate. *FEBS Lett.* **414**, 69–73.
82. Tai, P. K., Albers, M. W., Chang, H., Faber, L. E., and Schreiber, S. L. (1992). Association of a 59-kilodalton immunophilin with the glucocorticoid receptor complex. *Science* **256**, 1315–1318.
83. Ratajczak, T., Carrello, A., Mark, P. J., Warner, B. J., Simpson, R. J., Moritz, R. L., and House, A. K. (1993). The cyclophilin component of the unactivated estrogen receptor contains a tetratricopeptide repeat domain and shares identity with p59 (FKBP59). *J. Biol. Chem.* **268**, 13,187–13,192.
84. Tai, P. K., Albers, M. W., McDonnell, D. P., Chang, H., Schreiber, S. L., and Faber, L. E. (1994). Potentiation of progesterone receptor-mediated transcription by the immunosuppressant FK506. *Biochemistry* **33**, 10,666–10,671.
85. Ning, Y. M., and Sanchez, E. R. (1993). Potentiation of glucocorticoid receptor-mediated gene expression by the immunophilin ligands FK506 and rapamycin. *J. Biol. Chem.* **268**, 6073–6076.
86. Renoir, J. M., Mercier-Bodard, C., Hoffmann, K., Le Bihan, S., Ning, Y. M., Sanchez, E. R., Handschumacher, R. E., and Baulieu, E. E. (1995). Cyclosporin A potentiates the dexamethasone-induced mouse mammary tumor virus-chloramphenicol acetyltransferase activity in LMCAT cells: A possible role for different heat shock protein-binding immunophilins in glucocorticosteroid receptor-mediated gene expression. *Proc. Natl. Acad. Sci. USA* **92**, 4977–4981.
87. Pratt, W. B., Silverstein, A. M., and Galigniana, M. D. (1999). A model for the cytoplasmic trafficking of signalling proteins involving the hsp90-binding immunophilins and p50cdc37. *Cell Signal* **11**, 839–851.
88. Silverstein, A. M., Galigniana, M. D., Kanelakis, K. C., Radanyi, C., Renoir, J. M., and Pratt, W. B. (1999). Different regions of the immunophilin FKBP52 determine its association with the glucocorticoid receptor, hsp90, and cytoplasmic dynein. *J. Biol. Chem.* **274**, 36,980–36,986.

89. Carrello, A., Ingley, E., Minchin, R. F., Tsai, S., and Ratajczak, T. (1999). The common tetratricopeptide repeat acceptor site for steroid receptor-associated immunophilins and Hop is located in the dimerization domain of Hsp90. *J. Biol. Chem.* **274**, 2682–2689.
90. Taylor, P., Dornan, J., Carrello, A., Minchin, R. F., Ratajczak, T., and Walkinshaw, M. D. (2001). Two structures of cyclophilin 40. Folding and fidelity in the tpr domains. *Structure (Camb.)* **9**, 431–438.
91. Le Bihan, S., Marsaud, V., Mercier-Bodard, C., Baulieu, E. E., Mader, S., White, J. H., and Renoir, J. M. (1998). Calcium/calmodulin kinase inhibitors and immunosuppressant macrolides rapamycin and FK506 inhibit progestin- and glucocorticosteroid receptor-mediated transcription in human breast cancer T47D cells. *Mol. Endocrinol.* **12**, 986–1001.
92. Reynolds, P. D., Ruan, Y., Smith, D. F., and Scammell, J. G. (1999). Glucocorticoid resistance in the squirrel monkey is associated with overexpression of the immunophilin FKBP51. *J. Clin. Endocrinol. Metab.* **84**, 663–669.
93. Mark, P. J., Ward, B. K., Kumar, P., Lahooti, H., Minchin, R. F., and Ratajczak, T. (2001). Human cyclophilin 40 is a heat shock protein that exhibits altered intracellular localization following heat shock. *Cell Stress Chaperones* **6**, 59–70.
94. Kumar, P., Ward, B. K., Minchin, R. F., and Ratajczak, T. (2001). Regulation of the Hsp90-binding immunophilin, cyclophilin 40, is mediated by multiple sites for GA-binding protein (GABP). *Cell Stress Chaperones* **6**, 78–91.
95. Kumar, P., Mark, P. J., Ward, B. K., Minchin, R. F., and Ratajczak, T. (2001). Estradiol-regulated expression of the immunophilins cyclophilin 40 and FKBP52 in MCF-7 breast cancer cells. *Biochem. Biophys. Res. Commun.* **284**, 219–225.
96. Lebeau, M. C., Jung-Testas, I., and Baulieu, E. E. (1999). Intracellular distribution of a cytoplasmic progesterone receptor mutant and of immunophilins cyclophilin 40 and FKBP59: Effects of cyclosporin A, of various metabolic inhibitors and of several culture conditions. *J. Steroid Biochem. Mol. Biol.* **70**, 219–228.
97. Migita, K., Tanaka, H., Okamoto, K., Yoshikawa, N., Ichinose, Y., Urayama, S., Yamasaki, S., Ida, H., Kawabe, Y., Kawakami, A., and Eguchi, K. (2000). FK506 augments glucocorticoid-mediated cyclooxygenase-2 down-regulation in human rheumatoid synovial fibroblasts. *Lab. Invest.* **80**, 135–141.
98. Carrello, A., Ingley, E., Minchin, R. F., Tsai, S., and Ratajczak, T. (1999). The common tetratricopeptide repeat acceptor site for steroid receptor-associated immunophilins and Hop is located in the dimerization domain of Hsp90. *J. Biol. Chem.* **274**, 2682–2689.
99. Knoblauch, R., and Garabedian, M. J. (1999). Role for Hsp90-associated cochaperone p23 in estrogen receptor signal transduction. *Mol. Cell. Biol.* **19**, 3748–3759.
100. Chen, S., Sullivan, W. P., Toft, D. O., and Smith, D. F. (1998). Differential interactions of p23 and the TPR-containing proteins Hop, Cyp40, FKBP52 and FKBP51 with Hsp90 mutants. *Cell Stress Chaperones* **3**, 118–129.
101. Graumann, K., and Jungbauer, A. (2000). Quantitative assessment of complex formation of nuclear-receptor accessory proteins. *Biochem. J.* **345**, 627–636.
102. Miyata, Y., Chambraud, B., Radanyi, C., Leclerc, J., Lebeau, M. C., Renoir, J. M., Shirai, R., Catelli, M. G., Yahara, I., and Baulieu, E. E. (1997). Phosphorylation of the immunosuppressant FK506-binding protein FKBP52 by casein kinase II: Regulation of HSP90-binding activity of FKBP52. *Proc. Natl. Acad. Sci. USA* **94**, 14,500–14,505.
103. Wang, T., Donahoe, P. K., and Zervos, A. S. (1994). Specific interaction of type I receptors of the TGF-beta family with the immunophilin FKBP-12. *Science* **265**, 674–676.
104. Wang, T., Li, B. Y., Danielson, P. D., Shah, P. C., Rockwell, S., Lechleider, R. J., Martin, J., Manganaro, T., and Donahoe, P. K. (1996). The immunophilin FKBP12 functions as a common inhibitor of the TGF beta family type I receptors. *Cell* **86**, 435–444.
105. Huse, M., Chen, Y. G., Massague, J., and Kuriyan, J. (1999). Crystal structure of the cytoplasmic domain of the type I TGF beta receptor in complex with FKBP12. *Cell* **96**, 425–436.

106. Chen, Y. G., Liu, F., and Massague, J. (1997). Mechanism of TGF beta receptor inhibition by FKBP12. *EMBO J.* **16**, 3866–3876.
107. Razani, B., Zhang, X. L., Bitzer, M., von Gersdorff, G., Bottinger, E. P., and Lisanti, M. P. (2001). Caveolin-1 regulates transforming growth factor (TGF)-beta/SMAD signaling through an interaction with the TGF-beta type I receptor. *J. Biol. Chem.* **276**, 6727–6738.
108. Bassing, C. H., Shou, W., Muir, S., Heitman, J., Matzuk, M. M., and Wang, X. F. (1998). FKBP12 is not required for the modulation of transforming growth factor beta receptor I signaling activity in embryonic fibroblasts and thymocytes. *Cell Growth Differ.* **9**, 223–228.
109. Yao, D., Dore, J. J., and Leof, E. B. (2000). FKBP12 is a negative regulator of transforming growth factor-beta receptor internalization. *J. Biol. Chem.* **275**, 13,149–13,154.
110. Galat, A. (2000). Sequence diversification of the FK506-binding proteins in several different genomes. *Eur. J. Biochem.* **267**, 4945–4959.
111. Jin, Y. J., Albers, M. W., Lane, W. S., Bierer, B. E., Schreiber, S. L., and Burakoff, S. J. (1991). Molecular cloning of a membrane-associated human FK506- and rapamycin-binding protein, FKBP-13. *Proc. Natl. Acad. Sci. USA* **88**, 6677–6681.
112. Walensky, L. D., Gascard, P., Fields, M. E., Blackshaw, S., Conboy, J. G., Mohandas, N., and Snyder, S. H. (1998). The 13-kD FK506 binding protein, FKBP13, interacts with a novel homologue of the erythrocyte membrane cytoskeletal protein 4.1. *J. Cell Biol.* **141**, 143–153.
113. Nakamura, T., Yabe, D., Kanazawa, N., Tashiro, K., Sasayama, S., and Honjo, T. (1998). Molecular cloning, characterization, and chromosomal localization of FKBP23, a novel FK506-binding protein with Ca^{2+}-binding ability. *Genomics* **54**, 89–98.
114. Galat, A., Lane, W. S., Standaert, R. F., and Schreiber, S. L. (1992). A rapamycin-selective 25-kDa immunophilin. *Biochemistry* **31**, 2427–2434.
115. (a) Riviere, S., Menez, A., and Galat, A. (1993). On the localization of FKBP25 in T-lymphocytes. *FEBS Lett.* **315**, 247–251. (b) Jin, Y. J., and Burakoff, S. J. (1993). The 25-kDa FK506-binding protein is localized in the nucleus and associates with casein kinase II and nucleolin. *Proc. Natl. Acad. Sci. USA* **90**, 7769–7773.
116. Yang, W. M., Yao, Y. L., and Seto, E. (2001). The FK506-binding protein 25 functionally associates with histone deacetylases and with transcription factor YY1. *EMBO J.* **20**, 4814–4825.
117. Ahn, J., Murphy, M., Kratowicz, S., Wang, A., Levine, A. J., and George, D. L. (1999). Downregulation of the stathmin/Op18 and FKBP25 genes following p53 induction. *Oncogene* **18**, 5954–5958.
118. Johnson, K. L., and Lawen, A. (1999). Rapamycin inhibits didemnin B-induced apoptosis in human HL-60 cells: Evidence for the possible involvement of FK506-binding protein 25. *Immunol. Cell Biol.* **77**, 242–248.
119. Wiederrecht, G., Hung, S., Chan, H. K., Marcy, A., Martin, M., Calaycay, J., Boulton, D., Sigal, N., Kincaid, R. L., and Siekierka, J. J. (1992). Characterization of high molecular weight FK-506 binding activities reveals a novel FK-506-binding protein as well as a protein complex. *J. Biol. Chem.* **267**, 21,753–21,760.
120. Baughman, G., Wiederrecht, G. J., Campbell, N. F., Martin, M. M., and Bourgeois, S. (1995). FKBP51, a novel T-cell-specific immunophilin capable of calcineurin inhibition. *Mol. Cell. Biol.* **15**, 4395–4402.
121. Nair, S. C., Rimerman, R. A., Toran, E. J., Chen, S., Prapapanich, V., Butts, R. N., and Smith, D. F. (1997). Molecular cloning of human FKBP51 and comparisons of immunophilin interactions with Hsp90 and progesterone receptor. *Mol. Cell. Biol.* **17**, 594–603.
122. Peattie, D. A., Harding, M. W., Fleming, M. A., DeCenzo, M. T., Lippke, J. A., Livingston, D. J., and Benasutti, M. (1992). Expression and characterization of human FKBP52, an immunophilin that associates with the 90-kDa heat shock protein and is a component of steroid receptor complexes. *Proc. Natl. Acad. Sci. USA* **89**, 10,974–10,978.
123. Neye, H. (2001). Mutation of FKBP associated protein 48 (FAP48) at proline 219 disrupts the interaction with FKBP12 and FKBP52. *Regul. Pept.* **97**, 147–152.

124. Kieffer, L. J., Seng, T. W., Li, W., Osterman, D. G., Handschumacher, R. E., and Bayney, R. M. (1993). Cyclophilin-40, a protein with homology to the P59 component of the steroid receptor complex. Cloning of the cDNA and further characterization. *J. Biol. Chem.* **268**, 12,303–12,310.
125. Sananes, N., Baulieu, E. E., and Le Goascogne, C. (1998). Stage-specific expression of the immunophilin FKBP59 messenger ribonucleic acid and protein during differentiation of male germ cells in rabbits and rats. *Biol. Reprod.* **58**, 353–360.
126. Chambraud, B., Radanyi, C., Camonis, J. H., Rajkowski, K., Schumacher, M., and Baulieu, E. E. (1999). Immunophilins, Refsum disease, and lupus nephritis: The peroxisomal enzyme phytanoyl-COA alpha-hydroxylase is a new FKBP-associated protein. *Proc. Natl. Acad. Sci. USA* **96**, 2104–2109.
127. Shadidy, M., Caubit, X., Olsen, R., Seternes, O. M., Moens, U., and Krauss, S. (1999). Biochemical analysis of mouse FKBP60, a novel member of the FKBP family. *Biochim. Biophys. Acta* **1446**, 295–307.
128. Coss, M. C., Winterstein, D., Sowder, II, R. C., and Simek, S. L. (1995). Molecular cloning, DNA sequence analysis, and biochemical characterization of a novel 65-kDa FK506-binding protein (FKBP65). *J. Biol. Chem.* **270**, 29,336–29,341.
129. Coss, M. C., Stephens, R. M., Morrison, D. K., Winterstein, D., Smith, L. M., and Simek, S. L. (1998). The immunophilin FKBP65 forms an association with the serine/threonine kinase c-Raf-1. *Cell Growth Differ.* **9**, 41–48.
130. Davis, E. C., Broekelmann, T. J., Ozawa, Y., and Mecham, R. P. (1998). Identification of tropoelastin as a ligand for the 65-kD FK506-binding protein, FKBP65, in the secretory pathway. *J. Cell Biol.* **140**, 295–303.
131. Zeng, B., MacDonald, J. R., Bann, J. G., Beck, K., Gambee, J. E., Boswell, B. A., and Bachinger, H. P. (1998). Chicken FK506-binding protein, FKBP65, a member of the FKBP family of peptidyl prolyl *cis-trans* isomerases, is only partially inhibited by FK506. *Biochem. J.* **330**, 109–114.
132. Patterson, C. E., Schaub, T., Coleman, E. J., and Davis, E. C. (2000). Developmental regulation of FKBP65. An ER-localized extracellular matrix binding-protein. *Mol. Biol. Cell* **11**, 3925–3935.
133. Fukuda, K., Tanigawa, Y., Fujii, G., Yasugi, S., and Hirohashi, S. (1998). cFKBP/SMAP: A novel molecule involved in the regulation of smooth muscle differentiation. *Development* **125**, 3535–3542.
134. Cunningham, E. B. (1999). An inositolphosphate-binding immunophilin, IPBP12. *Blood* **94**, 2778–2789.
135. Milan, D., Griffith, J., Su, M., Price, E. R., and McKeon, F. (1994). The latch region of calcineurin B is involved in both immunosuppressant–immunophilin complex docking and phosphatase activation. *Cell* **79**, 437–447.
136. Altschuh, D., Vix, O., Rees, B., and Thierry, J.-C. A. (1992). Conformation of cyclosporin A in aqueous environment revealed by the X-ray structure of a cyclosporin–Fab complex. *Science* **256**, 92–94.
137. Schreiber, S. L. (1991). Chemistry and biology of the immunophilins and their immunosuppressive ligands. *Science* **251**, 283–287.
138. Kissinger, C. R., Parge, H. E., Knighton, D. R., Lewis, C. T., Pelletier, L. A., Tempczyk, A., Kalish, V. J., Tucker, K. D., Showalter, R. E., and Moomaw, E. W. (1995). Crystal structures of human calcineurin and the human FKBP12–FK506–calcineurin complex. *Nature* **378**, 641–644.
139. Ivery, M. T. (1999). A proposed molecular model for the interaction of calcineurin with the cyclosporin A–cyclophilin A complex. *Bioorg. Med. Chem.* **7**, 1389–1402.
140. Salowe, S. P., and Hermes, J. D. (1998). Competitive and slow-binding inhibition of calcineurin by drug x immunophilin complexes. *Arch. Biochem. Biophys.* **355**, 165–174.

141. Cardenas, M. E., Hemenway, C., Muir, R. S., Ye, R., Fiorentino, D., and Heitman, J. (1994) Immunophilins interact with calcineurin in the absence of exogenous immunosuppressive ligands. *EMBO J.* **13**, 5944–5957.
142. Lorenz, M. C., and Heitman, J. (1995). TOR mutations confer rapamycin resistance by preventing interaction with FKBP12-rapamycin. *J. Biol. Chem.* **270**, 27,531–27,537.
143. Sabers, C. J., Martin, M. M., Brunn, G. J., Williams, J. M., Dumont, F. J., Wiederrecht, G., and Abraham, R. T. (1995). Isolation of a protein target of the FKBP12-rapamycin complex in mammalian cells. *J. Biol. Chem.* **270**, 815–822.
144. Ye, X., Rivera, V. M., Zoltick, P., Cerasoli, F., Schnell, M. A., Gao, G., Hughes, J. V., Gilman, M., and Wilson, J. M. (1999). Regulated delivery of therapeutic proteins after *in vivo* somatic cell gene transfer. *Science* **283**, 88–91.
145. Vilella-Bach, M., Nuzzi, P., Fang, Y., and Chen, J. (1999). The FKBP12–rapamycin-binding domain is required for FKBP12–rapamycin-associated protein kinase activity and G1 progression. *J. Biol. Chem.* **274**, 4266–4272.
146. Zheng, X. F., Florentino, D., Chen, J., Crabtree, G. R., and Schreiber, S. L. (1995). TOR kinase domains are required for two distinct functions, only one of which is inhibited by rapamycin. *Cell* **82**, 121–130.
147. Burnett, P. E., Barrow, R. K., Cohen, N. A., Snyder, S. H., and Sabatini, D. M. (1998). RAFT1 phosphorylation of the translational regulators p70 S6 kinase and 4E-BP1. *Proc. Natl. Acad. Sci. USA* **95**, 1432–1437.
148. Kumar, V., Pandey, P., Sabatini, D., Kumar, M., Majumder, P. K., Bharti, A., Carmichael, G., Kufe, D., and Kharbanda, S. (2000). Functional interaction between RAFT1/FRAP/mTOR and protein kinase C delta in the regulation of cap-dependent initiation of translation. *EMBO J.* **19**, 1087–1097.
149. Parekh, D., Ziegler, W., Yonezawa, K., Hara, K., and Parker, P. J. (1999). Mammalian TOR controls one of two kinase pathways acting upon nPKC delta and nPKC epsilon. *J. Biol. Chem.* **274**, 34,758–34,764.
150. England, K., Watson, J., Beale, G., Warner, M., Cross, J., and Rumsby, M. (2001). Signalling pathways regulating the dephosphorylation of Ser729 in the hydrophobic domain of protein kinase C epsilon upon cell passage. *J. Biol. Chem.* **276**, 10,437–10,442.
151. Yokogami, K., Wakisaka, S., Avruch, J., and Reeves, S. A. (2000). Serine phosphorylation and maximal activation of STAT3 during CNTF signaling is mediated by the rapamycin target mTOR. *Curr. Biol.* **10**, 47–50.
152. Lee-Fruman, K. K., Kuo, C. J., Lippincott, J., Terada, N., and Blenis, J. (1999). Characterization of S6K2, a novel kinase homologous to S6K1. *Oncogene* **18**, 5108–5114.
153. Brunn, G. J., Fadden, P., Haystead, T. A., and Lawrence, Jr., J. C. (1997). The mammalian target of rapamycin phosphorylates sites having a (Ser/Thr)-Pro motif and is activated by antibodies to a region near its COOH terminus. *J. Biol. Chem.* **272**, 32,547–32,550.
154. Gingras, A. C., Gygi, S. P., Raught, B., Polakiewicz, R. D., Abraham, R. T., Hoekstra, M. F., Aebersold, R., and Sonenberg, N. (1999). Regulation of 4E-BP1 phosphorylation: A novel two-step mechanism. *Genes Dev.* **13**, 1422–1437.
155. Mothe-Satney, I., Yang, D., Fadden, P., Haystead, T. A., and Lawrence, Jr., J. C. (2000). Multiple mechanisms control phosphorylation of PHAS-I in five (S/T)P sites that govern translational repression. *Mol. Cell. Biol.* **20**, 3558–3567.
156. Mothe-Satney, I., Brunn, G. J., McMahon, L. P., Capaldo, C. T., Abraham, R. T., and Lawrence, Jr., J. C. (2000). Mammalian target of rapamycin-dependent phosphorylation of PHAS-I in four (S/T)P sites detected by phospho-specific antibodies. *J. Biol. Chem.* **275**, 33,836–33,843.
157. Sabatini, D. M., Barrow, R. K., Blackshaw, S., Burnett, P. E., Lai, M. M., Field, M. E., Bahr, B. A., Kirsch, J., Betz, H., and Snyder, S. H. (1999). Interaction of RAFT1 with gephyrin required for rapamycin-sensitive signaling. *Science* **284**, 1161–1164.

158. Denning, G., Jamieson, L., Maquat, L. E., Thompson, E. A., and Fields, A. P. (2001). Cloning of a novel phosphatidylinositol kinase-related kinase: Characterization of the human SMG-1 RNA surveillance protein. *J. Biol. Chem.* **276**, 22,709–22,714.
159. Nave, B. T., Ouwens, M., Withers, D. J., Alessi, D. R., and Shepherd, P. R. (1999). Mammalian target of rapamycin is a direct target for protein kinase B: Identification of a convergence point for opposing effects of insulin and amino-acid deficiency on protein translation. *Biochem. J.* **344**, Suppl. B, 427–431.
160. Kufe, D., Kharbanda, S., Kumar, V., Sabatini, D., Pandey, P., Gingras, A. C., Majumder, P. K., Kumar, M., Yuan, Z. M., Carmichael, G., Weichselbaum, R., and Sonenberg, N. (2000). Regulation of the rapamycin and FKBP-target 1/mammalian target of rapamycin and cap-dependent initiation of translation by the c-Abl protein-tyrosine kinase. *J. Biol. Chem.* **275**, 10,779–10,787.
161. Brunn, G. J., Williams, J., Sabers, C., Wiederrecht, G., Lawrence, Jr., J. C., and Abraham, R. T. (1996). Direct inhibition of the signaling functions of the mammalian target of rapamycin by the phosphoinositide 3-kinase inhibitors, wortmannin and LY294002. *EMBO J.* **15**, 5256–5267.
162. (a) Gingras, A. C., Kennedy, S. G., O'Leary, M. A., Sonenberg, N., and Hay, N. (1998). 4E-BP1, a repressor of mRNA translation, is phosphorylated and inactivated by the Akt(PKB) signaling pathway. *Genes Dev.* **12**, 502–513. (b) Peterson, R. T., Beal, P. A., Comb, M. J., and Schreiber, S. L. (2000). FKBP12–rapamycin-associated protein (FRAP) autophosphorylates at serine 2481 under translationally repressive conditions. *J. Biol. Chem.* **275**, 7416–7423.
163. Sekulic, A., Hudson, C. C., Homme, J. L., Yin, P., Otterness, D. M., Karnitz, L. M., and Abraham, R. T. (2000). A direct linkage between the phosphoinositide 3-kinase-AKT signaling pathway and the mammalian target of rapamycin in mitogen-stimulated and transformed cells. *Cancer Res.* **60**, 3504–3513.
164. Crabtree, G. R. (1989). Contingent genetic regulatory events in T lymphocyte activation. *Science* **243**, 355–361.
165. Emmel, E. A., Verweij, C. L., Durand, D. B., Higgins, K. M., Lacy, E., and Crabtree, G. R. (1989). Cyclosporin A specifically inhibits function of nuclear proteins involved in T cell activation. *Science* **246**, 1617–1620.
166. Brunner, T., Yoo, N. J., LaFace, D., Ware, C. F., and Green, D. R. (1996). Activation-induced cell death in murine T cell hybridomas. Differential regulation of Fas (CD95) versus Fas ligand expression by cyclosporin A and FK506. *Int. Immunol.* **8**, 1017–1026.
167. Tumlin, J. A. (1997). Expression and function of calcineurin in the mammalian nephron: Physiological roles, receptor signaling, and ion transport. *Am. J. Kidney Dis.* **30**, 884–895.
168. Jiang, H., Xiong, F., Kong, S., Ogawa, T., Kobayashi, M., and Liu, J. O. (1997). Distinct tissue and cellular distribution of two major isoforms of calcineurin. *Mol. Immunol.* **34**, 663–669.
169. Takahashi, N. (1998). The role of NF-AT-family transcription factor in T-cell activation, and their cellular localization [in Japanese]. *Clin. Immunol.* **30**, 98–104.
170. Kubo, M., Kincaid, R. L., and Ransom, J. T. (1994). Activation of the interleukin-4 gene is controlled by the unique calcineurin-dependent transcriptional factor NF(P). *J. Biol. Chem.* **269**, 19,441–19,446.
171. Kao, P. N., Chen, L., Brock, G., Ng, J., Kenny, J., Smith, A. J., and Corthesy, B. (1994). Cloning and expression of cyclosporin A- and FK506-sensitive nuclear factor of activated T-cells; NF45 and NF90. *J. Biol. Chem.* **269**, 20,691–20,699.
172. Hodge, M. R., Chun, H. J., Rengarajan, J., Alt, A., Lieberson, R., and Glimcher, L. H. (1996). NF-AT-driven interleukin-4 transcription potentiated by NIP45. *Science* **274**, 1903–1905.
173. Ranger, A. M., Grusby, M. J., Hodge, M. R., Gravallese, E. M., de la Brousse, F. C., Hoey, T., Mickanin, C., Baldwin, H. S., and Glimcher, L. H. (1998). The transcription factor NF-ATc is essential for cardiac valve formation. *Nature* **392**, 186–190.
174. de la Pompa, J. L., Timmerman, L. A., Takimoto, H., Yoshida, H., Elia, A. J., Samper, E., Potter, J., Wakeham, A., Marengere, L., Langille, B. L., Crabtree, G. R., and Mak, T. W. (1998). Role

of the NF-ATc transcription factor in morphogenesis of cardiac valves and septum. *Nature* **392**, 182–186.
175. Xanthoudakis, S., Viola, J. P., Shaw, K. T., Luo, C., Wallace, J. D., Bozza, P. T., Luk, D. C., Curran, T., and Rao, A. (1996). An enhanced immune response in mice lacking the transcription factor NFAT1. *Science* **272**, 892–895.
176. Nolan, G. P. (1998). Transcription and broken heart. *Nature* **362**, 129–130.
177. Barinaga, M. (1998). Signaling path may lead to better heart-failure therapies. *Science* **280**, 383.
178. Sussman, M. A., Lim, H. W., Gude, N., Taigen, T., Olson, E. N., Robbins, J., Colbert, M. C., Gualberto, A., Wieczorek, D. F., and Molkentin, J. D. (1998). Prevention of cardiac hypertrophy in mice by calcineurin inhibition. *Science* **281**, 1690–1693.
179. Zhang, B. W., Zimmer, G., Chen, J., Ladd, D., Li, E., Alt, F. W., Wiederrecht, G., Cryan, J., O'Neill, E. A., Seidman, C. E., Abbas, A. K., and Seidman, J. G. (1996). T cell responses in calcineurin A alpha-deficient mice. *J. Exp. Med.* **183**, 413–420.

Index

A

A2058, in 23S rRNA, mutation at, 459
A-66005, 120–121
A-177551, 131
A-184656, 135
A-201943, *see* ABT-773
A-202094, 131
A83543A, 209–210
ABT-773
    absorption and elimination, 348
    antibacterial and antimicrobial activity, 140–141
    chemical structure, 464
    distribution, 348–349
    inhibitory activity, 482
    metabolism, 349–350
    for pneumonia, 347
    synthetic process, 136
Accumulation, macrolide, decreased, 479–483
Acetylspiramycin, 150
ACP, *see* Acyl carrier protein
ACP domain, polyketide synthase, 292
Actinomycetes
    8- to 10-membered ring macrolides, 8–10
    12-membered ring macrolides, 10–11
    14- and 16-membered ring macrolides, 12–13
    15-, 17-, and 18-membered ring macrolides, 13–14
    24- to 48-membered ring macrolides, 14–17
    macrocyclic lactones, 59–60
    macrodiolides and macrotetrolides, 20–21
    macrolide lactams, 21
    polyene macrolides, 17–19
Actinoplanic acid A, 20–21
Acutiphycin, 271

4″-*O*-Acylation, tylosin, 151–152
Acyl carrier protein, component of polyketide synthases, 288
3″-*O*-Acyl derivatives, antibacterial activity, 147
Acyltransferase domain, polyketide synthase, 291–292
Adverse effects, *see also* Side effects
    erythromycin, 100–101, 327
    ivermectin, 412
    mebendazole, 409, 411
    thiabendazole, 408–409, 415
    topical tacrolimus, 436–437
Aerocavin, 23
Airway inflammation, erythromycin action points, 538–539
Alanin aminotransferase, 412–413
Albendazole, combined with ivermectin, 413–414
Albocycline, 12
Aldehyde group
    C-18, 16-membered macrolides, 147
    C-20, tylosin, 152–153
Aldol reaction
    asymmetric, 182–188
    Evans asymmetric, 212–213
Algae
    blue-green, 62–64
    dinoflagellates, 64–66
    macrolides from, 40–41
    red, 61–62
Alkaloids, macropolylide skeletons, 35–36
2′-*O*-(*N*-Alkyl succinamoyl)-erythromycin derivatives, 125
Alkyne metathesis, ring-closing, 196

623

Allylation
  asymmetric, 187
  Brown's asymmetric, 261
Almuheptolide A, 33
Alternation of function, polyketide synthase genes, 320
23-epi-Altohyrtin C, 257
Altohyrtins, 69, 183
  total synthesis, 243–257
    Evans's strategy, 243–248
    Kishi's strategy, 248–252
    Smith's strategy, 252–257
Ambrettolide, 33, 37, 43, 196–197
Aminoacyl-tRNA
  A-site-bound, 472
  formation, 459
  ribosome binding, 460
γ-Aminobutyric acid, gated Cl ion channel, 571–575
Aminoglycoside inactivating enzyme, 555–556
Amphidinolide J, 269
Amphidinolides, 66–67
Amphotericin B, 19, 201, 205–206, 318
Amphoterinolide B, 201–202, 205–206
Angiolam A, 26
2,3-Anhydroclarithromycin 11,12-carbazates, 124
Anticancer drugs
  clarithromycin antimetastatic effect, 563
  halichondrin B, 67–68
Antifungals
  callipeltoside A, 78
  goniodomin A, 64
  polyene macrolides, 17–19
  rhizoxin, 29
  soraphen $A_{1\alpha}$, 23
  tacrolimus, 435
Anti-inflammatory action
  erythromycin, 539–540
  macrolide antibiotics, 550–552
  14-membered ring macrolides, 557–558
Antimycin A_{1a}, 20
Antioxidant effect, erythromycin, 562
Apicularen A, 79
Aplidite A, 87
Aplyronine A, 81–82, 272
Aplysiatoxin, 62
Apoptolidin, 18
Archazolide A, 25
Arenolide, 77–78
Arisostatins, 60

Ascidiatrienolide A, 86–87
Aspartate aminotransferase, 412–413
Aspicilin, 33, 201
Atopic dermatitis, tacrolimus efficacy, 434–441
Aurantinins, 22
Aurenin, 19
Auriside A, 82–83
Autoimmune diseases
  pimecrolimus efficacy, 442
  rapamycin efficacy, 442
  tacrolimus efficacy, 431–442
Autoradiographic analysis, motilin receptor distribution, 523–524
Avermectin binding protein, cloning and structure, 572–575
Avermectins, 12–13, 183
  biocidal activities, 153–154
  biosynthetic genes, 301
  derivatives, 154, 156–161
  glycosidation, 202–205
  milbemycin derivatives, 161–164
  target of action, 571–572
  total synthesis, 232–243
    Danishefsky's strategy, 240–243
    Hanessian's strategy, 233
    Ley's strategy, 233–238
    White's strategy, 238–239
Azalides, 9a and 8a derivatives, 119–120
Azalomycin B, 206–207
Azamacrolides, produced by insects, 39
Azcarpine, 36
Azimine, 36
Azithromycin, 363–364
  absorption and elimination, 343
  antibacterial spectrum, 485–486
  anti-inflammatory action, 558
  biological properties, 116–117
  chemical modification, 110–112, 327–328
  chemical structure, 464
  clinical use, 372–380
  distribution, 343–345
  drug interactions, 353
  metabolism, 345

B

Bacteria
  erythromycin resistance, 473–474
  functioning, macrolide modulation of, 553–557

macrocyclic lactones from, 58–60
macrodiolides and macrotriolides, 25
macrolide lactams and oxazole-containing
  macrolides, 26–27
macrolide-resistant, 487
8- to 35-membered ring macrolides, 22–24
triene macrolides, 25
Bafilomycins, 13–14
Bancroftian filariasis, ivermectin for, 413–414
Bartanol, 30
BE-56384, 24
Biofilm, bacterial, macrolide effect, 554–555
Biological properties
  HMR-3647, 133–135
  newer macrolides, 115–117
Bistheonellide A, 70
BK223-B, 31
(S)-N-Boc-pipecolic acid, 213–214, 216
(S)-N-Boc-pipecolinal, 225
Bode and Carreira's total synthesis, epothilones, 267
Bone marrow transplantation, tacrolimus for, 429
Boromycin, 20–21, 25
Boron enolates, aldol reactions mediated by, 182–187
Borrelidin, 14
Botcinolide, 27
Brefeldin A, 28
Bronchitis, see also Diffuse panbronchiolitis
  dirithromycin effect, 369–370
Brown's asymmetric allylboration, 260–261
Bryostatin 1, 42, 80–81, 89
Bryostatin 7, 271
Bundlin B, 13
Buruli ulcer, *Mycobacterium*-caused, 22
2-*t*-Butyldimethylsilyl-1,3-dithiane, 252–254

C

Calcineurin
  complex with FK506 and cyclosporin A, 579
  neuron-specific functions, 581
  signaling pathway, intervention by FK506, 607–609
  target of FK506–FKBP12 complex, 599–604
Callipeltoside A, 78, 83
Callipeltoside aglycone, 270
Cancer, antimetastatic effect of macrolides, 563

11,12-Carbamate ketolides, 129–130
Carbomycin, 466
Carbonoide B, 203
β-Carbonyl reduction, in polyketide synthesis, 288
Ca^{2+}-releasing channel complex, stabilizing role of PPIases, 590–593
Carpaine, 36
Cerulenin, blocking synthesis of polyketides, 287
Chaksine, 36
Chalcomycin, 460
Chemical modification
  avermectin family, 153–164
  azithromycin, 327–328
  clarithromycin, 327–328
  erythromycin A, 508, 510
  erythromycins, 117–127
  ketolides, 127–141
  miokamycin, 148–149
  new-generation macrolides, 105–117
  rokitamycin, 149–150
  SN-41, 150–151
  spiramycin, 150–151
  tylosin, 151–153
Chemical modification (before 1984)
  16-membered ring macrolide antibiotics, 146–147
  14-membered ring macrolides, 101–105
Chivosazol A, 27
Chlamydia infections, roxithromycin effect, 371–372
Chloride channel
  GluCl-α and GluCl-β, 572–575
  inhibition of, 541–546
Chlorothricin, 12
Chondropsin A, 42, 79–80
Chondropsin B, 79–80
Chrohn's disease, tacrolimus for, 432, 441
Chronic airway infection, anti-inflammatory effect of erythromycin, 539–540
Cinachyrolide A, 69
Cisapride, effect on gastric emptying, 517–520
CJ-12,950, 28
Cladinose, erythromycin, 104
Clarithromycin, 105, 118–119, 125, 363–364
  absorption and elimination, 334–336
  anti-inflammatory action, 558
  antimetastatic effect, 563
  biological properties, 116–117

Clarithromycin (*Continued*)
  chemical modification, 107–110, 327–328
  clinical use, 365–369
  distribution, 336–340
  drug interactions, 352–353
  effect on sputum production, 545–546
  Helicobacter pylori resistant to, 477–478
  ketolide formed from, 128
  metabolism, 340–342
  in renal impairment, 342
Clindamycin, 465, 487
Clinical trials, motilides, 526–527
Clinical use
  azithromycin, 372–380
  clarithromycin, 365–369
  dirithromycin, 369–370
  erythromycin, 364–365
    efficacy for DPB, 535
  ivermectin
    bancroftian filariasis, 413–414
    onchocerciasis, 405–407
    scabies, 414
    strongyloidiasis, 407–413
  roxithromycin, 370–372
  tacrolimus, topical therapy, 425–426
Clonostachydiol, 30
Clostomicin B_1, 13
Coloradocin, 8
Concanamycin A, 14
Concanamycin F, 271
Contractility, GI tract, EM574 effect, 521
Copiamycin, 15
Corey method, macrolactonization, 191–192
Corticosteroids, topical, 435
Corymbi-7,13*E*-dienolide, 34
Coupled gating, through FKBP12, 590–591
Coupling reactions
  Stille, 197–199
  Wittig, 199–202
CP-225,600, 122
CP-544,372, 122–123
Crotalarine, 35
Crotylation, asymmetric, 187
Crotylboronation, Roush asymmetric, 227
(*Z*)-Crotyltriphenylsilane, 241
Cucujolides, 37
Cyanophytes, toxins from, 62–63
Cyclization, complex polyketides, 286
Cyclophilin A
  calcineurin A-binding, 603–604
  complex with CsA, 602
  initial target protein for CsA, 586
  role in protein folding, 587
Cyclophilins, 579
Cyclosporin A
  complex with cyclophilin A, 602
  cyclophilin A as initial target protein, 586
  immunosuppressive action, 577–579
Cytochalasins, produced by fungi, 30
Cytochrome *P*-450, erythromycin effect, 350
Cytokine/chemokine expression, macrolide effects, 546–552
Cytovaricin, 15, 208

D

Danishefsky's total synthesis
  avermectins, 240–243
  epothilones, 261–262
  rapamycin, 226–229
Dasypogalactone, 33
Debromoaplysiatoxin, 62
Decarestrictine D, 28
Decarestrictine M, 28
3-*O*-Decladinosyl erythromycin derivatives, 124–125
Defensive secretions, insect macrolides and azamacrolides, 38–39
Dehydratase domain, polyketide synthase, 292
Dental infections, roxithromycin effect, 370
9-Deoxo-9a-aza-9a-methyl-9a-homoerythromycin A, *see* Azithromycin
Deoxyaminosugar biosynthesis, 317–319
10-Deoxymethynolide, 201–202
18-Deoxynargenicin A_1, 269
6-Deoxy-3-*O*-methyl talose pyranoside, 59
2-Deoxysugar biosynthesis, 314–316
13-Deoxytedanolide, 68
Deposition binding, by rokitamycin, 458, 485
Dermatitis
  associated with neutrophil inflammation, 563
  atopic, tacrolimus efficacy, 434–441
Dermocanarins, 27–28
Dermostatins, 19
Desertomycin A, 16
D-Desosamine biosynthesis, 317–319
Didemnilactones, 86
Diethylcarbamazine, 405–406
Difficidin, 25

Diffuse panbronchiolitis
  erythromycin treatment, 534–535
  prognosis, 535
18-Dihydro compounds, aldehyde group at C-18, 147
Dihydroxylation, Sharpless asymmetric, 189
Dinoflagellates, macrocyclic lactones, 64–66
Dinophysistoxin 1, 83
1,3-Di-O-[3,4-bis-(3,4-dihydroxyphenyl)-cyclobutane-1,2-dicarbonyl]-4,5-di-O-caffeoylquinic acid, 34
1,3-Diol systems, asymmetric synthesis, 182–189
Dirithromycin, 113
  clinical use, 369–370
Discodermolide, 73
Disorazol A_1, 27
Divinylcarbinol, Sharpless asymmetric epoxidation, 215
22-Docosanolide, 38
(Z)-3-Dodecen-12-olide, 37
Dolabelide A, 82
Donovanosis, azithromycin effect, 374
Doramectin, 158–159
Doronenine, 36
Dotriacolide, 15
32-Dotriacontanolide, 38, 44
DPB, *see* Diffuse panbronchiolitis
Drug interactions
  azithromycin, 353
  clarithromycin, 352–353
  cytochrome *P*-450 in, 328
  primary mechanism, 350–352
  roxithromycin, 352
  telithromycin, 353

E

Elaiophylin, 20, 199, 206–207
Ellagitannins, 34
Emamectin, 159
Endectosides, 159–161
Enoyl reductase domain, polyketide synthase, 294
Enteritis
  azithromycin effect, 377
  rokitamycin effect, 380
Enterobactin, 25
Enzymatic inactivation, macrolide antibiotics, 484

Eosinophils, cytokine production, 550
Epilachnene, 39
Epithelial cells
  airway, cytokine production, 548
  bronchial, suppression of cytokine expression in, 550–551
Epothilone A, 23
Epothilones, 183
  total synthesis, 257–267
    macroaldolization strategy, 258–259
    macrolactonization strategy, 261–267
    ring-closing olefin metathesis strategy, 259–261
Epoxidation, asymmetric, 188–189
Eprinomectin, 159–160
ER42859, 121
Erythromycin A, 191
  chemical modification, 508, 510
  chemical structure, 12, 464
  effect on gastric emptying in humans, 526–527
  GI side effects, 503
  GMS activity, 507–508
  total synthesis, 207–208
Erythromycins
  accumulation, 481–482
  antioxidant effect, 562
  biosynthetic genes, 317
  clinical use, 364–365
  derivatives, 117–127, 558
  disadvantages, 100–101, 327
  effect on *Shigella* invasion into host cells, 556–557
  modifications of
    lactone ring, 104–105, 125, 127
    sugar moieties, 101, 104, 125, 127
  new-generation, 105–117
  9-oxime, 105–108, 110
  peptidyltransferase inhibition, 467
  pharmacokinetics, 329–330
  polyketide synthases, 295
  resistance to, 473–474
  ribosome binding, 455–456
  treatment of DPB, 534–535
    clinical efficacy, 535
    mechanism of action, 537–540
(9*S*)-9-Erythromycylamine, 105, 113
Escherichia coli O157, verotoxin production, 556
Ethyl *(S)*-malate, anti-selective methylation, 233

Etnangiene, 23
Evans's total synthesis
  altohyrtin, 243–248
  miyakolide, 270
  oleandolide, 268
  phorboxazoles, 270
  rutamycin B, 270
Evonine, 36
Exaltolide, 33

F

Fasted state
  motilide GMS effect, 510–517
  motilin effect, 504–506
Fatty acid synthase, reductive processing, 289–290
Fatty acid synthesis, compared to polyketide synthesis, 288
FD-895, 10
Fed state
  motilide GMS effect, 517–520
  motilin effect, 506
Ferrulactones, 37
Filariasis, bancroftian, 413–414
Filipin III, 19, 201
FK506, 21, 183
  complex with FKBP12, calcineurin as target of, 599–604
  dissociation of FKBP12 from RyR1, 591
  immunosuppressive action, 577–579
  initial cellular target: PPIases, 586–599
  intervention of calcineurin signaling pathway, 607–609
  mode of action, 585
  total synthesis, 210–219
    Ireland's strategy, 211, 218–219
    Merck's strategy, 211–214
    Schreiber's strategy, 211, 214–218
FK520, polyketide synthase, 309, 314
FKBP12
  binding to RyR1, 590–591
  complex with
    FK506: calcineurin as target, 599–604
    rapamycin, 582, 604–607
  isoforms, 599
  tacrolimus and rapamycin binding, 422–423
Flavofungin I, 19
Flavomycoin, 19
Fluorinated erythronolides, 114

Flurithromycin, 114–115
Fluvirucin B_1, 195
Foldase, role of PPIases as, 587
Forsyth's total synthesis, phorboxazoles, 270
Fourteen-membered ring macrolides
  actinomycete-produced, 12
  chemical modification before 1984, 101–105
  clinical use, 364–380
  effect on transcription factors, 551–552
  erythromycins, modifications to, 117–127
  first generation antibiotics, 485
  inhibition of Cl channel functions, 544–546
  ketolides, 127–142
  new generations, 105–117
  suppression of cytokine expression, 550–551
Fungi
  cytochalasins produced by, 30
  macropentolides, 30–31
  macrosphelides, 61
  8- to 16-membered ring macrolides, 27–29

G

Gain of function, polyketide synthase genes, 320
Galbonolide A, 12
Gastric emptying
  cisapride effect, 517–520
  motilin and erythromycin A effects, 526–527
Gastrointestinal motor stimulating activity
  erythromycin A, 507–508
  14-membered ring macrolides, 558
  motilides, 510–520
Genes
  in deoxyaminosugar biosynthesis, 317–319
  in 2-deoxysugar biosynthesis, 314–316
  encoding modular polyketide synthase, 295–314
  erm, 473–474, 476–477
  polyketide synthase
    alternation of function, 320
    loss of function, 319–320
  for resistance, 473
Gephyrin, interaction with mTOR, 606
Geraniin, 34
Geraniol, 188
Gloeosporone, 195–196
Glycosidation, to aglycon, 202–210
GMS, see Gastrointestinal motor stimulating activity
Goniodomin A, 64

Griseoviridin, 269
Growth hormone receptor complex, stabilizing role of PPIases, 597–598
Guanidylfungins, 15

H

Halichlorine, 84
Halichondramide, 71
Halichondrin B, 67–70, 272
C-2 Halogenated ketolides, 131–132
Hanessian's total synthesis, avermectins, 233
Haterumalide B, 88
Haterumalide NA, 88
Helicobacter infection
 azithromycin effect, 378
 clarithromycin effect, 368–369
 roxithromycin effect, 372
Helicobacter pylori, clarithomycin-resistant, 477–478
21-Henicosanolide, 38
Hennoxazole A, 75
17-Heptadecanolide, 33
26-Hexacosanolide, 38
(R)-15-Hexadecanolide, 33, 37–38
36-Hexatriacontanolide, 38
HMR-3647
 chemical structure, 464
 synthesis and biological properties, 133–135
Hoffmanniolide, 65
Homoepilachnene, 39
Horner–Wadsworth–Emmons reaction, 200–201
hsp90
 binding to TRP-containing proteins, 596
 in steroid receptor complexes, 594
Human T-lymphotropic virus type I, 407, 412
(14R)-14-Hydroxyclarithromycin, 109
4″-Hydroxyl group, chemical modification before 1984, 146–147
Hypersecretion
 airway, macrolide antibiotics effects, 541–542
 in DPB: erythromycin effect, 537

I

IB-96212, 60
20-Icosanolide, 38

Iejimalides, 86
Immunophilins
 complemented by parvulin, 588–589
 complex with immunosuppressant, 590
Immunosuppressive action
 FK506, 577–579
 pateamine, 76
 rapamycin, 422–423, 582
Infective endocarditis prophylaxis, 364–365, 379–380
Insecticides
 MK 244, 159
 spinosad, 11
 wilfordine, 36
Insects
 azamacrolides, 39
 8- to 19-membered ring macrolides, 33
 10- to 39-membered ring macrolides, 37–39
Interleukin-8, mRNA expression, 547
Intestinal infection, azithromycin effect, 379
Invertebrates
 macrocyclic lactones from, 80–88
 macrolides from, 40–41
Ireland's strategy, FK506 total synthesis, 211, 218–219
Islet microsomes, FKBP12.6 dissociation from, 592–593
Isostrictinin, 34
Ivermectin, 154, 156–157
 clinical use
  bancroftian filariasis, 413–414
  onchocerciasis, 405–407
  scabies, 414
  strongyloidiasis, 407–413
Izumenolide, 15

J

Jasmine ketolactone, 33
Jasminin, 34
Josamycin, 548
Juglorubin, 8

K

Kabiramide C, 71
Kallmerten's total synthesis, 18-deoxynargenicin A_1, 269
KAV-218, 161

Kende's total synthesis, lankacidin C, 270
β-Ketoacyl ACP synthase, in polyketide biosynthesis, 287–288
β-Ketoacyl synthase domain, polyketide synthase, 291
Ketolides
    ABT-773, 136, 140–141
    9-a-*N*, 130
    11,12-carbamate and 11,12-carbazate, 129–130
    C-2 halogenated, 131–132
    fourth generation macrolides, 486
    HMR-3647 synthesis, 133–135
    9-oxime, 128
    preparation attempts, 127–128
    6-*O*-substituted, 135–136
    tricyclic and tetracyclic, 130–131
β-Ketoreductase domain, polyketide synthase, 294
Kidney transplantation, tacrolimus for, 427–428
Kishi's total synthesis
    altohyrtin, 248–252
    halichondrin B, 272

L

L-783,277, 29
Labilomycin, 18
Lactimidomycin, 10
Lactone ring
    azithromycin, 111
    erythromycins, 104–105, 125, 127
    polyketide-derived, 314
    size, clinical usage based on, 363
Laingolide, 41, 64
Lankacidin A, 13
Lankacidin C, 270
Lasiodiplodin, 33, 195
Lasonolide A, 77
Latrunculins, 73–74
Laulimalide, 74
Lepicidin A, 209–210
Lepranthin, 33
Leucascandrolide A, 79
Leucomycin A_5, 464, 471
Leucomycins, 12, 148–150
Ley's total synthesis, avermectins, 233–238
Lichens, macrolides isolated from, 31–33
Lienomycin, 19
Lincomycin, 465

Lipiarmycin A_3, 13
Liver dysfunction, mebendazole-related, 409, 411
Liver transplantation, tacrolimus for, 426–427
LL-F-28249, 163–164
Lobatamide A, 25, 87
Lobophorins, 60
Loss of function, polyketide synthase genes, 319–320
Low-density lipoprotein receptor, CJ-12,950 effect, 28
Lucensomycin, 18
Luminamicin, 8
Lymphocytes
    cytokine production, 550
    erythromycin effect, 538
Lyngbyaloside, 63–64, 83

M

Macroaldolization strategy, total synthesis of epothilones, 258–259
Macrocyclic lactones, marine organism origin, 58–88
Macrodiolides
    actinomycete-produced, 20–21
    bacteria-produced, 25
    fungus-produced, 30–31
    invertebrate-produced, 41
    plant-produced, 34–36
Macrolactin A, 58, 198–199
Macrolactins, 24
Macrolactone
    macrolactonization, 191–193
    ring-closing olefin metathesis, 193–197
    Stille coupling reaction, 197–199
    synthetic methodology for, 190–202
    Wittig coupling reaction, 199–202
Macrolactonization strategy
    macrolactone synthesis, 191–193
    total synthesis of epothilones, 261–267
Macrolide antibiotics
    anti-inflammatory actions, 550–552
    effect on
        airway secretions, 541–542
        cytokine/chemokine expression, 546–552
        polypeptide synthesis, 459–468
    enzymatic inactivation, 484
    generations, 485–486

inhibition of
  biofilm formation, 554–555
  protein synthesis, 454–455
 interaction with peptidyltransferase, 468–472, 487
 mechanism of action, 467
 resistance mechanisms, 472–484
 ribosome binding, 455–459
 suppression of
  aminoglycoside inactivating enzyme, 555–556
  verotoxin production, 556
Macrolide lactams, 21
 actinomycete-produced, 21
 bacteria-produced, 26–27
 chondropsin A, 41
Macrolide synthesis: synthetic strategy
 asymmetric synthesis of 1,3-diol, 182–189
 glycosidation, 202–210
 macrolactone, 190–202
Macrolide synthesis: total synthesis
 acutiphycin, 271
 altohyrtin, 243–257
 amphidinolide J, 269
 aplyronine A, 272
 avermectins, 232–243
 bryostatin 7, 271
 callipeltoside aglycone, 270
 concanamycin F, 271
 18-deoxynargenicin A_1, 269
 epothilones, 257–267
 FK506, 210–219
 griseoviridin, 269
 halichondrin B, 272
 lankacidin C, 270
 miyakolide, 270
 oleandolide, 268
 pateamine A, 269
 phorboxazoles, 270
 rapamycin, 220–231
 rhizoxin, 269
 rutamycin B, 270
 sanglifehrin A, 268
 swinholide A, 268
 thiazinotrienomycin E, 269
Macropentolides, fungus-produced, 30–31
Macrophages
 alveolar, cytokine production, 550
 erythromycin effect, 538
Macrosphelide A, 30–31

Macrosphelide E, 61
Macrotetrolides
 actinomycete-produced, 20–21
 bacteria-produced, 25
Macrotriolides
 fungus-produced, 30–31
 plant-produced, 37
Madangolide, 41
Maduralide, 15, 59
Madurensine, 36
Malaria, azithromycin effect, 378–379
Malolactomycins, 15
Marginolactones, 16–17
Masamune method, macrolactonization, 192
Masamune's total synthesis, bryostatin 7, 271
Mebendazole, liver dysfunction due to, 409, 411
Megalomycin, 117
Merck's strategy, FK506 total synthesis, 211–214
Metabolism
 ABT-773, 349–350
 azithromycin, 345
 clarithromycin, 340–342
 roxithromycin, 333–334
 telithromycin, 347
Metal enolates, diastereoselective aldol reaction via, 184
Methotrexate, plus tacrolimus, 429
4″-o-(4-Methoxyphenyl)acetyltylosin, 465
Methylation, 23S rRNA domain V, 476–477
6-O-Methylerythromycin, see Clarithromycin
Methyl-α-D-glucopyranoside, 219
30-Methyl-31-hentriacontanolide, 44
(Z-Z)-36-Methyl-15,29-heptatriacontadien-37-olide, 44
28-Methyl-29-nonacosanolide, 44
(Z-Z)-34-Methyl-13,27-pentatriacontadien-35-olide, 44
12-Methyl-13-tridecanolide, 195
(Z)-32-Methyl-25-tritriaconten-33-olide, 44
Methymycin, 10
Meyer's total synthesis, griseoviridin, 269
Microfilarial effects, ivermectin, 406, 415
Microsomes, islet, FKBP12.6 dissociation from, 592–593
Midecamycin, clinical use, 381
Migrastatin, 12
Mikamycin B, 465
Milbemycins, 12–13
 derivatives, 161–164

Miokamycin, 148–149
  clinical use, 381–382
Misakinolide A, 42, 70–71
Mitsunobu method, macrolactonization, 192–193
Miyakolide, 75, 270
MK 244, 159
MK 397, 159–160
MLS antibiotics
  bacteria susceptible to, 454
  resistance, 474, 483–484
  target site, alteration in, 476–479
Mode of action
  avermectin, 571–575
  discodermolide, 73
  erythromycin in DPB, 537–540
  macrolide antibiotics
    effect on polypeptide synthesis, 459–468
    inhibition of protein synthesis, 454–455
    interaction with peptidyltransferase, 468–472
    ribosome binding, 455–459
  motilin, 504–507
  spinosad, 11
Molybdenum catalyst, for ring-closing olefin metathesis, 193–195
Monazomycin, 16
Monolactone macrolides
  invertebrate-produced, 41
  plant-produced, 31–32
Monorden, 29
Motilides
  autoradiographic analysis, 523–524
  clinical trials, 526–527
  conformational analysis, 524–526
  contractile activity, 521
  GMS activity *in vivo*, 510–520
  invention of, 507–510
  receptor binding *in vitro*, 522–523
Motilin, 502–503
  3-D structure, 524–526
  effect on gastric emptying in humans, 526–527
  endogenous release, EM574 effect, 512
  mode of action
    in vitro study, 506–507
    in vivo study, 504–506
Motilin receptors
  distribution analysis, 523–524
  EM574 binding, 522–523
  human, 528

Moxidectin, 163–164
mTOR
  inhibition by rapamycin, 582–583
  signaling pathway, intervention by rapamycin, 610–611
  target of rapamycin–FKBP12 complex, 604–607
Mukaiyama method, macrolactonization, 192–193
Multidrug efflux, transporters mediating, 479–483
Multidrug therapy, AIDS patients, 367–368
Mulzer's synthesis strategy, epothilones, 263–266
Mutation
  FKBP12, 591–592
  23S rRNA
    at A2058, 459
    domain II, 478–479
    domain V, 477–478
Mycalamide A, 76
Mycalolide A, 71–72, 81
D-Mycaminose biosynthesis, 317–319
L-Mycarose biosynthesis, 314–316
Mycinamicin IV, 206
Mycobacterium avium complex
  azithromycin effect, 376–377
  clarithromycin effect, 367–368
Mycofenolate mofetil, plus tacrolimus, 428
Mycolactones, 22
Myxobacteria
  apicularen A from, 79
  macrolides from, 22–27
Myxovirescins, 26–27

N

Nargenicin A_1, 8
Nemadectin, 163–164
Neoantimycin, 20–21
Neodidemnilactone, 86–87
Neospiramycins, 150
Neuroregenerative effects, FK506, 581
Neurotrophic action, tacrolimus, 433
Neutrophils
  accumulation in DPB, erythromycin effect, 537
  adhesion to endothelial cells, 561–562
  alveolar, cytokine production, 550
  clarithromycin uptake, 338

inflammation, dermatitis associated with, 563
NG-012, 31
Nicolaou's total synthesis
  epothilones, 260, 262–263
  rapamycin, 220–223
  sanglifehrin A, 268
  swinholide A, 268
Niddamycin, 296, 300, 466
Niphithricin A, 15
Nonactin, 20–21
Nongonococcal urethritis, azithromycin effect, 373–374
Nuclear factor of activated T cells, 607–609
Nuclear factors, NFκB, 551–552
Nystatin A_1, 18
Nystatins
  aglycon formation, 301
  biosynthetic genes, 318

O

Oasomycins, 16
28-Octacosanolide, 38, 44
(Z)-9-Octadecen-18-olide, 33, 37–38
Octalactins, 8, 59
38-Octatriacontanolide, 38
Ohno's total synthesis, rhizoxin, 269
Ointment, tacrolimus, 435–441
Okadaic acid
  shellfish-derived, 83
  from sponges, 67
L-Oleadrose biosynthesis, 314–316
Oleandolide, 206–207, 268
Oleandomycin, 206–207, 466
  GMS activity, 507–508
Oligomycin A, 15
Oligomycins, backbone formation, 301, 309
Oligomycin SC-2, 24
Onchocerciasis, ivermectin for, 157, 405–407
Ondansetron, effect on EM574 GMS, 514–515
Oocydin A, 88
Opportunistic infections, roxithromycin for, 372
Organostannanes, cyclization, 198
Organ rejection
  rapamycin prophylaxis, 430–431
  tacrolimus efficacy, 426–430, 443
Oscillariolide, 63
Ossamycin, 15
Otitis media, azithromycin effect, 373
Oxazole-containing macrolides
  bacteria-produced, 26–27
  invertebrate-produced, 41
9-Oxime erythromycin, 105–108, 110
9-Oxime ketolides, 128
5-Oxime milbemycin, 162–163
Oximidines, 23, 25

P

PA-46101, 10
Paclitaxel, synergy with discodermolide, 73
Palladium, Stille coupling reaction mediated by, 197
Pamamycin-607, 20
Parsonsine, 37
Partial macrolide and streptogramin B (PMS) resistance, 481–483
Parvulin, complementary to immunophilins, 588–589
Pateamine, 41, 43, 75–76
Pateamine A, 199–200, 269
Patellazoles, 85
Paterson's total synthesis
  callipeltoside aglycone, 270
  concanamycin F, 271
  oleandolide, 268
  swinholide A, 268
Patulolide A, 28
Pectenotoxins, 83–84
Pedunculagin, 34
Peloruside A, 76
Pelvic inflammatory disease, azithromycin effect, 374
Peptidyl prolyl cis-trans isomerases, see PPIases
Peptidyltransferase
  inhibition, 466–467
  interaction with macrolide antibiotics, 468–472, 487
  protein composition, 468, 470
Peptidyl-tRNA, dissociation from ribosomes, 467–468
Phagocytes, clarithromycin uptake, 338–339
Pharmacokinetics
  ABT-773, 347–350
  azithromycin, 343–345
  clarithromycin, 334–342
  erythromycin, 329–330
  roxithromycin, 330–334
  telithromycin, 346–347
  uptake process, 328–329

PHAS-I/4E-BP1, phosphorylation, 582–583, 605
Pheromones, 37–38
Phoracantholide I, 37
Phorboxazoles, 77, 270
Phthoramycin, 15
Pikromycin, biosynthetic genes, 300
Pimaricin
  backbone formation, 301
  biosynthetic genes, 318
Pimecrolimus, for autoimmune diseases, 442
Pin 1, substrate specificity, 588
Pinnaic acid, 84
Pinnatoxin A, 84
Pinolidoxin, 28
Plants
  macrodiolides and macrotriolides from, 34–37
  8- to 19-membered ring macrolides, 33
Plasmid pMS97, 484
Plecomacrolides, 14, 20
Pneumonia
  ABT-773 for, 347
  community-acquired
    azithromycin effect, 375
    clarithromycin effect, 365
Poisoning, red algae, 61–62
Polycavernoside A, 61–62, 210
Polyene macrolides
  biosynthetic genes, 318
  produced by actinomycetes, 17–19
Polyketides
  biosynthesis, reaction mechanism, 287–288
  complex, cyclization, 286
Polyketide synthase
  ACP domain, 292
  acyltransferase domain, 291–292
  dehydratase domain, 292
  deoxyaminosugar biosynthesis, 317–319
  2-deoxysugar biosynthesis, 314–316
  enoyl reductase domain, 294
  fatty acid biosynthetic pathway and, 289–290
  β-ketoacyl synthase domain, 291
  β-ketoreductase domain, 294
  thioesterase domain, 294–295
Polypeptide synthesis, effects of macrolide antibiotics, 459–468
PPIases
  FK506 and cyclosporin A binding, 579
  in highly differentiated cells, 587–588
  initial cellular target for FK506 and rapamycin, 586–599

  role as foldase in protein folding, 587
  stabilizing role in
    Ca^{2+}-releasing channel complexes, 590–593
    growth hormone receptor complexes, 597–598
    steroid hormone receptor complexes, 593–597
9-Propyl-10-azacyclododecan-12-olide, 39
Prorocentrolide, 40, 65
Protein folding, role of PPIases as foldase, 587
Protein L22, alteration, and erythromycin resistance, 468
Protein MsrA, regulation of, 480–481
Protein synthesis, inhibition by macrolide antibiotics, 454–455
Pseudomonas aeruginosa
  biofilm formation, 554–555
  susceptibility to serum, 554
  virulence factors, 555
Psoriasis, tacrolimus for, 432–433
Pulmonary fibrosis, bleomycin-induced, 561
Pulvomycin, 18
Puromycin reaction system, 466–467
Pycnocomolide, 34
Pyocyanine, 555
Pyrenophorin, 198–199
Pyrrolizidine alkaloids, 35–36

Q

Quality of life, tacrolimus ointment and, 440
Quinolidomicin A_1, 18
6-*O*-Quinolylallyl-11,12-carbamates, 136

R

Radicicol, 29
Rapamycin, 21, 183, 199
  for autoimmune diseases, 442
  chemical structure, 578
  complex with FKBP12, 422–423, 582, 604–607
  dissociation of FKBP12 from RyR1, 591
  initial cellular target: PPIases, 586–599
  intervention of mTOR signaling pathway, 610–611
  polyketide synthase, 309, 314
  prophylaxis of organ rejection, 430–431
  side effects, 582

total synthesis, 220–231
  Danishefsky's strategy, 226–229
  Nicolaou's strategy, 220–223
  Schreiber's strategy, 224–226
  Smith's strategy, 229–231
  triene segment, 581
Rapamycin binding protein, 217–218
Reductive processing, fatty acid synthase, 289–290
Renal impairment, clarithromycin in, 342
Renal insufficiency, roxithromycin in, 334
Rescue therapy, tacrolimus in, 428–429
Resistance mechanisms
  alteration of MLS target site, 476–479
  in clinical isolates, 473–476
  decreased macrolide accumulation, 479–483
  enzymatic inactivation of macrolide antibiotics, 484
  intrinsic and acquired, 472–473
  translational and transcriptional attenuation, 483–484
Respiratory infections
  clarithromycin effect, 365, 367
  erythromycin effect, 534–535
  miokamycin effect, 381–382
  roxithromycin effect, 371
Respiratory tract, clarithromycin in, 336–337
Rheumatoid arthritis, systemic tacrolimus for, 431–432
Rhizopodin, 27
Rhizoxin, 29, 269
Ribosomes, macrolide antibiotics binding to, 455–459
Ribozyme, 470–471
Ricinelaidic acid lactone, 195–196
Rifamycin, polyketide assembly, 309
Rimocidin, 18
Ring-closing olefin metathesis
  for macrolactone synthesis, 193–197
  total synthesis of epothilones, 259–261
Ring opening, epoxide, 188–189
Ripostatins, 23
River blindness, ivermectin for, 157
RNA
  IL-8 mRNA levels, 547
  peptidyltransferase, 470–472
  23S rRNA
    bacteria, 553
    domain V and domain II alterations, 476–479
Roflamycoin, 19, 201

Rokitamycin
  anti-inflammatory action, 558
  chemical modification, 149–150
  chemical structure, 464
  clinical use, 380
  deposition binding, 458, 485
Romo–Liu's total synthesis, pateamine A, 269
Roritoxin A, 30
Rosamicin, 13
Rosellichalasin, 30
Roxaticin, 19
Roxithromycin, 105–107
  absorption and elimination, 330–332
  clinical use, 370–372
  distribution, 332–333
  drug interactions, 352
  inhibition of cytokine production, 550
  metabolism, 333–334
  in renal insufficiency, 334
RP 63834, 15
Rustmicin, 12
Rutamycin B, 270
Ruthenium catalysts, for ring-closing olefin metathesis, 193–196
Ryanodine receptors, FKBP12-binding, 590–591

S

Saccharolidin A, 13
Saccharopolyspora spinosa, spinosyns isolated from, 11
Safety, topical tacrolimus, 435–441
Salicylihalamide A, 79, 87, 196
Salvifoliolide, 33
Sanglifehrin A, 21, 268
SANK 60,576, milbemycins produced by, 161
Scabies, ivermectin for, 414
Scapaundulin A, 34
SCH 23831, 13
Sch 38516, 195
Schreiber's total synthesis
  FK506, 211, 214–218
  rapamycin, 224–226
Scytophycin B, 63, 71
Sea hare, macrocyclic lactones from, 81–83
Sekothrixide, 12
Selamectin, 160–161
Seneciphylline, 35–36
Ser/Thr protein kinases, 588–589, 605

Serum, bactericidal activity, 554
Sharpless asymmetric epoxidation, 188–189
  divinylcarbinol, 215
Shellfish, macrocyclic lactones from, 83–84
Shibasaki group's total synthesis, epothilones, 267
Shigella, invasion into host cells: erythromycin effect, 556–557
Shizukaol B, 37
Side effects, *see also* Adverse effects
  erythromycin A, 503
  rapamycin, 582
  thiabendazole, 408–409, 415
Sin3-Rpd3 histone deacetylase, 589
Sinusitis
  clarithromycin effect, 365
  spiramycin effect, 381
Sirolimus, 21, 183, 220–231, 430–431, 581
Sixteen-membered ring macrolides
  actinomycete-produced, 12–13
  antibiotics
    chemical modification before 1984, 146–147
    leucomycin family, 148–150
    spiramycin family, 150–151
    tylosin family, 151–153
  clinical use, 380–382
  peptidyltransferase inhibitors, 466
  second generation drugs, 485
Skeletons
  macropolylide, alkaloids with, 35–36
  representative compounds, 5–7
Skin diseases, immune-mediated, 432–443
Smith's total synthesis
  acutiphycin, 271
  altohyrtin, 252–257
  rapamycin, 229–231
  thiazinotrienomycin E, 269
SN-41, 150–151
Sophorolipid lactone, 196–197
Sorangicin A, 25
Soraphen $A_{1\alpha}$, 23
Sphinxolide, 72, 78
Spinosad, mode of action, 11
Spinosyns, 11
Spiramycin, 466–467
  biosynthetic genes, 296, 300
  chemical modification, 150–151
  clinical use, 380–381
Sponges, macrocyclic lactones, 67–80
Spongistatins, 69, 77, 183, 243–257
Sporaviridins, 15–16

Sputum production, clinical studies, 545–546
Steroid hormone receptor complex, stabilizing role of PPIases, 593–597
Stille coupling reaction, 197–199
Streptococcal pharyngitis, miokamycin effect, 381
Streptogramin B, 465
  in PMS resistance, 481–483
Streptomyces, marine, octalactins produced by, 8
Stroke, tacrolimus effect, 433
Strongyloidiasis, ivermectin for, 407–413
Structure–activity relationship, 14-membered ring macrolides, 557–558
4″-*O*-Substituted erythromycin derivatives, 122–123
6-*O*-Substituted erythromycin derivatives, 117–118
9-*O*-Substituted erythromycin derivatives, 118–119
11-*O*-Substituted erythromycin derivatives, 120–122
13-Substituted erythromycin derivatives, 124
6-*O*-Substituted ketolides, 135–136
Sugar moieties, erythromycins, 101, 104, 125, 127
Superposing method, motilide structure, 524–526
Superstolide A, 42, 78
Suramin, 405
Swinholide A, 42, 69–71, 268
Synthetic strategy: macrolide synthesis
  asymmetric synthesis of 1,3-diol, 182–189
  glycosidation, 202–210
  macrolactone, 190–202

T

Tacrolimus, 21, 183
  in autoimmune diseases, 431–442
  combined with mycofenolate mofetil, 428
  developmental history, 424–425
  efficacy for
    bone marrow transplantation, 429
    kidney transplantation, 427–428
    liver transplantation, 426–427
  FKBP12 binding, 422
  immunosuppressive action, 577–579
  topical therapy, 425–426, 433–442
  total synthesis, 210–219

Tannins, macrodiolide, 34–35
Tartrolon B, 25
Tauropinnaic acid, 84
TE-802, 130–131
TEA-0769, 124–125
Tedanolide, 68
Telithromycin
  absorption and elimination, 346
  chemical structure, 464
  distribution and metabolism, 347
  drug interactions, 353
  inhibitory activity, 482
  synthesis and biological properties, 133–135
Tenuisine A, 36
24-Tetracosanolide, 38
(Z)-5-Tetradecen-14-olide, 33
4-Tetradecen-14-olide, 43
34-Tetratriacontanolide, 38
Tetratricopeptide repeat motifs, 594–595
Theonezolide A, 42–43, 75
Thiabendazole, side effects, 408–409, 415
Thiazinotrienomycin E, 269
Thiazole-containing macrolides, pateamine, 41
Thioesterase domain, polyketide synthase, 294–295
Tilmicosin, 153
Tolytoxin, 63, 72
Toshima's total synthesis, concanamycin F, 271
Toxoplasmosis, azithromycin effect, 379
Trachoma, azithromycin effect, 377–378
Transcription factors, of inflammatory cytokines, 551–552
Transforming growth factor β receptor, interaction with FKBP12, 597–598
Transporters, mediating multidrug efflux, 479–483
30-Triacontanolide, 38, 44
Trichothecenes, macrodiolide skeletons, 30
Tricyclic ketolides, 130–131
13-Tridecanolide, 33
Triene macrolides, produced by bacteria, 25
Tubelactomicin A, 13
Tuckolide, 191
Tunicates, macrocyclic lactones from, 84–88
Tylosin, 191
  biosynthetic genes, 318
  chemical modification, 151–153
  chemical structure, 465
  in polyketide synthases, 295
Typhoid fever, azithromycin effect, 377

U

Ulapualide A, 42–43, 71–72, 201
Urethritis, nongonococcal, azithromycin effect, 373–374
Urinary excretion, clarithromycin, 109–110

V

Vagus nerve, role in motilide-induced GI contractions, 515–516
Venturicidin A, 14
Verotoxin, produced by E. coli O157, 556
Verrucarin A, 30
Vertebrates, macrolides from, 43–45
Viridenomycin, 21
Virulence factors, P. aeruginosa, 555

W

Water secretions, erythromycin effect, 542–545
White's total synthesis
  avermectins, 238–239
  epothilones, 264–265
  rutamycin B, 270
Wilfordine, 36
Williams's total synthesis, amphidinolide J, 269
Wittig coupling reaction, 199–202, 255

Y

Yamada's total synthesis, aplyronine A, 272
Yamaguchi method, macrolactonization, 193
YM 133
  chemical structure, 465
  inhibition of S. aureus, 482–483
YM-75518A, 25
Yuzu lactone, 33, 196–197

Z

Zampanolide, 74
Zearalenone, 28–29, 195–196, 198
Zeranol, 29
Zhu and Park's total synthesis, epothilones, 266–267
Zooxanthellatoxin A, 65–66
Zooxanthellatoxins, 40–41

Macrolide Antibiotics:
Chemist

9780125264518.3

ISBN 0-12-526451-8